THE RUSSIAN FEDERATION,
CENTRAL ASIA, AND
THE TRANSCAUCASUS

EAST ASIA

SOUTHEAST
ASIA

AUSTRALIA,
NEW ZEALAND, AND
THE SOUTH PACIFIC

SOUTH ASIA

EUROPE

MIDDLE EAST
AND NORTH AFRICA

SUB-SAHARAN
AFRICA

World Regions in Global Context

Peoples, Places, and Environments

Sallie A. Marston
University of Arizona

Paul L. Knox
Virginia Tech

Diana M. Liverman
University of Arizona

Prentice
Hall

Upper Saddle River, New Jersey 07458

Pe.
Pears.

Library of Congress Cataloging-in-Publication Data

Marston, Sallie A.
    World regions in global context : peoples, places, and environments / Sallie A. Marston,
  Paul L. Knox, Diana M. Liverman.
        p. cm.
    Includes bibliographic references (p.).
    ISBN 0-13-022484-7
    1. Geography. I. Knox, Paul L. II. Liverman, Diana M. III. Title.

G116.M37 2002
910--dc21   2001052360

Executive Editor: Daniel Kaveney
Project Developmental Editor and Editor in Chief of Development: Ray Mullaney
Production Editor: Tim Flem/PublishWare
Vice President of Production and Manufacturing: David W. Riccardi
Executive Managing Editor: Kathleen Schiaparelli
Marketing Manager: Christine Henry
Manufacturing Manager: Trudy Pisciotti
Assistant Manufacturing Manager: Michael Bell
Art Editor: Adam Velthaus
Director of Creative Services: Paul Belfanti
Director of Design: Carole Anson
Art Directors: Joseph M. Sengotta, Heather Scott
Assistant to Art Directors: John Christiana
Interior Design: Tom Nery, Judith Matz-Coniglio
Managing Editor, Audio/Visual Assets: Grace Hazeldine
Editorial Assistant: Margaret Ziegler
Assistant Managing Editor, Science Media: Nicole Bush
Associate Editor: Amanda Griffith
Media Editor: Chris Rapp
Photo Research Administrator: Melinda Reo
Photo Researcher: Truitt and Marshall
Production Assistant: Nancy Bauer
Composition: PublishWare
Cartographer: MapQuest.com
Senior Manager, Artworks: Patty Burns
Production Manager, Artworks: Ronda Whitson
Manager, Production Technologies, Artworks: Matt Haas
Project Coordinator, Artworks: Connie Long
Illustrator, Artworks: Kathryn Anderson, Mark Landis, Jay McElroy
Cover images: Yann Arthus-Bertrand
Cover description: In the islands of the Samales group in the Sulu Archipelago, the southern Philippines, the Badjaos people, popularly known as "gypsies of the sea," live permanently on their boats, which are real floating homes, or in isolated villages built on stilts over the water. This village is located on a channel carved through a coral reef that gives the villagers' boats access to the open sea where the Badjaos make their living from harvesting shellfish and pearl oysters.

Excerpt from "Little Gidding" in *Four Quartets*. Copyright 1942 by T.S. Eliot and renewed 1970 by Esme Valerie Eliot, reprinted by permission of Harcourt, Inc.

© 2002 by Prentice-Hall, Inc.
Upper Saddle River, New Jersey 07458

ISBN 0-13-022484-7

Printed in the United States of America

10 9 8 7 6 5 4 3 2

Pearson Education Ltd., *London*
Pearson Education Australia Pty., Limited, *Sydney*
Pearson Education Singapore, Pte. Ltd
Pearson Education North Asia Ltd., *Hong Kong*
Pearson Education Canada, Ltd., *Toronto*
Pearson Educación de Mexico, S.A. de C.V.
    ͭson Education—Japan, *Tokyo*
        ͭn Education Malaysia, Pte. Ltd

Brief Contents

Contents

4 The Russian Federation, Central Asia, and the Transcaucasus   152

5 Middle East and North Africa   200

6 Sub-Saharan Africa   256

7 North America   314

Environment and Society in North America   316
Landforms and Landscapes   317
Climate   319
Environmental History   320

10 Southeast Asia   476

13 Future Regional Geographies 612

Preface

We shall not cease from exploration
And the end of all our exploring
Will be to arrive where we started
And know the place for the first time.

Excerpt from "Little Gidding"
in *Four Quartets*.
Copyright 1942 by T. S. Eliot
and renewed 1970 by Esme Valerie Eliot,
reprinted by permission of Harcourt, Inc.

Most people have an understanding of what their own lives are like and some knowledge of their own areas—their neighborhood, their city, their country. Yet, even as the countries and regions of the world become interconnected, most of us still know very little about the lives of people in other societies or about the ways in which the lives of those people connect to our own.

The lines from T. S. Eliot's poem remind us that learning about new places helps us to see familiar places in fresh and unexpected ways. This book provides an introduction to world regional geography that will make exotic places, landscapes, and environments accessible and will reveal the familiar in new ways. To study world regional geography, to put it simply, is to study the dynamic and complex relationships between people and the worlds they inhabit. Our book gives students the basic geographical tools and concepts needed to understand the complexity of regions and to appreciate the interconnections between their own lives and those of people in different parts of the world.

Objective and Approach

This book has two primary objectives. The first is to provide a body of knowledge about how natural, social, economic, political, and cultural phenomena come together to produce distinctive territories with distinctive landscapes and cultural attributes: that is, world regions. The second is to emphasize that although there is diversity among world regions, it is important for us to understand the increasing interdependencies that exist among and between regions in order to build any real understanding of the modern world.

In an attempt to achieve these objectives, we have taken a fresh approach to world geography, reflecting the major changes that have recently been impressed on the global, regional, and local landscapes. These changes include the global spread of new technologies, especially information technologies like the Internet, biotechnology such as genetically engineered seeds, and transportation technologies such as high-speed rail systems. They also include geopolitical shifts such as the formation of the European Union and the Free Trade Area of the Americas; economic trends, such as the growth of transnational corporations and the globalization of consumer culture; and environmental changes associated with increasing industrialization and global warming. The approach used in *World Regions in Global Context* provides access not only to the new ideas, concepts, and theories that address these changes but also to the fundamentals of geography: the principles, concepts, theoretical frameworks, and basic knowledge that are necessary to build a geographic understanding of today's world.

A distinctive feature of this approach is that it employs the concept of geographic scale and emphasizes the interdependence of places and processes at different scales. In overall terms, this approach is designed to provide an understanding of relationships between the global and the local and the outcomes of these relationships. It follows that one of the chief organizing principles is how globalization frames the social and cultural construction of particular places and regions at various scales.

This approach allows us to emphasize a number of important concepts.

- *Globalization and the links between global and local*—Throughout the book, we stress the increasing interconnectedness of different parts of the world through common processes of economic, environmental, political, and cultural change. We approach the processes of globalization through a world-systems framework based on ideas about geographic cores, peripheries, and semiperipheries. A world economy has in fact been in existence for several centuries, and it has been reorganized several times. Each time it has been reorganized, there have been major changes not only in world geography but also in the character and fortunes of individual regions. In this book, we look not only at world regions as they exist in modern times, but also at how each region has contributed to world history and has been affected by the role that it has played. This approach also helps us to point to the links between the global and the local. Recently there has been a pronounced change in both the pace and the nature of globalization. There has been an intensification of global connectedness, a major reorganization of the world economy, and a radical change in our relationships to other people and other places.

- *The unevenness of political and economic development*—We also explicitly recognize the underlying diversity of the world. While there are a range of processes that are likely to be common to most regions—urbanization, industrialization, and population distribution—the way these processes are manifested will vary from region to region and even within regions. In short, there are important variations within places and regions at every scale: for example, social well-being varies and there can be affluent enclaves in poor regions and pockets of poverty in rich regions.

- *Linking society and nature*—Inherent to the basic geographic concepts of landscape, place, and region are the interactions between people and the natural environment that shape landscapes and give places and regions their distinctive characteristics. In this book, we explore the nature-society and human-environment relationships that assist in our understanding of regional geography. We emphasize that human adaptation to Earth's physical environments has gone far beyond responses to natural constraints to produce significant modifications of environments and landscapes and widespread environmental degradation and pollution.

The Geography of World Regions

In this text we have divided the world into ten major regions—Europe; The Russian Federation, Central Asia, and the Transcaucasus; North America; Sub-Saharan Africa; the Middle East and North Africa; Latin America and the Caribbean; East Asia; Southeast Asia; South Asia; and Australia, New Zealand, and the South Pacific. There is no standard way of dividing the world into regions. Textbooks, international organizations, and regional studies groups within universities have chosen a variety of ways to divide up and make sense of the world. Although we review the distinctive characteristics of every region at the beginning of each chapter, the changing and sometimes controversial process of defining world regions merits some discussion here.

Early Greek geographers divided their known world into Europe, Africa, and Asia, with the boundaries defined by the Straits of Gibraltar (dividing Africa and Europe), the Red Sea (dividing Africa and Asia), and the Bosporus Strait (dividing Europe and Asia). As Europeans began to explore the world, new regions were associated with major landmasses or continents, with the Americas usually split into North and South America, and Australia and Antarctica added as the sixth and seventh continents. These divisions lumped together many different landscapes and cultures (especially in Asia) but served, in the minds of Europeans, to differentiate "us" from "them," and to provide a framework for organizing colonial exploration and administration. The colonial period produced many new nations and boundaries, and transformed cultures and landscapes in ways that produced more homogeneous regions. For example, 400 years of Spanish and Portuguese colonization of the region that stretches from Mexico to Argentina created a region of shared languages, religion, and political institutions that became known as Latin America. British colonization of what now comprises Sri Lanka, India, Bangladesh, Pakistan, and Nepal interacted with local culture to produce a region frequently known as South Asia. In the Middle East and North Africa, the persistence of Muslim religion and tradition gave these regions an identity that separated them from Asia and from Africa south of the Sahara.

In the twentieth century, new configurations of political power and economic alliances produced some reconfigurations of world regions. The most notable was the large block of Asia and eastern Europe associated with the socialist politics of the former Soviet Union centered on Russia, together with eastern European countries ranging from East Germany to Bulgaria.

In response to global conflicts and economic opportunities in the second half of the twentieth century, governments and universities established programs and centers that focused on specific world areas and their languages. For example, in the United States, the Department of Education established university centers that focused on apparently coherent regions such as Latin America, the Caribbean, the Pacific, Europe, Africa, the Soviet Union and Eastern Europe, the Middle East, and East, South, and Southeast Asia.

At the beginning of the twenty-first century these traditional divisions of the world into regions have been challenged by events, critics, and the latest phases of globalization. When the Soviet bloc disintegrated in 1989, some states reoriented toward western Europe and to the economic alliance of the European Community, whereas others remained closer to Russia or looked eastward to an identity with countries such as Afghanistan as part of central Asia. As we note in the relevant chapters, regionalizations have been criticized for being based on race or religion (for example, the Middle East and Sub-Saharan Africa), for being remnants of colonial thinking (for example, Latin America or Southeast Asia), or for being grounded only in physical proximity or environmental characteristics rather than on cultural or other human commonalities (for example, Australia, New Zealand, and the Pacific islands clustered in Oceania). We will also discuss a number of countries, such as Sudan, Cyprus, or Antarctica that do not fit easily into the traditional regions or that fit into more than one region. Some scholars and institutions have proposed a dramatic rethinking of world regions. They suggest, for example, that all Islamic or oil-producing countries be treated together, or that countries be grouped according to their level of economic development or integration into the global economy. For example, the World Bank commonly classifies nations into high-, middle-, and low-income countries, and this book identifies many regions and countries according to their relation to the core or periphery of the world system.

Our own division of the world tries to take into account some of these changing ideas about world regions without deviating too radically from other texts or course outlines, and by trying to create a manageable number and coherent set of chapters. Each chapter includes our rationale for treating the places in the chapter as a distinct region and a review of the limitations and debates about defining each region. In addition, each chapter emphasizes the links of the region under discussion to other regions and to processes of globalization that might be changing the nature and coherence of world regions.

Chapter Organization

Two of the central challenges to writing a world regional geography text appropriate for the modern world involve balancing an emphasis on globalization and global processes with the traditional and important emphasis on individual places, and in striking a balance between broad regional generalizations and overly divisive country-by-country regional descriptions. The internal structure of each of the regional chapters is critically important to achieving this balance. In order to strike such a balance, we divide each regional chapter into six standard categories.

Physical and Environmental Context: We begin each of the chapters with a concise discussion of the physical and environmental context of the region, ending this section with an explanation of the region's environmental history. Our aim here is to demonstrate the links between people and nature and how the environment is shaped by and shapes the region's inhabitants over time.

Region in the World: Consistent with our aim to highlight the enduring interdependence of the world's regions, we then provide a section that places each of the regions within the larger context of global history and geography.

Peoples of the Region: In this section we discuss, at various different scales, the people who live in the region.

Regional Change and Interdependence: This section outlines the contemporary role of the region in the global context. This material contrasts to the more historical material emphasized in the Region in the World section.

Core Regions and Key Cities: One of our approaches in the text is to demonstrate the ways in which core, periphery, and semiperiphery are unevenly distributed across geographical scales, in that a specific city or subregion in a peripheral region may share the characteristics of a core region. To illustrate this point we include a section on core regions and key cities that describes the politically and economically central subregions within each of the world regions we discuss.

Distinctive Regions and Landscapes: Our final section of each chapter, which is followed by the summary and conclusions, is devoted to exploring and understanding some of the distinctive regions and landscapes of each of the world's regions.

The organization of the book is pedagogically useful in several ways. First, the conceptual framework of the book is built on two opening chapters: Chapter 1 describes the basics of a regional perspective; Chapter 2 introduces the key concepts that are deployed throughout the remaining regional chapters, highlighting the importance of the globalization approach. Second, the concepts and conceptual framework that are laid out in Chapters 1 and 2 are explored and elaborated upon in the ten regional chapters that follow.

A third important aspect of the book is the distinctive ordering of the chapters. The sequencing of the chapters is a deliberate move to avoid privileging any one region over any other or to cluster the regions according to any economic or political categorization. Rather, because the key conceptual framework of the book is the globalization of the capitalist world-system, we begin with the European region (Chapter 3) because that is where contemporary capitalism and many of the impulses for the contemporary world map have their source. Following the initial appearance of this historically critical core region, however, we deliberately intersperse core, semiperipheral, and peripheral regions in order to signal the interdependence of each.

The final chapter provides a coherent summary of the main points discussed and illustrated in the preceding chapters through an elaboration of the possible futures of the world's regions. This chapter returns students to the conceptual foundations of the book and provides them with a sense of what the future of the globe—and the places and regions within it—might be like.

Features

This book takes a decidedly different approach to understanding world regions, and the features we use help to underscore that difference. The book employs an innovative cartography program, four different boxed features (Geography Matters, Sense of Place, A Day in the Life, and Geographies of Indulgence, Desire, and Addiction), as well as more familiar pedagogical devices such as end-of-chapter review questions and exercises and a listing of important films, music, and popular literature of each region.

Cartography: The signature projection is Buckminster Fuller's Dymaxion™ projection, which centers the globe on the Arctic Circle and arrays the continents around it. This projection helps illustrate the global theme of the book because no one region or continent commands a central position over and above any other. (The word *Dymaxion* and the Fuller Projection Dymaxion Map design are trademarks of the Buckminster Fuller Institute, Santa Barbara, CA, © 1938, 1967 & 1992. All rights reserved.) Each chapter includes a large number of regional and subregional maps that illustrate the geographical patterns and issues discussed in the text. While some of these maps are from existing sources, many were developed specifically for this text.

Geography Matters: This feature examines one of the key concepts of the chapter by providing an extended example of its meaning and implications through both visual illustration and text. The Geography Matters feature demonstrates to students that the focus of world regional geography is on real-world problems.

Sense of Place: This feature highlights specific places within the region with the intention of providing students with a more nuanced sense of what it is like to live in such a place. The Sense of Place feature draws students closer to the textures of a specific regional geography.

A Day in the Life: This feature brings the region into a more personal focus for students by introducing them to real individuals who live their daily lives within the region. The Day in the Life feature makes the abstract discussion of regions more concrete by exposing students to the people who inhabit and shape it, emphasizing the challenges and opportunities these people encounter in their daily lives.

Geographies of Indulgence, Desire, and Addiction: This feature links people in one world region to people throughout the world through a discussion of the local production and global consumption of one of the region's primary commodities. The Geographies of Indulgence, Desire, and Addiction feature helps students to appreciate the links between producers and consumers around the world, as well as between people and the natural world.

Instructional Package

In addition to the text itself, the authors and publisher have been pleased to work with a number of talented people to produce an excellent instructional package. This package includes the traditional supplements that students and professors have come to expect from authors and publishers, as well as new kinds of components that utilize electronic media.

For the Student

- *A Companion Web site* gives students the opportunity to use the Internet to explore topics presented in the book. The site contains numerous review exercises (from which students get immediate feedback), exercises to expand students' understanding of world geography, and resources for further exploration. This Web site provides an excellent platform from which to start using the Internet for the study of geography. Please visit the site at http://www.prenhall.com/marston

- *Science on the Internet: A Student's Guide* (0-13-028253-7): Written by Andrew T. Stull and Harry Nickla, this is a guide to the Internet specifically for students in the sciences. *Science on the Internet* is available at no cost to qualified adopters of *World Regions in Global Context*.

- *Study Guide* (0-13-091929-2): The study guide helps students identify the important points from the text, and then provides them with review exercises, study questions, self-check exercises, and vocabulary review.

For the Professor

- *Slides* (013-091928-4) and *Transparencies* (013-091922-5): All of the maps and figures in the book are featured on the slide and transparency sets. The images have been enlarged and edited for classroom presentation. In order to accommodate instructor preference, these images are available both on transparency acetates and 35-millimeter slides.

- *Digital Files* (013-091925-X): All of the maps and figures from the text, and some of the photographs, are available digitally as high-resolution JPEG's on a CD-ROM. These files are ideal for those professors who use PowerPoint or a comparable presentation software for their classes, or for professors who create text-specific Web sites for their students.

- *The New York Times Themes of the Times—Geography:* This unique newspaper-format supplement features recent articles about geography from the pages of *The New York Times*. This supplement, available at no extra charge from your local Prentice Hall representative, encourages students to make connections between the classroom and the world around them.

- *Instructor's Manual* (013-091926-8): The instructor's manual is intended as a resource for both new and experienced instructors. It includes a variety of lecture outlines, additional source materials, teaching tips, advice about how to integrate visual supplements (including the Web-based resources), and various other ideas for the classroom.

- *Test Item File* (0-13-091910-1): The test item file provides instructors with a wide variety of test questions for use in their exams.

- *PH Custom Test* (0-13-091920-9): Formatted for both Macintosh and IBM computers and based on the powerful testing technology developed by Renaissance Corporate Services, *Prentice Hall Custom Test* allows instructors to create and tailor exams to their own needs. With the online testing program, exams can also be administered online, and data can then be automatically transferred for evaluation. A comprehensive desk reference guide is included along with online assistance.

- *Course Management:* Prentice Hall is proud to partner with many of the leading course-management system providers on the market today. These partnerships enable us to combine our market-leading online content with the powerful course management tools Blackboard, WebCT, and our proprietary course management system, CourseCompass. Please visit our demo site, www.prenhall.com/demo, for more information, or contact your local Prentice Hall representative who can provide a live demonstration of these exciting tools.

Conclusion

As we wrote this text, we were trying to respond to recent major reforms in geographic education. One important outcome of these reforms was the inclusion of geography as a core subject in Goals 2000: Educate America Act (Public Law 103-227). Another was the publication of a set of national geography standards for K-12 education (*Geography for Life,* published by National Geographic Research and Education for the American Geographical Society, the Association of American Geographers, the National Council for Geographic Education, and the National Geographic Society).

More broadly, this book is the product of conversations among the authors, colleagues, and students about how best to teach a course on world regional geography. In preparing the text, we have tried to help students make sense of the world by connecting our conceptual materials to the most compelling current events. We have also been careful to represent the best ideas and concepts the discipline of geography has to offer by mixing cutting-edge and innovative theories and concepts with more classical and proven approaches and tools. Finally, we have also tried to make it clear that no textbook is the product of its authors alone but is instead built up from the intellectual and pedagogical toils and triumphs of thousands of colleagues around the world. Our aim has been to show how a geographical imagination is important, how it can lead to a greater understanding of the world and its constituent places and regions, and how it has practical relevance in our everyday and professional lives.

Acknowledgments

We are indebted to many people for their assistance, advice, and constructive criticism in the course of preparing this book. Among those who provided comments on various drafts of this book are the following professors:

Max Beavers, *University of Northern Colorado*
Richard Benfield, *Central Connecticut State University*
William H. Berentsen, *University of Connecticut*
Warren R. Bland, *California State University, Northridge*
Brian W. Blouet, *College of William and Mary*
Jean Ann Bowman, *Texas A & M University*
Stanley D. Brunn, *University of Kentucky*
Craig Campbell, *Youngstown State University*
David B. Cole, *University of Northern Colorado*
Lorraine Dowler, *Pennsylvania State University*
Ronald Foresta, *University of Tennessee*
Gary Gaile, *University of Colorado*
Roberto Garza, *University of Houston*
Mark Giordano, *Oregon State University*
Kris Jones, *Saddleback College*
Robert C. Larson, *Indiana State University*

Alan A. Lew, *Northern Arizona University*
Max Lu, *Kansas State University*
Donald Lyons, *University of North Texas*
Eugene McCann, *The Ohio State University*
Tom L. McKnight, *University of California, Los Angeles*
Sherry D. Moorea-Oakes, *University of Colorado, Denver*
Tim Oakes, *University of Colorado*
Nancy Obermeyer, *Indiana State University*
Jeffrey E. Popke, *East Carolina University*
Yda Schreuder, *University of Delaware*
Samuel Wallace, *West Chester University*
Gerald R. Webster, *University of Alabama*
Mark Welford, *Georgia Southern University*

Special thanks go to our editor, Dan Kaveney; to our development editor, Ray Mullaney; and to our production editor, Tim Flem. We would also like to thank our research assistants, Reasa Haggard and Lydia Breunig, Cathy Weppler, Administrative Secretary in the University of Arizona Geography and Regional Development Department, the staff of the Center for Latin American Studies at the University of Arizona, and Liz Roberson, Assistant to the Dean, College of Architecture and Urban Studies, Virginia Tech.

Finally, a number of colleagues gave generously of their time and expertise in guiding our thoughts, making valuable suggestions, and providing materials: Simon Batterbury (University of Arizona), Michael Bonine (University of Arizona), Jamey Essex (Syracuse University), John Krygier (Ohio Weslyan University), Maria Carmen Lemos (University of Arizona), Robert Merideth (University of Arizona), John Liverman (independent scholar), Ali Modarres (California State University, Los Angeles), Lynn Patterson (Georgia Tech), Farhang Rouhani (Mary Washington College), David Rain (U.S. Census Bureau), Dereka Rushbrook (University of Arizona), Joel Stillerman (University of Arizona), and Emily Young (San Diego Community Foundation).

Sallie A. Marston
Paul L. Knox
Diana M. Liverman

Sallie A. Marston

Sallie Marston received her Ph.D. in Geography from the University of Colorado, Boulder. She has been a faculty member at the University of Arizona since 1986. Her teaching focuses on the historical, social, and cultural aspects of American urbanization, with particular emphasis on race, class, gender, and ethnicity issues. She received the College of Social and Behavioral Sciences Outstanding Teaching Award in 1989. She is the author of numerous journal articles and book chapters and serves on the editorial board of several scientific journals. In 1994/1995 she served as interim director of Women's Studies and the Southwest Institute for Research on Women. She is currently a professor in, and serves as head of, the Department of Geography and Regional Development at the University of Arizona.

Paul L. Knox

Paul Knox received his Ph.D. in Geography from the University of Sheffield, England. After teaching in the United Kingdom for several years, he moved to the United States in 1985 to take a position as professor of urban affairs and planning at Virginia Tech. His teaching centers on urban and regional development, with an emphasis on comparative study. In 1989 he received a university award for teaching excellence. He has written several books on aspects of economic geography, social geography, and urbanization. He serves on the editorial board of several scientific journals and is co-editor on a series of books on world cities. In 1996 he was appointed to the position of University Distinguished Professor at Virginia Tech, where he currently serves as dean of the College of Architecture and Urban Studies.

Diana M. Liverman

Diana Liverman received her Ph.D. in Geography from the University of California at Los Angeles and also studied at the University of Toronto, Canada, and University College London, England. Born in Accra, Ghana, she is currently a professor of geography and regional development and the director of the Center for Latin American Studies at the University of Arizona. Her teaching focuses on global environmental issues and on Latin America; in 1993, she received a teaching award from Pennsylvania State University. Diana has served on several national and international advisory committees dealing with environmental issues, and has written recent journal articles and book chapters on topics such as natural disasters, climate change, and environmental policy in Mexico. She is an editor of the *Journal of Latin American Geography*.

A Global Focus With A Local View

Focus on Today's Global World

The modern world exhibits great diversity between and within world regions. Despite this diversity, recent changes in the global, regional and local landscapes have created increasing interdependencies between regions. These interdependencies must be thoroughly explored in order to understand current trends in world regional geography. This text carefully highlights these new interdependencies, as well as regional diversities, in order to trace their impact on today's global world.

Chapter Structure

In order to strike an appropriate balance between material about globalization and global processes, and material about individual places, six standard sub-sections appear in every chapter:
- Environment and Society in the Region
- Region in the World
- Peoples of the Region
- Regional Change and Interdependence
- Core Regions and Key Cities
- Distinctive Regions and Landscapes

Table of Contents

8. Latin America and the Caribbean

Environment and Society in Latin America and the Caribbean
Landforms and Landscapes
Climate
Environmental History

Latin America and the Caribbean in the World
The Colonial Experience in Latin America and the Caribbean
Independence Movements and the Export Boom
U.S. Dominance, Latin American Revolutions, and the Cold War
Import Substitution, the Debt Crisis, Neoliberalism, and NAFTA

The Peoples of Latin America and the Caribbean
History and Composition of the Peoples of Latin America and the Caribbean
Population Growth and Urbanization
Migration
The Latin American and Caribbean Diaspora
Language and Cultural Traditions
Religion

Regional Change and Interdependence
Green Revolution and Land Reform
Continuing Inequality Between Social Classes
Drugs in Latin America

Core Regions and Key Cities
Central Mexico
Southeastern Brazil
U.S.-Mexico Border Region
Central Chile

Distinctive Regions and Landscapes
Amazon Basin
The Andes
Caribbean Islands
Central America

Geography Matters: Hurricane Mitch in Honduras
Sense of Place: Lake Titicaca
Geographies of Indulgence, Desire, and Addiction: Sugar
Sense of Place: Havana
Geography Matters: The Economic and Environmental Effects of the North American Free Trade Agreement
A Day in the Life: Yesenia
Geography Matters: The Panama Canal

Summary and Conclusions
Key Terms
Review Questions
Further Reading
Film, Music, and Popular Literature

Geography Matters

***Geography Matters*
demonstrates to students
that the focus of world
geography is on the real
world.** In this feature, we
examine one of the key
concepts of each chapter,
providing an extended
example of its meaning and
implications through both
visual illustration and text.

Geography Matters

Kosovo

The Kosovo region of Yugoslavia provides a conspicuous and tragic example of the complex relationships between ethnicity, nationality, and territoriality in the Balkans. In the twentieth century this region of moderately prosperous farms, small villages, and ancient market towns was inhabited by a mixed population of Serbs, Roma, and Albanians, with Albanians representing the overwhelming majority. Serbians in modern Yugoslavia regard Kosovo with special significance as part of the heartland of Old Serbia, a region that was the platform for a Serbian empire that became, for a while, an important European power. Kosovo was where medieval kings were crowned and was the seat of the Serbian Orthodox Church, an institution synonymous with Serbs' self-identity as a nation. It was also where—at Kosovo Polje—the Serbs suffered a crucial defeat against the Turks in 1389, an event that was the subject of so much Serbian romantic literature and legend over the centuries that the military defeat was transformed into a moral victory. The Albanian majority in modern Kosovo are viewed by Serbs as latecomers and as Turkish surrogates who helped to drive Serbs from the region in the seventeenth and eighteenth centuries.

claim a strong affini-

Figure 1 War-damaged buildings, Kosovo *(Source: U.S. State ... Kosovo: An Accounting. December 1999. ...an_rights/kosovoii/*

Geography Matters

Globalization and Interdependence

Now that the world economy is much more globalized, patterns of local and regional economic development are much more open to external influences, much more interdependent with development processes elsewhere. The globalization of the world economy involves new patterns of regional economic specialization in association with the

a bowl of rice for his daily meal. He makes $2 a day and is hopeful for the future.*

These examples begin to reveal a complex and fast-changing interdependence that would have been unthinkable just 15 or 20 years ago. Joe lost his job because of ... as Maria, and now her ... from China. But Joe ... homy has gained from ... standard of living has

A Day in the Life

Yesenia

Yesenia is an 11-year-old girl who lives in Nogales, Sonora, Mexico, a city on the U.S.–Mexico border (**Figure 1**). When she was five years old, her family moved there from the agricultural state of Sinaloa, so her parents could find steady work.

Since the mid-1960s, when the Mexican government initiated an industrialization program to rehabilitate the northern economy, hundreds of maquiladoras—foreign-owned assembly plants—have located on the Mexican side of the border. Today about 80 factories operate in Nogales. In them, laborers assemble, among other things, suitcases, television remote controls, microchips, trombones, hospital supplies, auto parts, and computers for the world market.

While the early phase of maquiladora industry brought mostly single migrants to the border—mostly young women who ventured away from home and into the factories—today many neighborhoods in Nogales are filled with young families. Like Yesenia's parents, they came primarily to provide a better life for their children.

Yesenia's family, like other newcomers, built a house with whatever they could afford on land that was up for grabs. Yesenia lives with her parents, both of whom work in the maquiladoras, and her four younger brothers and sisters in a three-room house made of scrap wood and discarded shipping pallets. One cold Christmas Eve several years ago, Yesenia built a bonfire outside to try warm up. The flames grew too big too quickly and Yesenia was severely burned on the back of her right leg. Yesenia spent the holidays in the health clinic, where care is free for families of maquiladora workers. Before, when Yesenia's parents worked in the fields in Sinaloa, benefits like this were virtually nonexistent.

Figure 1   Yesenia's life Yesenia lives in Nogales, Mexico, adjacent to the border and to maquiladora manufacturing plants. She lives with her parents and her four younger brothers and sisters in a three-room house similar to those shown here. Many of these are constructed from scrap wood and discarded shipping pallets.

A Day in the Life

***A Day in the Life* makes
the abstract discussion of
regions more concrete by
exposing students to the
people who inhabit and
shape those regions.** By
introducing students to
individuals who live within
the region, this feature
brings the region into a
more personal focus.

Geographies of Indulgence, Desire, and Addiction

Geographies of Indulgence, Desire, and Addiction **helps students appreciate the linkages between producers and consumers around the world, as well as between people and the natural world.** This feature links people in one region to others throughout the world with a thumbnail discussion of the local production and global consumption of one of the region's primary commodities.

Geographies of Indulgence, Desire, and Addiction

Wine

The production and consumption of wine reflects the evolution of the world-system. Wine was one of the early luxury products that established the pattern of merchant trading within Europe. When Europeans branched out to incorporate more of the world into the orbit of their world-system, they began organizing the production of wine wherever climatic conditions were encouraging: in warm temperate zones, roughly between latitudes 30° and 50° north and south. Today, wine is one of the most widespread commodities of consumer indulgence, and fine wines are an important marker of affluence and distinction throughout the world's core regions and in the affluent enclaves of many of the metropolises of peripheral regions. In 1999, more than 26 million liters (5.72 million gallons) of wine were produced worldwide. Almost 15 million liters (3.3 million gallons) of this was consumed in Europe, compared with 2.2 million liters (484,000 gallons) in North America.

The original domestication of wine grapes (*Vitis vinifera*)

Wine had symbolic and ritual significance in early civilizations of the eastern Mediterranean, partly because its ability to intoxicate and engender a sense of "other-worldliness" provided a means through which people could feel that they could come into contact with their gods, and partly because of the apparent death of the vine in winter and its dramatic growth and rebirth in the spring. Greek civilization established viticulture—the cultivation of grape vines for winemaking—as one of the staples of the Mediterranean agrarian economy, along with wheat and olives. By the sixth century B.C., Greek wine was being traded as far as Egypt, the shores of the Black Sea, and the southern regions of France. Under the Roman empire, viticulture spread west along the north shores of the Mediterranean and along the valleys of navigable rivers in France and Spain, while the wine trade extended north, to the North Sea and the Baltic. By the first century A.D., wine had become a commodity of indulgence, desire, and—for some—addiction throughout Europe. Viticulture and the art of

Sense of Place

Sense of Place **gets students closer to the texture of a specific regional geography by highlighting specific places within the region.** Students will gain a more nuanced sense of what it's like to live in the area under study.

Sense of Place

Imperial St. Petersburg

In spite of seven decades of socialism and more than a decade of hardship and disorganization in the transition to a post-Soviet society, St. Petersburg remains an impressive and inspiring city. The city was home not only to Peter the Great but also to Dostoevsky, Nijinsky, and Lenin. The great composers Rimsky-Korsakov, Mussorgsky, Borodin, and Tchaikovsky are buried in the city's Tikhvin Cemetery. But it is the city's core of imperial architecture and urban design that provides its sense of place and symbolizes its sophistication. Often called "The Venice of the North" because of the opulence of its architecture and its canals, St. Petersburg was founded in 1703 by Peter the Great and was the tsars' imperial capital until the Bolshevik revolution of 1917. During that time, St. Petersburg was deliberately fashioned in the Grand Manner as a European-style capital city. The tsars' architects were able to lay out their work unrestricted by any legacy of old streets or buildings. Over two centuries, they collectively created a marvelous set piece of urban design, with imposing public buildings, imperial palaces, and churches in the baroque, rococo, and classical styles, all laid out around impressive plazas and along broad boulevards, all surrounded by large fashionable residences (**Figure 1**).

While Peter the Great founded the city, it was his daughter Empress Elizabeth (who ruled from 1741 to 1761) who comm... first grand wave of buildings, including the Winte... Subsequently... in a drive the G... to ma... miss...

(a)

Sense of Place

Lake Titicaca

At 3820 meters (12,580 feet) above sea level, Lake Titicaca, with an area of 9064 square kilometers (3500 square miles) and a depth of up to 150 meters (500 feet), reflects the luminescent blues of the Andean sky, framed by the towering

paca and llama. Together with the Uros people, who live on Lake Titicaca on floating islands made from reeds (**Figure 2**), the Aymara increasingly derive an income from tourism. People fish on the deep lake from canoelike boats also made

Superior Cartography

The book offers a rich, diverse cartographic program with hundreds of maps that help professors better teach their students the important spatial elements inherent to geography.

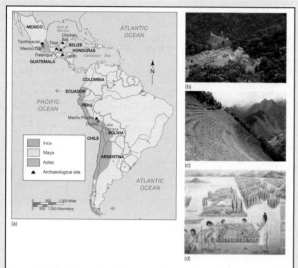

Figure 8.14   Maya, Aztec and Inca adaptations to environment (a) Map of the extent of Maya, Inca, and Aztec empires. (b) The Maya, Inca, and Aztec cultures adapted to environmental constraints in many resourceful ways. Around Mayan cities—such as Palenque, shown here with its pyramids in the Yucatán of Mexico—the forest was cleared using slash-and-burn agriculture, and elsewhere in the Yucatán flooded areas were farmed using raised fields. Palenque was abandoned sometime after A.D. 500 and is now a major tourist destination. (c) The Inca constructed terraces, such as these near Cuzco, Peru, so that they could create level surfaces for irrigating and growing crops and reduce frost risks by breaking the downslope flow of cold air. Constructing the terraces required considerable technical expertise and social organization, and the Inca rulers conscripted large numbers of laborers from local communities. Many terraces were abandoned as the native labor force was reduced due to the ravages of newly introduced European diseases and the need to shift laborers to the Spanish mines in the sixteenth century. (d) The Aztecs cultivated wetlands through the *chinampa* system of fields built from mud and vegetation and anchored to lake beds.

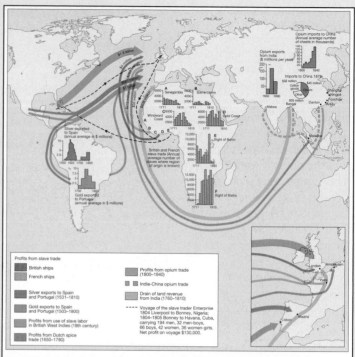

Figure 2.33   Profits from European global expansion, 1500–1800 This map illustrates the profits generated through European plunder of global minerals, spices, and human beings over a 300-year period. Silver and gold from the Americas, opium and tea from China, spices from the East Indies (Indonesia, Malaysia, and the Philippines), and slaves from Africa are represented. (*Source: Adapted from B. Crow and A. Thomas, Third World Atlas, Milton Keynes: Open University Press, 1982, p. 27.*)

Figure 1.5   Caricature of a New Yorker's mental map of the United States This representation of a New Yorker's image of the United States, with its exaggerated dominance of a small part of the city and its deliberate put-down of the rest of the country, was created by artist Saul Steinberg for the cover of *New Yorker* magazine.

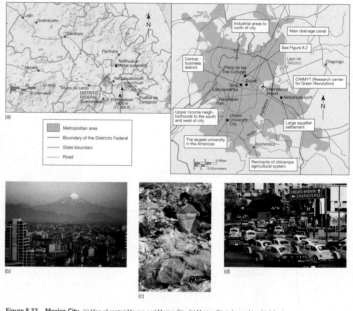

Figure 8.32   Mexico City (a) Map of central Mexico and Mexico City. (b) Mexico City is located in a high basin surrounded by mountains, including the volcanic pair of Popocatépetl on the left (5450 meters, 17,887 feet) and Iztaccihuatl on the right (5288 meters, 17,343 feet). (c) The growing city has inadequate garbage collection and the poor pick through dumps in search for food and items to sell in the informal sector. (d) The green taxis use nonleaded fuel as one of the environmental policies designed to reduce the air pollution that causes environmental health problems and obscures the view of the volcanoes for most of the year.

Figure 3.23 European Union regions eligible for aid, 2000 Just over half of the total population of the European Union lives in areas eligible for regional assistance from the European Regional Development Fund and Cohesion Fund. Most of the EU's regional aid is allocated to "lagging" regions (where per capita GDP is less than 75 percent of the Community average) and to declining industrial regions. (*Source:* EU Inforegio. European Commission, European Regional Development Fund and Cohesion Fund, 2000. Available at: http://www.inforegio.cec.eu.int/wbpro/prord/guide/euro_en.htm)

Legend:
- Lagging regions
- Declining industrial regions
- Distressed rural areas
- Areas with very low population density

Figure 1.6 Topographic maps Topographic maps are maps that represent the form of Earth's surface in both horizontal and vertical dimensions. This extract is from a British Ordnance Survey map of the area just to the south of Edinburgh, Scotland. The height of landforms is represented by contours (lines that connect points of equal vertical distance above sea level), which on this map are drawn every 50 feet, with contour values shown to the nearest meter. Features such as roads, power lines, and built-up areas are shown by stylized symbols. Note how the closely spaced contours of the hill slopes are able to represent the shape and form of the land. (*Source:* Extract from Sheet 66, 1:50,000 Series, Ordnance Survey of the United Kingdom. Copyright © Crown copyright 87375M.)

Buckminster Fulle

This projection helps ... he region or continent commands a central position over and ab... ploy this projection.

Figure 1.4 ... ons Development Programme's Gender Em... icipation in the labor force as administra... entage of parliamentary seats held by women ... with a perfect score would score 1.0, with ... he map reflects a broad core-periphery ... nomic prosperity and gender empowe... y require high levels of economic develop... Design, Santa Barbara, CA. The word *Dymaxio*... inster Fuller Institute, Santa Barbara, California ...

SOUTH AMERICA
NORTH AMERICA
ASIA
AUSTRALIA
EUROPE
AFRICA

Gender empowerment values
- 0.65 and above
- 0.45 to 0.64
- 0.25 to 0.44
- Less than 0.25
- No data

1 A World of Regions

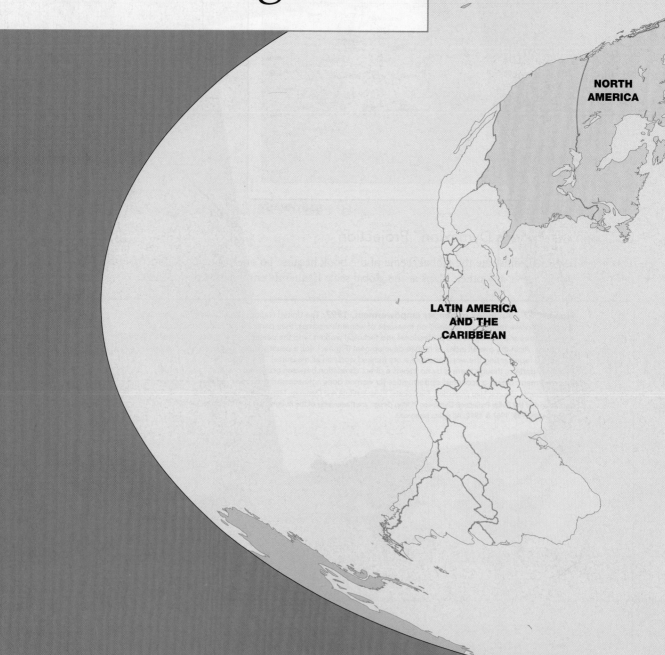

Figure 1.1 World regions World regions are large-scale geographic divisions based on continental and physiographic settings that contain major clusters of humankind with broadly similar cultural attributes. This map shows the world regions that are used in this book: each world region is the subject of a separate chapter.

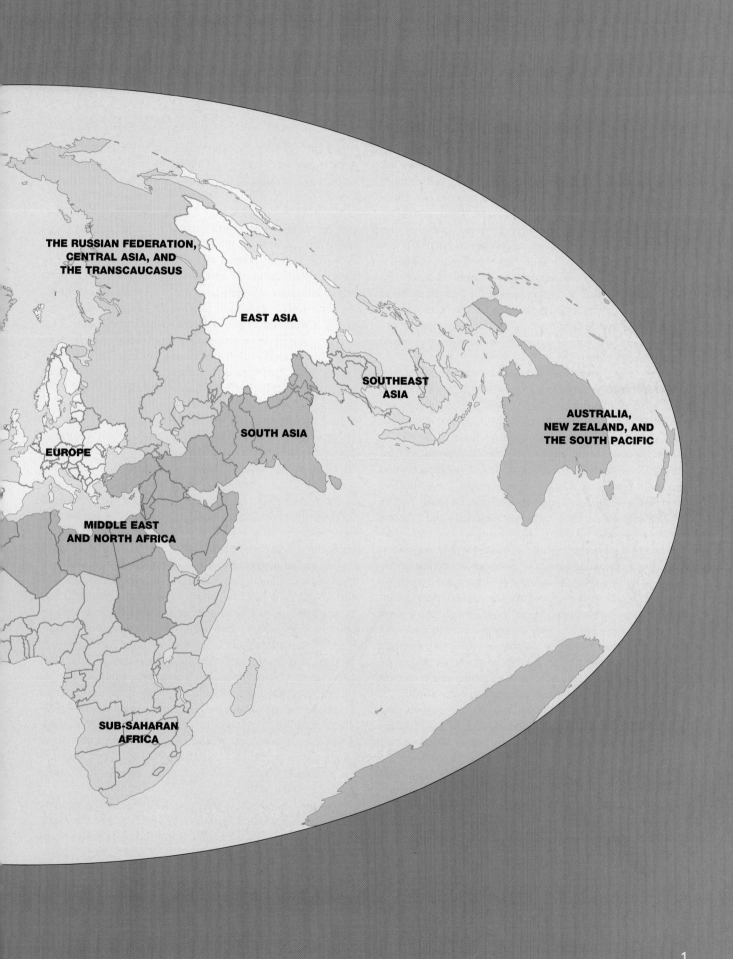

THE RUSSIAN FEDERATION,
CENTRAL ASIA, AND
THE TRANSCAUCASUS

EAST ASIA

SOUTHEAST
ASIA

AUSTRALIA,
NEW ZEALAND, AND
THE SOUTH PACIFIC

SOUTH ASIA

EUROPE

MIDDLE EAST
AND NORTH AFRICA

SUB-SAHARAN
AFRICA

One commonly held belief about today's information age is that instantaneous global telecommunications, satellite television, and the Internet will soon overthrow all but the last vestiges of geographical differentiation in human affairs. According to this view, companies will need no headquarters—they will be able to locate their activities almost anywhere in the world. Employees will work as effectively from home, car, or beach as they could in the offices that need no longer exist. Events halfway across the world will be seen, heard, and felt with the same immediacy as events across town. National differences and regional cultures will dissolve, as a global marketplace brings a uniform dispersion of people, tastes, and ideas.

Such developments are in fact highly unlikely. Even in the information age, geography will still matter and may well become more important than ever. Places and regions will undoubtedly change as a result of the new global context of the information age. But geography will still matter because of transport costs, differences in resource endowments and access to capital, fundamental principles of spatial organization, people's territorial impulses, the resilience of local cultures, and the legacy of the past.

Today, major world regions (**Figure 1.1**) are increasingly dependent on one another, and it is important to know something about regional geography and to understand how places and regions affect, and are affected by, one another. Consider, for example, some of the prominent news stories of 2000. At first glance, they were a mixture of achievements, disputes, and disasters that might seem to have little to do with geography, apart from the international flavor of the coverage. There was civil war in Sierra Leone; ethnic strife in Indonesia, Sri Lanka, Sudan, and Zimbabwe; simmering conflict between India and Pakistan; and the breakdown of peace talks between Israel and Palestine; but peace talks progressed in Ireland and Korea, and in Yugoslavia a "people's revolution" restored democratic governance. In Europe, there were mass rallies against genetically modified foods and demonstrations against international fast-food franchises; in the United States, there were protests against the policies of the International Monetary Fund and the World Bank. Off the front pages, we read of international agreements designed to combat global warming, of the impact of globalization on local cultures and local economic development prospects, and of the continuing struggle to control AIDS.

Most of these stories did, in fact, reflect important geographical dimensions. The diffusion of AIDS, for example, is a geographical as well as a social and cultural phenomenon. Thus, in the United States as in other countries, the AIDS epidemic has diffused through the country in a very distinctive geographical pattern. Behind some of the major news stories, geographical processes played a more central role. Stories about local economic development, local territorial disputes, the globalization of the economy, and global warming, for example, all involve a strong geographical element.

Regional geography is about understanding the variety and distinctiveness of places and regions, without losing sight of the interdependence among them. Geographers learn about the world by finding out where things are and why they are there and by analyzing the spatial patterns and distributions that underpin regional differentiation and regional change. In this chapter we introduce the basic tools and fundamental concepts that enable geographers to study the world in this way.

The Power of Geography

As a subject of scientific observation and study, geography has made important contributions both to the understanding of the world and to its development. As we move farther into the information age, geography continues to contribute to the understanding of a world that is more complex and fast-changing than ever before. With a good understanding of world regional geography, it is possible not only to appreciate the diversity and variety of the world's peoples and places but also to be aware of their relationships to one another and to be able to make positive contributions to regional, national, and global development.

The study of geography involves the study of Earth as created by natural forces and as modified by human action. This, of course, covers an enormous amount of subject matter. *Physical geography* deals with Earth's natural processes and their outcomes. It is concerned, for example, with climate, weather patterns, landforms, soil formation, and plant and animal ecology. *Human geography* deals with the spatial organization of human activity and with people's relationships with their environments. This focus necessarily involves looking at natural, physical environments insofar as they influence, and are influenced by, human activity. This means that the study of human geography must cover a wide variety of phenomena. These phenomena include, for example, agricultural production and food security, population change, the ecology of human diseases, resource management, environmental pollution, regional planning, and the symbolism of places and landscapes.

Regional geography, which combines elements of both physical and human geography, is concerned with the way that unique combinations of environmental and human factors produce territories with distinctive landscapes and cultural attributes. Geographers apply the concept of **region** to large-sized territories (such as counties, provinces, national states, or large sections of a national state, such as the U.S. Midwest) that encompass many **places**, all or most of which share a set of attributes that differ from the attributes of places that make up a different region. What is distinctive about the study of regional geography is not so much the phenomena that are studied as *the way they are approached*. The contribution of regional geography is to reveal how natural, social, economic, political, and cultural phenomena come together to produce distinctive geographic settings.

In the United States, a decade of debate about geography education has resulted in a widespread acceptance that being literate in geography is essential in equipping citizens to earn a decent living, to enjoy the richness of life, and to participate responsibly in local, national, and international affairs. In response to the inclusion of geography as a core subject in the *Goals 2000: Educate America Act* (Public Law 103-227), a major report on the goals of geographic education was produced jointly by the American Geographical Society, the Association of American Geographers, the National Council for Geographic Education, and the National Geographic Society.[1] Published in 1994, the report emphasized the importance of being geographically informed—that is, having an understanding that geography is the study of people, places, and environments from a spatial perspective and appreciating the interdependent worlds in which we live:

> The power and beauty of geography allow us to see, understand, and appreciate the web of relationships between people, places, and environments.
>
> At the everyday level, for example, a geographically informed person can appreciate the locational dynamics of street vendors and pedestrian traffic or fast-food outlets and automobile traffic; the routing strategies of school buses in urban areas and of backpackers in wilderness areas; the land-use strategies of farmers and of real estate developers.
>
> At a more expanded spatial scale, that same person can appreciate the dynamic links between severe storms and property damage or between summer thunderstorms and flash floods; the use of irrigation systems to compensate for lack of precipitation . . . ; the seasonal movement of migrant laborers in search of work and of vacationers in search of sunshine and warmth.
>
> At a global level, the geographically informed person can appreciate the connections between cyclical drought and human starvation in the Sahel or between the Chernobyl nuclear disaster and the long-term consequences to human health and economic activities throughout eastern and northwestern Europe; the restructuring of human migration and trade patterns as the European Union becomes increasingly integrated or as the Pacific rim nations develop a commonality of economic and political interests; and the uncertainties associated with the possible effects of global warming on human society or the destruction of tropical rain forests on global climate.[2]

The report cited four different reasons for being geographically informed:

- The Existential Reason. In 1977 the U.S. spacecraft *Voyager 1* set out on its epic journey to the outer solar system and beyond. When it had passed the most distant planet, its camera was turned back to photograph the solar system. Purely by chance, the camera recorded a pale blue dot in the vastness of space. Every human who has ever lived has lived on that blue dot—Earth (**Figure 1.2**). Humans want to understand the intrinsic nature of their home. Geography enables them to understand where they are, literally and figuratively.

- The Ethical Reason. Earth will continue to whirl through space for untold millennia, but it is not certain that it will

[1]Geography Education Standards Project, *Geography for Life. National Geography Standards 1994*. Washington, DC, National Geographic Research and Exploration, 1994.

[2]*Geography for Life*, p. 29.

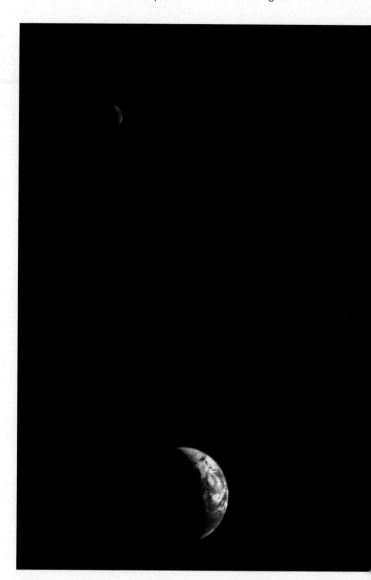

Figure 1.2 Earth from space This photograph of Earth was taken from the *Voyager I* satellite in 1977. It was an image that changed people's perceptions of our planet, literally providing them with a new perspective on the finite environment in which we live.

exist in a condition in which humans can thrive or even live. . . . Geography provides knowledge of Earth's physical and human systems and of the interdependency of living things and physical environments. That knowledge, in turn, provides a basis for humans to cooperate in the best interests of our planet.

- The Intellectual Reason. Geography captures the imagination. It stimulates curiosity about the world and the world's diverse inhabitants and places, as well as about local, regional, and global issues. By understanding our place in the world, humans can overcome parochialism and ethnocentrism. Geography focuses attention on exciting and interesting things, on fascinating peoples and places, on things worth knowing because they are absorbing and because knowing them makes humans better informed and, therefore, helps them make wiser decisions.

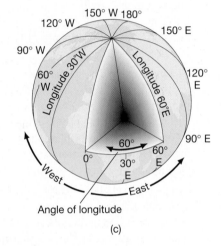

The prime meridian at the Royal Observatory in Greenwich, England. The observatory was founded by Charles II in 1675 with the task of setting standards for time, distance, latitude, and longitude—the key components of navigation.

Figure 1.3 Latitude and longitude Lines of latitude and longitude provide a grid that covers Earth, allowing any point on Earth's surface to be accurately referenced. Latitude is measured in terms of angular distance (that is, degrees and minutes) north or south of the equator, as shown in (a). Longitude is measured in the same way, but east and west from the prime meridian, a line around Earth's surface that passes through both poles (North and South) and the Royal Observatory in Greenwich, just to the east of central London, in England. Locations are always stated with latitudinal measurements first (c). The location of Paris, France, for example, is 48°51′ N and 2°20′ E, as shown in (b).

■ The Practical Reason. Geography has utilitarian value in the modern world. As the interconnectedness of the world accelerates, the practical need for geographic knowledge becomes more critical. Imagine a doctor who treats diseases without understanding the environment in which the diseases thrive and spread, or a manufacturer who is ignorant of world markets and resources, or a postal worker who cannot distinguish Guinea from Guyana. With a strong grasp of geography, people are better equipped to solve issues not only at the local level but also at the global level.[3]

The importance of geography as a subject of study has become more widely recognized in recent years as people everywhere have struggled to understand a world that is increasingly characterized by instant global communications, unfamiliar international relationships, unexpected local changes, and growing evidence of environmental degradation. Many more schools now require courses in geography than just a decade ago, and the Educational Testing Service's College Board has added the

subject to its Advanced Placement program. Between 1985–86 and 1996–97, the number of bachelor's degrees in geography awarded annually increased from 3056 to 4128. Meanwhile, many employers have come to realize the value of employees with expertise in geographical analysis and an understanding of the uniqueness, influence, and interdependence of places.

Some Fundamental Concepts

The study of regional geography draws on several fundamental concepts. In analyzing the spatial patterns and distributions that underpin regional differentiation, it is important to understand the ways in which it is possible to approach the apparently straightforward issue of location. Second, it is important to have an understanding of one of geographers' basic tools: maps. Third, it is important to understand the idea of interdependence between places and regions.

[3]*Geography for Life,* pp. 23–24.

Location

Often, location is *nominal,* or expressed solely in terms of the names given to regions and places. We speak, for example, of Washington, DC, or of Georgetown, a location within Washington, DC. Location can also be used as an *absolute* concept, whereby locations are fixed mathematically through coordinates of latitude and longitude (**Figure 1.3**). **Latitude** refers to the angular distance of a point on Earth's surface, measured in degrees, minutes, and seconds north or south from the equator, which is assigned a value of 0°. Lines of latitude around the globe run parallel to the equator, which is why they are sometimes referred to as *parallels.* **Longitude** refers to the angular distance of a point on Earth's surface, measured in degrees, minutes, and seconds east or west from the *prime meridian* (the line that passes through both poles and through Greenwich, England, and which is assigned a value of 0°). Lines of longitude, called *meridians,* always run from the North Pole (latitude 90° north) to the South Pole (latitude 90° south). Georgetown's coordinates are 38°55′ N, 77°00′ W.

Thanks to the Global Positioning System (GPS), it is very easy to determine the latitude and longitude of any given point. The **Global Positioning System** consists of 21 satellites (plus 3 spares) that orbit Earth on precisely predictable paths, broadcasting highly accurate time and locational information. The GPS is owned by the U.S. government, but the information transmitted by the satellites is freely available to everyone around the world. All that is needed is a GPS receiver. Basic receivers cost less than $100 and can relay latitude, longitude, and height to within 10 meters day or night, in all weather conditions, in any part of the world. The most precise GPS receivers, costing thousands of dollars, are accurate to within a centimeter. GPS has drastically increased the accuracy and efficiency of collecting spatial data. In combination with **geographic information systems** (GIS) and **remote sensing** (the collection of information about parts of Earth's surface by means of aerial photography or satellite imagery), GPS has revolutionized mapmaking and spatial analysis.

Location can also be *relative,* fixed in terms of site or situation. **Site** refers to the physical attributes of a location: its terrain, its soil, vegetation, and water sources, for example. **Situation** refers to the location of a place relative to other places and human activities: its accessibility to routeways, for example, or its nearness to population centers (**Figure 1.4**). Washington, DC, has a low-lying, riverbank site and is situated at the head of navigation of the Potomac River, on the eastern seaboard of the United States.

Finally, location also has a *cognitive* dimension, in that people have cognitive images of places and regions, compiled

Figure 1.4   The importance of site and situation The location of telecommunications activities in Denver, Colorado, provides a good example of the significance of the geographic concepts of site (the physical attributes of a location) and situation (the location of a place relative to other places and human activities). Denver has become a major center for cable television, with the headquarters of giant cable companies such as Tele-Communications and DirecTV, an industry-wide research lab, and a cluster of specialized support companies that together employ more than 3000 people. Denver's *site*, 1.6 kilometers (1 mile) above sea level, is important because it gives commercial transmitters and receivers a better "view" of communications satellites. Its *situation*, on the 105th meridian and equidistant between the telecommunications satellites that are in geostationary orbit over the Pacific and Atlantic oceans, allows it to send cable programming directly not just to the whole of the Americas but also to Europe, the Middle East, India, Japan, and Australia—to every continent, in fact, except Antarctica. This is important because it avoids "double-hop" transmission (in which a signal goes up to a satellite, then down, then up and down again), which increases costs and decreases picture quality. Places east or west of the 105th meridian would have to double-hop some of their transmissions because satellite dishes would not have a clear "view" of both the Pacific and Atlantic telecommunications satellites.

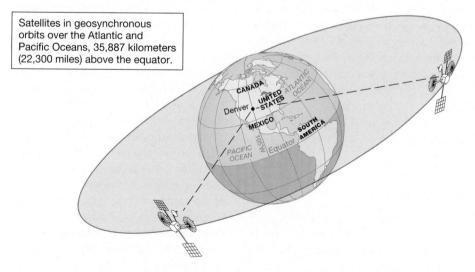

Satellites in geosynchronous orbits over the Atlantic and Pacific Oceans, 35,887 kilometers (22,300 miles) above the equator.

Figure 1.5   Caricature of a New Yorker's mental map of the United States   This representation of a New Yorker's image of the United States, with its exaggerated dominance of a small part of the city and its deliberate put-down of the rest of the country, was created by artist Saul Steinberg for the cover of *New Yorker* magazine.

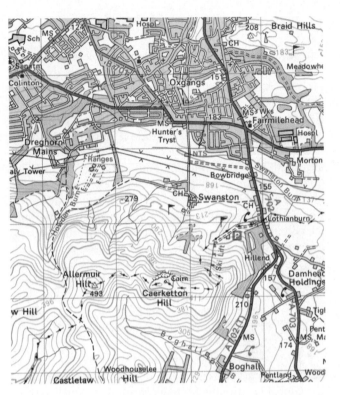

Figure 1.6   Topographic maps   Topographic maps are maps that represent the form of Earth's surface in both horizontal and vertical dimensions. This extract is from a British Ordnance Survey map of the area just to the south of Edinburgh, Scotland. The height of landforms is represented by contours (lines that connect points of equal vertical distance above sea level), which on this map are drawn every 50 feet, with contour values shown to the nearest meter. Features such as roads, power lines, and built-up areas are shown by stylized symbols. Note how the closely spaced contours of the hill slopes are able to represent the shape and form of the land. (*Source*: Extract from Sheet 66, 1:50,000 Series, Ordnance Survey of the United Kingdom. Copyright © Crown copyright 87375M.)

from their own knowledge, experience, and impressions. **Cognitive images** (sometimes referred to as *mental maps*) are psychological representations of locations that are made up from people's individual ideas and impressions of these locations. These representations can be based on people's direct experiences, on written or visual representations of actual locations, on hearsay, on people's imagination, or on a combination of these sources. Location in these cognitive images is fluid, depending on people's changing information and perceptions of the principal landmarks in their environment. Some things, indeed, may not be located in a person's cognitive image at all! **Figure 1.5** shows a caricature of a New Yorker's cognitive image of the United States.

Understanding Maps

Maps are representations of the world, and the body of practical and theoretical knowledge about mapmaking is known as **cartography.** Maps are usually two-dimensional, graphic rep-

resentations that use lines and symbols to convey information or ideas about spatial relationships. Maps that are designed to represent the *form* of Earth's surface and to show permanent (or, at least, long-standing) features such as buildings, highways, field boundaries, and political boundaries, are called *topographic maps* (see, for example, **Figure 1.6**). The usual device for representing the form of Earth's surface is the *contour,* a line that connects points of equal vertical distance above or below a zero data point, usually sea level.

Maps that are designed to represent the spatial dimensions of particular conditions, processes, or events are called *thematic maps.* These can be based on any one of a number of devices that allow cartographers or mapmakers to portray spatial variations or spatial relationships. One of these is the *isoline,* a line (similar to a contour) that connects places of equal data value (for example, air pollution, as in **Figure 1.7**). Maps based on isolines are known as *isopleth maps.* Another common device used in thematic maps is the *proportional symbol.* Thus, for example, circles, squares, spheres, cubes, or some other shape can be drawn in proportion to the frequency of occurrence of some particular phenomenon or event at a given lo-

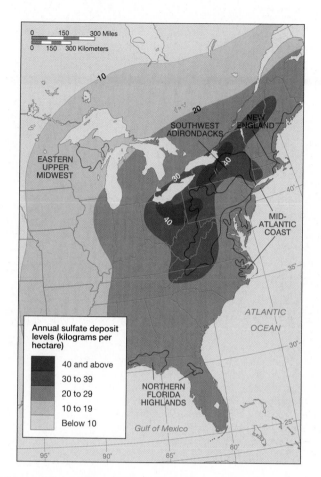

Figure 1.7   Isoline maps Isoline maps portray spatial information by connecting points of equal data value. Contours on topographic maps (see Figure 1.6) are isolines. This map shows air pollution in the eastern United States. (*Source:* Reprinted with permission of Prentice Hall, from J. M. Rubenstein, *The Cultural Landscape: An Introduction to Human Geography,* 1996, 584. Adapted from William K. Stevens, "Study of Acid Rain Uncovers Threat to Far Wider Area," *New York Times,* January 16, 1990, 21.)

cation. Symbols such as arrows or lines can also be drawn proportionally, in order to portray flows of things between particular places. **Figure 1.8** shows two examples of proportional symbols: flow lines and proportional circles. Simple distributions can be effectively portrayed through *dot maps,* in which a single dot or other symbol represents a specified number of occurrences of some particular phenomenon or event (**Figure 1.9**). Yet another device is the *choropleth map,* in which tonal shadings are graduated to reflect area variations in numbers, frequencies, or densities (**Figure 1.10**). Finally, thematic maps can be based on *located charts,* in which graphs or charts are located by place or region. In this way, a tremendous amount of information can be conveyed in a single map (**Figure 1.11**).

Map Scales   A *map scale* is simply the ratio between linear distance on a map and linear distance on Earth's surface. It is usually expressed in terms of corresponding lengths, as in "one centimeter equals one kilometer," or as a *representative fraction* (in this case, 1/100,000) or ratio (1:100,000). *Small-scale* maps are maps based on small representative fractions (for ex-

ample, 1/1,000,000 or 1/10,000,000). Small-scale maps cover a large part of Earth's surface on the printed page. A map drawn on this page to the scale of 1:10,000,000 would cover about half of the United States; a map drawn to the scale of 1:16,000,000 would easily cover the whole of Europe. *Large-scale* maps are maps based on larger representative fractions (for example, 1/25,000 or 1/10,000). A map drawn on this page to the scale 1:10,000 would cover a typical suburban subdivision; a map drawn to the scale of 1:1000 would cover just a block or two of it.

Map Projections   A map projection is a systematic rendering on a flat surface of the geographic coordinates of the features found on Earth's surface. Because Earth's surface is curved and it is not a perfect sphere, it is impossible to represent on a flat plane, sheet of paper, or monitor screen without some distortion. Cartographers have devised a number of different techniques of projecting latitude and longitude (see Figure 1.3) onto a flat surface, and the resulting representations of Earth each have advantages and disadvantages. None of them can represent distance correctly in all directions, though many can represent compass bearings or area without distortion. The choice of map projection depends largely on the purpose of the map.

Projections that allow distance to be represented as accurately as possible are called **equidistant projections.** These projections can represent distance accurately only in one direction (usually north-south), although they usually provide accurate scale in the perpendicular direction (which in most cases is the equator). Equidistant projections are often more aesthetically pleasing for representing Earth as a whole, or large portions of it. An example is the Polyconic projection (**Figure 1.12**).

Projections on which compass directions are rendered accurately are known as **conformal projections.** Another property of conformal projections is that the scale of the map is the same in any direction. The Mercator projection (Figure 1.12), for example, preserves directional relationships between places, and so the exact compass distance between any two points can be plotted as a straight line. As a result, it has been widely used in navigation. As Figure 1.12 shows, however, the Mercator projection distorts area more and more toward the poles—so much so that the poles cannot be shown as single points. Some projections are designed such that compass directions are correct only from one central point. These are known as **azimuthal projections.** They can be equidistant, as in the Azimuthal Equidistant projection (Figure 1.12), which is sometimes used to show air-route distances from a specific location, or equal-area, as in the Lambert Azimuthal Equal-Area projection.

Projections that portray areas on Earth's surface in their true proportions are known as **equal-area** or **equivalent projections.** Such projections are used where the cartographer wishes to compare and contrast distributions on Earth's surface: the relative area of different types of land use, for example. Examples of equal-area projections include the Eckert IV projection, Bartholomew's Nordic projection, and the Mollweide projection

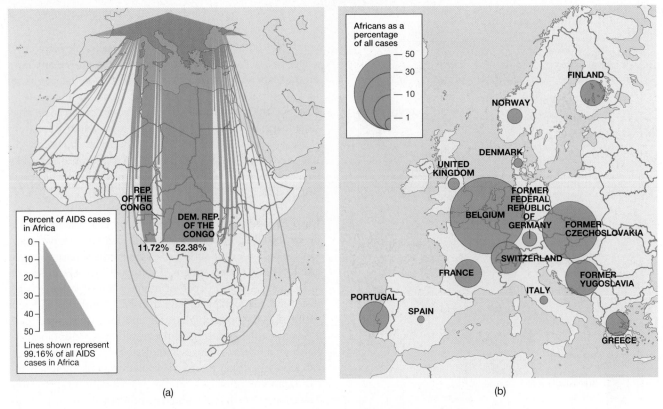

(a)   (b)

Figure 1.8   Two examples of proportional symbols in thematic mapping   (a) Flow lines: The country of origin of AIDS cases diagnosed in Europe. The magnitude of the flow from each country is expressed as a percentage of all AIDS cases from Africa in Europe. (b) Proportional circles: The percentage of AIDS patients originating from Africa as a proportion of national AIDS cases in Europe. (*Source:* M. Smallman-Raynor, A. Cliff, and P. Haggett, *London International Atlas of AIDS.* Oxford: Blackwell Reference, 1992, Figs 2.10(a) and 2.10(b).)

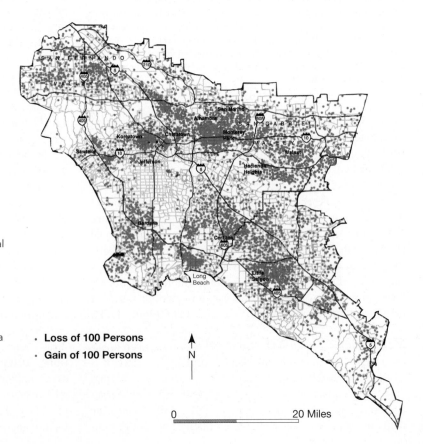

Figure 1.9   Dot maps   Dot maps show the spatial distribution of particular phenomena by means of simple, located symbols (usually dots or small circles, but they can be any shape). This map shows the changing distribution of the Asian population in Metropolitan Los Angeles and Orange counties between 1980 and 1990. Each blue dot symbolizes the loss of 100 Asian inhabitants from a census tract (a small areal unit used in collecting and reporting data from the U.S. Census of Population and Housing); each red dot shows an equivalent gain. (*Source:* E. Turner and J. P. Allen, *An Atlas of Population Patterns in Metropolitan Los Angeles and Orange Counties, 1990.* Department of Geography, California State University, Northridge, 1991, p. 6.)

• **Loss of 100 Persons**
· **Gain of 100 Persons**

0 ————————— 20 Miles

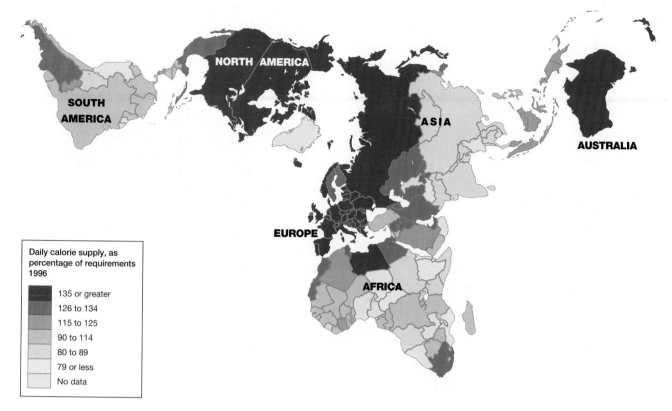

Figure 1.10 Choropleth maps Choropleth maps use tonal shadings to represent data based on areal units such as census tracts, counties, or regions. An important feature of such maps is that the tonal shading is proportional to the values of the data (that is, the higher the value, the darker the shading, or vice versa). This map shows international variations in people's food consumption in 1996. On this map, the darker the tone of the shaded area, the higher the country's food consumption, measured in terms of average daily calorie supply as a percentage of the minimum requirement. (*Source:* P. L. Knox and J. Agnew, *The Geography of the World Economy*, 3rd ed. London: Edward Arnold, 1998, p. 36. Map projection, Buckminster Fuller Institute and Dymaxion Map Design, Santa Barbara, CA. The word *Dymaxion* and the Fuller Projection Dymaxion™ Map design are trademarks of the Buckminster Fuller Institute, Santa Barbara, California, © 1938, 1967 & 1992. All rights reserved.)

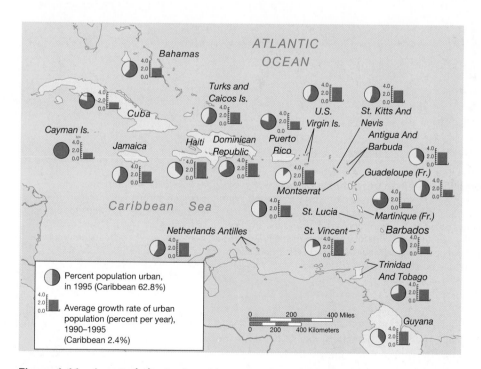

Figure 1.11 Located charts By combining graphs or charts with base maps, a great deal of information can be conveyed in a single figure. This example shows variations across the Caribbean in both the level (divided circles, or pie charts) and the rate (bar charts) of urbanization. (*Source:* R. Potter, "Urbanisation and Development in the Caribbean," *Geography*, 80 (1995), 336.)

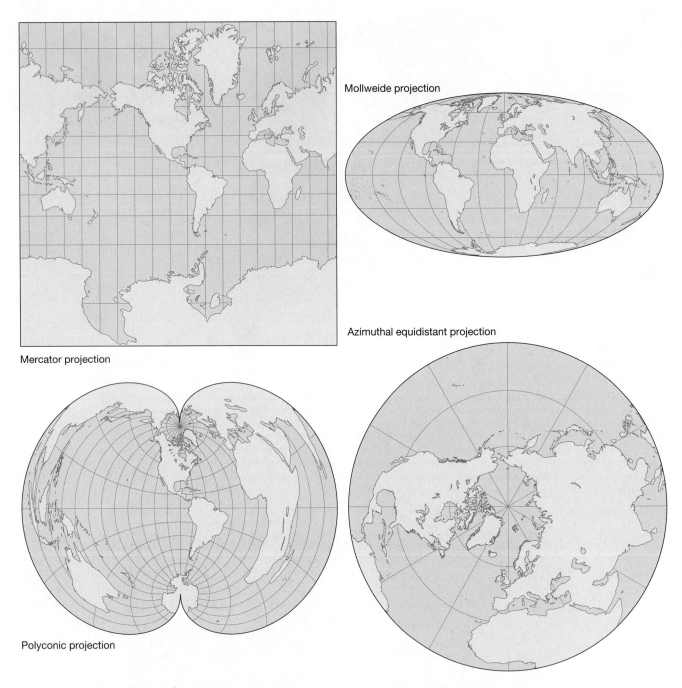

Figure 1.12   Comparison of map projections Different map projections have different properties. The Polyconic projection is true to scale along each east-west parallel and along the central north-south meridian. It is neither conformal nor equal-area, and it is only free of distortion along the central meridian. On the Mercator projection, compass directions between any two points are true and the shapes of land masses are true but their relative size is distorted. On the Azimuthal Equidistant projection, distances measured from the center of the map are true, but direction, area, and shape are increasingly distorted with distance from the center point. On the Mollweide projection, relative sizes are true but shapes are distorted.

(Figure 1.12). Equal-area projections such as the Mollweide projection are especially useful for thematic maps showing economic, demographic, or cultural data. Unfortunately, preserving accuracy in terms of area tends to result in world maps on which many locations appear squashed and have unsatisfactory outlines.

For some applications, aesthetic appearance is more important than conformality, equivalence, or equidistance, and so cartographers have devised a number of other projections. Examples include the Times projection, which is used in many world atlases, and the Robinson projection, which is used by the National Geographic Society in many of its publications.

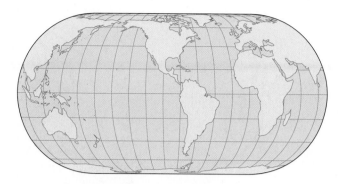

Figure 1.13   The Robinson projection On the Robinson projection, distance, direction, area, and shape are all distorted in an attempt to balance the properties of the map. It is designed purely for appearance and is best used for thematic and reference maps at the world scale. (*Source:* Reprinted with permission of Prentice Hall from E. F. Bergman, *Human Geography: Cultures, Connections, and Landscapes,* © 1995, p. 12.)

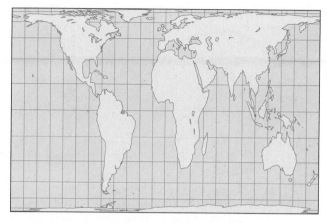

Figure 1.14   The Peters projection This equal-area projection was an attempt to offer an alternative to traditional projections which, Arno Peters argued, exaggerated the size and apparent importance of the higher latitudes—that is, the world's core regions—and so promoted the "Europeanization" of Earth. While it has been adopted by the World Council of Churches, the Lutheran Church of America, and various agencies of the United Nations and other international institutions, it has been criticized by cartographers in the United States on the grounds of esthetics: One of the consequences of equal-area projections is that they distort the shape of land masses. (*Source:* Reprinted with permission of Prentice Hall from E. F. Bergman, *Human Geography: Cultures, Connections, and Landscapes,* © 1995, p. 13.)

The Robinson projection (**Figure 1.13**) is a compromise projection that distorts both area and directional relationships but provides a general-purpose world map. There are also political considerations. Countries may appear larger and so more "important" on one projection rather than another. The Peters projection, for example (**Figure 1.14**), is a deliberate attempt to give prominence to the underdeveloped countries of the equatorial regions and the southern hemisphere. As such, it was officially adopted by the World Council of Churches and by numerous agencies of the United Nations and other international institutions. Its unusual shapes give it a shock value that gets people's attention. For some, however, its unusual shapes are ugly: It has been likened to laundry hung out to dry.

In this book, we shall sometimes use another striking projection, the Dymaxion projection devised by Buckminster Fuller (**Figure 1.15**). Fuller was a prominent modernist architect and industrial designer who wanted to produce a map of the world with no significant distortion to any of the major land masses. The Dymaxion projection does this, though it produces a world that, at first, may seem disorienting. This is not necessarily such a bad thing, for it can force us to take a fresh look at the world and at the relationships between places. Because Europe, North America, and Japan are all located toward the center of this map projection, it is particularly useful for illustrating two of the central themes of this book: the relationships among these prosperous regions, and the relationships between this prosperous group and the less prosperous countries of the world. On Fuller's projection, the economically peripheral countries of the world are shown as being cartographically peripheral, too.

One particular kind of map transformation that is sometimes used in small-scale thematic maps is the **cartogram.** In this kind of projection, space is transformed according to statistical factors, with the largest mapping units representing the greatest statistical values. **Figure 1.16 (a)** shows a cartogram of the world in which countries are represented as proportional to their population. This sort of transformation is par-

ticularly effective in helping to visualize relative inequalities among the world's populations. **Figure 1.16 (b)** shows a cartogram of the world in which the cost of telephone calls has been substituted for linear distance as the basis of the map. The deliberate distortion of the shapes of the continents in this sort of projection provides a dramatic way of emphasizing spatial variations.

Finally, the advent of computer graphics has made it possible for cartographers to move beyond the use of maps as two-dimensional representations of Earth's surface. Computer software that renders three-dimensional statistical data on to the flat surface of the monitor screen or a piece of paper facilitates the **visualization** of many aspects of regional geography in innovative and provocative ways (**Figure 1.17**).

Interdependence

Interdependence between places and regions can be sustained only through movement and flows. Geographers use the term *spatial interaction* as shorthand for all kinds of movement and flows that involve human activity. Freight shipments, commuting, shopping trips, migration, vacation travel, telecommunications, and electronic cash transfers are all examples of spatial interaction. In terms of regional change, one of the most important aspects of spatial interaction and interdependence is **spatial diffusion**—the way that things spread through space and over time. Disease outbreaks, technological innovations, political movements, and new musical fads all originate in specific places and subsequently spread to other places and regions

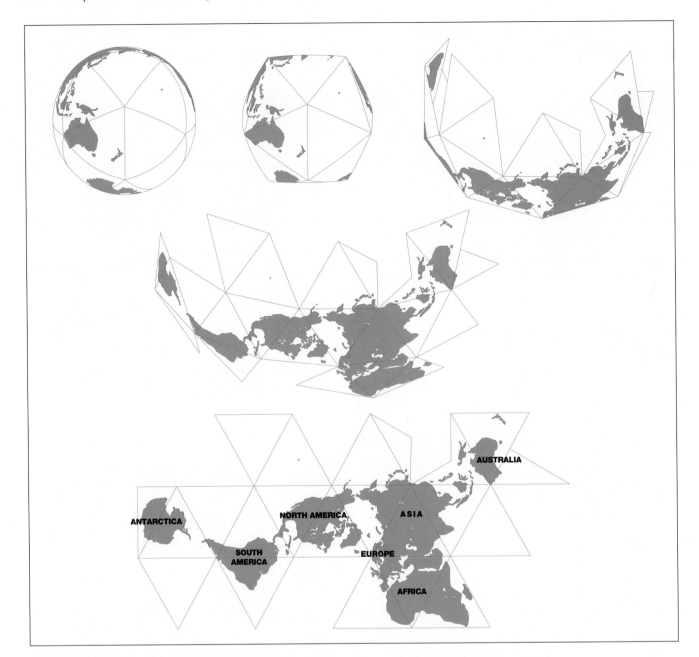

Figure 1.15   Fuller's Dymaxion projection This striking map projection was designed by Buckminster Fuller (1895–1983). As this figure shows, he achieved his objective of creating a map with the minimum of distortion to the shape of the world's major land masses by dividing the globe into triangular areas. Those areas not encompassing major land masses were cut away, allowing the remainder of the globe to be "unfolded" into a flat projection. (*Source:* Buckminster Fuller Institute and Dymaxion Map Design, Santa Barbara, CA. The word *Dymaxion* and the Fuller Projection Dymaxion™ Map design are trademarks of the Buckminster Fuller Institute, Santa Barbara, California, © 1938, 1967 & 1992. All rights reserved.)

through the process of spatial diffusion. Diffusion seldom occurs in an apparently random way, jumping unpredictably all over the map. Rather, it occurs in ways that can be predicted and modeled using statistical probability. The diffusion of a contagious disease, for example, is a function of the probability of physical contact, modified by variations in individual resistance to the disease. The diffusion of an agricultural innovation, such as genetically modified corn, is a function of the probability of information flowing between members of the farming community (itself partly a function of distance), modified by variations in

individual farmers' receptivity to innovative change and their ability to pay for the innovation.

In order for any kind of spatial interaction to occur between two places, there must be a demand in one place and a supply that matches, or *complements* it, in the other. This complementarity can occur as a result of several factors. One important factor is the variation in physical environments and resource endowments from place to place. For example, a heavy flow of vacation travel from often cool and damp Scottish cities to warm, sunny Mediterranean resorts is a largely

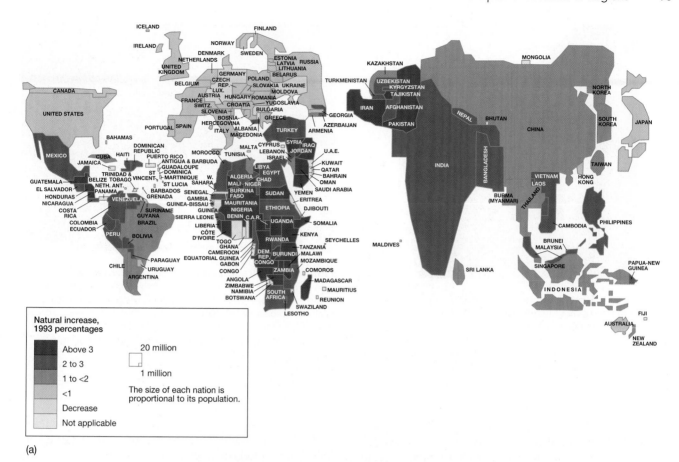

(a)

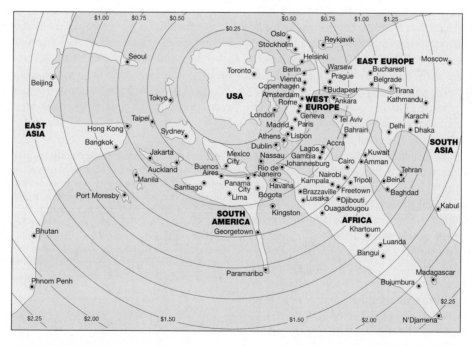

(b)

Figure 1.16   Examples of cartograms   In a cartogram, space is distorted in order to emphasize a particular attribute of places or regions. (a) In this example, the relative size of countries is based on their population rather than their area, while the designers of the cartogram have attempted to maintain the shape of each country as closely as possible in order to make the map easier to read. As this example shows, the use of population-based cartograms is very effective in demonstrating spatial inequality. (b) In this example, the cost of telephone calls has been substituted for linear distance as the basis of the map, thus deliberately distorting the shapes of the continents to dramatic effect. Countries are arranged around the United States according to the cost per minute of calls made from the United States in 1998. (*Source:* (a) M. Kidron and R. Segal (eds.), *The State of the World Atlas,* rev. 5th ed. London: Penguin Reference, 1995, pp. 28-29; (b) G. C. Staple (ed.), *TeleGeography 1999.* Washington DC: TeleGeography, 1999, p. 82.)

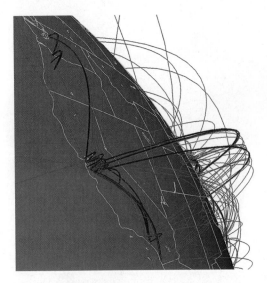

Figure 1.17   Visualization This example shows the spatial structure of the Internet MBone, the Internet's multicast backbone that is the most popular way of transmitting real-time video and audio streams. (*Source:* T. Munzner, E. Hoffman, K. Claffy, and B. Fenner, "Visualizing the Global Topology of the MBone," *Proceedings of the 1996 IEEE Symposium on Information Visualization,* 1996, 85-92, San Francisco, California.)

Figure 1.18   Trade as a result of complementarity
Japan's needs are complementary to those of Saudi Arabia. Japan's industrial economy requires vast amounts of crude oil, but Japan itself has no oil supplies. Saudi Arabia, on the other hand, is one of the world's biggest suppliers of crude oil but has to import most high-tech products and consumer goods. The result is a high degree of spatial interaction between the two countries. In 1995 Japan imported 1.7 billion barrels of crude oil, 79 percent of it from the Middle East.

a function of climatic complementarity. To take another example, the flow of crude oil from Saudi Arabia (with vast oil reserves) to Japan (with no oil reserves) is a function of complementarity in natural resource endowments (**Figure 1.18**).

A second factor that contributes to complementarity is the international division of labor that derives from the evolution of the world's economic systems. The more affluent and powerful countries of the world have sought to establish overseas suppliers for their food, raw materials, and exotic produce, allowing the more affluent countries to specialize in more profitable manufacturing and knowledge-based industries (see Chapter 2, p. 88). Over past centuries, less powerful countries have found themselves with economies that directly complement the needs of the more affluent countries. This has been partly through **colonialism,** the establishment and maintenance of political and legal domination by a **state** over a separate and alien society; partly through **imperialism,** the extension of the power of a state through direct or indirect control of the economic and political life of other territories; and partly through sheer economic dominance on the part of these more affluent and powerful states. Among the many flows resulting from this complementarity are shipments of sugar from Barbados to the United Kingdom, bananas from Costa Rica and Honduras to the United States, palm oil from Cameroon to France, automobiles from France to Algeria, school textbooks from the United Kingdom to Kenya, and investment capital from the United States to most other countries.

A third factor that contributes to complementarity is the operation of principles of specialization and economies of scale. Individual regions and countries can derive economic advantages from the efficiencies created through specializa-

tion, which allows for larger-scale operations. **Economies of scale** are cost advantages to manufacturers that accrue from high-volume production, since the average cost of production falls with increasing output (**Figure 1.19**). Among other things, fixed costs (for example, the cost of renting or buying factory space, which will be the same—fixed—whatever the level of output from the factory) can be spread over higher levels of output, so that the average cost of production falls. Economic specialization results in interdependence, which in turn contributes to patterns of spatial interaction. One exam-

Figure 1.19   Economies of scale In many manufacturing enterprises, the higher the volume of production, the lower the average cost of producing each unit. This is partly because high-volume production allows for specialization and division of labor, which can increase efficiency and hence lower costs. It is also partly because most manufacturing activities have significant fixed costs (such as product design and the cost of renting or buying factory space) that must be paid for irrespective of the volume of production, so that the larger the output, the lower the fixed cost per unit. These savings are known as *economies of scale.*

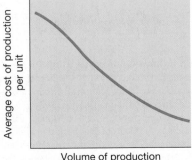

ple is the specialization of Israeli farmers in high-value fruit and vegetable crops for export to the European Union, which in return exports grains and root crops to Israel.

Spatial interdependence also varies over time as successive innovations in transport and communications technologies and successive waves of infrastructure development (canals, railways, harbor installations, roads, bridges, and so on) alter the geography of transport costs. New technologies and new or extended infrastructures have the effect of altering the costs of moving particular things between one place and another. As a result, the spatial organization of many different activities is continually changing and readjusting in order to take advantage of reduced transport costs. The consequent tendency toward a shrinking world gives rise to the concept of **time-space convergence**, the rate at which places move closer together in travel or communication time or costs. Time-space convergence results from a decrease in the friction of distance as new technologies and infrastructure improvements successively reduce travel and communication time between places. Such space-adjusting technologies have, in general, brought places "closer" together over time. Other important space-adjusting innovations include air travel and air cargo, telegraphic, telephonic, and satellite communications systems; state postal services, package delivery services, and fax machines; and modems, fiber-optic networks, and electronic-mail software.

What is most significant about the latest developments in transport and communication is that they are not only global in scope but also able to penetrate to local scales. As this is happening, some places that are distant in kilometers are becoming closer together, while some that are close in terms of absolute space are becoming more distant, in terms of their ability to reach one another electronically. Much depends on the mode of communication, the extent to which people in different places are "plugged in" to new technologies. Older wire cable can carry only small amounts of information; microwave channels are good for person-to-person communication, but depend on line-of-sight; telecommunications satellites are excellent for reaching remote areas but involve significant capital costs for users, while fiber-optic cable is excellent for areas of high population density but not feasible for remoter, rural areas.

A World of Regions

Places and regions are dynamic, with changing properties and fluid boundaries that are the product of the interplay of a wide variety of environmental and human factors. This dynamism and complexity is what makes travel so fascinating for many people. It is also what makes places and regions so important in shaping people's lives and in influencing the pace and direction of change.

Places and regions exert a strong influence, for better or worse, on people's physical well-being, their opportunities, and their lifestyle choices. Places and regions also contribute to peoples' collective memory and can become powerfully symbolic. Think, for example, of the evocative power for most Ameri-

Figure 1.20   Ordinary landscapes   Some ordinary cityscapes are powerfully symbolic of particular kinds of places. The New England village, the Main Street of Middle America, and California suburbia are in this category, so much so that they have been taken as being symbolic of the United States itself, part of the "iconography of nationhood," the symbolic landscapes that give the country a sense of identity, both at home and abroad. The New England townscape symbolizes for many a particular kind of community: intimate, family-centered, industrious, thrifty, democratic, and morally aware. Shown here is Stowe, Vermont.

cans of a region like rural New England. The stereotype of small-town New England (**Figure 1.20**), "marked by a steeple rising gracefully above a white wooden church which faces on a village green around which are arrayed large white clapboard houses which, like the church, show a simple elegance in form and trim" is widely taken to symbolize not just a certain type of regional architecture but the best that Americans have known "of an intimate, family-centered, Godfearing, morally conscious, industrious, thrifty, democratic community."[4] Meanwhile, for many people there are also ordinary geographic settings with special meaning: a childhood neighborhood, an ancestral homeland, or a family vacation area. This layering of meanings reflects the way that places are socially constructed—given different meanings by different groups for different purposes. Places exist, and are constructed, from a subjective point of view by their inhabitants. At the same time, though, the same places will likely be constructed rather differently by outsiders. Your own neighborhood, for example, centered on yourself and your home, is probably heavily laden with personal meaning and sentiment. But your neighborhood may well be viewed very differently, and perhaps unsympathetically, from an outsider's perspective. This distinction is useful in pointing to the importance of understanding places and regions from the point of view of the insider—the person who normally lives in and uses a particular place—as well as from the point of view of outsiders (including geographers).

Finally, places and regions can be crucibles for innovation, change, and resistance. The unique characteristics of specific places and regions can provide the preconditions for

[4]D. Meinig, "Symbolic Landscapes." In D. W. Meinig et al., eds., *The Interpretation of Ordinary Landscapes*. New York: Oxford University Press, 1979, 165.

new agricultural practices (for example, the development of seed agriculture and the use of plow and draft animals that sparked the first agricultural revolution in the Middle East in prehistoric times), for new modes of economic organization (for example, the Industrial Revolution that began in the English Midlands in the late eighteenth century), for new cultural practices (the hip-hop movement that began in New York's Bronx neighborhoods, for example), and for new lifestyles (for example, the "hippie" lifestyle that began in San Francisco in the late 1960s). It is in specific locales that important events happen, and it is from them that significant changes spread.

Nevertheless, the influence of places and regions is by no means limited to the occasional innovative change. Because of their distinctive characteristics, places and regions always modify and sometimes resist the imprint of even the broadest economic, cultural, and political trends. Consider, for example, the way that a global cultural trend—rock 'n' roll—interacted with an already complex Jamaican music to produce reggae, while in Iran and North Korea, rock 'n' roll has been resisted by the authorities, with the result that it has acquired an altogether different kind of value and meaning for the citizens of those countries.

The Regional Approach

At the heart of geographers' concern with understanding how combinations of environmental and human factors produce regions with distinctive landscapes and cultural attributes is the belief that this "regional approach" is one of the best ways of organizing knowledge about the world. There are, however, different methods of regional analysis and different ways of identifying and defining regions. Place making and regional differentiation are products of many factors, while the outcomes of place making and regional differentiation require careful consideration in terms of the attributes of landscape and sense of place.

Regionalization   Regionalization is the geographer's equivalent of scientific classification, with individual places or areal units being the objects of classification. The purpose of regionalization is to identify "regions" of one kind or another. There are several ways in which individual areal units can be assigned to classes (regions). One is that of *logical division,* or "classification from above." This involves partitioning a universal set of areal units into successively larger numbers of classes, using more specific criteria at every stage. Thus a world regional classification of national states might be achieved by first differentiating between rich and poor countries, then dividing both rich and poor countries into those countries that have a trade surplus and those that have a deficit, and so on. A second way in which individual areal units can be assigned to classes (regions) is that of *grouping,* or "classification from below." This involves searching for regularities or significant relationships among areal units and grouping them together in successively smaller numbers of classes, using a broader measure of similarity at each stage.

An implicit assumption in this type of classification of areal units into regions is that each unit is homogeneous with respect to the attribute or attributes under consideration. Where this assumption holds true, the result of regional classification is a set of formal regions. **Formal regions** are groups of areal units that have a high degree of homogeneity in terms of particular distinguishing features (such as religious adherence or household income). Few phenomena, however, exhibit such homogeneity over large areal units. For this reason, geographers also recognize **functional regions** (sometimes referred to as *nodal regions*)—regions that are defined and classified by patterns of spatial interaction or spatial organization. Functional regions are those within which, while there may be some variability in certain attributes (again, for example, religious adherence and income), there is an overall coherence to the structure and dynamics of economic, political, and social organization. The concept of functional regions allows us to recognize that the coherence and distinctive characteristics of a region are often, in reality, stronger in some places than in others. This point is illustrated by geographer Donald Meinig's *core-domain-sphere* model, which he set out in his classic essay on the Mormon region of the United States.[5] In the core of a region, the region's distinctive attributes are very clear; in the domain, they are dominant, but not to the point of exclusivity; in the sphere, they are present but not dominant.

In addition to these questions of classification, the art and science of the regional approach must consider questions of *geographic scale,* for we can (and must) see the world as, simultaneously, a mosaic of small-scale regions that exist within successively larger spatial frameworks. These frameworks are closely related, as both cause and effect, to the formal *boundaries* that have evolved (and that are continually challenged and amended) under national and international law. Finally, people's own conceptions of place, region, and identity may resonate with or against these boundaries to generate strong feelings of *regionalism* and *sectionalism* that feed back into the processes of place making and regional *differentiation*. **Regionalism** is a term used to describe situations where different religious or ethnic groups with distinctive identities coexist within the same state boundaries, often concentrated within a particular region and sharing strong feelings of collective identity. If such feelings develop into an extreme devotion to regional interests and customs, the condition is known as **sectionalism.**

Landscape   Geographers think of landscape as a comprehensive product of human action such that every landscape is a complex repository of society. It is a collection of evidence about our character and experience, our struggles and triumphs as humans. To understand better the meaning of landscape, geographers have developed different categories of landscape types based on the elements contained within them. **Ordinary landscapes** (or vernacular landscapes, as they are

---

[5]D. Meinig, "The Mormon Culture Region: Strategies and Patterns in the Geography of the American West," *Annals, Association of American Geographers,* 55 (1965), 191–220.

sometimes called) are the everyday landscapes that people create in the course of their lives together. From crowded city centers to leafy suburbs and quiet rural villages, these are landscapes that are lived in and changed and that in turn influence and change the perceptions, values, and behaviors of the people who live and work in them.

Symbolic landscapes, by contrast, stand as representations of particular values or aspirations that the builders and financiers of those landscapes want to impart to a larger public. For example, the neoclassical architecture of the buildings of the federal government in Washington, DC, along with the streets, parks, and monuments of the capital, constitute a symbolic landscape intended to communicate a sense of power, but also of democracy, in its imitation of the Greek city-state.

Geographers now recognize that there are many layers of meaning embodied in the landscape, meanings that can be expressed and understood differently by different social groups at different times. Landscapes reflect the lives of ordinary people as well as the more powerful, and they reflect their dreams and ideas as well as their material lives. The messages embedded in landscapes can be read as signs about values, beliefs, and practices, though not every reader will take the same message from a particular landscape (just as people may differ in their interpretation of a passage from a book). In short, landscapes both produce and communicate meaning, and one of our tasks as geographers is to interpret those meanings.

Sense of Place   The experience of everyday routines in familiar settings allows people to derive a pool of shared meanings. People become familiar with one another's vocabulary, speech patterns, gestures, and humor and with shared experiences of the physical environment such as vegetation and climate. Often, this carries over into people's attitudes and feelings about themselves and their locality. When this happens, the result is a self-conscious sense of place. The concept of a **sense of place** refers to the feelings evoked among people as a result of the experiences and memories that they associate with a place and to the symbolism they attach to that place. It can also refer to the character of a place as seen by outsiders: its unique or distinctive physical characteristics and/or its inhabitants.

For *insiders*, this sense of place develops through shared dress codes, speech patterns, public comportment, and so on. It also develops through familiarity with the history and symbolism of particular elements of the physical environment— a local mountain or lake, for example, the birthplace of someone notable, the location of some particularly well-known event, or the expression of community identity through community art (**Figure 1.21**). Sometimes, it is deliberately fostered by the construction of symbolic structures such as monuments and statues. Often, it is a natural outcome of people's familiarity with one another and their surroundings. Because of this consequent sense of place, insiders feel at home and "in place."

For *outsiders*, a sense of place can be evoked only if local landmarks, ways of life, and so on are distinctive enough to evoke a significant common meaning for people who have no direct experience of them. Central London, for example, is a setting that carries a strong sense of place to many outsiders, who can feel a sense of familiarity with the riverside panoramas, busy streets, and distinctive monuments and historic buildings that together symbolize the heart of the city.

Scale

Different aspects of regional differentiation are understood best, and most effectively analyzed, at different spatial scales. At the same time, these different aspects are interrelated and interdependent, so that geographers have to be able to relate things at one scale to things at another. The whole question can be problematic if we do not clarify what *scale* means.

It is useful to think of geographical scales as being materializations of real-world processes, not simply different levels of abstraction or convenient devices for zooming in and out from the global context to the detail of local settings. In this sense, scale represents a tangible partitioning of space within which different processes (economic, social, political, etc.) are played out. This partitioning, in turn, often consolidates the importance of particular patterns of geographical organization, at least until some major change occurs in the relationships among the forces of nature, culture, and human agency. The Industrial Revolution, for example, changed not only the character of economic development (from agrarian to manufacturing) but also the scales at which industrial production and consumption were organized (from local to national and international).

Figure 1.21 Community art Community art can provide an important element in the creation of a sense of place for members of local communities. This example is from the Mission district of San Francisco.

Figure 1.22 Spatial scales There are many scales at which geographic phenomena may be identified, analyzed, and understood. This diagram shows some of the principal scales that are commonly the focus of geographic research.

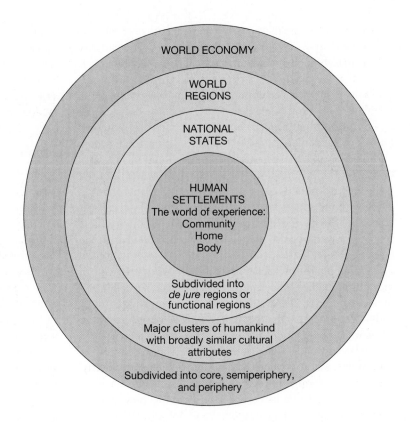

WORLD ECONOMY

WORLD REGIONS

NATIONAL STATES

HUMAN SETTLEMENTS
The world of experience:
Community
Home
Body

Subdivided into
de jure regions or
functional regions

Major clusters of humankind
with broadly similar cultural
attributes

Subdivided into core, semiperiphery,
and periphery

World Regions   At any particular moment we can thus identify a sequence of specific scales that represent significant confluences of geographical processes. In today's world, the large scale is represented by international regions, or world regions, large but relatively homogeneous territories with distinctive economic, cultural, and demographic characteristics (**Figure 1.22**). **World regions** are large-scale geographic divisions based on continental and physiographic settings that contain major clusters of humankind with broadly similar cultural attributes. Examples would be Europe, Latin America, and South Asia, as shown in Figure 1.1. These regions are constructed, unraveled, and reconstructed as the realities of natural resources and technologies form a framework of opportunities and constraints to which particular cultures and societies respond.

National States   Superimposed on these regions, sometimes with only an approximate fit, are the formal, *de jure* territories of national states. *De jure* simply means "legally recognized." Territories delimited by formal, legally recognized boundaries—national states, provinces, states, counties, municipalities, special districts, and so on—are known as *de jure* spaces or regions. Because of the inherent power of national governments, especially in relation to the flows of goods, money, and information that underpin "reality," national states represent a geographic scale that is often very important. National states tend to be established to "fit" economic reality as closely as possible at the time of their foundation. Once their boundaries are set, however, principles of national **sovereignty** mean that these boundaries tend increasingly to become regarded by their inhabitants as somehow natural or immutable.

National political boundaries are not, though, fixed and unchanging. When economic circumstances change, national states may feel the need to adjust their boundaries or seek other means of accommodating economic reality, such as joining supranational organizations. **Supranational organizations** are collections of individual states with a common goal that may be economic and/or political in nature and that diminish, to some extent, individual state sovereignty in favor of the group interests of the membership. Examples of supranational organizations include the European Union (EU), the North American Free Trade Agreement (NAFTA), and the Association of South East Asian Nations (ASEAN).

Within most national states and all international regions are smaller, functional regions. This geographical scale is constructed around specific resources and industries, with their networks of producers, suppliers, distributors, and ancillary activities; and their associated social, cultural, and political identities. These are the classic functional regions of traditional regional geography, and examples are the American Corn-Soybean Belt, the Argentine pampas, the Scottish coalfields, Japan's Pacific Corridor, and the Urals manufacturing region in Russia.

Community, Home, and Body   The realm of experience, for most people, is encompassed by the scale of human settlements. This is the scale that is constructed around the way people's lives are organized through their work, consumption, and recreation. It is also roughly coincident with another important scale of *de jure* territories: local municipalities that provide the framework for public administration and the means for "collective consumption" of certain goods and services (public transport, education, public housing, recreational amenities, and so forth).

But within the realm of experience there are other significant scales (Figure 1.22). Of these, the scale of community is the most important but also the most difficult to pin down. It is the scale of social interaction—of personal relationships and daily routine. It is a scale that depends a great deal on the economic, social, and cultural attributes of local populations. Much more sharply defined is the scale of the home, which is an important geographic site insofar as it constitutes the physical setting for the structure and dynamics of family and household. It also reflects, in its own spatial organization, the differential status accorded to men and women, and to the young and the elderly.

Finally, the body and the self represent the most detailed scale with which geographers have to deal. The body is of interest to geographers because it represents the scale at which difference and diversity are ultimately defined: not only through physical attributes (for example, racial characteristics) but also through the socially constructed attributes of the body, such as norms of personal space, preferred body styles, and acceptable uses of our bodies. Particularly important is the way that, in many cultures, the bodily scale is seen as less relevant to males. That is, males are regarded as being able to "transcend" the body, while females are regarded as being "trapped" by bodily attributes (for example, sensuality and nurturance) and functions (for example, menstruation, pregnancy, and childbirth). The result is that differential geographies are created and experienced by men and women—women's bodies and women's domestic spaces, but men's world(s). The self is of interest because it represents the operational scale for cognition, perception, imagination, free will, and individual behavior. The self has become an important scale of analysis for geographers because of the need to understand the interrelationships between nature, culture, and individual human agency in shaping places and regions.

Perhaps the most important conclusion that we can draw from this examination of scale is that, while certain phenomena can be identified and understood best at specific spatial scales, the reality of regional geographies is that they are very fluid phenomena, constantly being constructed, reinforced, undermined, and rebuilt. Similarly, although certain scales represent materializations of powerful real-world processes, the real world has to be understood, ultimately, as the product of interdependent phenomena at a variety of spatial scales. It follows that every world region should be seen in its diversity, comprising metropolitan cores and rural peripheries, each of which is part of a broader framework of interdependent places and regions within the global economy.

Boundaries and Frontiers

Boundaries are important phenomena because they allow claims on space to be defined and enforced and because they allow conflict and competition to be managed and channeled. The creation of boundaries is, therefore, an important

Figure 1.23   Boundary between the United States and Canada Most boundaries are established in order to regulate and control specific sets of people and resources within a given territory. Such boundaries need to be clearly identified but do not necessarily need to be fortified. This photograph shows part of the U.S./Canadian border, a good example of an "inclusionary" boundary.

element in region building and place making. Formal boundaries are normally inclusionary—that is, they are constructed in order to regulate and control specific sets of people and resources (**Figure 1.23**). Encompassed within a clearly defined area, all sorts of activity can be controlled and regulated—everything, in fact, from birth to death. The delimited area over which a state exercises control and that is recognized by other states is called **territory.** Such an area may include both land and water.

Formal boundaries can also be exclusionary, however. Again, this often fulfills the function of controlling people and resources. National boundaries, for example, can be used to control the flow of immigrants or the flow of imported goods (**Figure 1.24**). Municipal boundaries and land-use zoning boundaries can be used to regulate access to upscale residential neighborhoods, field boundaries can be used to regulate access to pasture, and so on (**Figure 1.25**).

The key point is that, once established, formal boundaries tend to reinforce regional differentiation. This is partly because of the outcomes of the operation of different sets of rules, both formal and informal, that apply within different regions and territories. It is also partly because boundaries often restrict contact between people and so foster the development of stereotypes of "others." This restricted contact, in turn, reinforces the role of boundaries in regulating and controlling conflict and competition between territorial groups.

Boundaries can be established in many different ways, however, and with differing degrees of permeability. At one extreme are informal, implied boundaries that are set by markers and symbols but never delineated on maps or set down in legal documents. Good examples are the "turf" of a city gang, the "territory" of an organized crime "family," and

Figure 1.24   Boundary between the United States and Mexico Some boundaries are designed to be exclusionary. The United States/Mexico border provides striking contrast to the one that separates the United States and Canada. The former is heavily patrolled and lined with barbed-wire chain-link fences along the highly urbanized parts. Aerial surveillance is also extensive along the U.S./Mexico border. In an effort to stem the flow of illegal immigration from Mexico, the U.S. government increased the Border Patrol from 5176 in 1996 to 10,000 in 2000. This photo shows the U.S./Mexico border along the Tijuana River estuary, with Southern California on the left.

Figure 1.25   Stone wall in Derbyshire, England Some boundaries, such as these stone walls in Derbyshire, England, contribute significantly to the character of places and regions. In a world of modernized agriculture and large-scale agribusiness, traditional field boundaries like these have become potent symbols of regional identity, visible links with past landscapes and past ways of life.

the range of a pastoral tribe. At the other extreme are formal boundaries established in international law, delimited on maps, demarcated on the ground, fortified, and aggressively defended—not only against the movement of people but also of goods, money, and even ideas. An extreme example of this sort of boundary is the one between North and South Korea (**Figure 1.26**). In between are formal boundaries that have some degree of permeability. The boundaries between the states of the European Union, for example, have become quite permeable, since people and goods from member states can now move freely between them, with no customs or passport controls.

Impermeability does not necessarily mean immutability, however. The boundary between East and West Germany, part of the "Iron Curtain" for more than 40 years (**Figure 1.27**), was as aggressively defended as the present boundary between North and South Korea, yet it is now dissolved, having been made obsolete by the unification of Germany. Similarly, the boundaries of the former Soviet Union have been entirely redrawn since 1989, allowing states such as Lithuania and Estonia to reappear.

Frontier Regions   Frontier regions occur where boundaries are very weakly developed or where population densities are especially low. They involve zones of underdeveloped human settlement, areas that are distinctive for their marginality rather than for their belonging. In the nineteenth century, for example, some vast frontier regions still existed, such as Australia, the American West, the Canadian North, and sub-Saharan

Figure 1.26   Border between North and South Korea Some boundaries are virtually impermeable. The border between North and South Korea is highly fortified and heavily patrolled. It was established at the conclusion of the Korean War (1950–53) between two states that still contest one another's territory.

Africa—major geographic realms that had not yet been conquered, understood, and settled by the states claiming jurisdiction over them (though there were, of course, indigenous groups with no apparatus of formal state boundaries). All of these frontier regions are now subject to formal occupancy at

Figure 1.27   Former border between East and West Germany The boundary between East and West Germany was virtually impermeable for more than 40 years. This photograph shows the Berlin Wall and "no-man's-land" looking east from West Berlin toward Potsdamer Platz, which was a busy hub in the days before the Second World War. In 1989, just after this photograph was taken, Germany was reunified, the wall was demolished, and the boundary erased.

various spatial scales (that is, from codified individual land ownership to local and national governmental jurisdiction). Only Antarctica, virtually unsettled, exists today as a frontier region in this sense.

There remain, nevertheless, many frontier regions that are still marginal (in that they have not been fully settled or do not have a recognized economic potential), even though national political boundaries and sovereignty are clear-cut. Examples are the Amazon River basin (see Chapter 8, p. 413), and the Sahelian region of Africa (Chapter 6, p. 306). Such regions often span national boundaries simply because they are inhospitable, inaccessible, and, at least at the moment, economically unimportant. Political boundaries sometimes get drawn through them because they represent the line of least territorial resistance.

At the local level, many examples of frontier regions exist. Although the residents of most towns and cities recognize a series of distinctive districts and neighborhoods, these are often separated by zones or spaces that are frontierlike. Not fully integrated into the territorial realm of any one sociocultural group, these "frontier" spaces are often transitional, with a relatively rapidly changing pattern of land use and an equally rapidly changing profile of residents.

Boundary Formation   Generally speaking, formal boundaries tend first to follow natural barriers such as mountain ranges, lakes, oceans, and, sometimes, rivers. A good example of countries with an important mountain-range boundary is that of India with China and Nepal, where the Himalayas act as a formidable barrier. Chile, though, provides the ultimate example: a cartographic freak, restricted by the Andes to a very long and relatively thin strip along the Pacific coast. An example of countries with a boundary partly shaped by major lakes is that of Canada and the United States (along the Great Lakes). Countries with boundaries formed by rivers include Laos and Thailand (the Mekong); Zambia and Zimbabwe (the Zambezi); and the United States and Mexico (Rio Grande, Rio Bravo). It should be noted, though, that while major rivers can form natural physical barriers, river basins often form the basis of functional economic regions. As a result, river basins are more often encompassed by formal boundaries than split along the course of the river itself.

Where no natural features occur, and usually before any heavy human settlement, formal boundaries tend to be fixed along the easiest and most practical cartographic device: a straight line. Examples include the boundaries between Egypt and Libya, between Syria and Iraq, and the western part of the boundary between the United States and Canada. Straight-line boundaries are also characteristic of formal boundaries that are established through colonization, which is the outcome of a particular form of territoriality. The reason, once again, is practicality. Straight lines are easy to survey and even easier to delimit on maps of territory that remains to be fully charted, claimed, and settled. Straight-line boundaries were established, for example, in many parts of Africa during European colonization in the nineteenth century.

In detail, however, formal boundaries often detour from straight lines and natural barriers in order to accommodate special needs and claims. Colombia's border, for instance, was established to contain the source of the River Orinoco; the border of the Democratic Republic of Congo (formerly Zaïre) was established to provide a corridor of access to the Atlantic Ocean; and Sudan's border detours in order to include a settlement, Wadi Halfa.

After primary divisions have been established, internal boundaries tend to evolve as smaller, secondary territories are demarcated. In general, the higher the population density, the smaller these secondary units tend to be. Their configuration tends to follow the same generalizations as for larger units, following physical features; accommodating special needs; and following straight lines where there are no appropriate natural features or where colonization has made straight lines expedient. This last reason, for example, explains the generally rectilinear pattern of administrative boundaries in the United States to the west of the Mississippi.

Historically, the world has evolved from a loose patchwork of territories (with few formally defined or delimited boundaries) to nested hierarchies (**Figure 1.28**) and overlapping systems of *de jure* territories. These *de jure* territories are often used as the basic units of analysis in regional geography, largely because they are both convenient and significant units of analysis. They are often, in fact, the only areal units for which reliable data are available. They are also important units of analysis in their own right, because of their importance as units of governance or administration. A lot of regional analysis and nearly all attempts at regionalization, therefore, are based on a framework of *de jure* spaces.

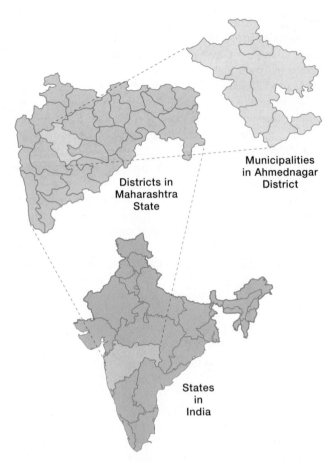

Figure 1.28   Nested hierarchy of *de jure* territories *De jure* territories are constructed at various spatial scales, depending on their origin and function. Administrative and governmental territories are often "nested," with one set of territories fitting within the larger framework of another, as in this example of states, districts, and municipalities in India.

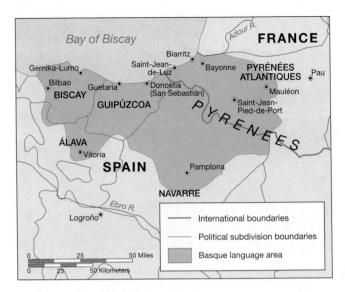

Figure 1.29   The Basque region The Basque regions of France and Spain have a strong sense of regional identity, to the point where there are strong grassroots movements in favor of political autonomy.

Regionalism and Sectionalism

Regionalism often involves ethnic groups whose aims include autonomy from a national state and the development of their own political power. A good example is Basque regionalism. Basque regionalism, or "Basquism," represents a regional movement that has roots back to industrialization and modernization beginning at the turn of the nineteenth century. The Basque people of northeastern Spain and the southern part of Aquitaine in southwestern France (**Figure 1.29**) feared that cultural forces accompanying industrialization would undermine Basque preindustrial traditions. Because of this, the Basque provinces of Spain and France have sought autonomy from those states for most of the twentieth century. Since the 1950s, agitation for political independence has occurred—especially for the Basques in Spain—through terrorist acts. For more than 25 years, the French, Spanish, and more recently the Basque regional police have attempted to undermine the Basque Homeland and Freedom movement through arrests and imprisonments. Not even the Spanish move to parliamentary democracy and the granting of autonomy to the Basque provinces, however, could slake the thirst for self-

determination among the Basques in Spain (**Figure 1.30**). Meanwhile, on the French side of the Pyrenees, although a Basque separatist movement does exist, it is neither as violent nor as active as the movement in Spain.

In certain cases, enclaves of ethnic minorities are claimed by the government of a country other than the one in which they reside. Such is the case, for example, of Serbian enclaves in Croatia, claimed by nationalist Serbs (see Chapter 3, p. 119). The assertion by the government of a country that a minority living outside its formal borders belongs to it historically and culturally is known as **irredentism**. In some circumstances, as with Serbia's claims on Serbian enclaves in Croatia in the early 1990s, irredentism can lead to war.

We need only look at the long list of territorially based conflicts that have emerged in the post–Cold War world to realize the extent to which territorially based ethnicity remains a potent force in regional geography. For example, the Kurds continue to fight for their own state separate from Turkey and Iraq (see Chapter 5, p. 235). A significant proportion of Quebec's French-speaking population, already accorded substantial autonomy, persists in advocating complete independence from Canada (see Chapter 7, p. 334). In the most recent electoral vote on the issue, the separatists were only very narrowly defeated. Consider also the former Yugoslavia, the geography of which has fractured along the lines of ethnicity (see Chapter 3, p. 120).

Not to be confused with regionalism or irredentism is the concept of sectionalism, an extreme devotion to regional interests and customs. Sectionalism has been identified as an overarching explanation for the U.S. Civil War. It was an attachment to the institution of slavery and the political and economic way of life that slavery enabled that prompted the southern states to secede from the Union. The Civil War was fought to ensure that sectional interests would not take priority over the unity of the

Figure 1.30   Basque independence poster This independence poster is plastered over the door of a shop in Donostia (San Sebastián) in one of the Basque provinces of Spain. The poster is a sign of the passionate opposition the Basques have adopted toward the central government in Madrid. Acts of terrorism continue to occur throughout Spain as the Basques maintain their desire for independence. What is most interesting about the sign is that it is written in neither Castillian Spanish (the national language) nor Euskadi (the Basque language). It appears as if this and declarations like it are directed at the tourists who have made Donostia a popular destination.

Places and Regions in a Changing World

Today, in a world that is experiencing rapid changes in economic, cultural, and political life, geographic knowledge is especially important and useful. In a fast-changing world, when our fortunes and our ideas are increasingly bound up with those of other peoples in other places, the study of geography provides an understanding of the crucial interdependencies that underpin everyone's life. One of the central themes throughout this book will be the interdependence of people, places, and regions.

The Interdependence of Regions

Places and regions have an importance of their own, yet at the same time they are interdependent, each filling specialized roles in complex and ever-changing geographies. Consider, for example, the way that the New York region operates as a specialized global center of corporate management, business, and financial services while relying on thousands of other places to satisfy its needs. For labor it draws on analysts and managers from the country's business schools; blue- and pink-collar workers from neighboring communities; and skilled professional immigrants from around the world. For food it draws on fruits and vegetables from Florida, dairy produce from upstate New York, specialty foods from Europe, the Caribbean, and Asia. For energy it draws on coal from southwest Virginia to fuel its power stations and on oil from the Middle East to run its transportation systems. For consumer goods it draws on specialized manufacturing settings all over the world.

This interdependence means that individual regions are tied in to wider processes of change that are reflected in broader geographical patterns. New York's attraction for business-school graduates, for example, is reflected in the overall pattern

whole; that is, that states' rights would not undermine the power of the federal government. Although the Civil War was waged around the real issue of permitting or prohibiting slavery, it was also fought at another level, a level that dealt with issues of the power of the state. As **Figure 1.31** shows, the election of Abraham Lincoln to the presidency in 1860 reflected the sectionalism that dominated the country: He received no support from slave states.

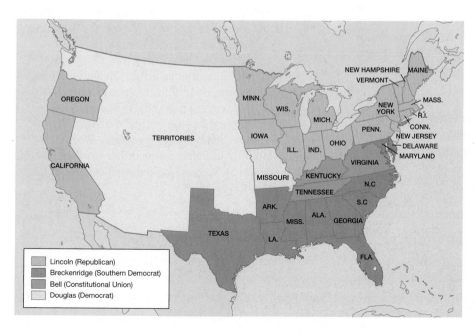

Lincoln (Republican)
Breckenridge (Southern Democrat)
Bell (Constitutional Union)
Douglas (Democrat)

Figure 1.31   The 1860 presidential election The U.S. presidential election of 1860 graphically illustrates the role of sectionalism in determining who receives votes from which geographical regions. Here, in a four-way race, we see that Abraham Lincoln failed to win the support of any of the slave states. (*Source: Presidential Elections Since 1789*, 4th ed. Washington, DC: Congressional Quarterly, Inc., 1987.)

of migration flows that, cumulatively, affects the size and composition of labor markets around the country: New York's gain is somewhere else's loss. An important issue for regional geographers—and a central theme of this book—is to recognize these wider processes and broad geographical patterns without losing sight of the individuality and uniqueness of specific places and regions.

This means that we have to recognize another kind of interdependence: the interdependence that exists *between different geographic scales*. In today's world, some of the most important aspects of the interdependence between geographic scales are provided by the relationships between the *global* and the *regional* scales. New York again can illustrate both ends of this relationship. In New York's stock exchanges and financial markets, brokers and clients must, in their own interest, take a global view of things. Their collective decisions influence stock prices, currency rates, and interest rates around the world, and these decisions often have very direct outcomes at the local level around the world. Factories in certain localities may be closed and workers laid off because changed currency rates make their product too expensive to export successfully; elsewhere, new jobs may be created because the same change in currency rates puts a different local economy at an advantage within the global marketplace. On the other hand, local events can reverberate through New York's stock exchanges and financial markets with global effects. Political instability in a region that produces a key commodity, for instance, can result in changes in global pricing. A striking example of this was provided in August 1990. Within 24 hours of Iraq's invasion of Kuwait, a major oil-producing country, gasoline prices in Europe and North America had risen by 10 percent. By the time United Nations forces interceded in January 1991, gasoline prices had increased by 36 percent, and the stock prices of many companies—especially those dependent on high inputs of oil or gasoline—had fallen significantly.

One of the most important tenets of regional geography is that places and regions are not just distinctive outcomes of geographical processes; they are part of the processes themselves. Regions are dynamic phenomena. They are created by people responding to the opportunities and constraints presented by their environments. As people live and work in particular geographic settings, they gradually impose themselves on their environment, modifying and adjusting it to suit their needs and express their values. At the same time, people gradually accommodate both to their physical environment and to the people around them. There is thus a *continuous two-way process*, in which people create and modify places while at the same time being influenced by the settings in which they live and work (**Figure 1.32**).

It is often useful to think of places and regions as representing the cumulative legacy of successive periods of change. For example, the present-day downtown of Edinburgh, Scotland, embodies elements of medieval, Georgian, Victorian, and modern urban fabric, while the regional landscape of Tuscany, in Italy, carries elements of Roman, medieval, Renaissance, and Modern development (**Figure 1.33**).

Following this approach, geographers look for superimposed layers of development. We can show how some patterns and relationships last, while others are modified or obliterated. We can show how different regions bear the imprint of different kinds of change, perhaps in different sequences, and with different outcomes.

Processes of geographic change are constantly modifying and reshaping places and regions, and the inhabitants of places and regions are constantly coping with change. It is important for geographers to be sensitive to this kind of interdependence without falling into the trap of overgeneralization or losing sight of the diversity and variety that constitute the heart of regional geography. It is equally important not to fall into the trap of treating places or regions as separate entities, the focus of study in and of themselves.

Globalization

Another important theme of this book will be globalization. **Globalization** involves the increasing interconnectedness of different parts of the world through common processes of economic, environmental, political, and cultural change. A world economy has been in existence for several centuries, and with it there has developed a comprehensive framework of sovereign national states and an international system of production and exchange. This system has been reorganized several times. Each time it has been reorganized, there have been major changes not only in world geography but also in the character and fortunes of individual regions.

The most recent round of reorganization has created a highly interdependent world. The World Bank has noted that "These are revolutionary times in the global economy."[6] A study by the Bank has shown how globalization has affected the lives of four very different people in very different places: a Vietnamese peasant, a Vietnamese city dweller, a Vietnamese immigrant to France, and a French garment worker.

> Duong is a Vietnamese peasant farmer who struggles to feed his family. He earns the equivalent of $10 a week for 38 hours of work in the rice fields, but he works full-time only six months of the year—during the off-season he can earn very little. His wife and four children work with him in the fields, but the family can afford to send only the two youngest to school. Duong's 11-year-old daughter stays at home to help with housework, while his 13-year-old son works as a street trader in town. By any standard Duong's family is living in poverty. Workers like Duong, laboring in family farms in low- and middle-income countries, account for about 40 percent of the world's labor force.

> Hoa is a young Vietnamese city dweller experiencing relative affluence for the first time. In Ho Chi Minh City she earns the equivalent of $30 a week working 48 hours in a garment factory—a joint venture with a French firm.

[6]World Bank, *World Development Report,* Washington, DC: The World Bank, 1995, p. 1.

Grape picker with chardonnay grapes to be used in making champagne.

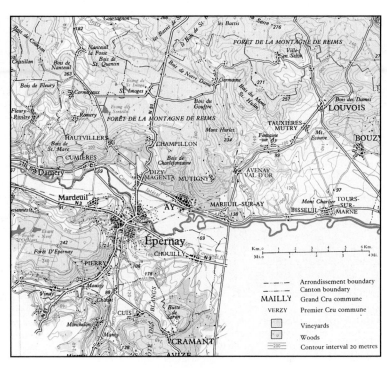

Champagne landscape, near Epernay.

It is man who reveals a country's individuality by moulding it to his own use. He establishes a connection between unrelated features, substituting for the random effects of local circumstances a systematic cooperation of forces. Only then does a country acquire a specific character, differentiating it from others, till at length it becomes, as it were, a medal struck in the likeness of a people.

Paul Vidal de la Blache
Tableau de la Geographie de la France,1903, p. viii

Figure 1.32   Place making People develop patterns of living that are attuned to the opportunities and constraints of the local physical environment. When this happens, distinctive regional landscapes are produced. These photographs and the map extract show part of Champagne, in northeastern France. In Champagne there has been a long interplay between humans and the natural environment. Champagne has been an agricultural society from the Middle Ages, its chalky soils and its temperate climate being particularly suited to viticulture, or vine-growing. The result was a distinctive landscape of small winemaking settlements surrounded by rolling slopes of carefully tended vineyards. (*Source:* Map: H. Johnson, *World Atlas of Wine*. London: Mitchell Beazley, 1971, p. 107.)

She works hard for her living and spends many hours looking after her three children as well; her husband works as a janitor. But Hoa's family has several times the standard of living of Duong's and, by Vietnamese standards, is relatively well off. There is every expectation that she and her children will continue to have a vastly better standard of living than her parents had. Wage employees such as Hoa, working in the formal sector in low- and middle-income countries, make up about 20 percent of the global labor force.

Françoise is an immigrant in France of Vietnamese origin who works long hours as a waitress to make ends meet. She takes home the equivalent of $220 a week, after taxes and including tips, for 50 hours work. By French standards she is poor. Legally, Françoise is a ca-

sual worker and has no job security, but she is much better off in France than she would have been in Vietnam. Her wage is almost eight times that earned by Hoa in Ho Chi Minh City. Françoise and other service workers in high-income countries account for about 9 percent of the global workforce.

Jean-Paul is a 50-year-old Frenchman whose employment prospects look bleak. For ten years he has worked in a garment factory in Toulouse, taking home the equivalent of $400 a week—twelve times the average in Vietnam's garment industry. But next month he will lose his job when the factory closes. Unemployment benefits will partly shield him from the shock, but his chances of matching his old salary in a new job are slim. Frenchmen of Jean-Paul's age who lose their jobs are likely to

Figure 1.34   The globalization of manufacturing industry
The globalization of the world economy has been made possible by the emergence of commercial corporations that are transnational in scope and that can take advantage of modern production technologies and computer-based information systems in order to keep track of materials and parts, inventories of finished products, and consumer demand.

Figure 1.33   Tuscany The landscapes of Tuscany are a cumulative record of changing economic, cultural, political, and environmental conditions. Shown here is the ancient hill town of San Gimignano, together with the rural landscape around the town.

stay unemployed for more than a year, and Jean-Paul is encouraging his son to work hard in school so he can go to college and study computer programming. Workers in industry in high-income countries, such as Jean-Paul, make up just 4 percent of the world's labor force.

These four families—two living in Vietnam, two in France—have vastly different standards of living and expectations for the future. Employment and wage prospects in Toulouse and Ho Chi Minh City are worlds apart, even when incomes are adjusted for differences in the cost of living. Françoise's poverty wage would clearly buy Hoa a vastly more affluent lifestyle. Much of the world's workforce, like Duong, works outside the wage sector, on family farms and in casual jobs, generally earning even lower incomes. But the lives of all workers in different parts of the world are increasingly intertwined. French consumers buy the product of Hoa's labor, and Jean-Paul believes it is Hoa's low wages that are taking his job, while immigrant workers such as Françoise feel the brunt of Jean-Paul's anger through Jean-Paul's support of right-wing, nationalist politicians. Meanwhile, Duong struggles to save so that his children can be educated and leave the countryside for the city, where foreign companies advertise new jobs at better wages (see also Geography Matters: Globalization and Interdependence, p. 27).

Recently there has been a pronounced change in both the pace and the nature of globalization. New telecommunication technologies, new corporate strategies, and new institutional frameworks have all combined to create a dynamic new framework for real-world geographies. New information technologies have helped create a frenetic international financial system, while transnational corporations are now able to transfer their production activities from one region of the world to another

in response to changing market conditions (**Figure 1.34**). This locational flexibility has meant that there is now a high degree of functional integration between economic activities that are increasingly dispersed, so that products, markets, and organizations are both spread and linked across the globe. Governments, in their attempts to adjust to this new situation, have sought new ways of dealing with the consequences of globalization, including new international political and economic alliances.

All this adds up to an intensification of global connectedness and the beginnings of the world as one place. Or, to be more precise, it adds up this way for the 800 million or so of the world's population (about 13 percent of the total) who are directly tied to global systems of production and consumption and who have access to global networks of communication and knowledge. All of us in this globalizing world are in the middle of a major reorganization of the world economy and a radical change in our relationships to other people and other places.

Nature-Society Interactions

Inherent to the basic geographical concepts of landscape, place, and region are the interactions between people and the natural environment that shape landscapes and give places and regions their distinctive characteristics. Geography has contributed to several important ideas about the nature-society or human-environment relationship that assist in our understanding of regional geography.

The idea of **environmental determinism** proposes that human activity and development is determined by the limits set by the physical environment. Early twentieth-century geographers such as Ellsworth Huntington and Ellen Churchill Semple argued that climate, landforms, and soils were the

Geography Matters

Globalization and Interdependence

Now that the world economy is much more globalized, patterns of local and regional economic development are much more open to external influences, much more interdependent with development processes elsewhere. The globalization of the world economy involves new patterns of regional economic specialization in association with the internationalization of finance, the deployment of new technologies such as robotics, telematics, and biotechnology, and the homogenization of consumer markets (see Chapter 2).

This new framework for economic geography has already left its mark on the world's economic landscapes. It has also meant that the lives of people in different parts of the world have become increasingly intertwined. Here are three more examples from the World Bank's *World Development Report*:

Joe lives in a small town in southern Texas. His old job as an accounts clerk in a textile firm, where he had worked for many years, was not very secure. He earned $50 a day, but promises of promotion never came through, and the firm eventually went out of business as cheap imports from Mexico forced textile prices down. Joe went back to college to study business administration and was recently hired by one of the new banks in the area. He enjoys a comfortable living even after making the monthly payments on his government-subsidized student loan.

Maria recently moved from her central Mexican village and now works in a U.S.-owned firm in Mexico's maquiladora sector. Her husband, Juan, runs a small car upholstery business and sometimes crosses the border during the harvest season to work illegally on farms in California. Maria, Juan, and their son have improved their standard of living since moving out of subsistence agriculture, but Maria's wage has not increased in years: she still earns about $10 a day. . . .

Xiao Zhi is an industrial worker in Shenzhen, a Special Economic Zone in China. After three difficult years on the road as part of China's floating population, fleeing the poverty of nearby Sichuan province, he has finally settled with a new firm from Hong Kong that produces garments for the U.S. market. He can now afford more than a bowl of rice for his daily meal. He makes $2 a day and is hopeful for the future.*

These examples begin to reveal a complex and fast-changing interdependence that would have been unthinkable just 15 or 20 years ago. Joe lost his job because of competition from poor Mexicans such as Maria, and now her wage is held down by cheaper exports from China. But Joe now has a better job, and the U.S. economy has gained from expanding exports to Mexico. Maria's standard of living has improved and her son can hope for a better future. Joe's pension fund is earning higher returns through investments in growing enterprises around the world, and Xiao Zhi is looking forward to higher wages and the chance to buy consumer goods. But not everyone has benefited, and the new international division of labor has come under attack by some in industrial countries where increasing wage inequality is making people feel less secure about the future. Some workers in industrialized countries are fearful of losing their jobs because of cheap exports from lower-cost producers. Others worry about companies relocating abroad in search of low wages and lax labor laws.

Most of the world's population now lives in countries that are either integrated into world markets for goods and finance or rapidly becoming so. As recently as the late 1970s, only a few less-developed countries had opened their borders to flows of trade and investment capital. About a third of the world's labor force lived in countries such as the Soviet Union and China with centrally planned economies, and at least another third lived in countries insulated from international markets by prohibitive trade barriers and currency controls. Today, three giant population blocs—China, the republics of the former Soviet Union, and India—with nearly half the world's labor force among them, are entering the global market. Many other countries, from Mexico to Thailand, have already become involved in deep linkages. According to World Bank estimates, fewer than 10 percent of the world's labor force remain isolated from the global economy. (The World Bank, properly called the International Bank for Reconstruction and Development, is a United Nations affiliate established in 1948 to finance productive projects that further the economic development of member nations.)

---

*World Bank, *World Development Report 1995. Workers in an Integrating World,* New York: Oxford University Press, 1995, p. 50.

major influences on patterns and levels of land use, economic development, and even human potential. They explained the lower levels of agricultural production and industrial development that they saw in the Tropics as a result of high temperatures, humidity, and diseases that limited ecological potential, human energy, and most controversially, human intelligence. Although their ideas were rejected for their racist implications and concept of an all-controlling nature, elements of environmental determinist thinking continue to emerge in explanations of regional and economic development problems.

Many critics of environmental determinism highlighted the ability of people to overcome the constraints of the natural environment using ingenuity, technology, and social organization. The concept of *adaptation* to the natural environment is associated with the geographical subfield of **cultural ecology** most closely associated with the work of Carl Sauer and his students. Cultural ecologists study how human society has adapted to environmental challenges such as aridity and steep slopes through technologies such as irrigation and terracing and the organization of people to construct and maintain these systems. These adaptations can be seen clearly in many traditional landscapes such as the rice terraces of Southeast Asia or the canals and reservoirs of the southwestern United States. More recent adaptations include the use of biotechnology and agricultural chemicals to increase agricultural production and the development of new pharmaceuticals to cope with diseases.

Human adaptation has gone far beyond responses to natural constraints to produce widespread modification of environment and landscapes. In some cases, the human use of nature has resulted in a decline in conditions commonly termed *environmental degradation or pollution*. For example, overcultivation of steep slopes can result in erosion of the soil needed for subsequent agricultural production, and the use of agricultural chemicals has caused the contamination of adjacent rivers and lakes by chemicals that are toxic to fish and humans. The Industrial Revolution produced a dramatic growth in the emissions of waste material to land, water, and atmosphere and resulted in serious air pollution and health problems in many cities.

The massive transformation of nature by human activity has led geographers such as Neil Smith and Margaret FitzSimmons to claim that we can no longer talk about "natural" environments or untouched wilderness. They use the concept of the *social production of nature* to refer to the refashioning of landscapes and species by human activity, especially capitalist production and labor processes.

Geographers have played a major role in highlighting the global scope of this transformation in their discussions of the *human dimensions of global environment change,* defined as the study of the social causes and consequences of changes in global environmental conditions. Of particular concern are global patterns of fossil fuel use and land use change that are producing serious changes in climate and biodiversity through carbon-dioxide-induced global warming or by deforestation.

Global climate change is causing sea levels to rise (as polar ice caps melt) and has increased the frequency of violent storms. Warmer oceans surrender greater quantities of water as evaporation. Warmer surface temperatures and more humid air masses intensify weather systems, resulting in fiercer cyclones and hurricanes. In 1998 worldwide flooding was the worst on record, with 96 floods in 55 countries, including the most serious flooding around China's Chang Jiang (Yangtze) River in half a century and the most long-lasting on record in Bangladesh. In 1999 the twin disasters associated with global warming—flooding and violent storms—came together in one place as a particularly violent cyclone hit the low-lying coast of northeastern India, pushing rivers backward and flooding much of the province of Orissa, killing an estimated 10,000 people and leaving hundreds of thousands homeless (**Figure 1.35**). During the twentieth century, global sea level rose by 20 centimeters (7.9 inches), and a 1999 report by Britain's Meteorological Office warned that flooding will increase more than ninefold over the next century, with four-fifths of the increase coming in South and Southeast Asia. Such a rise in sea level is potentially disastrous for some countries. About 70 percent of Bangladesh, for example, is at sea level, while much of Egypt's most fertile land, in the Nile delta, is also at sea level. Meanwhile, extensive regions of Africa, Asia, and Latin America are so marginal for agriculture that further drought could prove disastrous. In contrast, farmers in much of Europe and North America would welcome a local rise in mean temperatures, since it would extend their options for the kinds of crops that they could profitably raise.

The causes and consequences of these changes vary considerably by world region. For example, the industrial coun-

Figure 1.35 Flooding in Orissa Villager Panchu Behra, extreme right, and his family try to salvage good rice grain in the aftermath of the severe cyclone that devastated the Indian province of Orissa in 1999. The Behra's 20-year-old son was one of more than 10,000 who perished as a result of the cyclone, and they also lost their house and most of their rice crop.

tries have higher carbon dioxide emissions and poor, less-developed regions are experiencing rapid deforestation. In order to survive, the rural poor are often impelled to degrade and destroy their immediate environment, cutting down forests for fuelwood and exhausting soils with overuse. In order to meet their debt repayments, governments feel compelled to generate export earnings by encouraging the harvesting of natural resources.

We know enough about the growth of population and the changing geography of economic development to be able to calculate with some confidence that the air and water pollution generated by low-income countries will more than double in the next 15 years as they become more industrialized. We know, in short, that environmental problems will be inseparable from processes of demographic change, economic development, and human welfare. In addition, it is becoming clear that regional environmental problems are going to be increasingly enmeshed in matters of national security and regional conflict. The spatial interdependence of economic, environmental, and social problems means that some parts of the world are ecological time bombs. The prospect of civil unrest and mass migrations resulting from the pressures of rapidly growing populations, deforestation, soil erosion, water depletion, air pollution, disease epidemics, and intractable poverty is real.

The global nature of environmental changes has led to calls for global solutions including international agreements to reduce pollution and protect species. A more benign relationship between nature and society has been proposed under the principle of **sustainable development,** a term that is now widely used but vaguely defined. One definition is that of the World Commission on Environment and Development, chaired by the former prime minister of Norway, Gro Bruntland, stating that sustainable development is "development that meets the needs of the present without compromising the ability of future generations to meet their own needs."[7] This definition incorporates the ethic of intergenerational equity with its obligation to preserve resources and landscapes for future generations. Geographers such as William Adams and Timothy O'Riordan consider sustainable development to include ecological, economic, and social goals of preventing environmental degradation while promoting economic growth and social equality. Sustainable development means that economic growth and change should occur only when the impacts on the environment are benign or manageable and the impacts (both costs and benefits) on society are fairly distributed across classes and regions. This means finding less polluting technologies that use resources more efficiently and managing renewable resources (those that replenish themselves such as water, fish, and forests) so as to ensure replacement and continued yield. In practice, sustainable development policies of major international institutions, such as the World Bank, have promoted reforestation, energy efficiency and conservation, and birth control and poverty pro-grams to reduce the environmental impact of rural populations. At the same time, however, the expansion and globalization of the world economy has resulted in increases in resource use and inequality that contradict many of the goals of sustainable development.

The Fast World and the Slow World

As Ted Turner, former owner of CNN, observed in a 1999 United Nations report on international development, "It is as if globalization is in fast forward, and the world's ability to react to it is in slow motion."[8]

Ted Turner's observation points to an increasing division that now exists between the "fast world" and the "slow world." The **fast world** consists of people, places, and regions directly involved, as producers and consumers, in transnational industry, modern telecommunications, materialistic consumption, and international news and entertainment. The **slow world,** which accounts for about 85 percent of the world's population, consists of people, places, and regions whose participation in transnational industry, modern telecommunications, materialistic consumption, and international news and entertainment is limited. The slow world consists chiefly of less-developed countries, but it also includes many rural backwaters, declining manufacturing regions, and disadvantaged slums in affluent countries, all of them bypassed by this latest phase in the evolution of the modern world economy.

The center of gravity of the fast world is in the richer and more developed countries of the world. The United States, for example, with less than 5 percent of the world's population, accounts for more than 40 percent of the world's telephones. But the fast world also extends throughout the world to the more affluent regions, neighborhoods, and households that are "plugged in" to the contemporary world economy, whether as producers or consumers of its products and culture. The leading edge of the fast world is the Internet, the global web of computer networks that began in the United States in the 1970s as a decentralized communication system sponsored by the U.S. Department of Defense. Until the mid-1970s, there were fewer than 50 nodes (servers) in the whole system. Then, in the early 1980s, the original network (ARPANET) was linked with two important new networks: CSnet (funded by the National Science Foundation) and BITNET (funded by IBM). In July 1988 a high-speed backbone (NSFnet) was established in order to connect regional networks in the United States.

Today, these early networks have become absorbed into the Internet, a loose confederation of thousands of small, locally run computer networks for which there is no clear center of control or authority. The Internet has become the world's single most important mechanism for the transmission of scientific and academic knowledge. Roughly 50 percent of its traffic is electronic mail; the rest consists of scientific documents, data, bibliographies, electronic journals, bulletin

---

[7]World Commission on Environment and Development, *Our Common Future* (Brundtland Report), New York: Oxford University Press, 1987, p. 43.

[8]United Nations, *Human Development Report 1999,* New York: United Nations Development Programme, 1999, p. 100.

Geography Matters

The Digital Divide

The growth of the Internet has been phenomenal. In 1989 only 100,000 computers were connected to the Internet worldwide. In early 2001 the estimate was close to 90 million, with more than 380 million people around the world having access to the Internet, either at home or at work. By 2002 close to a billion people are expected to be using the Internet. The rapid spread of the Internet owes much to the tremendous advances made in computing power during the 1990s, combined with sharply falling costs. (If the automobile industry had enjoyed the kind of productivity growth the computer industry has experienced since 1990, the price of a family car in 2001 would be less than $10.)

This growth, together with the Internet's ability to bypass borders and interweave world cultures, has led many observers to hail a "digital revolution" that is going to shrink the globe. But geography still matters. The reality is that people in most places and regions throughout the world have little infrastructure and few computers through which to communicate digitally. The benefits of the digital revolution are being reaped by the world's affluent populations, leaving the poor even more marginalized than before.

The statistics on the basic building block of Internet connectedness—that is, telephone lines—are stark. Fully 65 percent of the world's households do not have a telephone. Cambodia and Haiti both had less than one telephone line for every 100 people in the late 1990s. Electricity is only available in urban areas in Haiti, and then for only a couple of hours per day. In Bangladesh the wait time for a new landline telephone exceeds six years. The Democratic Republic of Congo had the lowest telephone density in the world in the late 1990s, with one telephone line for 1318 people. In most of the world's less-developed countries, high per-minute telephone costs have rendered Internet use off-limits to all but the most affluent. Some less-developed countries, though, do have better communications situations than their economic status would suggest. Chile, for example, whose government has helped guide information technology development, has an entirely digital phone system, making quality connections easy and inexpensive.

It is not surprising, therefore, to find that there is a "digital divide" between countries. According to United Nations figures, the world's most affluent countries, with about 15 percent of the global population, account for nearly 90 percent of the Internet users. Not everyone in affluent countries is an Internet user, of course, so that in overall terms only 2 percent of the global population are online. The map of global Internet connectivity (**Figure 1**) shows very starkly the magnitude of the digital divide.

Yet access to the Internet is not limited simply by physical connectivity: Four-fifths of the world's Web sites in 2000 were in English. Meanwhile, the governments of Singapore, China, and Saudi Arabia all censor what can be accessed and sent on the Web. Syria's late President Hafez Assad went a step further and forbade Internet access to his citizens. Those who can afford it must pay long-distance telephone charges to access the Internet via an ISP (Internet Service Provider) in neighboring Beirut.

In spite of the digital divide, it is already clear that the Internet gives a voice to the politically powerless. For example, the Internet played a widely publicized role when used by pro-democracy students in Beijing in 1989 and in preventing a coup against then-Soviet leader Mikhail Gorbachev in 1991. In 1999 the first Internet center in a Palestine refugee camp was opened, allowing people living in the Dheisheh camp to speak to friends and relatives in Gaza and Lebanon. In Mexico City an organization called Mujer a Mujer (Woman to Woman) e-mailed contacts in California for assistance when plans for a new textile factory were announced in their community. The women went to meet the management armed with a bulky portfolio detailing the company's practices, profits, and ownership. For impoverished nations facing shortages of drinking water and food, the Internet can help bring desperately needed information about farming and health issues. The best-known example is HealthNet, a networked information service that supports healthcare workers in more than 30 countries, including 22 in Africa. Doctors in Central Africa used it to share information on the 2000 outbreak of the deadly Ebola virus, and malaria researchers at a remote site in northern Ghana use it to communicate daily with colleagues in the London School of Tropical Medicine.

The Digital Divide in the United States

Although the United States has a tremendous advantage in the global digital revolution, there is a serious digital divide within the country. Overall, in 2000 almost 60 percent of U.S. households had no access to the Internet at all. Once again, geography matters. The West Coast and eastern seaboard from New Hampshire to Virginia are at the forefront of the wired economy, while the Deep South and the upper Midwest lag far behind. A report by the U.S. Department of Commerce found that households with annual incomes above $75,000 are more than six times as likely to be Internet users as those earning less than $15,000.* Single-parent households are less than

---

Falling Through the Net: Toward Digital Inclusion. U.S. Department of Commerce, 2000 (www.ntia.doc.gov).

The number of Internet connections in Argentina more than doubled between 1999 and 2000, taking the total to 900,000. The typical Internet user in Argentina is male, in his thirties, with a college degree. E-mail is the most popular activity among Argentine Internet users, most of whom are reluctant to make online purchases: only 30 percent of those with access in 1999 had ever made an online purchase, compared to approximately 80 percent in the United States.

Japan has lagged behind other countries in its use of the Internet. Although there were more than 2 million Internet hosts in Japan in 2000 (compared with less than 2 million in both Britain and Germany), fewer than 25 percent of Japanese offices are computerized compared with over 70 percent in the United States. Similarly, only 20 percent of Japanese personal computers are hooked in to a network of some sort, compared to 70 percent in the United States.

China is connected to the Internet, but mostly through lines with only a very small capacity. In 2000 just under 17 million people in China had access to the Internet—less than 2 percent of the population.

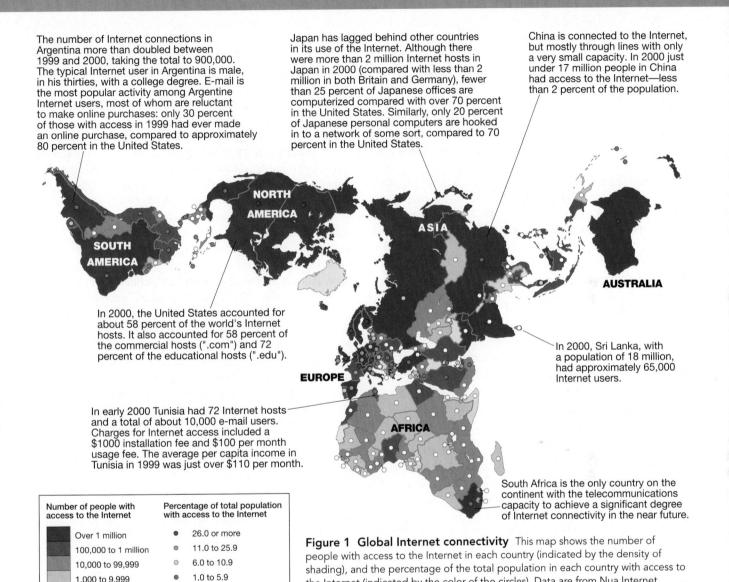

In 2000, the United States accounted for about 58 percent of the world's Internet hosts. It also accounted for 58 percent of the commercial hosts (".com") and 72 percent of the educational hosts (".edu").

In 2000, Sri Lanka, with a population of 18 million, had approximately 65,000 Internet users.

In early 2000 Tunisia had 72 Internet hosts and a total of about 10,000 e-mail users. Charges for Internet access included a $1000 installation fee and $100 per month usage fee. The average per capita income in Tunisia in 1999 was just over $110 per month.

South Africa is the only country on the continent with the telecommunications capacity to achieve a significant degree of Internet connectivity in the near future.

Number of people with access to the Internet

	Over 1 million
	100,000 to 1 million
	10,000 to 99,999
	1,000 to 9,999
	0 to 999
	No data

Percentage of total population with access to the Internet

●	26.0 or more
◉	11.0 to 25.9
◎	6.0 to 10.9
◉	1.0 to 5.9
○	Less than 1.0
	No data for countries shown without a dot

Figure 1 Global Internet connectivity This map shows the number of people with access to the Internet in each country (indicated by the density of shading), and the percentage of the total population in each country with access to the Internet (indicated by the color of the circles). Data are from Nua Internet surveys (http://www.nua.ie), taking the highest estimates from late 2000. (*Source*: Map projection, Buckminster Fuller Institute and Dymaxion Map Design, Santa Barbara, CA. The word *Dymaxion* and the Fuller Projection Dymaxion™ Map design are trademarks of the Buckminster Fuller Institute, Santa Barbara, California, © 1938, 1967 & 1992. All rights reserved.)

half as likely to be wired as two-parent families. The disparity is even greater among African-American families: Black children living with one parent are less than one-fourth as likely to have Internet access as those in two-parent households.

These aspects of the digital divide were reflected in variations by region and geographic setting. More than 50 percent of the adult population of Alaska, Colorado, Connecticut, Delaware, New Hampshire, and Oregon had access to the Internet in 2000, compared to less than 35 percent in Arkansas, Louisiana, Oklahoma, Mississippi, South Carolina, and West Virginia. The cities with the most Internet users included Atlanta, Austin, Boston, Dallas-Ft. Worth, Minneapolis-St. Paul, San Diego, San Francisco, San Jose, Seattle, and Washington, DC. Even within these cities, however, there exists a deep digital divide between the affluent suburbs and disadvantaged inner-city neighborhoods. In general, underserved groups such as the residents of poor inner-city neighborhoods and rural areas have fallen farther behind as the digital revolution has gathered pace. As at the global scale, the gap between the plugged-in and the shut out is reinforcing spatial inequalities, rather than reducing them.

boards, and a user interface to the Internet, the World Wide Web. In late 2000 more than 75 million Internet hosts existed in more than 150 countries; more than 380 million people had access to the Internet; and somewhere between 100 and 120 million people worldwide had Internet e-mail addresses. The Internet has been doubling in networks and users every year since 1990, but most Internet users are still in the world's affluent countries: In late 2000, about 42 percent were in North America and another 28 percent were in Europe. The rest are in Japan, Australia and New Zealand, and in the fragmentary outposts of the fast world that are embedded within the larger metropolitan areas of less-developed countries. Overall, 90 percent of all Internet traffic originates in, or is destined for, the United States (see Geography Matters: The Digital Divide, p. 30).

This division between fast and slow worlds is, of course, something of a caricature. In fact, the fast world encompasses almost every*where* but not every*body*. As a result, regional geography now has to contend with the apparent paradox of people whose everyday lives are lived part in one world, part in another. Consider, for example, the shantytown residents of Mexico City. With extremely low incomes, only makeshift housing, and little or no formal education, they somehow are knowledgeable about international soccer, music, film, and fashion and are even able to copy fast-world consumption through cast-offs and knock-offs. Much the same could be said about the impoverished residents of rural Appalachia (substitute NASCAR racing for international soccer) and, indeed, about most regions of the slow world. Very few regions remain largely untouched by globalization.

Globalization and Cultural Change   Anyone who has ever traveled between major world cities—or, for that matter, anyone who has been attentive to the backdrops of movies, television news stories, and magazine photojournalism—will have noticed the many familiar aspects of contemporary life in settings that, until recently, were thought of as being quite different from one another. Airports, offices, and international hotels have become notoriously alike, and their similarities of architecture and interior design have become reinforced by near-universal dress codes of the people who frequent them. The business suit, especially for males, has become the norm for office workers throughout much of the world. Jeans, T-shirts, and sneakers, meanwhile, have become the norm for both young people and those in lower-wage jobs. The same automobiles can be seen on the streets of cities throughout the world (though sometimes they are given different names by their manufacturers); the same popular music is played on local radio stations; and many of the movies shown in local theaters are the same ("Titanic," "South Park," and "Pokemon: the Movie," for example). Some of the TV programming is also the same—not just the music videos on MTV, but CNN's news, major international sports events, drama series such as "Baywatch" and "Ally McBeal" and comedy series such as "Friends." The same brand names also show up in stores and restaurants: Coca-Cola, Perrier, Carlsberg, Nestlè, Nike, Seiko, Sony, IBM, Nintendo, and Microsoft, to list just a few. Every-

where there is Chinese food, pita bread, pizza, classical music, rock music, and jazz.

It is these commonalities that provide a sense of familiarity among the inhabitants of the "fast world." From the point of view of cultural nationalism, the "lowest common denominator" of this familiarity is often seen as the culture of fast food and popular entertainment that emanates from the United States. Popular commentators have observed that cultures around the world are being Americanized, or "McDonaldized," which represents the beginnings of a single global culture that will be based on material consumption, with the English language as its medium (**Figure 1.36**).

There is certainly some evidence to support this point of view, not least in the sheer numbers of people around the world who view "Friends," drink Coca-Cola, and eat in McDonald's franchises or similar fast-food chains. Meanwhile, U.S. culture is increasingly embraced by local consumers and entrepreneurs around the world. It seems clear that U.S. products are consumed as much for their symbolism of a particular way of life as for their intrinsic value. McDonald's burgers, along with Coca-Cola, Hollywood movies, rock music, and NFL and NBA insignia, have become associated with a lifestyle package that features luxury, youth, fitness, beauty, and freedom.

The economic success of the U.S. entertainment industry has helped reinforce the idea of an emerging global culture based on Americanization. In 2000, the entertainment industry was a leading source of foreign income in the United States, with a trade surplus of $28 billion. Similarly, the United States transmits much more than it receives in terms of the sheer volume of cultural products. In 2000, the originals of more than half of all the books translated in the world (more than 20,000 titles) were written in English. In terms of international flows of everything from mail and phone calls to press-agency reports,

Figure 1.36   McDonald's in Poland   The American franchise restaurant McDonald's is becoming a fixture on the landscape of formerly communist eastern European countries such as Poland. Although menu prices are quite high by local standards, frequenting places like McDonald's is a sign of status and personal prosperity in Poland and other recently communist countries, such as Romania and Bulgaria.

television programs, radio shows, and movies, a disproportionately large share originates in the United States.

Neither the widespread consumption of U.S. and U.S.-style products nor the increasing familiarity of people around the world with global media and international brand names, however, adds up to the emergence of a single global culture. Rather, what is happening is that processes of globalization are exposing the inhabitants of both the fast world and the slow world to a common set of products, symbols, myths, memories, events, cult figures, landscapes, and traditions. People living in Tokyo or Tucson, Turin or Timbuktu, may be perfectly familiar with these commonalities without necessarily using or responding to them in uniform ways.

Equally, it is important to recognize that cultural flows take place in all directions, not just outward from the United States. Think, for example, of European fashions in U.S. stores; of Chinese, Indian, Italian, Mexican, and Thai restaurants in U.S. towns and cities; and of U.S. and European stores selling exotic craft goods from less-developed countries.

The answer to the question "Is there a global culture?" then, must be no. While an increasing familiarity exists with a common set of products, symbols, and events (many of which share their origins in U.S. culture of fast food and popular entertainment), these commonalties become configured in different ways in different places, rather than constitute a single global culture. The local interacts with the global, often producing hybrid cultures. Sometimes, traditional, local cultures become the subject of global consumption; sometimes it is the other way around.

The Increasing Significance of Places and Regions   At first glance, the emergence of the fast world, with its transnational architectural styles, dress codes, retail chains, and popular culture, and its ubiquitous immigrants, business visitors, and tourists, seems as if it might have brought a sense of placelessness and dislocation: a loss of territorial identity and an erosion of the distinctive sense of place associated with certain localities. Yet the common experiences associated with globalization are still modified by local geographies. The structures and flows of the fast world are variously embraced, resisted, subverted, and exploited as they make contact with specific places and specific communities. *In the process, places and regions are reconstructed rather than effaced.* Often, this involves deliberate attempts by the residents of a particular area to create or re-create territorial identity and a sense of place. Inhabitants of the fast world, in other words, still feel the need for enclaves of familiarity, centeredness, and identity. Regional geographies change, but they don't disappear.

At first glance, it might seem that globalization will render geography obsolete—especially in the more affluent regions of the world. High-tech communications and the global marketing of standardized products seem as if they might soon wash away the distinctiveness of people and places, permanently diminishing the importance of differences between places. Far from it. The new mobility of money, labor, products, and ideas actually increases the significance of place in some very real and important ways.

- The more universal the diffusion of material culture and lifestyles, the more valuable regional and ethnic identities become. One example of this is the way that the French government has actively resisted the Americanization of French language and culture by banning the use of English words and phrases and by subsidizing their domestic movie industry.

- The faster the information highway takes people into cyberspace, the more they feel the need for a subjective setting—a specific region or community—that they can call their own. Examples of this can be found in the private master-planned residential developments that have sprung up around every U.S. metropolitan area since the mid-1980s. Unlike most previous suburban developments, each of these master-planned projects has been carefully designed to create a sense of community and identity for their residents.

- The greater the reach of transnational corporations, the more easily they are able to respond to place-to-place variations in labor markets and consumer markets, and the more often and more radically that economic geography has to be reorganized. Athletic shoe manufacturers such as Nike, for example, frequently switch production from one country to another in response to the changing international geography of wage rates and currencies.

- The greater the success of transnational corporations and the more pervasive global consumer products and global culture become, the more likely it is that they will be actively resisted (see Geography Matters: Mobilization Against Globalization, p. 34).

- The greater the integration of transnational governments and institutions, the more sensitive people have become to localized cleavages of race, ethnicity, and religion. An example is the resurgence of nationalist movements, as in the near secession of Quebec from Canada in 1995 and the emergence of the Lega Nord party in Italy in the early 1990s. Lega Nord (the Northern League) is a federalist political party, whose supporters in northern Lombardy and rural northeastern Italy want to distance themselves from what they view as a distinctively different culture and society in the Italian South.

All in all, the reality is that globalization influences—and is influenced by—specific cultures and settings in very different ways. In the process, places and regions are modified, rather than being destroyed or homogenized.

The Global Context: Some Important Patterns

Beyond the very broad context that is provided by the concept of a global economy that is leading to an increasing degree of spatial interdependence, it is important to recognize the underlying *diversity* of the world. In this section, we examine several key elements of regional diversity: religion, language,

Geography Matters

Mobilization Against Globalization

Local mobilization against transnational business and the effects of economic globalization seems set to become an important cultural struggle in the early decades of the twenty-first century. Economic globalization is in many ways still in its beginning stages, but already it has brought a great deal of change to the economic, cultural, and political geography of places and regions throughout the world. A great deal of this change has been progressive, bringing increased overall levels of economic well-being, a strengthening of free enterprise and democracy, and an enriched flow of products, ideas, and culture among and between places and regions.

Inevitably though, as with previous epochs of economic change, economic globalization has also brought problems as some places and regions have experienced disinvestment in order that capital could be made available for more profitable investments elsewhere. Economic globalization has also undercut the power of national and local governments to regulate economic affairs and has erased a great deal of local diversity because of the economic success of global products: the "McDonaldization of everywhere."

Fundamental geographic differences—in climate, resources, culture, and so on—mean that economic globalization is variously embraced, modified, or resisted in different parts of the world. Indeed, there has emerged a counter-movement, a "mobilization against globalization," that could well affect the whole dynamic of economic globalization as it is played out over the next decade or two. One form of this mobilization is exemplified by the efforts of some local communities to cope with the negative effects of global economic change. In Sheffield, England, for example, the merger of British Steel and the Dutch steelmaker Hoogovens in 1998 put many jobs in Sheffield in jeopardy when the city already had more than a third of its jobless people on unemployment benefits for well more than a year. In response to the employment crisis, the city established a Sheffield Employment Bond and raised £1.9 million ($2.9 million) from local investors. Some of the money raised is being used to build new houses, creating new jobs for young trainee construction workers taken from the pool of local unemployed. The remainder of the fund is being used as seed money to finance small businesses such as fair-trade shops, bicycle workshops, mobile hairdressers, and neighborhood cafes and bookshops.

Another form of mobilization is exemplified by activists who use the legal system in order to resist what they see as the undesirable local outcomes of transnational business practices. In the United States, activists have resurrected a 1789 alien tort law that was originally designed to provide redress for foreigners against sea pirates and slavers. The law had already been used to track down individuals, such as Ferdinand Marcos. In the late 1990s, activists pursued transnational corporations, accusing them of helping to suppress human rights. For example, four U.S. retailers and clothing manufacturers charged with unethical labor practices in a $1 billion alien-tort suit that had been filed on behalf of some 50,000 garment workers in Saipan agreed to settle in 1999. While admitting no liability, Nordstrom, Gymboree, Cutter & Buck, and J. Crew agreed to pay $1.25 million into a fund to support the independent monitoring of their overseas suppliers.

A third form of mobilization is old-fashioned popular protest. The most vivid examples are provided by French farmers who, in protest over trade-liberalization policies, regularly take to tactics such as blocking streets with tractors, with produce, with farmyard manure, or with farm animals (**Figure 1**).

Finally, and perhaps most significantly in terms of future cultural struggles between local interests and transnational business interests, mobilization can be organized by coalitions of nongovernmental agencies (NGOs). This form of mobilization against globalization became much more powerful in the 1990s as a result of the Internet. Groups such as Kenya's Consumers' Information Network, Ecuador's Accion Ecologica, and Trinidad and Tobago's Caribbean Association for Feminist Research and Action are linked through scores of Web sites, list servers, and discussion groups to U.S., European, and Asian counterparts. NGOs set the agenda for the Earth Summit in Rio in 1992 and lobbied governments to attend; they publicized the Chiapas rebellion in Mexico in 1994, thereby preventing the Mexican government from suppressing it violently. In 1997 a loose alliance of 350 NGOs from 23 countries set out to ban land mines; they soon persuaded 122 nations to sign on to a treaty. In 1998 another NGO alliance, this time reckoned to number 600 groups in nearly 70 countries, blocked a painstakingly negotiated treaty on international multilateral investment. In 1999 more than 775 NGOs registered with the World Trade Organization (WTO) and took more than 2000 observers to the

---

population distribution, urbanization, economic development, and social well-being. In Chapter 2 we describe the ways in which the human and environmental diversity of the world have unfolded, emphasizing the legacies of dependence and interdependence among world regions.

Religion

Although religious affiliation is on the decline in some parts of the world's more affluent regions, it still acts as a powerful shaper of daily life in much of the world, influencing every-

Figure 1 French farmers protest Globalization and transnational business often means the downward convergence of wages and environmental standards, an undermining of democratic governance, and a general recoding of nearly all aspects of life on Earth to the language and logic of global markets. French farmers have been especially militant in protesting such outcomes, particularly when their livelihood and traditional farming practices have been threatened.

were designed to ensure minimal levels of pollution. The WTO ruled in favor of Venezuela. The EPA subsequently had to change its regulations, leaving it with a weakened ability to enforce federal air-quality standards. When the European Union banned U.S. beef raised with the assistance of hormone injections in 1998, the United States took the case to the WTO. The WTO ruled that the European Union's act was illegal under international trading rules, even if the Europeans were distrustful of the hormone-injected meat. Another WTO ruling in 1998 undermined the U.S. Endangered Species Act. The United States has attempted to protect endangered sea turtles from extinction by requiring that shrimp-fishing boats install devices that allow the turtles to escape their nets. The law applied to all shrimp sold in the United States, but the WTO ruled that this was a restraint on other countries' ability to trade freely.

Figure 2 Seattle protests In November 1999 representatives of the World Trade Organization (WTO) met in Seattle, Washington, to discuss a new round of agreements on liberalizing international trade. An estimated 70,000 protesters converged on the city to demonstrate against the WTO and its policies. "No Globalization Without Representation" read the placards in the protest march, and "Hey, hey, ho, ho, WTO has got to go." "Whose world? Our world! Whose streets? Our streets!" chanted the crowd. Riot police fired tear gas, pepper spray, and rubber bullets at peaceful protesters blockading the WTO meeting, while roving gangs of anarchists smashed windows, overturned newspaper stands, and attacked cars. When demonstrators refused to disperse, the city declared a state of civil emergency. The WTO meetings themselves, delayed by protests, ended in collapse.

WTO summit in Seattle, Washington. They also helped to organize some 70,000 protesters who took part in the most extensive teach-ins and demonstrations in the United States since the Vietnam War (**Figure 2**).

These demonstrations against the WTO highlight some of the central issues that surround economic globalization. Economic globalization depends on free trade, but should the abolition of economic protectionism be accompanied by the abolition of social and environmental protection? The WTO's mandate is the "harmonization" of safety and environmental standards among member nations as well as the removal of tariffs and other barriers to free trade. Most people support free trade, but not if it harms public health and not if it is based on child labor.

There have been several examples of how the free trade principles embodied in the WTO can erode national environmental standards. In 1997 Venezuela and Brazil, on behalf of their gasoline producers, challenged U.S. Environmental Protection Agency (EPA) regulations on gasoline quality, which

thing from eating habits and dress codes to coming-of-age rituals and death ceremonies. Religious beliefs and practices change as new interpretations are advanced or new spiritual influences are adopted. From the onset of globalization in the fifteenth century, religious missionizing—propagandizing and

persuasion—has been a key element. In the 500 years since the onset of the Columbian Exchange, conversion of all sorts has escalated throughout the globe. The **Columbian Exchange** refers to the interdependence between the Old World and the New World, originating with the voyages of Columbus. In fact,

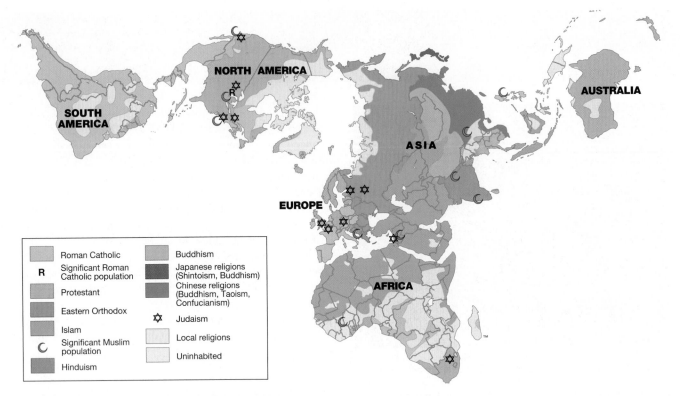

Figure 1.37 World distribution of major religions This map shows a generalized picture of the world's major religions. Most of the world's peoples are members of one of these religions. Not evident on this map are the local variations in practices, as well as the many other different religions that are practiced worldwide. (*Source:* Map projection, Buckminster Fuller Institute and Dymaxion Map Design, Santa Barbara, CA. The word *Dymaxion* and the Fuller Projection Dymaxion™ Map design are trademarks of the Buckminster Fuller Institute, Santa Barbara, California, © 1938, 1967 & 1992. All rights reserved.)

since 1492, traditional religions have become dramatically dislocated from their sites of origin.

The processes of global political and economic change that led to the massive movement of the world's populations over the last five centuries has also meant the dislodging and spread of the world's many religions from their traditional sites of practice. Religious practices have become so spatially mixed that it is a challenge to present a map of the contemporary global distribution of religion that reveals more than it obscures. This is because the global scale is too gross a level of resolution to portray the wide variation that exists among and within religious practices. **Figure 1.37** identifies the contemporary distribution of what are considered by religious scholars to be the world's "major religions" because they contain the largest number of practitioners globally. As with other global-scale representations, the map is useful in that it helps to present a generalized picture. **Figure 1.38** identifies the source areas of four of the world's major religions and their diffusion from those sites over time. The map illustrates the fact that the world's major religions originated and diffused from two fairly small areas of the globe. The first, where Hinduism and Buddhism (as well as Sikhism) originated, is an area of the lowlands of the subcontinent of India drained by the Indus (Punjab on the map) and Ganga rivers. The second, where Christianity and Islam (as well as Judaism) originated, is in the deserts of the Middle East.

Language

The distribution and diffusion of languages tells much about changing geographies and the impact of globalization on culture. **Figure 1.39** shows the distribution of the world's indigenous **language families.** The geography of language has been significantly affected by globalization. The plethora of languages and dialects in many regions has made communication and commerce among the different language speakers difficult. These problems often lead governments to impose standard languages (also known as official languages because they are maintained by offices of government, such as education and the courts). Where official languages are put into place, indigenous languages may eventually be lost. Yet the actual unfolding of globalizing forces—such as official languages—works differently in different places and in different times. The overall trend appears to be toward the loss of indigenous language (and other forms of culture). It is also important to recognize, however, that language and other forms of cultural identity can also be used as a means of challenging the political, economic, cultural, and social forces of globalization as they occur in France, Spain (the Basque Separatist Movement—see page 22), Canada (the Quebeçois Movement—see Chapter 7, p. 334), and other countries.

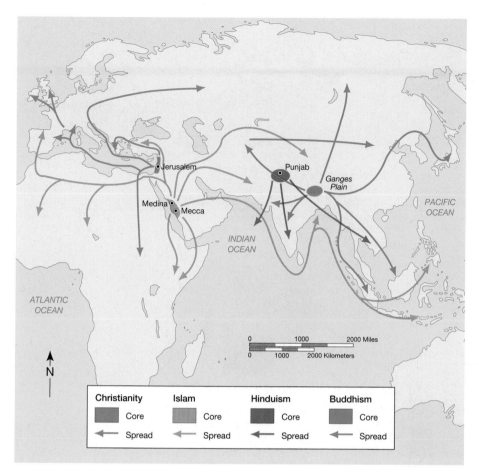

Figure 1.38 Origin areas and diffusion of four major religions The world's major religions originated in a fairly small region of the world. Judaism and Christianity began in present-day Israel and Jordan. Islam emerged from western Arabia. Buddhism originated in India, and Hinduism in the Indus Valley of Pakistan. The source areas of the world's major religions are also the cultural hearth areas of agriculture, urbanization, and other key aspects of human development.

Population

As the world population density map demonstrates (**Figure 1.40**), some areas of the world are very heavily inhabited, while others only sparsely. Almost all of the world's inhabitants live on 10 percent of the land. Most live near the edges of land masses, near the oceans, seas, or along rivers with easy access to a navigable waterway. Approximately 90 percent live north of the equator, where the largest proportion of the total land area (63 percent) is located. Finally, most of the world's population live in temperate, low-lying areas with fertile soils. In mid-2000 the world contained just over 6 billion people. The United Nations Population Fund projects that the world's population will increase by 90 million per year to mid-century. This means that by the year 2050, the world is projected to contain nearly 10 billion. Compare this figure to the fact that over the course of the entire nineteenth century less than a billion people were added to the population. The geography of this projected population growth is noteworthy. Over the next century, population growth is predicted to occur almost exclusively in Africa, Asia, and Latin America, while Europe and North America will experience very low and in some cases zero population growth.

The history of demographic change in industrialized countries has prompted some analysts to suggest that many of the economic, political, social, and technological transforma-tions associated with industrialization and urbanization lead to a demographic transition. The **demographic transition** is a model of population change when high birth and death rates are replaced by low birth and death rates. Once a society has moved from a preindustrial economic base to an industrial one, population growth slows. According to the demographic transition model, the slowing of population growth is attributable to improved economic production and higher standards of living brought about by changes in medicine, education, and sanitation.

As **Figure 1.41** illustrates, the high birth and death rates of the preindustrial phase (Phase 1) are replaced by the low birth and death rates of the industrial phase (Phase 4) only after passing through the critical transitional phase (Phase 2) and then more moderate rates (Phase 3) of natural increase and growth. This transitional phase of rapid growth is the di-rect result of early and steep declines in mortality while fertil-ity remains at high, pre-industrial levels.

The model suggests that it is inevitable that countries will be stalled for a while in the transitional high-growth phase, which has been called a "demographic trap." The rea-soning for this is that, while new and more effective meth-ods for fighting infectious diseases have been advanced, social attitudes about the desirability of large families have only recently begun to be affected. It should be emphasized, though, that the demographic transition model is based on

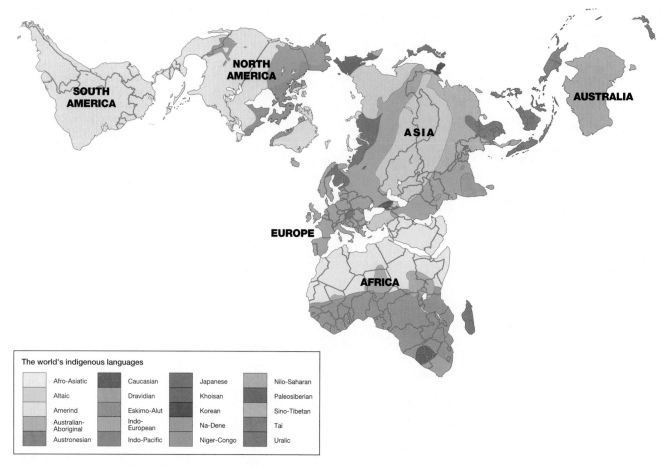

Figure 1.39   World distribution of major language families This map provides insights into linkages among seemingly disparate cultures widely separated in space and time and helps us to begin to understand something about the consequences of population movements across broad expanses of time and space. (*Sources:* Reprinted with permission from Prentice Hall, E. F. Bergman, *Human Geography: Cultures, Connections, and Landscapes,* © 1995, p. 240. Western Hemisphere adapted from *Language in the Americas* by Joseph H. Greenberg with the permission of the publishers, Stanford University Press, 1987, by the Board of Trustees of the Leland Stanford Junior University. Eastern Hemisphere adapted with permission from David Crystal, *Encyclopedia of Language.* New York: Cambridge University Press, 1987.)

the actual experience of developed countries and is thought by many experts to be less useful in explaining the demographic trends affecting less-developed countries and regions, whose entire development experience, as we shall see in Chapter 2, is quite different from that of industrialized countries.

Contemporary concerns about excessive population growth, especially in the world's poorest countries, have led to the development of international and national policies and programs. Most of the international population policies of the last two decades have been directed at reducing the number of births worldwide through family-planning programs—offering free contraceptives and family-planning counseling; authorizing the minimum age of marriage; offering incentives to couples who have only one child; and mandating disincentives to couples who have larger families. It is now broadly recognized that the history, social and cultural practices, development level and goals, and political structures for countries and even regions within countries are highly variable and

that one rigid and overarching policy to limit fertility will not work for all.

It is also now widely accepted among policymakers that a close relationship exists between women's status and fertility. Women who have access to education and employment tend to have fewer children since they have less of a need for the economic security and social recognition that children are thought to provide. In Botswana, for instance, women with no formal education have, on average, 5.9 children, while those with four to six years of school have just 3.1 children. In Senegal, women with no education give birth to an average of seven children. In contrast, the average number of children born to a woman with 10 years of education drops to 3.6. The numbers are comparable for Asia and South America. Success at damping population growth in less-developed countries appears to be very much tied to enhancing the possibility for a good quality of life and to empowering people, especially women, to make informed choices.

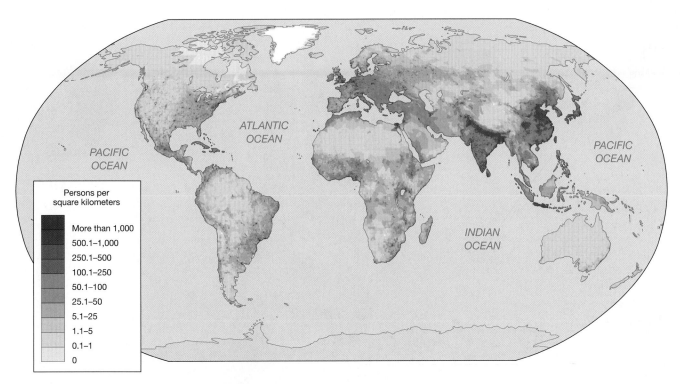

Figure 1.40   World population density 1995 As this map shows, the world's population is not uniformly distributed across the globe. Such maps are useful in understanding the relationship between population distribution and the national contexts in which they occur. Note the extremely high densities of population in China and India. (*Source:* Center for International Earth Science Information Network (CIESIN), Columbia University; International Food Policy Research Institute (IFPRI); and World Resources Institute (WRI), 2000. Gridded Population of the World (GPW), Version 2. Palisades, NY: CIESIN, Columbia University. Available at http://sedac.ciesin.org/plue/gpw/index.html?main.html&2)

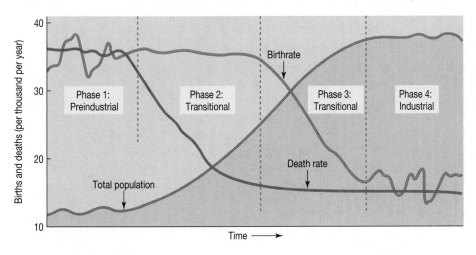

Figure 1.41   Demographic transition The transition from a stable population based on high birth and death rates to one based on low birth and death rates tends to progress in clearly defined stages, as illustrated by this graph. Population experts disagree about the usefulness of the model, however. Many insist that it is applicable only to the demographic history of industrialized countries.

Urbanization

In 1950 only 29.7 percent of the world's population was urbanized. In that year there were only 83 metropolitan areas of a million or more, and only 8 of five million or more existed. In 2000 there were approximately 372 metropolitan areas of a million or more people and 45 with more than 5 million. Cities now account for almost half the world's population. Much of the developed world has become almost completely urbanized (**Figure 1.42**), while in many less-developed regions the current *rate* of urbanization is without precedent (**Figure 1.43**). North America is the most urbanized continent in the world, with more than 77 percent of its population living in urban areas. In contrast, Africa and Asia are less than 40 percent urban. Urbanization on this scale is a remarkable geographical phenomenon—one of the most important sets of processes shaping the world's landscapes.

The single most important aspect of world urbanization, from a geographical perspective, is the striking difference in trends and projections between affluent and less-affluent

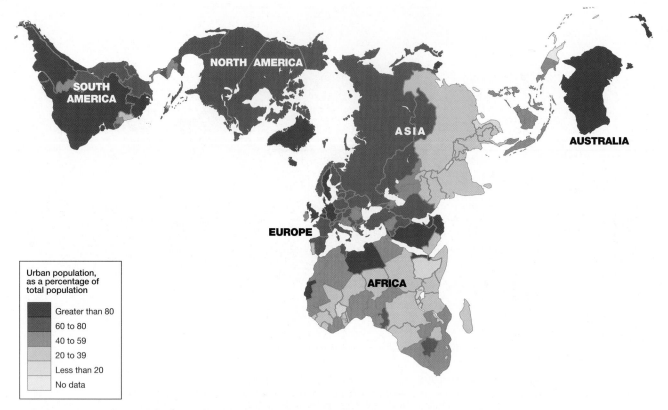

Figure 1.42   The percentage of each country's population living in urban settlements, 2000 The lowest levels of urbanization are found in the African countries of Rwanda and Burundi, where only 7 percent of the population lived in urban settlements in 2000. Most of the core countries are highly urbanized, with between 60 and 95 percent of their populations living in urban settlements. (*Source:* Map projection, Buckminster Fuller Institute and Dymaxion Map Design, Santa Barbara, CA. The word *Dymaxion* and the Fuller Projection Dymaxion™ Map design are trademarks of the Buckminster Fuller Institute, Santa Barbara, California, © 1938, 1967 & 1992. All rights reserved. Data: United Nations, *World Urbanization Prospects,* New York: U.N. Department of Economic and Social Affairs, 1998.)

countries. In 1950, two-thirds of the world's urban population was concentrated in the more affluent countries of Europe and North America. Since then, the world's urban population has increased threefold, the bulk of the growth having taken place in the less-developed countries of other world regions. Asia provides some of the most dramatic examples of this trend. From a region of villages, Asia is fast becoming a region of cities and towns. Between 1950 and 1985, for example, its urban population rose nearly fourfold to 480 million people. By 2020, about two-thirds of Asia's population will be living in urban areas.

The reasons for this urban growth vary. Wars in Liberia and Sierra Leone have pushed hundreds of thousands of people into their capitals, Monrovia and Freetown. In Mauritania, Niger, and other countries bordering the Sahara, deforestation and overgrazing have allowed the desert to expand and swallow up villages, forcing people toward cities. For the most part, though, urban growth in less-developed countries is a consequence of the onset of the demographic transition, which has produced fast-growing rural populations in regions that face increasing problems of agricultural development. As a response, many people in these regions migrate to urban areas seeking a better life.

In many countries, a single city dominates economic affairs to such an extent that their population is several times larger than the next-largest city. Geographers call this condition **primacy.** In Argentina, for example, Buenos Aires is more than 10 times the size of Rosario, the second-largest city. In the United Kingdom, London is more than nine times the size of Birmingham, the second-largest city. In France, Paris is more than 8 times the size of Marseilles, France's second-largest city. In Brazil, both Rio de Janeiro and São Paulo are five times the size of Belo Horizonte, the third-largest city. Primacy is a result of the roles played by particular cities within their own national urban systems. Primacy in less-developed countries is usually a consequence of primate cities' early roles as gateway cities. In more developed countries it is usually a consequence of primate cities' roles as imperial capitals and centers of administration, politics, and trade for an urban system that extends beyond their own national borders.

Ever since the sixteenth century, certain cities known as **world cities** (sometimes referred to as *global cities*) have played key roles in organizing space beyond their own national boundaries. Initially, these roles involved the organization of trade and the execution of colonial, imperial, and geopolitical strategies. The world cities of the seventeenth century were

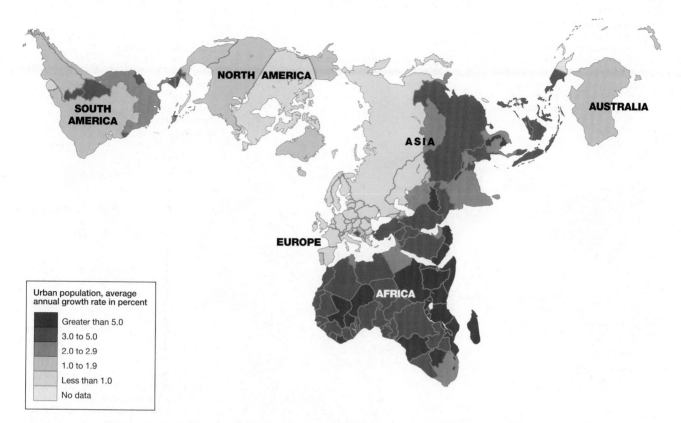

Figure 1.43   Rates of growth in urbanization, 1995–2000   This map shows the annual average growth rate between 1995 and 2000 in the proportion of people in each country living in urban settlements. Core countries, already highly urbanized, grew quite slowly. The urban populations of peripheral countries, such as Afghanistan, Botswana, Burundi, Liberia, Rwanda, and Yemen, on the other hand, grew by more than 6 percent each year, creating tremendous pressure on cities' capacity to provide jobs, housing, and public services. (*Source:* Map projection, Buckminster Fuller Institute and Dymaxion Map Design, Santa Barbara, CA. The word *Dymaxion* and the Fuller Projection Dymaxion™ Map design are trademarks of the Buckminster Fuller Institute, Santa Barbara, California, © 1938, 1967 & 1992. All rights reserved. Data: United Nations, *World Urbanization Prospects*, New York: U.N. Department of Economic and Social Affairs, 1998.)

London, Amsterdam, Antwerp, Genoa, Lisbon, and Venice. In the eighteenth century, Paris, Rome, and Vienna also became world cities, while Antwerp and Genoa became less influential. In the nineteenth century, Berlin, Chicago, Manchester, New York, and St. Petersburg became world cities, while Venice became less influential. Today, with the globalization of the economy, the key roles of world cities are concerned less with the deployment of imperial power and the orchestration of trade and more with transnational corporate organization, international banking and finance, supranational government, and the work of international agencies. World cities have become the control centers for the flows of information, cultural products, and finance that collectively sustain the economic and cultural globalization of the world. World cities also provide an interface between the global and the local. They contain the economic, cultural, and institutional apparatus that channels national and provincial resources into the global economy and that transmits the impulses of globalization back to national and provincial centers.

Today, the global urban system is dominated by three world cities whose influence is truly global: London, New York, and Tokyo. The second tier of the system consists of world cities with influence over large regions of the world economic system. These include, for example, Brussels, Frankfurt, Los Angeles, Paris, Singapore, and Zürich. A third tier consists of important international cities with more limited or more specialized international functions (including Amsterdam, Madrid, Miami, Mexico City, Seoul, and Sydney). A fourth tier exists of cities of national importance and with some transnational functions (including Barcelona, Dallas, Manchester, Munich, Melbourne, and Philadelphia).

Economic Development and Social Well-Being

Patterns of economic development are the result of many different factors. One of the most important is the availability of key resources such as cultivable land, energy sources, and valuable minerals. Key resources are unevenly distributed across the world. Just as important, the *combinations* of energy and minerals that are crucial to economic development are especially uneven in their distribution (**Figure 1.44**). A lack

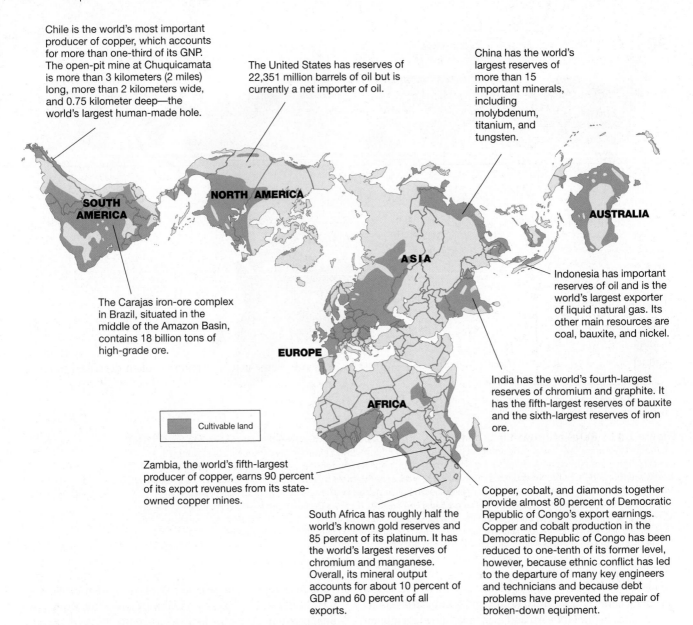

Chile is the world's most important producer of copper, which accounts for more than one-third of its GNP. The open-pit mine at Chuquicamata is more than 3 kilometers (2 miles) long, more than 2 kilometers wide, and 0.75 kilometer deep—the world's largest human-made hole.

The United States has reserves of 22,351 million barrels of oil but is currently a net importer of oil.

China has the world's largest reserves of more than 15 important minerals, including molybdenum, titanium, and tungsten.

Indonesia has important reserves of oil and is the world's largest exporter of liquid natural gas. Its other main resources are coal, bauxite, and nickel.

The Carajas iron-ore complex in Brazil, situated in the middle of the Amazon Basin, contains 18 billion tons of high-grade ore.

India has the world's fourth-largest reserves of chromium and graphite. It has the fifth-largest reserves of bauxite and the sixth-largest reserves of iron ore.

Cultivable land

Zambia, the world's fifth-largest producer of copper, earns 90 percent of its export revenues from its state-owned copper mines.

South Africa has roughly half the world's known gold reserves and 85 percent of its platinum. It has the world's largest reserves of chromium and manganese. Overall, its mineral output accounts for about 10 percent of GDP and 60 percent of all exports.

Copper, cobalt, and diamonds together provide almost 80 percent of Democratic Republic of Congo's export earnings. Copper and cobalt production in the Democratic Republic of Congo has been reduced to one-tenth of its former level, however, because ethnic conflict has led to the departure of many key engineers and technicians and because debt problems have prevented the repair of broken-down equipment.

Figure 1.44   The uneven distribution of the world's known resources Some countries, such as the United States, are fortunate in having a broad resource base of energy, minerals, and cultivable land, which allows for many options in economic development. Many countries have a much narrower resource base, and must rely on the exploitation of one major resource as a means to economic development. Countries that have few natural resources have to pursue pathways to development that are based on manufacturing or services, relying on the profits from exports to pay for imported energy resources, foodstuffs, and raw materials. (*Source*: Map projection, Buckminster Fuller Institute and Dymaxion Map Design, Santa Barbara, CA. The word *Dymaxion* and the Fuller Projection Dymaxion™ Map design are trademarks of the Buckminster Fuller Institute, Santa Barbara, California, © 1938, 1967 & 1992. All rights reserved.)

of natural resources can, of course, be remedied through international trade (Japan's success is a prime example of this), but for most countries the resource base remains an important determinant of development.

A high proportion of the world's key industrial resources—basic raw materials and sources of energy—are concentrated in Russia, the United States, China, Canada, South Africa, and Australia. The biggest single exception to the concentration of key resources in these countries is presented by

the vast oilfields of the Middle East. It is an exception that has enabled formerly less-developed countries such as Saudi Arabia to become wealthy and that has made the region especially important in international politics.

The concentration of known resources in just a few countries is largely a result of geology, but it is also partly a function of countries' political and economic development. Political instability in much of post-colonial Africa, Asia, and Latin America has seriously hindered the exploration and exploitation of resources. In contrast, the relative affluence and strong political stability of the United States has led to a much more intensive exploration of resources. We should bear in mind, therefore, that Figure 1.44 reflects only the currently *known* resource base.

We should also bear in mind that the significance of particular resources is often tied to particular technologies. As technologies change, so resource requirements change, and the geography of economic development is "rewritten." One important example of this was the switch in industrial energy sources from coal to oil, gas, and electricity early in the twentieth century. When this happened, coalfield areas like central Appalachia found their prospects for economic development on indefinite hold, while oilfield areas such as west Texas suddenly had potential. Another example was the switch in the manufacture of mass-produced textiles from natural fibers such as wool and cotton to synthetic fibers in the 1950s and 1960s. When this happened, many farmers in the American South, for example, had to switch from cotton to other crops. Regions and countries that are heavily dependent on one particular resource are vulnerable to the consequences of technological change. They are also vulnerable to fluctuations in the price set for their product on the world market. These vulnerabilities are particularly important for countries such as Bolivia, Chile, Guyana, Liberia, Mauritania, Sierra Leone, Surinam, and Zambia, whose economies are especially dependent on nonfuel minerals.

Resources and Technology   Technological innovations in power and energy, transportation, and manufacturing processes have been important catalysts for changes in the pattern of economic development. They have allowed a succession of expansions of economic activity in time and space; as a result, many existing industrial regions have grown bigger and more productive. Industrial development has also spread to new regions, whose growth has become interdependent with the fortunes of others through a complex web of production and trade. Each major cluster of technological innovations tends to create new requirements in terms of natural resources as well as labor forces and markets. The result is that each major cluster of technological innovations—called *technology systems*—has tended to favor different regions, and different kinds of places. **Technology systems** are clusters of interrelated energy technologies, transportation technologies, and production technologies that dominate economic activity for several decades at a time, until a new cluster of improved technologies evolves. What is especially remarkable about these technology systems is that they have come along at about 50-year intervals. Since the beginning of the Industrial Revolution, we can identify four of them:

1790–1840: Early mechanization based on water-power and steam engines, the development of cotton textiles and ironworking, and the development of river transport systems, canals, and turnpike roads

1840–1890: The exploitation of coal-powered steam engines, steel products, railroads, world shipping, and machine tools

1890–1950: The exploitation of the internal combustion engine, oil and plastics, electrical and heavy engineering, aircraft, radio and telecommunications

1950–: The exploitation of nuclear power, aerospace industries, and electronics and petrochemicals; and the development of limited-access highways and global air routes

A fifth technology system, still incomplete, began to take shape in the 1980s with a series of innovations that are now being commercially exploited:

1990–: The exploitation of solar energy, robotics, microelectronics, biotechnology, advanced materials (fine chemicals, thermoplastics, for example), and information technology (digital telecommunications and geographic information systems, for example)

Each of these technology systems has rewritten the geography of development as it has shifted the balance of advantages between regions. The contemporary economic structure of a country or region is often described in terms of the relative share of primary, secondary, tertiary, and quaternary economic activities. **Primary activities** are those that are concerned directly with natural resources of any kind. They include agriculture, mining, fishing, and forestry. **Secondary activities** are those concerned with manufacturing or processing. They involve the processing, transforming, fabricating, or assembling the raw materials derived from primary activities, or the reassembling, refinishing, or packaging of manufactured goods and include, for example, steel making, food processing, furniture making, textile manufacturing, and garment manufacturing. **Tertiary activities** are those that involve the sale and exchange of goods and services. They include warehousing, retail stores, personal services such as hairdressers, commercial services such as accounting and advertising, and entertainment. **Quaternary activities** are those that deal with handling and processing knowledge and information. Examples include data processing, information retrieval, education, and research and development (R & D).

The economic structure of much of the world is dominated by the primary sector (that is, primary activities such as agriculture, mining, fishing, and forestry). In much of Africa and Asia, between 50 and 75 percent of the labor force is engaged in primary-sector activities. In contrast, the primary sector of the world's affluent countries is typically small, occupying only 5 or 10 percent of the labor force. The secondary sector is much larger in the developed countries, where the world's specialized manufacturing regions are located. The tertiary and quaternary sectors are significant only in the most affluent countries. In the United States, for example, the primary sector in 1999 accounted for less than 4 percent of the labor force, the secondary sector for about 22 percent, the tertiary sector for just over 50 percent, and the quaternary sector for 24 percent of the labor force.

Figure 1.45 **Gross domestic product (GDP) per capita** GDP per capita is one of the best single measures of economic development. This map, based on 1998 data, shows the tremendous gulf in affluence between the core countries of the world economy—such as the United States, Norway, and Switzerland, with an annual per capita GDP (in PPP "international" dollars) of more than $25,000—and peripheral countries such as Haiti, India, and Mali, where annual per capita GDP was less than $2500. In semiperipheral countries such as South Korea, Brazil, and Mexico, per capita GDP ranged between $5000 and $15,000. The global average per capita GDP in 1997 was $6256. (*Source:* Map projection, Buckminster Fuller Institute and Dymaxion Map Design, Santa Barbara, CA. The word *Dymaxion* and the Fuller Projection Dymaxion™ Map design are trademarks of the Buckminster Fuller Institute, Santa Barbara, California, © 1938, 1967 & 1992. All rights reserved.)

Measuring Economic Development Understanding the structure of the world's economies, however, tells us only part of the story of their level of development. At the global scale, levels of economic development are usually measured by economic indicators such as gross domestic product and gross national product. **Gross domestic product** (GDP) is an estimate of the total value of all materials, foodstuffs, goods, and services that are produced by a country in a particular year. To standardize for countries' varying sizes, the statistic is normally divided by total population, which gives an indicator, *per capita* GDP, that provides a good yardstick of relative levels of economic development. **Gross national product** (GNP) includes the value of income from abroad—flows of profits or losses from overseas investments, for example. In making international comparisons, GDP and GNP can be problematic because they are based on each nation's currency. Recently, it has become possible to compare national currencies based on *purchasing power parity* (PPP). In effect, PPP measures how much of a common "market basket" of goods and services each currency can purchase locally, including goods and services that are not traded internationally. Using PPP-based currency values to compare levels of economic prosperity usually produces lower GDP figures in wealthy countries and higher

GDP figures in poorer countries, compared with market-based exchange rates. Nevertheless, even with this compression between rich and poor, economic prosperity is very unevenly distributed across countries.

As **Figure 1.45** shows, most of the highest levels of economic development are to be found in northern latitudes (very roughly, north of 30° N), which has given rise to another popular shorthand for the world's economic geography: the division between the "North" and the "South." In almost all of the more developed countries of North America, northwestern Europe, and Japan, annual per capita GDP (in PPP) in 1998 exceeded $20,000. The only other countries that matched these levels were Australia and Singapore, where annual per capita GDP in 1998 was $22,452 and $24,210, respectively.

In the rest of the world, annual per capita GDP (in PPP) typically ranges between $1000 and $7000. The gap between the highest per capita GDPs ($33,505 in Luxembourg, $29,605 in the United States, and $26,342 in Norway) and the lowest ($458 in Sierra Leone, $480 in Tanzania) is huge. The gap between the world's rich and poor is also getting wider rather than narrower. In 1970 the average GDP per capita of the 10 poorest countries in the world was just one-fiftieth of the average GDP per capita of the 10 most prosperous coun-

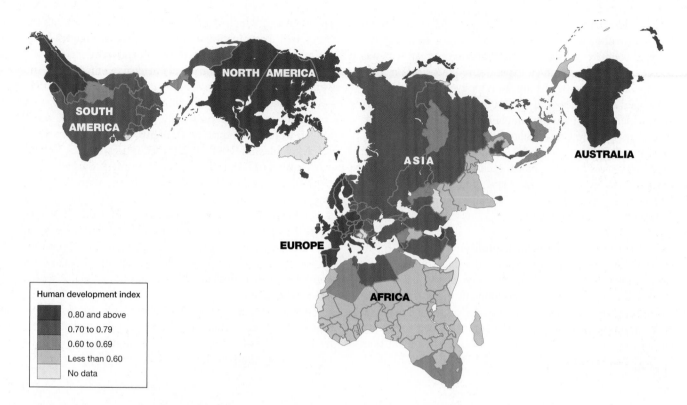

Figure 1.46   An index of human development, 1998 This index, calculated by the United Nations Development Programme, is based on measures of life expectancy, educational attainment, and personal income. A country that had the best scores among all of the countries in the world on all three measures would have a perfect index score of 1.0, while a country that ranked worst in the world on all three indicators would have an index score of 0. Most of the affluent core countries have index scores of 0.9 or more, while the worst scores—those less than 0.4—are concentrated in Africa. (*Source*: Map projection, Buckminster Fuller Institute and Dymaxion Map Design, Santa Barbara, CA. The word *Dymaxion* and the Fuller Projection Dymaxion™ Map design are trademarks of the Buckminster Fuller Institute, Santa Barbara, California, © 1938, 1967 & 1992. All rights reserved.)

tries. By 1990 the relative gap had doubled, and by 2000, the average of the bottom 10 was approaching one two-hundredth of the average of the top 10.

Patterns of Social Well-Being This inequality is reflected—and reinforced—by many aspects of human well-being. Patterns of infant mortality, a reliable indicator of social well-being, show the same steep North-South gradient. For adults in the industrial countries, life expectancy is high and continues to increase. Life expectancy at birth in Australia in 1998 was 78.3 years, in Canada it was 79.1, and in the United States it was 76.8. In contrast, life expectancy in the poorest countries is dramatically shorter. In 1998 in Namibia life expectancy at birth was 50.1 years; in Ethiopia it was 43.4, and in Sierra Leone it was 37.9. In most African countries, only 60 to 75 percent of the population can expect to survive to age 40.

The United Nations Development Programme (UNDP) has devised an overall index of human development based on measures of life expectancy, educational attainment, and personal income. The index is calculated in such a way that a country that had the best scores among all of the countries in the world on all three indicators would have a perfect index score of 1.0, while a country that ranked worst in the world

on all three indicators would have an index score of 0. **Figure 1.46** shows the international map of human development in 1998. Australia, Belgium, Canada, Iceland, the Netherlands, Norway, Sweden, and the United States had the highest overall levels of human development (0.93), while Burkina Faso, Ethiopia, Niger (all at 0.30) and Sierra Leone (0.25) had the lowest levels. The same fundamental pattern is repeated in terms of the entire array of indicators of human development: adult literacy, poverty, malnutrition, access to physicians, public expenditure on higher education, telephone lines, Internet users, and so on. Inequality on this scale poses the most pressing, as well as the most intractable questions of national and international policymaking. It also raises important questions of **spatial justice**—the fairness of geographical variations in people's levels of affluence and well-being, given people's needs and their contributions to the production of wealth and social well-being.

These questions are underscored by some simple comparisons between the needs of people in less-developed countries and the spending patterns of those in the world's most affluent countries. The UNDP has calculated that the annual cost of providing a basic education for all children in less-developed countries would be in the region of $6 billion, which is less than the annual sales of cosmetics in the United States. Providing water and sanitation for everyone in less-developed countries is estimated at $9 billion per year, which is less than Europeans' annual expenditure on ice cream. Providing for basic health and nutrition for everyone in the less-developed countries would cost an estimated $13 billion per year, which is less than the annual expenditure on pet foods in Europe and the United States. Reducing the military expenditures of affluent countries (in the region of $500 billion per year) by less than 10 percent each year would pay for the costs of basic education, water and sanitation, basic health and nutrition, and reproductive health programs for everyone in less-developed countries.

Development and Gender Equality North-South patterns are also reflected in indicators that measure economic development in terms of *gender equality*. The UNDP has established a gender-sensitive development index that adjusts the overall human development index for gender inequality in life expectancy, educational attainment, and income. According to this index, in no country are women better off than men. Perhaps most revealing is the UNDP's Gender Empow-

erment Index, which is based on measures of women's incomes, their participation in the labor force as administrators and managers, professional and technical workers, and the percentage of parliamentary seats held by women. As in the overall index of human development, a country with a perfect score (ranked the best in the world on all measures) would score 1.0, with zero representing the worst possible score (ranked worst on all measures). **Figure 1.47** shows the actual index values for 1998. The top countries were Scandinavian: Norway (0.83), Iceland (0.80), Denmark, Sweden (both at 0.79), and Finland (0.76). The countries with the worst gender empowerment index score were Niger (0.12), Jordan (0.22), Egypt (0.27), and Bangladesh (0.30). Nevertheless, as Figure 1.47 demonstrates, high levels of economic development are not a prerequisite for creating economic opportunities for women. Costa Rica and Trinidad and Tobago both scored better than Italy and Japan, and the Bahamas scored better than Ireland.

Women are, in fact, playing a central and increasing role in processes of development and change in the global economy. In many less-developed countries, women constitute the majority of workers in the manufacturing sector created by

Figure 1.47   An index of gender empowerment, 1997 The United Nations Development Programme's Gender Empowerment Index is based on measures of women's incomes, their participation in the labor force as administrators and managers and professional and technical workers, and the percentage of parliamentary seats held by women. As in the overall index of human development (Figure 1.46), a country with a perfect score would score 1.0, with 0 representing the worst possible score (ranked worst on all measures). The map reflects a broad core-periphery pattern, though there is by no means a direct correlation between economic prosperity and gender empowerment: creating economic opportunities for women does not necessarily require high levels of economic development. (*Source*: Map projection, Buckminster Fuller Institute and Dymaxion Map Design, Santa Barbara, CA. The word *Dymaxion* and the Fuller Projection Dymaxion™ Map design are trademarks of the Buckminster Fuller Institute, Santa Barbara, California, © 1938, 1967 & 1992. All rights reserved.)

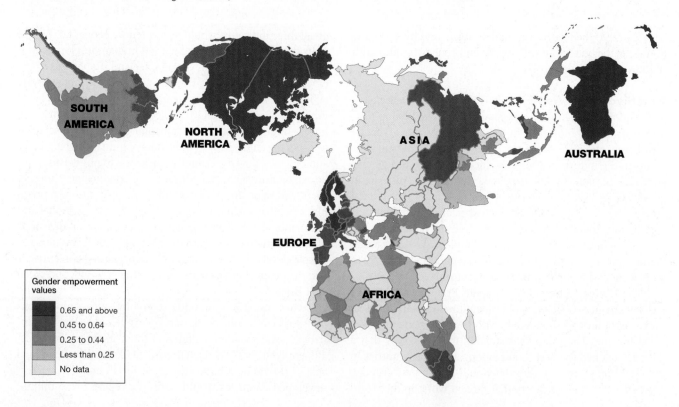

Gender empowerment values

- 0.65 and above
- 0.45 to 0.64
- 0.25 to 0.44
- Less than 0.25
- No data

the new international division of labor. In others, it is women who keep households afloat in a world economy that has resulted in localized recession and intensified poverty. On average, women earn 30 to 40 percent less than men for the same work. They also tend to work longer hours than men: 12 to 13 hours a week more (counting both paid and unpaid work) in Africa and Asia.

Globalization appears to lead to increasing levels of participation by women in the formal labor force. Large firms producing for export tend to employ women in assembly-line jobs because they can be hired for lower wages than men. But increasing participation does not always mean less discrimi-

nation. Women constitute a large share of workers in informal subcontracting—often in the garment industry—at low wages and under poor conditions. Globalization is also associated with increasing levels of home work, tele-work and part-time work. In the United Kingdom, the share of workers in such positions rose from 17 percent in 1965 to 40 percent in 1991. Similar changes have taken place in many other countries, and in most of them women constitute 70 or 80 percent of home, tele-, and part-time workers. This is a mixed blessing. Informal work arrangements can accommodate women's care obligations in the family, but such jobs are typically precarious and underpaid.

Summary and Conclusions

Regional geography combines elements of both physical and human geography and is concerned with the way that unique combinations of environmental and human factors produce territories with distinctive landscapes and cultural attributes. What is distinctive about the study of regional geography is not so much the phenomena that are studied as the *way* they are approached. The study of regional geography draws on several fundamental concepts. Understanding maps and concepts of location, spatial interdependence, sense of place, and landscape allows geographers to analyze the spatial patterns and distributions that underpin regional differentiation and regional change.

Geography matters because it is in specific places that people learn who and what they are and how they should think and behave. Places are also a strong influence, for better or worse, on people's physical well-being, their opportunities, and their lifestyle choices. Places also contribute to peoples' collective memory and become powerful emotional and cultural symbols. Places are the sites of innovation and change, of resistance and conflict.

We must, however, be able to frame our investigations of specific places within the compass of the entire globe. This is important for two reasons. First, the world consists of a complex mosaic of places and regions that are interrelated and interdependent in many ways. Second, place-making forces—especially economic, cultural, and political forces that influence the distribution of human activities and the character of places and regions—are increasingly operating at global and international scales. The interdependence of regions means that individual places are tied in to wider

processes of change that are reflected in broader geographical patterns. An important issue for regional geographers is to recognize these wider processes and broad geographical patterns without losing sight of the individuality and uniqueness of specific places and regions.

While the regional approach provides a rich and intuitively appealing way of organizing knowledge about the world, there are, as we have noted, various methods of regional analysis and different ways of identifying and defining regions. In this book, we emphasize the interdependence of regions, explaining and analyzing them as the outcomes, in different physical environments, of successive eras of human activity that have been organized on the basis of different economic, cultural, and political systems and successive phases of economic and technological development.

In the next chapter, we describe the evolution of this world of interdependent regions, emphasizing the broad divisions of global space in terms of physical environments and in terms of the functional organization of economic and political regions into core, periphery, and semiperiphery. These broad divisions provide the "big picture"—an essential context for understanding the nature and evolution of specific places and regions. Subsequent chapters are organized in terms of major world regions. Within each of these chapters, we describe the region's changing roles in the context of globalization and explore the consequences of these changes for internal regional differentiation and interdependence. Against this background, we examine systematically the core subregions, key cities, and distinctive representative landscapes of the region.

Key Terms

azimuthal projection (p. 7)

cartography (p. 6)

cartogram (p. 11)

cognitive image (p. 6)

colonialism (p. 14)

Columbian Exchange (p. 35)

conformal projection (p. 7)

cultural ecology (p. 28)

demographic transition (p. 37)

economies of scale (p. 14)

environmental determinism (p. 26)

equal-area (equivalent) projection (p. 7)

equidistant projection (p. 7)

fast world (p. 29)

formal region (p. 16)
functional region (p. 16)
geographic information
  systems (p. 5)
Global Positioning System
  (p. 5)
globalization (p. 24)
gross domestic product
  (GDP) (p. 44)
gross national product
  (GNP) (p. 44)
imperialism (p. 14)
irredentism (p. 22)

language family (p. 36)
latitude (p. 5)
longitude (p. 5)
ordinary landscapes (p. 16)
place (p. 2)
primacy (p. 40)
primary activity (p. 43)
quaternary activity (p. 43)
region (p. 2)
regional geography (p. 2)
regionalism (p. 16)
remote sensing (p. 5)
secondary activity (p. 43)

sectionalism (p. 16)
sense of place (p. 17)
site (p. 5)
situation (p. 5)
slow world (p. 29)
sovereignty (p. 18)
spatial diffusion (p. 11)
spatial justice (p. 45)
state (p. 14)
supranational organization
  (p. 18)
sustainable development
  (p. 29)

symbolic landscape (p. 17)
technology system (p. 43)
territory (p. 19)
tertiary activity (p. 43)
time-space convergence
  (p. 15)
vizualization (p. 11)
world city (p. 40)
world regions (p. 18)

Review Questions

Testing Your Understanding

1. What distinguishes regional geography from physical geography and human geography?
2. Define the following forms of *location*: nominal, absolute, relative, cognitive.
3. What is the difference between site and situation?
4. Why is it important to understand different kinds of map projection?
5. How would you differentiate between colonialism and imperialism?
6. Why might a national state join a supranational organization? Give two examples.
7. In a few sentences each, discuss Basque regionalism, Serbian irredentism, and American sectionalism in the former Confederate states.
8. What processes are involved in globalization? Does globalization always result in the reduction of the distinctiveness of places and regions?
9. Does a "digital divide" exist between the fast world and the slow world? Does it exist anywhere else? Explain.
10. What is the relationship between a woman's access to education, employment, and/or healthcare and the number of children she is likely to have?
11. In what ways can the existence of natural resources affect social well-being in a region?
12. Distinguish among the following economic sectors: primary activities, secondary activities, tertiary activities, and quaternary activities.
13. Define gross domestic product (GDP), gross national product (GNP), purchasing power parity (PPP), and human development index (HDI).
14. What are some techniques used to mobilize against globalization? Are they effective? What would you recommend to ordinary people who are not activists?

Thinking Geographically

1. As the global marketplace becomes more accessible to more people worldwide, why will regional geography remain relevant? List at least five reasons why people will maintain their sense of place in an increasingly global economy.
2. According to the 1994 publication *Geography for Life,* which is discussed in this chapter, there are four main reasons for being geographically informed. Briefly, what are they and how are they relevant to your own life?
3. It has been said that "Good fences make good neighbors." If so, then is it a good idea to remove trade or travel barriers between neighboring countries?
4. How realistic is the concept of sustainable development? Please provide two arguments for and against it in your community.
5. Sketch out what happens to a society's birthrate, death rate, and technological level as it moves along the demographic transition from a pre-industrial phase to an industrial phase. Where does a population explosion occur? Why is this a "demographic trap" for so many developing countries?
6. While the concept of primacy may apply to many cities around the world, only a very few can ever be called world cities. How have New York, London, and Tokyo become world cities? Could the status of any of these cities change over time? Could any other cities eventually become world cities? Which ones and why?
7. Issues of spatial justice and gender equity are becoming increasingly salient around the world. Please discuss at least one example of each, indicating how public perceptions of the issues have evolved over the past decade.

Further Reading

Agnew, J., Livingstone, D. N., and Rogers, A., *Human Geography: An Essential Anthology.* Oxford: Blackwell, 1996.

Claval, P., *An Introduction to Regional Geography.* Oxford: Blackwell, 1998.

Dickinson, R. E., *The Regional Concept.* London: Routledge & Kegan Paul, 1976.

Entrikin, N., *The Betweenness of Place.* Baltimore: Johns Hopkins University Press, 1991.

Giddens, A., *Runaway World: How Globalization Is Reshaping Our Lives.* New York: Routledge, 2000.

Gilbert, A., "The New Regional Geography in English- and French-speaking Countries." *Progress in Human Geography,* 12(1988), 208–28.

Gregory, D., *Geographical Imaginations.* Oxford: Blackwell, 1994.

Haggett, P., *The Geographer's Art.* Oxford: Blackwell, 1990.

Harvey, D., *Explanation in Geography.* London: Edward Arnold, 1969.

Knox, P. L., and Marston, S. A., *Human Geography: Places and Regions in Global Context,* 2nd ed. Upper Saddle River, NJ: Prentice Hall, 2001.

Pudup, M. B., "Arguments Within Regional Geography." *Progress in Human Geography,* 12(1988), 369–90.

Rogers, A., Viles, H., and Goudie, A. (eds.), *The Student's Companion to Geography.* Cambridge, MA: Blackwell, 1992.

Tuan, Y-F., "Space and Place: A Humanistic Perspective." *Progress in Human Geography,* 6(1974), 211–52.

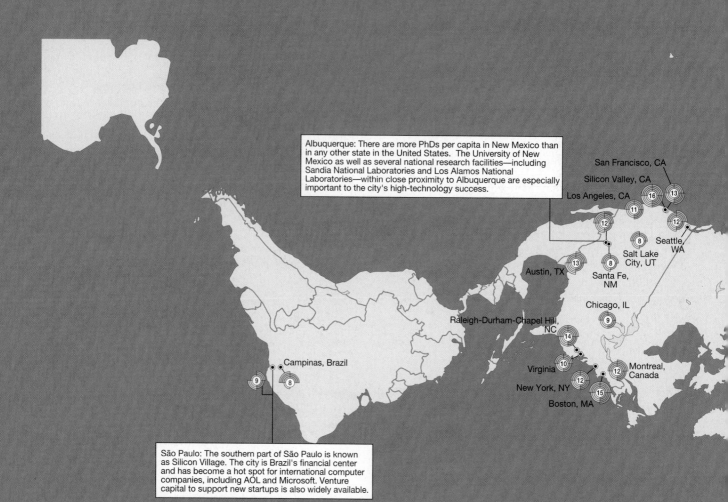

Albuquerque: There are more PhDs per capita in New Mexico than in any other state in the United States. The University of New Mexico as well as several national research facilities—including Sandia National Laboratories and Los Alamos National Laboratories—within close proximity to Albuquerque are especially important to the city's high-technology success.

San Francisco, CA
Silicon Valley, CA
Los Angeles, CA
Seattle, WA
Salt Lake City, UT
Austin, TX
Santa Fe, NM
Chicago, IL
Raleigh-Durham-Chapel Hill, NC
Virginia
Montreal, Canada
New York, NY
Boston, MA
Campinas, Brazil

São Paulo: The southern part of São Paulo is known as Silicon Village. The city is Brazil's financial center and has become a hot spot for international computer companies, including AOL and Microsoft. Venture capital to support new startups is also widely available.

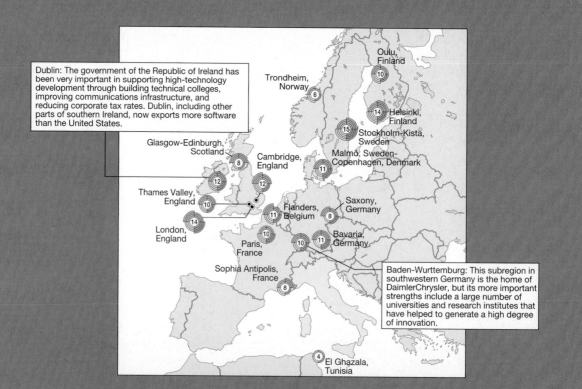

Dublin: The government of the Republic of Ireland has been very important in supporting high-technology development through building technical colleges, improving communications infrastructure, and reducing corporate tax rates. Dublin, including other parts of southern Ireland, now exports more software than the United States.

Oulu, Finland
Trondheim, Norway
Helsinki, Finland
Stockholm-Kista, Sweden
Glasgow-Edinburgh, Scotland
Cambridge, England
Malmö, Sweden-Copenhagen, Denmark
Thames Valley, England
Saxony, Germany
London, England
Flanders, Belgium
Paris, France
Bavaria, Germany
Sophia Antipolis, France

Baden-Wurttemburg: This subregion in southwestern Germany is the home of DaimlerChrysler, but its more important strengths include a large number of universities and research institutes that have helped to generate a high degree of innovation.

El Ghazala, Tunisia

2 The Foundations of World Regions

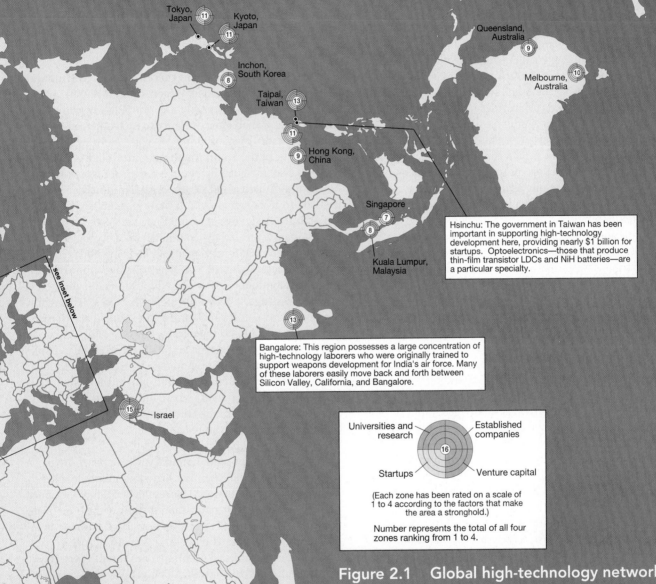

Tokyo, Japan ⑪
Kyoto, Japan ⑪
Inchon, South Korea ⑧
Taipai, Taiwan ⑬
⑪
Hong Kong, China ⑨
Queensland, Australia ⑨
Melbourne, Australia ⑩
Singapore ⑦
⑧
Kuala Lumpur, Malaysia

Hsinchu: The government in Taiwan has been important in supporting high-technology development here, providing nearly $1 billion for startups. Optoelectronics—those that produce thin-film transistor LDCs and NiH batteries—are a particular specialty.

⑬
Bangalore: This region possesses a large concentration of high-technology laborers who were originally trained to support weapons development for India's air force. Many of these laborers easily move back and forth between Silicon Valley, California, and Bangalore.

see inset below

⑮ Israel

④ Gauteng, South Africa

Universities and research — Established companies
⑯
Startups — Venture capital

(Each zone has been rated on a scale of 1 to 4 according to the factors that make the area a stronghold.)

Number represents the total of all four zones ranking from 1 to 4.

Figure 2.1   Global high-technology network

This map shows 46 locations around the globe that are the anchors of the high- technology network. *Wired* magazine consulted with local sources in government, industry, and the media to identify the locations and then rated each on a scale of 1 to 4 in four areas: universities and research, established companies, startups, and venture capital. On the map, the number in the center of the symbol is the location's total score, which represents the sum of the number of concentric rings in each of the four categories. (*Source:* Modified from J. Hillner, "Venture Capitals," *Wired Magazine*, July 2000, pp. 258–59.)

The Changing World

An essential foundation for an informed regional geography is an ability to understand places and regions as components of a constantly changing global system. In this sense, all regional geography is historical geography. Built into every place and each region is the legacy of a sequence of major changes in world geography. One of the key features of this aspect of human geography, however, is that the sequence of changes has not been the same everywhere. We can best understand these changes and their consequences for different places and regions by thinking in terms of the world as a changing, competitive, political-economic system.

We need also to think of the changing world as dependent upon how we alter our physical environment through the economic, political, and sociocultural systems we create. The environment that surrounds us provides us with what we need to produce and enjoy the worlds in which we live. Understanding the limits and resources that the natural world furnishes is just as important as understanding the strengths and weaknesses of the social systems we have created to organize our lives. Human geography as a discipline emphasizes the fact that humans are part of nature, and our practices as sophisticated developers and users of technology and culture have implications not only for the economy and politics but for the air, land, and water upon which we depend. Thus the world of regions we currently operate within is as dynamic, in the physical environmental sense, as it is in the political and economic one (**Figure 2.1**).

The Earth System

The physical and biological characteristics of places make distinctive contributions to our understanding of regional geography because they provide constraints on, and opportunities for, human activities. Many geographers view physical and environmental conditions as highly dynamic and best understood if one thinks of Earth as a system, in which humans play an important role. **Earth system science** helps us understand global patterns of geology, climate, and ecosystems and how they have changed over time and space, producing a physical geography that is dynamically shaped by both natural forces and human impacts. Physical geographers work with other Earth scientists to understand the functioning of the Earth system and with human geographers to interpret the interactions between the Earth system and social, cultural, economic, and political circumstances.

The fundamental Earth system processes that shape world regions are plate tectonics, atmospheric circulation, and ecosystem functioning. Plate tectonics explains the formation of continents and mountain ranges, atmospheric circulation shapes the pattern of world climates, and ecosystem characteristics affect the geography of vegetation, the cycling of key minerals, and the location of animals and other key organisms.

Plate Tectonics   The theory of **plate tectonics** explains how Earth's outer layer, or crust, is structured and has changed over time. This layer, which is from 50 to 100 kilometers (31 to 62 miles) thick, is composed of about a dozen large "plates" of solid rocks floating on a layer of molten material. These plates move very slowly over the more fluid deeper layer and interact at their boundaries, where the resulting tensions are responsible for most of the world's volcanic, earthquake, and mountain-building activity. The continents sit on the plates and emerge from the oceans where Earth's crust is thicker or where the rocks are lighter and therefore more buoyant. At the core of each continent is a region of old (500 million years or older) crystalline rocks called a **continental shield**. Shield areas often contain many important minerals.

Plate tectonics builds on theories about **continental drift**, associated with Alfred Wegener (1880–1930), a German geophysicist who nearly a century ago hypothesized that landmasses had moved relative to each other, and across Earth's surface, over millions of years. He noted the remarkable fit of the South American and African continents and was also intrigued by plant and animal fossils found in both South America and Africa, organisms that he thought unlikely to have swum or to have been transported across the Atlantic Ocean by some other means. According to Wegener's theory, the earliest continent was a single landmass called Pangaea that formed where molten lava emerged from the ocean floor about 225 million years ago (**Figure 2.2**). This supercontinent slowly shifted position, and about 200 million years ago it broke up into two pieces, Laurasia to the north and Gondwanaland to the south. Laurasia was eventually split by narrow ocean gaps into North America and the landmass of Europe and Asia. Gondwanaland split into South America, Africa, Antarctica, India, and Australia. As parts of Gondwanaland drifted north, they connected with Laurasia, most notably where India collided with Asia, causing the uplifting of the Himalayas. The permanent connection between North and South America occurred around 5 million years ago.

At a **convergent plate boundary**, where plates are moving toward each other, one plate may sink under the other in a process called **subduction**. This results in uplift of the upper plate, thus creating mountains, and melting of the other plate as it moves down into Earth, with associated volcanic pressures and earthquakes (**Figure 2.3**). Usually an ocean plate moves under a continental plate, as it does along the west coast of South America. At convergent boundaries, plates may also collide, pushing the surface into steep folds associated with mountain ranges such as the Himalayas and Alps. Collision and compression can also result in the fracturing of rocks in a process known as **faulting**, displacing rocks horizontally and vertically.

At a **divergent plate boundary**, where plates are moving apart, molten rock from deep within Earth may reach the surface through cracks to form volcanoes, and the tearing of Earth's crust may result in earthquakes along what are called *fault zones*. Much of this activity occurs on the ocean floors, but

225 million years ago

200 million years ago

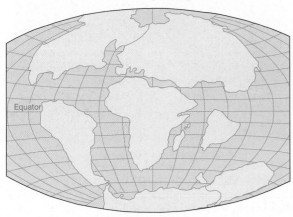

65 million years ago

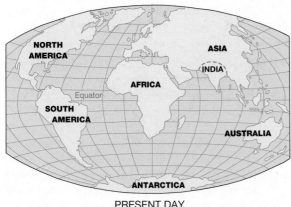

PRESENT DAY

plates can diverge within continents also. On continents, the spreading of the plates may cause blocks of crust to drop, creating a steep-sided **rift valley**, with high, flat plateaus on either side. One example of this is the Rift Valley of East Africa (**Figure 2.4**). Plates may also slide past each other horizontally at a **transform boundary**, creating a zone of earthquakes, an example of which is the San Andreas fault in California. The current configuration of the plates is shown in **Figure 2.5**, which also shows the major zones of fracturing and faulting and of mountain building (also called *orogeny*). This map helps to explain the major landforms of different world regions and highlights the key zones of earthquake and volcanic activity. For example, we can see where the sinking (or subduction) of the Pacific, Cocos, and Nazca plates under the North and South American plates has caused the uplift and folding of the Rocky Mountains and Andes and has created zones of active volcanoes and earthquakes. In a similar manner, the collision of the Eurasian and Indian plates has led to the formation of the Himalaya Mountains (**Figure 2.6**).

Within the continents, other more local processes—erosion, weathering, and sedimentation—create important regional landforms. *Erosion* occurs when water and wind move across the land surface, picking up material and transporting it to other locations. In some cases, heat and the characteristics of water or rocks cause chemical changes and breakdown of material in the process called *weathering*. Erosion has affected many of the world's great mountain ranges, moving material to lower regions and depositing it in a process called *sedimentation*. Extensive areas of deposited sediment occur in the large river basins, such as the Amazon, and across some of the vast plains, such as the North American prairies. Erosion and weathering have been critical in the development of better soils and, over the longer term, in the formation of layers of *sedimentary rocks*. Over time, organic material in sedimentary basins can compress and form reserves of oil and coal. Igneous rock is formed when hot, molten material approaches Earth's surface and solidifies and crystallizes. When existing sedimentary or igneous rock undergoes physical or chemical change under conditions of high temperature and pressure, metamorphic rocks, which often contain valuable minerals, are formed.

Weather and Climate   Earth scientists have combined an understanding of basic physics with information about global patterns of temperature and precipitation to provide explanations of atmospheric circulation, the global movement of air that transports heat and moisture and explains the climates of different regions. Whereas **weather** is defined as the atmospheric condition (for example, a rainy or freezing day) at a particular time and place, **climate** is the typical or average condition over

Figure 2.2   Plate tectonics and continental drift According to the continental drift theory, the supercontinent Pangaea began to break up about 225–200 million years ago, eventually fragmenting into the continents as we know them today.

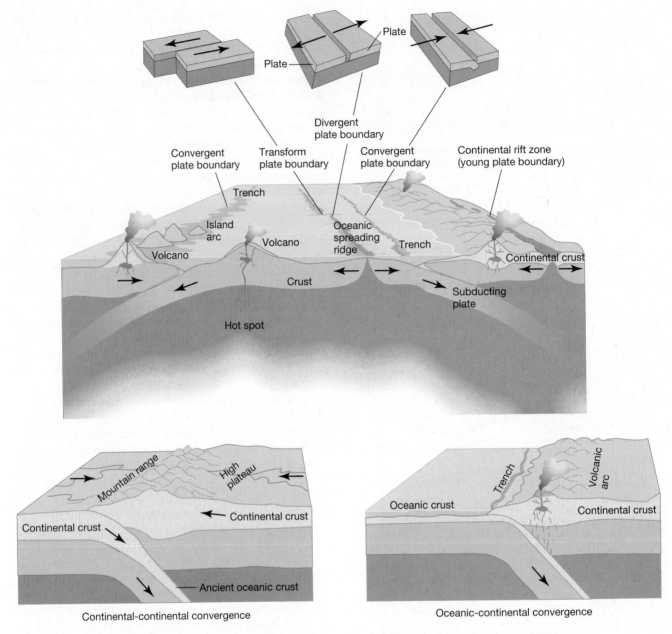

Figure 2.3   Plate convergence and divergence There are four types of plate boundaries illustrated in this figure: (1) Divergent boundaries, where new crust is generated as the plates pull away from each other. (2) Convergent boundaries, where crust is destroyed as one plate dives under another. (3) Transform boundaries, where crust is neither produced nor destroyed as the plates slide horizontally past each other. (4) Plate boundary zones—broad belts in which boundaries are not well defined and the effects of plate interaction are unclear.

a much longer time period and at different seasons (for example, a place with wet, cool winters and hot, dry summers).

A simple model of atmospheric circulation is based on variations in the input of energy from the Sun and the configuration of the major continents and mountain ranges. The spherical shape of Earth, the tilt of its axis, and its revolution around the Sun mean that the sunlight does not hit all parts of Earth's surface at the same angle (**Figure 2.7**). As Earth moves around the Sun, the angle at which sunlight hits Earth varies according to the seasons. The Sun's rays hit Earth most directly and focus the greatest solar energy and heat at the

equator in March and October, at the latitude of the tropic of Cancer (23.5° N) in the Northern Hemisphere in June, and at the latitude of the tropic of Capricorn (23.5° S) in the Southern Hemisphere in December.

The constant high inputs of solar radiation at the equator produce warm temperatures throughout the year, and this warmer air has a tendency to rise into the atmosphere, creating low pressure at ground level, and cooling and condensing into clouds that eventually generate heavy rainfall. This process is called *convectional precipitation* and is typical of the equatorial climate with high temperatures and rainfall year-round.

Figure 2.4   East African rift valley   The East African rift valley formed when a block of crust dropped down between two diverging plates, much as the keystone in an arch will fall if the walls of the arch move apart. This process is responsible for the relatively symmetrical cross sections of most parts of the East African rift system, where the valley floor lies 1000 meters (3280 feet) or more below the higher plateaus of Ethiopia and Kenya. The rift has been forming for some 30 million years (as Africa and the Arabian Peninsula separated) and has been accompanied by extensive volcanism along parts of its length, producing such peaks as Mount Kilimanjaro, at 5950 meters (19,340 feet) the highest mountain in Africa.

The cooler air that rises high into the atmosphere moves out from the equator toward the poles and eventually sinks over tropical latitudes (23.5° N and S), creating a zone of high pressure (**Figure 2.8**). As the air moves toward the surface, it becomes warmer and drier, holding so little moisture by the time it reaches ground level that these regions are characterized by the very low rainfall, sparse vegetation, and warm, dry conditions of desert climates.

When the sinking air reaches ground level, it diverges and some of the air flows back toward the equator, where it converges with the heated air and rises again. This vertical circulation of air from the equator to the Tropics is called the *Hadley cell*, and the zone of convergence near the equator is called the **intertropical convergence zone (ITCZ)** (see Figure 2.8).

The spin of Earth drags air flowing back from the tropical latitudes to the equator into a more east-to-west flow and creates a major wind belt called the *trade winds* that blow from east to west between the dry Tropics and the equator. Air moving from the Tropics toward the poles is similarly pulled by Earth's spin into a major west-to-east flow called the *westerly winds*. Within each of these major wind belts, more complex processes produce high-speed jet streams that can meander across the continents, driving weather systems and the paths of major storms. Where atmospheric circulation brings contrasting air masses with different temperatures and other characteristics into contact along what is called a *front*, the collision and movement of one air mass over another often produces cyclonic precipitation.

The seasonal variation associated with the tilt of Earth's axis and associated changing orientation of the Northern and

Figure 2.5   Major tectonic plates   The crust of the Earth we live on is broken into a dozen or so rigid slabs or tectonic plates that are moving relative to one another. Each arrow represents 20 million years of movement and the direction of movement. The longer arrows indicate that the Pacific and Nazca plates are moving more rapidly than others. The square jagged boundaries indicate spreading ridges offset by transform faults, and the boundaries with small triangles are subduction zones.

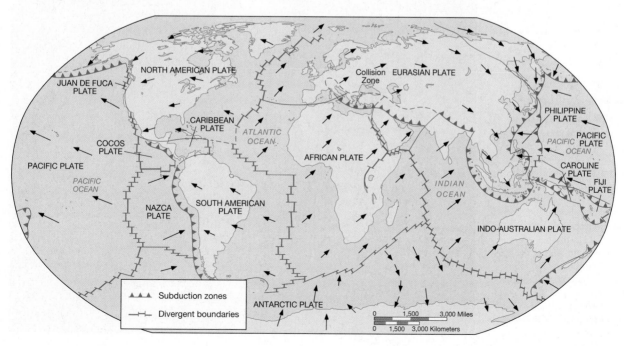

Figure 2.6   Creation of the Himalayas (a) The collision between the Indian and Eurasian plates has pushed up the Himalayas and the Tibetan Plateau. (b) These cross-sections show the orientation of the two plates before and after their collision. The reference points (small squares) show the amount of uplift of an imaginary point in Earth's crust during this mountain-building process. (c) The Himalayas are among the highest mountains in the world resulting from very active mountain building in the region.

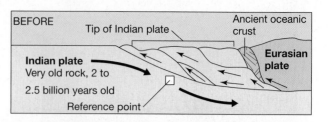

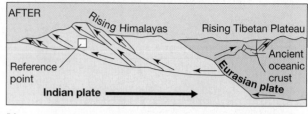

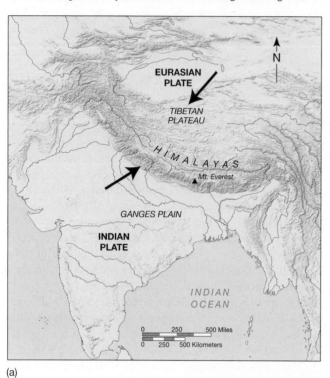

(a)

(b)

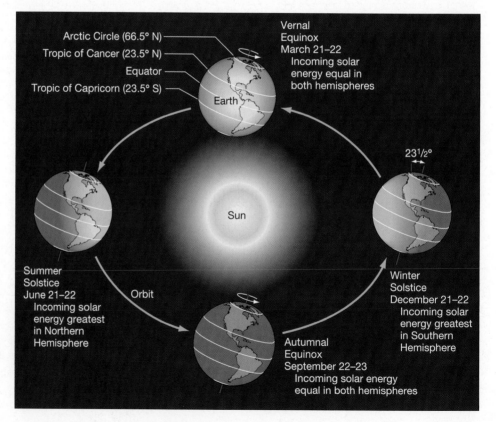

(c)

Figure 2.7   Seasonal incidence of Sun's rays by latitude As Earth moves around the Sun, the angle at which sunlight hits Earth varies according to the seasons. The Sun's rays hit Earth most directly and focus the greatest solar energy and heat at the equator in March and October, at the latitude of the tropic of Cancer (23.5° N) in the Northern Hemisphere in June, and at the latitude of the tropic of Capricorn (23.5° S) in the Southern Hemisphere in December. (*Source:* E. Aguado, and J. E. Burt, *Understanding Weather and Climate*, 2nd ed. Upper Saddle River, NJ: Prentice Hall, 2001.)

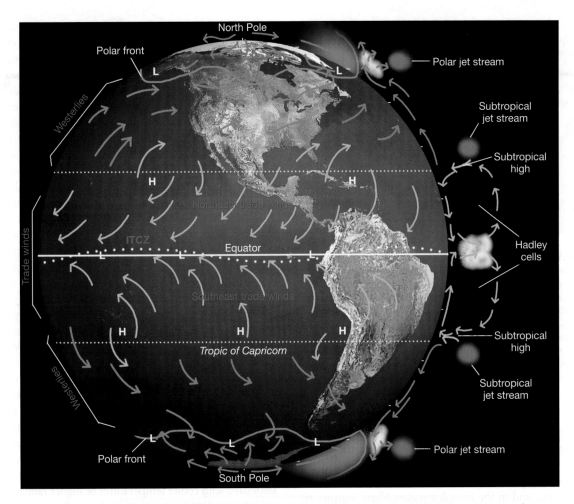

Figure 2.8   Atmospheric circulation The general circulation of the atmosphere is based on air moving from regions of high to low pressure and on the vertical lift of air in the regions of highest heating over the equator at the ITCZ. Air that is heated by the Sun's direct rays at the equator rises and then moves north and south toward the poles high in the atmosphere, sinking in a high pressure zone at about 30° latitude north and south and returning at the surface to the equator in a system called the *Hadley cell*. The stability of the air in the high-pressure zones creates a zone of little surface wind and dry conditions called the *horse latitudes*. Rising air at the equator cools and moisture condenses to cause heavy rainfall. Air moves from high pressure at about latitude 30° north and south toward low pressure at the equator and toward low pressure at the polar front at about 60° north and south. Earth's spin turns these winds clockwise in the Northern Hemisphere and anticlockwise in the Southern Hemisphere (called the *Coriolis effect*) to create the trade winds (or easterlies) blowing from east to west between the equator and 30°, and the westerlies blowing from west to east between 30° and 60°. These major wind and pressure belts shift north and south with the seasons (north when the Sun is strongest in the Northern Hemisphere in June and south when the Sun is strongest in the Southern Hemisphere in December). (*Source:* R. W. Christopherson, *Geosystems: An Introduction to Physical Geography,* 4th ed. Upper Saddle River, NJ: Prentice Hall, 2000.)

Southern Hemispheres toward the Sun (the Northern Hemisphere facing the Sun more directly in June than in December) means that the zones of rising and sinking air, and the major wind belts, move northward in June and southward in December, with corresponding shifts in the zones of rainfall and dry conditions.

When winds blow across warmer oceans, they tend to pick up moisture, and when moisture-laden air encounters a landmass, especially coastal mountains, it condenses into rainfall or snow. Precipitation associated with mountains, called *orographic precipitation,* may result in the formation of a dry rainshadow region on the inland, or lee, side of the mountains, where sinking air that has lost its moisture becomes even drier. The trade winds flow across the oceans in tropical lati-

tudes and frequently produce rain on east-facing coasts in what is sometimes called the *trade wind climates* (**Figure 2.9**).

Similarly the westerly winds bring rain as they blow from the oceans onto western coasts. The regions on the margins of the trades, and of the equatorial rainfall zone, have highly seasonal climates with a distinct rainy season. Seasonal shifts in pressure and wind belts mean that the westerlies move nearer the equator in December and to the poles in June, resulting in distinct wet and dry seasons on the margins of the westerly circulation. When the global circulation shifts southward in December, storms spinning out of the Northern Hemisphere westerlies bring rain to the poleward margins of drier regions in the Northern Hemisphere.

Figure 2.18 Terraced landscapes This landscape in Sikkim is the legacy of a hydraulic society—a world-empire in which despotic rulers once organized labor-intensive irrigation schemes that allowed for significant increases in agricultural productivity. Irrigation schemes such as this were widely copied and, having been maintained for generations, have become the basis for local economies and ways of life in a number of subregions in Asia and Latin America. The photo shows rice cultivation in Sikkim, formerly a kingdom but now an Indian state.

colonizers and colonies. It was also important in establishing hierarchies of settlements and creating improved transportation networks. The military underpinnings of colonization also meant that new towns and cities now came to be carefully sited for strategic and defensive reasons.

The legacy of these important changes is still apparent in today's landscapes. The clearest examples are in Europe, where the Roman world-empire colonized an extensive territory that was controlled through a highly developed system of towns and connecting roads (see Chapter 3, p. 104). Some world-empires were exceptional in that they were based on a particularly strong central state, with totalitarian rulers who were able to organize large-scale, communal land improvement schemes using forced labor. These world-empires were found in China, India, the Middle East, MesoAmerica, and the Andean region of South America. Their dependency on large-scale land improvement schemes (particularly irrigation and drainage schemes) as the basis for agricultural productivity has led some scholars to characterize them as *hydraulic societies*. Today, their legacy can be seen in the landscapes of terraced fields that have been maintained for generations in places such as Sikkim, India, and East Java, in Indonesia (**Figure 2.18**).

The Geographic Foundations of the Modern World

Figure 2.19 shows the generalized framework of human geographies in the Old World as they existed around A.D. 1400. The following characteristics of this period are important. First, harsher environments in continental interiors were still peopled by isolated, subsistence-level, kin-ordered hunting and gathering minisystems. Second, the dry belt of steppes and

desert margins stretching across the Old World from the western Sahara to Mongolia was a continuous zone of kin-ordered pastoral minisystems. Third, the hearths of sedentary agricultural production extended in a discontinuous arc from Morocco to China, with two main outliers: in the central Andes and in MesoAmerica. The dominant centers of global civilization were China, northern India (both of them hydraulic variants of world-empires), and the Ottoman Empire of the eastern Mediterranean. Other important world-empires were based in Southeast Asia, in Muslim city-states of coastal North Africa, in the grasslands of West Africa, around the gold and copper mines of East Africa, and in the feudal kingdoms and merchant towns of Europe.

These more-developed realms were interconnected through trade, which meant that there were several emerging centers of capitalism. Port cities were particularly important, and among the leading centers were the city-state of Venice, the Hanseatic League of independent city-states in northwestern Europe (including Bergen, Bremen, Danzig, Hamburg, Lübeck, Riga, Stockholm, and Tallinn, and affiliated trading outposts in other cities, including Antwerp, Bruges, London, Turku, and Novgorod), Cairo, Calicut, Canton, Malacca, Lisbon, Madrida, and Sofala. Traders in these port cities began to organize the production of agricultural specialties, textiles, and craft products in their respective hinterlands. The **hinterland** of a town or city is its sphere of economic influence—the tributary area from which it collects products to be exported and through which it distributes imports. By the fifteenth century there were several regions of budding capitalism: in northern Italy, Flanders; southern England; the Baltic; the Nile Valley; Malabar, Coromandel, and Bengal in India; as well as northern Java and southeast coastal China. (See Chapter 3, p. 104, for an extended discussion of the role of Europe in globalizing capitalism.)

Organizing the Core

The transformation of much of Europe, and later the United States, as core regions beginning in the fifteenth century was predicated on complex innovations and institutions that stabilized and enabled capitalist political and economic organization to flourish, particularly those central to trade and later industrialization. While all of these institutions and innovations emerged prior to or during the mercantile period, they were stabilized and extended geographically during the industrial period in order to support the requirements of capitalist economic organization. **Mercantilism** was an economic policy prevailing in Europe during the sixteenth, seventeenth, and eighteenth centuries, when the idea of governmental control over industry and trade was first introduced. The following are some of the primary factors that played critical roles in organizing and consolidating the core from the fifteenth century onward:

1. The division of labor that enabled increased productivity
2. The standardization of time, space, measure, value, and money that allowed for more predictability and consistency in commerce and manufacture

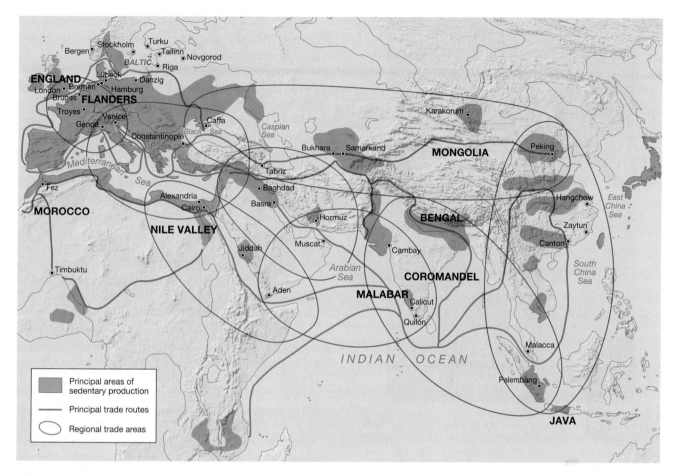

Figure 2.19 The precapitalist Old World, circa A.D. 1400 Principal areas of sedentary agricultural production are shaded. Some long-distance trade took place from one region to another, but for the most part it was limited to a series of overlapping regional circuits of trade. (*Source:* After R. Peet, *Global Capitalism: Theories of Societal Development.* New York: Routledge, 1991; J. Abu-Lughod, *Before European Hegemony: The World-System A.D. 1200-1350.* New York: Oxford University Press, 1989; and E. R. Wolf, *Europe and the People Without History.* Berkeley: University of California Press, 1983.)

3. The forging of national identities that affirmed the emergence of a unified and powerful state

4. The controlling and commodifying of nature

5. The development of internal physical infrastructures, such as railroads, canals, and communications systems, which improved the movement of goods, people, and ideas

The Division of Labor

All contemporary societies are organized around a division of labor, whether it is a complex or simple one. Furthermore, the division of labor has been a feature of human societies from the earliest times. A **division of labor** is the separation of productive processes into individual operations, each performed by different workers or groups of workers. For example, in many pre-industrial societies, men hunt, trap, and fish while women manage household gardens and tend to children. When the separation of productive processes is based on gender, it is known as a **gender division of labor** (**Figure 2.20**). Other divisions of labor might be based on skill or vocation, as it was in the me-

dieval period when the system of social and economic organization was based on different groups of artisans and merchants organized into guilds.

For the world-system core, as it concentrated its role as the dominant economic force around the globe through industrialization, a technical division of labor premised on factory production enabled remarkably greater productivity at the same time that it increasingly divided tasks into many discrete processes and laborers into numerous skill groups. Although an important innovation, the division of labor is not unequivocally beneficial. At one level, the division of labor is ingenious because it enables a substantial increase in individual and collective efficiency due to the increase in skill that specialization provides. At another level, however, the division of labor separates workers from the product of their labor so that they contribute only a small fragment to the whole and may come to feel meaningless in the larger scheme of production. The technical division of labor is a central institutional feature of the economic organization of the core. At the same time that it has enabled the core to be efficient in productive activities in an unprecedented way, it

Figure 2.20   The gender division of labor, Lake Turkana, Kenya Throughout the world there are gender divisions of labor, in some places more apparent than in others. In core countries, men tend to occupy the highest paid occupations and, although this is changing, women are more likely to be the primary caretakers of children. Pictured here is a gender division of labor in Kenya where a man is seen handling the herd while women have dug water holes in a dry river bed and are lifting water from it to carry back to their homes, often many kilometers away.

has also narrowed and segmented individuals' social identities, virtually equating a person's social identity with his or her role in the division of labor.

Standardization of Time, Space, Measure, Value, and Money

Standardization of timekeeping, measurement, currency, and value also became a central requirement for conducting transactions in the daily life of the core as well as in an increasingly interconnected world. At the level of everyday life, standardization of time, for instance, ensured that workers arrived each morning at the factory gate at the same time. Whereas previously national, regional, and local time standards, or just general daily and seasonal sensitivity to light and dark, shaped human activity patterns, the standardization of time meant that sharply defined work schedules and deadlines became part of factory life and daily habits. With the increasing interconnectedness of the globe, the standardization of space—a framework for determining relative location and distance—was also critical (see Geography Matters: Globalization and the Standardization of Space, p. 72).

Other standardizations besides time, however, were also necessary. The standardization of measure, for instance, ensured that replacement parts for a machine would fit all the same machines in any factory anywhere in the world. Furthermore, the prices for goods could be fixed more precisely and consistently through an agreed-upon system of weights and measures. Moreover, value—how much a worker was paid and how much a product was worth—came to be defined in terms of money or currency, and the standardization of money simplified and extended the transactions of market exchange. The stabilization of credit—the ability to borrow money—was also an important accomplishment of core industrialization.

Forging of National Identities and Construction of States

Another important aspect of the organization of the core was the increasing political significance of a national population unified around a strong state. A state is an independent political unit with territorial boundaries that are internationally recognized by other states; a **nation** is a community of people often sharing common elements of culture, such as religion or language, or a history or political identity.

Following the overthrow or decline of monarchies in Europe in the late eighteenth to mid-nineteenth centuries, a number of new republics were created. Republican government, as distinct from monarchy, requires the democratic participation and support of its population. Monarchical political power is derived from force and subjugation; republican political power derives from the support of the governed. By creating a sense of nationhood, the newly emerging states of Europe were attempting to homogenize their multiple and sometimes conflicting constituencies so that they could govern with their active cooperation according to a sense of a common purpose.

France provides an excellent illustration of the process of nation building and its advantages to the state. As early as the sixteenth century, the highly centralized French state began actively discouraging the use of regional languages and dialects in official transactions. Following the overthrow of the monarchy in France in the late eighteenth century, the new French republic advanced a policy intended to establish unity among the various provinces by suppressing the regional languages. The multiplicity of languages was seen as a barrier to stable democracy and egalitarianism. The argument for such a policy was that free people—that is, people who were no longer subjects of the monarch—must speak the same language (north-central, or Parisian, French) in order to unify France and promote democracy as a way of life (**Figure 2.21**). After all, how could the people create and operate a government if they could not speak to each other? As a result, the regional languages and dialects of France went into a decline, hastened by official government policies spanning an extended period from the time of Napoleon Bonaparte (emperor from 1810 to 1814) to that of Charles de Gaulle (president from 1959 to 1969).

In addition to enabling the creation of a stable democracy, the construction of a nation also enables the organization

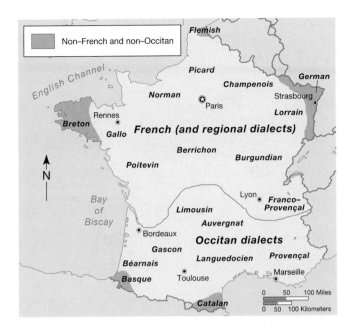

Figure 2.21   The languages of France in 1789 On the eve of the French Revolution, language diversity in France was not so dissimilar from other European regions that were consolidating into states. Whereas a multiplicity of local languages and dialects prevailed before the emergence of a strong central state, many governments created policies to eliminate them. Local languages made it difficult for states to collect taxes, enforce laws, and teach new citizens. (*Source:* D. Bell, "Lingua Populi, Lingua Dei," *American Historical Review,* 1995, p. 1406.)

of a more extensive and coherent market where buyers and sellers all speak the same language and all have an investment in the success of the economic enterprise. A national identity is built not only upon a common language but also upon a common sense of history and purpose such that individuals feel compelled to defend the nation and to further the objectives of the state.

Diaspora is the term used to characterize the spatial dispersion of a previously homogeneous group. Because of the increasing interconnectedness of the world, there has been more opportunity for population mobility, which has resulted in the widespread dispersal of Earth's peoples. As such, contemporary members of a nation recognize a common identity, but they need not reside within a common geographical area. For example, the Jewish nation refers to members of the Jewish culture and faith throughout the world regardless of their place of origin. The term **nation-state** is an ideal form consisting of a homogeneous group of people governed by their own state. In a true nation-state, no significant group exists that is not part of the nation.

In fact, few pure or true nation-states exist today. Rather, multinational states exist—states composed of more than one regional or ethnic group. Spain is such a multinational state (composed of Catalans, Basques, Gallegos, and Castilians), as is France, Kenya, the United States, and Bolivia. Multinational states are far more typical than homogeneous nation-states. Since the First World War, it has become increasingly common for groups of people sharing an identity different from the majority, yet living within the same political unit, to agitate to form their own state separate from the existing one. The existence of nationalist conflict demonstrates that the project of the state to homogenize its citizens has not always been entirely successful. This has been the case with the Québécois in Canada and the Basques in Spain, as discussed in Chapter 1, p. 22. It is out of this desire for autonomy that the term *nationalism* emerges. **Nationalism** is the feeling of belonging to a nation as well as the belief that a nation has a natural right to determine its own affairs (**Figure 2.22**).

Controlling and Commodifying Nature

Perhaps the most widespread conception of nature that informed the imperialist practices of the core—and one that has persisted under different labels for thousands of years—is that humans are the center of all creation and that nature in all its wildness was meant to be dominated by humans. Judeo-Christian belief insists that Man (as an ideal type), made in the image of God, was set apart from nature and must be encouraged to control it.

While early Christianity held that nature was to be dominated, that idea existed more in the religious and spiritual realm than in the political and social realm. In terms of the conduct of everyday life, it was not until the sixteenth century that Christian theology was coaxed from its isolation and conscripted to aid the goals of science. Before 1500 in Europe there existed a widely held image of Earth as a living entity such that human beings conducted their daily lives in an intimate relationship with the natural order of things. The prevailing metaphor was that of the organism, which emphasized interdependence among human beings and between them and Earth (**Figure 2.23**). Yet even within this organic idea of nature, there are two opposing conceptions. One was of a nurturing Earth that provided for human needs in a beneficent way; the other was of a violent and uncontrollable nature that could cause general chaos in human lives. In both of these views, Earth and nature were regarded as female.

Francis Bacon (1561–1626) and Thomas Hobbes (1588–1679) were English philosophers who, as prominent promoters of science and technology, were influential in changing the prevailing organic view of nature. Borrowing from the Christian ideology, they advanced a view of nature as something subordinate to Man. Bacon and Hobbes sought to rationalize benevolent nature as well as to dominate disorderly and chaotic nature.

As feminist environmental historian Carolyn Merchant writes:

> The change in controlling imagery was directly related to changes in human attitudes and behavior toward the earth. Whereas the nurturing earth image can be viewed as a cultural constraint restricting the types of socially and morally sanctioned human actions allowable with respect to the earth, the new images of mastery and

Figure 2.22   Nationalist conflict around the globe This map depicts the locations around the globe where nationalism has provoked serious disagreement or conflict. Some conflicts have lasted for years, sometimes even decades, whereas others have flared and subsided. The conflict between India and Pakistan over Kashmir is more than 50 years old and is based largely on religious differences. Many of the conflicts ongoing in the Democratic Republic of Congo date back to independence in 1960. Here four major ethnic groups and more than 200 smaller groups, or tribes, have home territories that extend beyond the national borders that act as artificial barriers to the interactions of homogenous groups. More recent conflict has occurred in Guatemala. Here peasants are fighting against government forces.

domination functioned as cultural sanctions for the denudation of nature. Society needed these new images as it continued the process of commercialization and industrialization, which depended on activities directly altering the earth—mining, drainage, deforestation, and assarting [grubbing up stumps to clear fields]. The new activities used new technologies—lift and force pumps, cranes, windmills, geared wheels, flap valves, chains, pistons, treadmills, under- and overshot watermills, fulling mills, flywheels, bellows, excavators, bucket chains, rollers, geared and wheeled bridges, cranks, elaborate block and tackle systems, worm spur, crown, and lantern gears, cams and eccentrics, ratchets, wrenches, presses, and screws in magnificent variation and combination.[1]

Merchant shows that by the sixteenth and seventeenth centuries the power of science was too great for the organic idea of nature. Subsequently, a view that nature was the instrument of Man became dominant in Western culture. This view of nature underlay the "age of discovery" that propelled Europeans to claim Africa, North and South America, and parts of Asia for their own commercial and political uses.

The view that nature was to be controlled by humans was accompanied by the parallel idea that nature was a commod-

ity to be exploited and produced. The word *commodity* is of medieval English origin, and it means anything that has a use value, which means that it has some usefulness to someone. Sheep, rivers, and trees were all referred to as commodities because they could be used to sustain and shelter life. With the rise of capitalism, the popular understanding of use value was replaced by that of exchange value, and a commodity came to mean anything useful that *could be bought or sold*. Production for sale in the marketplace is what makes things—whether foodstuffs, minerals, animals, even human beings—commodities. Capitalism has made it possible for everything and anything to be a commodity under certain conditions. For example, recent environmental agreements coordinated in Kyoto, Japan, have made the right to pollute the air a commodity that can be bought and sold.

The commodifying of nature that accompanied the rise of capitalism was feverishly pursued during the age of discovery as European explorers found new resources to exploit in new places. Unfortunately, the history of the capitalist pursuit of producing commodities often includes the destruction and substantial dislocation of whole social systems so that these commodities could be marketed and made profitable. Important for the development of the interconnectedness of world regions is the fact that the turning of nature into commodities also has a history that is part of the history of the globalization of capitalism. In brief, any commodity has a history that links it to an origin and then traces its diffusion out-

[1]C. Merchant, *The Death of Nature*, San Francisco: Harper & Row, 1979, pp. 2–3.

Figure 2.23 Commodifying and controlling nature This painting by Jacopo Zucchi, completed in the mid-sixteenth century, depicts gold mining. The painting carries a wide range of messages about society during this period and, in particular, the relationships between rich and poor. Notice that the men who are digging, carrying, and extracting the ore from the rock are depicted as extremely muscular and well matched to the rugged Earth with which they must contend.

Figure 2.24 Canal St. Martin, Paris The canal systems that opened up the interiors of Europe and North America in the eighteenth century were initially dependent on horse power. Later, barges were able to utilize steam- and oil-powered engines, giving new life to larger canals and navigable rivers. The three-mile-long canal pictured here opened in 1825 and provided a shortcut between meanders in the Seine River. The boat pictured in this photograph is a pleasure boat, but barges are still an important mode of commercial transport for bulky goods.

ward to other parts of the globe. A commodity is linked to the societies that first recognized its use value, then understood how to enhance that value by cultivating or processing it, and finally began to trade it. But the history of any commodity must also take into account the links that were created to connect it economically, politically, socially, and culturally to other parts of the world. Finally, the history of a commodity is not only about the commodity itself, such as wine, tobacco, sugar, oil, or diamonds—commodities whose history will be discussed in the following chapters—but it is also the history of the billions of people who have produced, desired, and consumed the commodities.

In order to market commodities, links must be created between producers and consumers: Commodities must be transported from where they are harvested or produced to where they will be consumed. The development of the transportation infrastructure for moving commodities from one part of the world to another or from one side of a city to another is also a significant aspect of the organization of the core of the world-system.

Development of Internal Infrastructures

Within the world's core regions, the transformation of regional geographies hinged on successive innovations in transport technology. These innovations opened up continental interi-

ors for commercial agriculture and intensified inter-regional trading networks. Farmers were able to mechanize their equipment, while manufacturing companies were able to take over more resources and more markets.

Canals and the Growth of Industrial Regions The first phase of this internal geographic expansion and regional integration was in fact based on an old technology: the canal (**Figure 2.24**). Merchant trade and the beginnings of industrialization in both Britain and France were underpinned by extensive navigation systems that joined one river system to another. By 1790, France had just over 1000 kilometers (620 miles) of canals and canalized rivers; Britain had nearly 3600 kilometers (2230 miles). The Industrial Revolution provided both the need and the capital for a spate of additional canal building that began to integrate and extend emerging industrial regions.

In Britain, 2000 more kilometers (1240 miles) of canals were built between 1790 and 1810. In France, which did not industrialize as early as Britain, 1600 additional kilometers (990 miles) were built between 1830 and 1850. In the United States the landmark was the opening of the Erie Canal in 1825. This breakthrough enabled New York, a colonial gateway port, to reorient itself toward the nation's growing interior. The Erie Canal was so profitable that it set off a "canal fever" that resulted in the construction of some 2000 kilometers (1240 miles) of navigable waterways in the next 25 years. This canal system helped to bind the emergent Manufacturing Belt together.

Steamboats, Railroads, and Internal Development The scale of the United States was such that a network of canals was a viable proposition only in more densely settled

Geography Matters

Globalization and the Standardization of Space

Great Britain and France, two countries at the center of the industrialization of the core, were instrumental in getting other countries around the world to accept the key standards that enabled increasing global interconnectedness and independence. Great Britain established and garnered worldwide acceptance of the standardization of time (Greenwich Mean Time) and the standardization of space (Greenwich prime meridian). France convinced most of the world and all of the scientific community to accept the metric system as the standard for measurement.

The standardization of space was important to exploration and the opening up of trading opportunities. Although the concepts of geographical longitude and latitude as defining positions on Earth's surface likely were formulated by the ancient Greeks, they were not then conceived as they are today. The Greeks used latitude and longitude to measure *time,* particularly with respect to the numbers of hours of daylight at a particular location at a particular time of the year. Today, we think of latitude as measuring *distance*—in degrees, north or south of the equator. We think of longitude as also measuring distance, but east and west of the prime meridian. Although the Greeks understood the equator as their 0-degree latitude line, they used an arbitrary line as their 0-degree longitude line and placed it in the westernmost part of their known world, the "Fortunate Isles"—now known as the Canary and Madeira islands off the northwest coast of Africa. The major difference between the ancient Greeks and us is that we *equate time with distance* in establishing longitude. In the mid-eighteenth century the determination of longitude was finally standardized through the painstaking labor of a British clockmaker who was able to build

a reliable and precise method of keeping marine time. Dava Sobel, who has published an account of the eighteenth-century quest for a practical means of determining longitude, writes:

> To learn one's longitude at sea, one needs to know what time it is aboard ship and also the time at the home port or another place of known longitude—at the very same moment. The two clock times enable the navigator to convert the hour difference into a geographical separation. Since the Earth takes twenty-four hours to complete one full revolution of three hundred and sixty degrees, one hour marks one twenty-fourth of a spin, or fifteen degrees. And so each hour's time difference marks a progress of fifteen degrees of longitude east or west. . . . Those same fifteen degrees of longitude correspond to distance traveled.[*]

For ships traveling before the late eighteenth century, there was no precise way to establish the time in two different places because clocks were unreliable timekeepers, particularly because they were pendulum clocks, and the rocking of the ship, temperature changes, changes in barometric pressure, and ambient moisture from the sea caused them to run down, speed up, or stop completely.

In 1714 the British Parliament passed the Longitude Act. The act established a prize, worth several million dollars in today's currency, for anyone who could develop a "Practicable and Useful" means of determining longitude. (A French government official offered a series of related, though less lu-

---

[*]D. Sobel, *The Illustrated Longitude,* New York: Walker, 1995, p. 46.

---

areas. The effective colonization of the interior could not take place until the development of steam-powered transportation: first riverboats, and then railroads. The first steamboats were developed in the early 1800s, offering the possibility of opening up the vast interior by way of the Mississippi River and its tributaries. By 1830 the technology and design of steamboats had been perfected, and navigable channels had been established. The heyday of the river steamboat was between 1830 and 1850. During this period, vast acreage of the U.S. interior was opened to commercial, industrialized agriculture—especially cotton production for export to British textile manufacturers. At the same time, river ports such as New Orleans, St. Louis, Cincinnati, and Louisville grew rapidly, extending the frontier of industrialization and modernization.

By 1860 the railroads had taken over the task of internal development, further extending the frontier of settlement and industrialization and intensifying the use of previously developed regions (see Geography Matters: Railroads and Geographic Change, p. 74). The railroad originated in Britain, where George Stephenson engineered the world's first commercial railroad, a 20-kilometer (12.4-mile) line between Stockton and Darlington that opened in 1825. The famous *Rocket,* the first-ever locomotive for commercial passenger trains, was designed mainly by Stephenson's son Robert for the Liverpool & Manchester line that opened four years later. The economic success of this line sparked two railroad-building booms that eventually created a highly efficient transportation network for Britain's manufacturing industry. In other core countries, where there was sufficient capital to license (or

crative, prizes as well.) Shipping disasters, particularly the loss of life and cargo despite the best available charts and compasses, were behind the announcement of prizes.

John Harrison was the clockmaker who worked nearly a lifetime to build a precise and portable clock, in the form of a nautical chronometer, that would determine longitude. Over the years, Harrison refined his nautical timekeepers through four prototypes known as H1, H2, H3, and the final successful one, H4 (**Figures 1, 2**). Nearly 60 years after the Longitude Act was announced, Harrison's marine timekeeper was presented to a panel of judges—made up of scientists, naval officers, and government officials—and accepted as possessing the ability to determine longitude accurately. Ultimately, the solution to the longitude problem linked science and technology for the purpose of promoting more effective and dependable trade. With an accurate scientific understanding and an elegant technological application, the precise determination of longitude ensured that ships would not get lost and ships' cargo would come to port.

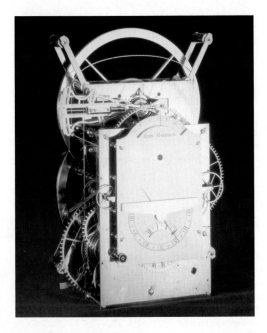

Figure 1 H3 timepiece Being able to measure time accurately and consistently was the key to establishing longitude. British clockmaker John Harrison was eventually able to accomplish this difficult feat through the development of a series of timepieces, each of which was more dependable, accurate, and portable than the last. H3 pictured here is housed in the museum at Greenwich, England.

Figure 2 H4 timepiece The H4 timepiece was not only accurate, it was also portable. Designed to be carried in a mariner's pocket, the piece is clearly elegant as well as functional.

copy) the locomotive technology and install the track, railroad systems led to the first full stage of economic and political integration.

While the railroads integrated the economies of entire countries and allowed vast territories to be colonized, they also brought some important regional and local restructuring and differentiation. In the United States, for example, the railroads led to the consolidation of the Manufacturing Belt. They also contributed to the mushrooming of Chicago as the focal point for railroads that extended the Manufacturing Belt's dominance over the West and South. This reorientation of the nation's transportation system effectively ended the role of the cotton regions of the South as outliers of the British trading system. Instead, they became outliers of the U.S. Manufacturing Belt, supplying factories in New England and the Mid-Atlantic Piedmont. This left New Orleans, which had thrived on cotton exports, to cope with an abrupt end to its phenomenal growth.

Tractors, Trucks, Road Building, and Spatial Reorganization

In the twentieth century the internal combustion engine powered further rounds of internal development, integration, and intensification (**Figure 2.25**). The replacement of horse-drawn farm implements with lightweight tractors powered by internal combustion engines, beginning in the 1910s, amounted to a major revolution in agriculture. Productivity was increased, the frontiers of cultivable land were extended, and vast amounts of labor were released for industrial work in cities. The result was a parallel revolution in the geographies of both rural and urban areas.

Railroads and Geographic Change

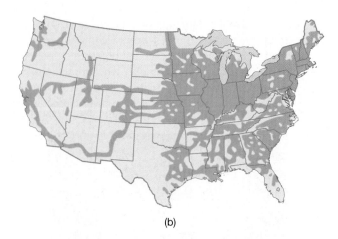

(b)

Railroad service in (a) 1860 and (b) 1880 Areas shown in gray are no more than 24 kilometers (15 miles) from a railroad line. (*Source:* P. Hugill, *World Trade Since 1431*. Baltimore: Johns Hopkins University Press, 1993, p. 179.)

(a)

Railroad construction was labor-intensive but required little more than common farm equipment applied to a graded surface. Most of the labor force consisted of immigrants: European immigrants on railroads driven westwards from the East Coast, and Chinese immigrants on those driven eastwards from the West Coast.

Chicago's railyards at their peak, in 1942.

The first passenger train services averaged little more than 20 to 35 kilometers per hour (15–20 m.p.h.), but locomotive technology changed rapidly and made it easier, faster, and cheaper to conquer the vast interior distances of the United States. Between 1830 and 1845, the United States created the world's largest rail system—some 5458 kilometers (3688 miles), compared to Britain's 3083 kilometers (2069 miles), Germany's 2956 kilometers (1997 miles), France's 817 kilometers (552 miles), and Belgium's 508 kilometers (343 miles).

Grand Central Station, New York City, with the Hotel Commodore, behind.

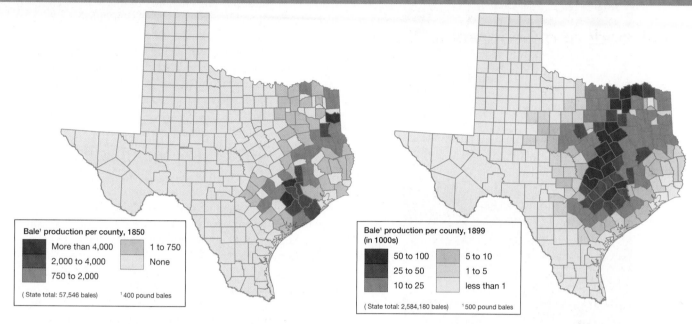

Bale[1] production per county, 1850

More than 4,000	1 to 750
2,000 to 4,000	None
750 to 2,000	

(State total: 57,546 bales) [1] 400 pound bales

Bale[1] production per county, 1899 (in 1000s)

50 to 100	5 to 10
25 to 50	1 to 5
10 to 25	less than 1

(State total: 2,584,180 bales) [1] 500 pound bales

The railroads reorganized the geography of America's cities and opened up its interior regions. Railway stations reordered land-use patterns in central business districts, where new hotels and department stores competed for sites near the station's entrance. Farther away, the railroad lines cut swathes through the urban fabric, separating neighborhoods from one another and establishing the basis for a radial framework for the physical development of the city.

Before the railroads reached Texas, cotton production was concentrated in a few areas along short sections of navigable rivers. The railroads, which reached the interior of Texas after 1850, allowed cotton production to extend inland onto the high plains. (*Source:* P. Hugill, "The Macro-Landscape of the Wallerstein World Economy," *Geoscience and Man,* 25, 1988, figs. 1 and 5, pp. 78 and 80, Department of Geography and Anthropology, Louisiana State University.)

The Illinois Central Railroad issued a promotional poster in the 1860s (left), celebrating the role it envisaged—that its trunk line from Chicago to New Orleans would significantly affect the settling of the American interior. Note the deliberate juxtaposition of the high technology of the railroad locomotive and the telegraph line, contrasting with the obsolescent technologies of the stagecoach (inset top right) and barge (inset bottom right). Even the oceangoing ship is shown with masts and sails, to diminish its status in comparison to the railroad locomotive. The poster issued 50 years later (right), on the other hand, shows the railroad locomotive in the company of several important new technologies: aircraft, automobiles and trucks, and electric streetlights.

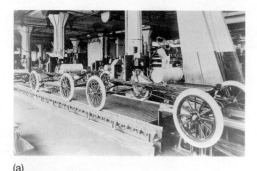

(a)

(b)

Figure 2.25 The impacts of the internal combustion engine The internal combustion engine revolutionized the geography of the more affluent and developed parts of the world between 1920 and 1970. (a) The production of trucks in the United States increased from 74,000 in 1915 to 750,000 in 1940 and 1.75 million in 1965. Automobile production increased from 896,000 in 1915 to 3.7 million in 1940 and 9.3 million in 1965. (b) Meanwhile, road building opened up interior regions and opened out metropolitan regions. Between 1945 and 1965, federal highways increased from a total of 456,936 kilometers (308,741 miles) to 1,344,908 kilometers (908,722 miles). Later, limited-access highways (autobahns in Germany, autostrada in Italy, interstate highways in the United States, and motorways in Britain) provided such an efficient and flexible means of long-distance passenger and freight movement that railways were eclipsed as the principal framework for the transportation geography of industrial countries. With this eclipse, the spatial organization of industries and land uses was radically reorganized, and a new round of geographical change took place.

The development of trucks in the 1910s and 1920s suddenly released factories from locations tied to railroads, canals, and waterfronts. Trucking allowed goods to be moved farther, faster, and cheaper than before. As a result, trucking made it feasible to locate factories on inexpensive land on city fringes and in smaller towns and peripheral regions where labor was cheaper. It also increased the market area of individual factories and reduced the need for large product inventories. This decentralization of industry, in conjunction with the availability of buses, private automobiles, and massive road-building programs, brought about another phase of spatial reorganization. The outcomes of this phase were the specialized and highly integrated regions and urban systems of the modern core of the world-system. This integration was not simply a question of their being interconnected through highway systems. It also involved close economic linkages between manufacturers, suppliers, and distributors, linkages that enabled places and regions to specialize and develop economic advantages.

Organizing the Periphery

Parallel with the internal development of core regions were changes in the geographies of the periphery of the world-system. Indeed, the growth and internal development of the core regions simply could not have taken place without the foodstuffs, raw materials, and markets provided by the colonization of the periphery and the incorporation of more and more territory into the sphere of industrial capitalism.

As soon as the Industrial Revolution had gathered momentum in the early nineteenth century, the industrial core nations embarked on the inland penetration of the world's midcontinental grassland zones in order to exploit them for grain or stock production. This led to the settlement, through the emigration of European peoples, of the temperate prairies and pampas of the Americas, the veld in southern Africa, the Murray-Darling Plain in Australia, and the Canterbury Plain in New Zealand. At the same time, as the demand for tropical plantation products increased, most of the tropical world came under the political and economic control—direct or indirect—of one or another of the industrial core nations. In the second half of the nineteenth century, and especially after 1870, there was a vast increase in the number of colonies and the number of people under colonial rule.

The colonization and imperialism that accompanied the expansion of the world-system was closely tied to the evolution of world leadership cycles. **Leadership cycles** are periods of international power established by individual states through economic, political, and military competition. In the long term, success in the world-system depends on economic strength and competitiveness, which brings political influence and pays for military strength. With a combination of economic, political, and military power, individual states can dominate the world-system, setting the terms for many economic and cultural practices and imposing their particular ideology by virtue of their preeminence. This kind of dominance is known as *hegemony*. **Hegemony** refers to domination over the world economy, exercised—through a combination of economic, military, financial, and cultural means—by one national state in a particular historical epoch. Over the long run, the costs of maintaining this kind of power and influence tend to weaken the hegemon. This phase of the cycle is known as *imperial overstretch*. It is followed by another period of competitive struggle, which brings the possibility of a new dominant world power.

Imperialism and Colonialism: Imposing New Geographies on the World

The incorporation of the external arena into the periphery was motivated by several factors, among them the basic logic of free trade, investment, and the desire for new territories. Although Britain was the leading world economic and military hegemon in the late nineteenth century, several other European countries (notably Germany, France, and the Netherlands), together with the United States—and later Japan—were competing for global influence. This competition developed into a scramble for territorial and commercial domination. The core countries engaged in preemptive geographic expansionism in order to protect their established interests and to limit the opportunities of others. They also wanted to secure as much of the world as possible—through a combination of military oversight, administrative control, and economic regulations—in order to ensure stable and profitable environments for their traders and investors. This combination of circumstances defined a new era of imperialism. Over the last 500 years, imperialism has resulted in the political or economic domination of strong core states over the weaker states of the periphery. Imperialism involves some form of authoritative control of one state over another (**Figure 2.26**).

Africa, more than any other peripheral region, was given an entirely new geography under imperialism and colonialism. As previously mentioned, colonialism is the establishment and maintenance of political and legal domination by a state over a separate and alien society. Virtually the entire continent of Africa was carved up into a patchwork of European colonies and protectorates in just over 30 years, between 1880 and 1912, with little regard for either physical geography or the preexisting human geographies of minisystems and world-empires. At the same time, the major powers jostled and squabbled over small Pacific islands that had suddenly become valuable as strategic provisioning stations for their navies and merchant fleets. Resistance from indigenous peoples was quickly brushed aside by imperial navies with iron steamers and high-explosive guns and by troops with rifles and cannon. European weaponry was so superior that Otto von Bismarck, the founder and first Chancellor (1871–90) of the German empire, referred to these conflicts as "sporting wars." Between 1870 and 1900, European countries added almost 22 million square kilometers (10 million square miles) and 150 million people to their spheres of control: 20 percent of Earth's land surface and 10 percent of its population.

The imprint of imperialism and colonization on the geographies of the newly incorporated peripheries of the world-system was immediate and profound. The periphery became almost entirely dependent on European and North American capital, shipping, managerial expertise, financial services, and news and communications. As a consequence of this, it also came to depend on European cultural products: language, education, science, religion, architecture, and planning. All of this came to be etched into the landscapes of the periphery in a variety of ways as new places were created, old places were remade, and regions were reorganized.

One of the most striking changes in the periphery was the establishment and growth of externally oriented port cities through which commodity exports and manufactured imports were channeled. Often, these major ports were also colonial administrative and political capitals, so that they became overwhelmingly important, growing rapidly to sizes far in excess of other settlements. Good examples include Georgetown (Guyana), Lagos (Nigeria), Luanda (Angola), Karachi (Pakistan), and Rangoon (Burma). Meanwhile, the interior geography of peripheral countries was restructured as smaller settlements were given new functions: colonial administration and commercial marketing. As with the interior development of the core countries, transport networks were vital to this process. Railroads provided the principal means of spatial reorganization, and in the colonies of Africa, Central America, and South and Southeast Asia, railroad lines evolved into linear patterns with simple feeder routes and limited interconnections that focused almost exclusively on major port cities.

An International Division of Labor

The fundamental logic behind imperialism and colonization was economic: the need for an extended arena for trade, an arena that could supply foodstuffs and raw materials in return for the industrial goods of the core. The outcome was an international division of labor, driven by the needs of the core and imposed through its economic and military strength. This **international division of labor** involved the specialization of different people, regions, and countries in certain kinds of economic activities. In particular, colonies began to specialize in the production of those foodstuffs and raw materials:

- For which there was an established demand in the industrial core (for example, foodstuffs, industrial raw materials)
- For which colonies held a **comparative advantage**, in that their productivity was higher than for other possible specializations
- That did not duplicate or compete with domestic suppliers within core countries (tropical agricultural products such as cocoa and rubber, for example, simply could not be grown in core countries)

The result was that colonial economies were founded on narrow specializations that were oriented to, and dependent upon, the needs of core countries. Examples of these specializations are many: bananas in Central America; cotton in India; coffee in Brazil, Java, and Kenya; copper in Chile; cocoa in Ghana; jute in East Pakistan (now Bangladesh); palm oil in west Africa; rubber in Malaya (now Malaysia) and Sumatra; sugar in the Caribbean islands; tea in Ceylon (now Sri Lanka); tin in Bolivia; and bauxite in Guyana and Surinam. Most of these specializations have continued through to the present. Thus, for example, 48 of the 55 countries in Sub-Saharan Africa still depend on just three products—tea, cocoa, and coffee—for more than half of their export earnings.

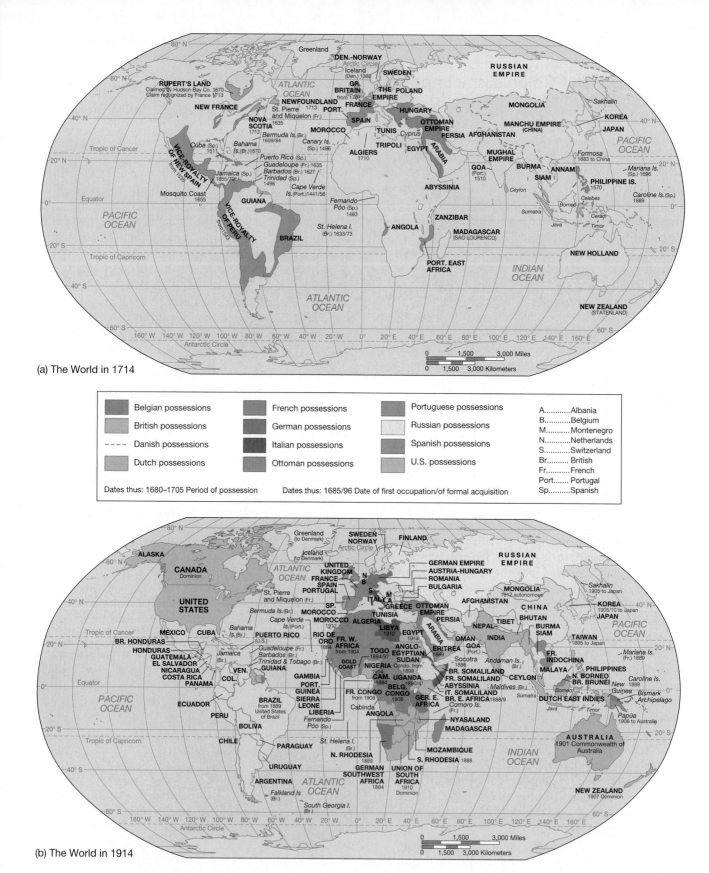

(a) The World in 1714

▢ Belgian possessions	▢ French possessions	▢ Portuguese possessions
▢ British possessions	▢ German possessions	▢ Russian possessions
- - - Danish possessions	▢ Italian possessions	▢ Spanish possessions
▢ Dutch possessions	▢ Ottoman possessions	▢ U.S. possessions

A............Albania	
B............Belgium	
M...........Montenegro	
N............Netherlands	
S............Switzerland	
Br..........British	
Fr...........French	
Port.......Portugal	
Sp..........Spanish	

Dates thus: 1680–1705 Period of possession   Dates thus: 1685/96 Date of first occupation/of formal acquisition

(b) The World in 1914

Figure 2.26   The world in 1714 and 1914 These two maps illustrate the transformation in the political geography of the globe that occurred during the 200-year period that marks the most intense period of European global imperialism and colonialism. The maps show that imperialism in the Americas was effectively over by the nineteenth century, whereas imperialism in Africa and Asia was in full swing. The Spanish and Portugese empires expanded early and for an extended period of time, though the populations ruled by them were relatively small. By the mid-eighteenth century, more and more of the global population was being brought under European control, with the expansion of the British empire being the most dramatic. (a) In 1714, European possessions amounted to less than 10 percent of the world's land area and only 2 percent of the world's population. (b) By 1914, European colonies amounted to more than 55 percent of the world's land area and 34 percent of the world's population. (*Source:* Redrawn from B. Crow and A. Thomas, *Third World Atlas.* Milton Keynes: Open University Press, 1982, pp. 37, 41).

This new world economic geography took some time to establish, and the details of its pattern and timing were heavily influenced by technological innovations. The incorporation of the temperate grasslands into the commercial orbit of the core countries, for example, involved successive changes in regional landscapes as critical innovations such as barbed wire, the railroad, and refrigeration were introduced. But the single most important innovation behind the international division of labor was the development of metal-hulled, ocean-going steamships. This development was in fact cumulative, with improvements in engines, boilers, transmission systems, fuel systems, and construction materials adding up to produce dramatic improvements in carrying capacity, speed, range, and reliability. The construction of the Suez Canal (opened in 1869) and the Panama Canal (opened in 1914) was also critical, providing shorter and less hazardous routes between core countries and colonial ports of call. By the eve of the First World War, the world economy was effectively integrated by a system of regularly scheduled steamship trading routes (**Figure 2.27**). This integration in turn was supported by the second most important innovation behind the international division of labor: a network of telegraph communications (**Figure 2.28**) that enabled businesses to monitor and coordinate supply and demand across vast distances on an hourly basis.

The international division of labor brought about a substantial increase in trade and a huge surge in the overall size of the capitalist world economy. The peripheral regions of the world contributed a great deal to this growth. By 1913 Africa and Asia provided more *exports* to the world economy than either North America or the British Isles. Asia alone was *importing* almost as much, by value, as North America. The industrializing countries of the core bought increasing amounts of foodstuffs and raw materials from the periphery, financed by

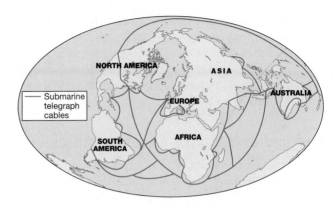

Figure 2.28   The international telegraph network, 1900
For Britain, submarine telegraph cables were the nervous system of its empire. Of the global network of 246,000 kilometers (152,860 miles) of submarine cable, Britain had laid 169,000 kilometers (105,015 miles).

profits from the export of machinery and manufactured goods. Britain, the leading world power of the period, drew on a trading empire that was truly global (**Figure 2.29**).

Patterns of international trade and interdependence became increasingly complex. Britain used its capital to invest not just in peripheral regions but also in profitable industries in other core countries, especially the United States. At the same time, these other core countries were able to export cheap manufactured goods to Britain. Britain financed the purchase of these goods, together with imports of food from its dominion states (Canada, South Africa, Australia,

Figure 2.29   The British empire, late 1800s Protected by the all-powerful Royal Navy, the British merchant navy established a web of commerce that collected food for British industrial workers and raw materials for its industries, much of it from colonies and dependencies appropriated by imperial might and developed by British capital. So successful was the trading empire that Britain also became the hub of trade for other states. (*Source:* After P. Hugill, *World Trade Since 1431.* Baltimore: Johns Hopkins University Press, 1993, p. 136.)

Figure 2.27   Principal steamship routes, 1920 The shipping routes reflect (1) the transatlantic trade between the two core regions of the world-system at the time; and (2) the colonial and imperial relations between the world's core economies and the periphery. Transoceanic shipping boomed with the development of steam-turbine engines for merchant vessels and with the construction of shipping canals, such as the Kiel Canal, the Suez Canal, the St. Lawrence Seaway, and the Panama Canal. When the 82 kilometers (51 miles) of the Panama Canal opened in 1914, shipping could move between the Atlantic and the Pacific without having to go around South America, saving thousands of kilometers of steaming.

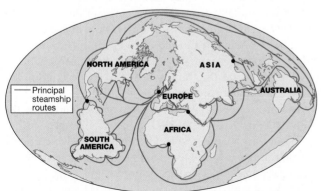

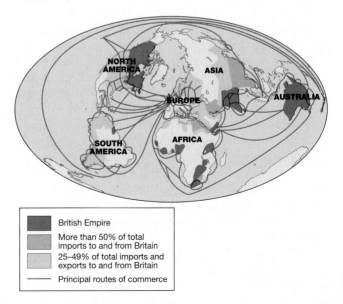

and New Zealand) and colonies, through the export of its own manufactured goods to peripheral countries. India and China, with large domestic markets, were especially important. Thus there developed a widening circle of exchange and dependence, with constantly switching patterns of trade and investment.

Political and Cultural Geographies

Keep in mind that the impact of colonialism was not uniform throughout the globe. The local expressions of colonialism were immensely complex. For example, the impact of the British on what we now call India was substantially different from the impact of the French of what we now call Algeria. Moreover, not only did colonialism affect different places differently because the domestic political, economic, cultural, and social systems of the various colonial powers differed markedly and came into contact with dramatically different existing systems, but also the deployment and the impact of colonialism was experienced differently by different social groups in the colonies. Men and women, young and old, elites and peasants, rural people and urban dwellers, these and others each negotiated and were shaped by colonialism in distinct ways. In short, **culture**, a shared set of meanings that are lived through the material and symbolic practices of everyday life, is a central mediating element in the way groups experience change. Yet, despite the enormous variety that characterizes the impact of colonialism on colonized peoples and places, there are at least two broad repercussions that can be identified.

First, the territorial divisions that reflected colonial boundaries were seldom consistent with preexisting political or environmental geographies. Second, core social institutions—such as legal systems of property ownership—that were transplanted to the colonies disrupted or severely dislocated existing ones. As a result of both, political and environmental geographies as well as ethnic geographies were fragmented or reconfigured, increasing the potential for conflict or problems.

As mentioned previously, the nation-state was a European invention the aim of which was to encourage culturally, geographically, politically, and socially distinct groups to imagine themselves as possessing significant, binding connections to each other. In order to achieve this sense of "imagined community," the new states of Europe erased and replaced fiefdoms and other forms of feudal political organization through a whole range of practices from the extermination of local languages to the organizing of a national army. In colonizing the periphery, European powers acted in similar though perhaps less conscious ways. As national aggrandizement through the formal incorporation of new territory was often a central aim of imperial efforts, areas were occupied and bounded often without any heed to the political, social, cultural, and environmental relationships that existed before colonization.

For instance, in central Africa, both the British and the French vied for the same territory. In establishing the southern boundary of Niger and the northern boundary of Nigeria,

France and Britain fragmented the Hausa city-state that had been established in the eleventh century. Those Hausa people to the north of the boundary became French subjects; those to the south, British. In each case their experience of colonialism was different. Although culturally homogenous when the colonial powers arrived, Hausaland ceased to exist as a political entity at the beginning of the twentieth century when it was divided between the French and the British. While the new political geographies imposed by colonialism had severely dislocating impacts on numerous cultural groups around the globe, contemporary Hausa people on both sides of the present borders of Nigeria/Niger have maintained a great deal of interaction, and most of them continue to farm, as they have for many generations, growing millet and sorghum in the sandy soil of the Sahelian desert.

A second broadly consistent impact of colonization for the colonized was the imposition of core social institutions that reshaped or undermined existing ones. Arguably, the most profoundly dislocating social institution to be introduced through colonization were European systems of property rights with respect to land. Throughout the periphery, distinctive land tenure systems existed that determined access to land, largely for farming, herding, or hunting, but also for other sorts of settlement uses. It is unwise to generalize about the types of land tenure systems that existed throughout the periphery as they were far too numerous and varied. Before the British began colonization in the late eighteenth century in present-day India, several major types—and lots of local variations—of tenure systems were in operation. Many of these had been the legacy of previous colonizers stretching back to the Greeks, who came to India in the third century B.C. European legal frameworks for acquiring, holding, and selling land ran up against a whole host of existing systems that were based not on ownership rights through monetary purchase but use rights through kinship or communal affiliation.

European land tenure systems and legal frameworks for establishing property rights introduced to North America clashed calamitously with the various Native American ones. American Indian groups believed that land was free to be used and enjoyed by all and that it belonged to or was owned by no one. Indians had no concept of private property rights. Europeans, however, brought with them an arsenal of ideological and legal understandings of land that contradicted existing Indian understandings. There were three central ways in which Europeans formally acquired land: through the doctrine of discovery, treaty negotiation, and the theory of higher use. The doctrine of discovery allowed early European explorers in North America to regard native peoples as conquered subjects of the Crown, and thus any land discovered reverted to the Crown through preemption. Under the doctrine of discovery, Native Americans retained the right to occupy and enjoy the fruits of the land.

Treaty negotiations were a second way in which Europeans acquired land in North America. Treaties were elaborate legal frameworks that recognized Native Americans as independent nations with the ability to negotiate land transfers, often through exchanges for other non-land objects. Representatives of the Crown were empowered to negotiate treaties with Native American representatives while the United States and Canada were still colonies. Finally, the theory of higher land use allowed the rational purchase and acquisition of Native American lands that were not being put to their highest and best use in the eyes of the Crown and later the U.S. government. As American Indian groups increasingly came to understand the implications of their loss of access to land for hunting, agriculture, and the fundamentals of their way of life, they resisted external acquisition through wars and other forms of hostility and conflict. European frameworks for acquiring land were largely incomprehensible to native groups, as the very notion of land ownership was so fundamentally contrary to their own. In recent years, North American Indian groups have recognized that they have legal recourse to challenge the taking of their lands, and they have begun to raise significant and legally compelling challenges based on treaties that continue to have the power of legal documents defensible in a court of law (**Figure 2.30**).

Exploration and Exploitation

The scramble to incorporate the periphery into the world-system in the late eighteenth and early nineteenth centuries was aimed at acquiring territory and cultivating commercial opportunities. Imperialism and colonialism were operationalized through the processes of exploration and exploitation. Exploration is generally understood to mean the growth of knowledge of the globe that occurred as a result of voyages of discovery and scientific expeditions. It should be noted, however, that an alternative view is to see the encounter between the core and the periphery that took place during this period as, more accurately, one of invasion and conquest. Whichever way one looks at the moral issues behind the historical meeting of the core and periphery, it is unquestionable that exploration and scientific discovery went hand-in-hand and that geography as a discipline contributed substantially to both (**Figure 2.31**).

The experience of the encounter between the core and the periphery was certainly complicated. It is uncontestable

Figure 2.30 Diminishing tribal lands, 1790–2000 Two hundred years ago American Indians controlled three-quarters of the present-day United States. Today the 358 federally recognized Indian reservations cover less than 2 percent of the United States. (*Source:* Bureau of Indian Affairs; Smithsonian Institution; *New York Times*, "Mending a Trail of Broken Treaties," June 25, 2000, Section WK, p. 3.)

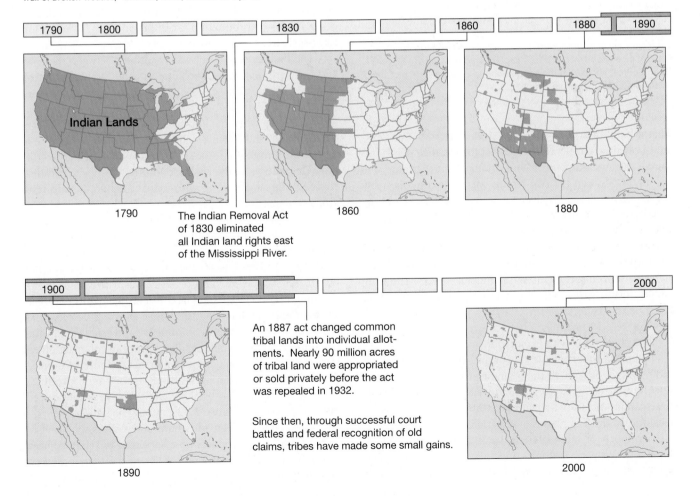

Figure 2.31   Captains Speke and Grant addressing the Royal Geographical Society, London By the mid-nineteenth century, thriving geographical societies had been established in a number of cities, including Berlin, London, Frankfurt, Moscow, New York, and Paris. By 1899 there were 62 geographical societies worldwide, and university chairs of geography had been created in many of the most prestigious universities around the world. It has to be said, however, that geography was seen at first in narrow terms, as the discipline of exploration. Because the importance of geography was linked so clearly to European commercial and political ambitions, ways of geographic thinking also changed. Places and regions tended to be portrayed from a distinctly European point of view and from the perspective of particular national, commercial, and religious interests. Geography mattered, but mainly as an instrument of colonialism. Pictured here are two of Britain's most well-known African explorers addressing the Royal Geographical Society on their expedition to find the source of the Nile River. (See Chapter 6, p. 278.)

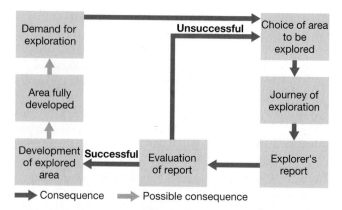

Figure 2.32   Principal elements in the process of exploration This diagram illustrates the main elements in the process of exploration, beginning with a need in the home country that prompts a desire to look outward to satisfy that need. Geographers have figured prominently in the process of exploration by identifying areas to be explored as well as actually traveling to these far-flung places, cataloging resources and people. Nineteenth-century geography textbooks are records of these explorations and the way in which geographers conceptualized the worlds they encountered. Exploration is one aspect of the process of imperialism; colonization is another. (*Source:* J. D. Overton, "A Theory of Exploration," *Journal of Historical Geography* 7, 1981, p. 57.)

that the core countries—effectively those of western Europe—transformed existing lands and peoples to meet their own commercial, evangelical, and colonial motives. Furthermore, the encounter between the two worlds was framed by the view that Western civilization was superior to the barbaric rest of the world. Thus, the engagement between "the West" and "the rest" was as much a moral event as anything else. As a result, the core often attempted to "master" the periphery and ended up marginalizing, if not completely destroying, existing and highly developed social, cultural, and moral systems. The same thing was largely true of the core's impact on the natural world of the periphery.

Figure 2.32 provides a theorization of the process of imperialism beginning with exploration and culminating in development via either colonization or the exploitation of people and resources or both. At the beginning of the process, a state perceives a need for exploration. This need is often the result of a scarcity or lack of a critical natural resource. Broadly speaking, in the first phases of imperialism, the core exploits the periphery for raw materials. Later, as the periphery becomes developed, colonization may occur, and cash economies are introduced where none has previously existed.

The periphery may also become a market for the manufactured goods of the core. Eventually, though not always, the periphery—because of the availability of cheap labor, land, and other inputs to production—can become a new arena for large-scale capital investment.

The first cases of sustained encounter between the core and the periphery usually resulted in the establishment of trading relations, sometimes even trading settlements, as was the case with European contact with Africa during the fifteenth century, when the Portugese explored the West African coastline and established a chain of trading settlements. These new opportunities for trade with Europe often dislocated the ongoing patterns of internal trade in the periphery and disrupted political life along with existing economic and religious systems.

After trading links were established, core states would often increase exploration and accelerate exploitation. For instance, by the late eighteenth century European countries had substantial and well-established trade with Africa. Later, between 1880 and 1912, most of Africa was partitioned among Belgium, France, Germany, Great Britain, Italy, and Portugal. At the Berlin Conference (1884–85), Europe's colonial nations defined their spheres of influence and established rules for future occupation of the globe. European colonization efforts in Africa, Asia, and South America expanded as domestic demand for certain agricultural and mineral products increased. European technology and crops were introduced into their colonies (**Figure 2.33**); mining operations were established (**Figure 2.34**); other natural resources such as fish, animals, and wood products were harvested and shipped back to European markets; and an exchange economy based on money evolved.

The classic institution of colonial agriculture was the plantation—an extensive, European owned, operated, and financed enterprise where single crops were produced by local or imported labor for a world market. The first plantations are likely to have been started in the Caribbean only two decades after the first voyage of Columbus in the late fifteenth century. Sugar was the first plantation crop with other crops—both edible and nonedible—following, including coffee plantations in India and rubber plantations in Sri Lanka. Over time, some plantation crops were replaced by others as competition over the production of particular crops between colonies increased or crops in some locations became increasingly susceptible to pests.

Development of Internal Infrastructures

The agents of the physical development of the colonies were explorers, traders, soldiers, missionaries, and settlers. Throughout the colonial periphery, these agents, with the financial, intellectual, and military backing of their home countries, constructed transportation systems, water-delivery systems, and communication systems to facilitate commerce. In addition, they instituted tax systems that enabled them to undertake these projects. The building of infrastructure was certainly not something that peripheral regions had the capital (or perhaps even the inclination) to undertake, so it is unlikely that such projects would have been accomplished without the intervention of the colonizers. It is also true, however, that many of the systems that were established reflected the needs of the colonial powers and not those of the local people. In Zimbabwe, for example, the principal transportation links reflect the main areas of European settlement. Transportation in the communal areas of the country continues to be inferior (**Figure 2.35**).

The most significant legacy of colonialism in terms of infrastructural development is the promotion of urbanization that resulted, following independence, in **overurbanization** which is a condition in which cities grow more rapidly than the jobs and housing can sustain. During the colonial period, ports, as has already been pointed out, were critical connective points to the evolving global economy. Ports encouraged population migration and settlement as opportunities for work acted as attractors to rural people. In colonial cities, the colonial powers have left perhaps their most significant physical legacy in the form of government buildings, roads, bridges, schools, and other public structures as well as substantial elite housing and recreational developments and parks. Many of the colonial ports have become the primate cities of independent countries, exhibiting all the problems of large cities throughout the periphery in terms of unemployment, environmental problems, and inadequate housing and public welfare provisions.

Globalization and Economic Development

The imperial world order began to disintegrate shortly after the Second World War. The United States emerged as the dominant state within the world-system core. This core came to be called

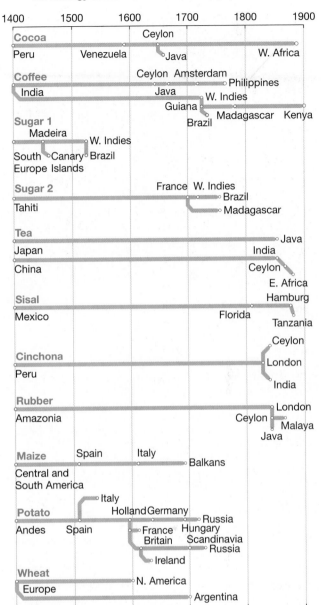

Chronology of the transfer of cash crops, 1400–1900

Figure 2.33   Global transfer of crops, 1400–1900 This diagram depicts the timing of the transfer of crops as they were introduced around the globe. Perhaps the most significant impact of European imperialism beginning in the sixteenth century was the gradual incorporation of agriculture into one comprehensive world system. The diagram illustrates the very important point that production, which began in one location, was moved to other locations where the climate was suitable, labor could be acquired, and production and trade managed. Thus agriculture became a globally organized system where crops were produced in numerous locations to feed world markets. An important point to note about this diagram is the relevance of the colonizing countries to the diffusion of plants. Cinchona, a South American tree whose dried bark contains the anti-malarial agent quinine, is a good example of this. Cinchona seeds were brought from Peru, propagated in Kew Gardens, the national botanical gardens of Britain, and then sent off to British plantations in India and Ceylon (present-day Sri Lanka), where the bark was harvested and processed into commercial quinine. As the diagram illustrates, botanical gardens all over Europe, as both scientific and commercial enterprises, were important to the global diffusion of commercial agriculture. (*Source:* B. Crow and A. Thomas, *Third World Atlas*, Milton Keynes: Open University Press, 1982, p. 29.)

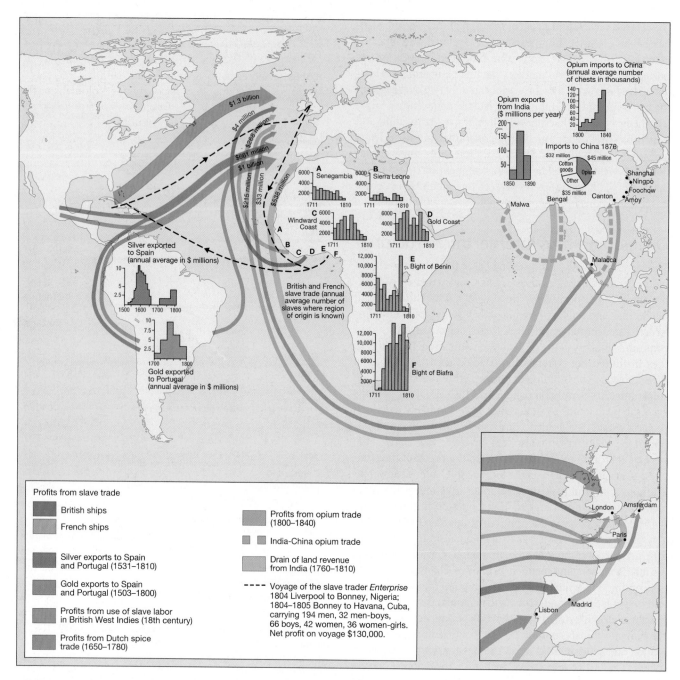

Figure 2.34   Profits from European global expansion, 1500–1800 This map illustrates the profits generated through European plunder of global minerals, spices, and human beings over a 300-year period. Silver and gold from the Americas, opium and tea from China, spices from the East Indies (Indonesia, Malaysia, and the Philippines), and slaves from Africa are represented. (*Source:* Adapted from B. Crow and A. Thomas, *Third World Atlas*, Milton Keynes: Open University Press, 1982, p. 27.)

the "First World." The Soviet Union and China, opting for alternative paths of development for themselves and their satellite countries, were seen as a "Second World," withdrawn from the capitalist world economy. Their pursuit of alternative political economies was based on radically different values.

By the 1950s many of the old European colonies began to seek political independence. Some of the early independence struggles were very bloody, because the colonial powers were

initially reluctant to withdraw from colonies where strategic resources or large numbers of European settlers were involved. In Kenya, for example, a militant nationalist movement known as the Mau Mau launched a campaign of terrorism, sabotage, and assassination against British colonists in the early 1950s. Their actions killed more than 2000 white settlers between 1952 and 1956; in return, 11,000 Mau Mau rebels were killed by the colonial army and 20,000 put into detention camps by

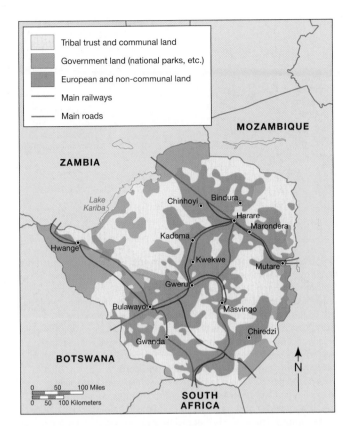

Figure 2.35 Main roads and railway lines in Zimbabwe
This map depicts the main transportation lines constructed by the British in Zimbabwe. It also shows the locations of communal land occupied by various tribal groups as well as land owned by the government and European settlers. The transportation lines clearly link the European settlements and ignore the communal ones. Transportation networks in the communal areas are far inferior, and, as a result, development is hindered. (*Source:* Redrawn from R. Potter, T. Binns, J. Elliott, and D. Smith, *Geographies of Development.* Harlow: Addison Wesley Longman, 1999, p. 39.)

the colonial administration. But by the early 1960s the process of decolonization had become relatively smooth. (In Kenya, Jomo Kenyatta, who had been jailed as a Mau Mau leader in 1953, became prime minister of the newly independent country in 1962.) The periphery of the world-system now consisted of a "Third World" of politically independent states, some of which adopted a policy of nonalignment vis-à-vis the geopolitics of the First and Second worlds. They were, nevertheless, still highly dependent, in economic terms, on the world's core countries.

As newly independent peripheral states struggled to be free of their economic dependence through industrialization, modernization, and trade from the 1960s onward, so the capitalist world-system became increasingly integrated and interdependent. The old imperial patterns of international trade broke down and were replaced by more complex patterns. Nevertheless, the newly independent states were still influenced by many of the old colonial links and legacies that remained intact. The result was a neocolonial pattern of international development. **Neocolonialism** refers to economic and political strategies by which powerful states in core

economies indirectly maintain or extend their influence over other areas or people. Instead of formal, direct rule (colonialism), controls are exerted through such strategies as international financial regulations, commercial relations, and covert intelligence operations. Because of this neocolonialism, the human geographies of peripheral countries continued to be heavily shaped by the linguistic, cultural, political, and institutional influence of ex-colonial powers and by the investment and trading activities of their firms.

Deploying and Encountering Development

The rationale whereby the postwar world became more fully integrated was a *set* of approaches to economic and political transformation known as *development theory.* **Development theory** is an analysis of social change that assesses the economic progress of individual countries. Development theory first came into being as the justification for the rebuilding of Europe and as a weapon against the Cold War threat of communism. It was soon after applied to the situation of newly emerging states in Africa, Asia, and South America struggling under the legacies of colonialism and post-colonialism. Numerous authors, world leaders, and economic policies have been associated with development theory and the development project, but the centerpiece of all mainstream approaches to development is the aim to replicate the prosperity of the core in the periphery by encouraging economic growth through industrialization and modernization. President Truman expressed this view in his inaugural address in 1949: "More than half the people of the world are living in conditions approaching misery. . . . I believe that we should make available to peace-loving peoples the benefits of our store of technical knowledge in order to help them realize their aspirations for a better life." The two underlying assumptions of this quote and the core's attitude about development are: (1) that the periphery should aim to be like the core in its pathway to development, and (2) that the economic problems of development in the periphery are poverty and backwardness.

This overall relationship between the economy and levels of prosperity makes it possible to interpret economic development in terms of distinctive *stages.* Each region or country, in other words, might be thought of as progressing from the early stages of development, with a heavy reliance on primary activities (and relatively low levels of prosperity), through a phase of industrialization and on to a "mature" stage of postindustrial development (with a diversified economic structure and relatively high levels of prosperity). This, in fact, is a commonly held view of economic development, conceptualized by a prominent economist, W. W. Rostow (**Figure 2.36**). Most development theorists and practitioners hew to this perspective and view the situation of peripheral and semiperipheral countries as one in which they must simply be helped to move along a clear path so that eventually they will "catch up" economically.

In such a model, an economy is understood to consist of a modern sector(s) and a nonmodern sector(s) with the latter

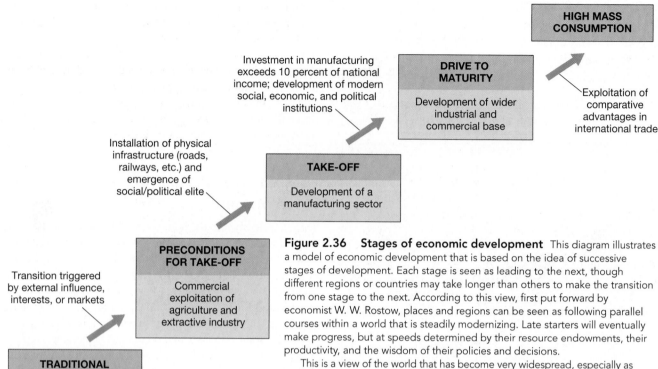

Figure 2.36 **Stages of economic development** This diagram illustrates a model of economic development that is based on the idea of successive stages of development. Each stage is seen as leading to the next, though different regions or countries may take longer than others to make the transition from one stage to the next. According to this view, first put forward by economist W. W. Rostow, places and regions can be seen as following parallel courses within a world that is steadily modernizing. Late starters will eventually make progress, but at speeds determined by their resource endowments, their productivity, and the wisdom of their policies and decisions.

This is a view of the world that has become very widespread, especially as applied to the experience and prospects of different countries, as intended by Rostow's model. It is, however, too simplistic to be of much help in understanding world regional geography. The reality is that places and regions are interdependent. The fortunes of any given place are tied up with those of many others, and increasingly so. Rostow's model perpetuates the myth of "developmentalism": the idea that every country and region will eventually make economic progress toward "high mass consumption" provided that they compete to the best of their ability within the world economy. But the main weakness of developmentalism is that it is simply not fair to compare the prospects of late starters to the experience of those places, regions, and countries that were among the early starters. For these early starters, the horizons were clear: free of effective competition, free of obstacles, and free of precedents. For the late starters, the situation is entirely different. Today's less-developed regions must compete in a crowded field while facing numerous barriers that are a direct consequence of the success of some of the early starters.

characterized by low labor productivity. If the country or region with these conflicting economic sectors is to develop, then the non-modern sector must be transformed or modernized. In short, economic development occurs when investment rates enable higher levels of industrialization, thus raising labor productivity and increasing the GDP per capita levels. The ultimate goal in this staged economic development model is an end period of high mass consumption.

The modernization version of development theory was not without its detractors. Chief among the critics were scholars and policy makers from the periphery who became part of the dependency school of development. The dependency theorists argued that it was not labor inefficiency or absence of modernization that was the root cause of the poverty in the periphery but rather that the very nature of the core-periphery relationship was such that the core, through economic exploitation, had created a state of *underdevelopment* in the periphery. Thus the pathway to development was not the same for the periphery as it was for the core. The progression of the core from undeveloped to developed status was enabled at the direct expense of the periphery. *The core had expressly underdeveloped the periphery by taking the economic surplus generated there and repatriating it to the core.* Moreover, the process of underdevelopment in the periphery had rendered it dependent on the core for inputs of capital and technology. The solution, the dependency school theorists argued, was for the periphery to break off from the capitalist world-system in order to escape active underdevelopment and the status of economic dependence. One way in which the periphery attempted to accomplish this was through *import substitution,* a practice that is discussed in greater detail in later chapters.

Another critique of development theory was launched by feminists in the 1970s. Their objection to mainstream development theory and policy was that it left out women, either by assuming that women had no role to play in development or by failing to recognize that women might experience development differently from men. This criticism came to be widely recognized as a legitimate one, and development theory and policies were revised not only to take account of the differential impact of development on women but to incorporate women directly into development projects. More recently, however, this "women in development" approach has also been criticized by a new generation of feminists, who see it as failing to recognize that women's identities, and access to power and resources are also shaped by other important identities, such as generation, ethnicity, religion, sexuality, and marital status, just to name a few.

Development theory, like most economic theories, rests on certain simplifying assumptions about the world. The real world, however, is highly differentiated, not just in terms of natural resources, but in terms of demographics, culture, and politics. The assumptions in Rostow's model, for instance, fit the experience of some parts of the world, but by no means all of it. In reality, there are a variety of pathways to development, and a variety of different processes and outcomes of development. Indeed, this narrow construction of the pathway toward, and the existing conditions necessary for, development ultimately created serious challenges for modernization theory and its policy proponents. At the same time, dependency theory was also seen as seriously flawed. The result has been a reconsideration and restructuring of modernization theory in the form of neoliberalism, the emergence of the idea of sustainable development (see Chapter 1, p. 29), and alternative and critical theories of development advanced more recently by scholars from the periphery.

Neoliberal policies of development have emanated from the core and have been very much associated with the increasing influence of the World Bank and the International Monetary Fund (IMF) over the last 25 years. The **World Bank** is a development bank and the largest source of development assistance in the world. Its goal is to help countries strengthen and sustain the fundamental conditions that will attract and retain private investment. The **International Monetary Fund**, in contrast, provides loans to governments throughout the world. In order to obtain these loans, governments must submit to IMF conditions. This often means rewriting laws so that they are more favorable to foreign investment. Along with the World Bank, the IMF aims to help countries strengthen their banking system. There have been a number of means by which the World Bank and the IMF have attempted to shape and assist economic development in the periphery. These include neoliberal policies that emphasize privatization, export production, and limited restrictions on imports. The term neoliberal (new liberal) refers to the revival of ideas popular at the end of the nineteenth century promoting free trade and economic integration. **Neoliberalism** promotes a reduction in the role and budget of government, including reduced subsidies and the privatization of formerly publicly owned and operated concerns such as utilities. Like modernization theory, the goal of neoliberal development policies is to enable peripheral countries to achieve core economic standards of wealth and prosperity while recognizing that preexisting conditions will have to be taken into account to construct a place-specific development path. For the most part, neoliberal development is premised on policies of structural adjustment,

focused on market-led economic growth, with the aim of duplicating the economic and political organization of the core.

In response to the revival of modernization theory in the guise of neoliberalism, critics have emerged who question whether the kind of development promulgated by neoliberalism is environmentally and socially sustainable. In practice, sustainable development policies argue for using renewable natural resources in a manner that does not eliminate or degrade them—by making greater use, for example, of solar and geothermal energy and by greater use of recycled materials. It means *managing* economic systems so that all resources—physical and human—are used optimally. It means *regulating* economic systems so that the benefits of development are distributed more equitably (if only to prevent poverty from causing environmental degradation). And it means *organizing* societies so that improved education, health care, and social welfare can contribute to environmental awareness and sensitivity and an improved quality of life. A final and more radical aspect of sustainable development is a move away from wholesale globalization toward increased "localization": a desire to return to a more locally based economy where production, consumption, and decision making can be oriented to local needs and conditions.

Since the 1980s, the apparent failure of existing development theory and policy to modernize the periphery has generated radical criticism by feminist, postcolonial, and other perspectives to reconceptualize development as violent imposition of core economic values upon the periphery and a means for the core to exercise control over the periphery. As geographers Philip Porter and Eric Sheppard have written, development in the periphery has been predicated on the initial "violence of colonialism, gunboat diplomacy and wars between superpowers; impoverishment; external control over domestic affairs; the dissolution of indigenous institutions and cultures; and environmental deterioration. In short, they [the periphery] have encountered rather than propagated development. . . ."[2] As a result of this radical refiguring of development theory and policy, contemporary theorists have called into question the whole grand notion of development and argue instead for indigenous alternatives that empower grassroots movements and promote local knowledge and that can repair the damage done by core development projects. They have shown how development is just one way to think about a very complicated reality, but one that has come to dominate as the only solution to social, cultural, economic, and political crises in the periphery.

The contemporary radical critique of mainstream development theory challenges the core's position that it knows best how to solve the problems of the periphery and further argues that the neoliberal goal of economic development for all states and nations of the world might even be called into question. These challenges are part of a broader critique that argues for the need to recognize the importance of social and cultural factors in the process of economic and political development.

[2]P. Porter and E. Sheppard, *A World of Difference: Society, Nature, Development.* New York: Guilford Press, 1998, pp. 97.

Focusing on women and other underrepresented groups, the radical critique of development theory contends that different groups in different places have different access to the power and resources that shape their daily lives.

Five Key Factors of Globalization

The increasing integration of the world-system over the last 25 years has been informed by the theories and policies of development discussed above. The level of globalization that has resulted has been caused by dramatic changes in all aspects of economic life from production to consumption. We treat the five key factors in the following five sections. They include a new international division of labor, an internationalization of finance, a new technology system, the homogenization of international consumer markets, and the proliferation of the transnational corporation.

A New International Division of Labor The new international division of labor that has accompanied and enabled the most recent round of globalization has resulted in three main changes. First, the United States has declined as an industrial producer, relative to the spectacular growth of Japan and the resurgence of Europe as industrial producers. Second, the **new international division of labor** has involved the decentralization of manufacturing production from all of these core regions to some semiperipheral and peripheral countries. In 1995 U.S.-based companies employed about 5.5 million workers overseas, 80 percent of whom were in manufacturing jobs. An important reason for this trend has been the prospect of keeping production costs low by exploiting the huge differential in wage rates around the world.

A third result of the new international division of labor is that new specializations have emerged within the core regions of the world-system: high-tech manufacturing and producer services (that is, services such as information services, insurance, and market research that enhance the productivity or efficiency of other firms' activities or that enable them to maintain specialized roles). The most significant reflection of this new international division of labor is that global trade has grown much more rapidly over the past 25 years than has global production, a clear indication of the increased economic integration of the world-system.

The Internationalization of Finance The second factor contributing to today's globalization is the internationalization of finance: the emergence of global banking and globally integrated financial markets. These changes are of course tied in to the new international division of labor. In particular, they are a consequence of massive increases in levels of international direct investment. Between 1988 and 1996, the flow of investment capital from core to semiperipheral and peripheral countries increased twentyfold. These increases include transnational investments by individuals and businesses as well as cross-border investments undertaken within the internal structures of transnational corporations. In addition, the capacity of computers and information systems to deal very quickly with changing international conditions has added

a speculative component to the internationalization of finance. All in all, about $100 billion worth of currencies are traded every day. The volume of international investment and financial trading created a need for banks and financial institutions that could handle investments on a large scale, across great distances, quickly and efficiently. The nerve centers of the new system are located in just a few places—London, Frankfurt, New York, and Tokyo, in particular. Their activities are interconnected around the clock (**Figure 2.37**), and their networks penetrate every corner of the globe.

A New Technology System The third factor contributing to globalization is a new technology system based on a combination of innovations, including solar energy, robotics, microelectronics, biotechnology, digital telecommunications, and computerized information systems. This new technology system has required the geographical reorganization of the core economies. It has also extended the global reach of finance and industry and made for a more flexible approach to investment and trade. Especially important in this regard have been new and improved technologies in transport and communications: the integration of shipping, railroad, and highway systems through containerization, the introduction of wide-bodied cargo jets, and the development of fax machines, fiber-optic networks, communications satellites, and electronic mail and information-retrieval systems. Finally, many of these telecommunications technologies have also introduced a wider geographical scope and faster pace to many aspects of political, social, and cultural change (**Figure 2.38**).

Homogenization of International Consumer Markets A fourth factor in globalization has been the growth of consumer markets. Among the more affluent populations of the world, similar trends in consumer taste have been created by similar social processes. A new and materialistic international culture has taken root, in which people save less, borrow more, defer parenthood, and indulge in affordable luxuries that are marketed as symbols of style and distinctiveness. This culture is easily transmitted through the new telecommunications media, and it has been an important basis for transnational corporations' global marketing of "world products" (German luxury automobiles, Swiss watches, British raincoats, French wines, American soft drinks, Italian shoes and designer clothes, and Japanese consumer electronics, for example). It is also a culture that has been easily reinforced through other aspects of globalization, including the internationalization of television—especially CNN, MTV, Star Television—and the syndication of TV movies and light entertainment series (**Figure 2.39**).

The Transnational Corporation As we discussed in a previous section, the division of labor is one way to differentiate the workforce in order to introduce new efficiencies. The introduction of a division of labor during the mercantile period fostered a new form of commercial organization, which would enable the combination of resources and labor on a much larger scale than was possible with the individual or even partnerships. This new form of commercial organization

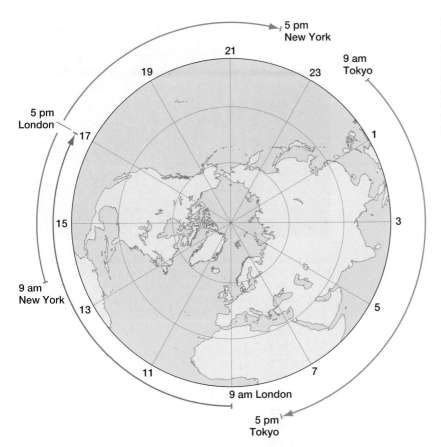

Figure 2.37 Twenty-four-hour trading between major financial markets Office hours in the most important financial centers—New York, London, and Tokyo—overlap one another because the three cities are situated in broadly separated time zones. This means that, between them, they span the globe with 24-hour trading in currencies, stocks, and other financial instruments.

was the corporation with limited liability. Although first appearing during the fifteenth century, limited liability corporations really took off during the early days of the American republic when the developing nation had no banks or other established institutions to support economic growth. Groups of individuals incorporated in order to construct roads, create trading or mining groups, or build and operate factories.

The corporation as a form of business organization has a number of advantages. First, it exists independently of its owners (the stockholders). Second, in U.S. law, as well as in most other countries, the corporation is recognized as a legal person with many of the same rights that individuals have, such as the right to buy and sell property and to enter into contracts. Third, the corporation is an excellent device for raising vast amounts of business capital by pooling the financial resources of thousands of individuals, at the same time spreading the risks of a new venture among many people. Finally—and this is where the notion of *limited liability* comes in—the owners of a corporation are not liable for its debts beyond their investment.

As former colonies gained their independence and neocolonialism emerged as a new form of exploitation of the periphery by the core, a new form of imperialism was also emerging in the form of the giant corporation. These corporations had grown within the core countries through the elimination of smaller firms by mergers and takeovers. By the 1960s quite a few of them had become so big that they were

transnational in scope, having established overseas subsidiaries, taken over foreign competitors, or simply bought into profitable foreign businesses.

A **transnational corporation** (TNC) has investments and activities that span international boundaries, with subsidiary companies, factories, offices, or facilities in several countries. In 1999 there were 60,000 parent transnational corporations with 500,000 foreign affiliates around the world. In 1997 the top 100 TNCs together held $1.8 trillion in foreign assets, sold products worth $2.1 trillion abroad, and employed more than six million people in their foreign affiliates (**Table 2.1**). Transnational corporations have been portrayed as imperialist by some geographers because of their ability and willingness to exercise their considerable power in ways that adversely affect peripheral states. They have certainly been central to a major new phase of geographical restructuring that has been under way for the last 25 years or so. This phase has been distinctive because an unprecedented amount of economic, political, social, and cultural activity has spilled beyond the geographic and institutional boundaries of states. It is a phase of *globalization*, a much fuller integration of the economies of the worldwide system of states and a much greater interdependence of individual places and regions from every part of the world-system. And the increasing integration of the world economy is increasingly predicated on nongovernmental corporate enterprises that operate at a worldwide scale.

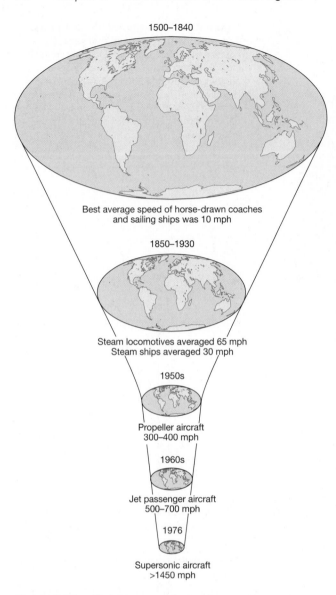

1500–1840

Best average speed of horse-drawn coaches and sailing ships was 10 mph

1850–1930

Steam locomotives averaged 65 mph
Steam ships averaged 30 mph

1950s

Propeller aircraft
300–400 mph

1960s

Jet passenger aircraft
500–700 mph

1976

Supersonic aircraft
>1450 mph

Figure 2.38   Shrinking world This diagram illustrates how the interconnectedness of transportation systems around the globe has enabled travel times to shrink dramatically over the last 500 years of capitalist development. (*Source:* Redrawn from P. Dicken, *Global Shift*. London: Paul Chapman, 1998.)

Figure 2.39   Global marketing of television programming The globalization of culture has been facilitated more than anything else by television broadcasting via satellite and by the sales of popular television programs to markets around the world. This photograph shows a television fair held in Miami Beach, Florida.

Transnational Economic Integration

The five key factors of globalization described above have helped to usher in a contemporary transnational world economy highly integrated politically and economically. It is important to keep in mind, however, that the dominant processes both in rivalry within the core as well as in the struggle by the periphery to escape from dependency have been dominated by conflict and competition. Economic nationalism, whether drawing on practical examples (for example, eighteenth-century Britain; see Chapter 3, p. 110), political ideology (for example, Juan Perón in Argentina and Getúlio Vargas in Brazil in the 1940s and 1950s; see Chap-

ter 8, p. 394), or development theory (for example, neoliberalism, as described in this chapter), continues to dominate global economics and geopolitics.

At the same time that we consider the salience of economic nationalism, we must not underestimate the relevance of the long-term trend among the various national economies toward the progressive integration of local, regional, and national economic systems. What has happened is that the logic of the world economy has in many ways transcended, and in some ways undermined, the scale of nation-states. The logic and apparatus of statehood is not conducive to transnational integration, whether economic or political, but the outcomes of the new international division of labor have forced many states to explore cooperative strategies of various kinds. As a result, the world's economic landscapes now bear the imprint, in a variety of ways, of *transnational* economic and political integration.

Types and Levels of Economic Integration   Figure 2.40 summarizes some of the history and geography of transnational economic integration since 1945. In practice, integration can be pursued in a variety of ways and at different levels. It can be *formal,* involving an institutionalized set of rules and procedures—for example, the United Nations Organization or the General Agreement on Tariffs and Trade (GATT)—or *informal,* involving coalitions of interests, such as UN voting blocs. It can be *trans*national, involving attempts to foster integration between nation-states—the North Atlantic Treaty Organization (NATO), the Organization of Asian Unity (OAU), or the World Trade Organization (WTO)—or *supra*national, involving a commitment to an institutionalized body with certain powers over member states—for example, the European Union (EU). It can be *economic* (WTO or the European Free Trade Association [EFTA]), *strategic* (NATO or the Warsaw Pact), *political* (UN

| Table 2.1 | The World's Top Twenty Transnational Corporations, 1997 (Billions of Dollars and Number of Employees) | | | | | | | | |

Ranking	Corporation	Country	Industry	Assets		World Sales		World Employment	
				Foreign	Total	Foreign	Total	Foreign	Total
1	General Electric	United States	Electronics	97.4	304.0	24.5	90.8	111,000	276,000
2	Ford Motor Company	United States	Automotive	72.5	275.4	48.0	153.6	174,105	363,892
3	Royal Dutch/ Shell Group	Netherlands/ United Kingdom	Petroleum expl./ ref./distr.*	70.0	115.0	69.0	128.0	65,000	105,000
4	General Motors	United States	Automotive	0.0	228.9	51.0	178.2	–	608,000
5	Exxon Corporation	United States	Petroleum expl./ ref./distr.	54.6	96.1	104.8	120.3	–	80,000
6	Toyota	Japan	Automotive	41.8	105.0	50.4	88.5	–	159,035
7	IBM	United States	Computers	39.9	81.5	48.9	78.5	134,815	269,465
8	Volkswagen Group	Germany	Automotive	–	57.0	42.7	65.0	133,906	279,892
9	Nestlé SA	Switzerland	Food and beverages	31.6	37.7	47.6	48.3	219,442	225,808
10	Daimler-Benz AG	Germany	Automotive	30.9	76.2	46.1	69.0	74,802	300,068
11	Mobil Corporation	United States	Petroleum expl./ ref./distr.	30.4	43.6	36.8	64.3	22,220	42,700
12	FIAT Spa	Italy	Automotive	30.0	69.1	20.2	50.6	94,877	242,322
13	Hoechst AG	Germany	Chemicals	29.0	34.0	24.3	30.0	–	137,374
14	Asea Brown Boveri (ABB)	Switzerland	Electrical equipment	–	29.8	30.4	31.3	200,574	213,057
15	Bayer AG	Germany	Chemicals	–	30.3	–	32.0	–	137,374
16	Elf Aquitaine SA	France	Petroleum expl./ ref./distr.	26.7	42.0	25.6	42.3	40,500	83,700
17	Nissan Motor Co., Ltd.	Japan	Automotive	26.5	57.6	27.8	49.7	–	137,201
18	Unilever	Netherlands/ United Kingdom	Food and beverages	25.6	30.8	44.8	46.4	262,840	269,315
19	Siemens AG	Germany	Electronics	25.6	67.1	40.0	60.6	201,141	386,000
20	Roche Holding AG	Switzerland	Pharmaceuticals	–	37.6	12.7	12.9	41,832	51,643

* Exploration, refining, and distribution

Source: United Nations, *World Investment Report: Foreign Direct Investment and the Challenge of Development.* New York and Geneva: United Nations, 1999, p. 3.)

voting blocs), *sociocultural* (the United Nations Educational, Scientific, and Cultural Organization [UNESCO]), or *mixed* (the EU or the OAU).

The GATT Framework and the WTO   Within the contemporary world-system, economically oriented integration schemes have had to conform to the rules of GATT, a transnational association of most of the world's trading nations formed in the aftermath of the Second World War to promote global free trade and ease the complex trade restrictions that had accumulated. The original GATT agreement (in 1947) reduced the average tariff on goods from more than 40 percent to less than 30 percent. Subsequent rounds of renegotiation have brought the average tariff level down to about 5 percent.

Unfortunately, GATT became the victim of its own success. In short, trade issues have also become increasingly complex as more countries have joined the agreement and the world economy has become increasingly globalized and interdependent. The original agreement was written to deal primarily with trade in manufactured goods among core countries,

yet by 1990 only about 60 percent of world export earnings came from manufacturing. Instead, services accounted for an increasing share of world trade; many of the newly industrializing countries (or NICs), such as Indonesia or Taiwan, were not fully subject to GATT rules. And although tariffs on manufactured goods were successfully reduced through the GATT, substantial nontariff barriers (for example, import quotas, import licenses, exchange rate manipulation, government subsidies to domestic industries, special labeling and packaging regulations, and so forth) remained a problem. As a result of the increasing complexity of the trading landscape, GATT renegotiation is an increasingly protracted process. The most recent round of renegotiation, the Uruguay Round, began in 1986 and was not concluded until December 1993. The chief obstacle was disagreement between the United States and the European Union over nontariff barriers in the form of subsidies to farmers. The crowning achievement of the Uruguay round was the creation of the WTO as a replacement for GATT (which had become labeled by wags as the "General Agreement to Talk and Talk").

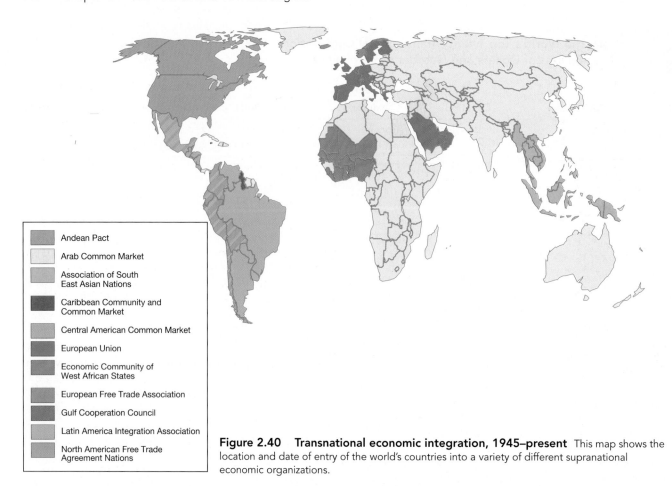

Andean Pact

Arab Common Market

Association of South
East Asian Nations

Caribbean Community and
Common Market

Central American Common Market

European Union

Economic Community of
West African States

European Free Trade Association

Gulf Cooperation Council

Latin America Integration Association

North American Free Trade
Agreement Nations

Figure 2.40   Transnational economic integration, 1945–present This map shows the location and date of entry of the world's countries into a variety of different supranational economic organizations.

Whereas GATT had little ability to enforce its decisions, the WTO is a global body with both judicial and regulatory power. Its framework is a series of agreements that are some 24,000 pages long. These agreements go beyond trade in manufactured goods to cover investment, services, and intellectual property rights. In the words of the organization's former director-general, Renato Ruggiero, the WTO "is writing the constitution of a single global economy." A significant step toward this was taken in February 1997 when the 68 original members of the WTO signed an agreement to free up their markets to international competition in telecommunications. Currently, the WTO has a membership of nations, with a waiting list of others, including Russia and China.

Many of the WTO's agreements are derived from GATT rulings, including the provision that each member state shall extend most-favored status to all other member countries. (Thus, if the United States were to lower its import duty on textile products from Canada, it would immediately have to extend that same reduced rate to every other WTO member.) There is, however, an exception to this principle for free trade associations and customs unions, members of which may reduce their tariffs against each other without extending such concessions to remaining WTO members. It is this exception that has provided the basis for regional economic integration within the globalizing world economy. As was discussed in Chapter 1, the emergence and growing power of supranational

organizations such as the WTO has not occurred without serious opposition (see Geography Matters: Mobilization Against Globalization, on p. 34 in Chapter 1).

The Institutional Forms of Integration   In a **free trade association**, member countries eliminate tariff and quota barriers against trade from other member states, but each individual member continues to charge its regular duties on materials and products coming from outside the association. The only significant free trade association has been the European Free Trade Association (EFTA), whose membership now comprises only Iceland, Liechtenstein, Norway, and Switzerland.

Another way to eliminate tariffs between member states, but with a common protective wall against nonmembers, is a customs union. A **customs union** is an international association organized to eliminate customs restrictions on goods exchanged between member nations and to establish a uniform tariff policy toward nonmember nations. Where internal restrictions on the movement of capital, labor, and enterprise are also removed from the basic framework of a customs union, the result is a **common market**. Most customs unions have gone at least some way toward common-market status. Some examples include:

■   The European Union: Austria, Belgium, Denmark, Finland, France, Germany, Greece, Ireland, Italy, Luxembourg, Netherlands, Portugal, Spain, Sweden, and the United Kingdom

- The Central American Common Market: Costa Rica, El Salvador, Guatemala, Honduras, Nicaragua
- The Arab Common Market: Egypt, Iraq, Syria, Jordan
- The Andean Pact: Bolivia, Colombia, Ecuador, Peru, Venezuela
- The Caribbean Community and Common Market: Antigua and Barbuda, the Bahamas, Barbados, Belize, Dominica, Grenada, Guyana, Jamaica, Montserrat, St. Kitts and Nevis, St. Lucia, St. Vincent and the Grenadines, Suriname, Trinidad and Tobago
- Economic Community of Central African States: Congo, Central African Republic, Democratic Republic of the Congo, Gabon, Rwanda, Burundi, and the island republic of São Tomé and Príncipe
- Economic Community of West African States: Benin, Côte d'Ivoire, Gambia, Ghana, Guinea, Guinea-Bissau, Liberia, Mali, Niger, Nigeria, Senegal, Sierra Leone, Togo, Burkina Faso, Cape Verde

A form of integration higher than the common market is the economic union. In addition to the characteristics of a common market, the **economic union** provides for integrated economic policies among member states, as in the EU. The highest form of integration possible would have to involve some form of supranational political union, with a single monetary system and a central bank, a unified fiscal system, a common foreign economic policy, and a supranational authority with executive, judicial, and legislative branches.

Although the EU appears to be a success, many other free-trade associations and common markets have been less so. Some of the most difficult obstacles to overcome have been the result of memberships that include nations at very different levels of development and that involve enormous distances and poorly developed transportation networks. It was in response to such problems that GATT authorized, in 1971, the waiver of the Article 1 most-favored-nation provision for peripheral countries offering concessions to other peripheral countries. As a result, Mexico, for example, could offer to reduce its duty on a product from Bolivia without having to extend the same lower rate to the United States. As a result of the waiver of Article 1, many peripheral countries were free to experiment with a variety of integration models without incorporating internal free trade as a legally binding obligation. The result has been the emergence of a series of trade-preference associations, such as the Association of Southeast Asian Nations, or ASEAN (Brunei, Burma, Cambodia, Indonesia, Laos, Malaysia, the Philippines, Singapore, Thailand, Vietnam), and the Latin American Integration Association (LAIA), formerly the Latin American Free Trade Association (Argentina, Bolivia, Brazil, Chile, Colombia, Ecuador, Mexico, Paraguay, Peru, Uruguay, Venezuela).

The increasing globalization of the world economy has broadened, deepened, and accelerated the trend toward regional economic integration. In 1990, for example, the Andean Pact decided to establish an Andean Common Market

by 1996. In the following year, the pact approved more trade liberalization reforms than it had in the previous 22 years of the pact's existence. In 1989 the ASEAN countries joined with Australia, Canada, China, Hong Kong, Japan, New Zealand, South Korea, Taiwan, and the United States to form the Asia Pacific Economic Co-operation group. Their objective was to promote the liberalization of trade as well as cooperation in trade and investment around the Pacific Rim. In 1992 EFTA and the EU agreed to establish a unified free trade zone, the European Economic Area (EEA), with a combined market size of 379 million people.

Meanwhile, Canada, Mexico, and the United States established a trading zone in 1992 with the completion of the North American Free Trade Agreement (NAFTA). This was not only an unprecedented economic integration of core countries with a semiperipheral country but also the first instrument of economic integration to liberalize trade in services. Over a 15-year period, NAFTA is scheduled to phase out tariffs and other trade and investment barriers between the three countries. After nearly a decade, NAFTA has had a significant impact on the reorganization of the economic geography of North America. At the same time that thousands of manufacturing jobs have been transplanted from Canada and the United States to Mexico, the growing Mexican market has become open to U.S. and Canadian automobiles and automobile parts, telecommunications, and financial services. Eventually, there will be total access in agricultural markets, which will rewrite the agricultural geography of Mexico and significantly modify agricultural patterns in the southwestern United States and possibly elsewhere.

Spatial Outcomes of Economic Integration   The expansion of markets and the removal of artificial barriers to trade predictably result in a reorganization of economic geography with respect to two main sets of effects that result in either creating or impairing trade. As national economies become more integrated across political boundaries, *the removal of trade barriers should lead to a more pronounced regional division of labor,* with each region in the larger association tending to specialize in those activities in which it has the greatest comparative advantage. More particularly, production is reallocated from high-cost to low-cost settings—from core to periphery—and a great deal of trade is generated within the association. An additional outcome is that lower costs can, theoretically, be passed on to consumers, thus contributing to improved levels of living. These effects of transnational integration are generally referred to as **trade creation effects.**

Countries that are not a part of the association, however, tend to lose trade. This is because the external tariff wall prevents outsiders from competing effectively with higher-cost inside producers whose output is able to circulate duty-free within the association. To the extent that the old producers were more efficient producers than the new ones, **trade diversion effects** will have taken place, with the result that consumption is shifted away from lower-cost external sources to higher-cost internal sources. Consumers outside the bounds of the association will have to pay more for certain goods, and levels of living may be depressed.

Geography Matters

How Politics and Culture Modify the Economics of Development

Within the modern world-system, the territorial framework of political states provides a *competitive* economic system within which each state attempts to insulate itself as much as possible from the rigors of the world market, while attempting to turn the market to its own advantage. In terms of world-systems theory, the most important feature of this competitive system is that it has ensured that the modern world has not become a global world-empire: that is, that no single core area has been able permanently to dominate the rest of the world. In terms of human geography, the significance of the competitive system lies in the way that states influence the trajectory of change of individual countries, regions, and places within the international division of labor. States attempt to consolidate their advantages within the world market (and to overcome their disadvantages) in three main ways:

- Through their ability to organize territorial expansion and control, either through diplomacy or through military strength. This organization provides a simple and direct means of accessing and controlling resources. Where territorial expansion and control are contested, however, conflict and war may result.

- Through their ability to distort markets to the advantage of domestic producers and/or consumers. This is routinely achieved through (1) tariffs (taxes on imports) and trade quotas that restrict the access of foreign producers to domestic markets; (2) financial incentives (tax breaks, for example) to stimulate and encourage domestic firms; and (3) control of currency exchange rates, domestic interest rates, and money supply in order to manipulate production and consumption.

- Through their ability to develop an infrastructure that helps to mobilize national resources. This includes both the physical infrastructure of roads, railways, harbors, airports, dams, and so on, and the social infrastructure of legal codes, educational systems, public health care, and social welfare services.

Another important characteristic of the world-system is that it is inherently *uneven*, not just in its structures of economic and political power but also in the degree to which individual places and regions are incorporated into the system. The periphery, for example, is not uniform, even though its regions do share some important attributes of economic and political subordination. Individual countries, regions, and places have their own histories, their own specific attributes, and their own integrity as geographical entities.

Much of this unevenness is a result of the history and geography of the modern world-system. Three main factors have been important. First, differences in local and regional cultures have resulted in different reactions to the economic and political logic of the world-system. These reactions range from acceptance and cooperation to resistance and insurrection. A great deal, of course, depends on cultural compatibility with the values of materialism, individualism, competitiveness, popular democracy, and so on, that are attached to the market system that drives the core. Thus, for example, Islamicists in Iran see the capitalism of the world-system as decadent and corrupt. As a result, Iran has been only tenuously incorporated into the world-system since 1979, when a popular revolution installed a Shi'a Islamic government dedicated to economic and cultural insularity.

The second factor at work has been the geographical spread of the modern world-system, which took place under the auspices of nation-states whose strengths and objectives have varied. As a result, patterns of incorporation into the world-system have been equally varied. In some instances (for example, North America and Australia), colonization almost completely displaced indigenous societies. In others (Brazil and South Africa, for example), colonization was imposed on subjugated peoples. In some cases (Egypt and India) incorporation was deep and widespread; in others (Cambodia and Libya) it was patchy and superficial. In some instances (Brazil and Malaysia) incorporation was motivated by economic opportunity; in others it was motivated simply by political territoriality—preempting a rival power or securing a militarily strategic but economically unimportant location. Somalia's location, for example, made it strategically important for controlling access to the Indian Ocean and the Persian Gulf, with the result that the United States and the former Soviet Union vied for influence in this country that was otherwise unimportant to core countries.

The third factor modifying economic development has been that some states have responded to the unevenness and inequalities inherent to the modern world-system by pursuing alternatives to trade-based capitalism. The most important of these have been the communist political economy of the People's Republic of China and the socialist political economies of the former Soviet Union, its eastern European satellites, and Cuba. These alternatives have been presented to the world as being based on political beliefs—an opting-out of the world-system in favor of the pursuit of egalitarian sociocultural ideals. They can also be interpreted, however, as tactical alternatives—independent attempts to maintain sovereignty over territory, to distort and manipulate markets in order to achieve domestic economic efficiency, and to develop a modern infrastructure as a means of catching up with (and perhaps eventually joining) the core.

The result of transnational economic integration has been to reinforce the dominant core-periphery patterns in the world's economic landscapes at the macro scale. Patterns of trade between core economies, for instance, are already so strong that integration is able to draw on a good deal of momentum. Furthermore, it is relatively easy for core states to meet the political, social, and cultural prerequisites for successful economic integration.

In the case of peripheral economies, existing patterns of trade offer little realistic scope for the reallocation of output following the removal of trade barriers in trade preference organizations, common markets, or free trade associations. Most peripheral nations produce primary commodities that are exported to the core rather than to each other. Also, most are so short of capital that even pooled resources will not enable them to break free from their functional dependency on trade with core economies. Experience has shown, meanwhile, that it is difficult for peripheral states to meet the political, social, and cultural preconditions for successful economic integration. For instance, cultural differences are too great or very passionately held and political institutions are underdeveloped or often corrupted.

One important response to such problems has been the so-called "North-South dialogue." The most important influence on this dialogue has been the United Nations Conference on Trade and Development (UNCTAD), launched in Geneva in 1964. By the end of the Geneva meetings, a degree of political solidarity had emerged among peripheral countries.

By 1987 the Group of 77 had nearly 130 members and had succeeded in articulating demands for a "New International Economic Order" (not to be confused with the new international division of labor). Central to the new order envisioned by the Group of 77 are demands for fundamental changes in the marketing conditions of world trade in primary commodities. These changes would require a variety of measures, including price and production agreements among producer countries, the creation of international buffer stocks of commodities financed by a common fund, multilateral long-term supply contracts, and the indexing of prices of primary commodities against the price of manufactured goods. Such changes have been at the center of discussions in a series of UNCTAD conferences, special sessions of the United Nations General Assembly, meetings of a specially convened Conference on International Economic Cooperation, and successive meetings of the heads of state of the British Commonwealth. Throughout these discussions, however, the core countries in general and the United States in particular have been reluctant to do more than agree to general statements about the desirability of a new international economic order. As a result, it remains true that so far the North-South dialogue has been a failure, and the New International Economic Order has yet to begin.

In practice, therefore, there have been two dominant sets of spatial outcomes of transnational economic integration. One has simply been the reinforcement of the dominant core-periphery structure of the world economy because of the relative success of economic integration between core states. The second has been the imprint of this success on particular regions. This imprint can be discerned in terms of (1) the effects of trade creation, trade diversion, and socio-spatial tensions within core associations; and (2) the dislocations experienced within nonmember states. It is important to recognize, however, that other factors besides economics play key roles in the shaping the prospects for success of any region or subregion. (See Geography Matters: How Politics and Culture Modify the Economics of Development, p. 94.)

Summary and Conclusions

Places and regions everywhere carry the legacy of a sequence of major changes in world geography. The evolution of world geography can be traced from the prehistoric hearths of agricultural development and human settlement, through the trading systems of the precapitalist, preindustrial world, to the foundations of the geography of modern world. These foundations were cast through industrialization, the colonization of the world, and the spread of an international market economy. Today, these foundations can be seen in the geography of the information age, a geography that now provides a new, global context for places and regions.

The world today is highly integrated. Places and regions have become increasingly interdependent, connected through complex and rapidly changing linkages that are orchestrated by transnational corporations. Using new technology systems that allow for instantaneous global telecommunications and flexible patterns of investment and production, these corporations span the fast world of the core and the slow world of the periphery. This integration does tend to blur some national and regional differences as the global marketplace brings a dispersion of people, tastes, and ideas. The overall result, though, has been an intensification of the differences between the core and the periphery.

Within this new global context, local differences in resource endowments remain, and people's territorial impulses endure. Many local cultures continue to be resilient or adaptive. Fundamental principles of spatial organization also continue to operate. All this ensures that, even as the world-system becomes more and more integrated, places and regions continue to be made and remade. The new global context is filled with local variety that is constantly changing, just as the global context itself is constantly responding to local developments.

Key Terms

biogeography (p. 58)
biome (p. 58)
climate (p. 53)
colonialism (p. 62)
comparative advantage
  (p. 77)
common market (p. 92)
continental drift (p. 52)
continental shield (p. 52)
convergent plate boundary
  (p. 52)
core regions (p. 62)
culture (p. 80)
customs union (p. 92)
development theory (p. 85)
diaspora (p. 69)
divergent plate boundary
  (p. 52)

division of labor (p. 67)
Earth system science (p. 52)
economic union (p. 93)
ecosystem (p. 58)
faulting (p. 52)
free trade association
  (p. 92)
gender division of labor
  (p. 67)
global warming (p. 60)
greenhouse effect (p. 60)
hegemony (p. 76)
hinterland (p. 66)
international division
  of labor (p. 77)
International Monetary
  Fund (p. 87)

intertropical convergence
  zone (p. 55)
island biogeography (p. 60)
law of diminishing returns
  (p. 65)
leadership cycles (p. 76)
mercantilism (p. 66)
minisystem (p. 65)
nation (p. 68)
nationalism (p. 69)
nation-state (p. 69)
neocolonialism (p. 85)
neoliberalism (p. 87)
new international division
  of labor (p. 88)
overurbanization (p. 83)
peripheral region (p. 62)

plate tectonics (p. 52)
rift valley (p. 53)
semiperipheral region
  (p. 62)
subduction (p. 52)
trade creation effects (p. 93)
trade diversion effects
  (p. 93)
transform boundary (p. 53)
transnational corporation
  (p. 89)
weather (p. 53)
World Bank (p. 87)
world-empire (p. 65)
world-system (p. 61)

Review Questions

Testing Your Understanding

1. How does the theory of plate tectonics explain the current configurations of continents, mountains, and volcanically active zones? Explain the three main types of plate boundaries and give an example of a specific place where they are important.

2. What are major elements of the atmospheric circulation (winds, rising and sinking air) and how do they explain the geography of global precipitation? What are the main factors that explain the pattern of average temperature? What features of the climate explain rainforest, desert, and tundra vegetation?

3. How do weather and climate differ and what are the major causes of climate change? What is global warming and what is the evidence that it may already be happening?

4. In today's world-system, what are the main characteristics that differentiate the core regions, the semiperiphery, and the periphery?

5. What is colonialism? How is it different from imperialism?

6. What is the relationship between minisystems and world-empires? The development of world-empires resulted in the emergence of what two important trends?

7. What is hegemony? How did dominance move from one state to another during the European Colonial period? Which states were hegemons when, and why?

8. According to the concept of the international division of labor, why do colonies specialize in production of certain products? Give an example of international specialization and explain where specialization occurs (for example, the cocaine trade).

9. What were the three main means by which European colonizers formally acquired land in North America? Why are these differences of significance?

10. What is development theory? Describe the main elements of Rostow model of economic development. What are some arguments against development theory?

11. How do the World Bank and the International Monetary Fund (IMF) support neoliberal policies of development?

12. What are the possible spatial outcomes of continued global economic integration?

Thinking Geographically

1. Over one month, collect news articles and provide a summary of articles that you believe provide evidence of global warming.

2. How did the international slave trade of the 1600s help to establish the current world-system?

3. Why are some regions considered peripheral while others are semiperipheral? Provide and explain at least two examples of each.

4. What is urban infrastructure? Discuss the dynamic relationship between a city's wealth, population level, and ability to provide desired services. Use a specific city as a case study if possible.

5. As cultures evolve over time or vary from place to place, divisions of labor change as well. For three separate locations around the world today, please discuss the division of labor with regard to gender, technical skill, and social class.

Further Reading

Cardoso, F. H., and Faletto, R., *Dependency and Development*. Berkeley, CA: University of California Press, 1979.

Crow, B. M., Thorpe, M., et al., *Survival and Change in the Third World*. Cambridge: Open University Press, 1988.

Escobar, A., *Encountering Development*. Princeton, NJ: Princeton University Press, 1995.

Frank, A. G., *World Accumulation, 1492–1789*. New York: Monthly Review Press, 1978.

Kidron, M., and Segal, R., *The State of the World Atlas*, 5th ed. London: Penguin, 1995.

Kidron, M., and Smith, D., *The New State of War and Peace*. Hammersmith: HarperCollins, 1991.

Porter, P. W., and Sheppard, E., *A World of Difference: Society, Nature, Development*. New York: Guilford Press, 1998.

Potter, R. B., Binns, T., Elliott, J. A., and Smith, D., *Geographies of Development*. Harlow: Addison Wesley Longman, 1999.

Preston, P. W., *Development Theory: An Introduction*. Oxford: Blackwell, 1996.

Robbins, J. T., and Hite, A. (eds.), *From Modernization to Globalization: Perspectives on Development and Social Change*. Malden, MA: Blackwell, 2000.

Schaeffer, R. K., *Understanding Globalization: The Social Consequences of Political, Economic, and Environmental Change*. Lanham, MD: Rowman & Littlefield, 1997.

United Nations Conference on Trade and Development, *World Investment Report: Foreign Direct Investment and the Challenge of Development*. New York and Geneva: United Nations, 1999.

Wallerstein, I., *The Capitalist World Economy*. Cambridge, UK: Cambridge University Press, 1979.

Wallerstein, I., *The Politics of the World Economy*. Cambridge, UK: Cambridge University Press, 1984.

ARCTIC OCEAN

ICELAND
Reykjavik

More than 5 million
1–5 million
Fewer than 1 million
Capital cities are underlined

ATLANTIC
OCEAN

SWEDEN

NORWAY
Bergen

Turku

Oslo
Stockholm
Eskilstuna
Karlskoga

Aberdeen
Glasgow
Edinburgh

North
Sea

Göteborg
Huskvarna

Belfast
Newcastle
upon Tyne

DENMARK

Baltic Sea

Dublin
**UNITED
KINGDOM**
Leeds
IRELAND
Liverpool
Manchester
Birmingham

Århus
Copenhagen
Malmö

RUSSIA
Kaliningrad

Hamburg
Hannover
Gdansk

Amsterdam
NETH.
Rotterdam
Bielefeld
Berlin
POLAND

Thames
London
Portsmouth
Antwerp
Bruges
Rye
Brussels
Lille
BELGIUM
Aachen
Essen
Düsseldorf
Köln
Manheim
GERMANY
Frankfurt

Warsaw
Lodz
Katowice

Paris
LUX.
Luxembourg
Nürnberg
Prague
CZECH REPUBLIC
Vistula

Loire
Seine
**BLACK
FOREST**
Stuttgart
Munich
Vienna
SLOVAKIA
Bratislava

FRANCE
JURA
Bern
Zürich
SWITZ.
LIECH.
AUSTRIA
Budapest
HUNGARY

Bordeaux
Garonne
Lyon
ALPS
DOLOMITES
Ljubljana
SLOVENIA

Como
Turin
Milan
Venice
Verona
Zagreb

PORTUGAL
Ebro
PYRENEES
Lugano
Po
Genoa
Florence
Belgrade
**BOS. &
HERZ.**
DINARIC ALPS
CROATIA

Rhône
Nice
Pisa
CROATIA
Sarajevo

SPAIN
Madrid
ANDORRA
Marseille
MONACO
Siena
**SAN
MARINO**
Adriatic Sea

Lisbon
Toledo
Barcelona
Corsica
Tiber
APENNINES
YUGOSLAVIA

Valencia
Rome
Tiran

Seville
Benidorm
Murcia
Balearic Islands
Naples
ITALY
ALBANIA

Sardinia

GIBRALTAR

M e d i t e r r a n e a n   S e a

Sicily

0   200   400 Miles
0   200   400 Kilometers

MALTA

30°W   20°W   10°W   0°   10°E

60°N

50°N

40°N

Arctic Circle

Bremen/Weser

Meuse

Elbe

Oder

3 Europe

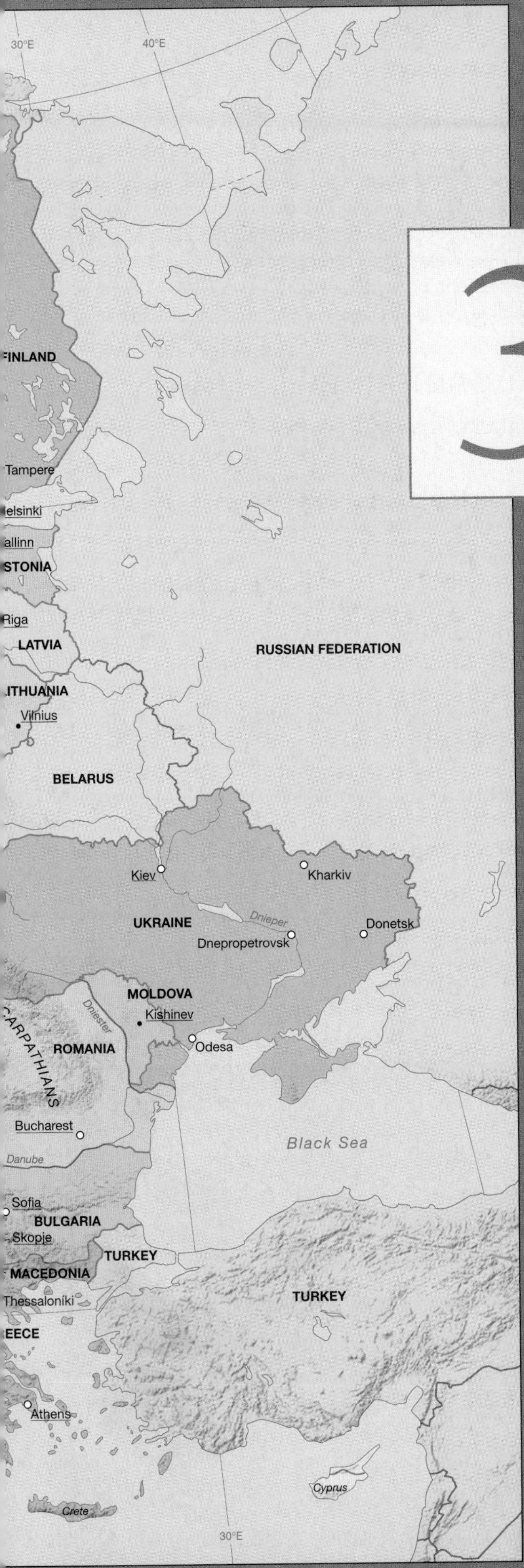

Figure 3.1

urope is the most intensively settled of the major world regions, with a population of 728 million, 75 percent of whom live in Europe's cities, 142 of which have populations of more than 500,000 (**Figure 3.1**). The western, northern, and southern limits of Europe as a whole are quite clearly defined, since they are formed by the Atlantic Ocean on the west, the Arctic Ocean on the north, and the Mediterranean Sea on the south. The eastern edge of Europe merges into the vastness of Asia and is less easily defined. The upland ranges of the Urals are sometimes used by geographers to mark the boundary between Europe and Asia, but the most significant factors separating Europe from Asia are human and relate to race, language, and a common set of ethical values that stem from Roman Catholic, Protestant, and Orthodox forms of Christianity. As a result, the eastern boundary of Europe is often demarcated through political and administrative boundaries rather than physical features. For a few decades, between 1945 and 1989, there was a significant geopolitical division within Europe into eastern and western Europe. These two subregions share a great deal in terms of physical geography, racial characteristics, and cultural values, but for a while they were divided by the Cold War territorial boundary that separated the capitalist democracies of western Europe from Soviet **state socialism,** a form of economy based on principles of collective ownership and administration of the means of production and distribution of goods, dominated and directed by state bureaucracies. The result was what Winston Churchill called an "Iron Curtain" along the western frontier of Soviet-dominated territory: a militarized frontier zone across which Soviet and East European authorities allowed the absolute minimum movement of people, goods, and information. Soviet influence to the east of the Iron Curtain created significantly different economic, social, and cultual conditions for a long enough period to have resulted in some modifications to the geography of eastern Europe. Since the collapse of the Soviet Union in 1989, however, its former **satellite states** in eastern Europe have reoriented themselves as part of the broader European world region. Meanwhile, most European states have joined together in the European Union, creating an extremely powerful economic and political force in world affairs. Europe as a whole still bears the legacy of a complex history of social and political development, but it has reemerged since World War II as a crucible of technological, economic, and cultural innovation.

Environment and Society in Europe

Two aspects of Europe's physical geography have been fundamental to its evolution as a world region and have been important influences on the evolution of regional geographies within Europe itself. First, as a world region, Europe is situated between the Americas, Africa, and the Middle East (see Figure 1.1). Second, as a satellite photograph of Europe reveals, Europe consists mainly of a collection of peninsulas and islands at the western extremity of the great Eurasian landmass (**Figure 3.2**).

The largest of the European peninsulas is the Scandinavian Peninsula, whose prominent western mountains separate Atlantic-oriented Norway from continental-oriented Sweden. Equally striking are the Iberian Peninsula, whose square mass projects into the Atlantic, and the boot-shaped Italian Peninsula. In the southeast is the broad triangle of the Balkan Peninsula, which projects into the Mediterranean, terminating in

Figure 3.2 Europe from space This image underlines one of the key features of Europe: the many arms of the seas that penetrate deep into the western extremity of the great Eurasian landmass. [*Source:* (inset) Map projection, Buckminster Fuller Institute and Dymaxion Map Design, Santa Barbara, CA. The word *Dymaxion* and the Fuller Projection Dymaxion™ Map design are trademarks of the Buckminster Fuller Institute, Santa Barbara, California, © 1938, 1967 & 1992. All rights reserved.]

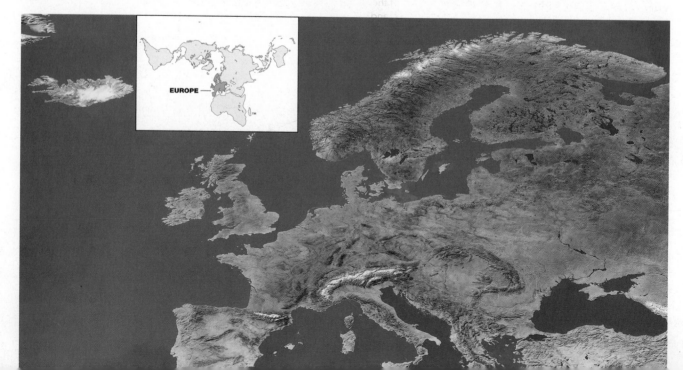

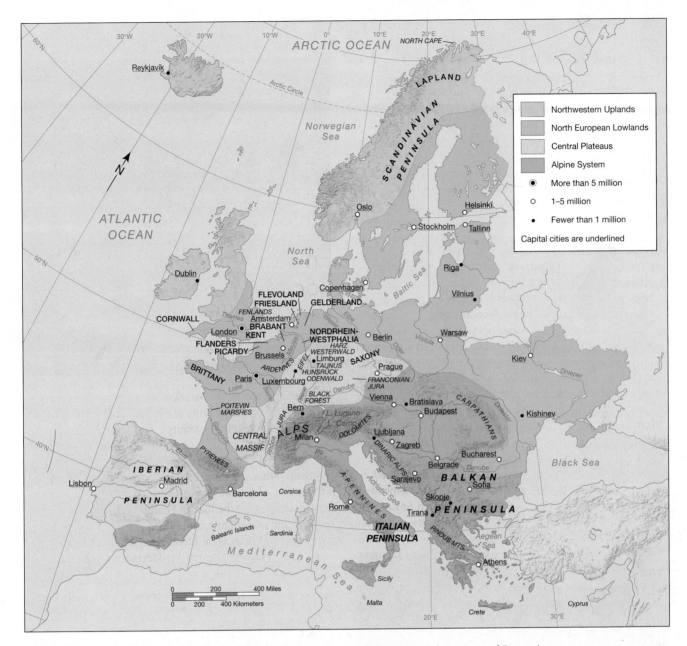

Figure 3.3 Europe's physiographic regions Each of the four principal physiographic regions of Europe has a broad coherence in terms of geology, relief, landforms, soils, and vegetation. (Adapted from R. Mellor and E. A. Smith, *Europe: A Geographical Survey of the Continent*. London: Macmillan, 1979.)

the intricate coastlines of the Greek peninsulas and islands. In the northwest are Europe's two largest islands, Britain and Ireland. The overall effect is that tongues of shallow seas penetrate deep into the European landmass. This was especially important in the pre-Modern period, when the only means of transporting goods were by sailing vessel and wagon. The Mediterranean and North seas, in particular, provided relatively sheltered sea lanes, fostering seafaring traditions in the peoples all around their coasts. The penetration of the seas deep into the European landmass provided numerous short land routes across the major peninsulas, making it easier for trade and communications to take place in the days of sail and wagon. As we shall see, Europeans' relationship to the surrounding seas has been a crucial factor in the evolution of European—and, indeed, world—geography.

Europe's navigable rivers were also important in shaping the human geography of the region. Although small by comparison with major rivers in other regions of the world, some of the principal rivers of Europe—the Danube, the Dneiper, the Elbe, the Rhine, the Seine, and the Thames—played key roles as routeways, and the low-lying watersheds between the major rivers of Europe's plains allowed canal building to take place relatively easily, thereby increasing the mobility of river traffic.

Landforms and Landscapes

The physical environments of Europe are complex and varied. It is impossible to travel far without encountering significant changes in physical landscapes. There is, however, a broad pattern to this variability, and it is based on four principal **physiographic regions** that are characterized by broad coherence of geology, relief, landforms, soils, and vegetation. These regions are the Northwestern Uplands, the Alpine System, the Central Plateaus, and the North European Lowlands (**Figure 3.3**).

The Northwestern Uplands are composed of the most ancient rocks in Europe, the product of the Caledonian mountain-building episode about 400 million years ago. Included in this region are the mountains of Norway and Scotland and the uplands of Iceland, Ireland, Wales, Cornwall (in England) and Brittany (in France). The original Caledonian mountain system was eroded and uplifted several times, and following the most recent uplift the northwestern uplands have been worn down again, molded by ice sheets and glaciers. Many valleys were deepened and straightened by ice, leaving spectacular glaciated landscapes. Since the last glaciation, sea levels have risen, forming **fjords** wherever these glaciated valleys have been flooded by the rising sea (**Figure 3.4**).

The Alpine System is the product of the most recent of Europe's mountain-building episodes, which occurred about 50 million years ago. The Alpine System stretches west to east across the southern part of Europe from the Pyrenees, which mark the border between Spain and France, through the Alps and the Dolomites and on to the Carpathians, the Dinaric Alps, and some ranges in the Balkan Peninsula. The Apennines of Italy and the Pindus Mountains of Greece are also part of the Alpine System. The Alpine landscape is characterized by jagged mountains with high, pyramidal peaks and deeply glaciated valleys (**Figure 3.5**). Seven of the peaks in the Western Alps exceed 4000 meters (13,123 ft) in height.

Figure 3.5   The Alps   Jagged peaks and glaciated valleys are typical of the Alpine ranges. This photograph is of the western Alps, near Chamonix, France.

Figure 3.4   Fjord landscape   Norway's western coastline is dominated by deeply eroded glacial valleys that have been drowned by the sea, creating distinctive fjord landscapes such as this one at Hestenesoyra, Norway.

Between the Alpine System and the Northwestern Uplands are the landscapes of the Central Plateaus and the North European Lowlands. The Central Plateaus are formed from 250- to 300-million-year-old rocks that have been eroded down to broad tracts of uplands. Beneath the forest-clad slopes and fertile valleys of these plateaus lie many of Europe's most important coalfields. For the most part, the plateaus reach between 500 and 800 meters (1640 and 2625 feet) in height, though they rise to more than 1800 meters (5905 feet) in the Central Massif of France. The Central Plateaus were generally too low to have been glaciated and too far south to have been covered by the great northern ice sheets of the last Ice Age. Rather, their landscape is characterized by rolling hills, steep slopes and dipping vales, and deeply carved river valleys (**Figure 3.6**).

The North European Lowlands sweep in a broad crescent from southern France, through Belgium, the Netherlands, and southeastern England and into northern Germany, Denmark, and the southern tip of Sweden. Continuing eastward, they broaden into the immense European plain that extends through Poland, the Czech Republic, Slovakia, and Hungary, all the way into Russia. Coal is found in quantity under the lowlands of England, France, Germany, and Poland and in smaller deposits in Belgium and the Netherlands. Oil and natural gas deposits are found beneath the North Sea and under the lowlands of southern England, the Netherlands, and northern Germany. Nearly all of this area lies below 200 meters (656 feet) in elevation, and the topography everywhere is flat or gently undulating. As a result, the region has been particularly attractive to farming and settlement (**Figure 3.7**). There is a great deal of variation, however, in the fertility of the soils, so that settlement patterns are uneven and agriculture is finely tuned to the limits and opportunities of local soils, landscape, and climate.

Figure 3.6 The Central Plateau The river systems of central Europe have cut into the rolling hills of Europe's Central Plateaus, creating deep gorges and steep-sided valleys whose slopes are often suited to viticulture. This photograph shows the Mosel Valley, near Bremm.

Figure 3.7 The North European Lowlands The rolling plains of the North European Lowlands provide many fertile and easily tilled soils that have given rise to lush agricultural landscapes. Shown here is the lowland farming landscape near Cantal, France.

Climate

The seas that surround Europe have an important influence on the region's climate. In winter, seas cool more slowly than the land, while in summer they warm up more slowly than the land. As a result, the seas provide a warming effect in winter and a cooling effect in summer. Europe's arrangement of islands and peninsulas means that this moderating effect is particularly marked, contributing to an overall climate that does not have great seasonal extremes of heat and cold. This moderating effect is intensified by the North Atlantic Drift, which carries great quantities of warm water from the tropical Gulf Stream as far as the British Isles and the North Sea. Given its latitude (Paris, at almost 49° N, is the same latitude as Vancouver and Newfoundland in Canada), most of Europe is remarkably warm. It is continually crossed by moist, warm air masses that drift in from the Atlantic. The effects of these warm, wet, westerly winds are most pronounced in northwestern Europe, where squalls and showers accompany the passage of the succession of eastward-moving weather systems. Weather in northwestern Europe tends to be unpredictable, partly because of the swirling movement of air masses as they pass over the Atlantic and partly because of the complex effects of the widely varying temperatures of interpenetrating bodies of land and water. Further east, in continental Europe, seasonal weather tends to be more settled, with more pronounced extremes of summer heat and winter cold. In these interior regions, local variations in weather are influenced a great deal by the direction in which a particular slope or land surface faces and its elevation above sea level.

The Mediterranean Basin has a different and quite distinctive climate. In winter, low-pressure systems along the northern Mediterranean draw in rain-bearing weather fronts from the Atlantic. When low pressure over the northern Mediterranean coincides with high pressure over continental Europe, southerly airflows spill over mountain ranges and down valleys, bringing cold blasts of air. These events have local names: the *mistral,* for example, which blows down the Rhône Valley in southern France, and the *bora,* which blows over the eastern Alps toward the Adriatic region of Italy. In summer, hot, dry air masses from Asia and Africa dominate the Mediterranean Basin, producing dry, sunny conditions.

Environmental History

Temperate forests originally covered about 95 percent of Europe, with a natural ecosystem dominated by oak, together with elm, beech, and lime. By the end of the medieval period, Europe's forest cover had been reduced to about 20 percent; and today it is around 5 percent. The first clearing of natural forests took place as Europe's small prehistoric populations began to cultivate small patches of land under the "slash-and-burn" or **swidden** system. Under this system, people would create clearings in the forest, cultivate the land for several years, and then abandon it once fertility had declined, allowing the area to revert to secondary growth. As population slowly increased over many centuries, permanent fields were established, usually through clearing the woodland around small villages in small increments.

Then, between A.D. 1000 and A.D. 1300, a period of warmer climate, together with advances in agricultural knowledge and practices, led to a significant transformation of the European landscape. The population more than doubled, from around 36 million to more than 80 million, and a vast amount of land was brought under cultivation for the first time. By about 1200 most of the best soils of western Europe had been cleared of forest and new settlements were increasingly forced into the more marginal areas of heavy clays or thin sandy soils on higher ground and heathlands. Many parts of Europe undertook large-scale drainage projects in order to reclaim marshlands. The Romans had already demonstrated the effectiveness of drainage schemes, reclaiming parts of Italy and northwestern Europe. In the twelfth and thirteenth centuries there were important and extensive drainage and resettlement schemes in Italy's Po Valley, in the Poitevin marshes of France, and in the Fenlands of eastern England. In eastern Europe, clearances were organized by agents acting for various princes and bishops who controlled extensive tracts of land. The agents would arrange finance for settlers and develop villages and towns, often to standardized designs.

This great medieval colonization came to a halt nearly everywhere around 1300. One important factor was the so-called Little Ice Age, a period of cooler climate during which the growing season was reduced significantly—perhaps by as much as five weeks. Another factor was the catastrophic loss of population in the Black Death (1347–51) and the periodic recurrences of the plague that continued for the rest of the fourteenth century. The Little Ice Age lasted until the early sixteenth century, by which time many villages, and much of the more marginal land, had been abandoned.

The resurgence of European economies from the sixteenth century onward coincided with overseas exploration and trade, but domestic landscapes were significantly impacted by repopulation, by land tenure and reforms, and by advances in science and technology that changed agricultural practices, allowing for a more intensive use of the land. In the Netherlands, a steadily growing population and the consequent requirement for more agricultural land led to increased efforts to reclaim land from the sea and to drain coastal marshlands. Hundreds of small estuarine and coastal barrier islands were slowly joined into larger units, and sea defense walls were constructed in order to protect low-lying land, which was drained by windmill-powered water pumps, the excess water being carried off into a web of drainage ditches and canals. The resulting **polder** landscape provided excellent, flat, fertile, and stone-free soil. Between 1550 and 1650, 165,000 hectares (407,715 acres) of polderland was established in the Netherlands, and the sophisticated techniques developed by the Dutch began to be applied elsewhere in Europe—including eastern England and the Rhône estuary in southern France. While most of these schemes resulted in improved farmland, the environmental consequences were often serious. In addition to the vulnerability of the polderlands to inundation by the sea, large-scale drainage schemes devastated the wetland habitat of many species, while some ill-conceived schemes simply ended in widespread flooding.

These environmental problems, though, were but a prelude to the environmental changes and ecological disasters that accompanied the industrialization of Europe, beginning in the eighteenth century. Mining—especially coal mining—created derelict landscapes of spoil heaps; urbanization encroached on rural landscapes and generated unprecedented amounts and concentrations of human, domestic, and industrial waste; and manufacturing, unregulated at first, resulted in extremely unhealthy levels of air pollution and in the devastating pollution of rivers and streams.

Europe in the World

The foundations of Europe's human geography were laid by the Greek and Roman empires. Beginning around 750 B.C., the ancient Greeks developed a series of fortified city-states (called *poleis*) along the Mediterranean coast, and by 550 B.C. there were about 250 of them. **Figure 3.8** shows the location of the largest of these trading colonies, some of which subsequently grew into important cities (for example, Athens and Corinth), while others remain as isolated ruins or as archaeological sites (for example, Delphi and Olympia). The Roman Republic was established in 509 B.C. and took almost 300 years to establish control over the Italian Peninsula. By A.D. 14, however, the Romans had conquered much of Europe, together with parts of North Africa and Asia Minor. Most European cities that are important today had their origin as Roman settlements. In quite a few of these cities, it is possible to trace the original Roman street layouts. In some, it is possible to glimpse remnants of defensive city walls and of paved streets, aqueducts, viaducts, arenas, sewage systems, baths, and public buildings. In the modern countryside, the legacy of the Roman empire is represented by arrow-straight roads, built by their engineers and maintained and improved by successive generations.

The decline of the Roman empire, beginning in the fourth century A.D., was accompanied by a long period of rural reorganization and consolidation under feudal systems, a period that is often characterized as uneventful and stagnant. In fact, the roots of European regional differentiation can be traced to this long feudal era of slow change. **Feudal systems** were almost wholly agricultural, with 80 or 90 percent of the workforce engaged in farming and most of the rest occupied in basic craft work. Most production was for people's immediate needs, with very little of a community's output ever finding its way to wider markets.

By A.D. 1000 the countryside of most of Europe had been consolidated into a regional patchwork of feudal agricultural subsystems, each of which was more or less self-sufficient. For a long time, towns were small and relatively unimportant, their existence tied mainly to the castles, palaces, churches, and cathedrals of the upper ranks of the feudal hierarchy. These economic landscapes—inflexible, slow-motion and introverted—nevertheless contained the essential preconditions for the rise of Europe as the dynamic hub of the world economy.

Trade and the Age of Discovery

The rise of Europe as a major world region had its origins in the emergence of a system of merchant capitalism in the fifteenth century. Long-standing trading patterns that had been developed from the twelfth century by the merchants of Venice, Pisa, Genoa, Florence, Bruges, Antwerp, and the Hanseatic League (a federation of city-states around the North Sea and Baltic coasts that included Bremen, Hamburg, Lübeck, Rostock, and Danzig, as shown in **Figure 3.9**) provided the foundations for a trading system of immense complexity that soon came to span Europe.

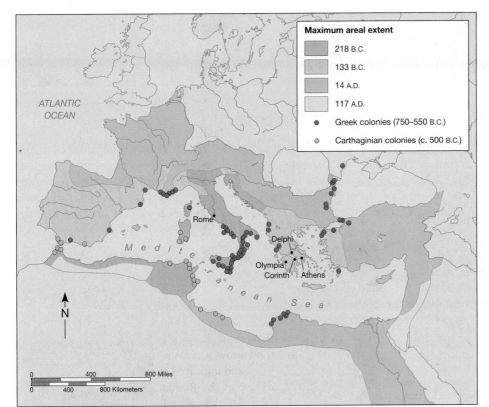

Figure 3.8   Greek colonies and extent of the Roman empire This map shows the distribution of Greek *poleis* (city-states) and Carthaginian colonies, and the spread of the Roman empire from 218 B.C. to A.D. 117. (*Source:* Redrawn from R. King et al., *The Mediterranean.* London: Arnold, 1997, pp. 59 and 64.)

In the fifteenth and sixteenth centuries a series of innovations in business and technology contributed to the consolidation of Europe's new merchant capitalist economy. These included several key innovations in the organization of business and finance: banking, loan systems, credit transfers, company partnerships, shares in stock, speculation in commodity futures, commercial insurance, and courier/news services. Meanwhile, technological innovations began to further strengthen Europe's economic advantages. Some of these innovations were adaptations and improvements of Oriental discoveries—the windmill, spinning wheels, paper manufacture, gunpowder, and the compass, for example. In Europe, however, there was a real passion for the mechanization of the manufacturing process. Key engineering breakthroughs included the more efficient use of energy in water mills and blast furnaces, the design of reliable clocks and firearms, and the introduction of new methods of processing metals and manufacturing glass.

It was the combination of innovations in shipbuilding, navigation, and naval ordnance, however, that had the most far-reaching consequences for Europe's role in the world economy. In the course of the fifteenth century, the full-rigged sailing ship was developed, enabling faster voyages in larger and more maneuverable vessels that were less dependent on favorable winds. Meanwhile, navigational tools such as the quadrant (1450) and the astrolabe (1480) were developed, and a systematic knowledge of Atlantic winds had been acquired. By the mid-sixteenth century, armorers in England, Holland, and Sweden had perfected the technique of casting iron guns, making it possible to replace bronze cannons with larger numbers of more effective guns at lower expense. Together, these

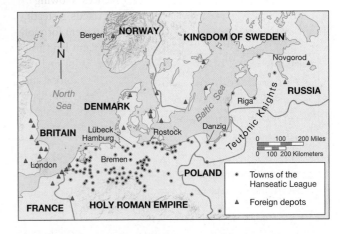

Figure 3.9   The Hanseatic League The Hanseatic League was a federation of city-states founded in the thirteenth century by north German towns and affiliated German merchant groups abroad to defend their mutual trading interests. The League, which remained an influential economic and political force until the fifteenth century, laid the foundations for the subsequent growth of merchant trade throughout Europe. (*Source:* Redrawn from P. Hugill, *World Trade Since 1431.* Baltimore: Johns Hopkins University Press, 1994, p. 50.)

advances made it possible for the merchants of Europe to establish the basis of a worldwide economy in the space of less than a hundred years (see Geography Matters: European Exploration and the Age of Discovery, p. 106).

Gold and silver from the Americas provided the first major economic transformation of Europe, allowing Spain and Portugal, in particular, to live well above their means. The bullion

Geography Matters

European Exploration and the Age of Discovery

During the fifteenth century, the Portuguese, under the sponsorship of Dom Henrique (known as "Prince Henry the Navigator"), established a school of navigation and cartography and began to explore the Atlantic and the coast of Africa. *Cartography* is the name given to the system of practi-

cal and theoretical knowledge about making distinctive visual representations of Earth's surface in the form of maps. **Figure 1** shows the key voyages of discovery. Portuguese explorer Bartholomeu Dias reached the Cape of Good Hope (the southern tip of Africa) in 1488. In 1492 Cristóbal Colón (Christo-

Figure 1 The European Age of Discovery The European voyages of discovery can be traced to Portugal's Prince Henry the Navigator (1394–1460), who set up a school of navigation and financed numerous expeditions with the objective of circumnavigating Africa in order to establish a profitable sea route for spices from India. The knowledge of winds, ocean currents, natural harbors, and watering places built up by Henry's captains was an essential foundation for the subsequent voyages of Cristóbal Colón (Columbus), Vasco da Gama, Ferdinand de Magalhães (Magellan), and others. The end of the European Age of Discovery was marked by Captain James Cook's voyages to the Pacific.

This colored engraving shows Henry Hudson, with his ship the *Half Moon* anchored in what is today the Hudson River in New York. Hudson's 1609 voyage up the Hudson in search of an outlet to the Pacific Ocean took him as far as present-day Albany, where he gave up his search.

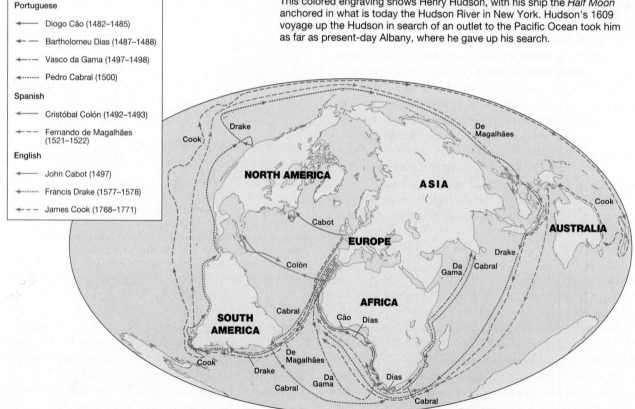

Portuguese

←—— Diogo Cão (1482–1485)

←— — Bartholomeu Dias (1487–1488)

←– – – Vasco da Gama (1497–1498)

←······· Pedro Cabral (1500)

Spanish

←—— Cristóbal Colón (1492–1493)

←— — Fernando de Magalhães (1521–1522)

English

←—— John Cabot (1497)

←······· Francis Drake (1577–1578)

←– – – James Cook (1768–1771)

pher Columbus), from Genoa, Italy, arrived in Hispaniola (the island that is now Haiti and the Dominican Republic) under the sponsorship of the Castillian (Spanish) monarchy. Six years later, Vasco da Gama reached India; two years after that, Pedro Cabral crossed the Atlantic from Portugal to Brazil. A small fleet of Portuguese ships reached China in 1513, and the first circumnavigation of the globe was completed in 1522 by Juan Sebastián del Cano, a survivor of the expedition led by Portuguese navigator Fernando de Magalhães (better known as Magellan). Portuguese successes inspired other countries to attempt their own voyages of discovery, all of them in pursuit of commercial advantage and economic gain. Between them, these explorations led to an invaluable body of knowledge about ocean currents, wind patterns, coastlines, peoples, and resources.

Geographical knowledge acquired during this Age of Discovery was crucial to the expansion of European political and economic power in the sixteenth century. In societies that were becoming more and more commercially oriented and profit-conscious, geographical knowledge became a valuable commodity in itself. Information about overseas regions was a first step to control and have influence over them; this in turn was an important step to wealth and power. At the same time, every region began to open up to the influence of other regions because of the economic and political competition that was unleashed by geographical discovery. Not only was the New World affected by European colonists, missionaries, and adventurers; but also the countries of the Old World found themselves pitched into competition with one another for overseas resources. Meanwhile, new crops such as corn and potatoes, introduced to Europe from the New World, had a profound impact on local economies and ways of life.

The growth of a commercial world economy meant that objectivity in cartography and geographical writing became essential. Navigation, political boundaries, property rights, and rights of movement all depended on accuracy and impartiality. Success in commerce depended on clarity and reliability in describing the opportunities and dangers presented by one region or another. International rivalries required sophisticated understandings of the relationships between countries, regions, and places. Geography became a key area of knowledge. The historical period in Europe known as the Renaissance (from the mid-fourteenth to the mid-seventeenth centuries) saw an explosion of systematic mapmaking and geographical description (**Figure 2**).

Figure 2 Vermeer's geographer Dutch master Johannes Vermeer's painting *The Geographer* (1668–69). In Renaissance Europe, the study of geography not only contributed to the growth of scientific knowledge but also helped to support European overseas expansion. Vermeer's geographer is surrounded with accurately rendered cartographic objects, including a wall chart of the sea coasts of Europe, published by Willem Blau in 1658, and a globe made by Jodocus Hondius in 1618.

created an effective demand for consumer and capital goods of all kinds—textiles, furniture, weapons, ships, food, and wine (see Geographies of Indulgence, Desire and Addiction: Wine, p. 108)—thus stimulating production throughout Europe and creating the basis for a Golden Age of prosperity for most of the sixteenth century. Meanwhile, overseas exploration and expansion made available a variety of new and unusual products—cocoa, beans, maize, potatoes, tomatoes, sugarcane, tobacco, and vanilla from the Americas, tea and spices from the Orient—that opened up large new markets to enterprising merchants. It is also important to note that the emergence of a worldwide system of exploration and trade helped to establish the foundations of modern academic geography.

The demands of overseas expansion stimulated Europeans to achieve new developments in nautical mapmaking, naval artillery, shipbuilding, and the use of sails. Maritime insurance emerged as one of a growing number of sophisticated business and financial services industries; and the whole experience of overseas expansion provided Europeans with a great practical school of entrepreneurship and investment. Just as important

Geographies of Indulgence, Desire, and Addiction

Wine

The production and consumption of wine reflects the evolution of the world-system. Wine was one of the early luxury products that established the pattern of merchant trading within Europe. When Europeans branched out to incorporate more of the world into the orbit of their world-system, they began organizing the production of wine wherever climatic conditions were encouraging: in warm temperate zones, roughly between latitudes 30° and 50° north and south. Today, wine is one of the most widespread commodities of consumer indulgence, and fine wines are an important marker of affluence and distinction throughout the world's core regions and in the affluent enclaves of many of the metropolises of peripheral regions. In 1999, more than 26 million liters (5.72 million gallons) of wine were produced worldwide. Almost 15 million liters (3.3 million gallons) of this was consumed in Europe, compared with 2.2 million liters (484,000 gallons) in North America.

The original domestication of wine grapes (*Vitis vinifera*) seems to have taken place as early as 8000 B.C. along the mountain slopes of Georgia, eastern Turkey, and western Iran.

Wine had symbolic and ritual significance in early civilizations of the eastern Mediterranean, partly because its ability to intoxicate and engender a sense of "other-worldliness" provided a means through which people could feel that they could come into contact with their gods, and partly because of the apparent death of the vine in winter and its dramatic growth and rebirth in the spring. Greek civilization established viticulture—the cultivation of grape vines for winemaking—as one of the staples of the Mediterranean agrarian economy, along with wheat and olives. By the sixth century B.C., Greek wine was being traded as far as Egypt, the shores of the Black Sea, and the southern regions of France. Under the Roman empire, viticulture spread west along the north shores of the Mediterranean and along the valleys of navigable rivers in France and Spain, while the wine trade extended north, to the North Sea and the Baltic. By the first century A.D., wine had become a commodity of indulgence, desire, and—for some—addiction throughout Europe. Viticulture and the art of winemaking survived (but did not prosper) through the Middle Ages, even in those parts of Mediterranean Europe that

Figure 1   The global distribution of viticulture   In general, the best areas for viticulture lie between the 10°C and 20°C annual isotherms, equating approximately to the warm temperate zones between latitudes 30° and 50° north and south. In detail, the geography of viticulture is heavily influenced by soils and micro-climatic conditions. Where grapes are produced successfully nearer the equator, as in parts of Bolivia and Tanzania, it is usually because they are grown at higher altitudes. (*Source:* Adapted from T. Unwin, *Wine and the Vine.* New York: Routledge, 1996, pp. 35 and 219.)

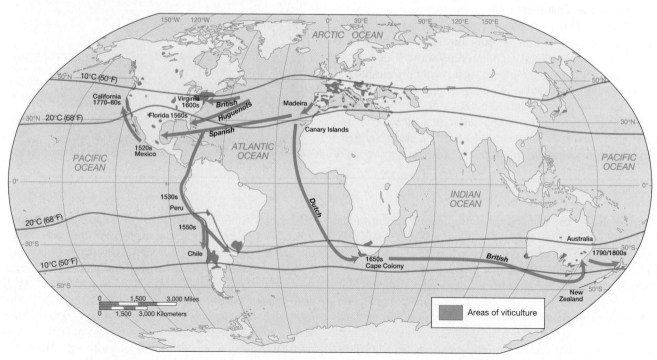

came under Islamic rule, where the consumption of alcohol was, theoretically, prohibited. Then, in the late Medieval period, the growth of towns provided a large and increasingly affluent consumer market that led to the development of large, commercially oriented vineyards. Commercial viticulture spread to the southeast-facing slopes of the major river valleys in northern France and Germany, and merchant traders, drawing on innovations in finance, banking, and credit, facilitated the movement of vast quantities of wine, together with spices, perfumes, and silks, from the Mediterranean to England, Flanders, Scandinavia, and the Baltic. These northern European regions, meanwhile, paid for their luxury imports with the proceeds of exports of furs, fish, dairy produce, timber, and wool.

In the sixteenth century, Spanish and Portuguese overseas expansion saw the introduction of viticulture to the New World—to Mexico in the 1520s, Peru in the 1530s, Chile in the 1550s, and Florida in the 1560s. The British introduced viticulture to Virginia in the 1600s and the Dutch established vineyards in the Cape Colony of southern Africa in the 1650s. The first vineyards in California were established by Franciscan missions in the 1770s, in southeastern Australia in the 1790s, and in New Zealand in the early 1800s. Meanwhile, in Europe, demographic growth and increasing prosperity rapidly expanded the market for wine. Winemakers developed new types of wine (including champagne, claret, and port), began to specialize in particular varieties of grapes (red grapes such as Cabernet Sauvignon, Nebbiolo, and Pinot Noir, and white grapes such as Chardonnay, Riesling, and Sauvignon Blanc), and found ways of storing wine, so that especially good quality wines could be aged without spoiling. Vintage wines, carefully aged and stored, acquired special value for connoisseurs and so added an important new dimension to the production and consumption of wine.

Disaster hit European winemakers in the 1860s in the form of an aphid, *phylloxera*, that had somehow been brought to Europe on American vines. Though American vines were immune to *phylloxera*, the aphid killed European species. Before it was discovered, in 1881, that grafting European vines onto American rootstock would produce high-quality and phylloxera-resistant plants, many European vineyards were devastated. On recovering from the *phylloxera* episode, however, the makers (and consumers) of fine wines faced other problems: unscrupulous merchants and foreign competitors sought to pass off lesser wines as prestigious wines, limited supplies of good wines were watered to stretch limited supplies, and poor wines were adulterated with chemicals to improve their color or their shelf life. In response, the exclusivity of wines was protected by new systems of regulation. In France, for example, the *Appellation*

Contrôlée system was introduced in order to guarantee the authenticity of wines, district by district. In Germany, the classification system was based not on geographic origin but on levels of quality and degrees of sweetness. Such regulations have been important in determining the nature of wine as a commodity of indulgence and desire. Nevertheless, the globalization of viticulture and the more widespread consumption of wine have meant that branding remains an issue. Japanese supermarket shelves are lined with locally produced bottles with French-language labels that allude to nonexistent châteaus, while California wineries produce "Chablis," "Champagne," and "Burgundy."

Two of the most important changes in relation to wine production date from the mid-twentieth century and have contributed a great deal to the globalization of wine as a global commodity of indulgence and desire. First, the deployment of scientific approaches and new technologies, combined with large-scale capital investment in the industry, allowed the development of first-class wines in North America, Australia, New Zealand, and elsewhere. Second, the hedonistic cultural shift of the 1960s in Western societies brought the consumption of wine firmly into the routine practices of the middle classes. The result is that the production and retailing of wine is now a significant component of the activities of large, conglomerate, and transnational corporations. The most successful of these are able to exploit and manipulate changing patterns of consumption, introducing profitable new products such as wine coolers in order to broaden the market.

Figure 2 The globalization of wine consumption Store assistant Zhang Weigang holds a bottle of imported wine at a store in Beijing. China's prosperous middle classes are rapidly developing a taste for wine, with most imports coming from Spain and France.

was the way that the profits from overseas colonies and trading provided capital for investment in domestic agriculture, mining, and, most important of all, manufacturing.

These changes also had a profound effect on the geography of Europe itself. Before the mid-fifteenth century, Europe was organized around two subregional maritime economies—one based on the Mediterranean and the other on the Baltic. The overseas expansions pioneered first by the Portuguese and then by the Spanish, Dutch, English, and French reoriented Europe's geography toward the Atlantic. The river basins of the Rhine, the Seine, and the Thames rapidly became the focus of a thriving network of **entrepôt** seaports, intermediary centers of trade and transshipment, that transformed Europe. These three river basins, backed by the increasingly powerful states in which they were embedded—the Netherlands, France, and Britain, respectively—then became engaged in a struggle for economic and political hegemony. Although the Rhine was the principal natural routeway into the heart of Europe, the convoluted politics of the Netherlands allowed Britain and France to become the dominant powers by the late 1600s. Subsequently, France, under Napoleon, made the military error of attempting to pursue both maritime and continental power at once, allowing Britain to become the undisputed hegemonic power of the industrial era.

Industrialization and Imperialism

Europe's regional geographies were comprehensively recast once more by the new production and transportation technologies that marked the onset of the Industrial Revolution (from the late 1700s). New production technologies, based on more efficient energy sources, helped to raise levels of productivity and to create new and better products that stimulated demand, increased profits, and created a pool of capital for further investment. New transportation technologies enabled successive phases of geographic expansion that completely reorganized the geography of Europe. As the application of new technologies altered the margins of profitability in different kinds of enterprise, so the fortunes of particular places and regions shifted.

There was in fact not a sudden, single Industrial Revolution but three distinctive transitional waves of industrialization, each having a different degree of impact on different regions and countries. The first, between about 1790 and 1850, was based on a cluster of early industrial technologies (steam engines, cotton textiles, and ironworking) and was highly localized. It was limited to a few regions in Britain where industrial entrepreneurs and workforces had first exploited key innovations and the availability of key resources (coal, iron ore, and water). These included north Cornwall, eastern Shropshire (where Abraham Darby II built the world's first iron bridge—at Ironbridge, on the River Severn), south Staffordshire, south Lancashire, southwestern Yorkshire, Tyneside, and Clydeside (**Figure 3.10**).

The second wave, between about 1850 and 1870, involved the diffusion of industrialization to most of the rest of Britain and to parts of northwest Europe, particularly the coalfield areas of northern France, Belgium, and Germany (see

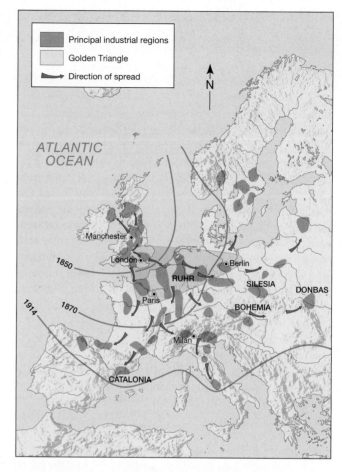

Figure 3.10   The spread of industrialization in Europe
European industrialization began with the emergence of small industrial regions in several different parts of Britain, where early industrialization drew on local mineral resources, water power, and early industrial technologies. As new rounds of industrial and transportation technologies emerged, industrialization spread to other regions with the right locational attributes: access to raw materials and energy sources, good communications, and large labor markets.

Figure 3.10). New opportunities were created as railroads and steamships made more places accessible, bringing their resources and their markets into the sphere of industrialization. New materials and new technologies (steel, machine tools) created opportunities to manufacture and market new products. These new activities brought some significant changes in the logic of industrial location. The importance of railway networks, for example, attracted industry away from smaller towns on the canal systems and toward larger towns with good rail connections. The importance of steamships for coastal and international trade attracted industry to larger ports. At the same time, the importance of steel produced concentrations of heavy industry in places with nearby supplies of coal, iron ore, and limestone.

The third wave of industrialization saw a further reorganization of the geography of Europe as yet another cluster of technologies (including electricity, electrical engineering, and telecommunications) brought different needs and created new opportunities. During this period, industrialization spread for the first time to remoter parts of the United Kingdom, France, and Germany and to most of the Netherlands, southern Scandinavia, northern Italy, eastern Austria, Bohemia (in what was then Czechoslovakia), Silesia (in Poland), Catalonia (in Spain), and the Donbas region of the Ukraine, then in Russia. The overall result was to create the foundations of a core-periphery structure within Europe (see Figure 3.10), with the heart of the core centered on the "Golden Triangle" stretching between London, Paris, and Berlin. The peripheral territories of Europe—most of the Iberian peninsula, northern Scandinavia, Ireland, southern Italy, the Balkans, and east-central Europe—were slowly penetrated by industrialization over the next 50 years. The environmental impacts of these changes were profound. Much of Europe's forest cover was cleared, while remaining forests and woodlands suffered from the **acid rain** resulting from heavy doses of atmospheric pollution. Many streams and rivers also became polluted, while the landscape everywhere was scarred with quarries, pits, cuttings, dumps, and waste heaps.

Meanwhile, several of the most powerful and heavily industrialized European countries (notably the United Kingdom, Germany, France, and the Netherlands) were by now competing for influence on a global scale. This competition developed into a scramble for territorial and commercial domination through **imperialism**—the deliberate exercise of military power and economic influence by core states in order to advance and secure their national interests. European countries engaged in pre-emptive geographic expansionism in order to protect their established interests and to limit the opportunities of others (**Figure 3.11**). They also wanted to secure as much of the world as possible—through a combination of military oversight, administrative control, and economic regulations—in order to ensure stable and profitable environments for their traders and investors. This combination of circumstances defined a new era of imperialism.

During the first half of the twentieth century, the economic development of the whole of Europe was disrupted twice by major wars. The devastation of the First World War

Figure 3.11   Colonial powers in 1914   The main era of imperialism was the period from 1880 to 1914, when many European powers sought to gain territories in Africa and Asia. The principal motivation was economic, and territories were taken by force before being subject to rule by the imperial power. It was not until the mid-twentieth century that colonialism came to be generally regarded as illegitimate.   (*Source:* Redrawn from B. Crow and A. Thomas, *Third World Atlas.* Milton Keynes: Open University Press, 1982, p. 41).

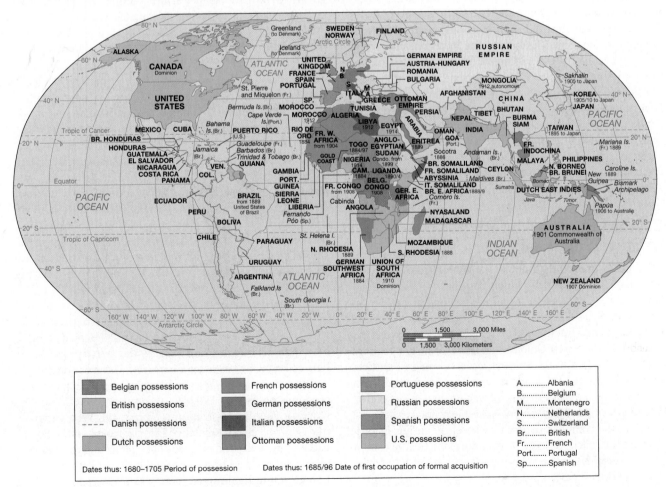

was immense. The overall loss of life, including the victims of influenza epidemics and border conflicts that followed the war, amounted to between 50 and 60 million. About half as many again were permanently disabled. For some countries, this meant a loss of between 10 and 15 percent of the male workforce.

Just as European economies had adjusted to these dislocations, the Great Depression created a further phase of economic damage and reorganization throughout Europe. World War II resulted in yet another round of destruction and dislocation. The total loss of life in Europe this time was 42 million, two-thirds of whom were civilian casualties. Systematic German persecution of Jews—the Holocaust—resulted in approximately 4 million Jews being put to death in extermination camps such as Auschwitz and Treblinka, with up to 2 million more being exterminated elsewhere, along with gypsies and others. The German occupation of continental Europe also involved ruthless economic exploitation. By the end of the war, France was depressed to below 50 percent of her pre-war standard of living and had lost 8 percent of its industrial assets. The United Kingdom lost 18 percent of its industrial assets (including overseas holdings), and the Soviet Union lost 25 percent. Germany lost 13 percent of its assets and ended the war with a level of income per capita that was less than 25 percent of the prewar figure. In addition to the millions killed and disabled during the Second World War, approximately 46 million people were displaced between 1938 and 1948 through flight, evacuation, resettlement, or forced labor. Some of these movements were temporary, but most were not.

After the war, the Cold War rift between eastern and western Europe resulted in a further erosion of the coherence of the European economy and, indeed, of its economic geography. Ironically, this rift helped to speed economic recovery in western Europe. The United States, whose leaders believed that poverty and economic chaos in Western Europe would foster communism, embarked on a massive program of economic aid under the **Marshall Plan**. This pump-priming action, together with the backlog of demand in almost every sphere of production, provided the basis for a remarkable recovery. Meanwhile, eastern Europe began an interlude of state socialism.

Eastern Europe's Interlude of State Socialism

After the Second World War, the leaders of the Soviet Union felt compelled to establish a **buffer zone** between their homeland and the major Western powers in Europe. The Soviet Union rapidly established its dominance throughout eastern Europe: Estonia, Moldova, Latvia, and Lithuania were absorbed into the Soviet Union itself, and Soviet-style regimes were installed in Albania, Bulgaria, Czechoslovakia, East Germany, Hungary, Poland, Romania, and Yugoslavia. In addition to the installation in 1947 of the Iron Curtain, which severed most economic linkages with the West, this intervention resulted in the complete nationalization of the means of production, the collectivization of agriculture, and the imposition of rigid social and economic controls within the East European **satellite states.**

It should be emphasized at the outset that the economies of the former Soviet Union and its satellites were *not* based on true socialist or communist principles in which the working class had democratic control over the processes of production, distribution, and development. Rather, they evolved as something of a hybrid, in which state power was used by a bureaucratic class to create **command economies** in the pursuit of modernization and economic development. The Communist Council for Mutual Economic Assistance (CMEA, better known as COMECON) was established to reorganize eastern European economies in the Soviet mold—with individual members, each pursuing independent, centralized plans designed to produce economic self-sufficiency. This quickly proved unsuccessful, however, and in 1958 COMECON was reorganized. The goal of economic self-sufficiency was abandoned, mutual trade among the *Soviet bloc*—the Soviet Union plus its eastern European satellite states—was fostered, and some trade with western Europe was permitted. Meanwhile, Albania withdrew from the Soviet bloc in pursuit of a more authoritarian form of communism inspired by the Chinese revolution of 1949 (see Chapter 9); and Yugoslavia was expelled from the Soviet bloc (because of ideological differences over the interpretation of socialism) and allowed to pursue a more liberal, independent form of state socialism.

The experience of the east European countries under state socialism varied considerably, but, in general, rates of industrial growth were high. As in western Europe, industrialization brought about radical changes in economic geography. In practice, however, the command economies of eastern Europe did not result in any really distinctive forms of spatial organization. As in the industrial regions of the West, the industrialized landscapes of eastern Europe came to be dominated by the localization of manufacturing activity, by regional specialization, and by center-periphery contrasts in levels of economic development. The geography of industrial development under state socialism, as in democratic capitalism, was in practice heavily influenced by the unevenness of the distribution of natural resources and by the economic logic of initial advantage, specialization, and **agglomeration economies.** The most distinctive landscapes of state socialism were those of urban residential areas, where mass-produced, system-built apartment blocks allowed impressive progress in eliminating urban slums and providing the physical framework for an egalitarian proletariat—though at the price of uniformly modest dwellings and strikingly sterile cityscapes (**Figure 3.12**).

Eventually, the economic and social constraints imposed by excessive state control and the dissent that resulted from the lack of democracy under state socialism combined to bring the experiment to a sudden halt. By the time that the Soviet bloc collapsed in 1989 (see Chapter 4), Poland and Hungary had already accomplished a modest degree of democratic and economic reform. By 1992, East Germany (the German Democratic Republic) had been reunited with West Germany (the German Federal Republic); Estonia, Latvia, and Lithuania had become independent states once more; and the whole of eastern Europe had begun to be reintegrated with the rest of Europe. Only Kaliningrad, a small province on the Baltic

Figure 3.12 Socialist housing The socialist countries of eastern Europe eradicated a great deal of substandard housing in the three decades following the Second World War, re-housing the population in mass-produced, system-built apartment blocks. Although this new housing provided adequate shelter and basic utilities at very low rents, space standards were extremely low and housing projects were uniformly drab. This example is from Budapest, Hungary.

between Poland and Lithuania, remains as part of the Russian Federation, retained as a Russian exclave because of its important warm-water naval port.

The Peoples of Europe

A distinctive characteristic of Europe as a whole is the size and relative density of its population. With less than 7 percent of Earth's land surface, Europe contains about 13 percent of its population at an overall density of nearly 100 persons per square kilometer (km^2; 260 per square mile). Within Europe, the highest national densities (344 per km^2, or 894 per mi^2, in the Netherlands, 325 per km^2, or 865 per mi^2, in England and Wales, and 322 per km^2, or 837 per mi^2, in Belgium) match those of Asian countries such as Japan, the Republic of Korea, and Sri Lanka. On the other hand, population density in Finland, Norway, and Sweden stands at about 15 persons per km^2: the same as in Kansas and Oklahoma (**Figure 3.13**). This reflects a fundamental feature of the human geography of Europe: the existence of a densely populated core and a sparsely populated periphery. We have already noted the economic roots of this core-periphery contrast.

Figure 3.13 Population density in Europe 1995 The distribution of population in Europe reflects the region's economic history, with the highest densities in the "Golden Triangle," the newer industrial regions of northern Italy, and the richer agricultural regions of the North European Lowlands. For a view of worldwide population density and a guide to reading population-density maps, see Figure 1.40. (*Source:* Center for International Earth Science Information Network (CIESIN), Columbia University; International Food Policy Research Institute (IFPRI); and World Resources Institute (WRI). 2000. *Gridded Population of the World* (GPW), Version 2. Palisades, NY: CIESIN, Columbia University. Available at http://sedac.ciesin.org/plue/gpw)

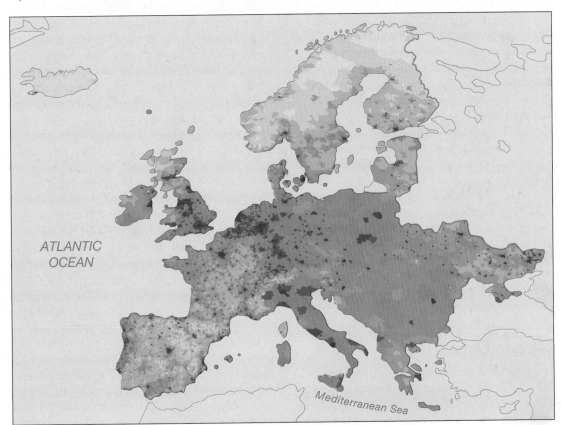

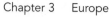

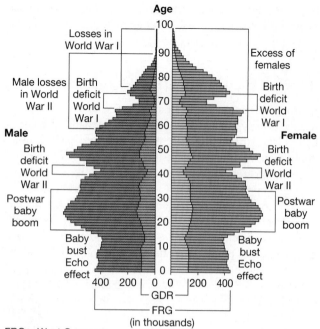

FRG = West Germany
GDR = East Germany

Figure 3.14   Population of Germany, by age and sex, 1989 Germany's population profile is that of a wealthy country that has passed through the postwar baby boom and currently possesses a low birthrate. It is also the profile of a country whose population has experienced the ravages of two world wars. (*Source:* J. McFalls, Jr., "Population: A Lively Introduction," *Population Bulletin*, 46 (2), 1991.)

While the population of the world as a whole is increasing fast, the population of Europe is roughly stable. Europe's population boom coincided roughly with the Industrial Revolution of the late eighteenth to late nineteenth centuries. Today, Europe's population is growing slowly in some regions, while declining slightly in others. The main reason for Europe's slow population growth is a general decline in birthrates (though certain subgroups, especially immigrant groups, are an exception to this trend). It seems that conditions of family life in Europe, including readily available contraception, have led to a widespread fall in birthrates. The average size of families has dropped well below the rate needed for replacement of the population (about 2.1 children per family), to about 1.75 per family. A "baby boom" after World War II has been followed by a "baby bust." Meanwhile, life expectancy has increased, due to improved health care, medical knowledge, and healthier lifestyles. The effect is not sufficient to outweigh falling birthrates, but it has meant a dramatic increase in the proportion of people over the age of 65, from 9 percent in 1950 to nearly 17 percent in 2000. Germany's population (**Figure 3.14**) reflects these trends, and shows the impact of two world wars.

The European Diaspora

The upheavals associated with the transition to industrial societies, together with the opportunities presented by colonialism and imperialism and the dislocations of two world wars, have dispersed Europe's population around the globe.

Beginning with the colonization of the Americas, vast numbers of people have left Europe for overseas destinations. The full flood of emigration began in the early nineteenth century, partly in response to population pressure during the early phases of the demographic transition, and partly in response to the poverty and squalor of the early phases of the Industrial Revolution. The main stream of migration was to the Americas, with people from northwestern and central Europe heading for North America and southern Europeans heading for destinations throughout the Americas. In addition, large numbers of British left for Australia and New Zealand and eastern and southern Africa. French and Italian emigrants traveled to North Africa, Ethiopia, and Eritrea, and the Dutch went to southern Africa and Indonesia. The final surge of emigration occurred just after the Second World War, when various relief agencies helped homeless and displaced persons to move to Australia and New Zealand, North America, and South Africa, and large numbers of Jews settled in Israel.

Migration Within Europe

Industrialization and geopolitical conflict has also resulted in a great deal of population movement within Europe. With the onset of industrialization, the regional redistribution of population within Europe followed the pattern described above, with three major waves of industrial development drawing migrants from less-prosperous rural areas to a succession of industrial growth areas around coalfields.

As industrial capitalism evolved, it was the diversified economies of national metropolitan centers that offered the most opportunities and the highest wages, thus prompting a further redistribution of population. In Britain this involved a drift of population from manufacturing towns southward to London and the southeast; in France, migration to Paris from towns all around France resulted in a polarization between Paris and the rest of the country. Some countries, developing an industrial base after the "coalfield" stage, experienced a more straightforward shift of population, directly from peripheral rural areas to prosperous metropolitan regions. In this way, Barcelona, Copenhagen, Madrid, Milan, Oslo, Stockholm, and Turin all emerged as regionally dominant metropolitan areas.

Wars and political crises have also led to significant redistributions of population within Europe. The First World War forced about 7.7 million people to move. Another major transfer of population took place in the early 1920s, when more than 1 million Greeks were repatriated from Turkey in the aftermath of an unsuccessful Greek attempt to gain control over the eastern coast of the Aegean Sea. Soon afterward, more people were on the move, this time in the cause of ethnic and ideological purity, as the policies of Nazi Germany and fascist Italy began to bite. Jews, in particular, were squeezed out of Germany, Italy, and Spain. With the Second World War, there occurred further forced migrations involving approximately 46 million people. These migrations left large parts of West and Central Europe with significantly fewer ethnic minorities than before the Second World War. In Poland, for example, minorities constituted 32 percent of the population before the war but only 3 percent after the war. Similar changes occurred in Czechoslovakia—from 33 percent to 15 percent—and in Romania—from 28 percent to 12 percent. Southeast Europe did not experience such large-scale transfers, and as a result many ethnic minorities remained surrounded and isolated, as in the republic of Yugoslavia. The geopolitical division of Europe after the war also resulted in significant transfers of population: West Germany, for example, had absorbed nearly 11 million refugees from eastern Europe by 1961, when the Berlin Wall was built.

More recently, the main currents of migration within Europe have been a consequence of patterns of economic development. Rural-urban migration has continued to empty the countryside of Mediterranean Europe as metropolitan regions have become increasingly prosperous. Meanwhile, most metropolitan regions themselves have experienced a decentralization of population as factories, offices, and housing developments have moved out of congested central areas. Another stream of migration has involved better-off retired persons, who have tended to congregate in spas, coastal resorts, and picturesque rural regions. The most striking of all recent streams of migration within Europe, however, have been those of migrant workers (**Figure 3.15**). These population movements were initially the result of Western Europe's postwar economic boom in the 1960s and early 1970s, which created labor shortages in western Europe's industrial centers. The demand for labor represented welcome opportunities to many of the unemployed and poorly paid workers of Mediterranean Europe and of former European colonies. By the mid-1970s these migration streams had become an important early component of the globalization of the world economy. By 1975, between 12 and 14 million immigrants had arrived in northwestern Europe. The majority came from Mediterranean countries—Spain, Portugal, southern Italy, Greece, Yugoslavia, Turkey, Morocco, Algeria, and Tunisia. In Britain and France the majority of immigrants came from former colonies in Africa, the Caribbean, and Asia. In the Netherlands most came from former colonies in Indonesia. Most of these immigrants have stayed on, adding a striking new ethnic dimension to many of Europe's cities and regions.

Finally, it is estimated that more than 18 million people moved within Europe during the 1980s and 1990s as refugees from war and persecution or in flight from economic collapse in Russia and eastern Europe. Civil war in the former Yugoslavia displaced more than 4 million people in the early 1990s, and by 2000 another 2.5 million had been displaced as a result of continuing conflict in the region (see p. 119).

European Cultural Traditions

The foundations of European culture were established by the ancient Greeks, who, between 600 B.C. and 200 B.C. built up an intellectual tradition of rational inquiry into the causes and explanations of everything, along with a belief that individuals are free, self-understanding, and valuable in themselves. The Romans took over this intellectual tradition, added Roman law and a tradition of disciplined participation in the state as a central tenet of citizenship, and spread the resulting culture throughout their empire. From the Near East came the Hebrew tradition, which in conjunction with Greek thought produced Judaism and Christianity, religions in which the individual spirit is seen as having its own responsibility and destiny within the creation. At the heart of European culture, then, are the curiosity, open-mindedness, and rationality of the Greeks; the civic responsibility and political individualism of both Greeks

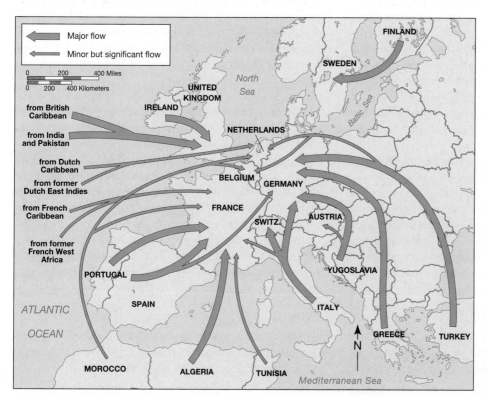

Figure 3.15 International labor migration This map shows the main international labor migration flows to European countries between 1945 and 1973. (*Source*: D. Pinder (ed.), *The New Europe: Economy, Society, and Environment.* New York: John Wiley & Sons, 1998, p. 265.)

and Romans; and the sense of the significance of the free individual spirit that is found in the main tradition of Christianity.

When Europeans pushed out into the rest of the world in their colonial and imperial ventures, they took these values with them, imposing them onto some cultures, grafting them into others. By the eighteenth century they were also the carriers of a new dimension of their own culture. This was the idea of **Modernity,** an idea whose genesis was in the changing world geography of the Age of Discovery. As Europeans tried to make sense of their own ideas and values in the context of those they encountered in the East, in Africa, in Islamic regions, and among native Americans during the sixteenth and seventeenth centuries, many of the old certainties of traditional thinking were cracked open. In the eighteenth century this ferment of ideas culminated in the **Enlightenment** movement, which was based on the conviction that all of nature, as well as human beings and their societies, could be understood as a rational system. Politically, the Enlightenment reinforced the idea of human rights and of democratic forms of government and society. Expanded into the fields of economics, social philosophy, art, and music, the Enlightenment gave rise to the cultural sensibility of Modernity, which emphasizes innovation over tradition, rationality over mysticism, and utopianism over fatalism.

In the late twentieth century, after the decline of heavy industry and repeated episodes of economic recession; after two terrible world wars; after interludes of fascist dictatorships in Germany, Italy, Greece, Portugal, and Spain; after a protracted period of being in the front line of a Cold War that divided European geography in two; and after intermittent episodes of regional and ethnic conflict, it is not surprising that the culture of Europe is a culture of doubt and criticism, heavily influenced by a search for radical rethinking. In this search, Europeans have not only established a new cultural sensibility for themselves but also generated some powerful new ideas and philosophies that have begun to influence other cultures around the world. Dismay with the side effects of laissez-faire industrial capitalism and, later, horror at the results of fascism and Nazism gave a strong impetus to left-wing critiques that have been powerful enough to reshape entire national and regional cultures and, with them, some important dimensions of regional geographies.

The most profound influence of all was Karl Marx, whose penetrating critique of industrial capitalism (written in London, and drawing heavily on descriptions of conditions in Manchester, England, supplied by his colleague Friedrich Engels) inspired both a socialist political economy in Russia and a fascist countermovement in Germany. After the Second World War, Western European left-wing critique portrayed both fascism and Soviet-style socialism as essentially imperialist, while American-style capitalism was critiqued as being intrinsically exploitative in privileging the individual and property over the community and the public good. Another powerful post-war movement that was deeply critical of the dominant structures of capitalist society was feminism, built on the ideas of Simone de Beauvoir in her book *The Second Sex,* published in 1949.

Subsequently, Europe has generated a succession of cultural impulses that, together, provide contemporary Europe with a distinctive cultural cast. Intellectual debate—about the role of culture itself; about whether people's thoughts and lives should be understood in terms of the dynamics of the cultures in which they are embedded (*structuralism*) or in terms of individual consciousness (*existentialism*); and about whether any kind of single-viewpoint, big-picture understanding of the world is really possible (*postmodernism*)—has spilled over into European literature, cinema, television, magazines, and newspapers, with the result that contemporary European culture is marked by a critical awareness of the role of culture itself (see Film, Music, and Popular Literature, p. 151).

Culture and Ethnicity, Nations and States

While European culture is distinctive at the global scale, it is also characterized by some sharp internal regional variations. In the broadest of terms, there is a significant north-south cultural divide. Southern Europe has always been more traditional in its religious affiliations—not just in terms of the dominance of Roman Catholicism over Protestantism, but of the prevalence of the conservative and more mystical forms of Catholicism. The Roman Catholic Church, still one of the most widespread within Europe, emerged in the fourth century under the bishop of Rome and spread quickly through the weakening Roman Empire. Missionaries helped spread not only the gospel but also the use of the Latin alphabet throughout most of Europe. The Eastern Orthodox Church, under the auspices of the Byzantine Empire centered in Constantinople, dominated the eastern margins of Europe and much of the Balkans, while Islamic influence spread into parts of the Balkans (present-day Albania, the European part of Turkey, and parts of Bosnia-Herzegovina) and, for a while, southern Spain. With the religious upheavals of the sixteenth and seventeenth centuries, Protestant Christianity came to dominate much of northern Europe (see Figure 1.37).

Another distinctive aspect of the cultural makeup of southern Europe is the continuing importance attached to traditional patterns of family life, with larger, close-knit families that tend to stick together as a buffer against unemployment and poverty in societies with relatively underdeveloped social welfare systems. The western part of southern Europe also shares the same family of languages: the Romance languages whose development was fostered by the spread of the Roman Empire. A second major group of languages, Germanic languages, occupy northwestern Europe, extending as far south as the Alps (**Figure 3.16**). English is one of the Germanic family of languages, an amalgam of Anglo-Saxon and Norman French, with Scandinavian and Celtic traces. A third major language group consists of Slavic languages, which dominate eastern Europe.

These broad geographic divisions of religion, language, and family life are reflected in other cultural traits: folk art, traditional costume, music, and folklore. Thus there is a Scandinavian cultural subregion with a collection of related languages (except Finnish), a uniformity of Protestant denominations,

Figure 3.16 Major languages in Europe Although three main language groups—Romance, Germanic, and Slav—dominate Europe, differences between specific languages are significant. These differences have contributed a great deal to the cultural diversity of Europe but have also contributed a great deal to ethnic and geopolitical tensions. (*Source:* Redrawn from R. Mellor and E. A. Smith, *Europe: A Geographical Survey of the Continent.* London: Macmillan, 1979, p. 22.)

and a strong cultural affinity in art and music that reaches back to the Viking age and even to pre-Christian myths. A second distinctive cultural subregion is constituted by the sphere of Romance languages in the south and west. A third is constituted by the British Isles, bound by language, history, art forms, and folk music, but with a religious divide between the Protestant Anglo-Saxon and Catholic Celtic spheres. A fourth clear cultural subregion is the Germanic sphere of central Europe, again with mixed religious patterns—Lutheran Protestantism in the north, Roman Catholicism in the south—but with a common bond of language, folklore, art, and music. The Slavic subregion of eastern and southeastern Europe forms another broad cultural subregion, though beyond the commonalities of related languages and certain physical traits among the general population, there is considerable diversity.

It is, in fact, the cultural and ethnic diversity of Europe's peoples and their languages, religions, and cultures that is one of the most significant aspects of its geography. Europe's cultural diversity has made it vital and attractive; it has contributed in large measure to the modern ideal of nation-states; and it has also made it the theater of innumerable wars, including two world wars within a single generation.

Ethnicity and National Identity Europe's broad cultural subregions underpin complex patterns of national, regional, and ethnic identity. It is important to recognize that many of the

countries of Europe are relatively new creations and that the political boundaries of many of them have been changed quite often. The whole idea of national states, in fact, can be traced to the Enlightenment in Europe, when the ferment of ideas about human rights and democracy, together with widening horizons of literacy and communication, combined to create new perspectives on allegiance, communality, and identity. In 1648 the Treaty of Westphalia, signed by most European powers, brought an end to Europe's seemingly interminable religious wars by making national states the principal actors in international politics and establishing the principle that no state has the right to interfere in the internal politics of any other state.

Gradually, these new perspectives began to undermine the dominance of the great European continental empires controlled by family dynasties—the Bourbons, the Hapsburgs, the Hohenzollerns, the House of Savoy, and so on. After the French Revolution (1789–93) and the kaleidoscopic changes of the Napoleonic Wars (1800–15), Europe was reordered, in 1815, to be set in a pattern of modern state**s**. Denmark, France, Portugal, Spain, and the United Kingdom had long existed as separate, independent states. The nineteenth century saw the unification of Italy (1861–70) and of Germany (1871) and the creation of Belgium, Greece, Luxembourg, the Netherlands, and Switzerland as independent national states. Early in the twentieth century they were joined by Norway, Sweden, and then Finland. Austria was created in its present form in the aftermath of the First World War, as part of the carve-up of the German and Austro-Hungarian empires. In 1921 long-standing religious cleavages in Ireland resulted in the creation of the Roman Catholic Free State (now known as the Republic of Ireland), with the six Protestant counties of Ulster remaining in the United Kingdom.

The European concept of the nation-state has been of immense importance in the modern world. As we saw in Chapter 2 (p. 69), the idea of a nation-state is based on the concept of a homogeneous group of people governed by their own state. In a true nation-state, no significant group exists that is not part of the nation. In practice, most European states were established around the concept of a nation-state but with territorial boundaries that did in fact encompass substantial ethnic minorities (**Figure 3.17**), and the result has been that the geography of Europe has been characterized by **regionalism** and **irredentism** throughout the twentieth century and into the twenty-first.

We have already cited the example of Basque regionalism in Spain and France (see p. 22 in Chapter 1). Other examples of regionalism include regional independence movements in Catalonia (within Spain), Scotland (within the United Kingdom), and the Turkish Cypriots' determination to secede from Cyprus. Examples of irredentism include the Irish Republic's claim on Northern Ireland (renounced in 1999), the claims of Nazi Germany on Austria and the German parts of Czechoslovakia and Poland, and the claims of Serbia and Croatia on various parts of Bosnia-Herzegovina. Some cases of regionalism have led to violence, social disorder, or even civil war, as in Cyprus. For the most part, however, regional ethnic separatism has been pursued within

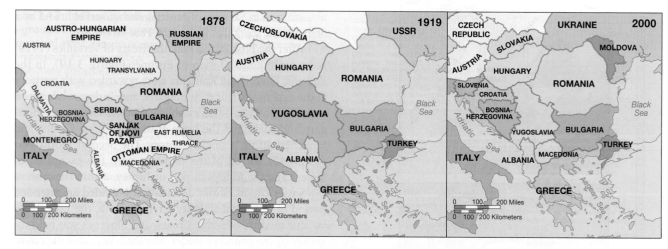

Figure 3.19 Changing political boundaries in the Balkans, 1878, 1919, and 2000 Nineteenth-century empires were dismantled after the First World War, creating in the Balkans an entirely new political geography that survived until the 1990s, when the breakup of Yugoslavia marked the end of the Great Powers' attempt to unite Serbs, Croats, and Slovenes within a single territory.

remain the focus of continued or potential hostility. In Romania, for example, there are more than 1.6 million Hungarians, while in Bulgaria there are more than 800,000 Turks (**Table 3.1**). It is Serbian nationalism, however, that has provided the principal catalyst for violence and conflict in the region. Within Yugoslavia, Serbian nationalism has led to attempts at **ethnic cleansing**: the forcible expulsion of non-Serb minorities from certain areas, accompanied by rape, murder, looting, and torture. The most extreme example of this has been in the Kosovo region, where the systematic persecution of the majority Albanian population led to the intervention of NATO forces in 1999 (see Geography Matters: Kosovo, p. 124).

A "European" Identity? Ethnic tensions and feelings of nationalism throughout Europe have been intensified by globalization. Globalization has heightened people's awareness of cultural heritage and ethnic identities. As we saw in Chapter 1, the more universal the diffusion of material culture and lifestyles, the more valuable regional and ethnic identities tend to become. Globalization has also brought large numbers of foreign immigrants to some European countries, and their presence has heightened still further people's awareness of cultural identities. In the more affluent countries of northwestern Europe, immigration has emerged as one of the most controversial issues since the end of the Cold War. Although the economic benefits of immigration far outweigh any additional demands that may be made on a country's health or welfare system, fears that unrestrained immigration might lead to cultural fragmentation and political tension have provoked some governments to propose new legislation to restrict immigration from the former communist states of eastern Europe and from outside Europe. The same fears have been responsible for a resurgence of popular **xenophobia** in some countries. In Germany, for example, right-wing nationalistic groups have attacked hostels housing immigrant families, while citizenship laws have prevented second-generation *Gastarbeiter* ("guest worker") families from obtaining

German citizenship (**Figure 3.20**). In France, claims that immigrants from North Africa are a threat to the traditional French way of life have led to some success for the National Front Party. Similar xenophobic and nationalistic impulses were reflected in election results in both Austria and Switzerland in 1999.

All this raises the question, "How 'European' *are* the populations of Europe?" There can be no doubt that European history and ethnicity have resulted in a collection of national prides, prejudices, and stereotypes that are strongly resistant to the forces of cultural globalization. Germans continue to be seen by most other Europeans as a little overserious, preoccupied by work, and inclined to arrogance. Scots continue to carry the popular image of a dour, unimaginative, ginger-haired people who love bagpipe and accordion music, dress in kilts and sporrans, live on whisky and porridge, and generally spend as little as possible. The English are seen as a nation of lager-swilling hooligans, well-meaning middle classes, and out-of-touch aristocrats. Norwegians and Danes continue to resent the Swedes' "neutrality" during the Second World War, and so on. In reality, such stereotypes are of course exaggerations that stem from the behaviors of a relative minority, and opinion surveys show that these stereotypes, prejudices, and identities are steadily being countered by a growing sense of European identity, especially among younger and better-educated persons. Much of this can be attributed to the growing importance and influence of the European Union, which is discussed in the next section.

Regional Change and Interdependence

Contemporary Europe is a cornerstone of the world economy with a complex, multilayered, and multifaceted regional geography. In overall terms, Europe, with about 13 percent of the world's population, accounts for almost 35 percent of the

Table 3.1	Ethnicity in Selected Southeast European Countries

Albania

Albanians	3,118,700
Greeks	572,800
Others	63,600
Total population	*3,755,100*

Bosnia-Herzegovina

Muslims	1,902,900
Serbs	1,370,400
Croats	755,000
Others	336,000
Total population	*4,364,300*

Croatia

Croats	3,736,500
Serbs	583,700
Others	464,100
Total population	*4,784,300*

Macedonia

Macedonians	1,313,900
Albanians	427,100
Turks	97,600
Roma	54,900
Serbs	42,700
Others	97,600
Total population	*2,033,800*

Romania

Romanians	20,347,800
Hungarians	1,616,000
Roma	409,700
Germans	113,800
Others	273,100
Total population	*22,760,400*

Serbia

Serbs	7,003,900
Albanians	1,727,700
Hungarians	341,400
Muslims	331,100
Roma	134,500
Croats	113,800
Others	693,100
Total population	*10,224,500*

Source: D. Hall and D. Danta, *Reconstructing the Balkans.* New York: John Wiley & Sons, 1996, pp. 10–11.

Figure 3.20 Anti-immigrant violence A woman looks at a burned house, site of the slayings of five Turkish immigrants by neo-Nazis in Solingen, Germany.

world's exports, almost 43 percent of the world's imports, and 33 percent of the world's aggregate GNP. Europe's inhabitants, on average, now consume about twice the quantity of goods and commercial services than they did in 1970. Purchasing power has risen everywhere to the extent that basic items of food and clothing now account for only about 30 percent of household expenditure, leaving more resources for leisure and consumer durables. Levels of material consumption in much of Europe approach those of households in the United States (**Figure 3.21**). The development of European welfare states has helped to maintain households' purchasing power during periods of recession and to ensure at least a tolerable level of living for most groups at all times. Levels of personal taxation are high, but all citizens receive a wide array of services and benefits in return. The most striking of these services are high-quality medical care, public transport systems, social housing, schools, and universities. The most important benefits are pensions and unemployment benefits.

Contemporary Europe is a very dynamic region that embodies a great deal of change. Because of the legacies of European history and culture, and because modern regional development in Europe was so closely tied to the technology systems of the Industrial Revolution, the challenges and opportunities presented to places and regions within Europe by economic globalization and by the new, high-tech, information-based technology

Figure 3.21   A German family with their material possessions The Pfitzner family from Köln, Germany, photographed with their possessions outside their home in the mid-1990s, represent a statistically average German family in terms of family size, residence, and income.

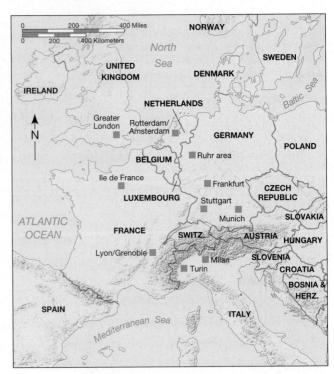

Figure 3.22   Regions of science-based growth The principal clusters of science-based commercial innovation in Europe are relatively small, urban-based networks of research and development (R & D) laboratories and enterprises devoted to new product development and processes of production. (*Source:* European Commission, *Competitiveness and Cohesion: Trends in the Regions.* Luxembourg: Office for Official Publications of the European Communities, 1994.)

system are immense. Formerly prosperous industrial regions have suffered economic decline, while some places and regions have reinvented themselves in order to take advantage of new paths to economic development. Meanwhile, the former Soviet satellite states have been reintegrated into the European world region; and much of Europe has joined together in the European Union, a supranational organization founded in order to recapture prosperity and power through economic and political integration (see p. 123).

Growth, Deindustrialization, and Reinvestment

Europe provides a classic example of the way that longer-term shifts in technology systems tend to lead to regional economic change (see Chapter 2). The innovations associated with new technology systems generate new industries that are not yet tied down by enormous investments in factories or tied to existing industrial agglomerations. Combined with innovations in transport and communications, this creates windows of locational opportunity that can result in new industrial dis-

tricts and in some towns and cities growing into dominant metropolitan areas through new rounds of investment. Within Europe the regions that have prospered most through the onset of a new, high-tech, science- and information-based technology system are the Thames Valley to the west of London, the Île de France region around Paris, the Ruhr valley in northwestern Germany, and the metropolitan regions of Lyon–Grenoble (France), Amsterdam–Rotterdam (Netherlands), Milan and Turin (Italy), and Frankfurt, Munich, and Stuttgart (Germany) (**Figure 3.22**).

Just as these high-tech industries and regions have grown, the profitability of old, established industries in established regions has declined. Wherever the differential in profitability has been large enough, disinvestment has taken place in the less-profitable industries and regions. **Disinvestment** means selling off assets such as factories and equipment. Widespread disinvestment leads to deindustrialization in formerly prosperous industrial regions. **Deindustrialization** involves a relative decline (and in extreme cases an absolute decline) in industrial employment in core regions as firms scale back their activities in response to lower levels of profitability. This is what happened to the industrial regions of northern England, South Wales, and central Scotland in the early part of the twentieth century; and it is what happened to the industrial region of Alsace-Lorraine, in France, and to many other traditional manufacturing towns and regions within Europe in the 1960s and 1970s.

The European Union

The European Union (EU) had its origins in the political and economic climate following the Second World War. The idea behind the EU was to ensure European autonomy from the United States and to recapture the prosperity that Europe had forfeited as a result of the war. The first stage in the evolution of the EU was the creation in the 1950s of three institutions that were set up in order to promote economic efficiency through integration: Euratom, the European Coal and Steel Community (ECSC), and the European Economic Community (EEC). These were amalgamated into the European Community (EC) in 1967, and in 1992 the Maastricht Treaty extended the scope of the EC, creating the European Union (EU).

In 1972 the EC expanded from its six original members—Belgium, France, Italy, Luxembourg, the Netherlands, and West Germany—to include Denmark, the Republic of Ireland, and the United Kingdom. Greece was added in 1981; Portugal and Spain in 1984; and Austria, Finland, and Sweden in 1995. The European Union now boasts a population of more than 370 million, with a combined gross domestic product (GDP) 10 percent larger than that of the United States. The EU has developed into a sophisticated and powerful institution with a pervasive influence on patterns of economic and social well-being within its member states. It also has a significant impact on certain aspects of economic development within some non-member countries.

The cornerstone of the EEC was a compromise worked out between the strongest two of the original six members. West Germany wanted a larger but protected market for its industrial goods, while France wanted to continue to protect its highly inefficient but large and politically important agricultural sector from overseas competition. The result was the creation of a tariff-free market within the Community, the creation of a unified external tariff, and a Common Agricultural Policy (CAP) to bolster the EEC's agricultural sector. Some of the most striking changes in the regional geography of the EU have been related to the operation of the CAP. Although agriculture accounts for less than 3 percent of the EU workforce, the CAP has dominated the EU budget from the beginning. For a long time, it accounted for more than 70 percent of the EU's total expenditures, and it still accounts for more than 45 percent. Its operation has had a significant impact on rural economies, rural landscapes, and rural standards of living, and it has even influenced urban living through its effects on food prices.

The Impacts of EU Agricultural Policy

The basis of the CAP is a system of EU support of wholesale prices for agricultural produce. This support has the dual effect of stabilizing the price of agricultural products and of subsidizing farmers' incomes. The CAP was originally designed to encourage farm modernization by securing higher incomes for farmers. An additional attraction of the policy, though, is that stable, guaranteed prices provide consumers with security and continuity of food supplies. Stable markets also allow trends in product specialization and concentration by farm, region, and country to proceed in an ordered and predictable fashion.

The overall result has been a realignment of agricultural production patterns, with a general withdrawal from mixed farming. Ireland, the United Kingdom, and Denmark, for example, have increased their specialization in the production of wheat, barley, poultry, and milk, while France and Germany have increased their specialization in the production of barley, maize, and sugar beets. It is at regional and subregional scales that these changes have been most striking. CAP support for oilseeds, for example, made rapeseed (canola) a profitable crop in cereal-producing regions of the United Kingdom, where the bright yellow flowers of the crop bring a remarkable change to the summer landscapes of the countryside.

The reorganization of Europe's agricultural geography under the CAP also brought some unwanted side effects, however. First, environmental problems have occurred as a result of the speed and scale of farm modernization, combined with farmers' desire to take advantage of generous levels of guaranteed prices for arable crops. Moorlands, woodlands, wetlands, and hedgerows have come under threat, and some traditional mixed-farming landscapes have been replaced by the prairie-style settings of specialized agribusiness.

A second serious problem concerns the large surpluses fostered by the price-support system. Prices that were set to give a reasonable return to producers on small farms were so favorable to the modernized sector of European agriculture that mountains of beef, butter, wheat, sugar, and milk powder and lakes of olive oil and wine had to be sold off at a loss to neighboring countries, dumped on world markets, or denatured (rendered unfit for human consumption) at a considerable cost.

A third set of problems arose from the indirect transfers of income that were brought about by CAP policies. The CAP's price-support mechanisms involve an indirect transfer of income from taxpayers to producers and from consumers to producers. These transfers are regressive (that is, the benefits disproportionately favor those who are better off). Food expenditures generally account for a larger proportion of disposable income in poorer households than in better-off households. Furthermore, because farmers benefit from price-support policies in proportion to their total production, the larger and more prosperous farmers receive a disproportionate share of the benefits. Regional inequity arises because countries or regions that are major producers of price-supported products receive the major share of the benefits, while the costs of price support are shared among member nations according to the overall size of their agricultural sector. For a long time the CAP pricing system made no concessions to the variety of agricultural systems practiced on farms of different sizes and in different regions. As a result, areas with particularly large and/or intensive or specialized farm units (such as northern France and the Netherlands) benefited most, together with regions specializing in the most strongly supported crops (cereals, sugar beets, and dairy products). Effectively, this has meant that the most prosperous agricultural regions have benefited most from the CAP, so that farm income differentials within member countries have been maintained, if not reinforced.

By 1983 the budgetary cost of the CAP had become burdensome, and the CAP had become a source of serious

Geography Matters

Kosovo

The Kosovo region of Yugoslavia provides a conspicuous and tragic example of the complex relationships between ethnicity, nationality, and territoriality in the Balkans. In the twentieth century this region of moderately prosperous farms, small villages, and ancient market towns was inhabited by a mixed population of Serbs, Roma, and Albanians, with Albanians representing the overwhelming majority. Serbians in modern Yugoslavia regard Kosovo with special significance as part of the heartland of Old Serbia, a region that was the platform for a Serbian empire that became, for a while, an important European power. Kosovo was where medieval kings were crowned and was the seat of the Serbian Orthodox Church, an institution synonymous with Serbs' self-identity as a nation. It was also where—at Kosovo Polje—the Serbs suffered a crucial defeat against the Turks in 1389, an event that was the subject of so much Serbian romantic literature and legend over the centuries that the military defeat was transformed into a moral victory. The Albanian majority in modern Kosovo are viewed by Serbs as latecomers and as Turkish surrogates who helped to drive Serbs from the region in the seventeenth and eighteenth centuries.

Albanians in modern Yugoslavia also claim a strong affinity to Kosovo, pointing out that their Illyrian ancestors had inhabited Kosovo at least 300 years before Slavic tribes displaced them from the region. Much later, in the 1800s, Kosovo became the cradle of an Albanian rebirth, the seat of Albanian literary inspiration and political will to pursue national freedom after 500 years of Turkish domination.

In March 1998 Yugoslavia's Serbian leader, Slobodan Milosevic, initiated a brutal, premeditated, and systematic campaign of ethnic cleansing that was aimed at expelling Kosovar Albanians from what had become their homeland. More than 1.5 million Kosovar Albanians—at least 90 percent of the estimated 1998 Kosovar Albanian population of Kosovo—were

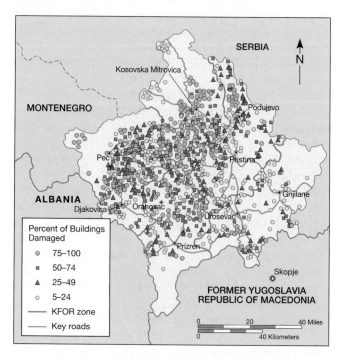

Figure 1 War-damaged buildings, Kosovo (*Source:* U.S. State Department, *Ethnic Cleansing in Kosovo: An Accounting.* December 1999. Available at: http://www.state.gov/www/global/human_rights/kosovoii/homepage.html)

forcibly expelled from their homes. Tens of thousands of homes in at least 1200 cities, towns, and villages were damaged or destroyed (**Figures 1** and **2**). According to a report by the U.S. State Department, Serbian forces expelled Kosovar Albanians at gunpoint from villages and larger towns, looted and burned their homes, organized the systematic rape of young Kosovar Albanian women, and used Kosovar Albanians

---

disharmony, particularly in the United Kingdom, where the CAP brought a higher and regressive system of food prices without any compensatory benefits. Peasant farming and inefficient agricultural practices had been purged from the UK economy long before. Together with increasing awareness of the unwanted side effects of the CAP, these considerations eventually led to a major reform of the CAP, with drastic reductions in guaranteed cereal and beef prices through the 1990s, along with sharply reduced production quotas for dairy farmers.

The EU and Regional Interdependence   By the early 1980s the U.S. and Japanese economies had become increasingly interdependent and prosperous on the basis of globalized

producer services and new, high-tech industries. In response, the European Community relaunched itself as the European Union. The Treaty of European Union of 1992 (the Maastricht Treaty) gave to the EU most of the major functions of a sovereign national state, including the creation of a single currency (the euro, launched in 1999), the coordination of economic policies, the redistribution of wealth among regions, and the management of a common external policy covering foreign relations and defense. This relaunching was an impressive achievement, particularly since it was undertaken at a time of major distractions: coping with the reunification of Germany and the breakup of the former Soviet empire in Eastern Europe and, not least, having to deal with a resurgence of nationalism within Europe.

Figure 2   War damage in Kosovo An elderly ethnic Albanian woman grieves in front of her destroyed home in Pec, Kosovo.

as human shields to escort Serbian military convoys.* In addition, Serbian forces carried out widespread summary executions, dumping bodies in mass graves in an attempt to destroy evidence, and systematically stripped Kosovar Albanians of identity and property documents, including passports, land titles, automobile license plates, identity cards, and other forms of documentation. Finally, by systematically destroying schools, places of worship, and hospitals, Serbian

---

*Ethnic Cleansing in Kosovo: An Accounting. Washington, DC: U.S. Department, December 1999.

---

forces sought to destroy the social identity and fabric of Kosovar Albanian society.

International outrage at these human rights violations led to the declaration of war against Yugoslavia by NATO in March 1999. NATO bombing, however, seemed only to spur the Serbs to greater ruthlessness. More than 800,000 ethnic Albanians were displaced in eight days, and between 3000 and 5000 were massacred. Eleven weeks later, after more than 11,000 NATO air strikes, NATO peacekeeping forces—49,000 troops and 1,800 international police—entered Kosovo. The war against Yugoslavia was significant in that it represented the first internationally recognized refutation of the central principle of the 1648 Peace of Westphalia—that no state has the right to interfere in the internal politics of another state. Solobodan Milosevic and other Serbian leaders were indicted in the International Court of Justice for their roles in human rights violations, though NATO stopped short of pursuing them inside Yugoslavia.

Meanwhile, most Kosovar Albanian survivors had returned to their villages and towns. By early 2000, 810,000 Albanian refugees had returned to their homes in Kosovo. Their return, however, triggered a mass exodus of Serbian civilians who—despite NATO efforts to protect them—were fearful of retribution from returning Kosovar Albanians. The complex relationships between ethnicity, nationality, and territoriality in the Balkans thus continue to fuel spasms of violence. In the most recent phase, it was Kosovar Serbs who were subjected to kidnapping, murder, arson, grenade attacks, shootings, and a variety of other intimidation tactics. International agencies also documented abuses against Serb patients in hospitals in Kosovo and, according to Serbian Orthodox Church officials, more than 40 Serbian Orthodox churches and monasteries were damaged or destroyed in the first months of 2000. In October 2000, Milosevic was ousted from office in a "people's revolution" following disputed election results. The new President, Vojislav Kostunica, though a Serbian nationalist, was not implicated in the ethnic cleansing.

The overall economic benefits of EU membership are widely recognized and have been apparent for a long time. Not surprisingly, there is a growing list of countries seeking membership. In 2001 the official EU list of candidates numbered 11: Bulgaria, the Czech Republic, Estonia, Hungary, Latvia, Lithuania, Poland, Romania, Slovakia, Slovenia, and Turkey. Of these, the most likely to be admitted in the near future are the Czech Republic, Hungary, Poland, Slovakia, and Slovenia. Yet EU membership brings regional stresses as well as the prospect of overall economic gain. Existing member countries have found that the removal of internal barriers to labor, capital, and trade has worked to the clear disadvantage of peripheral regions and in particular to the disadvantage of

those farthest from the Golden Triangle (the region centered on the triangle defined by London, Paris, and Berlin) that is increasingly the European center of gravity in terms of both production and consumption.

This was recognized by the Single European Act (SEA) of 1985, which included "economic and social cohesion" as a major policy. The SEA doubled the funding—in real terms—for grants for regional development assistance and established a Cohesion Fund to help Greece, Ireland, Portugal, and Spain achieve levels of economic development comparable to those of the rest of the EU. Regions that are eligible for these funds are shown in **Figure 3.23**. In its 1994–99 budget cycle, the EU spent $159.5 billion on funding projects and policies designed

Figure 3.23 European Union regions eligible for aid, 2000
Just over half of the total population of the European Union lives in areas eligible for regional assistance from the European Regional Development Fund and Cohesion Fund. Most of the EU's regional aid is allocated to "lagging" regions (where per capita GDP is less than 75 percent of the Community average) and to declining industrial regions. (*Source:* EU Inforegio. European Commission, European Regional Development Fund and Cohesion Fund, 2000. Available at: http://www.inforegio.cec.eu.int/wbpro/prord/guide/euro_en.htm)

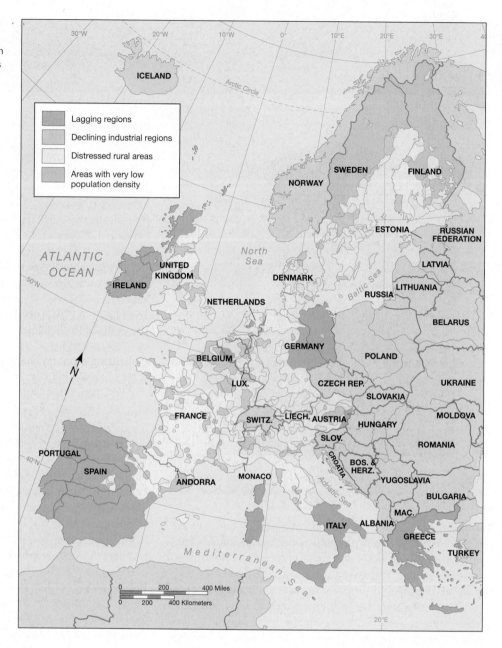

to improve economic and social cohesion within and among its member countries. The 2000–06 budget cycle includes an allocation of $339 billion, about one-third of the total EU budget.

The scale of the EU and its maintenance of a strongly protectionist agricultural policy has inevitably had a significant impact on nonmember countries. The EU does extend favorable trading privileges to a large group of countries in Africa, the Caribbean, and the Pacific, most of them former colonial territories of EU member states. Nevertheless, sensitive products (that is, those that compete directly with EU agricultural and industrial products) are excluded from preferential treatment or are subject to seasonal restrictions.

Because the whole idea of the EU is based on improving Europe's competitiveness with the United States and Japan, it is not surprising that the EU's trade relations with those countries have been fractious. Other countries, especially those

with strong traditional ties to European markets, have found themselves excluded by the EU's external tariff wall. New Zealand is a good example. The United Kingdom used to take nearly all of New Zealand's butter, cheese, and lamb. But after the United Kingdom joined the EU, New Zealand no longer had access to markets in the United Kingdom. As a result, New Zealand's agriculture had to be restructured, new products had to be developed—a notable success here being the kiwi fruit—and new markets had to be developed in Latin America, India, and Japan (see Chapter 10).

The Reintegration of Eastern Europe

Between 1989 and 1992, all of the former satellite states of the Soviet Union in eastern Europe turned away from state socialism with command economies and began the process of establishing democracies with capitalist economies. In 1991

COMECON was abolished, and one by one eastern European countries began a complex series of reforms. These included the abolition of controls on prices and wages, the removal of restrictions on trade and investment, the creation of the financial infrastructure to handle private investment, the creation of government fiscal systems to balance taxation and spending, and the privatization of state-owned industries and enterprises. After more than 40 years of state socialism, such reforms were both difficult and painful. Indeed, the reforms are by no means complete in any of the countries, and economic and social dislocation is a continuing fact of life.

Nevertheless, the reintegration of eastern Europe has added a potentially dynamic market of 344 million consumers to the European economy. Within a capitalist framework, eastern Europe has the comparative advantage of relatively cheap land and labor. This has attracted a great deal of foreign investment, particularly from transnational corporations and from German firms and investors, many of whom have historic ties with parts of eastern Europe. In some ways, the transition toward market economies has been remarkably swift. It did not take long for Western-style consumerism to appear on the streets and in many of the stores in larger eastern European cities. On the flip side, it did not take long for inflation, unemployment, and homelessness to appear either. Overall, eastern Europe is increasingly reintegrated with the rest of Europe, but for the most part as a set of economically peripheral regions, with agriculture still geared to local markets and former COMECON trading opportunities and industry still geared more to heavy industry and standardized products than to competitive consumer products. The service sector in general and knowledge-based industries in particular remain very weakly developed.

In detail, the pace and degree of reintegration varies considerably across eastern Europe. Ethnic conflict in Bosnia-Herzegovina, Croatia, and Yugoslavia has severely retarded reform and reintegration, while Albania, Bulgaria, Macedonia, Moldova, and Romania suffer from the combined disadvantages of having relatively poor resource bases, weakly developed communications and transportation infrastructures, and political regimes with little ability or inclination to press for economic and social reform. In Ukraine, which has a much better infrastructure, a significant industrial base, and the capacity for extensive trade in grain exports and in advanced technology, reintegration has been retarded by a combination of geographical isolation from western Europe, continuing economic and political ties with Russia, and a surviving political elite that has little interest in economic and social reform.

The Baltic states of Estonia, Latvia, and Lithuania have been more successful in reintegrating with the rest of Europe. Their small size and relatively high levels of education have made them attractive as production subcontracting centers for Western European high-technology industries. They are reviving old ties with neighboring Nordic countries, and in this regard Estonia is particularly well placed because its language belongs to the Finnish family, and it was part of the Kingdom of Sweden when it was annexed by the Russians in 1710. The best-integrated states of eastern Europe are the Czech Republic, Hungary, Poland, and Slovenia. All have a relatively strong industrial base, and Hungary has a productive agricultural sector. Poland and Hungary have been especially open to foreign direct investment and have been swift and vigorous in pursuit of economic and institutional reform. They are among the principal contenders for membership in an expanded EU.

Reinventing Europe: Place Marketing

Throughout Europe, disinvestment, deindustrialization, and economic restructuring following economic integration mean that many places and regions are increasingly seeking to attract new forms of investment and new jobs. As a result, places are increasingly being reinterpreted, reimagined, designed, packaged, and marketed, in order to influence the ways in which they are perceived by tourists, businesses, media firms, and consumers. Sense of place has become a valuable commodity, and culture has become an important basis for economic activity. Furthermore, culture has become a significant factor in the ability of places to attract and retain other kinds of economic activity. Seeking to be competitive within the globalizing economy, many places in Europe have sponsored extensive makeovers of themselves that include the creation of pedestrian plazas, cosmopolitan cultural facilities, festivals, and sports and media events.

Central to place marketing is the deliberate manipulation of material and visual culture in an effort to enhance the appeal of places to key groups. These groups include the upper-level management of large corporations, the higher-skilled and better-educated personnel sought by expanding high-technology industries, wealthy tourists, and the organizers of business and professional conferences and other events that bring money into the local economy. In part, this manipulation of culture depends on promoting traditions, lifestyles, and arts that are locally rooted; in part, it depends on being able to tap into globalizing culture through new cultural amenities and specially organized events and exhibitions. Some of the most widely adopted strategies for the manipulation and exploitation of culture include investments in facilities for the arts, investment in public spaces, the re-creation and refurbishment of distinctive settings such as waterfronts and historic districts, the expansion and improvement of museums (especially with blockbuster exhibitions of spectacular cultural products that attract large crowds and can be marketed with commercial tie-ins), and the designation and conservation of historic landmarks.

The Dutch city Amsterdam provides a good example of how investment in the arts can provide a catalyst for economic development. The construction in the late 1980s of an arts and civic complex, the Stopera, in a declining neighborhood in the eastern part of the central area of the city led to the recovery of the whole area within less than a decade. By the early 1990s, nearly 2000 people were working in the Stopera itself, and its presence had attracted bookstores, record and magazine stores, restaurants, cafes, and specialized food stores. Altogether, the neighborhood experienced a 60 percent increase in the number of shops in the area between 1988 and 1998. A number of small businesses have also been

Geography Matters

New Infrastructure for an Integrated Europe

The economic integration of Europe, following fundamental geographic principles, leans heavily on policies designed to increase accessibility and spatial interaction. In 1996 the European Union (EU) approved a far-reaching plan for a series of trans-European networks (TENs) to weld together Europe's patchwork of national transport systems. The plan includes a $400 billion network of new or upgraded roads, bridges, waterways, rail lines, and airports to be completed by 2010. The goal is to link states on the EU's margins with the center, boosting economic efficiency, reinforcing the social and political cohesiveness of the EU, and in the process creating up to 1 million new jobs by 2030.

Improvements to Europe's railway infrastructure account for 8 of the EU's 14 priority projects and about 60 percent of the budget. With its relatively short distances between major cities, Europe is ideally suited for rail travel and less suited, because of population densities and traffic congestion around airports, to air traffic. Allowing for check-in times and accessibility to terminals, it is already quicker to travel between many major European cities by rail than by air. The high-speed London–Paris rail service—in direct competition with the airlines—has already captured 60 percent of intercity traffic since opening in 1994. The EU plans to coordinate and subsidize a $250 billion investment in 30,000 kilometers (almost 20,000 miles) of high-speed track, to be phased in through 2012. The heart of the system will be the "PBKAL web," which will con-

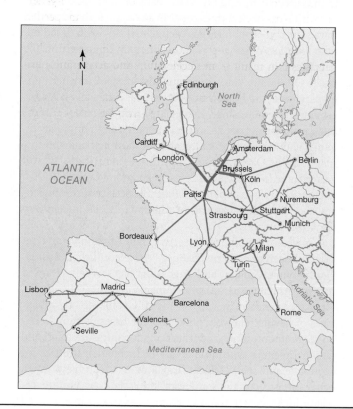

Figure 1 High-speed rail in Europe Europe, with its relatively short distances between major cities, is ideally suited for rail travel and less suited, because of population densities and traffic congestion around airports, to air traffic. Allowing for check-in times and accessibility to terminals, it is already quicker to travel between many major European cities by rail than by air. The European Union plans to coordinate and subsidize a $250 billion investment in 30,000 kilometers (almost 20,000 miles) of high-speed track, to be phased through 2012. The heart of the system will be the "PBKAL web," which will connect Paris, Brussels, Cologne (Köln), Amsterdam, and London and which will be completed by 2003. This web will lead to some restructuring of the geography of Europe. High-speed rail routes will have only a few scheduled stops, because the time penalties resulting from deceleration and acceleration undermine the advantages of high-speed travel. Cities with no scheduled stops will be less accessible and, then, less attractive for economic development.

attracted into the area because of its new atmosphere, and increasing numbers of tourists, who previously avoided the area, now seek it out.

Core Regions and Key Cities of Europe

The economic and political integration of Europe has intensified core-periphery differences within the region. The removal of internal barriers to flows of labor, capital, and trade has worked

to the clear disadvantage of geographically peripheral regions within the EU, while core metropolitan regions have benefited. The principal core region within Europe is the Golden Triangle, which centers on the area between London, Paris, and Berlin. A secondary emergent core is developing along a north-south crescent that straddles the Alps, stretching from Frankfurt, just to the south of the Golden Triangle, through Stuttgart, Zürich, and Munich, to Milan and Turin. Both cores are linked by an important new development in European transportation: a high-speed rail system (see Geography Matters: New Infrastructure for an Integrated Europe, above).

nect Paris, Brussels, Cologne (Köln), Amsterdam, and London and which will be completed by 2007 (**Figure 1**). This will be the cause of some restructuring of the geography of Europe. Improved locomotive technologies and specially engineered tracks and rolling stock will make it possible to offer passenger rail services at speeds of 275 to 350 kilometers per hour (180 to 250 miles per hour). New tilt-technology railway cars, which are designed to negotiate tight curves by tilting the train body into turns in order to counteract the effects of centrifugal force, are being introduced in many parts of Europe in order to raise maximum speeds on conventional rail tracks. German Railways (DB), for example, introduced third-generation ICE (inter-city express) trains with a maximum speed of 330 kilometers per hour (205 mph) in 2000. High-speed rail routes will have only a few time-tabled stops because the time penalties that result from deceleration and acceleration undermine the advantages of high-speed travel. The geographic implications of these systems are significant. Places that do not have scheduled stops will be less accessible and, then, less attractive for economic development. Places that are linked to them will be well situated to grow in future rounds of economic development.

Other trans-European network projects feature new tunnels through the Alps and the Pyrenees and under the city of Antwerp; new motorways across Greece (connecting Athens to the Bulgarian border), and between Lisbon (Portugal) and Valladolid (Spain); and improvements to the connections between Milan's Malpensa airport, the city, and the railway network. There are also numerous energy-related infrastructure projects aimed at increasing the efficiency and capacity of power stations and extending natural gas pipelines and electricity grids across member countries—and even into neighboring countries that have petitioned for EU membership, in order to help boost their economic efficiency before they join the EU. The most striking project of all is the Øresund Fixed Link (**Figure 2**), a 16-kilometer bridge and tunnel that opened in 2000,

Figure 2 The Øresund Fixed Link Opened in July 2000, this bridge, which cost $2.7 billion, connected Sweden to continental Europe for the first time in 7000 years. The fixed link consists of a four-lane motorway and double-track railway that connects the Danish capital, Copenhagen, to the heart of Sweden's third-largest city, Mälmo, via the bridge over the 15-kilometer (9.3-mile) Øresund gap. The fixed link is expected to regenerate southern Sweden, creating a dynamic economic area of 3.5 million people that will encompass 11 universities and 130,000 students.

connecting Sweden and Denmark for the first time since prehistoric times (before the land link was breached by rising sea levels). The two sides of the Øresund Sound, hitherto entirely separate, will soon develop into a single functional economic region that will be well placed to form an important commercial hub.

Europe's Golden Triangle

The character and relative prosperity of Europe's chief core region, the Golden Triangle, stems from four advantages:

1. Its geographic situation provides access to southern and central Europe by way of the Rhine and Rhône river systems and access to the sea lanes of the Baltic, the North Sea, and, by way of the English Channel, the Atlantic.

2. Within it are found the capital cities of the major former imperial powers of Europe.

3. It includes the industrial heartlands of central England, northeastern France, and the Ruhr district of Germany.

4. Its concentrated population provides both a skilled labor force and an affluent consumer market.

These advantages have been reinforced by the integrative policies of the European Union, whose administrative headquarters are situated squarely in the heart of the Golden Triangle, in Brussels, Belgium. They have also been reinforced by the emergence of Berlin, Paris, and, especially, London, as **world cities**—cities in which a disproportionate share of the world's most important business—economic, political, and cultural—is conducted. As these world cities have come to play an increasingly important role in the world economy, so they have become home to a vast web of sophisticated financial, legal, marketing, and communications services. These services, in turn, have added to the wealth and cosmopolitanism of the region. Agriculture within the Golden Triangle tends to be highly intensive and geared toward supplying the highly urbanized population with fresh dairy produce, vegetables, and flowers. Industry that remains within the Golden Triangle tends to be rather technical, drawing on the highly skilled and well-educated workforce; most heavy industry and large-scale, routine manufacturing has relocated from the region in favor of cheaper land and labor found elsewhere in Europe or beyond.

The landscape of the Golden Triangle is highly urbanized, with a tightly knit network of towns and cities linked by an elaborate infrastructure of canals, railways, and highways. The region has reached saturation levels of urbanization, as exemplified in Randstad Holland (see Sense of Place: Randstad Holland, p. 132). Nevertheless, the region still contains fragments of attractive rural landscapes, together with some unspoiled villages and small towns. These landscape elements have survived partly because of market forces: They are very attractive to affluent commuters. Equally important to their survival, though, has been the relatively strong role of environmental, land use, and conservation planning in European countries.

Surrounding the advanced city regions of the Golden Triangle are the fruit orchards and hop-growing fields of Kent, in southeastern England; the bulb fields of the Netherlands; the dikes and rectangular fields of the reclaimed marshland (polders) of North-Holland, Flevoland, and Friesland along the Dutch coastal plain; the pastures, woodlands, and forests of the upland plateaus of the Ardennes, the Eifel, the Westerwald, and the Harz; and the meadows and cultivated fields separated by hedgerows and patches of woodland that characterize most of the remaining countryside: Picardy in France; Flanders and Brabant in Belgium; Limburg, Nordrhein-Westphalia, and Saxony in Germany; and Gelderland in the Netherlands, for example. Embedded among these distinctive landscapes are hundreds of villages and scores of market towns. Those villages and market towns within the orbit of the metropolises and advanced city regions have lost a great deal of their character. A few, like Rye, in southeastern England, have been bypassed both by industrialization and by the post-industrial economy and have retained much of their traditional character (**Figure 3.24**). Some, like Bruges in Belgium, have traded on their legacy to become important tourist stops. It is the metropolises and advanced city regions, however, that both define and dominate the Golden Triangle.

Berlin If it were not for the geopolitical aftermath of the Second World War, Berlin would probably be a world city to rival Paris and London. Berlin is situated on a natural east-west commercial axis on the north European plain, with a favorable location on the river system that provides connections to the Elbe and Oder rivers. It is at once the most westerly city of eastern Europe and the most easterly city of western

(a)

(b)

Figure 3.24 Landscapes of the Golden Triangle Though the Golden Triangle is highly urbanized, the landscapes surrounding the cities are distinctly rural, a result of a combination of market forces and strict planning controls. (a) Bocage landscape, Cornwall, England; (b) Rye, in southeastern England.

Europe. As such, it enjoys an excellent strategic location within continental Europe. At the beginning of the eighteenth century, Berlin became the capital of the Hohenzollern dynasty of the Prussian monarchy. By the late eighteenth century it had a population of about 150,000, with soldiers and their families accounting for one in every five inhabitants. Berlin's position was further enhanced by the formation of the German Customs Union in 1834 and by the creation of the German Empire after the Franco-Prussian War (1871). The city is still graced by the monumental architecture of this period, although much was destroyed during the Second World War. With the Industrial Revolution, Berlin became an important producer of machinery, chemicals, textiles, electrical goods, electronics, and clothing, an important hub in the central European transportation system, and a major banking center. At its peak, in 1939, Greater Berlin had a population of 4.3 million. In physical terms, the building blocks of Berlin were *Meitskaserne* ("rental barracks"), four- or five-story apartment houses arranged around a courtyard and often extended by a series of rear courtyards, with access to the street only from the first court. With a new building ordinance in 1925, the classic *Meitskaserne* was effectively outlawed. The term continues to be used, however, for large-scale working-class housing developments.

Berlin was an important cultural and intellectual center in the early part of the twentieth century. It was a seedbed of avant-garde theater, film, cabaret, art, and architecture. The Second World War changed everything. By the end of the war, 34 percent of Berlin's housing had been destroyed, and another 54 percent was damaged. The city found itself embedded within the eastern, socialist part of a partitioned nation and was itself partitioned into eastern and western sectors. As a result, Berlin's dynamism was seriously disrupted. Economic hinterlands for both halves of the city were truncated. West Berlin had to develop a new **central business district** (CBD), the old one having fallen within the eastern sector. In 1961 the division between the two half-cities was physically reinforced when East Germany built the Berlin Wall (see Figure 1.27) in order to stem the flow of migrants to West Berlin. After the wall went up, both East and West Berlin remained highly militarized, with troops and their equipment in a very visible part of the urban landscape. Both also redeveloped their industry and refurbished their housing. East Berlin, with a population of just under 1.5 million by the late 1980s, was a showcase for the German Democratic Republic but was nevertheless dominated by bleak modernist architecture. West Berlin, with a population of well more than 2 million (of whom more than a quarter of a million were foreign workers and their families), developed a significant youth counterculture, partly because of its many institutions of higher education (which between them account for more than 120,000 students), and partly because its residents were not required to perform military service in the army of the Federal Republic of Germany.

With the reunification of Germany in 1989, Berlin reassumed its pre-war role as a national political and cultural center. The city experienced a surge of construction as the two parts of the city were reconnected, wired, and plumbed together again, and as the federal government and investors raced to install new infrastructure, department stores, office blocks, hotels, and entertainment centers in keeping with the city's restored position in the world (**Figure 3.25**). Federal offices have been moved from Bonn to Berlin, and Potsdamer Platz, once a no-man's-land of barbed wire, tank traps, and mines, has been redeveloped with 111,000 square meters (about 1.18 million square feet) of apartment space, 310,000 square meters (about 3.3 million square feet) of office space, 57,000 square meters (about 613,500 square feet) of retail shops and restaurants, plus two Imax theaters, eight cinemas, a concert stage, an underground train station, and a shopping arcade.

London   London is a vast, sprawling city that covers more than 3900 square kilometers (about 1500 square miles) of continuously built-up area. It has a total population of just over 7 million (13 million including the metropolitan fringes). In the nineteenth and early twentieth centuries it was the center of global economic and geopolitical power. It dominates the economic and political life of the whole of the United Kingdom and remains the single most cosmopolitan city in Europe.

Figure 3.25   Berlin Since the reunification of Germany in 1989, Berlin has experienced a major construction boom, reflecting the city's restored role as a world city. Shown here is Potsdamer Platz. For more than 28 years Potsdamer Platz was a no-man's land in the heart of the city, lit at night by powerful lamps to prevent East German citizens from escaping over the Berlin Wall to the West. After the reunification of Germany, the wall was very quickly demolished and Potsdamer Platz was redeveloped with movie theaters, shops, galleries, and restaurants.

London grew up around two core areas: a commercial core centered on its port and trading functions and an institutional core centered on its religious and governmental functions. The commercial core has Roman roots: *Londinium* was the fifth largest Roman city north of the Alps, a major trading center that enjoyed the advantages of a deep-water port (on the River Thames) and a key situation facing the continental North Sea and Baltic ports. These same advantages helped London prosper with the resurgence of trade in the medieval period, when the wealth accumulated by wool merchants provided the economic foundation for future growth. From this commercial nucleus grew an extensive merchant and financial quarter (**Figure 3.26**). The docks spread eastward from the financial precinct

Figure 3.26   London's financial core London's original river port trade gave rise to a commercial core that developed into a major financial hub in the district of London known as the City. Though the skyline of this financial precinct is still dominated by the dome of St. Paul's Cathedral, the skyline of the City has acquired a few high-rise buildings.

Sense of Place

Randstad Holland

Randstad ("rim city") is the name that has been given to the densely settled, horseshoe-shaped region in the western Netherlands. It stretches from Dordrecht in the southeast through—clockwise—Rotterdam, Delft, The Hague, Leiden, Haarlem, Amsterdam, Hilversum, and Utrecht (**Figure 1**). Randstad is not a city; it is not a place but an area. It is an example of the urban agglomerations that have arisen in western Europe as saturated urbanization has led to the coalescing of neighboring towns and cities.

In 1998 the Randstad contained more than 6 million inhabitants. Until about 1970 the population of the Randstad increased fairly rapidly, but since then the overall rate of growth has declined sharply, mainly because the net loss through migration to other parts of the Netherlands has exceeded the net gain from foreign immigration. The fundamental reason for the high degree of urbanization in this region is its location at the "Golden Delta" of the Rhine, Meuse, and Scheldt river systems, a nodal point of European trade routes for centuries. The particular shape of the region is a product of physical geography. When the sea level rose after the last Ice Age, the major part of what is now Randstad was transformed into a marshy lagoon. The best sites for human settlement were higher places on the edge of the swamps, such as the dunes in the west (the location of Haarlem and The Hague), the Pleistocene sandy terrain in the east (Hilversum), and the levees along the rivers (Amsterdam, Leiden, Rotterdam, and Utrecht). Later, as the swampy area was drained and colonized for agriculture, it became the "Green Heart" of the emerging Randstad region. In recent years, the name *Randstad* has come to be used increasingly for the whole region including the Green Heart, where the growth of smaller towns and villages has been striking.

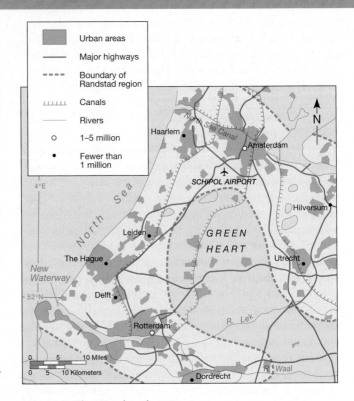

Figure 1 The Randstad region

Within the Randstad, each town and city has retained its own identity, although there is certainly a common visual element in the dominance of the vernacular style of scalloped red roofing tiles. Administrative, industrial, commercial, and service functions are distributed throughout the agglomeration, though each of the larger cities has its own specialization. The Hague, for example, is the national seat of government, while Utrecht, situated in the center of the country, forms a node of

---

(the City), and specialized market areas grew to the north and east. Today, this area remains the commercial core of the city. It contains the Stock Exchange, the Bank of England, the Royal Exchange, the Guildhall, and the Central Criminal Courts, as well as specialized commercial areas, such as Fleet Street (the press/media precinct), Lincoln's Inn Fields (the legal precinct), and the sites of old marketplaces, such as Billingsgate (fish) and Smithfield (meat). Older docks, such as St. Katherine's next to Tower Bridge, have meanwhile been renovated and now boast yacht basins, hotels, a trade center, upscale pubs, bistros, specialty retail stores, galleries, and condominiums.

London's institutional core developed around Westminster Abbey, some 3.2 kilometers (2 miles) upstream from the commercial core (**Figure 3.27**). Early meetings of Parliament were held in the Abbey's Chapter House; the present Houses of Parliament date only from the nineteenth century. St. James's

Palace was built as a London residence for the monarchy in the sixteenth century; Buckingham Palace, built for the Duke of Buckingham in 1703 and purchased by George III in 1762, became the royal residence during Queen Victoria's reign. To the north of the Houses of Parliament, along Whitehall, are government offices; between these and Buckingham Palace is St. James's Park; and to the north of the park are the palaces and mansions of the nobility, centers of culture (the Royal Academy, the National Gallery, and the Royal Opera House), the exclusive shops that cater to the city's elite, and the squares and townhouses of the rich and powerful.

London's population grew sixfold during the nineteenth century, reaching more than 6 million by 1900. As it grew, the two core areas merged together as part of a huge central business district. Down river from the CBD, new docks and manufacturing industry attracted concentrations of low-income

road and rail communications and derives from this a function as a center for trade fairs and conferences. The two largest cities, the anchors of the agglomeration, are Rotterdam (1.08 million in 2000) and Amsterdam (1.15 million in 2000). Both owe their origin and growth to their geographical situation at points of entry to the European hinterland, and both have developed mainly as industrial and commercial centers. They are, however, completely different in character. Amsterdam is striking for its rings of seventeenth-century canals, its countless bridges, and its old merchants' houses (**Figure 2**). Within Amsterdam's historic core, nearly 7000 structures have been designated as monuments, the maintenance of which requires large sums of public money. Because the center is so completely unsuited to modern traffic, many businesses have moved out to the suburbs or to fringe municipalities such as Amstelveen, or to the area around Schipol airport, where more than 400 firms employ a total of 30,000 people. In contrast, the historic core of Rotterdam is striking for its modern, large-scale business and entertainment center, with wide traffic streets and spacious pedestrian promenades. The reconstruction of Rotterdam city center after the Second World War was seen at the time as an opportunity to create a modern, functional city center. It was not long, however, before the result became unpopular with Rotterdam residents, despite attempts to soften the modernist architecture and planning by planting trees, laying out café terraces, and inserting small shops. In both cities, housing is now being built on filled sites in the central area, particularly around old dock basins that have lost their original function because of the seaward shift of port activities. Old warehouses, for the same reason, have meanwhile been transformed into up-market spaces.

Many of the urban problems of the Randstad stem from the extremely high population densities. Housing shortages are acute in the larger cities, and there is a general problem of congestion. There is also a shortage of recreational space in and around the Randstad. The Green Heart is mainly agricultural and does not lend itself to mass recreation. The Green Heart is also seriously under threat from suburban sprawl, which is spreading the water, air, soil, and noise pollution that have reached serious levels in parts of the Randstad.

Figure 2 Merchants' houses in Amsterdam

housing in the city's East End, while from around 1840 the railways triggered a process of suburbanization that created a mosaic of neighborhoods of high-density terrace housing. These inner-city neighborhoods are now mostly obsolescent, and a good deal of the original housing has disappeared, having been replaced by high-density social housing projects. In addition, the Dockland Development Corporation, established in 1981, has regenerated large tracts of formerly derelict docks and slums in the East End (see Sense of Place: London's Docklands, p. 136). Extensive areas of poor housing remain in much of inner London, however—a mixture of older terraced housing and newer, but run-down, municipal housing.

London's outer suburbs are the product of the extension of the city's underground rapid-transit system, the establishment of a suburban railway network, and the diffusion of the private automobile. London's outer suburbs are relatively affluent and conservative, characterized by semidetached "villas" that form a broad ring about 11 kilometers (6.8 miles) deep, punctuated only by neighborhood shopping streets and industrial parks. London's outer suburbs stopped suddenly at the point they reached in 1947, when a strategic plan for the city established a greenbelt that was designed to halt suburban sprawl, protect valuable agricultural land, and provide an amenity for the city's population. Inside the greenbelt, which covers some 2330 square kilometers (about 900 square miles) in a zone between 8 and 16 kilometers (5 and 10 miles) wide, development has been strictly policed by city planners, with the result that villages and small market towns have a picture-postcard quality that is much sought after by affluent commuter households. The major problem for outer London is traffic, the whole region being swamped by a density of automobiles that is several times the capacity of the road system.

Sense of Place

London's Docklands

Between the late eighteenth and the mid-twentieth centuries, London's docklands became the trading heart of Britain's empire, occupying the north bank of the River Thames for several miles downstream of Tower Bridge (**Figure 1**). The docksides and wharves developed into distinctive settings. Immediately downstream from Tower Bridge were elegant multistory Georgian warehouses for high-value goods linked to the commodity trades of the imperial capital: ivory, teas, furs, tobacco, plant and flower oils, spices, and other exotic imports. Farther downstream there developed facilities for handling and storing bulkier cargoes of tropical fruit and vegetables, coal, cattle feed, chemicals, cement, paper, and so

on. Farther still, newer, larger docks and huge, refrigerated warehouses were added, providing a modern infrastructure for global trade. This commerce required an extensive labor force, which was crammed into cheap and often substandard housing in the neighborhoods surrounding the docks. The distinctive environment of the working docklands—grimy, bustling, utilitarian, and hard-edged—evoked a strong sense of place and an even stronger sense of community.

Relatively little of either the traditional sense of place or sense of community now remains, though there is enough to give flavor to the contemporary scene. The docklands' new image is a result of the largest single urban redevelopment project in the

Figure 1 London's docklands Once the commercial heart of Britain's empire, employing more than 30,000 dockyard laborers, London's extensive docklands fell into a sharp decline in the 1960s because of competition from specialized ports using new container technologies. In 1981 the London Docklands Development Corporation was created by the central government and given extensive powers to redevelop the derelict dock areas. The docklands are now recognized as the largest urban redevelopment scheme in the world, with millions of square feet of office and retail space and substantial amounts of new housing. (*Source:* P. Knox and S. Marston, *Human Geography, Second Edition.* Upper Saddle River, NJ: Prentice Hall, 2001.)

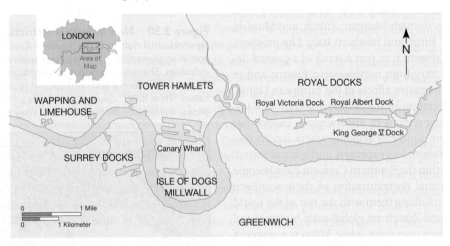

innovative, flexible, and high-quality manufacturers make products that include textiles, knitwear, jewelry, shoes, ceramics, machinery, machine tools, and furniture. Other examples of flexible production regions based on a similar mixture of design- and labor-intensive industries include the Baden-Württemberg region around Stuttgart in Germany (textiles, machine tools, auto parts, and clothing) and the Rhône-Alpes region around Lyon in France.

The northern end of the Southern Crescent is dominated by the pastures, woodlands, and forests of the upland plateaus of the Hunsrück, the Taunus, the Odenwald, and the Fran-

conian Jura. To the south is the imposing landscape of the Black Forest, or Schwartzwald, with its silent, solid mass of fir forests, edged with cheerful farms set amid small fields and meadows. All of these uplands are crossed by rivers whose valleys contain prosperous, manicured landscapes of mixed farming, vineyards, and orchards (**Figure 3.31**). Farther south still are the rich meadows and dairy farms of the Alpine fringe, followed quickly by the spectacular mountains and lakes of the Alps themselves. Beyond the Alps is a belt of **moraines** that dam a series of sub-Alpine lakes (Maggiore, Lugano, Como, Iseo, and Garda) and form the hills upon which north-world. The docklands fell into a steep decline in the 1960s and 1970s as a result of a combination of factors: labor problems, competition from Rotterdam and other European ports, and the construction of container port facilities at Tilbury, farther downstream. The disused and derelict docks, so near to the heart of the city of London, represented both an embarrassment to Britain and a huge potential property asset. In 1979 the government, taking up an idea by geographer Peter Hall, established an experimental "enterprise zone" in the Isle of Dogs, in the heart of the docklands, in an attempt to attract new businesses to the area by suspending certain taxes and regulations. The following year, the government created the London Docklands Development Corporation and charged it with planning the economic regeneration of the docklands.

It was at this point that London's financial markets began to respond to globalization. Automation and information-based dealing required large, modern office units, flexible in plan, with deep floor plates to accommodate state-of-the-art technology in suspended ceilings and underfloor cabling. London's central financial district had little such space and few opportunities to create it. Then, in 1985, the world's largest property development company, Canada-based Olympia and York, put together an ambitious redevelopment scheme for the docklands that would provide several million square feet of new office space. The government saw the opportunity to present a new image to the world (and expunge the old one) and quickly committed to furnish a new transportation infrastructure: a driverless light railway line and a $4.2 billion extension to the underground subway system. Although the redevelopment scheme ran into financial difficulties for a while (and Olympia and York fell into disastrous debt), the docklands have been wholly regenerated (**Figure 2**).

The docklands' new sense of place centers around the sleek office tower of Canary Wharf and its associated office complexes that are populated by day not by a bustling working class but by intense Euro-yuppies working for international financial and publishing companies. There is a powerful

Figure 2 The new docklands Recently linked to the London's underground railway system, new office developments have become the nucleus of the economic regeneration of London's docklands.

new sense of place that is evoked by the sheer scale and quality of the redevelopment—a huge tableau of polished granite, plate glass, stainless steel, fountains, and sculptures, all surrounded by cleaned-up waterfronts and restored warehouses that have been converted to expensive condominiums and retail complexes (Figure 2). However, there is very little sense of community in the new docklands: Few of the office workers actually live in the docklands, while the residual working-class population, having lost its ties with the river, has lost a great deal of its self-identity and cohesion.

ern Italy's distinctive hill towns (such as Bergamo and Brescia) stand. The landscape then flattens out into a zone of low terraces that descend to the rich, broad floodplain of the River Po before rising again toward the hills of the Apennines.

Milan Milan provides a good example of the prosperity of the Southern Crescent. Although Rome is the Italian capital and Naples is perhaps better known, Milan is the leading city of the country: the richest, the most fashionable, the most innovative, and the best governed. With only 7 percent of the country's population, it accounts for 28 percent of Italy's na-

tional income. If Milan were still an independent city-state, it would be richer than Switzerland. It has a diversified economic base that has expanded in recent years as, along with other cities in the European sunbelt, it has attracted a significant amount of foreign investment. The most important industries include automobile assembly, aircraft, motorcycles, electrical appliances, railroad materials, metal trades, chemical production, and graphic and publishing industries. Although parts of the industrial sector have declined—Pirelli, for example, closed its huge tire factory in the northern suburbs of the city in the 1980s—the service sector is growing

(a)

(b)

Figure 3.31   Landscapes of the Southern Crescent   There is great variety in the landscapes of the Southern Crescent, the greatest contrasts being between those to the north and those to the south of the Alps. (a) Valley of the River Altmuhl, Bavaria, Germany; (b) Lake Como, Italy.

(a)

(b)

Figure 3.32   Milan   Milan has long been a major regional center and today has developed into a prosperous city of global importance in finance, fashion, and industrial design. (a) Catwalk models at a Milan fashion show; (b) the Galleria Vittorio Emanuele.

rapidly. For every manufacturing firm that closed down or moved out during the 1990s, three service firms were created. Services now account for 70 percent of the local economy, and Milan has gained international significance as a financial center and as a center for design—both industrial and fashion. The Milanese are hard-working and entrepreneurial, with a strong progressive streak that is vigorously pursued by the (traditionally socialist) city government. The city exudes an exhilarating combination of the historic and the modern, of industry and culture, and of the fast track and laid-back urbanity.

The center of the city owes its street pattern to the pattern of successive defensive walls. It is dominated by the Duomo (cathedral), begun in the fourteenth century and completed in the nineteenth, and the Castello Sforzesco, a fifteenth-century castle. They look out on a low-rise city center in which are tucked away such landmarks as the Piazza Mercanti (the center of activity in medieval times), the theater of La Scala facing the Piazza della Scala with its statue of Leonardo da Vinci, the church of Santa Maria Delle Grazie with its

mural of the Last Supper by Leonardo da Vinci, and the magnificent indoor shopping arcade of the Galleria Vittorio Emanuele (**Figure 3.32**). The narrowness of the streets in the center has precluded the development of much industry; it is the suburbs, particularly to the north/northeast and south/southwest, that are the setting for Milan's industry. These suburbs are for the most part well built and well served. The pressure of population growth has, however, resulted in the expansion of slum housing and the appearance of shanty-towns such as Brianza on the urban fringe.

A Day in the Life

Anne-Lise Bamberg

Anne-Lise Bamberg is a 28-year-old Swiss banker who moved to the Italian-speaking Ticino region of Switzerland with her parents from German-speaking northern Switzerland when she was seven years old. Like many other middle-class Swiss citizens, she is fluent in English as well as French, German, and Italian, the three principal languages of Switzerland. Anne-Lise lives in the small Swiss village of Capolago, just north of the Italian border on the southern shore of Lake Lugano, in a small apartment that she rents from the owner, who runs a pharmacy on the street level of the building.

Every weekday morning, Anne-Lise walks along the shore of Lake Lugano toward the small modern train station in Capolago and takes a commuter train to work in the city of Lugano. Arriving at the main railway station high on the hillside, she walks down the steep winding road that leads to the central piazza. On the way down, she stops at one of the market stalls or bakeries to buy something to eat during her lunch break. She works in one of the big regional banks that are located near to the fashionable shopping street of the Via Nassa. On fine days in spring, summer, and fall, she is able to take her lunch to a bench by the lakeside. After work, she takes the funicular (a hillside cable railway system in which an ascending car counterbalances a descending car) up to the railway station to catch the 5:50 train that will get her back in Capolago in time to stop at the local co-operative store to get milk, wine, and groceries before it closes at 7:00 P.M.

In the evenings, she splits her time between studying international banking law and touring the back roads on her new sports motorcycle. On weekends, Anne-Lise often takes the train to Milan, just 50 minutes or so across the nearby border with Italy, to visit her friend Paulo, who is a clothes designer. Anne-Lise met him when he was studying in Paris and she was there visiting her sister, who works as an interpreter at UNESCO.

Distinctive Regions and Landscapes of Europe

Beyond Europe's core regions, major metropolitan areas, and specialized industrial districts, a mosaic of different landscapes has developed around the broad physiographic regions described at the beginning of this chapter. In detail, these landscapes are a product of centuries of human adaptation to climate, soils, altitude, and aspect, and to changing economic and political circumstances. Farming practices, field patterns, settlement types, traditional building styles, and ways of life have all become attuned to the opportunities and constraints of regional physical environments, with the result that distinctive regional landscapes have been produced.

The landscapes of mountain regions are the most natural. These are essentially physiographic subregions, with just a scattering of isolated settlement. The farmers in these mountain subregions depend on **pastoralism,** a system of farming and way of life based on keeping herds of grazing animals, eked out with a little produce grown on arable land on the valley floors. In the less mountainous parts of Scandinavia, as in much of Baltic Europe, with their short growing season and cold, acid soils, agriculture supports only a low density of settlement. Landscapes reflect a mixed farming regime of oats, rye, potatoes, and flax, with hay for cattle.

The more humid and temperate regions of Atlantic Europe (Belgium, France, Ireland, the Netherlands, and the United Kingdom) are dominated by dairy farming on meadowland, sheep farming on exposed uplands, and arable farming (mainly wheat, oats, potatoes, and barley) on drier lowland areas. Settlement density tends to vary according to the productivity of the land, and the traditional settlement form is mostly the *Haufendorf* (**Figure 3.33**). The exceptions are those areas where the initial colonization was by Celtic peoples and where environmental conditions support only low population densities (much of Brittany, Ireland, Scotland, and Wales, for example).

Figure 3.33 Rural settlement The typical settlement form in the rural areas of the more humid and temperate regions of Europe is the nucleated village, or *Haufendorf*. (*Source:* R. Mellor and E. A. Smith, *Europe: A Geographical Survey of the Continent.* London: Macmillan, 1979, p. 67.)

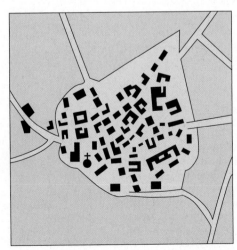

In these areas, dispersed settlement, in the form of hamlets and scattered farms, is characteristic.

The traditional form of settlement throughout most of Mediterranean Europe is the very large nucleated village. This form of settlement is partly a product of traditional forms of land holding and partly a product of past needs for people to cluster together for reasons of defense. The dry, warm climate of the Mediterranean is reflected in its landscapes, with agricultural regimes based on pastoralism, extensive arable farming (mainly hard wheat), and tree crops (especially olives).

It should be stressed that, within these broad divisions, marked variations exist. The mosaic of regions and landscapes within Europe is both rich and detailed. Physical differences are encountered over quite short distances, and there are numerous specialized farming regions where agricultural conditions have influenced local ways of life to produce distinctive landscapes. Within these landscapes are towns and cities of regional importance whose characteristics owe a great deal to the life and traditions of their region. The following sections describe the attributes of Europe's most distinctive regions, together with representative examples of major settlements.

Alpine Europe

The Alps occupy a vast area of Europe, stretching eastward for nearly 1290 kilometers (about 800 miles) from Montélimar, France, to Vienna, Austria, and with a maximum width of approximately 195 kilometers (about 120 miles) (**Figure 3.34**). The Alps are a recently created range of mountains, dating from about 20 to 100 million years ago. Their relative youth explains the sharpness of their outlines and the boldness of their peaks. The highest peak, Mont Blanc, reaches 4810 meters (15,781 feet). Most of the rest of the mountains are between 2500 and 3600 meters (about 8200 and 11,800 feet) in height. Although the Alps pose a formidable barrier between northwestern Europe and Italy and the Adriatic, a series of great passes—including the

Brenner Pass, the Simplon Pass, the Saint Gotthard Pass, and the Great Saint Bernard Pass—and longitudinal valleys have always provided transalpine routeways.

The dominant direction of the Alps and their parallel valleys is roughly southwest to northeast. The major Alpine valleys thus have one sunny, fully exposed slope that is suitable for vine growing and a shaded side rich with orchards, woods, and meadows. The mountains and valleys of the Alps proper are surrounded by an Alpine foreland that provides rich farmland on glacial outwash deposits. The limestone of the Alpine region is economically important for quarrying and cement, while mineral deposits—lead, copper, and iron—and small deposits of coal and salt have been locally important throughout the region. In addition, the Alps are a valuable source of hydroelectric power: about 65 billion watt-hours in Switzerland (60 percent of the country's electricity consumption), about 72 billion watt-hours in France (15 percent of the country's electricity consumption), and about 45 billion watt-hours in Austria (85 percent of the country's electricity consumption).

The traditional staple of the economy, however, has been agriculture, and it has been farming that has given the Alpine region its distinctive human landscape (**Figure 3.35**). The Alpine rural landscape is a patchwork of fields, orchards, vineyards, deciduous woodlands, pine groves, and meadows on the lower slopes of the valleys, with broad alpine pasture above. In these pastures, which are dotted with wooden haylofts and summer chalets, dairy cattle wander far and wide. Farmers attach bells around the necks of their animals in order to be able to locate them, and the consequent effect is a resonant pastoral "soundscape" of clanking cowbells. Farms and hamlets tend to cling to lower elevations, the chalet-style architecture drawing on timber or rough-cast stone construction, with overhanging eaves, tiers of windows, and painted ornamentation. The landscapes of the Alpine foreland are more lush. Lavender and fruit have been introduced to enrich and give variety to the mixed farming regime of the foreland, which features vine and wheat

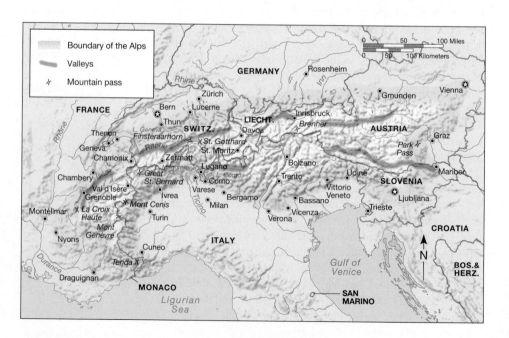

Figure 3.34 Alpine Europe
The Alps pose a formidable barrier between northwestern Europe and Italy and the Adriatic, penetrated only by a series of mountain passes. (*Source*: P. Deffontaines (ed.), *Larousse Encyclopedia of Geography: Europe.* London: Paul Hamlyn, 1961, p. 190.)

Figure 3.35 Alpine landscapes The distinctive landscapes of the Alps juxtapose lush meadows and prosperous villages against jagged mountains. (a) Wolkenstein, an alpine village in the Italian Dolomites; (b) Steinen, Switzerland.

growing, along with dairy cattle. The higher slopes, which receive more rainfall, provide lush pastures that have made the region famous for its rich cheeses, such as Gruyère.

The principal industry of the Alpine region, however, is not farming but tourism. Attractive rural landscapes, together with magnificent mountain scenery, beautiful lakes, and first-class winter sports facilities, have attracted tourists to this region since the 1800s. Lakeside resorts such as Lucerne and Lugano, Switzerland; mountain resorts such as Chamonix, France, and Innsbruck, Austria; and winter sports resorts such as Val d'Isere, France, Davos, St. Moritz, and Zermatt, Switzerland, are all well established, with an affluent clientele from across Europe. With the growth of the global tourist industry, these resorts, along with many others, have attracted an important new clientele from North America and Japan.

Nordic Europe

The term *Nordic* refers to the entire northern part of Europe: Scandinavia (Norway, Denmark, and Sweden), plus Finland and Iceland. The whole region is sometimes referred to as "Norden." The three Scandinavian countries have been closely interlinked since 1397, when they were united under the Danish crown. In 1523 Sweden (incorporating the southwest part of modern Finland) separated from Denmark, and in the early nineteenth century lost Finland to Russia but meanwhile gained control over Norway. Norway gained independence in 1905 and Finland in 1917. All are closely allied by ties of soil, custom, and common interest, and all except Finland have a close linguistic affinity.

The Nordic countries are also tied by their peoples' orientation to the sea. Surrounded variously by the North Sea, the Atlantic Ocean, Arctic Ocean, and the Baltic Sea, the peoples of Nordic Europe have a deep-rooted seafaring tradition. This orientation is reinforced by the inhospitable (though spectacularly beautiful) nature of much of the land. The region experienced several successive phases of glaciation, which have left a distinctive imprint on the landscape. There are *cirques* (deep, bowl-shaped basins on mountainsides, shaped by ice action), glaciated valleys, and fjords (some as deep as 1200 meters—about 3900 feet) in Norway; countless lakes (nearly 100,000 of them in Sweden alone); lines of moraines that mark the ice sheet's final recession from Denmark to Sweden, Norway, and Lapland; extensive deposits of sand and gravel from ancient glacial deltas, and vast expanses of peat bogs that lie on the granite shield that forms the physiographic foundation of the region.

This formidable environment is rendered even more forbidding as a result of climatic conditions. In the far north, the sun shines for 57 days without setting, but the winter nights are interminable. Helsinki sees a *total* of only 17 hours of sunshine in the whole month of December. Rivers and lakes in the north are frozen from mid-October on, while even farther south in Norway, Sweden, and Finland, they freeze at the end of November and remain frozen until May. Snow, which is permanent in parts of Iceland and Lapland, begins to fall toward the middle of September and it covers much of the landscape from October through early April.

The single most dominant feature of the landscapes of Nordic Europe is the forests (**Figure 3.36**). There are fine forests of beech trees, with their smooth majestic trunks, in parts of Denmark and southwestern Sweden. Farther north, more and more conifers (mostly evergreen trees such as pine, spruce, and fir) are mixed with birches and other deciduous trees, until the conifers become entirely dominant. In the far north, toward the North Cape, the forest gives way to desolate treeless stretches of the **tundra**, with its gray lichens and dwarf willows and birches. Overall, forests cover more than half of the land. Not surprisingly, forestry is a major industry. The forests supply timber for domestic building and fuel and produce the woods that are in greatest demand on world markets: pine and spruce for timber and for the paper industry; birch for plywood and cabinetmaking; aspen for making matches. Sweden and Finland are second only to Canada as exporters of paper and wood products.

The forests give way to agriculture along the coasts, on valley floors, and across the gentler and more fertile outwash plains of southeastern Norway, southern Sweden, and most of Denmark. In Denmark about 60 percent of the land is farmed,

Figure 3.36   Nordic landscapes The landscapes of Nordic Europe are dominated by forests, lakes, moorland, and the sea. Across most of the region only 5 percent of the land is farmed. Denmark, with its low topography, better soils, and gentler climate is the major exception. (a) The high plateau of Hardangervidda, Norway; (b) fishing village, Lofoten Islands, Norway.

Elsewhere, with less rigid traditional landholding systems and with no history of invasions between the Middle Ages and the twentieth century, farms and hamlets are casually situated on any habitable site, their buildings often widely scattered. The countryside is dotted with trim wooden houses, roofed with slate, tiles, shingles, or even sods of turf. Often, the buildings of a particular district are distinguished by some special stylistic feature. In some places, the old weather-board houses that were used as shelters in bad weather still exist. They date back to the Middle Ages, as does the custom of storing reserve stocks of food or hay in a special isolated building, the *stabbur*, decorated with beautifully carved woodwork. The predominant color of the buildings is gray, the natural color that wood acquires with age. Nearer to towns, the houses are painted brighter colors: light brown, light yellow, or white in Finland; yellow, dark red, and mid-blue in Norway and Sweden.

The larger towns and cities of Nordic Europe, as elsewhere in Europe, are the nodal points of the economy. Some owe their existence to mining and metallurgy (Karlskoga and Eskilstuna in Sweden, for example), and some to textile manufacture (Tampere and Turku, in Finland, for example). Most of the important towns and cities are ports, such as Bergen (**Figure 3.37**): shipbuilding and repair, fishing, and shipping are key industries. Some cities have a significant manufacturing base. Sweden is a major producer of automobiles, trucks, and aircraft and has significant steel, chemical, plastics, pharmaceutical, and biotechnology industries; cities such as Malmö and Huskvarna are primarily industrial cities. It was upon their productivity that Sweden developed a prosperous economy that

Figure 3.37   Bergen Bergen's prosperity as a trading port led to its emergence as the capital of Norway in the twelfth and thirteenth centuries and its affiliation with the Hanseatic League (see Figure 3.9). Today, Bergen is still an active trading port and manufacturing center. The most striking and distinctive aspect of Bergen's economy is its role as one of the two major operational centers (Aberdeen, in Scotland, is the other) for the North Sea oil and gas fields that were opened up in the 1970s. The city's roots are most visible in the turn-of-the-century burgher houses, fishmarket, and commercial berths around the historic Vågen harbor.

with another 5 percent under permanent pasture. Throughout the rest of Norden, however, only about 5 percent of the land is farmed, with less than 1 percent under permanent pasture. More than two-thirds of the area under cultivation outside Denmark produces food for livestock. Oats, the largest single crop, often has to be harvested while it is still green. More than half of the milk from dairy cattle is used to produce butter and cheese. In Denmark, agricultural cooperatives have established a highly efficient and very productive industry that includes dairy farming, pig-raising, and the cultivation of wheat, oats, barley, and root crops. The dependence of Danish agriculture on exports led the Danes to become early members of the European Union (in 1973); in contrast, Norway remains outside the European Union, largely because its farmers, subsidized by Norway's North Sea oil revenues, could not survive in open competition with farmers elsewhere in Europe.

The contrast between Denmark and the rest of Norden carries over into settlement patterns. In Denmark a feudal system (until 1788) led to the development of large farms and estates, with settlement concentrated into compact villages.

has been able to sustain a model welfare state. Sweden's prosperity was also helped by its avoidance of involvement in both world wars (Sweden's industries, in fact, profited from both wars). Meanwhile, the prosperity of both Norway and Denmark was boosted by the discovery, in the late 1960s, of extensive oil and natural gas deposits in the North Sea. The most significant economic change, however, has been a more recent shift toward service-sector employment, which now accounts for more than 65 percent of employment throughout Norden, with most of the advanced service jobs concentrated in the major cities of Århus (population 280,000), Bergen (224,000), Copenhagen (1,362,000), Göteborg (454,000), Helsinki (532,000), Oslo (494,000), and Stockholm (718,000).

The Danubian Plains

The Danube River is one of the most important waterways in Europe, stretching 2859 kilometers (1776 miles) from Donaueschingen, in southwestern Germany, to the Black Sea. Over the second half of its course, it passes through two broad lowland plains, each surrounded by complexes of mountains and uplands. The Danube pierces these uplands at a gorge known as the Iron Gates, which separates the mid-Danubian Alföld, Vojvodina, and Pannonian Plains from the Wallachian Plains of the lower Danube (**Figure 3.38**). These Danubian Plains share some fundamental physiographic features—lowland river basins with a rolling cover of **loess** and sandy river deposits—and they developed a considerable unity in terms of landscape and human settlement as a result of the influence of the Ottoman and Austro-Hungarian empires (see Figure 3.18). The constrained economies of Hungary, Yugoslavia, Romania, and Bulgaria during the second half of the twentieth century, under state socialism, meant that the rural and small-town landscapes of the Danubian Plains have remained largely unaltered for decades.

Figure 3.38   The Danubian Plains Reference map showing principal physical features, political boundaries, and major cities of the Danubian Plains.

The Danube enters its first basin of broad plains by swirling through a series of picturesque gorges, surrounded by hills that, to the west, are covered with oak and beech forests and, to the east, are volcanic upland with fertile soils that are favorable for market gardening and vineyards that produce, for example, the famous Bull's Blood wines of the Eger region in Hungary. The city of Budapest straddles the river and commands the gateway to the the mid-Danubian Plains. These lands were depopulated during the Ottoman occupation of the sixteenth and seventeenth centuries. The advances of the Turkish army in its campaigns were marked by swaths of smoking hamlets, and its withdrawals were followed by long trains of captives destined for the slave markets of Anatolia. When the Turks took control of the region, people flocked to town-sized villages, where they enjoyed a measure of protection. The country between the towns was left empty except for scattered huts (*tanyas*) in which the menfolk spent the summers tending the fields. The vast town-like villages, and a few of the *tanyas,* survive as distinctive features of the region's heritage.

When the Ottoman Empire was rolled back in the eighteenth century, the landed gentry of the Austro-Hungarian Empire attracted Serbs, Hungarians, and Germans who built villages, repopulated the tributary valleys of the Danube, drained marshes, and tilled the plains. This gave rise to the *Alföld,* or "cultivated plain." Before the Second World War, the Alföld was dominated by huge estates worked by thousands of peasant-servants. The Esterhazy family, for instance, owned more than 129,600 hectares (320,000 acres), which included 164 villages. Much of the region consists of loess plateaus where rainfall is very irregular and averages around 40 centimeters—approximately 16 inches—a year; there are no woods and irrigation is often necessary to sustain the typical 2-year rotation of corn and wheat. The region's tradition of large, townlike villages was continued by the Serbs, Hungarians, and Germans who resettled the plains, building *varos,* or trading villages with an open, rectilinear central area surrounded by administrative buildings, inns, and shops, and roads radiating from this central nucleus like spokes from a wheel, bordered for kilometers by straggling farms (**Figure 3.39**). Some villages have grown to several tens of thousands of inhabitants. The whole way of life is inherently rural, though a few centers, such as Pécs, Debrecen, and Szeged, have acquired something of an urban character as a result of their regional administrative functions (see Sense of Place: Szeged, p. 144).

In the Hortobágy area, west of Debrecen, there is a residual region of **steppe**, semiarid, treeless, grassland plains known as the *Puszta.* This land, which is too dry or too marshy to have invited cultivation, was once the domain of virtually wild horses, cattle, and pigs, but today huge flocks of sheep find pasture there. This is the landscape of old, traditional picturesque Hungary, associated in the national consciousness with shepherds, poets, gypsies, fine horses, and colorful costumes.

In contrast, the southern part of the mid-Danubian Plains, to the north of the River Sava and extending eastward beyond Belgrade, has long been famous as prosperous cereal-growing areas. These are the Pannonian Plains and the

Sense of Place

Szeged

Szeged, one of the larger towns of the Danubian Plains, had a population of 189,000 in 2000. Situated at an important regional crossroads at the junction of the Tisza and Maros rivers, Szeged began as a Roman trading post, developed into a medieval fortress town, and was occupied by the Turks between 1543 and 1686. Ironically, the Turkish occupation resulted in the town's first significant growth phase, as the Turks laid waste to surrounding villages and people retreated into Szeged for safety. As the surrounding countryside was repopulated by Hungarians and Serbs in the eighteenth century, Szeged became the hub of the farming community. It remains so today and is noted for its paprika processing, salami manufacture, and fruit canning. Prosperity as an agricultural market town brought a cathedral, a university, fine churches, and a neo-Baroque town hall. In the nineteenth century, steamer and railway services connected Szeged to a wider region and increased its prosperity still further. In 1879, however, a disastrous flood destroyed all but the most substantial of the town's buildings. Rebuilding took place very quickly, using a new pattern of radial streets with circular boulevards, along the lines then fashionable in western Europe. Rebuilding was assisted by financial aid from around Europe, and sections of the outer boulevard are still named in honor of Brussels, London, Moscow, Paris, Rome, and Vienna. During the reconstruction, the town developed a high bourgeois appearance (**Figure 1**), with public parks and statuary complementing the ostentatious homes of wealthy burghers.

During the socialist period after 1945, these parts of the town changed little, though patterns of ownership did change, and some of the major streets were renamed (after Marx, Lenin, and other revolutionary heroes). The principal change to the city was the addition of suburban residential estates with the sort of system-built apartment blocks that came to characterize much of eastern Europe. Today, after more than a decade of readjustment to a market economy, Szeged remains a prosperous agricultural market center with all of the hallmarks of a traditional central European town but with the increasing imprint of globalization. Fast-food restaurants and stores selling international brands of clothing and consumer goods have found their way into the fabric of the town, while an 11-hectare (27.2-acre) development project with the ambitious name Vilag Utcaja, or Street of the World, is planned just outside the town as a regional business and commercial center, which will include a 70,000-square-meter (753,200-square-foot) shopping center, an 18,000-square-meter (193,680-square-foot) conference building, a 500-bed hotel, and an open-air exhibition area. In a very different reflection of contemporary times, approximately 30,000 displaced Yugoslav citizens have arrived in Szeged, most of them ethnic Hungarians from the northern Serbian province of Vojvodina who are staying with relatives, friends, or people they know who resettled there earlier.

Figure 1 Central Szeged

Vojvodina (**Figure 3.40**), where several meters of loess and rich, sandy soil rest on the rocky substratum. Stone and trees are so scarce that houses are built of *pisé*, a kind of rammed-earth brick. Storks' response to the scarcity of trees is to build nests atop chimneys and telegraph poles. The large villages are less compact than farther north, and some are strung out as street-villages, along major routeways. The rich soils produce high yields of wheat and corn, together with hops, sugar beets, and forage crops for livestock.

Beyond the Iron Gates, where a canal allows shipping to avoid the river's turbulent passage over rocky ledges, the Wal-lachian Plains offer a landscape that is infinitely monotonous. These also are fertile lands, with rich soils washed down by the Danube's tributaries. As in the rest of the Danubian Plains, wheat and corn are the principal crops. The climate, though, is harsh, with seasonal extremes of burning hot and freezing cold, the winter easterlies blowing down from mid-continent Russia. Population densities are low, and there are few villages. At the eastern extremity of the region is the Danube delta, where the river meets two tributaries—the Siret and the Prut. Here, low, marshy country, with patches of oak forest and both freshwater and saltwater lagoons, offers a rich habitat for a

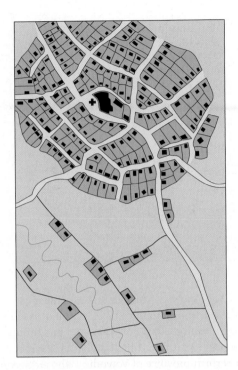

Figure 3.39   Rural settlement Large nucleated villages are the typical settlement form in the Danubian Plains. (*Source:* R. Mellor and E. A. Smith, *Europe: A Geographical Survey of the Continent.* London: Macmillan, 1979, p. 69.)

great diversity of birds and wildlife, now recognized as a UNESCO World Heritage Site.

Mediterranean Europe

Mediterranean Europe is an extensive region that stretches along the coastline of the Mediterranean Sea from southern Spain to eastern Greece, a distance of some 2414 kilometers (about 1500 miles). The distinctiveness of the region derives not only from its ties with the sea but also from its climate and vegetation and its long tradition of urban life. The watershed of rivers that drain into the Mediterranean provides a good approximation of the extent of the region (**Figure 3.41**).

The Mediterranean climate is such that winters are cool, with an Atlantic air stream that brings overcast skies and intermittent rain—though snow is unusual. In spring the temperature rises rapidly and rainfall is more abundant. Then summer bursts forth suddenly as dry, hot, Saharan air brings three months of hot, sunny weather. There is no rain save an occasional storm; the soil cracks and splits and is easily washed away in the occasional downpours. In October the temperature drops, and deluges of rain show that Atlantic air prevails once more.

In such conditions, delicate plants cannot survive. The Mediterranean climate precludes all plant species that cannot support cold as well as heat and drought as well as wet conditions. The result is a distinctive natural landscape of dry terrain that is dotted with cypress trees, holm-oaks, cork oaks, parasol pines, and eucalyptus trees (**Figure 3.42**), or that is covered with a low scrub of asphodel, cistus, lentiscus, and myrtle (known as *maquis* or, in its more sparse version, as *garrigue*). These same conditions make agriculture a challenge. The crops that prosper best include olives, figs, almonds, vines, oranges, lemons, wheat, and barley, with sheep and goats on dry pastureland and stubblefields. Irrigation is often necessary, and in some localities it sustains high yields of fruit, vegetables, and rice, as in the *huertas* (derived from the Latin *hortus,* a garden) of southeastern Spain.

In a few subregions, conditions are naturally more favorable, making for rich rural landscapes. Examples include the coastal plains of the Tyrrhenian (western) side of the Apennines in Italy, where volcanic soils contribute to productive agricultural regimes. Tuscany, in particular, has evolved a regional landscape that reflects an intensive and carefully developed regime that is closely adapted to the land. On the better and well-watered soils of the valleys, there are artificial meadows that favor stock-breeding, along with fields of wheat, mulberries, and corn. On the hills around the scattered farms and villas, elegant cypress trees stand out against the silvery-green of olive trees, and in the fields there is a rich mixture of cereals, vegetables, fruit trees, and vines (for chianti wine). This classical Tuscan landscape (see Figure 3.42) developed over centuries, became emblematic of Italy itself with

Figure 3.40   Landscapes of the Danubian Plains The Danubian Plains are monotonous, but they provide distinctive habitat for wildlife. (a) Corn harvest on the Danubian Plain in Romania; (b) Stork's nest.

(a)

(b)

Figure 3.41   The Mediterranean region
Reference map showing principal physical features, political boundaries, and major cities of the Mediterranean region.

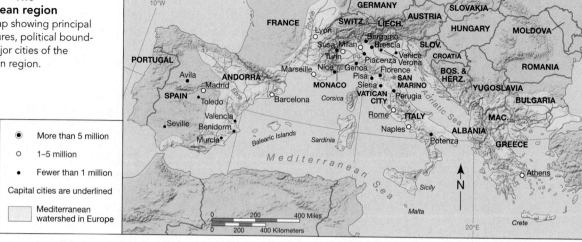

the creation of modern Italy and the *Risorgimento* ("revival through unification," 1815–61), and has been the subject of landscape painters, romantic poets, and novelists ever since.

The landscapes of the Mediterranean reflect the imprint of successive cultures over a very long history. The fields themselves are a good example. Under Roman colonization, land was often subdivided into a checkerboard pattern of rectilinear fields. This highly ordered system was known as *centuriation,* and the pattern can still be seen in some districts today—in parts of the Po valley, for example. Elsewhere, across large tracts of the Mediterranean, the soil can be cultivated only on a large scale, and the poor quality of pastureland necessitates vast untilled areas being left for flocks and herds. In these areas, successive conquerors, from the Greeks, Phoenicians, and Carthaginians, to the Ottoman Turks and Christian Crusaders, carved out huge estates, known as **latifundia,** on which they set peasants to work. Land that did not belong to these big estates was often subdivided by independent peasant farmers into very small, intensively cultivated lots, or **minifundia,** most of which are barely able to support a family.

Settlement patterns also reflect these influences. Both *latifundia* and *minifundia* systems tend to result in clustered settlements. In more productive districts, these can be quite large, with "villages" of 10,000–12,000 people, as in the huerta districts of southeastern Spain and the picturesque whitewashed villages of the Greek islands. The history of the Mediterranean, however, has made it above all a land of towns. Fears of invasion encouraged people to cluster together in easily defended sites: typically on steep-sided hills, as in Avila and Toledo in Spain; in Siena, Perugia, and Potenza in Italy. Other towns naturally emerged at strategic locations: at the foot of a pass (for example, Susa, Italy); at the entrance to a valley (for example, Murcia, Spain, and Verona, Italy); and at bridging points (for example, Piacenza, Italy, and Seville, Spain).

More recently, the urban landscapes of the Mediterranean have been transformed by tourism and retirement migration. The classic landscapes, picturesque hill towns, and ancient cities of the Mediterranean have attracted affluent tourists and retirees from the rest of Europe since the late 1700s, but the advent of mass-market tourism after the Second World War has created resort towns and brought tourist amenities—hotels, restaurants, night clubs, bars, and so on—to much of the region's coastline. Benidorm, on Spain's Costa del Sol, was

Figure 3.42   Mediterranean landscapes   The dry climate of the Mediterranean has led to distinctive landscapes. (a) Landscape near Siena, Tuscany; (b) olive groves, near Jaen, Andalucia, Spain.

(a)   (b)

a fishing village of just 1500 inhabitants in the early 1950s. Today, it is a mass-market, package-tour resort with more than 30,000 hotel beds and 100,000 more in rental apartments. Similar transformations have taken place elsewhere, as millions of vacationers and retirees from the colder, industrialized regions of northern Europe have made Mediterranean beaches their destination (**Figure 3.43**). The impact on local economies is markedly seasonal, and the physical results are not always pleasant, as noted by travel writer Paul Theroux as he passed through Spain's Costa del Sol one winter:

> It was a sort of cut-price colonization, this stretch of coast, bungaloid in the extreme—bungalows and twee little chalets and monstrosities in all stages of construction, from earthworks and geometrically excavated foundations filled with mud puddles to brick and stucco condos and huts and houses. There were cheap hotels, and golf courses, and marinas and rain-sodden tennis courts and stagnant swimming pools at Estepona, where "Prices Slashed" was a frequent sign on housing developments in partially built clusters with names such as "Port Paradise" and "The Castles" and "Royal Palms"— no people on the beach, no people on the road, no golfers, no sign of life at all, only suggestions here and there that the place was known to English-speaking people: "English Video Club" was one, and another that was hardly out of view from Gibraltar to the French frontier at Port-Bou: "Fish and Chips."[1]

Athens   Athens, like Rome, was once one of the centers of the world economy. Like Rome, it is now semiperipheral within Europe. Having given birth to Western civilization through the development of philosophy, political ideals, literature, and architecture, Classical Athens was eclipsed in the second century B.C. and fell into decline and neglect for almost two thousand years. In 1833, when the modern nation-state of Greece gained independence, Athens was no more than a village of some 4000 inhabitants. Thus, although it is semiperipheral, its immediate past has been one of growth. In 1921 the resettlement of Greeks from Turkey increased the city's population to 750,000, intensifying the pressure on land and housing caused by rural-urban migration and leading to the appearance of unauthorized suburban sprawl and shantytowns. Before the city could recover, it went through a period of brutal Nazi occupation, followed by a violent communist rebellion. Under repressive fascist governments of the 1950s, 1960s, and early 1970s, Athens resumed its economic and demographic growth, but the speed of growth led to acute problems of traffic congestion, water supply, land price inflation, and air pollution that have carried over into the democratic period of the past 15 years.

Modern Athens is a sprawling metropolis of more than 1.5 million, an administrative, financial, publishing, and commercial center with a broad manufacturing base, and an exporter—through its port city of Piraeus—of marble, minerals, and agricultural produce. Classical Athens is very much in evidence in the city center, where the Parthenon looks down

Figure 3.43   Benidorm   Seafront apartment blocks and hotels.

from the Acropolis onto the restored Agora (**Figure 3.44**). "Old" Athens is also present in neighborhoods such as Pláka (on the northern slopes of the Acropolis), where small, whitewashed houses cluster along narrow streets and around small squares with taverns and vine-covered arbors. For the most part, however, Athens is a modern, sprawling city with a rather characterless and unfinished appearance that derives from speculative building in concrete and stucco. Migrants from the Greek countryside and immigrants and refugees from the Balkans have to endure harsh conditions in comparison with the majority of western Europeans (see A Day in the Life: Daniela Stefani, p. 149).

Rome   Rome is one of the most beautiful and exciting cities in the whole Western world. Known as the Eternal City, it is associated with some of the pinnacles of humanity's artistic and intellectual achievement. It is also one of the most congested,

Figure 3.44   Athens   This aerial view of the city shows the remnants of the classical city in the midst of a modern metropolis of more than 1.5 million.

[1]P. Theroux, *The Pillars of Hercules.* New York: Putnam, 1995, p. 28.

(a)   (b)   (c)

Figure 3.45   **Rome**   Central Rome is a dense mixture of historic monuments, narrow streets, and exclusive shops. (a) Piazza Navona; (b) shoppers on the Via Condotti; (c) houses on the Piazza della Rotonda.

noisy, and debt-ridden cities in Europe. Located on a defensible hilltop site dominating the last high-banked crossing point of the river Tiber in central Italy, Rome was on an important trade route between Tuscany to the north and Greek colonies in southern Italy. By the second century A.D., it had become the capital of an empire of more than 100 million people, its own population reaching almost 2 million by the third century. After the transfer of the capital of the empire to Constantinople (present-day Istanbul) in A.D. 330, Rome was sacked by barbarians and there followed a thousand years of decline and stagnation. Earthquakes, invasions, and the pillaging of old buildings for quarry stone left much of the classical city in ruins, and it was only with the return of the papacy to the city from Avignon in 1378 that it began to grow again. Rome became the center of Renaissance culture in the early sixteenth century, only to be sacked by the armies of Charles V (Holy Roman Emperor) in 1527, when thousands of churches, palaces, and houses were destroyed in just over a week. This led, in turn, to one of the more important redevelopments of the city in the 1580s, when Pope Sixtus V and his architect Domenico Fontana undertook a vast program of city planning, laying out new streets and squares, building palaces, and repairing walls, bridges, and aqueducts. Yet the influence of the church was not always conducive to successful economic development, and it was not until the unification of Italian states in 1870 and Rome's selection as the national capital that it began to grow and prosper as a modern city. By 1910 the city's population had reached 500,000, and by 1930 it reached 1 million, expanding for the first time beyond the city walls (the Aurelian Wall) that had been built in A.D. 270.

It is within the compass of these walls that most of the glory of Rome is to be found. Within this area, which extends for little more than 1.5 kilometers (about 1 mile) in every direction from the Colosseum, are the ruins of the Roman Forum, the Pantheon, the Capitoline, the column of Marcus

Aurelius, the Piazza del Populo, the Via del Corso, Bernini's fountains, the Scala di Spagna (the Spanish Steps), the Piazza dei Cavelieri di Malta, the Piazza Navona (**Figure 3.45**), and hundreds of other historic structures and monuments. Embedded within this historic fabric are some of the world's most exclusive shops and restaurants, together with a tremendous mixture of other shops, offices, and workshops, providing a diversity that is emphasized by the variety of people—including artists, students, and movie and fashion people, as well as tourists from all over the world—who pack the central spaces.

The major employers in modern Rome are the building, tourism, and movie industries, and the government. While not an industrial city of international or national significance, Rome does have quite a broad industrial base in which the most prominent activities are engineering, electronics, chemicals, printing, clothing, and food processing. The modern city's growth has been reflected by an increase in population (mostly migrants from poorer regions of southern Italy) from 1.7 million in 1960 to 2.7 million in 2000.

The city's extensive historic fabric has made it difficult to cope with this growth. The most immediate problems are traffic congestion and pollution. Traffic fumes and vibrations are so bad that the city's monuments have deteriorated rapidly in the past 25 years. These, however, represent only part of the problem for Rome's inhabitants. Acute housing shortages, rampant land speculation, uncontrolled building, deteriorating slums, inadequate infrastructure, and overstretched social services are endemic, adding up to a city that is sharply polarized and acutely distressed. Although Rome has had a long history of urban planning, the modern planning system is complex, slow-moving, and ineffective. Plans can take so long to be adopted that the growth they are intended to regulate has already taken place, while patronage and influence have resulted in thousands of exceptions to planning regulations.

A Day in the Life

Daniela Stefani

Daniela Stefani has only two luxuries in life: a white card and a color television. The former grants her permission to work for a living, while the latter allows her to escape the drudgeries of life.

At the age of 45, after laboring for more than half her life, Daniela has little to show for her troubles. For 20 years she worked as a music teacher in Korqd, her native town in Albania, but now she makes more money cleaning people's houses in Greece. Like many of her compatriots, she fled across the border in pursuit of pecuniary improvement and peace.

Of course there are disadvantages to living and working in Athens, and sacrifices have to be made constantly. The worst is the discrimination and racism that she encounters at work and on the streets. Daniela doesn't complain; she is used to making sacrifices. She has been making them since she was born, the citizen of a poverty-stricken, unstable country.

Before moving to Greece three years ago, she and her husband shared a small house with her brother and his family. The couple's combined salary barely kept them in essentials. Along with their two children Kristos, aged nine, and Alda, aged six, they now live in a 20-square-meter (215-square-foot) windowless room in the southern suburbs of the Greek capital. Apart from the obvious logistical problems of crowding four people (it was five before her mother died last year) into

such a tight space, they have to endure humidity in the winter, oppressive heat in the summer, and unfriendly neighbors all year round.

Her husband is out of work and joins the groups of other Albanians forlornly loitering on street corners in the hope of being hired for the day. He does not help in the house, although he will look after the children when they return from school. Daniela is thankful for small mercies.

She sits in her dark, dank room, fingering her cigarette, her thinning hair and lined face testimony to an eternity of hardships. Kristos lies beside her, curled on the sofa that doubles as a bed. He is a star pupil "with an excellent character," she assures me, glancing down at him. "He has a good ear and should study music," she adds wistfully.

He's like her, I realize: sensitive, easily hurt, and somewhat withdrawn, as if he has understood at such a young age what a battle life will be. Alda, on the other hand, is the opposite: energetic, outspoken, and undaunted. She wants to be a ballerina. She has manufactured her own sparkling world of spotlights and tutus, and Daniela does not want to disillusion her yet.

Source: Letter from Athens: "The luxury of dreams," by Lucie Warrillow. *The Guardian,* February 24, 1999. Available at http://www.guardianunlimited.co.uk/Archive/Article/0,4273,3826945,00.html

Summary and Conclusions

Contemporary Europe is highly urbanized and is a cornerstone of the world economy with a complex, multilayered, and multifaceted regional geography. In overall terms, Europe accounts for almost two-fifths of world trade and about one-third of the world's aggregate GNP.

The rise of Europe as a major world region had its origins in the emergence of a system of merchant capitalism in the fifteenth century, when advances in business practices, technology, and navigation made it possible for the merchants of Europe to establish the basis of a worldwide economy in the space of less than 100 years. These changes also had a profound effect on the geography of Europe itself, reorienting the region toward the Atlantic and away from the subregional maritime economies of the Mediterranean and the Baltic. Since then, Europe's regional geographies have been comprehensively recast several times: by the new production and transportation technologies that marked the onset of the Industrial Revolution, by two world wars, and by the Cold War rift between eastern and western Europe.

The European Union emerged after the Second World War as a major factor in reestablishing Europe's role in the

world. The EU is now a sophisticated and powerful institution with a pervasive influence on patterns of economic and social well-being within its member states. It has a population of more than 370 million, with a combined GDP 10 percent larger than that of the United States. The reintegration of eastern Europe has added a potentially dynamic market of 344 million consumers to the European economy. Overall, eastern Europe functions as a set of economically peripheral regions, with agriculture still geared to local markets and former COMECON trading opportunities, and industry still geared more to heavy industry and standardized products than to competitive consumer products.

The principal core region within Europe is the Golden Triangle, which stretches between London, Paris, and Berlin. A secondary, emergent, core is developing along a north-south crescent that straddles the Alps, stretching from Frankfurt, just to the south of the Golden Triangle, through Stuttgart, Zürich, and Munich, to Milan and Turin.

Beyond Europe's core regions, major metropolitan areas, and specialized industrial districts, a mosaic of different landscapes has developed around Europe's broad physiographic regions. In detail, these landscapes are a product of centuries of human adaptation to climate, soils, altitude, and aspect, and to changing economic and political circumstances. Farming practices, field patterns, settlement types, local architecture, and ways of life have all become attuned to the opportunities and constraints of regional physical environments, with the result that distinctive regional landscapes have been produced.

Key Terms

acid rain (p. 111)
agglomeration economies
  (p. 112)
balkanization (p. 119)
buffer zone (p. 112)
canton (p. 119)
central business district
  (CBD) (p. 131)
command economy (p. 112)
deindustrialization (p. 122)

disinvestment (p. 122)
enclave (p. 119)
Enlightenment (p. 116)
entrepôt (p. 110)
ethnic cleansing (p. 120)
exclave (p. 119)
feudal system (p. 104)
fjord (p. 102)
flexible production region
  (p. 135)

latifundia (p. 145)
loess (p. 142)
Marshall Plan (p. 112)
minifundia (p. 145)
Modernity (p. 116)
moraine (p. 136)
pastoralism (p. 139)
physiographic region
  (p. 101)
polder (p. 104)

satellite state (p. 100)
state socialism (p. 100)
steppe (p. 143)
swidden (p. 103)
world city (p. 129)
xenophobia (p. 120)

Review Questions

Testing Your Understanding

1. How has Europe benefited from its location and its major physical features?
2. Many European cultures have a strong history of seafaring. How did that become a crucial factor in European and world geography?
3. As the Roman Empire spread westward, what modifications did the ancient Romans make to the European landscape? How did people living in medieval feudal systems affect the landscape of Europe?
4. What key inventions during the period from 1400 to 1600 helped European merchants establish the basis of today's global economy? Why?
5. Which imports from the American colonies helped transform Europe? Focus on natural resources and new crops.
6. What was an entrepôt seaport? How have entrepôt functions affected a city like London?
7. What factors led to the end of the European colonial era?
8. What was the Cold War? Why did the Soviet Union establish a buffer zone in eastern Europe? Which two eastern European countries left the Soviet bloc to pursue alternate forms of state socialism?
9. How did the European Union (EU) develop? Why is the EU's Common Agricultural Policy (CAP) so important? What are the three unwanted side effects of CAP implementation?

10. What are the main characteristics of Europe's two core regions? Please explain where these core regions are and why they are prosperous.

Thinking Geographically

1. Why was geographic knowledge and accurate mapmaking crucial to the growth of European power during the 1500s?
2. How did aid from the Marshall Plan and COMECON help rebuild Europe after the Second World War? Which regions or economies benefited first?
3. What migration patterns characterized Europe during the nineteenth and twentieth centuries? Consider movement within Europe as well as movement to and from Europe.
4. From Greece to Portugal, many European countries pride themselves on their wine production and rely on income from their wine exports. How did the history of wine as a commodity reflect the history of Europe from classical Greece to the present day?
5. Summarize and discuss the ongoing situation in Kosovo. Is Kosovo an independent country? While ethnic Albanians suffered greatly during the late 1990s, are they justified in retaliating with violence against ethnic Serbs today?

Further Reading

Carter, F. W., and Norris, H. T. (eds.), *The Changing Shape of the Balkans*. Boulder: Westview Press, 1996.

Champion, A., Monnesland, J., and Vandermotten, C., "The New Regional Map of Europe." *Progress in Planning*, 46(1996), 1–89.

Foster, J., *Docklands: Cultures in Conflict, Worlds in Collision*. Philadelphia: UCL Press, 1999.

Gibb, R., and Wise, M., *The European Union: Challenges of Economic and Social Cohesion*. London: Arnold, 2000.

Gowland, D., O'Neill, B., and Reid, A. (eds.), *The European Mosaic. Contemporary Politics, Economics, and Culture*, 2nd ed. London: Longman, 1999.

Graham, B., (ed.), *Modern Europe: Place, Culture, Identity*. London: Arnold, 1998.

Grove, A. T., and Rackham, O., *The Nature of Mediterranean Europe: An Ecological History*. New Haven: Yale University Press, 2001.

Hall, D., and Danta, D. (eds.), *Reconstructing the Balkans*. New York: John Wiley & Sons, 1996.

Hebbert, M. *London*. Chichester: John Wiley & Sons, 1998.

King, R., Proudfoot, L., and Smith, B. (eds.), *The Mediterranean: Environment and Society*. London: Arnold, 1997.

Livi-Bacci, M., *The Population of Europe*. Trans. by Carl Ipsen. Oxford: Blackwell, 2000.

McDonald, J. R., *The European Scene: A Geographic Perspective*, 2nd ed. Upper Saddle River, NJ: Prentice Hall, 1997.

Noin, D., and White, P., *Paris*. Chichester: John Wiley & Sons, 1997.

Petrakos, G., *Integration and Transition in Europe*. New York: Routledge, 2000.

Piening, C., *Global Europe: The European Union in World Affairs*. Boulder: Lynne Rienner Publishers, 1997.

Pinder, D. (ed.), *The New Europe: Economy, Society and Environment*. New York: John Wiley & Sons, 1998.

Tickle, A., and Welsh, I. (eds.), *Environment and Society in Eastern Europe*. London: Addison Wesley Longman, 1998.

Tilly, C., "The Geography of European Statemaking and Capitalism Since 1500." In E. D. Genovese and L. Hochberg (eds.), *Geographic Perspectives in History* (pp. 158–81). Oxford: Blackwell, 1989).

Townsend, A. R., *Making a Living in Europe: Geographies of Economic Change*. New York: Routledge, 1997.

Turnock, D. (ed.) *Eastern Europe and the Former Soviet Union*. London: Arnold, 2000.

Unwin, T., *Wine and the Vine*. New York: Routledge, 1996.

Unwin, T. (ed.), *A European Geography*. London: Longman, 1998.

Film, Music, and Popular Literature

Film

The Back Roads of Europe. PBS Home Video series, 1998. A three-tape travelogue through rural Europe.

Bosnia: Peace Without Honor. BBC Production, 1999. This documentary traces the roots of the Bosnian conflict through the efforts of U.S. and British diplomats.

Diamonds in the Dark. Directed by Olivia Carrescia, 1999. This documentary tells the stories of 10 Romanian women: how they lived under the old regime and how they are confronting the new problems of the post-communist era.

The Grand Canal. PBS Home Video Great Streets series, 2000. An intimate view of Venice, Italy, focusing on the Grand Canal.

In Search of Tuscany with John Guerrasio. PBS Home Video series, 1999. A tour of Tuscany with food critic and author John Guerrasio.

Jonah, Who Will Be 25 in the Year 2000 (Jonas qui aura 25 ans en l'an 2000). Directed by Alain Tanner, 1976. This movie captures the liberal aspirations of the European baby-boom population who had become 30-something by the mid-1970s.

One World: The Baltic States. PBS documentary by Ward Television Corporation, 1997. A documentary on the post-Soviet Baltic States, presenting multiple views on current issues, the Soviet heritage, ethnic minorities, economic development.

Rome: Power and Glory. PBS Home Video series, 1997. A six-part documentary on the Roman Empire.

Music

Bocelli, Andrea. *Sogno*. Philips, 1999.

Bulgarian Women's Choir. *Voices of Life*. Globe Music Media Arts, 2000.

Chao, Manu. *Clandestino*. Uni/Ark, 1998.

Chumbawamba. *Tubthumper*. Universal Records, 1997.

Hardy, Francoise. *Ma Jeunesse Fout Le Camp*. Virgin, 2000.

Lien, Annbjorg. *Baba Yaga*. Northside, 2000.

Madredeus. *Antologia*. Blue Note, 2000.

Various Artists. *Paris Is Sleeping: Respect Is Burning*. Caroline, 1998.

Vasen. *Whirled*. Northside, 1997.

Popular Literature

Bryson, Bill. *Notes from a Small Island*. New York: William Morrow, 1996. Britain and the British closely observed by an American writer who lived in Britain for 20 years.

Carhart, Thaddeus. *The Piano Shop on the Left Bank: Discovering a Forgotten Passion in a Paris Atelier*. New York: Random House, 2001. A novel set in present-day Paris, describing an American's search for the ideal piano for his small apartment.

Davies, Norman. *Europe: A History*. New York: HarperCollins, 1998. A well-written and comprehensive history of Europe.

Doyle, Roddy. *The Barrytown Trilogy. (The Commitments; The Snapper; The Van.)* New York: Penguin USA, 1995. Three novels that depict working-class life in contemporary Dublin with humor and compassion. *The Commitments* was made into a hit movie (directed by Alan Parker, 1999).

Epstein, Alan. *As the Romans Do: The Delights, Dramas, and Daily Diversions of Life in the Eternal City*. New York: William Morrow, 2000. An evocative book on the people, places, and everyday life of Rome.

Kaplan, Robert D. *Balkan Ghosts: A Journey Through History*. New York: Vintage, 1994. A political travelogue that deciphers the Balkan landscapes and the people's ancient passions and intractable hatreds for outsiders.

Mayle, Peter. *Encore Provence*. New York: Vintage Books, 2000. A vivid and intimate description of one of France's most popular regions.

Mazower, Mark. *Dark Continent: Europe's Twentieth Century*. New York: Vintage Books, 2000. Explores the conflicts that dominated Europe in the twentieth century and the social value systems that informed them.

Rackham, Oliver. *The Illustrated History of the Countryside*. New York: Sterling Publishing, 1994. An historical ecology of the British countryside, showing how everyday landscapes reflect past activities.

4 The Russian Federation, Central Asia, and the Transcaucasus

Figure 4.1

N

ARCTIC
OCEAN

Bering Strait

Providyeniya

170°W

70°N

60°N

180°

Kolyma R.

170°E

80°N

50°N

Verkhoyansk

Lena R.

160°E

Sea of Okhotsk

RUSSIAN FEDERATION

PACIFIC
OCEAN

Lena R.

150°E

Bratsk

Lake
Baykal

YABLONOVYY MTS.

Amur R.

40°N

Krasnoyarsk

KHINGAN MTS.

Novokuznetsk

Angarsk   Irkutsk

SAYAN MTS.

Vladivostok

MOUNTAINS

30°N

◉   More than 5 million

○   1–5 million

•   Fewer than 1 million

Capital cities are underlined

0   250   500 Miles

0   250   500 Kilometers

130°E

The Russian Federation (the principal successor state to the Soviet Union), together with neighboring Belarus and the former Soviet satellite states in Central Asia (Kazakhstan, Kyrgyzstan, Tajikistan, Turkmenistan, and Uzbekistan) and the Transcaucasus (Armenia, Azerbaijan, and Georgia), constitute a vast world region (**Figure 4.1**). The Russian Federation alone stretches across 11 time zones, from St. Petersburg (just across the Gulf of Finland from Helsinki) in the west to Provideniya (just across the Bering Strait from Nome, Alaska) in the east. Altogether the Russian Federation amounts to 17,075,400 square kilometers (6,591,100 square miles)—roughly twice the size of the United States. The region as a whole amounts to a landmass of 21,024,210 square kilometers (8,115,340 square miles). It is the most sparsely settled of the major world regions, with a population in 2000 of 232 million, 69 percent of whom live in urban areas, 43 of which have populations that exceed 500,000. The region is bounded to the north by the icy seas of the Arctic and to the south by a mountain wall that stretches from the Elburz Mountains of northern Iran through the Pamir Mountains and Tien Shan ("Mountains of Heaven") along the southern borders of the Central Asian countries, to the Altay and Sayan Mountains that separate Siberia from Mongolia, and the Yablonovyy and Khingan ranges that separate southeastern Siberia from northern China. The boundary to the east is the Pacific Ocean, while to the west, as we saw in Chapter 3, the boundary is political rather than physical, the former Soviet republics in the Baltic and the former satellite states of eastern Europe having returned to their European orientation. The vast area covered by this world region is relatively sparsely settled, mainly because of the harsh climate, poor soils, and difficult terrain that characterize much of the region. The bulk of the population, in fact, is concentrated in the southern parts of European Russia, to the west of the Ural Mountains, which constitutes the heartland of the old Russian empire.

Environment and Society in the Russian Federation, Central Asia, and the Transcaucasus

A satellite photograph of the territory (**Figure 4.2**) suggests several of the fundamental attributes of this vast world region. First, its sheer size: more than 10,000 kilometers (6200 miles) east–west and more than 2500 kilometers (1550 miles) north–south at its broadest. It takes a full week to traverse the region by train from Vladivostok in the east to St. Petersburg in the west. Second, its northerliness: Nearly half of the territory of the Russian Federation is north of 60° N, while Moscow is approximately the same latitude as Juneau, Alaska, and Tbilisi, Georgia—one of the southernmost cities of the region—is approximately the same latitude as Chicago (42° N). A third important attribute suggested by Figure 4.2 is how restricted is the region's access to the world's seas. The northerliness of most of the region means that most ports are icebound dur-

Figure 4.2   The Russian Federation, Central Asia, and the Transcaucasus from space   This image underlines one of the key features of the Russian Federation, Central Asia, and the Transcaucasus: its sheer size. [*Source:* (inset) Map projection, Buckminster Fuller Institute and Dymaxion Map Design, Santa Barbara, CA. The word *Dymaxion* and the Fuller Projection Dymaxion™ Map design are trademarks of the Buckminster Fuller Institute, Santa Barbara, California, © 1938, 1967 & 1992. All rights reserved.]

THE RUSSIAN FEDERATION, CENTRAL ASIA, AND THE TRANSCAUCASUS

ing the long winter. Murmansk, on the Kola Peninsula in the far north, is a major exception. It benefits from its location near the tail end of the warm Gulf Stream and is open year-round. Some ports, such as Vladivostok, on the Sea of Japan in the far east, can be kept open by icebreakers. At least the Russian Federation has some warm-water ports (including Kaliningrad, a small province on the Baltic between Poland and Lithuania, retained as an exclave by the Russian Federation because of its important naval port). All of the other countries in the region are landlocked, with the exception of Georgia, which has access to international sea lanes by way of its Black Sea ports.

The Black Sea itself is an inland sea, connected to the Aegean Sea and the Mediterranean by way of the Bosporus, a narrow strait, the Sea of Marmara, and then another narrow strait, the Dardanelles. The many rivers that empty into the Black Sea give its surface waters a low salinity, but it is almost tideless, and below about 80 fathoms it is stagnant and lifeless. The Caspian Sea is the largest inland sea in the world, at 371,000 square kilometers (143,205 square miles—roughly the size of Germany). With the Black Sea and the Aral Sea, it once formed part of a much greater inland sea. Though perhaps dwarfed by the vastness of the region, there are several inland lakes of significant size, including Lake Balkhash (17,400 square kilometers; 6715 square miles) and Lake Baykal (30,500 square kilometers; 11,775 square miles), which, with a depth of 1615 meters (5300 feet), is the deepest lake in the world.

The northerliness of the region limits the use of its rivers for navigation and for the generation of hydroelectric power. Many rivers are frozen for much of the year, while the mouths of some remain frozen through the spring, causing backed-up meltwater to flood extensive areas of wetlands. In May 2001, when the worst floods in a century caused tens of thousands of people along the Lena River to become homeless, SU-24 supersonic bombers had to be deployed to break up a 30-kilometer (18-mile) ice floe on the river with 250-kilogram (550-pound) bombs. Nevertheless, the sheer size of the territory sustains several rivers of considerable size, all of which were historically important as transport routes, allowing for conquest, colonization, and trade. It takes some of the longest rivers on Earth to drain the huge Siberian landmass. Rising in the southern mountains of Central Asia, the Lena, Kolyma, Ob', and Yenisey flow north to the Arctic Ocean, while the Amur flows north to the Pacific. In the western part of the Russian Federation, the most important rivers flow south from the Central Region occupied by Moscow and Nizhniy Novgorod. The Dnieper flows south to the Black Sea, the Don to the Sea of Azov (which in turn connects to the Black Sea), and the Volga to the Caspian Sea. In the modern period, the Volga has become particularly important as a navigable waterway and source of hydropower, with huge reservoirs built during the Soviet era to regulate the flow of the river, conserving spring floodwaters for the dry summer months.

Landforms and Landscapes

The most striking feature of the entire region is the monotony of its plains over thousands and thousands of kilometers. The reason for this monotony lies in the geological structure of the region. Two large and geologically ancient and stable shields of highly resistant crystalline rocks provide platforms for extensive plains of sedimentary material and glacial debris. In the west is the first shield, the Russian Plain, an extension of the Central European Plain that extends from Belarus in the west to the Ural Mountains in the east, and from the Kola Peninsula in the north to the Black Sea in the south (**Figure 4.3**). East of the Urals is a second shield that extends as far as the Lena River. It is so vast that it is conventionally divided into several physiographic subregions: the West Siberian Plain, the Central Siberian Plateau, and the plateaus of Central Asia. The Urals themselves represent a third distinctive physiographic region, and a fourth is the mountain wall that runs along the southern and eastern margins of the two shields.

The Russian Plain (**Figure 4.4**) has a gently rolling topography, the hard crystalline rock shield providing a flat platform that is covered by several meters or more of sedimentary deposits. Much of the Russian Plain is poorly drained, boggy, and marshy, though the major rivers that drain the Russian Plain—the Dnieper, the Don, and the Volga—have eroded the sedimentary layer in places, resulting in more varied and attractive topography. The West Siberian Plain is even flatter and contains still more extensive wetlands and tens of thousands of small lakes. Poorly drained by the slow-moving Ob' and Irtysh rivers (**Figure 4.5**), the West Siberian Plain (**Figure 4.6**) is mostly inhospitable for settlement and agriculture, though it contains significant oil and natural gas reserves. The West Siberian Plain is distinctive for its absolute flatness: across the whole broad expanse—more than 1800 kilometers (1116 miles) in each direction—relief is no more than 400 meters (1312 feet). The monotony of the landscape is captured in this quote from Russian writer A. Bitov, describing a train journey:

> Once I was travelling through the Western Siberian Lowlands. I woke up and glanced out of the window—sparse woods, a swamp, level terrain. A cow standing knee-deep in the swamp and chewing, levelly moving her jaw. I fell asleep, woke up—sparse woods, a swamp, a cow chewing, knee deep. I woke up the second day—a swamp, a cow. . . ."[1]

The Yenisey River marks the eastern boundary of this flat condition and the beginning of the Central Siberian Plateau, where the rock shield, having been uplifted by geological movements, averages about 700 meters (2297 feet) in elevation. Stretching between 800 and 1900 kilometers (496 to 1178 miles) west–east, the Central Siberian Plateau (**Figure 4.7**) has been dissected by rivers into a hilly upland topography with occasional deep river gorges.

The Urals (**Figure 4.8**) consist of a once-great mountain range of ancient rocks that have been worn down over the ages. For the most part only 600 to 700 meters (1969 to

---

[1]A. Bitov, *A Captive of the Caucasus.* Cambridge: Cambridge University Press, 1993, p. 50. Quoted in A. Novikov, "Between Space and Race: Rediscovering Russian Cultural Geography," in M. J. Bradshaw (ed.), *Geography and Transition in the Post-Soviet Republics,* Chichester: John Wiley & Sons, 1997, p. 45.

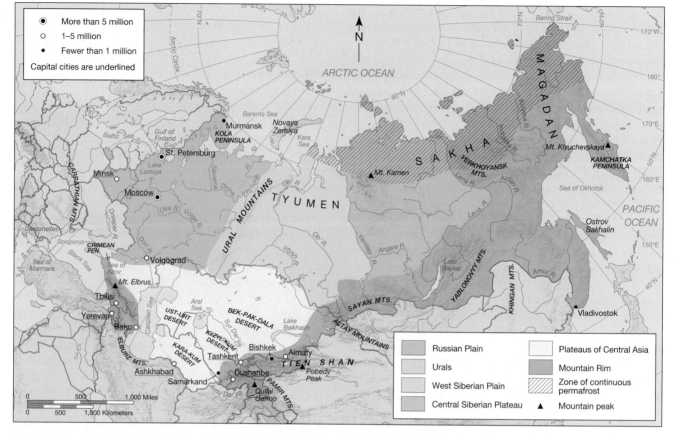

Figure 4.3   Physiographic regions of the Russian Federation, Central Asia, and the Transcaucasus The framework for the physical geography of the region consists of two stable shields on either side of the ancient mountains of the Urals, with a wall of young mountains that runs along the southern and eastern margins of the shields.

Figure 4.4   The Russian Plain The Russian Plain is an extension of the Central European Plain. It is one of the most densely settled parts of the Russian Federation, with some of the best farmland.

Figure 4.5   River Ob' After spring thaws, the boggy plains of Siberia drain into its great rivers, which move slowly toward the Arctic Ocean. In winter most Siberian rivers are frozen, and the mouths of some remain frozen through the spring, causing backed-up meltwater to flood extensive areas of bogs and marshlands. Shown here is the River Ob' near Novosibirsk, in summer.

2297 feet) above sea level, and only in a few places rising above 2000 meters (6562 feet) in elevation, the Urals are penetrated by several broad valleys, and so they do not constitute a major barrier to transport. The Urals stretch for more than

3000 kilometers (1864 miles) from the northern frontier of Kazakhstan to the Arctic coast of the Russian Federation, the range reappearing across the Kara Sea in the form of the islands of Novaya Zemlya. The rocks of the Urals are heavily miner-

Figure 4.6   The West Siberian Plain This photograph of marshland near Primorye, western Siberia, shows very clearly the difficult, boggy conditions that prevail in much of western Siberia in summer.

Figure 4.8   Urals landscape The mountains of the Urals are ancient and have been worn down to rounded landforms with thin soils that support small farms. Shown here is a small farming village near Kungur.

Figure 4.7   The Central Siberian Plateau The rock shield of the Central Siberian Plateau has been uplifted by geological movements, with the result that the land has been dissected by rivers into a hilly upland with occasional deep river gorges.

Figure 4.9   The Pamir Mountains The Pamirs stand at the western end of the Mountain Rim, a region of high plateaus, peaks, and glaciers that forms the northwestern limb of the Himalayas.

alized and contain significant quantities of chromite, copper, gold, graphite, iron ore, nickel, titanium, tungsten, and vanadium. As a result, a number of significant industrial cities, including Chelyabinsk, Magnitogorsk, Perm', Ufa, and Yekaterinburg have developed in the Urals, together forming a major industrial region (see p. 187).

The mountain wall that runs along the southern and eastern margins of the two stable shields on either side of the Urals is the product of geological instability. Younger, sedimentary rocks have been pushed up against the older and more stable shields in successive episodes of mountain-building, forming a series of mountain ranges of varying height, composition, and complexity. The highest ranges are those of the Caucasus (where Mt. Elbrus reaches 5642 meters, or 18,510 feet) and the Pamirs (**Figure 4.9**) and Tien Shan ranges along the borders with Iran, Afghanistan, and China, where many peaks reach 5000 to 6000 meters, and two—Pobedy and Qullai Garmo (formerly Communism Peak)—exceed 7400 meters

(24,278 feet). In the far east, the ranges of the Kamchatka Peninsula contain numerous active volcanoes, including Mt. Klyuchevskaya (4750 meters; 15,584 feet) and Mt. Kamen (4632 meters; 15,197 feet). Only in the western part of Central Asia does the mountain wall fall outside the region, running along the southern side of Turkmenistan's border with Iran and Afghanistan. North of this border, extending through Uzbekistan into Kazakhstan and beyond, is a huge geosyncline, a geological depression of sedimentary rocks. This syncline is of special importance as a source of energy resources, with oil reserves equivalent to between 15 and 31 billion barrels—about 2.7 percent of the world's proven reserves—plus significant deposits of coal and about 7 percent of the world's proven reserves of natural gas.

Climate

The northerliness and vast size of the region exert strong influences on its climate. The absence of mountainous terrain, except in the far south and east, and the lack of any significant moderating influence of oceans and seas means that the prevailing climatic pattern is relatively simple. The region is dominated by a severe continental climate, with long, cold winters and relatively short, warm summers. The cold winters become colder eastward, as one moves away from the weak marine influence that carries over from the westerly weather systems that cross Europe from the Atlantic. Pronounced high-pressure systems develop over Siberia in winter, bringing clear skies and calm air. Average January temperatures in Verkhoyansk, a mining center in the middle of this high-pressure area, are in the region of –50°C (–58°F). The northerliness of the region, with its long and intense winters, means that the subsoil is permanently frozen—a condition known as **permafrost**—in more than two-thirds of the Russian Federation. In the extreme northeast, winter conditions can last for 10 months of the year.

Summer comes quickly over most of Belarus and the Russian Federation, spring being a brief interlude of dirty snow and much mud. Because many rural roads remain unpaved, they are typically impassable for a period in the spring, before the summer heat bakes the mud. As the landmass warms, low-pressure systems develop, drawing in moist air across the western Russian Federation from Atlantic Europe and resulting in moderate summer rains. In late summer, the Chinese **monsoon** brings heavy rains to the southeastern corner of the far east. Across much of Siberia, though, summer rainfall is quite low. The summers become hotter southward, and drought is a frequent problem in the southwestern and southern parts of the Russian Federation. In Central Asia, aridity is a severe problem, with desert and semidesert covering much of Kazakhstan, Uzbekistan, and Turkmenistan.

In the Transcaucasus, climatic patterns are distinctive, mainly as a result of the presence of massive mountain ranges to the north and south and substantial bodies of water to both east and west. A lot of precipitation falls on the windward side of the mountains, though the Transcaucasus is also influenced by the warm, dry air masses that originate over the deserts of Central Asia. The most distinctive feature of climatic patterns in the Transcaucasus, however, is the subtropical niche of western Georgia, on the shores of the Black Sea—a unique and striking feature in a region of otherwise severe climatic regimes.

Environmental History

The natural landscapes of the Russian Federation, Central Asia, and the Transcaucasus follow a strikingly straightforward pattern of seven long, latitudinal zones that run roughly from west to east. These zones are very closely related to climate, glacial geomorphology, and soil type and remain easily recognizable to the modern traveler, despite centuries (or, in places, millennia) of human interference and modification. The northernmost zone is that of the tundra, which fringes the entire Arctic Ocean coastline and part of the Pacific (see Figure 4.3). The **tundra** (**Figure 4.10**) is an arctic wilderness where the climate precludes any agriculture or forestry. Permafrost and very short summers mean that the natural vegetation consists of mosses, lichens, and certain hardy grasses. The tundra is not wholly hostile to life, however. It supports reindeer on its lichen, waterfowl on its many summer swamps and pools, and

Figure 4.10 Tundra landscape The tundra landscape is bleak, with sparse vegetation and a surface strewn with rocks that in places have been arranged into geometric patterns through frost-heave action. This photograph was taken near the Kolyma River Delta in Siberia.

fish and walruses in its neighboring seas—resources that have been exploited for centuries by indigenous peoples and, more recently, Russians.

South of the tundra is the most extensive zone of all—a belt of coniferous forest known as the *taiga*. The term **taiga** originally referred to trackless or virgin forest, though it is now used to describe the entire zone of boreal coniferous forest (spuce, fir, and pines, for example) that stretches from the Gulf of Finland to the Kamchatka Peninsula. Spruce is the dominant tree in the west, while larch trees do better on the poorly drained soils of Siberia and the far east (**Figure 4.11**). The indigenous inhabitants of the taiga were hunters and gatherers, not farmers. Where the forest is cleared, some cultivation of hardy crops such as potatoes, beets, and cabbage is possible, but the poor, swampy soils and short growing season make agriculture chancy. More recently the taiga has become commercially important for the fur-bearing animals whose luxuriant pelts are well adapted to the bitter cold, and for the forest itself, whose timber is now methodically exploited and exported.

The next zone is a continuation of the mixed forests of central Europe that extend through Belarus and into the Russian Federation as far as the Urals, with discontinuous patches in Siberia and the far east. Here, firs, pines, and larches are mixed with stands of birch and oak. It quickly shades into another relatively narrow zone, of wooded steppe. In this zone the grasslands of the steppe are interspersed with less extensive stands of mixed woodland, mostly in valley bottoms. Both the mixed forest and the wooded steppe were cleared and cultivated early in Russian history, providing both an agricultural heartland for the emerging Russian empire and a corridor along which Russian traders and colonists pushed eastward in the sixteenth and seventeenth centuries—through the middle Volga region to the southern Urals and eventually to the Pacific coast via the mixed forests of the Amur valley.

Figure 4.11   Taiga landscape The taiga is a zone of boreal coniferous forest that stretches from the Gulf of Finland to the Kamchatka Peninsula. In much of the Siberian taiga, as in the example shown here, the dominant tree species is the larch.

Figure 4.12   Steppe landscape This typical steppe landscape is from the southern part of the West Siberian Plain, near the Kazakhstan border.

The wooded steppe, in turn, quickly shades into the **steppe** proper: flat, treeless, and dominated by tall and luxuriant feather-grass, whose matted roots are able to trap whatever moisture as is available in this rather arid region (**Figure 4.12**). The accumulated and decayed debris of these grasses has produced a rich dark soil, known as black earth, or **chernozem.** These soils, along with related brown and chestnut soils, have high natural fertility, but when they are plowed they are vulnerable to the aridity of the region and can easily degenerate into wind-driven dustbowl conditions.

South of the steppe are zones of semidesert and desert. They are largely a feature of Central Asia, and they continue south of Siberia into Chinese and Mongolian territory. The semidesert is characterized by boulder-strewn wastes and saltpans (areas where salt has been deposited as water evaporated from short-lived lakes and ponds created by runoff from surrounding hills) and patches of rough vegetation used by nomadic pastoralists. The desert proper is characterized by bare rock and extensive sand dunes, though there are occasional oases and fertile river valleys. Finally, there is a relatively tiny area of subtropical forest in western Georgia, adjoining the Black Sea coast.

The Russian Federation, Central Asia, and the Transcaucasus in the World

It should be clear from the previous section that huge tracts of this world region are decidedly marginal. Agriculture and settlement have been greatly restricted by severe climatic conditions, highly acidic soils, poor drainage, and mountainous terrain. Even in the zone of rich chernozem soils, low and irregular rainfall rendered agriculture and settlement marginal until large-scale irrigation schemes were introduced in the twentieth century. Only in the mixed forest and the wooded steppe west of the Urals were conditions suitable for the emergence of a more prosperous and densely settled population. It was this area, in fact—from Smolensk in the west to Nizhniy Novgorod in the east, and from Tula in the south to Vologda and Velikiy Ustyug in the north—that was the Russian "homeland" that developed around the principality of Muscovy from late medieval times.

Muscovy and the Russian Empire

In the mid-fifteenth century, Muscovy was a principality of approximately 5790 square kilometers (2235 square miles) centered on the city of Moscow. Over a 400-year period, the Muscovite state expanded at a rate of about 135 square kilometers (52 square miles) per day so that by 1914, on the eve of the Russian Revolution, the empire occupied more than 22 million square kilometers (roughly 8.5 million square miles), or one-seventh of the land surface of Earth (**Figure 4.13**). At first, Muscovy formed part of the Mongol-Tatar empire whose

armies were known as the Golden Horde, and Russian princes were obliged to pay homage to the Khan, the leader of the Golden Horde. In 1552, under Ivan the Terrible, the Muscovites defeated the Tatars at the battle of Kazan'—a victory that prompted the commissioning of the construction of St. Basil's Cathedral in Moscow.

Desirous of more forest resources—especially furs—Muscovy expanded into Siberia (see Geographies of Indulgence, Desire and Addiction: Furs, p. 162). Gradually, more and more territory was colonized. By the mid-seventeenth century, the eastern and central parts of Ukraine had been wrested from Poland. The steppe regions, though, remained very much a frontier region of the Russian empire because of the constant threat of attack by nomads. Early in the eighteenth century, Peter the Great (1682–1725) founded St. Petersburg and developed it as the planned capital of Russia. Beyond the wealth and grandeur of a few cities, however, the Russian empire was very much a rural, peasant economy. In the latter part of the eighteenth century, under Catherine the Great (1762–1796), Russia secured the territory of what would eventually become southern Latvia, Lithuania, Belarus, and western Ukraine. Then, with the defeat of the Crimean Tatars in the late eighteenth century, the steppes were opened to colonization by Russians and

Figure 4.13   Territorial growth of the Muscovite/Russian state The Muscovite empire was vast and was conquered over the same period (fifteenth century to the late twentieth) that corresponds to the globalization of the world economy. What makes the Russian case different is that the lands conquered were adjacent ones and not overseas. When the Bolsheviks came to power at the beginning of the twentieth century, some of the territory was lost. Eventually, however, the Bolsheviks were able to control most of the territories formerly held by the tsars, and it was upon this that they also built the Soviet state. (*Source:* Redrawn from D. J. B. Shaw, *Russia in the Modern World.* Oxford: Blackwell, 1999, p. 7.)

by ethnic and religious minorities—including Mennonites and Hutterites—from the Russian heartland. It was during this period that Russia ousted the Ottoman Turks from the Crimean Peninsula and gained the warm-water port city of Odessa on the Black Sea.

Russia's imperial expansion followed the same impulses as other European empires. The factors behind expansion were the drive for more territorial resources (especially a warm-water port) and additional subjects. Different for Russia, however, was that vast stretches of adjacent land on the Eurasian continent were annexed, whereas other empires established new territories overseas.

The final phases of expansion of the Russian state occurred in the late eighteenth and nineteenth centuries. Finland was acquired from Sweden in 1809 and given the status of Grand Duchy. In the Transcaucasus, Georgians and Armenians were "rescued" from the Turks and Persians. In Central Asia, the Moslem Khanates fell one by one under Russian control: the city of Tashkent in 1865, the city of Samarkand in 1868, the Emirate of Bukhara in 1868, and the Khanate of Khiva in 1873. Meanwhile, in the Far East, the weakening of the Manchu dynasty, which had ruled China since 1664, prompted the Russian annexation of Chinese territory, where colonization and settlement was aided by the construction of the Trans-Siberian Railroad in the final years of the nineteenth century. By 1904, when defeat in Manchuria by Japan brought a halt to Russian territorial expansion, the Russian empire contained about 130 million persons, only 56 million of whom were Russian. Of the rest, which included more than 170 distinct ethnic groups, some 23 million were Ukrainian, 6 million were Belorussian, more than 4 million were Kazakh or Kyrgyz, nearly 4 million were Jews, and nearly 3 million were Uzbek.

To meet the challenge of different ethnicities under one state, Russia needed to apply binding policies and practices. Russia's strategies to bind together the 100-plus "nationalities" (non-Russian ethnic peoples) into a unified Russian state were oftentimes punitive and not at all successful. Non-Russian nations were simply expected to conform to Russian cultural norms. Those that did not were more or less persecuted. The result was opposition and, sometimes, rebellion and stubborn refusal to bow to Russian cultural dominance.

Meanwhile, ever since the time of Peter the Great, tsarist Russia had been seeking to modernize. By 1861, when Tsar Alexander II decreed the abolition of serfdom, Russia had built up an internal core with a large bureaucracy, a substantial intelligentsia, and a sizable group of skilled workers. The abolition of feudal serfdom was designed to accelerate the industrialization of the economy by compelling the peasantry to raise crops on a commercial basis, the idea being that the profits from exporting grain would be used to import foreign technology and machinery. In many ways, the strategy seems to have been successful: grain exports increased fivefold between 1860 and 1900, while manufacturing activity expanded rapidly. Further measures in 1906, known as the Stolypin Agricultural Reform, helped to establish large, consolidated farms in place of some of the many small-scale peasant holdings. The consequent flood of dispossessed peasants to the cities created acute problems as housing conditions deteriorated and urban labor markets became inundated.

These problems, to which the tsars remained indifferent despite the petitions of desperate city governments, nourished deep discontent among the population. At the turn of the twentieth century, Russia was in the grip of a severe economic recession. Inflation, with high prices for food and other basic commodities, led to famine and widespread hardship, but there was no real mechanism for legitimately voicing the concerns and aspirations of the majority of the population. Unions were illegal, as were strikes. Nevertheless, riots spread across the countryside and, in 1905, after the embarrassing military defeat by the Japanese in Manchuria the previous year, there was a revolutionary outbreak of strikes and mass demonstrations. A network of grassroots councils of workers—called *soviets*—emerged spontaneously in order not only to coordinate strikes but also to help maintain public order. The unrest was eventually subdued by brute force, and the soviets were abolished. But the discontent continued, intensified if anything by the flood of dispossessed peasants to cities after the Stolypin Agricultural Reform of 1906. The First World War intensified the discontent of the population, as casualties mounted and the government's handling of both the armed forces and the domestic economy led to the socialist revolution of 1917.

The Soviet Empire

From the beginning, the **state socialism** of the Soviet Union was based on a new kind of social contract between the state and the people. In exchange for people's compliance with the system, their housing, education, and health care were to be provided by state agencies at little or no cost. This new social contract, though, had its roots in the traditional Russian traits of collectivism and authoritarianism. It was not the exploited peasantry or the oppressed industrial proletariat that emerged from the chaos of revolution to take control of this new system. It was the Bolsheviks, a dissatisfied element drawn from the former middle classes, whose orientation from the beginning favored a strategy of economic development in which the intelligentsia and more highly skilled industrial workers would play the key roles.

In the early years of the Soviet Union, the government took control over production, but the ravages of war and the upheavals of revolution made planned economic reorganization of any kind impossible. There were strict state controls on the economy, but these resulted as much from the need for national and political survival as from ideological beliefs. Similarly, it was rampant inflation that led to the virtual abolition of money, not revolutionary purism. By 1920, industrial production was only 20 percent of the prewar level, agricultural production was only 44 percent of the prewar level, and per capita national income stood at less than 40 percent of the prewar level.

In 1921 a New Economic Policy was introduced in an attempt to catch up. Central control of key industries, foreign

Geographies of Indulgence, Desire, and Addiction

Furs

From earliest times, fur has been a prized commodity. In cold regions, fur coats, hats, and boots are valued for their warmth in harsh winter weather but they are regarded as a practical investment that is also a portable form of wealth. In Russia and Europe, fur has long had royal and aristocratic connotations and, as a result, became a status symbol for all who could afford it. In the world of women's fashion, furs have become synonymous with *haute couture,* and in much of the world fur is seen by status-conscious consumers as a fashionable luxury good, a clear marker of material wealth. Furs have, however, become a controversial luxury item. In Europe and North America in particular, the market for furs has been significantly affected by people's concern that certain animal species might be threatened with extinction and realization that fur trapping and fur farming can both involve unnecessary cruelty to animals.

European merchants began trading fur in the Middle Ages, and fur was the commodity that drew Russian trappers and traders to Siberia in the sixteenth century. In 1581, under the sponsorship of the rich merchant family of Stroganov, a military expedition opened a routeway to Siberia for *promyshlenniks* (fur hunters), who were drawn eastward in search of sable and sea otter. The pelts of these animals were exchanged for Chinese and Indian goods, and the tax revenues from the trade were the mainstay of the Russian imperial treasury for the next 300 years. It became government policy to encourage the fur trade and to support the *promyshlenniks* in their ruthless displacement of indigenous peoples and their sustainable economies. By the reign of Peter the Great (1682–1725) the *promyshlenniks* had reached the Sea of Okhotsk, and fur hunting was beginning to reach saturation point. In response, Peter the Great sponsored maritime expeditions to the Kamchatka Peninsula and to the offshore islands of the northeast. After Vitus Bering's expedition to the Northern Pacific in 1741–42 established that the islands had abundant populations of sea otter, foxes, seals, and walruses, there was a "fur rush" that drew *promyshlenniks* all the way across the Bering Sea to Alaska.

Meanwhile, trade in fur pelts (beavers, muskrats, minks, and martens) had attracted Europeans' initial interest in North America. Beaver, trapped by Native Americans, was a main source of barter at trading posts that later grew into such cities as Chicago, Detroit, Montréal, New Orleans, Québec, St. Louis, St. Paul, and Spokane. The Hudson's Bay Company,

Figure 1   Russian fur farm

trade, and banking was codified under *Gosplan,* the central economic planning commission. But in other spheres—and in agriculture in particular—a substantial degree of freedom was restored, with heavy reliance on market mechanisms operated by "bourgeois specialists" from the old intelligentsia. Improvement in national economic performance was immediate and sustained, with the result that recovery to prewar levels of production was reached in 1926 for agriculture and in 1927 for industry.

By the early 1920s, Nikolai Lenin, whose real name was Vladimir Ilich Ulyanov, the revolutionary leader and head of state, was also able to focus attention on the more idealistic aspects of state socialism. The Bolsheviks were internationalists, believing in equal rights for all nations and wanting to break down national barriers and end ethnic rivalries. Lenin's solution was recognition of the many nationalities through the newly formed Union of Soviet Socialist Republics (USSR). Lenin believed that a *federal system,* with *federal units* delimited according to the geographic extent of ethno-national communities, would ensure political equality among at least the major nations in the new state. A **federal state** allocates power to units of local government within the country. Federal states can be contrasted with a **unitary state,** in which power is concentrated in the central government. The Russian state under the tsar had been a unitary state. Federation was also a way of bringing reluctant areas of the former Russian empire into the Soviet fold. The new federal arrangement recognized the different nationalities and provided them a measure of independence. Each of the ethno-national territories that comprised the Soviet Union had specific rights in a hierarchical political-

founded in England in the mid-seventeenth century to trade skins for guns, knives, and kettles, gained almost total control of the North American fur trade. For the company's first 200 years, its business consisted entirely of trading in furs. The company abandoned fur sales in 1991 but in 1997 started them again, largely because of a surge in demand from China and the other rapidly growing capitalist economies of Asia. Today, fur farming (raising animals in captivity under controlled conditions), rather than trapping, is the principal source of furs for the world market. Fur farming was started in Canada in 1887 on Prince Edward Island. Animals with unique characteristics of size, color, or texture can pass those characteristics on to their offspring through controlled breeding. The silver fox, developed from the red fox, was the first fur so produced.

Overall, the Scandinavian countries produce about 45 percent of the world supply of pelts, Russia 30 percent, the United States 10 percent, and Canada 3 percent. Retail sales of furs in the United States grew from less than $400 million in the early 1970s to $1.5 billion by the mid-1980s and have since stagnated at between $1.8 billion and $2 billion annually. The most important reason for this leveling-off in sales during a sustained economic boom is a shift in attitudes toward furs, led by animal rights activists. Anti-fur demonstrations targeting designers, led by pop icons such as Chrissie Hynde, the B-52s, and kd lang, have captured widespread attention. Calvin Klein, Giorgio Armani, Oleg Cassini, and Bill Blass all dropped furs from their fashion lines in the mid-1990s. Nevertheless, a 1999 survey by a British consumer group found that some 180 of the world's leading clothes designers were using fur in some of their product lines.

In the late 1990s, Russia's economic recession led to a number of well-publicized cases of maltreatment on fur farms, further reinforcing the case of Western anti-fur activists. Some Russian fur farmers, faced with a combination of falling consumer demand because of economic recession, higher taxes, and widespread corruption, let their animals go hungry. Western visitors to Siberian fur farms found starving animals in tiny cages with no bedding, no protection against the elements, and no veterinary care. Many Russian fur farmers have slaughtered most of their animals rather than watch them starve. Of the 200 fur farms in Russia in the early 1990s, only 30 were still operating in 2000. The others have closed or were gradually phasing out production. Nevertheless, Russian consumers have not been affected by Western activists' concerns. Many middle-class Russians own fur coats, and most consider them necessities in the harsh winter, even though few can afford to purchase new furs while the recession persists.

Figure 2 Anti-fur protest

administrative structure. The 15 Soviet Socialist Republics that stood at the apex of the system had, in theory, all of the rights of independent states. Below them in the hierarchy were 20 autonomous Soviet Socialist Republics, 8 autonomous *oblasts* (regions), and 10 autonomous *okrugs* (areas).

Lenin was optimistic that once international inequalities were diminished, and once the many nationalities became united as one Soviet people, the federated state would no longer be needed: Nationalism would be replaced by communism. Lenin's vision was short-lived, and, following his death in 1924, the federal ideal faded. After eliminating several rivals, Joseph Stalin came to power in 1928 and enforced a new nationality policy, the aim of which was to construct a unified Soviet people whose interests transcended nationality. Although the federal framework remained in place, nations increasingly lost their independence and by the 1930s were punished for displays of nationalism. **Figure 4.14** shows the administrative units and nationalities that were part of the USSR during Stalin's tenure as premier (1928–53). Figure 4.14 also shows how, during and immediately after the Second World War, Stalin expanded the power of the Soviet state westward to include Albania, Bulgaria, Czechoslovakia, the German Democratic Republic, Hungary, Poland, Romania, and Yugoslavia.

Meanwhile, Soviet aspirations for an egalitarian society required the reshaping of the country's geography at every scale. Under Lenin, a number of visionary, utopian (and often impractical), architectural, and city-planning schemes emerged, together with hundreds of new standards and norms that were designed to ensure an equitable allocation of resources. It was decided, for example, that there were to be 35 cinema seats

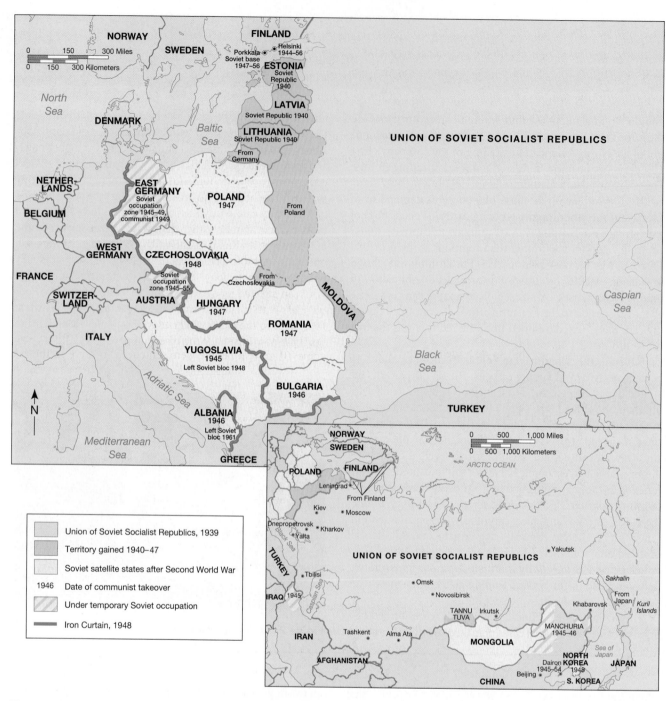

Figure 4.14 Soviet state expansionism, 1940s and 1950s The Second World War gave the Soviet state the opportunity to move westward for additional territories. Insisting that these countries would never again be used as a base for aggression against the USSR, Stalin retained control over Poland, East Germany, Czechoslovakia, Hungary, Romania, Bulgaria, Albania, Yugoslavia, and eastern Austria. In 1945 Stalin promised democratic elections in these territories. After 1946, however, Soviet control over eastern and central Europe became complete as noncommunist parties were dissolved and Stalinist governments installed. (*Source: Atlas of Twentieth Century World History.* New York: HarperCollins Cartographic, 1991, pp. 86–87.)

per 1000 urban inhabitants; that the maximum journey-to-work time should be no more than 40 minutes; and that the optimal size of a city would be between 50,000 and 60,000 persons. Cities were to be planned in such a way as to provide a basic range of services to everyone, while at the same time engendering a sense of neighborliness and collective re-sponsibility. The key organizational unit was to be the ***mikro-rayon,*** a planned development with a radius of 300 to 400 meters (328 to 437 yards), accommodating 8000 to 12,000 people with a representative mix of the city's socioeconomic and ethnic groups, ample green space, perimeter thorough-fares with public transportation, day care, schools, sports and recreation facilities, and health services. Within each *mikrorayon*, the community would be organized into superblocks of 1000 to 1500 people living in standardized housing, with an allowance of 9 square meters of living space (97 square feet) per person.

While many of the utopian and visionary ideals of the 1920s were left on the drawing boards of Soviet planners, the *mikrorayon* concept came to inscribe a distinctive stamp on the character of Soviet cities: by the late 1980s, about one half of the Soviet urban population lived in a *mikrorayon*. The standardization of housing construction also left a very distinctive mark on Soviet urban landscapes. One of the great achievements of the Soviet era was the accommodation of the bulk of the population, formerly immiserated in substandard dwellings, in decent new sanitary housing. Nevertheless, the new prefabricated housing (**Figure 4.15**) was drab, uniform, and cramped, the norm of 9 square meters of living space per person remaining unchanged throughout the Soviet era.

Collectivization and Industrialization
Under Stalin's leadership, there occurred a major shift in power within the Soviet Union. This power shift swept aside both the New Economic Policy and its "bourgeois specialists." They were replaced by a much more centralized allocation of resources: a **command economy** operated by a new breed of engineers, managers, and *apparatchiks* (state bureaucrats) drawn from the new intelligentsia that had developed among the membership of the Communist Party. With this shift there came also a more explicit strategy for industrial development. In giving national economic and political independence the highest priority, the Soviet Union chose to withdraw from the capitalist world economy as far as possible, relying on the capacity of its vast territories to produce the raw materials needed for rapid industrialization. The foundation of Stalin's industrialization drive was the collectivization of agriculture. The capital for creating manufacturing capacity and the required infrastructure and educational improvements was to be extracted from the agricultural sector. This involved the compulsory relocation of peasants into state or collective farms, where their labor was expected to produce bigger yields. The state would then purchase the harvest at relatively low prices so that, in effect, the collectivized peasant was to pay for industrialization by "gifts" of labor.

In the event, the Soviet peasantry proved reluctant to make these gifts, and it proved very difficult to organize them. Government requisitioning parties and inspectors were met with violence, passive resistance, and the slaughter of animals. At this juncture Stalin employed police terror to compel the peasantry to comply with the requirements of the Five-Year Plans that provided the framework for his industrialization drive. Severe exploitation required severe repression. Dissidents, along with enemies of the state uncovered by purges of the army, the bureaucracy, and the Communist Party, provided convict (*zek*) labor for infrastructure projects. Altogether, some 10 million people were sentenced to serve in the *zek* workforce, to be imprisoned, or to be shot. The barbarization of Soviet society was the price paid for the modernization of the Soviet economy.

The Soviet economy *did* modernize, however. Between 1928 and 1940 the rate of industrial growth increased steadily, reaching levels of more than 10 percent per year in the late 1930s: growth rates that had never before been achieved and that have been equaled since only by Japan (in the 1960s) and China (in the 1990s). The annual production of steel had increased from 4.3 million tons to 18.3 million tons; coal production had increased nearly five times; and the annual production of metal-cutting machine tools had increased from 2,000 to 58,400. An industrial revolution in the Western sense had been passed through in one decade. When the Germans attacked the Soviet Union in 1941, they took on an economy that in absolute terms (though not *per capita*) had industrial output figures comparable with their own.

The Second World War cost the Soviet Union 25 million dead, the devastation of 1700 towns and cities and 84,000 villages, and the loss of more than 60 percent of all industrial installations. In the aftermath, the Soviet Union gave first priority to national security. The *cordon sanitaire* of independent eastern European nation-states that had been set up by the Western nations after the First World War was appropriated as a buffer zone by the Soviet Union. Because this buffer zone happened to be relatively well developed and populous, it also provided the basis of a Soviet empire—the Soviet bloc—as an alternative to the capitalist world economy, thus providing economic as well as military security.

But the Soviet Union felt vulnerable to the growing influence and participation of the United States in world economic and political affairs, and in 1947 Stalin felt compelled to intervene more thoroughly in eastern Europe. In addition to the installation of the "iron curtain" that severed most remaining economic linkages with the West, this intervention

Figure 4.15   Soviet-era prefabricated housing   One of the Soviet era's great achievements was the construction of an adequate supply of sound housing for Soviet citizens. Most of the housing, however, came from a very restricted range of prefabricated apartment block designs, with the result that places throughout the Soviet empire acquired the same drab appearance.

resulted in the complete nationalization of the means of production, the collectivization of agriculture, and the imposition of rigid social and economic controls in all of the eastern European **satellite states.** The Communist Council for Mutual Economic Assistance (CMEA, better known as COMECON) was also established to reorganize the eastern European economies in the Stalinist mold—with individual members, each pursuing independent, centralized plans for economic self-sufficiency. This proved unsuccessful, however, and in 1958 COMECON was reorganized by Stalin's successor, Nikita Kruschev. The goal of economic self-sufficiency was abandoned, mutual trade among the Soviet bloc was fostered, and some trade with western Europe was permitted.

Meanwhile, the whole Soviet bloc gave high priority to industrialization. Between 1950 and 1955, output in the Soviet Union grew at nearly 10 percent per year, though it subsequently fell away to more modest levels. In addition to their desire for rapid growth, Soviet economic planners sought to follow three broad criteria in shaping the economic geography of state socialism. First was the idea of technical optimization. Without free markets to provide competitive cost-minimization strategies, Soviet planners had to organize industry in ways that ensured both internal and external economies. Perhaps the most striking result of this was the development of **territorial production complexes,** regional groupings of production facilities based on local resources that were suited to clusters of interdependent industries: petrochemical complexes, for example, or iron-and-steel complexes (**Figure 4.16**). Second was the idea of fostering industrialization in economically less-developed subregions, such as Central Asia and the Transcaucasus. A third consideration was secrecy and security from external military attack. This criterion led to some military-industrial development in Siberia and to the creation of scores of so-called secret cities—closed cities, where even the inhabitants' contacts with relatives and friends were strictly controlled because of the presence of military research and production facilities.

Soviet regional economic planners also sought to ameliorate many of the country's marginal environments through ambitious infrastructure schemes. Stalin insisted that it must be feasible to harness and transform nature through the collective will and effort of the people. As a result, plans were drawn up to reverse the flow of major rivers and divert them to feed irrigation schemes. There were also plans to ameliorate local climatic conditions in steppe regions through vast plantings of trees in shelter-belts. The prohibitive cost of these grandiose schemes kept most of them on the drawing board, but nevertheless the tendency to undertake civil engineering projects of heroic scale lasted through most of the Soviet era, resulting in some dramatic examples of the mismanagement of natural resources (see Environmental Challenges, p. 179).

Figure 4.16   Industrial regions of the Soviet Union Soviet planners gave a high priority to industrialization and sought to take advantage of agglomeration economies by establishing huge regional concentrations of heavy industry. (*Source:* Redrawn from P. L. Knox and J. Agnew, *The Geography of the World Economy,* 3rd ed. London: Arnold, 1998, p. 168.)

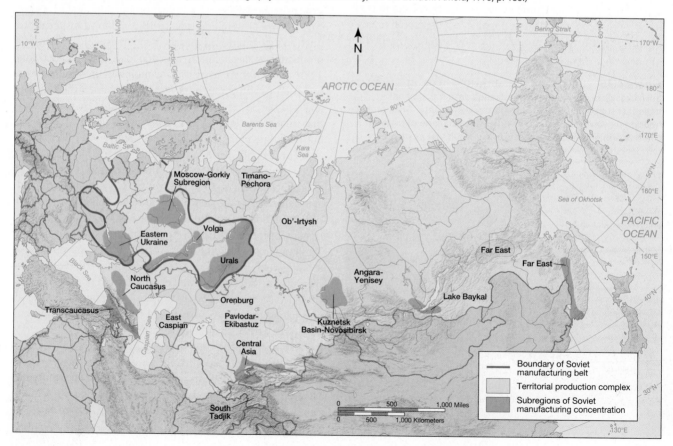

By the 1960s, the Soviet Union had clearly demonstrated its technological capabilities with its manned space program and the production of some of the world's most sophisticated military hardware. These successes were paralleled by the Soviet Union's geopolitical influence. The Soviet Union not only had an extensive nuclear arsenal but also an ideological alternative to the capitalist and imperialist ideology that had created peripheral regions throughout much of the world. Armed with these, the Soviet Union posed a very real threat to U.S. hegemony, waging a Cold War that between 1950 and 1989 provided the principal framework for world affairs. In that period, Soviet influence caused significant tension and a succession of geopolitical crises in many regions of the world, including Cuba, much of the Middle East, South Asia, East Asia, Southeast Asia, and parts of South America (Chile), Central America (Nicaragua and Panama), and Africa (Angola, Libya, and Egypt).

Yet throughout most of the Soviet Union itself, millions of peasants worked with primitive and obsolete equipment as they toiled to meet centrally planned production targets. Most nonmilitary industrial productivity was also constrained by technological backwardness and by cumbersome and bureaucratic management systems. A second, informal or shadow economy of private production, distribution, and sale emerged, and it was largely tolerated by the government—mainly because without it the formal economy would not have been able to function as well as it did. By the 1970s the Soviet economic system was steadily being enveloped by an era of stagnation.

The Breakup of the Soviet Empire   By the 1980s the Soviet system was in crisis. In part, the crisis resulted from a failure to deliver consumer goods to a population that had become increasingly well informed about the consumer societies of their foreign enemies. Persistent regional inequalities also contributed to a loss of confidence in the Soviet system as an alternative mode of economic development. The cynical manipulation of power for personal gain by ruling elites and the drain on national resources from the arms race with the United States also undoubtedly played some role in undermining the Soviet model. The critical economic failure, however, was state socialism's inherent inflexibility and its consequent inability to take advantage of the new computerized information technologies that were emerging elsewhere.

Surprising even the most astute observers, the Soviet system unraveled rapidly between 1989 and 1991, leaving 15 independent countries as successors to the former USSR. As we saw in Chapter 3, the former states of Yugoslavia and Czechoslovakia were broken into smaller entities; East Germany was absorbed into Germany; Hungary, Poland, and the Baltic states (Estonia, Latvia, and Lithuania) were drawn rapidly into the European Union's sphere of influence; and Moldova and Ukraine began to show signs of a Western orientation. Now, Belarus, the Russian Federation, and the states of Central Asia and the Transcaucasus are now experiencing somewhat chaotic transitions, at different speeds, toward market economies. In the process, all local and regional economies have been disrupted, leaving many

people to survive by supplementing their income with informal activities, such as street trading and domestic service (see A Day in the Life: Valeri Novikov, p. 168).

New Realities

For the Russian Federation, the principal successor state to the Soviet Union, these changes have greatly weakened its position in the world. The Russian Federation is an elaborate hierarchy of administrative units (**Figure 4.17**). There are 21 republics (successors to the former autonomous republics and autonomous *oblasts* of the Soviet Union), 10 autonomous *okrugs,* and 58 regional administrative districts (*oblasts*) and metropolitan districts (*krays*). Separate treaties of federation have resulted in a complex federal system in which these different administrative units have different sets of rights and privileges. The republics have elected presidents and written constitutions and have their own legislatures. Some republics have won special tax concessions and have come to agreements with the federal authorities on language and cultural rights. Nevertheless, the Russian Federation is still highly centralized in terms of real political power.

Although still a nuclear power with a large standing army and a vast territory containing a rich array of natural resources, the Russian Federation is economically weak and internally disorganized. The latter years of the Soviet system left industry in the Russian Federation with obsolete technology and low-grade product lines, epitomized by its automobiles and civilian aircraft. Similarly, the infrastructure inherited by the Russian Federation's economy is poorly developed, shoddy, and often downright dangerous, as witnessed by the Chernobyl disaster of 1986, when a nuclear power plant in Ukraine exploded, causing a runaway nuclear reaction and widespread radiation pollution (see p. 180). Investment in the development of computers and new information networks was deliberately suppressed by Soviet authorities because, like photocopiers and fax machines, computers were seen as a threat to central control. As a result, the Russian Federation's economy now faces a massive task of modernization before it can approach its full potential. Overall, the economy of the Russian Federation shrank by between 12 and 15 percent each year between 1991 and 1995, and by between 5 and 10 percent each year between 1996 and 2000. Foreign capital has flowed into the Russian Federation, but it has been targeted mainly at the fuel and energy sector, natural resources, and raw materials (which now account for about half of the Russian Federation's total exports) rather than manufacturing industry. By the end of the 1990s, the Russian Federation's economy was in crisis. About one half of the government's budget revenue was being absorbed by the cost of repaying debts to creditor nations; the rate of inflation had reached 100 percent; and economic output had plunged to about one-half that of 1989.

At the same time that there looms the equally massive task of establishing the institutions of business and democracy after 70 years of state socialism, the Russian Federation has been unable to create some of the essential pillars of a

A Day in the Life

Valeri Novikov

When Valeri Novikov and his wife, Galina, were youngsters and went to work at the "Ivanovo Order of Lenin Blended Yarn Mill, Named for Konstantin Frolov" (a local revolutionary hero), they were proud to belong to a flagship of the Soviet textile industry. But, after a year in which they were paid for only five months by a company struggling to survive, their attitudes are markedly less enthusiastic. . . .

Toward the end of last year, the Ivanovo Blended Yarn Mill Ltd. (it is now a joint stock company) began falling behind in paying wages. . . . In fact, Galina has not been paid since May, when she was given her February pay packet. Misha, their eldest son, is in the same boat. Valeri was paid his July wages last week and told that his June pay had been "frozen." Precisely what that officially means was not explained.

What it means to the family's everyday life, though, is clear from a look inside their refrigerator. Although Galina insists they eat properly ("that is why we can't afford any new clothes") the shelves are austere—a jar of preserved cabbage, a small enamel churn of milk, a single frankfurter, a pot of borscht, a plate of margarine, a lump of smoked pork fat, a saucepan of potatoes. . . . Everyday fare is . . . basic—porridge, vegetable soup, fried potatoes, bread, and milk, with the occasional egg for variety. And much of what the family eats, they grow themselves on a small plot of land an hour's bicycle ride away.

The kitchen windowsill is crammed with green tomatoes that will never ripen now, as autumn sets in. Valeri and Galina's bedroom reeks of the onions and garlic that fill two sacks under the single bedside chair. Lyuba, their teenage daughter, has to clamber around a mountain of apples—it has been a good year for apples—to get into bed. Under the balcony outside the sitting room, Valeri has constructed a chamber where carrots will keep through the winter without freezing. . . .

But even relative self-sufficiency and frugality do not see a family of five through four months without a kopeck in wages. . . . It is moonlighting money that keeps them fed. Their eldest daughter, Natasha, who is married, has a job as a night watchwoman at a local kindergarten. Her husband does not like her to work alone at night, however, and earns enough himself to

take care of his family. So Natasha's parents do her job, and Natasha hands them her $75 paycheck each month. . . .

And even then, the money they earn scarcely pays for the staple two loaves of bread and three liters of milk that Galina buys each morning. Lyuba is wearing the same clothes that her big sister wore 10 years ago; the last time Valeri and Galina bought anything for the house was in 1991, when grandpa, a World War II veteran, gave them his ration card to buy a sofa and two armchairs. . . .

Galina doesn't feel able to go into the petty trading business, setting up a stall on the street, which is how many unpaid Russians make a living today. "I can't even sell my apples, because I'd feel ashamed to stand on the pavement and sell things," Galina explains. "And anyway, I'm no good at mental arithmetic; I'd be cheated in no time." Nor is Valeri attracted to commerce. "I'm a worker; that's not my kind of life, and I wouldn't be any good at it," he says. Nor is there much prospect of a job anywhere else in Ivanovo, one of the most depressed regions in Russia.

Source: Adapted from P. Ford, "No Paycheck Means Apples Under the Bed," *Christian Science Monitor,* 88 (1996), p. 1.

Figure 1 Valeri's workplace This photograph shows the textile mill in Ivanovo in its heyday as a flagship of Soviet industry.

market economy. The institutional framework for the legal enforcement of private contracts and effective competition is still rudimentary. Another major weakness has been public finances. A system of fair and efficient tax collection has yet to be put in place, while the relationship between federal and state taxes and spending has remained obscure. Meanwhile, by allowing itself to fall into arrears on its own debts,

the government has contributed to the growth of an arrears culture across all sectors of the economy. Widespread theft of state property and the collapse of constitutional order have undermined respect for the law, and organized crime has flourished amid the factionalism and ideological confusion of the government. Meanwhile, real wages for most people have already fallen to 1950s levels.

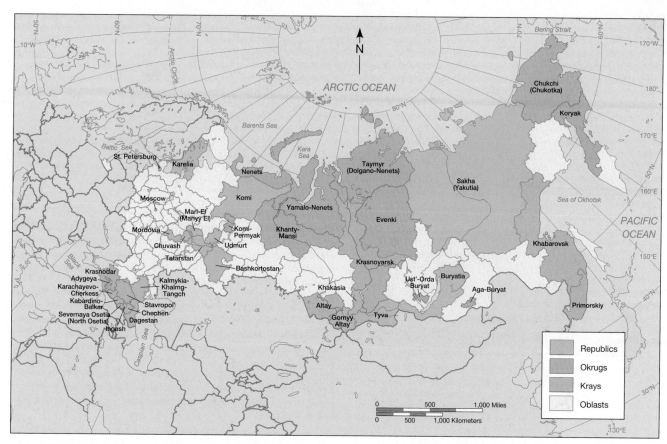

Figure 4.17   Administrative units of the Russian Federation   The complex administrative fabric of the Russian Federation is derived in large measure from the hierarchical framework of ethno-national territories established by the Bolsheviks, but the present system is lacking in hierarchical clarity and cohesiveness.

In an attempt to counter some of the economic disruption caused by the political disintegration of the Soviet Union, several of the successor states agreed to form a loose association, known as the Commonwealth of Independent States (CIS). The CIS was designed to provide a forum for the discussion of the management of economic and political problems, including defense issues, cooperation in transport and communications, the creation of regional trade agreements, and environmental protection. The founder members were the Russian Federation, Belarus, and Ukraine, and they were soon joined by the Central Asian and some Transcaucasus states. Meanwhile, however, the reorientation of the Baltic and eastern European states toward Europe and the imminent prospect of European Union and NATO membership for some of these states has not only undercut the economic prospects of the CIS (which has never really blossomed) but has also weakened the geopolitical security of the Russian Federation. In response, the leadership of the Russian Federation has asserted that country's claims to a special sphere of influence in what it calls the **Near Abroad:** the former components of the Soviet Union, particularly those countries that contain a large number of ethnic Russians. The Russian Federation is clearly finding it problematic to adjust to a new role in the world. But although embarrassed by the disintegration of the Soviet Union and bankrupt by the subsequent dislocation to economic de-

velopment, the Russian Federation is still accorded a great deal of influence in international affairs and may yet eventually reemerge as a major contender for world power.

The Peoples of the Russian Federation, Central Asia, and the Transcaucasus

A distinctive characteristic of this world region as a whole is the relatively low density of its population (**Figure 4.18a**). With a total population of some 232 million and almost 14 percent of Earth's land surface, the Russian Federation, Central Asia, and the Transcaucasus contains about 3.8 percent of Earth's population at an overall density of only 11 persons per square kilometer (28 per square mile). The highest national densities—123 per square kilometer (km^2) in Armenia, 87 per km^2 in Azerbaijan, and 79 per km^2 in Georgia—approximate the population densities of Colorado, Kansas, and Maine. Within the Russian Federation there is a core of relatively high population density (between 40 and 60 per km^2) that corresponds to the region of mixed forest and the wooded steppe west of the Urals. In contrast, population density in much of

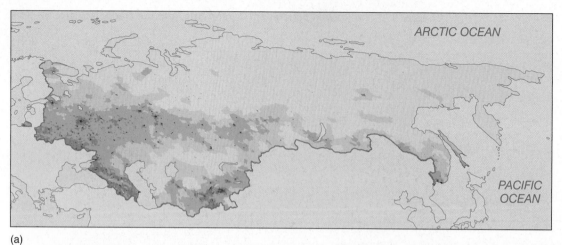

(a)

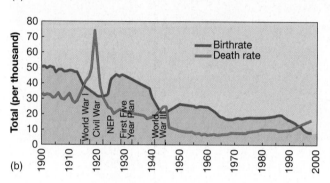

(b)

Figure 4.18 Population density and vital rates in the Russian Federation, Central Asia, and the Transcaucasus 1995
(a) The distribution of population in the Russian Federation, Central Asia, and the Transcaucasus reflects the region's economic history, with the highest densities in the industrial regions of the western parts of the Russian Federation and the richer agricultural regions of the Transcaucasus. (b) This graph shows the dramatic drop in the birthrate in Russia that characterized the 1960s and 1990s. Note also the sharp rise in death rates since 1990. For a view of worldwide population density and a guide to reading population-density maps, see Figure 1.40.
[*Sources:* (a) Center for International Earth Science Information Network (CIESIN), Columbia University; International Food Policy Research Institute (IFPRI); and World Resources Institute (WRI). 2000. *Gridded Population of the World (GPW), Version 2.* Palisades, NY: CIESIN, Columbia University. Available at http://sedac.ciesin.org/plue/gpw; (b) Updated from J. H. Bater, *Russia and the Post-Soviet Scene.* London: Arnold, 1996, p. 93.]

the far north, Siberia, and the far east stands at less than one person per km², about the same as in the far north of Canada. Levels of urbanization reflect this same broad pattern. Most of the large cities are in the European part of the Russian Federation and in the Urals. These include Moscow, Nizhniy Novgorod, St. Petersburg, Volgograd, and Yekaterinburg (**Table 4.1**). Most of the other cities of the Russian Federation that are of any significant size are found in southern Siberia, on or near the Trans-Siberian Railway. Overall, both Belarus and the Russian Federation are quite highly urbanized, with 72 and 76 percent of their total populations living in cities, according to their respective census counts in the mid-1990s. The populations of the Transcaucasus are moderately urbanized (56 to 69 percent living in cities), while those of Central Asia are more rural (only 30 to 50 percent living in cities).

Overall, this is a world region with a relatively slow-growing population. Throughout the twentieth century there was a general decline in both birth and death rates (**Figure 4.18b**). In the 1990s the population began to register a decline as a result of an excess of deaths over births. Viewed in greater detail, it is clear that this is a trend that masks some important regional differences. In Belarus and the Russian Federation, population growth has for a long time been relatively modest, and it is in these countries that recent declines have been most pronounced. In contrast, in Central Asia and the Transcaucasus, birthrates have historically been relatively high, and rates of natural increase remain at a level comparable with those in South Asia and Southeast Asia.

Both the First World War and the Second World War resulted in huge population losses that are still reflected in the

age-sex profile of the Russian Federation (**Figure 4.19**). It was not until the 1960s, however, that rates of natural population increase in the Soviet empire began to decrease significantly on a long-term basis. At the beginning of the 1960s, birthrates fell sharply as a result of a combination of the legalization of abortion, a greater propensity to divorce, planned deferral of marriage among the rapidly expanding urban population, and a growing preference to trade off parenthood for higher levels of material consumption. At about the same time, there began a steady rise in death rates, which increased sharply after the breakup of the Soviet empire. The reasons for this increase in death rates are several. Deteriorating healthcare systems and the worsening health of mothers have contributed to an escalation of infant mortality rates. Meanwhile, public health standards have generally deteriorated, environmental degradation has intensified, and the rate of industrial accidents and alcohol-related illnesses has increased. By 1999 the average life expectancy of those born in the Russian Federation had slipped from the mid-1980s peak of 70.1 years to 66.5.

Languages and Ethnic Groups

The last census count of the Soviet Union in 1989 acknowledged 92 distinct ethnic groups. Dominant today throughout Belarus and the Russian Federation are Slavic peoples, among whom Russians represent one particular ethnic group. The hearth area of the original Slav tribes was in the Danubian lands of present-day Hungary and Bulgaria. During the mid-

Table 4.1	Major Cities of the Russian Federation, Central Asia, and the Transcaucasus (1995 or Later Estimate)	
		Population
Armenia		
Yerevan		1,248,700
Azerbaijan		
Baku		1,739,900
Belarus		
Minsk		1,700,000
Georgia		
Tbilisi		1,253,100
Kazakhstan		
Almaty		1,150,500
Kyrgyzstan		
Bishkek		589,800
Russian Federation		
Chelyabinsk		1,100,000
Kazan'		1,100,000
Moscow		8,400,000
Nizhniy Novgorod		1,400,000
Novosibirsk		1,400,000
Omsk		1,200,000
Perm'		1,000,000
Rostov-on-Don		1,000,000
St. Petersburg		4,200,000
Samara		1,200,000
Ufa		1,100,000
Volgograd		1,003,000
Yekaterinburg		1,300,000
Tajikistan		
Dushanbe		524,000
Turkmenistan		
Ashkhabad		536,000
Uzbekistan		
Tashkent		2,107,000

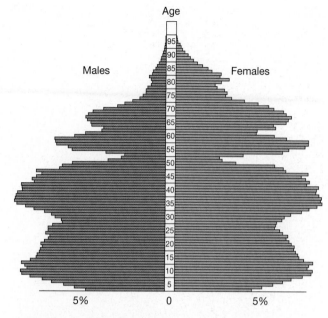

Figure 4.19   Age-sex pyramid for the Russian Federation This profile of the Russian Federation's population in the mid-1990s shows very clearly the effects of the Second World War (the relative lack of men and women in their early 50s and the reduced number of men aged 70 and older) and the reduced birthrates of the 1960s and 1990s. (*Source:* Updated from J. H. Bater, *Russia and the Post-Soviet Scene.* London: Arnold, 1996, p. 101.)

The Slavs are fundamentally defined by linguistic commonalities rather than territorial, racial, or other attributes. The Slavonic group of languages forms one of the major components of the great Indo-European language family, whose speakers range from north India (Hindi and Urdu) through Iran (Farsi) and parts of Middle Asia (Tajik) to virtually the whole of Europe. Written language came late to the Slavs, and when it did it was the deliberate effort of two missionaries—Constantine (later, as a monk, called Cyril) and Methodius—who were sent by the ninth-century Byzantine emperor Michael III to the Slavic nation of Greater Moravia (which occupied much of present-day Hungary, Germany, Slovakia, and the Czech Republic) in order to spread the Scriptures. The new alphabet that Constantine/Cyril devised in order to accommodate Slavonic speech sounds became known as the Cyrillic alphabet. As **Figure 4.20** shows, Slavonic-speaking peoples correspond to the most densely settled parts of the region, extending eastward along the zone of wooded steppe and steppe to the far east.

A second important language group is that of Turkic languages, which belong to the Altaic family of languages. These are spoken by the peoples of Central Asia and parts of the Transcaucasus and were spread into Russia itself through the Tatar invasion and period of rule (c. A.D. 1240–1480). Much of northern and eastern Siberia is occupied by peoples who speak other branches of the Altaic language group, while in the far east are peoples whose languages are part of the Paleo-Siberian language family, including Gilyak and Koryak. Finally, there are several smaller

dle centuries of the first millennium A.D., these tribes spread outward to occupy a vast swathe of the continent, extending to the Elbe River in Germany, to the Baltic and Adriatic seas, and the Gulf of Corinth, and eastward into the mixed forest and wooded steppe of present-day Ukraine, Belarus, and the Russian Federation. In the course of these great waves of migration, the Slavs split into three main branches: western (including Czechs, Poles, and Slovaks), southern (including Serbians, Serbo-Croats, and Slovenes), and eastern (the East Slavs, subdividing only in the late Middle Ages into Belarussians, Russians, and Ukrainians).

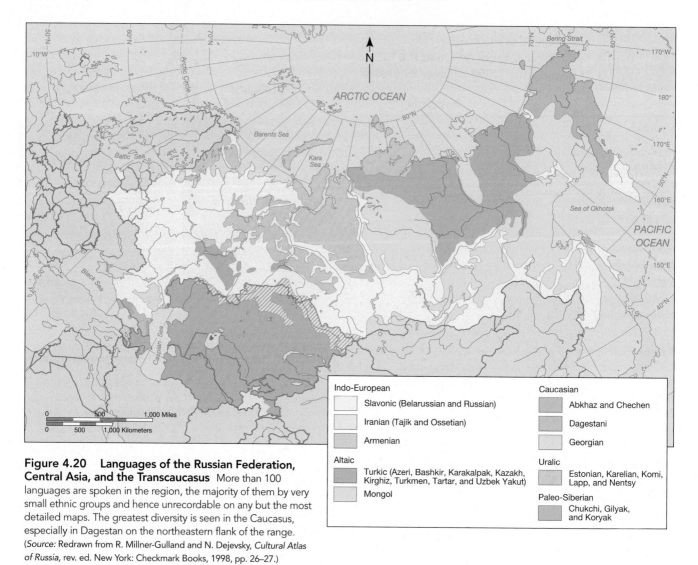

Figure 4.20 Languages of the Russian Federation, Central Asia, and the Transcaucasus More than 100 languages are spoken in the region, the majority of them by very small ethnic groups and hence unrecordable on any but the most detailed maps. The greatest diversity is seen in the Caucasus, especially in Dagestan on the northeastern flank of the range. (*Source:* Redrawn from R. Millner-Gulland and N. Dejevsky, *Cultural Atlas of Russia,* rev. ed. New York: Checkmark Books, 1998, pp. 26–27.)

areas of Caucasian languages: Abkhaz and Chechen on the northern slopes of the Caucasus, and Georgian and Dagestani languages in the Transcaucasus.

The Russian Diaspora and Migration Streams

The spread of the Russian empire from its hearth in Muscovy took Russian colonists and traders to the Baltic, Finland, Ukraine, most of Siberia, the Far East, and parts of Central Asia and the Transcaucasus. In the late nineteenth and early twentieth centuries many Russians joined the stream of emigrants headed toward North America. Concentrations of Russian immigrants developed in Chicago, New York, and San Francisco. They were joined by others who fled the civil war and Bolshevik revolution of 1917. More recently, in the first five years after the breakup of the Soviet Union, the United States resettled nearly 250,000 refugees from the former Soviet Union, mostly from Russia. Over a quarter of these immigrants have settled in New York City, the majority in Brooklyn, where distinctive Russian exclaves, such as the Brighton

Beach neighborhood of southern Brooklyn, have emerged as vital nodes in the Russian global diaspora.

With the rise of the Soviet empire, many Russians were directed and encouraged to settle in the Baltic, Ukraine, Siberia, the Far East, Central Asia, and the Transcaucasus—partly in order to further the Stalinist ideal of a transcendant Soviet people and partly to provide the workers needed to run the mines, farms, and factories required by Soviet economic, strategic, and regional planners. By the time of the breakup of the Soviet Union, 80 percent or more of the population of Siberia and the far east were Russian, and the Russian diaspora had become very pronounced in most of the Soviet Union's successor states beyond the borders of the Russian Federation.

In 1989, without any sense of ever having emigrated from their homeland, some 25 million Russians suddenly found themselves to be ethnic minorities in newly independent countries (**Table 4.2**). The largest number was in Ukraine, where more than 11.3 million Russians made up 22 percent of the population of the new state. In Kazakhstan, Russians represented nearly 38 percent of the population. Overall, the sudden collapse of the Soviet Union created

Table 4.2	The Russian Diaspora	
Republic	Number of Russians	Russians as % of Total Population of Republic
Ukraine	11,356,000	22.1
Belarus	1,342,000	13.2
Estonia	475,000	30.3
Latvia	906,000	34.0
Lithuania	344,000	9.4
Moldova	562,000	13.0
Georgia	341,000	6.3
Armenia	51,600	1.6
Azerbaijan	392,000	5.6
Kazakhstan	6,228,000	37.8
Uzbekistan	1,653,000	8.3
Kyrgyzstan	917,000	21.5
Turkmenistan	334,000	9.5
Tajikistan	388,000	7.6

Source: D. B. Shaw, *Russia in the Modern World.* Malden, MA: Blackwell, 1999, p. 256.

havoc in the lives of many families, who suddenly found themselves living "abroad." During the 1990s, a good number of them decided to migrate back to the Russian Federation. In the Transcaucasus, where the proportion of Russians was generally lower than elsewhere, strongly nationalistic governments of the successor states quickly enacted policies that encouraged Russians to leave: reducing the number of Russian-language schools, for example. In Central Asia too, nationalistic policies were enacted with similar effect. Kyrgyzstan, Turkmenistan, and Uzbekistan dropped the use of the Cyrillic alphabet, deliberately creating institutional barriers for Russian speakers. Civil war in Tajikistan led to the departure of 80 percent of that country's Russian-speaking population within just three years of its independence from the Soviet Union. About 17 percent of the Russian population of Kyrgyzstan departed in that same period, mainly because of the withdrawal of the Russian Federation's defense industry enterprises and military installations. Altogether, almost 4 million ethnic Russians migrated to the Russian Federation from the other Soviet successor states between 1989 and 1999.

Meanwhile, an even greater number of people emigrated from the Russian Federation and the other successor states to countries elsewhere in the world. The annual loss, at about 100,000 per year, is not particularly significant in terms of raw numbers. What is significant, however, is the fact that the most are well-educated individuals, and some are among the most talented. The countries of the former Soviet Union have thus been suffering something of a "brain drain," with the principal beneficiaries being Germany, Israel, and the United States.

Nationalisms

The prelude to the breakup of the Soviet Union involved not only a massive restructuring of the Soviet economy through radical economic and governmental reforms (*perestroiyka*) but also the direct democratic participation of the republics in shaping these reforms through open discussions, freer dissemination of information, and independent elections (*glasnost*). Both *perestroiyka* and *glasnost* were initiated by Mikhail Gorbachev when he became the Soviet leader in 1985. *Glasnost* resulted in the removal of restrictions that had been placed on the legal formation of national identity by Stalin. By 1987, grassroots national movements were already emerging, first in the Baltic republics and later in the Transcaucasus, Ukraine, and Central Asia. In 1989 *perestroiyka* and *glasnost* together culminated in the breakup of the Soviet Union. The Soviet Union's federated structure enabled the relatively peaceful breakup of the country, but the demise of a strong central government and the exhaustion of state socialism as an ideology opened the way for a reemergence of nationalist political identities based upon ethnic divisions. At the same time, the end of the Cold War meant that localized territorial disputes between ethnic groups no longer had to be suppressed for fear that they might spark a world war.

Within the Russian Federation, there are approximately 27 million non-Russians. This number encompasses 92 different ethno-national groups (though 25 of these groups include minority peoples of the north, who together number less than 200,000). Although most of the larger ethno-national groups enjoy a fair degree of administrative territorial autonomy within the Russian Federation, secessionist and irredentist claims are numerous (**Figure 4.21**). One of the most troubled regions is the North Caucasus, a complex mosaic of mountain peoples with strong territorial and ethnic identities. Soon after the breakup of the Soviet Union, Ingushetia broke away from the Chechen-Ingush Republic, and Chechnya promptly declared independence from the Russian Federation (see Geography Matters: Chechnya, p. 176). The Ingush themselves have irredentist claims to parts of neighboring North Ossetia; while the autonomous *oblast* of South Ossetia, in Georgia, declared its intent to secede from Georgia and unite with North Ossetia in the Russian Federation—a move that resulted in a brief civil war in 1992. Beyond the North Caucasus, the two most powerful nationalisms are in Tatarstan (where Tartars have irredentist claims on neighboring Bashkortostan) and in the area around Lake Baykal (where ethnic Buryats have called for the reunification of the Ust'-Ordin and Agin Buryat autonomous *okrugs* with the Republic of Buryatia).

In the Transcaucasus, the big trouble spot is the region of Nagorno-Karabakh, in Azerbaijan. For many years, this region was dominated by Armenians. At one time the region's population had been about 90 percent Armenian. By the mid-1980s the population of Nagorno-Karabakh was still more than 75 percent Armenian, and *glasnost* brought the opportunity for them to formally petition for secession from Azerbaijan. When the petition was refused, pent-up anger was unleashed in both Azerbaijan and Armenia against ethnic minorities from the

Chechnya

Of the many ethno-national movements that surfaced with *glasnost* and the subsequent breakup of the Soviet Union, the Chechen independence movement has been the most bloody. In this region of the North Caucasus (**Figure 1**), clans, not territory, had been the traditional form of political organization. From the time that imperial Russia began its territorial expansion into the northern Caucasus in the late 1700s, the Sunni Muslim Chechens put up strong resistance, periodically waging holy wars against Christian Russia. When revolution came in 1917, Chechens scarcely looked upon the Bolsheviks as a liberating force, not least because of the formal adoption of scientific atheism as the state religion of the newly created Soviet Union. Following a brief, failed attempt on the part of the peoples of the North Caucasus and Transcaucasus to resist Soviet domination, the Soviet strategy was to divide and conquer by creating administrative regions that encompassed a mixture of clans and ethnic groups. The anti-Soviet Chechens were put in the same region as the Ingush peoples.

The Chechens remained defiant, but paid a terrible price for doing so. In the late 1930s, tens of thousands of Chechens were liquidated by Stalin in his purges against all suspected anti-Soviet elements. Then, in 1944, after invading German forces had been forced to retreat from the North Caucasus, Stalin accused the Chechens of having collaborated with the Nazis and ordered the entire Chechen population—then numbering about 700,000—to be exiled to Kazakhstan and Siberia. Brutal treatment during this mass deportation led to the death of more than 200,000 Chechens. In 1957 Nikita Kruschev embarked on a program of de-Stalinization that included the rehabilitation of Chechens. But when Chechens returned, they found that newcomers had taken over many of their homes and possessions. Over the next 30 years, many of these newcomers withdrew, while the Chechen population consolidated and grew to almost 1 million. When Mikhail Gorbachev initiated his policy of *glasnost* in 1985, Chechens finally saw the possibility for self-determination, and with the breakup of the Soviet empire in 1989, Chechens wasted no time in unilaterally declaring their complete independence. Ingushetia decided to separate from Chechnya in 1992 and signed the Treaty of Federation. The Russian Federation chose at first to ignore Chechnya's declaration of independence but could not tolerate the possibility of the loss of the region, particularly since the area around Grozny is one of the Russian Federation's major oil-refining centers and has significant natural gas reserves. In December 1994, Russian troops invaded Chechnya. The ensuing conflict brought terrible suffering to the Chechen population and resulted in mass migrations away from the scene of the fighting. Chechen resistance continued, with increased popular support because of the invasion. Russian forces were disgraced in the fighting, and in 1996 the Russian Federation settled for peace, leaving Chechnya with *de facto* independence. For 3 years there were protracted

Figure 1 The Northern Caucasus Reference map showing principal physical features, political boundaries, and major cities of the northern Caucasus.

negotiations over the nature of the peace settlement. Then, in the summer of 1999, after Chechen rebels had taken the fight to the neighboring republic of Dagestan and to the Russian heartland with a series of terrorist bombings of apartment blocks, the Russian military effort was renewed. After bitter and intense fighting, during which the Russian army suffered more than 400 deaths and nearly 1500 wounded while hundreds of thousands of Chechens were made homeless and several thousand were dead or missing, Russian troops took the capital, Grozny, in February 2000. By that time, Grozny was virtually uninhabitable (**Figure 2**). Since early 2000, Russian Federation troops have maintained control of Grozny, though they continue to be harassed by Chechen rebel guerillas.

Figure 2 Grozny Refugees leaving Grozny, capital of Chechnya, in December 1999, after Russian troops advised residents to leave the city before it was destroyed.

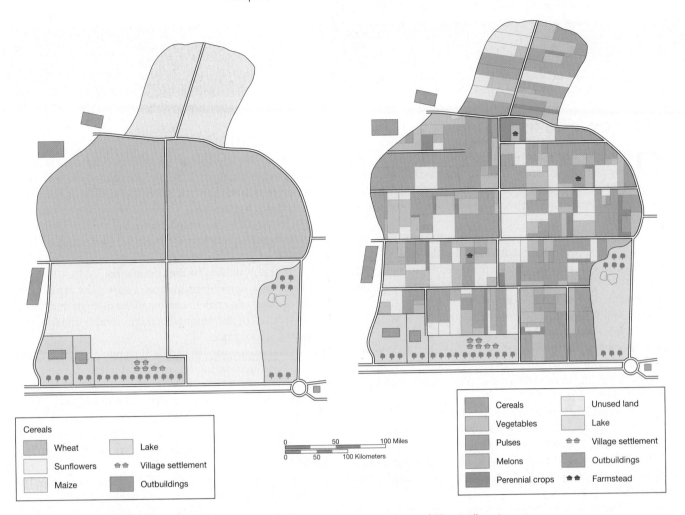

Figure 4.22   The landscape of decollectivization This figure shows the layout of Chocti village in eastern Georgia (a) before and (b) after the decollectivization of the early 1990s. (*Source:* Bradshaw, M. J. (ed.), *Geography and Transition in the Post-Soviet Republics.* New York: John Wiley & Sons, 1997, p. 114.)

In practice, the vouchers were promptly devalued as millions of citizens sought to sell their allocation. Banks and a relatively small number of investors accumulated the bulk of the vouchers and used them to purchase those state enterprises with the greatest profit potential. As a result, the privatization of industry has been very uneven. There remain significant pockets of collectively owned and state-run enterprises, many of them inefficient and undercapitalized. Meanwhile, the growing private sector in retailing and services has been only loosely regulated, resulting in some rapacious aspects of everyday life (see Geography Matters: Moscow's Taxis, p. 178).

In the countryside, market reform has meant dismantling the collective and state farms, and in some regions large farms have been decollectivized and split into a multitude of small holdings of just a few hectares each, leading to significant changes in rural landscapes. A good example is provided by the countryside surrounding the village of Chocti, in Georgia (**Figure 4.22**). In general, however, rural reform has been slow. Most of the land remains under some form of collective ownership, and there have been relatively few progressive changes in farming practice. Shortages of machinery are an obstacle to the modernization of agriculture: It has been estimated that

fully half of the tractors, combine harvesters, and plows in the Russian Federation were out of order in 1999. Output of agricultural products has fallen everywhere, as people have turned to semisubsistence forms of farming as an insurance against the risks of post-Soviet transition. A 1999 report issued by the Russian Federation's ministry of agriculture showed that the extent of farmland shrank by 35.2 million hectares, to 83.5 percent of the 1990 level. Years of out-migration by the young and enterprising have left behind a population that is frequently elderly and conservative, used to the Soviet way of doing things and unwilling or unable to take the risk of establishing private farms. Both politically and socially the countryside remains very conservative and is still dominated by farm managers and officials who derive from the old regime. Meanwhile, decision making in those enterprises that have been privatized tends to occur within a rather vague institutional and legal environment, with limited accountability either to shareholders or to government. As in other sectors of the economy, this has provided enormous scope for crime and corruption (see Geography Matters: Crime and Corruption, p. 181).

The overall state of affairs, then, is more chaotic than transitional. Price inflation, caused by market reform in conjunction

Geography Matters

Moscow's Taxis

On the mad, bad roads of Moscow, there's no point taking a registered taxi. Back in communist times, the city's taxis were state-owned, all 2000 of them. Just 150 of them still are; the rest have been "privatized" and then rented out to the drivers.

But since the ruble went through the floor in 1998, destroying savings overnight, thousands of males have taken to cruising Moscow, picking up fares in an attempt to make ends meet. Two dollars will take you halfway across the city. Two seconds by the side of the road will get you a cab.

An ambulance stopped for me the other day. A police car the week before.

Alyosha, a charming champion Dagestani kick-boxer, picked me up at 2 A.M., eager to reminisce about his glory days competing in Miami and Melbourne. "Ever been to New York? What a place."

Or Sergei, burly in a sheepskin, eking out his pension of £17 ($25.50) a month after 28 years as an Aeroflot pilot.

Or Sasha, a police captain with a wife and baby to support. He drives all night and sleeps all day at the police station, where he earns £30 ($45) a month. "On a good night, I can make that," he said.

Driving in Moscow is like dodgems for grown-ups, very dangerous, thoroughly criminalized, but free of road rage since it is assumed that every Muscovite male is entitled to drive like a maniac. He loves his car. A recent poll put the car as the Moscow male's number one desire, ahead of vodka and sex.

Time was when the lane down the middle of the broad thoroughfares was reserved for the communist elite's Chaika and Zil limousines. They may be gone, but the perks multiply. The lane remains reserved for bigshots, these days in Mercedes, BMWs, and Landcruisers, on a list of approved pampered drivers published by the government.

Red plates signify foreign diplomats, yellow plates identify foreigners, and endless combinations of digits and letters are deciphered by the ubiquitous traffic cops to translate into perks and "indulgences."

Special passes, papers, and flashing lights legalizing driving like a madman without fear of police interference are theoretically reserved for state officials. But *nouveau riche* Russians snap them up from the traffic police for up to $10,000. "It just takes a bit of money and imagination," said a Moscow banker.

A gang has taken to ramming yellow-plated foreigners on the roads and then instantly extorting large sums of cash "for repairs."

The police advice to their victims? Park your car in the middle of the road and ask a passing driver for help.

Source: Adapted from I. Traynor, "Capital Letters." *The Guardian*, January 15, 2000. Copyright © 2000 The Guardian. Available at http://www.guardianunlimited.co.uk/Archive/Article/0,4273,3951030,00.html

---

with declining productivity and unsound fiscal policies, has caused real hardship to millions and resulted in a serious economic crisis for the Russian Federation. A two-year, $23 billion international financing package was provided to the Russian Federation by the International Monetary Fund (IMF) in July 1998, but the Russian Federation's gross domestic product plunged by 9 percent in the final quarter of 1998, while inflation soared to 100 percent. The following year, the Russian Federation had to relaunch negotiations with the IMF to help overcome the crisis. Russian Federation currency was devalued, extensive restrictions were imposed on the foreign exchange market, the government deficit was increasingly financed by direct borrowing from the central bank, and exporters were forced to surrender 75 percent of their export earnings in order to help pay the interest on the IMF loan.

Today, the modest levels of material welfare to which citizens had become accustomed under state socialism are increasingly difficult to sustain (**Figure 4.23**). For most people, real income has fallen significantly, and personal savings have been eroded or disappeared altogether. For many, employment has become a part-time or informal affair. Meanwhile, as we have seen, there has been a deterioration of public health standards, an increase in the rate of industrial accidents and alcohol-related illnesses, a steady rise in death rates, and a rapid escalation of infant mortality rates. The transition to market economies has also intensified two important geographic problems: regional inequality and environmental degradation.

Regional Inequality   Regional inequality was part of the legacy of the Soviet era. Soviet patterns of economic development came to be characterized by regional specialization and by center-periphery contrasts. One reason for this was that principles of scientific rationality and the primacy of national economic growth took precedence over ideological principles of spatial equality. As a result, Soviet planners revised upward the optimal size of cities and applied the logic of agglomeration economies to regional planning, developing territorial production complexes. Another reason was that centralized economic planning was unable to redress the resulting regional inequalities because of conservatism and compartmentalization throughout the Soviet economic system. Regional resource allocations were guided not by principles of equity or efficiency but by *incrementalism,* whereby successive rounds of budgeting were based on previous patterns of funding.

The transformation to market economies has intensified the unevenness of patterns of regional economic development. Market forces have introduced a much greater disparity between the economic well-being of regional winners and losers while at the same time allowing for the more volatile spatial effects of the ebbs and flows of investment capital. After just a decade of transition, many of the regional winners are the same as under state socialism. This is partly because of the natural advantages of certain regions and partly because of the initial advantage of economic development inherited from Soviet-era regional planning. Four different kinds of regions have prospered through the transformation to date: gateway regions, natural resource regions, rich farming regions, and established high-tech manufacturing regions.

Gateway regions are those that are centered on metropolitan areas that have inherited good transportation and communication links not only within the former Soviet empire but also with Europe and beyond. Moscow and St. Petersburg are the most important, but most of the regions around the metropolitan areas listed in Table 4.1 are also in this category, as are Vladivostok, on the Pacific coast, and Kaliningrad, on the Baltic. Prosperous natural resource regions include the oil-rich geosyncline of Central Asia, Magadan (northeastern Siberia), Sakha (Yakutia), and Tyumen' (western Siberia). The best-endowed agricultural regions have also prospered, partly because of their natural advantages of better soils and climate and partly because these attributes have attracted the greatest levels of privatization and capital investment. As we have seen, these regions are mostly in the mixed forest and the wooded steppe west of the Urals and in the niche of subtropical farming in western Georgia, on the shores of the Black Sea. Finally, the established high-tech manufacturing regions that have prospered most are those with major research institutes and other facilities associated with the Soviet military-industrial complex. Many of these were located in the gateway regions of Moscow, St. Petersburg, and Nizhniy Novgorod, but others include Chelyabinsk, Samara, Saratov, and Voronezh.

In contrast, three different kinds of regions have experienced decreasing levels of prosperity through the transformation to date. The first consists of regions of armed territorial conflict (such as North Ossetia, Ingushetia, and Chechnya in the North Caucasus; Nagorno-Karabakh; and Tajikistan). A second consists of resource-poor peripheral regions—mainly in the European north, Siberia, and the far east. The third consists of "smokestack" regions of declining heavy industry. These include many of the core industrial regions of the old command economy, including much of the Central Region—the area that extends about 400 kilometers (248 miles) from Moscow in all directions—Volga-Vyatka (the Volga region) the Urals, and southern Siberia.

Environmental Challenges   The more open discourse that followed *glasnost* and the creation of democratic societies revealed a legacy of serious environmental challenges that stem from the mismanagement of natural resources and the failure

Figure 4.23   A Russian family with their material possessions   The Kapralov family from Suzdal, Russia, photographed with their possessions outside their home in the mid-1990s, represent a statistically average Russian family in terms of family size, residence, and income. Less than a month after Eugeny Kapralov posed for this photograph with his family and possessions, he was beaten to death by unknown assailants who smashed the windows of his car in what was presumed to be a robbery.

to control pollution during the Soviet era. Soviet central planning placed strong emphasis on industrial output, with very little regard for environmental protection. Stalin had propagated the view that it would be feasible to harness and transform nature through the collective will and effort of the people. Nature, it was asserted, is a dangerous force that needs to be subdued and transformed, and natural resources have no value in a socialist society until people's labor has been applied to them. As a result of this way of thinking, there was a tendency during the Soviet era to squander natural resources and to "take on" and "conquer" nature through ambitious civil engineering projects. Problems of pollution and environmental degradation were seen by the authorities as an inevitable cost of modernization and industrialization, and the people most affected—the general public—had no political power or means of voicing environmental concerns.

The resulting legacy of environmental problems includes overcutting of forests, widespread overuse of pesticides, heavy pollution of many rivers and lakes, extensive problems of acid rain and soil erosion, and serious levels of air pollution in industrial towns and cities (**Figure 4.24**). Fragile environments at the margins of human settlement and on the peripheries of the Soviet empire have been among the worst affected. Across the far north, for example, air pollution produces a phenomenon

Figure 4.24 Industrial pollution Soviet industry was not only undercapitalized and obsolescent in most cases, but also poorly regulated in terms of its environmental impacts. As a result, much of the Russian Federation, Central Asia, and the Transcaucasus has inherited a legacy of continuing industrial pollution.

known as "Arctic haze," seriously reducing sunlight and so destroying delicate vegetation complexes that underpin fragile ecosystems. In the semi-arid and arid regions of the south, the diversion of rivers for irrigation schemes aimed at boosting agricultural productivity have depleted water resources in some areas and led to widespread soil erosion and desertification.

Contamination by radioactivity is seen by many to epitomize the consequences of Soviet attitudes toward the environment. Both the large-scale civilian nuclear energy program and the military nuclear capability of the Soviet Union were developed in ways that have resulted in an alarming incidence of radioactive pollution. Up to half a million people are believed to have been exposed to harmful doses of radiation in the southern Urals as a result of a series of events and leakages, the most dramatic of which was the "Kyshtym incident" at a nuclear reprocessing center in 1957. A nuclear waste tank exploded, severely contaminating an area 8 kilometers (4.96 miles) wide and 100 kilometers (62 miles) long, requiring the permanent evacuation of more than 10,000 people and the bulldozing of 23 villages. Other sites associated with serious radioactive pollution include reprocessing and waste storage facilities at Tomsk-7 and Krasnoyarsk-26, nuclear dumps on the island of Novaya Zemlya in the Arctic and the adjacent Kara and Barents seas, Lake Ladoga, and parts of Primorsky Kray in the far east.

It was, though, the disaster at Chernobyl, in the former Soviet republic of Ukraine, that became emblematic of the

Soviet nuclear legacy. Geographer Peter Gould, in a small but important book called *Fire in the Rain,* provided a dramatic account of the greatest nuclear disaster the world has ever known. As he put it:

> At 1 hour, 23 minutes, and 43 seconds after midnight on April 26, 1986, Reactor 4 at Chernobyl went into a soaring and uncontrollable chain reaction. Two seconds later the resulting steam explosion to the concrete housing blew the thousand-ton "safety" cover off the top of the reactor, and spewed radioactive materials high into the night sky equal to all the atomic tests ever conducted above ground.[2]

While Gould tells, in gripping detail, the sequence of events that led to the meltdown of Reactor 4, what is more central to our discussion is the impact that this technological accident had on people and the environment in the region immediately surrounding the nuclear facility and in the former Soviet Union more generally. The stories of radiation sickness and eventual ghastly deaths of the facility workers, firefighters, medical personnel, and other volunteers filled the international newspapers for weeks following the meltdown. Less graphic and less well remembered are the invisible and enduring impacts on the population of the area surrounding the power plant as well as the natural environment. Radiation particles entered the soil, the vegetation, the human population, and the rivers, effectively contaminating the entire food chain of the region. Secondary radiation continues to be a problem. In the immediate area surrounding Chernobyl, all the trees were contaminated. As the trees slowly die, rot, and decay, radioactive material enters the physical system as "hot" nutrients. More than 3000 square kilometers (1161 square miles) of trees were turned brown from radiation immediately following the accident, and it is still not clear how to decontaminate such a large area. The town of Chernobyl, as well as the surrounding area, will not be inhabitable for decades, if then. Fifteen years after the accident, it is clear that the most extensive legacy of pollution is in the newly independent republic of Belarus. Here, 25 percent of the land is considered uninhabitable, and thousands of villages have been abandoned. The government of Belarus has spent 15 percent of its gross national product—more than $235 billion—over the last decade on paying medical bills and resettling tens of thousands of people affected by the disaster.

Yet the Chernobyl event is only one of several major environmental disasters that have left an enduring legacy to the Soviet Union's successor states. The overexploitation of the Amu Dar'ya and Syr Dar'ya rivers for irrigation schemes to support intensive cotton monoculture in southern Central Asia resulted in a dramatic draw-down of the Aral Sea. The level of the Aral Sea has already dropped by more than 10 meters (33 feet) and the dessication of the former seabed, now littered with stranded ships (**Figure 4.25**), generates a constant series of dust storms that are thought to be the cause

[2]P. Gould, *Fire in the Rain: The Democratic Consequences of Chernobyl.* Cambridge: Polity Press, 1990, p. 2.

Geography Matters

Crime and Corruption

As the Soviet Union began to break up, crime and corruption began to appear on a scale that was unprecedented. *Perestroika* and *glasnost* allowed new criminal networks to emerge amid the rapidly changing political and social environment. With the breakup of the Soviet Union, there was a proliferation of ethnic *mafiyas*—Chechens, Azeris, Georgians, and so on—and a rapid spread of corruption that has become ingrained in the business world and the political system. As a result, the Russian Federation has joined the ranks of the world's kleptocracies (as in *kleptomania,* an irresistible desire to steal): Democratic governance has been subverted by organized crime and widespread corruption, which have also affected the nature of economic and social development in many places and regions.

Privatization and market reform created unprecedented opportunities for corruption and organized crime. In the institutional vacuum that followed the breakup of the Soviet Union, there was no accepted code of business behavior, no civil code, no effective bank system, no effective accounting system, and no procedures for declaring bankruptcy. Security agencies were disorganized, bureaucratic lines of command were blurred, and border controls between the new post-Soviet states were nonexistent. Before long, virtually all small private businesses were paying tribute to criminal groups, while an estimated 70 to 80 percent of larger firms and commercial banks had criminal connections. The U.S. Drug Enforcement Agency has estimated that up to 25 percent of commercial banks in Moscow were controlled by organized crime in 2000. Organized crime did not stop at the traditional activities of prostitution, drugs, auto theft, and protection rackets but moved straight into the more lucrative business of illegal traffic in weapons, nuclear materials, rare metals, oil, natural resources, and currency. In Chechnya, revenues from the sales of petroleum from more than 300 small but illegal oil refineries were being used in 2000 to finance Chechen rebels and Chechen *mafiya*. Levels of criminal violence escalated in parallel with rising levels of crime and corruption, and by the mid-1990s, contract killings had become a way of life in the Russian business world.

In 1999, U.S. investigators discovered that billions of dollars were being laundered through Bank of New York accounts, apparently shunted through a maze of companies traced to Russia. At least $4.2 billion and as much as $10 billion may have been laundered between October 1997 and March 1999. This was done through Bank of New York accounts largely in the name of Benex Worldwide Ltd., a firm that was said to be controlled by Semyon Mogilevich, allegedly a vicious individual known as the "brainy don." It is not clear exactly from where the money came: It may simply have been capital fleeing from the wreckage of the Russian economy; it may have been looted from International Monetary Fund loans or from revenues from state assets such as oil or aluminum; or it may have been from organized crime.

One result of all this is that crime and corruption have pillaged a significant proportion of resources that would otherwise have gone into restructuring the economy. Another is that crime and corruption have stifled the emergence of a civil society with a democratic base. A third is that the scale of crime and corruption in the Russian Federation has become an issue of geopolitical importance. Russian *mafiyas* have extensive connections to international organized crime, and so much money has been laundered through the international financial system by corrupt politicians and businessmen that the ramifications are truly global in scope. Global organized crime, particularly the Sicilian Mafia and the Colombian drug cartels, seized the chance to link up with Russian *mafiyas* to launder huge sums of money, to circulate counterfeit dollars by the millions, and to establish smuggling networks. The Russian *mafiyas,* in turn, seized the chance to extend their operations to the rest of the world. Today, the scope of Russian *mafiya* organizations ranges from prostitution rings in major tourist hubs outside Bangkok, Thailand, to heroin trafficking between Afghanistan and Europe by way of Central Asia, the Russian Federation, Belarus, and the Baltic states.

---

of unusually high levels of respiratory ailments among the people of the region. Another notorious example is provided by the case of Lake Baykal, whose unique ecosystem has been threatened by industrial pollution (see Geography Matters: Lake Baykal, p. 183).

Today, serious environmental degradation affects all parts of the region (**Figure 4.26**), a legacy of problems that in many ways have been intensified by the transition to market economies. The ubiquitous corruption that has come to characterize the region during the post-Soviet transition means that environmental regulations are easily ignored or circumvented. In addition, the 1998 economic crisis in the Russian Federation has clearly limited the country's ability to address its environmental problems. In fact, economic stress has led the Russian Federation to agree to become a dumping ground for other countries' nuclear waste. All over the industrialized world, atomic power plant construction is significantly slowing down because of unresolved safety considerations and the failure to develop a safe, permanent means of disposing long-lived nuclear waste. In 2000 the Russian Federation volunteered to

Figure 4.25   The Aral Sea   Like this area on the southern shore, some 24,000 square kilometers (11,000 square miles) of former seabed in the Aral Sea have become a desert of sand and salt.

store 2000 metric tons (about 4.5 million pounds) of highly radioactive nuclear waste from Switzerland over the next 30 years for roughly $2 billion. Several more countries, including Germany, France, South Korea, Taiwan, and Japan, are likely to take advantage of the relatively low price charged by the Russian Federation for nuclear reprocessing. The risks, though, are appallingly high, given the Russian Federation's industrial inefficiency, corruption, and organized crime.

Core Regions and Key Cities of the Russian Federation, Central Asia, and the Transcaucasus

As we have seen, the breakup of the Soviet Union and the transition toward market economies is beginning to modify patterns of regional development. Nevertheless, the core regions and key cities of this world region remain broadly the same, for the moment, as under state socialism and, before that, Imperial Russia.

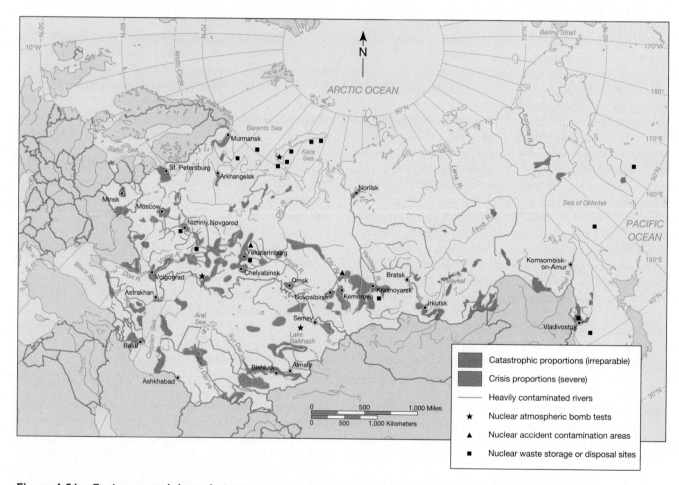

Figure 4.26   Environmental degradation   Serious environmental degradation afflicts all parts of the Russian Federation, Central Asia, and the Transcaucasus. The range of problems is enormous, and the intensity of many is such that severe problems are likely to persist for several generations.   (*Source:* Adapted from J. H. Bater, *Russia and the Post-Soviet Scene.* London: Arnold, 1996, p. 314.)

Geography Matters

Lake Baykal

Lake Baykal is the world's deepest lake at 1615 meters (5300 feet—more than a mile) and contains about 20 percent of all the fresh water on Earth—more than North America's five Great Lakes combined. It is also an unusually ancient lake. Most of the world's large lakes are less than 20,000 years old, but evidence from the 7-kilometer-thick sediment at the bottom of the lake shows it to have been in existence for at least 25 million years, perhaps even 50 million. Lake Tanganyika in Africa, the world's second-oldest lake, is 2 million years old. Lake Tahoe in the United States is a mere 10,000 years old. Most lakes fill with sediment in about 20,000 years, but Baykal has survived because it is located on a geological rift that grows nearly 2.5 centimeters (about an inch) a year—enough to accommodate the silt and animal remains that drift to the bottom. It has a unique ecology, with more than 2500 recorded plant and animal species, 75 percent of which are found nowhere else. These include the nerpa, Baykal's freshwater seal, separated by more than 3000 kilometers (1864 miles) from its nearest relative, the Arctic ringed seal. Ecologists have no understanding of how seals ever got to the lake or how they adapted to freshwater. Lake Baykal is also a place of incredible beauty—"The Pearl of Siberia"—that has become emblematic of the pristine wilderness of the region. In summer, the water's surface can be as smooth as glass, reflecting the snow-capped mountains, taiga forest, and little settlements of wood cabins that ring the lake. In places, the water is clear enough to see 30 meters (98 feet) deep. During a storm, the lake can be as rough as the ocean, with waves big enough to capsize a large boat. In winter, Baykal's surface freezes to a depth of a meter or more, forming a natural highway that temporarily provides a 636-kilometer (395-mile) north–south route for cars and trucks.

But the lake's purity and unique ecosystem have been compromised by environmental mismanagement. The first evidence of this was in the 1950s, when the lake's commercial fish populations nose-dived, partly as a result of overfishing and partly as a result of the construction of the Irkutsk dam (which raised the level of the lake and destroyed many of the shallow-water feeding grounds used by the fish). Then, in the 1960s, increasing levels of pollution were carried into the lake by the Selenga River, which supplies about half of the water that flows into the lake. The Selenga rises in mountain ranges to the south, but collects human and industrial waste, much of it untreated, from several large cities before entering Lake Baykal. The Selenga and other rivers also began to carry increasing amounts of agricultural chemicals such as DDT and PCB into the lake. Meanwhile, the purity of the lake's waters caught the attention of Soviet eco-nomic planners, who began to see the lake as a good location for factories that needed plentiful supplies of pure water (**Figure 1**). When the huge Baikalsk Pulp & Paper Mill, which produces high-quality cellulose for the Russian defense industry, was opened in the early 1960s, Lake Baykal became the birthplace of Russia's environmental movement. The mill pumps 140,000 tons of waste—including deadly dioxins—into the lake every day, along with 23 tons of pollutants into the atmosphere. It is estimated that over the past 38 years the mill has spewed more than a billion tons of waste into the lake.

As a result of unprecedented expressions of public concern, the Soviet government ordered the mill closed in 1986. But the government collapsed before the closure took effect, and since 1989 the mill has been partially privatized and now makes pulp for low-quality paper rather than cellulose. The mill has managed to stay afloat and pay its workers on time but, with its obsolete equipment, it is nearing the end of its life span. No money has been invested in a decade. From time to time, the mill owners pay symbolic fines of 500 to 1000 rubles ($85 to $170) for exceeding pollution limits, but if required to pay the hundreds of thousands of dollars they owe for environmental damage, the mill would go bankrupt. When thousands of the lake's freshwater seals began dying in 1997, the lake's fragile ecology came under international scrutiny, and in 1998 the lake was designated a World Heritage Site by UNESCO, the U.N. cultural agency. Nevertheless, it remains to be seen whether Russia can solve its environmental problems at a time when its economy is in disarray.

Figure 1 Lake Baykal A chemical plant on the shores of Lake Baykal.

The principal core region is the Central Region that extends for a radius of approximately 400 kilometers (248 miles) around Moscow. Secondary core areas, developed around long-standing industrial subregions, exist in the lower Volga region and the Urals. Between them, these core regions contain 11 of the 13 largest Russian Federation cities listed in Table 4.1 and about half of the total population of the Russian Federation.

The Central Region

The Central Region was the hearth of the Muscovite state, the base from which its growing power thrust along the rivers in all directions toward distant seas. Many of the ancient towns have preserved their beautiful medieval *kremlins* (citadels) and churches, as in Rostov-on-Don, Suzdal, Yaroslavl', and, of course, Moscow. The industrial roots of the region go back to the 1600s, when various early industries, drawing on local resources of flax, hemp, hides, wool, and bog-iron, developed to serve the needs of the growing capital Moscow. By the 1700s, the region had developed a specialization in textiles; and with the onset of Russia's industrial revolution in the 1800s, the textile industry expanded and was joined by a broad range of engineering and manufacturing. Although the Central Region has no significant sources of energy (apart from low-grade lignite and peat deposits that can be converted into electricity in power stations), Soviet economic planners regarded the region as pivotal to their industrialization policies. Significant imports of coal, oil, gas, and electricity were used to develop a broad economic base. The region was also key to the Soviet Union's drive for technological supremacy, and a considerable proportion of the country's leading scientific research and development institutes were established as part of the Central Region's massive military-industrial complex.

Today, the Central Region is highly urbanized, with about 85 percent of the population living in towns and cities. Moscow (population 8,400,000) dominates the entire region, but other significant centers include Dzerzhinsk (285,000), Ivanovo (474,000), Nizhniy Novgorod (1,400,000), Ryazan' (536,000), Serpukhov (139,000), Smolensk (355,000), Tula (532,000), Vladimir (339,000), and Yaroslavl' (629,000). The region accounts for about 80 percent of the Russian Federation's textile manufactures. Cotton textiles are most important and are produced mainly in towns along the Klyazma valley between Moscow and Nizhniy Novgorod and at Ivonovo. Woolens are manufactured in and around Moscow; synthetic fibers are produced in Kalinin, Serpukhov, and Vladimir; and the ancient linen industry survives at Kostroma. Engineering, automobile and truck manufacture, machine tools, chemicals, electrical equipment, and food processing are also important. Overall the Central Region accounts for about 20 percent of the Russian Federation's industrial production.

In spite of the high degree of urbanization and industrialization, much of the region has a rural flavor (**Figure 4.27**). The rural landscape is varied, the effects of glacial action having produced a complex topography with relatively poor quality soils except to the south, which borders on the fertile zone of chernozem soils. About 25 percent of the Central Region

remains forested, and there are numerous lakes and marshy areas. The traditional staple crop of the region was rye, but when the railways made it possible to import cheaper grain, the farmers of the region turned to industrial crops, such as flax, and to potatoes, sugar beets, fodder crops, dairying, and market gardening. Around rural settlements, there are orchards of apples, cherries, pears, and plums.

Beyond these traditional rural landscapes, the geography of the Central Region is being significantly rewritten by the transition to an open, internationalized market economy. The effects of the transition are very uneven. On the positive side, the gateway situation of the Moscow region, with good transport and communication links to other regions and countries, is attracting a good deal of foreign direct investment, while the region's high-tech labor force and research institutes have also been attractive to investors. On the negative side, reduced domestic demand and competition from cheap imports have had severe adverse effects on the region's smokestack industries, especially textiles, machine building, and engineering. Among the places worst affected by **deindustrialization** are the eastern parts of the Moscow region, Bransk, Kostroma, Vladimir, and Yaroslavl'.

Moscow   Moscow is situated at the center of the vast Russian Plain, on the Moskva River, a tributary of the Oka, which in turn leads to the Volga. The city's growth is reflected in its layout by a series of ring roads: the Boulevard Ring, the Garden Ring (both following the line of former fortifications), the Greater Moscow Ring Railway, and the Moscow Circular Beltway. Radial boulevards penetrate these ring roads, converging toward the ancient hub of the city, the Kremlin, built as a fortified palace complex in the fourteenth century and subsequently used as the seat of government for Russia, the Soviet Union, and, now, the Russian Federation. Next to the Kremlin is the famous Red Square, the ceremonial center of the capital that is anchored at its southern end by the Cathedral of St. Basil the Blessed, built between 1554 and 1560 by Ivan the Terrible to commemorate the defeat of the Tatars (**Figure 4.28**).

Moscow's inner city differs from large European and North American cities in that it does not contain any significant slums, nor does it contain a modern central business district. Rather, it contains a mixture of buildings representative of every period of the city's development: churches and institutional buildings from the fifteenth through the nineteenth centuries, interspersed with Soviet-era offices, apartment buildings, squares, and boulevards that gradually replaced the city's slums and dilapidated buildings from the 1920s onward. Beyond the Garden Ring as far as the Ring Railway is a zone of eighteenth- and nineteenth-century development—including the principal railway stations and freight yards, factories, and associated housing—that has been the target of extensive urban renewal projects. Here also are many of the larger institutional buildings of the Soviet era, including most of the distinctive "wedding-cake" skyscrapers of the Stalinist period. The outer zones of the city, beyond the Ring Railway, are almost entirely the product of post-Second World War growth, when the built-up area of the city increased more than tenfold. Immediately

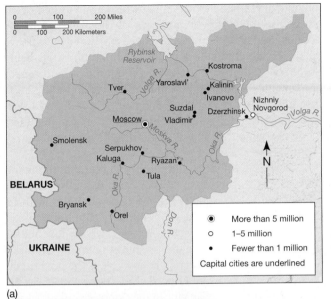

Figure 4.27   The Central Region
(a) Map of the region; (b) farmland in the Central Region, to the Northeast of Moscow; (c) rural housing; and (d) cottages on the outskirts of Suzdal.

beyond the Ring Railway are the *mikrorayoni* of the 1950s, dominated by five- to nine-story apartment buildings of yellowish brick. Farther out, larger factories and standardized high-rise apartment blocks of precast concrete dominate the cityscape, while in the outermost zones are scattered a series of satellite industrial towns amid open land and forest.

The transition to a market economy has already left its mark on Moscow in several ways. The combination of a newly emerging wealthy class, together with a current chaotic planning situation, has sparked a spate of uncontrolled housing construction. Some of this new housing has taken place within the borders of the city but a good deal of the recent growth has been in the forest protection belt, where speculative developments, mostly funded by foreign companies, have sprung up, providing expensive housing in community-style developments with tight security. New office buildings for

Figure 4.28   Moscow   (a) Red Square; and (b) Moscow cityscape.

transnational companies and new, Western-style stores have begun to appear in the center of the city, consuming many of the green sites created by Soviet planners. Meanwhile, a sharp rise in the number of private automobiles (from approximately 0.5 million vehicles in 1985 to more than 1.7 million by 2000) has led to an equally sharp rise in traffic congestion and unprecedented strain on the existing road infrastructure.

The greatest impacts of the transition to a market economy, however, are the social and economic consequences. Formerly a leading industrial center, by 2000, Moscow had slipped to fourteenth place among the Russian Federation's economic regions in terms of industrial production. The decline of the traditional industrial sector in Moscow has to a large extent been compensated by the rise of new sectors in the economy, particularly in tourism, retailing, and banking. Moscow has quickly developed as the Russian Federation's principal center for financial and business services, which in turn has attracted a good deal of foreign investment and many joint ventures. Moscow's new status as a **world city** has also led to the emergence of new culture and entertainment industries, while the overall climate of change has fostered a proliferation of small, private enterprises. The pace of change, however, has far outstripped the capacity of the city's authorities to regulate and control it, so that the positive aspects of transition have been accompanied by dramatic increases in social polarization and crime and by a sudden backlog in the provision of an adequate infrastructure for ground and air transportation and telecommunications.

St. Petersburg   St. Petersburg is located some 650 kilometers (403 miles) from Moscow, at the eastern end of the Gulf of Finland. As such, it falls well outside the Central Region proper. Nevertheless, it must be considered as an extension of the Central Region, an industrial, cultural, and administrative metropolis whose communications and fortunes have always been tied to Moscow and the Central Region. Most of the surrounding northwest region forms part of the original Russian homeland, though the area bordering on the Gulf of Finland was long disputed. When it was finally annexed to Russia by Peter the Great at the beginning of the eighteenth century, Peter decided to build a new capital city—St. Petersburg—on a swampy site at the mouth of the River Neva. Like Venice, Italy, St. Petersburg rests on countless wooden piles to prevent it from subsiding into its marshes. It was built at the cost of thousands of human lives, but the tsar was determined to create an imperial capital to rival those of continental Europe. He also wanted St. Petersburg to be Russia's "window on Europe," exposing Russia to new ideas and technology.

Canal and, later, railway connections to Moscow allowed St. Petersburg to flourish as Russia's chief **entrepôt**, and the city quickly became an important cultural and intellectual center. With the Soviet revolution of 1917, however, Moscow was reinstated as the capital city. Anti-German sentiment, meanwhile, had caused St. Petersburg to be renamed as Petrograd. The city's entrepôt function withered under state socialism's doctrine of national economic self-sufficiency, but Soviet planners

quickly developed Petrograd—renamed again as Leningrad in 1924—into a key component of the Soviet military-industrial complex. During the Soviet era, the city was above all a manufacturing center, the principal industries being electrical and power machinery, shipbuilding and repair, armaments, electronics, chemicals, and high-quality engineering. The city suffered terribly in the Second World War, with an estimated 1 million of its residents dying from hunger or disease while the city was under siege by the German army for 872 days.

Today, the city's imperial past is very visible in the Grand Design of the core area on the south bank of the River Neva—the Palace Square, the Admiralty building with its landmark elegant spire, and the Winter Palace (see Sense of Place: Imperial St. Petersburg, p. 188), which now houses the Hermitage Museum, a treasure house of fine art of worldwide significance that originated in 1764 as the private collection of Empress Catherine II. The Soviet past is visible, as in every other Russian city, in the extensive industrial and residential suburbs of standardized high-rise apartment blocks of precast concrete. Renamed once more as St. Petersburg, the city has faced a tough period of readjustment as its older military-industrial base experienced a sharp decline in fortune. Crime and corruption also emerged as striking features of the post-Soviet city. Nevertheless, St. Petersburg is once again poised to take advantage of its gateway situation, while its imperial legacy makes it an attractive international tourist destination. Already, the city handles about 35 percent of the Russian Federation's imports and about 30 percent of its exports. Although the city's infrastructure badly needs upgrading, its history and its European ambience are beginning to prove attractive not only to tourists but also to Western investors.

The Volga Region

This region (**Figure 4.29**) is known in the Russian Federation as Povolzhe, which means "along the Volga River." It has been the main artery of the Russian heartland for more than a thousand years, with Slav and Norse adventurers trading along the Volga en route between the Baltic and Byzantium. In modern times the lower Volga region has become a core economic region, with a series of industrial cities along the river—from Kazan in the north to Astrakhan, situated on the Volga delta as it enters the Caspian Sea—and extensive areas of productive agricultural land. The region has a long tradition of crafts and engineering, a strong supply of natural resources, and a good strategic location between the industries of the Central Region and the raw materials and natural resources of the Urals. In the 1920s oil was discovered in the northern part of the Volga Region (part of the Volga-Ural oilfields that extend northeastward to the western foothills of the Urals), and natural gas was found near Astrakhan. The Volga itself was exploited as a major resource as well as a routeway in the 1930s, when a string of monster hydroelectric projects was installed along the middle reaches of the slow-moving river.

Figure 4.29   The Volga region Reference map showing principal physical features, political boundaries, and major cities of this region.

These resources helped to sustain the industries of the Central Region as well as those of the Volga Region. The principal industries of the cities of the Volga Region are chemicals and petrochemicals, engineering, aerospace, and automotive manufacture and assembly. The Volga Automobile Plant at Togliatti produces two-thirds of the Russian Federation's passenger cars, while the Kama Automobile Plant at Naberezhnye Chelny and the Ul'yanovsk Automobile Plant are also important. The surrounding agricultural lands, meanwhile, are among the Russian Federation's most productive, accounting for almost 20 percent of the country's total arable land.

Volgograd is the largest city in the region (population 1,003,000 in 1998). Formerly called Stalingrad, the city was almost totally destroyed between July 1942 and February 1943 in fighting between the German and Soviet armies. The Soviet victory proved to be a major turning point in the Second World War. Completely rebuilt since 1945, Volgograd became a "hero-city" of the Soviet Union, an important river port and manufacturing center producing iron and steel, machine tools, tractors, oilfield machinery, petrochemicals, railway equipment, footwear, and clothing.

Overall, the Volga Region contains almost 17 million people and remains one of the most prosperous regions within the Russian Federation. Its relative prosperity is, however, threatened by nationalist tensions. Tatarstan, in the northern part of the region, declared its autonomy from the Russian Federation in 1990, though in the end it did not secede. Another shadow on the region's prosperity is the serious pollution that has been allowed to affect the Volga. Huge amounts of sewage and

industrial wastes enter at every town and city on the river, whose flow has become much reduced by the numerous dams and reservoirs. As a result, large stretches of the Volga are stagnant and algae-covered, in danger of becoming dead zones. One notable casualty of this pollution has been the sturgeon fisheries of the lower Volga. The roe, or raw eggs, of the sturgeon are the source of caviar. Production of this highly prized commodity has fallen from 2000 metric tons a year in 1978 to just over 150 metric tons in 1998, partly as a result of pollution and partly as a result of illegal poaching of immature fish, prompted by the shortages caused in the first place by pollution.

The Urals

The Urals have been economically important to Russia since the early 1700s, when Peter the Great established iron-smelting works in Yekaterinburg and Nizhniy Tagil using local ore and charcoal. In addition to iron ore, the Urals contain a wealth of mineral resources, including asbestos, bauxite, chromium, copper, magnesium and potassium salts, and zinc. The energy resources of the southern Urals include extensive low-grade bituminous coal and lignite deposits, the Volga-Urals oilfield (which extends southwestward from the western foothills of the Urals to the Saratov area in the Volga Region), and a major natural gas field at the southern end of the mountains, near Orenburg (**Figure 4.30**).

The rise of the Urals as an economic core region can be traced to Stalin's industrialization drive of the late 1920s, when

Figure 4.30   The Southern Urals Reference map showing principal physical features, political boundaries, and major cities of the southern Urals.

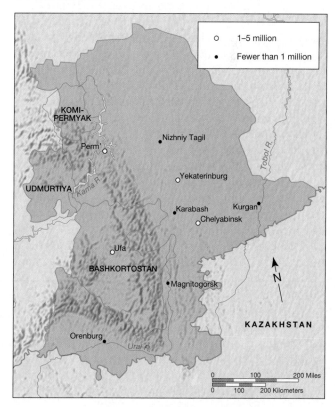

Imperial St. Petersburg

In spite of seven decades of socialism and more than a decade of hardship and disorganization in the transition to a post-Soviet society, St. Petersburg remains an impressive and inspiring city. The city was home not only to Peter the Great but also to Dostoevsky, Nijinsky, and Lenin. The great composers Rimsky-Korsakov, Mussorgsky, Borodin, and Tchaikovsky are buried in the city's Tikhvin Cemetery. But it is the city's core of imperial architecture and urban design that provides its sense of place and symbolizes its sophistication. Often called "The Venice of the North" because of the opulence of its architecture and its canals, St. Petersburg was founded in 1703 by Peter the Great and was the tsars' imperial capital until the Bolshevik revolution of 1917. During that time, St. Petersburg was deliberately fashioned in the Grand Manner as a European-style capital city. The tsars' architects were able to lay out their work unrestricted by any legacy of old streets or buildings. Over two centuries, they collectively created a marvelous set piece of urban design, with imposing public buildings, imperial palaces, and churches in the baroque, rococo, and classical styles, all laid out around impressive plazas and along broad boulevards, all surrounded by large fashionable residences (**Figure 1**).

While Peter the Great founded the city, it was his daughter Empress Elizabeth (who ruled from 1741 to 1761) who commissioned the first grand wave of buildings, including the Winter Palace and Smolny Cathedral. Subsequently, Catherine the Great (1762–96) and Alexander I (1801–25), in a drive to make St. Petersburg Europe's most imposing capital, commissioned dozens of elaborate projects. Among the most notable of these are the Russian empire-style Admiralty buildings, the baroque Winter Palace (commissioned by Empress Elizabeth but remodeled and extended by Catherine the Great and her successors into a palace complex that now houses the glittering national art collection of the State Hermitage), and the extravagant St. Isaac's Cathedral.

Taken together, these projects were intended to be the proscenium for the lifeworld of the imperial court and its attendant diplomats, artists, and society. Today, the built environment of imperial St. Petersburg still carries a strong sense of place, though the feeling of the city varies strikingly with the seasons. St. Petersburg's latitude (60° N) is level with Cape Farewell, Greenland, and its long summer days and "white nights" (when, although the sun sets, it stays light for

(a)

(b)

Figure 1 St. Petersburg (a) Palace Square and the General Staff Headquarters and (b) frozen canal in Central St. Petersburg.

hours) see the imperial city packed with citizens and with tourists from around the world, strolling the great boulevards and along the banks of the Neva River. On summer evenings, with the setting sun glinting gold off the canals and intensifying the rich colors of the churches and major buildings, the city has a sophisticated and romantic air. In winter, when the Neva is frozen and temperatures rarely exceed freezing, the low daytime sun and long nights conspire to reinforce a muted, overly spacious atmosphere.

Soviet economic planners established a major industrial complex focused on mining, ferrous and nonferrous metallurgy, and heavy engineering. The discovery of the Volga-Urals oilfield in the 1930s added chemicals and petrochemicals, as well as a new source of energy. At about the same time, Soviet planners decided to bring in high-grade coking coal from the

Kuznetsk basin in Central Siberia, 2000 kilometers (1240 miles) to the east, in order to fully exploit the vast reserves of iron ore in the Urals. Trains returning to the Kuznetsk coalfields were loaded with iron ore from the Urals, thus facilitating the creation of an iron and steel complex in the Kuznetsk basin. This arrangement was known as the Urals-Kuznetsk

Combine. During the Second World War, a substantial amount of additional industry was moved to the Urals from European Russia in order to be safely out of range of German air attack. After the Second World War, the Urals were further developed as a key region of the Soviet military-industrial complex, and most of the region was closed to foreigners until 1990. Five of the 10 secret cities of the Soviet Union's "nuclear archipelago" for weapons research and production were located in the Urals: Sverdlovsk-44, Sverdlovsk-45, Chelyabinsk-65, Chelyabinsk-70, and Zlatoust-36 (the numbers are their postal codes; Sverdlovsk was the Soviet name for Yekaterinburg, after Yakov Sverdlov, who arranged the murders of Tsar Nicholas II and his family in the wake of the Soviet revolution).

Today, the southern Urals remain a key component of the Russian Federation's economy. The region is highly urbanized: 75 percent of its 20 million people live in cities, the largest of which is Yekaterinburg (population 1.3 million), an important railway, metallurgical, and heavy engineering center. Perm' (population 1 million) has a similar economic profile, while Ufa (1.1 million) is a center of oil refining, chemical manufacture, and heavy engineering. Chelyabinsk (1.1 million), Nizhniy Tagil (409,000), and Magnitogorsk (427,000) are all important iron and steel centers with associated heavy engineering complexes. Many of these industries, however, have fared poorly in the transition from state socialism to a capitalist economy. Nonferrous metallurgy, petrochemicals, and the oil industry have attracted foreign investment and provided an important source of export revenues for the Russian Federation, but the iron and steel and heavy engineering industries, with their obsolescent plants, have been noncompetitive and unattractive. Many of the defense-related industries, meanwhile, have suffered as a result of the inability of the Russian Federation to fund them at a level comparable to that of the Soviet era.

In many communities, the economic hardship resulting from the decline of smokestack industries has been compounded by the legacy of pollution. Take, for example, the small town Karabash (population 15,000 in 2000), about two hours south of Yekaterinburg in the foothills of the Urals. The Karabash Copper Smelting Works, established in the 1940s to produce copper for ammunition, went on to produce copper for electrical equipment until 1987, when it was closed by the government because the plant's productivity was too low and its hazardous emissions too high. But the plant was reopened in 1998, when the town's residents had become desperate for jobs, and the government had become desperate for tax revenues from the region's copper. Air pollution has returned, compounding the environmental hazards from the 15-meter (49-foot) heaps of slag that line the main road, encircle the factory, and spill over into backyards. Slag, laced with lead, arsenic, and cadmium, is the waste from the copper smelting furnaces after raw copper is extracted from ore. Two-thirds of the children in Karabash suffer from lead, arsenic, or cadmium poisoning, and a 1995 government study found that they also suffer from at least twice the rates of congenital defects, disorders of the central nervous system, and diseases of the blood, glands, and the immune and metabolic systems compared with children in nearby towns.

Distinctive Regions and Landscapes of the Russian Federation, Central Asia, and the Transcaucasus

In comparison with other world regions, the distinctive regions and landscapes of the Russian Federation and Central Asia (though not the Transcaucasus) are vast and seemingly endless. Distinctive mid-sized regions comparable to, say, Alpine Europe or the American Midwest, have not emerged, while landscapes run virtually unchanged over thousands of kilometers across the two large and geologically stable shields of highly resistant crystalline rocks that provide platforms for extensive plains of sedimentary material and glacial debris. Nevertheless, there are several broad regions with distinctive natural and cultural landscapes. In the far north there are the zones of tundra and taiga. Farther south, the steppes have always been identified as a distinguishing feature of Russian geography, while in Central Asia it is the deserts that frame the regional identity of the area and provide the settings for distinctive natural and cultural landscapes.

The Tundra

The tundra zone extends along the northern shores of the Russian Federation and includes its Arctic islands (**Figure 4.31**). Altogether, it amounts to 2.16 million square kilometers (833,760 square miles), representing almost 13 percent of the country. Almost all of it was sculpted by one or another of the great ice sheets of the Quaternary ice ages (between 1.7 million and 10,000 years ago). These, wrote geographer W. H. Parker, "scraped, polished, grooved, crushed or sheared the rocks in their advance, and dropped boulders and stones haphazardly on their retreat."[3] Frost action is still the main modifier of the landscape, since running water is almost entirely absent from the tundra zone. For nine months or more, the landscape is locked up by ice and covered by snow, while during the brief summer, much of the melted snow and ice is trapped in ponds, lakes, boggy depressions, and swamps by permafrost that extends, in places, to a recorded depth of 1450 meters (4757 feet). Water seeps slowly to streams that drain into the slow-moving rivers that cross the tundra to drain into the Arctic.

During the winter, the tundra landscape has a uniformity that derives from the snow cover and the somber effect of long nights and weak daylight, the sun remaining low in the sky. There is little sign of life, apart from herds of reindeer or occasional polar bear or fox. In summer, vegetation bursts into life, and animals, birds, and insects appear (**Figure 4.32**). The days are long—in June the sun circles the horizon and there is no night at all. Mosses and tiny flowering plants provide color to the landscape, contrasting with the black peaty soils and the luminous bright skies. Wildfowl—swans, geese, ducks, and snipe—arrive on lakes and wetlands for their breeding season, as do seabirds and seals along the coast. Wolves enter

---

[3]W. H. Parker, *The Soviet Union*. Chicago: Aldine, 1969, p. 42.

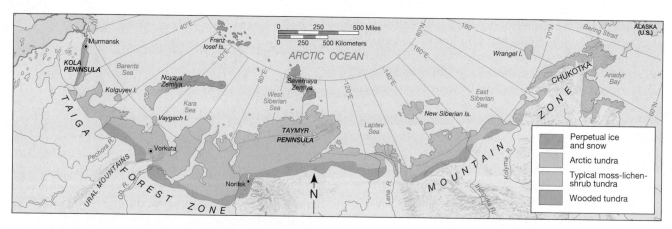

Figure 4.31   The tundra region Reference map showing principal physical features, political boundaries, and major cities. (*Source:* After W. H. Parker, *The Soviet Union.* London: Longman, 1969, pp. 46–47.)

the tundra from the forests to the south in search of prey that includes Alpine hare and lemmings (small, fat rodents that can produce five or six litters annually). Everywhere there are swarms of gnats and mosquitoes. Summer also brings a zonal differentiation to the region, with shrubs—blackberry, crowberry, and cowberry—and dwarf trees—birch, spruce, and willow—becoming more frequent as one moves south toward the taiga.

The indigenous peoples of the region were Sami in the west, Samoyeds or Nenets in the center, and Chukchi and Inuit in the east. Nomadic peoples who subsisted on reindeer, along with a variety of fish, eggs, and fowl, they probably never amounted to more than 30,000 or 40,000 individuals in total. Between the thirteenth and sixteenth centuries, they were joined by a few Russians who had pushed north in order to trade with them for furs and fish. Eventually, between 15,000 and 30,000 Russians settled in small fortified trading posts along the coasts of the Kola Peninsula and the lower estuaries of the larger rivers, supporting themselves by fishing, trapping, and hunting. Today, tundra landscapes contain very little evidence of either nomadic peoples or frontier settlers. The

human imprint on the landscape derives mainly from Soviet-era development.

Soviet interest in the tundra was driven initially by a desire to exploit the natural resources of the region. In addition, Stalin found use for the region as a setting for the "gulag archipelago" of prisons and labor camps that were described so tellingly by novelist Aleksandr Solzhenitsyn. Then the Cold War gave strategic significance to the region, with the result that a number of important military installations were developed. By 1989 the tundra was, remarkably, home to more than a million people.

The natural resources that attracted Soviet economic planners include coal, natural gas, petroleum, copper, gold, nickel, platinum, and tin. Fish-processing plants and fat-rendering furnaces were established along the Arctic coast, while a fleet of icebreakers was established in order to create a northern sea route that saves more than 7000 kilometers (4350 miles) in travel by sea between the Far East and the Baltic. These developments have not transformed the landscapes of the region as a whole but have introduced towns and cities that stand in stark contrast to their surroundings. Murmansk (**Figure 4.33**), the largest city and a major naval port, has a population of

Figure 4.32   Tundra landscape A herd of reindeer on the Siberian Tundra, Chukchi Peninsula.

Figure 4.33   Murmansk The world's largest city north of the Arctic Circle, Murmansk was founded in 1916 on an estuary of the Kola River, where the North Atlantic Drift keeps the waters ice-free all year-round.

407,000; Norilsk, the second largest, is a mining and smelting center and has a population of 159,000. The population of the entire region, though, is now declining rapidly. The economic crisis that has afflicted the Russian Federation throughout its transition to capitalism has made it increasingly difficult to support the towns and cities strung across the Arctic. Many of the smaller settlements, in particular, have become hopelessly expensive to maintain, and life for their inhabitants is hard. Unable to make a living, many people have been moving out. The Chukotka region, for example, which embraces the Russian Federation's northeastern tip, just across the Bering Strait from Alaska, lost fully half of its population in the 1990s. Twenty-six villages were abandoned, Russians and Ukrainians leaving the huge expanse of tundra to the native Chukchi people, whose traditional way of life had been nearly wiped out by a combination of Soviet policy and alcohol abuse.

The Central Siberian Taiga

The boreal (northern) forests of central Siberia cover a vast territory—more than 3.4 million square kilometers (1.3 million square miles)—from the Yenisey River to the Verkhoyansk mountain range beyond the Lena River. The underlying rock shield, having been steadily lifted by geological movements, averages about 700 meters (2296 feet) in elevation and has been dissected by rivers into a hilly upland topography with deep river gorges that are cut 300 meters (984 feet) or more below the general level of the plateau. The steep sides of the valleys are notched by numerous terraces, marking successive stages in the uplift of the ancient crystalline rock plateau. This topography, though, is given a striking uniformity by the distinctive forest cover of the region. The characteristic forest is made up of larches: hardy, flat-rooted trees that can establish themselves above the permafrost. Slow-growing and long-lived, they are able to counter the upward encroachment of moss and peat by putting out fresh roots above the base of the trunk. The dominant larches can grow to 18 meters (59 feet) or more, allowing an undergrowth of dwarf willow, juniper, dog-rose, and whortleberry.

The indigenous peoples of the region were Tungus, who lived by hunting and fishing; and Yakuts, a seminomadic, pastoral people who moved in from the south, creating pastureland of their own by burning down patches of forest. These peoples were largely displaced in the sixteenth century, when Russians, having defeated the Tatars, moved into the region to exploit the bark, timber, pitch, and fur-bearing animals of the forests. Wherever Russians settled, trees were cut down to make forts, stores, churches, and cabins. Logs were also laid on streets and roads in order to make them passable in the muddy season. Over time, more of the region's resources were discovered—fish and hydro power from the rivers, fossil fuels, and minerals in the rocks—but the region has remained one of the most sparsely settled in the world because of its formidable natural difficulties: the coldest winters on Earth, summers that bring plagues of mosquitoes, boggy and infertile soils, vast distances, and topography that makes the building

Figure 4.34   Siberian village Most settlements in the Central Siberian Taiga, like Uruscha, shown here, are small and isolated.

of roads and railways difficult and expensive. The settlements of the region are characteristically small and isolated villages (**Figure 4.34**).

The region's physical isolation and remoteness have led to one of its distinctive cultural attributes—as a region of voluntary escape and forced exile. On the one hand, central Siberia became an asylum. Peasants moved to the forests of Siberia to escape the serfdom of European Russia, while some better-off families moved to escape the Orthodox religious doctrine of pre-revolutionary Russia. On the other hand, both the tsars and the Soviets made central Siberia a place of exile and punishment. Siberia has never been seen as a region to be developed and settled, so much as an obstacle on the way to the far east. The famous Trans-Siberian Railway was not aimed at opening up Siberia but at securing communications between European Russia and the far east. Nevertheless, it was the railway line, running across the southern part of the region, that attracted the most settlement and economic development. To further facilitate the development of Siberia, the Soviets were encouraged to build the 3204-kilometer (1990-mile) Baykal-Amur Mainline (BAM) in the 1980s. The principal cities of the region are strung along the Trans-Siberian Railway. These include Krasnoyarsk (population 869,000 in 1998), Irkutsk (585,000), Novokuznetsk (572,000), and Angarsk (267,000). Bratsk (257,000) is the principal settlement along the BAM.

These cities support a variety of smelting, engineering, metalworking, and food-processing industries, but it is the forest that remains the economic and cultural staple of the region. The cities of the region rely on the forests for woodworking, pulp and paper industries, and wood chemicals. Their citizens rely on the forests not simply for recreation but for a sense of identity (see Sense of Place: Krasnoyarsk, p. 192). The central Siberian taiga is one of the richest timber regions in the world. Overall, close to 90 percent of the territory is covered with forest. More than a quarter of the

Sense of Place

Krasnoyarsk

It is a frigid, ice-blue morning as we wait for a train out of town. The sun has not yet risen, and distant mountains stand black against the horizon. This is Krasnoyarsk (**Figure 1**), crumbling metropolis of 1 million on the banks of the mighty Yenisey River. Away from the belching smokestacks of its industrial left bank, it is almost a handsome city. On broad, tree-lined avenues, ornate wooden mansions remind one of a Cossack past, and gaudy 1950s blocks of flats mingle with the fortress-like towers of the newly rich.

For most people, though, life is hard. Wage arrears mount month on month. Pensions are late and inadequate. Soaring inflation means lean pickings for the majority. Crime, unemployment, and homelessness add to the rising tide of despair. It's a bleak world if you let it get to you, which is why even on the coldest winter weekends people throng station platforms for an out-of-town break. "It's what keeps us sane," explains Albina Aleksandrovna, a feisty grandmother of six. A librarian at the city's institute of technology, she has not been paid for months. "The snow and the sun and the freedom of the forest are what sustain me from week to week." Forty minutes out of town, the train carriages have almost emptied. There's time for a communal cup of tea fortified with illicit home-brewed samagon before we skiers bundle out at a broken-down siding.

Half an hour along our cross-country route we are well strung out along the forest path. Lungs pounding out torrents of frozen breath, we speed down narrow forest corridors, then burst into dazzling clearings. Thick layers of snow blanket every horizontal surface. Trees slumber under their voluptuous covers. The only movement is the flutter of finches, and our kick-glide, kick-glide as we speed exhilaratingly past.

Twelve o'clock, and a wooden sign tells us we've come halfway. It's time for the Russian winter sportsman's brew-up. Soon a healthy blaze is going and great, battered boiling cans emerge from rucksacks. There are hunks of black sourdough bread, pickled cucumbers, salted fish, steaming potato soup, and weighty slabs of smoked pork fat that melt deliciously in the mouth.

Figure 1 Krasnoyarsk

In high spirits we strike out again into the forest. By early afternoon the forest thins, and we catch glimpses of the sluggish Yenisey, heaving its way north to the Arctic Ocean. Our destination is Divnogorsk (Marvellous Mountain), a beleaguered 1960s township built for workers on Krasnoyarsk's vast hydroelectric dam. As we skim along high, wooded banks we catch the tiny sounds of a crooning Russian diva. The unmistakable aroma of kebabs wafts on the evening air, signaling civilization and our journey's end.

Under flickering lights at the grimy bus station other groups converge in a clatter of skis. On the 10-ruble bus ride home, approaching the industrial glare of the night city, we retreat into contemplation. I think of sons at war, of jobless husbands, brothers, wives, of bright children with dreary futures, crowded homes with empty shelves, bank balances "lost," and wages due. But next to me, Sasha Alexeiyevich, Exalted Mountaineer of the USSR, is beaming. "Isn't new snow in the forest the most beautiful thing in the world?"

Source: Adapted from G. Mocatta, "Letter from Siberia," *Guardian Weekly,* February 17–23, 2000, p. 22.

Russian Federation's lumber production comes from the region. Poor forest resource management, however, gives cause for concern. After decades of relentless Soviet exploitation, the taiga, once dense and practically impassable, has been cleared throughout much of the southern parts of the region and around all the larger towns and cities. Loggers are now moving farther and farther north to cut down century-old pines. The post-Soviet transition has intensified concern over forest resource management as privatization has attracted

U.S., Korean, and Japanese transnational corporations to invest in "slash-for-cash" logging operations in a loosely regulated and increasingly corrupt business environment. The future of the central Siberian taiga has become an issue of international concern because the taiga accounts for a significant fraction of the world's temperate forests, which absorb huge amounts of carbon dioxide gas in the process of photosynthesis, thereby removing from the atmosphere a gas that is a main contributor to global warming.

The Steppes

The steppe belt stretches about 4000 kilometers (2486 miles) from the Carpathians to the Altay Mountains (see Figure 4.3), covering a total area of more than 3.25 million square kilometers (1.25 million square miles). The topography of the region is strikingly flat: yellow **loess** (a fine-grained, extremely fertile soil), several meters thick, has blanketed the underlying geology, creating a rolling landscape of unbounded horizons, punctuated here and there by incised streams and river valleys. The natural landscape of feather grasses and steppe fescue has generated the characteristic rich chernozem soils of the region. Trees and shrubs are restricted to valleys, where broadleaved woods of oak, ash, elm, and maple have established themselves, with pinewoods and a low scrub of blackthorn, laburnum, dwarf cherry, and Siberian pea-trees in drier locations.

For centuries, the steppe region was the realm of nomadic peoples, including Pechenegs, Kazakhs, Scythians, and Tatars. It was not until the late 1700s, when the Turkish empire's hold on the steppes was broken, that large numbers of colonists began to enter the western steppe. Along with Russians and Ukrainians came Armenians, Bulgarians, French, Germans, Greeks, Montenegrins, Serbs, and Swiss. Wheat-growing rapidly expanded wherever transportation was good enough to get the grain to the expanding world market, but large flocks of sheep dominated most of the colonized steppe until the railway arrived. With the arrival of the railway, German Mennonites began mixed farming on the rich soils, Greeks began tobacco farming, and Armenians specialized in business and commerce. Further east, the flat steppe of northern Kazakhstan remained largely untouched until the nineteenth century, save for Kazakh nomads and their herds and a few Russian forts and trading posts. During the nineteenth century, settlers came from the west in increasing numbers—a million or more by 1900—displacing the Kazakh nomads, thousands of whom died in famines or in unsuccessful uprisings against the Russians.

The Soviet period brought significant modifications to the region and its landscape (**Figure 4.35**). The colonists' small farms were merged into collectives, linear shelter belts of trees were planted in an attempt to modify climatic conditions, and rivers were dammed to provide hydroelectric power and irrigation for extensive farming of wheat, corn, and cotton. In the 1950s, Nikita Kruschev announced that the "virgin and idle lands" of the eastern steppe would be plowed up and farmed for wheat. State farms were organized, and large villages of new wooden houses were built to receive an army of 350,000 immigrants from European Russia. The eastern steppe was quickly transformed by modern machinery, fertilizers, and pesticides, initially producing as much wheat, on average, as the annual wheat harvest of Canada or France. This extensive wheat farming quickly led to dust-bowl conditions, however. In response, dry-farming techniques and irrigation have been introduced. **Dry-farming** techniques are those that allow the cultivation of crops without irrigation in regions of limited moisture

Figure 4.35   Extensive agriculture on the steppes
Beginning in the 1950s, the eastern steppes were transformed into an extensive wheat farming region through massive investments of modern machinery, fertilizers, and pesticides.

(50 centimeters, or 20 inches, of rain per year). Such techniques include keeping the land free from weeds and leaving stubble in the fields after harvest to trap snow. Together with irrigation schemes, dry-farming techniques now allow for the cultivation not only of wheat, but also millet and sunflowers, together with silage corn and fodder crops to support livestock.

Today, the steppes are an important agricultural region. The North Caucasus, which receives more rainfall than elsewhere in the steppes, is particularly productive. Nevertheless, the North Caucasus has traditionally been one of Russia's least economically developed regions. There are several large cities within or bordering the North Caucasus steppe, including Rostov-on-Don (population 1 million in 1998), Krasnodar (646,000), Grozny (364,000), and Stavropol' (342,000), but their industrial base is tied narrowly to agricultural processing and distribution, while the whole subregion has suffered acutely from the political instability and conflicts of the post-Soviet period. The economic vitality of the eastern steppe of northern Kazakhstan is similarly clouded. Aqmola (population 302,000 in 2000) is the largest city of the eastern steppe, having become the capital of Kruschev's "virgin lands" scheme. In 1997 it replaced Almaty as the capital of Kazakhstan and is enjoying a small boom as a result. Elsewhere, though, the towns of the eastern steppe are bleak and impoverished. The area near Semey was the Soviet Union's chief nuclear-testing ground, the site of more than

Sense of Place

Khiva

Khiva is ancient, perhaps as old as 6000 years. Legend has it that the settlement was founded by Shem, one of the sons of Noah. Khiva means "sweet water." The town grew around an oasis in the Kyzylkum Desert, and became a fortified trading post on a branch of the Silk Road. Khiva came into its own after the decline of Timur's Central Asian empire in the sixteenth century, when it became the capital of an Uzbek khanate. The wealth of the town increased significantly, thanks largely to its busy slave market. Between 1740 and 1873, Khiva came under the control of the Persian empire, and its slave market grew to be the largest in Central Asia. After several unsuccessful attempts, the Russians finally took Khiva in 1873, ending the slave trade and consigning the town to the role of a small and economically unimportant speck in the vast Russian empire. The town's fortunes changed little throughout the Soviet era, and Khiva might well have crumbled away to a picturesque ruin. But in the 1970s the Soviet government decided that it should be preserved as a living museum. The Soviet restoration was so thorough that the town acquired something of a clinical and sanitized feeling. Nevertheless, Khiva today is an impressive spectacle of mud walls, mosques, medressas (Islamic colleges), and minarets (**Figure 1**).

Khiva's castellated walls rise sheer out of the desert sands, hard against the sky. Its streets are still dominated by mosques, minarets, medressas, and palaces, and the detail on many of the buildings is breathtaking: mosaics of spiralling blue flowers on a white background and carved wooden ceilings delicately painted with red and ochre. Yet despite the strong Islamic flavor of the architecture, Islam is a cultural identifier for the townspeople, rather than an ideology. In Khiva, as in the rest of Central Asia, Islamic law has not replaced civil law, and society in general is tolerant of a range of behavior and practices not permitted in other Islamic regions. As a result, one can enter an attractive medressa, only to discover a bar serving Budweiser, with a Pretenders tape playing on a tinny Japanese stereo. Many of the restored buildings now house museums and tourist boutiques, and the city is poised to be "discovered" by global tourism. Outside the restored walls of the old city, Khiva is largely the product of Soviet-era development, with rather shabby, dusty, low-rise buildings that ac-

Figure 1   Khiva (a) Khiva city walls; (b) Alley in the historic district of Khiva.

commodate small apartments, interspersed with a few workshops, garages, and cafés. The town does boast a single department store, but the bazaar that adjoins the east gate of the city is still the principal place to shop.

450 underground nuclear detonations between 1949 and 1989. Most of the towns and cities of the eastern steppe are experiencing significant rates of out-migration, as Russians and other Slavs, seeing little likelihood of local economic development, are emigrating back to European Russia in search of better prospects.

The Central Asian Deserts

The desert and semidesert region of Central Asia covers an extensive area that includes most of Kazakhstan, Turkmenistan, and Uzbekistan (see Figure 4.3). The climate is harsh: Total annual precipitation in the deserts is less than 18 centimeters

(7 inches), shade temperatures can reach 50°C (122°F), and ground surfaces can heat up to 80°C (176°F). The scorching heat is aggravated by strong drying winds, called *sukhovey,* which blow on more than half the days of summer, often causing dust storms. In late summer and fall, the increasing temperature range between the hot days and the longer, cooler nights becomes so extreme that rocks exfoliate, or "peel," leaving the debris to be blown away by the wind. The landscapes of the northern zone of the desert region are arid plateaus with rocky outcrops, hillocks, and shallow depressions that have become crusted with salty deposits as a result of the evaporation of runoff from surrounding hills. The two principal deserts here are the Ust-Urt and Bek-pak-Dala. Farther south is a zone of sandy deserts: the Kara-Kum (Black Sands) and the Kyzyl-Kum (Red Sands). Here, the landscape is dominated by long ridges of sand in crescent-shaped dunes called *barchans,* and by vast plains of level sand punctuated by patches of sand hills and by isolated remnants of worn-down mountain ranges called *inselbergs.* Several important rivers—including the Amu Dar'ya, the Syr Dar'ya, and the Zeravshan—drain from the Tien Shan and Pamir mountain ranges across these deserts toward the Aral Sea. Dotted along their valleys are irrigated oases, while *tugay*—impenetrable thickets of hardy trees and thorny bushes—thrives in the rather salty soils of the valley floors.

This unpromising environment supported some of the world's most advanced civilizations and sophisticated cities of the premodern period. Until the eleventh century, sedentary agriculturalists prospered in the fertile oases, buffered from outside intrusion by the surrounding deserts. During this period the region was an important and lucrative link in the **Silk Road** along which east–west trade passed between Europe and China from Roman times until Portuguese navigators found their way around Africa and established the seaborne trade routes that exist to this day. The Silk Road was a shifting trail of caravan tracks that facilitated the exchange of silk, spices, and porcelain from the East and gold, precious stones, and Venetian glass from the West. The ancient cities of Samarkand, Bukhara, and Khiva stood along the Silk Road (see Figure 9.9), places of glory and wealth that astonished Western travelers such as Marco Polo in the thirteenth century. These cities were east–west meeting places for philosophies, knowledge, and religion, and in their prime they were known for their leaders in mathematics, music, architecture, and astronomy: scholars such as Al Khoresm (780–847), Al Biruni (973–1048), and Ibn Sind (980–1037). The cities' prosperity was marked by impressive feats of Islamic architecture (see Sense of Place: Khiva, p. 194). Their civilization was overcome by Mongol Tatar horsemen, who ruled until the fourteenth century. Timur (Tamerlane), one of Genghis Khan's descendants and a convert to Islam, subsequently built up a vast Central Asian empire stretching from northern India to Syria, with its capital in Samarkand. The decline of Timur's empire in the sixteenth century saw the rise of nomadic peoples, who established three khanates, or kingdoms: in Bukhara, Khiva, and Kokand. They prospered as trading posts on the trans-desert caravan routes until the late nineteenth century, when the three khanates fell to Russian troops.

In the twentieth century, Soviet modernization programs brought large-scale irrigation schemes, such as the Kara-Kum Canal, a 770-kilometer (478-mile) irrigation canal that diverts water from the Amu Dar'ya and waters 1.5 million hectares (3.7 million acres) of arable land and 5 million hectares (12.4 million acres) of pasture as it trails across southern Turkmenistan. Cotton was the dominant crop in these irrigated lands and remains so. In Turkmenistan, for example, over half the arable land is devoted to cotton monoculture. Uzbekistan is the world's fifth-largest producer of raw cotton and third-largest exporter of cotton. In many ways, however, irrigated cotton cultivation has been harmful. Yields remain comparatively low, in spite of irrigation, due to soil exhaustion and salinization. **Salinization** is caused when water evaporates from the surface of the land and leaves behind salts that it has drawn up from the subsoil. An excess of salt in the soil seriously affects the yield of most crops. In addition to salt, residues of the huge doses of defoliants, pesticides, and fertilizers used on the cotton fields have found their way into the drinking-water systems of the region. Meanwhile, cotton monoculture has rendered the countries of the region heavily dependent on food imports. The worst consequence of the Soviet program of irrigated cotton cultivation, however, has been the effects of excessive withdrawals of water from the main rivers that drain into the Aral Sea. The Kara-Kum Canal alone took away almost one-quarter of the Aral Sea's annual supply of water. Overall, the Aral Sea has shrunk by more than 40 percent in surface area, and it continues to shrink. As we have seen (see Figure 4.25), the result has been the acute dessication of the Aral Sea region, devastating its fishing industry and leaving ports such as Aralsk and Moynaq stranded more than 40 kilometers (25 miles) from the retreating lakeshore in the midst of a new "White Desert" of former lakebed sands.

Today, much remains of traditional crafts (**Figure 4.36**) and ways of life, though the lifestyles of younger people

Figure 4.36   Craft market Traditional crafts are still an important part of local economies in the Central Asian deserts. Shown here is part of a carpet market in Ashkhabad, Turkmenistan.

have departed dramatically from those of their parents and grandparents. The Central Asian geosyncline, with its oil, gas, and coal reserves, is one of the region's greatest economic assets. Proven oil reserves around and beneath the Caspian Sea amount to between 15 and 31 billion barrels, but estimates of the potential reserves run between 60 and 140 billion barrels. Only the oilfields of the Persian Gulf states (see Chapter 5) and Siberia are larger. This represents a tremendous economic asset for the Central Asian states. Exploiting these assets is beset with difficulties, however. Most of the states involved are effectively landlocked, which means that expensive pipelines have to be constructed before the oilfields can be fully developed. The problem here is that political tensions and instability in the region, along with the competing claims and interests of consumers and investors from different geographic markets—principally the Russian Federation vis-à-vis Europe and North America—make pipeline construction and routing both risky and contentious. Within the desert region considered here, Kazakhstan has the bulk of the oil reserves, while the natural gas fields are mainly to the south, in Turkmenistan and Uzbekistan.

Summary and Conclusions

The Russian Federation, Central Asia, and the Transcaucasus is a world region that is very much in transition. After more than seven decades under the Soviet system, the Russian Federation, Belarus, and the Soviet Union's other successor states in Central Asia and the Transcaucasus are now experiencing somewhat disordered transitions to new forms of economic organization and new ways of life. These transitions are taking place at different speeds in different places, and with rather uncertain outcomes. It is clear, though, that the process is having a significant impact on local economies and ways of life throughout the region. The region itself, meanwhile, is still struggling to find its new place in the world economy. Some of the old ties among the countries of the region have been weakened or reorganized, as have the interdependencies with Ukraine, Moldova, and the Baltic states. All of the new, post-Soviet states have joined the capitalist **world system** in semi-peripheral roles, and all of them have to find markets for uncompetitive products while at the same time engaging in domestic economic reform. Inevitably, patterns of regional interdependence have been disrupted and destabilized.

The Russian Federation, as the principal successor state to the Soviet Union, remains a nuclear power with a large standing army, but its future geopolitical standing remains uncertain. The Russian Federation has a formidable arsenal of sophisticated weaponry; a large, talented, and discontented population; a huge wealth of natural resources; and a pivotal strategic location in the center of the Eurasian landmass. Now freed from the economic constraints of state socialism, the Russian Federation stands to benefit a great deal by establishing economic linkages with the expanding world economy. Similarly, the collapse of the Communist Party has removed a major barrier to domestic economic and political development. The Russian Federation also has an ample labor force and a domestic market large enough to form the basis of a formidable economy. At present, the Russian Federation's economy is shrinking as it withdraws from the centrally planned model. Yet, although embarrassed by the disintegration of the Soviet Union and bankrupt by the subsequent dislocation to economic development, the Russian Federation is still accorded a great deal of influence in international affairs. The embarrassment and insolvency may also prove in the long run to be the spur that pushes the Russian Federation once again to contend for great-power status.

Nevertheless, it will be some time before the Russian Federation can once again be a contender for world-power status. The latter years of the Soviet system left the Russian Federation's industry with obsolete technology, low-grade product lines, and a shoddy infrastructure. The Russian Federation's economy now faces a massive task of modernization before it can approach its full potential. The Russian Federation must also renew civil society and the institutions of business and democracy after 70 years of state socialism. The partial breakdown of constitutional order has undermined respect for the law, and organized crime has flourished amid the factionalism and ideological confusion of the government. These problems could have serious implications for the future world order: A weak Russian Federation invites geopolitical instability.

Key Terms

chernozem (p. 159)	**federal state** (p. 162)	**salinization** (p. 195)	**territorial production**
civil society (p. 175)	**loess** (p. 193)	**satellite state** (p. 166)	**complex** (p. 166)
command economy (p. 165)	*mikrorayon* (p. 164)	**Silk Road** (p. 195)	**tundra** (p. 158)
deindustrialization (p. 184)	**monsoon** (p. 158)	**state socialism** (p. 161)	**unitary state** (p. 162)
dry farming (p. 193)	**Near Abroad** (p. 169)	**steppe** (p. 159)	**world city** (p. 186)
entrepôt (p. 186)	**permafrost** (p. 158)	**taiga** (p. 159)	**world system** (p. 196)

Review Questions

Review and Summary

1. How does the Russian Federation suffer from its location, physical features, and climate? What is unique about the Transcaucasus area in terms of climate?
2. Define: permafrost; tundra; taiga; steppe; chernozem.
3. Why is fur far more than an indulgence in Russia? What role did the fur trade play in the expansion of Russia?
4. Define: soviet; Bolshevik; *perestroika; glasnost.*
5. Under Stalin's rule, how did peasant farmers pay for Russia's industrialization and modernization?
6. How did the establishment of the Soviet bloc aid development of the Soviet Union following the Second World War? Please discuss with regard to technical optimization, industrialization, and military security.
7. What factors led to the breakup of the Soviet empire?
8. What is the Near Abroad? Please list significant areas.
9. What challenges are currently facing the Russian Federation?
10. Please explain the tensions in Nagorno-Karabakh. What is Russia's role there today?
11. What is a civil society and what does it need in order to flourish?
12. How do crime and corruption hinder the economic development of Russia in the twenty-first century?
13. Which flaws in Soviet economic-development practices led to regional inequality? How do market forces reinforce regional inequality today? Which regions benefit as a result?
14. In 1986, what happened at Chernobyl? Today, what policy does the Russian Federation have regarding the storage of nuclear waste?
15. During their transition to a market economy, how are Moscow and St. Petersburg changing in terms of city planning, economic sector activity, and cultural life?

Thinking Geographically

1. How have climate and physical geographic features spurred Russia's imperial expansion?
2. Why was the loss of Ukraine in the west and Georgia in the south so devastating for the Russian Federation? How is the Russian Federation responding to fill its agricultural needs without them? What are the implications for the ongoing conflict in Chechnya?
3. During the Soviet era, natural resources were devalued and therefore resource mismanagement was rampant. Please discuss the environmental degradation of Lake Baykal and the Aral Sea.
4. How do population dynamics in the Transcaucasus or Central Asian areas of the former Soviet Union compare to those in the Russian Federation? What cultural factors may explain the disparities? What role does voluntary migration play? What might occur to the Russian Federation's dominance of the region if current trends continue?
5. What are some of the economic and social problems facing post-Soviet Russian Federation? Using the examples from the textbook, please discuss:
   a. The changing roles between Ivan Gorod, the Russian Federation, and Narva, Estonia
   b. The failure of vouchers to assist equitable privatization
   c. Decollectivization
   d. International Monetary Fund (IMF) loans
6. Contrast and compare the Central Region, the Volga Region, and the Urals Region.
7. During the Soviet era, the human population of Siberia's tundra and taiga rose sharply. Currently people are leaving the area. Please discuss migration in and out of Siberia with regard to natural resource development; industrialization; forced labor; military strategy; free market forces.
8. What role does oil and related fossil fuels play in the development of Central Asia today? What physical and cultural factors affect the development of the oil industry in Central Asia?
9. Ethnic diversity in Central Asia contributed to the breakup of the Soviet Union. How have national identities been asserted in the decade since the Central Asian republics became independent countries? What cultural factors serve to unify or separate the states in this region?
10. How has organized crime spread through the Russian Federation, Central Asia, and the Transcaucasus region since the breakup of the Soviet Union in 1991? How have Russian *mafiyas* interacted with the global crime community?

Further Reading

Andrle, V., *A Social History of Twentieth-Century Russia.* London: Arnold, 1994.

Bater, J. H., *The Soviet City: Ideal and Reality.* London: Arnold, 1980.

Bater, J. H., *Russia and the Post-Soviet Scene: A Geographical Perspective.* London: Arnold, 1996.

Bradshaw, M. J. (ed.), *Geography and Transition in the Post-Soviet Republics.* New York: John Wiley & Sons, 1997.

Brown, A., Kaser, M., and Smith, G. S. (eds.), *The Cambridge Encyclopedia of Russia and the Former Soviet Union.* Cambridge: Cambridge University Press, 1993.

Castells, M., "The Crisis of Statism and the Collapse of the Soviet Union." In *End of Millenium.* Oxford: Blackwell, 1998.

Hanson, P., *Regions, Local Power and Economic Change in Russia.* London: Royal Institute of International Affairs, 1994.

Hunter, S. T., *The Transcaucasus in Transition: Nation Building and Conflict*. Washington, DC: Center for Strategic and International Studies, 1994.

Marples, D. R., *Belarus: A Denationalized Nation*. London: Harwood Academic, 1999.

Milner-Gulland, R., and Dejevsky, N., *Cultural Atlas of Russia and the Former Soviet Union*. New York: Checkmark Books, 1998.

Oldfield, J., "The Environmental Impact of Transition—A Case Study of Moscow City." *Geographical Journal* 165 (1999), 222–31.

Parker, W. H., *The World's Landscapes. The Soviet Union*. Chicago: Aldine, 1969.

Pryde, P. R. (ed.), *Environmental Resources and Constraints in the Former Soviet Republics*. Boulder: Westview Press, 1995.

Shaw, D. J. B., *Russia in the Modern World: A New Geography*. Oxford: Blackwell, 1999.

Smith, G. (ed.), *The Nationalities Question in the Post-Soviet States*. London: Longman, 1996.

Smith, G., *The Post-Soviet States: Mapping the Politics of Transition*. London: Arnold, 1999.

Stenning, A., and Bradshaw, M. J., "Globalization and Transformation: The Changing Geography of the Post-Socialist World." In J. Bryson, N. Henry, D. Keeble, and R. Martin (eds.), *The Economic Geography Reader* (pp. 97–107). New York: John Wiley & Sons, 1999.

Stewart, J. M. (ed.), *The Soviet Environment: Problems, Policies and Politics*. Cambridge: Cambridge University Press, 1992.

Turnock, D. (ed.), *Eastern Europe and the Former Soviet Union*. London: Arnold, 2000.

Van Selm, B., "Economic Performance in Russia's Regions." *Europe-Asia Studies* (1998), 603–18.

Film, Music, and Popular Literature

Film

Anna Karenina. Directed by Bernard Rose, 1997. Dramatization of Tolstoy's classic novel.

Baikal—Blue Eye of Siberia. Directed by Iuriii Beliankin for Channel Four (U.K.), 1992. Describes the lake and its environmental problems.

Chechnya: A Russian Nightmare. Produced by Jenny Bristow, for Assignment, BBC, 1995. An account of the early years of ethnic strife in Chechnya.

Dealing with the Devil. Produced by Alex Holmes for World in Action, ITV (U.K.), 1995. British businessmen in St. Petersburg and the *mafiya*.

Dirty Money. Produced by Peter Molloy for the BBC, 1994. Documentary on Russia's gangster capitalism.

Dr. Zhivago. Directed by David Lean, 1965. Set just before and during the years following the Bolshevik Revolution in Russia, the film follows the life of Dr. Zhivago as he marries, raises a family, has his life totally disrupted, first by the First World War, and then by the revolution.

Journey to Hell. Directed by Basile Grigoriev, for Dispatches, Channel Four (U.K.), 1995. Documentary on Krasnoyarsk 26 and plutonium pollution.

Kazakhstan. Directed by Mark Kidel and Tony Harrison for Channel Four (U.K.), 1994. Describes the land and peoples of Kazakhstan.

Peter the Great. Directed by Marvin Chomsky, 1986. The story of Peter I, tsar of Russia from 1682, and his efforts in transforming Russia into a "European" country, importing scientists, costumes, technology and military tactics.

The Cold War. BBC, 1998. Deals with the geopolitics of the Cold War period, 1950–89.

Where the Sky Meets the Land. Directed by Frank Müller (2000). Daily life and landscapes in Kyrgyzstan.

Music

Black Pearl. *Air Mail Music: Russian Gypsy Music*. Playasound, 2000.

Ensemble Kolkheti. *Batonebo*. Pan, 1996.

Inna and the Farlanders. *The Dream of Endless Nights*. Shanachie, 1999.

Kino. *Istoria Etogo Mir*. Musicrama, 2000.

Tuva. *Voices from the Center of Asia*. Smithsonian Folkways, 1990.

Usmanova, Yulduz. *The Selection Album*. Blue Flame, 1997.

Popular Literature

Hopkirk, P. *The Great Game. The Struggle for Empire in Central Asia*. New York: Kodansha International, 1992. Tells the story of the nineteenth-century imperial struggle between agents of Victorian Britain and Tsarist Russia for strategic and economic supremacy over an area stretching from the Caucasus to China.

Kaplan, Robert. *Eastward to Tartary: Travels in the Balkans, the Middle East, and the Caucasus*. New York: Random House, 2000. Geopolitical journalism from one of the best writers in the genre, who is adept at conveying both a sense of place and of the history of the places he describes.

Legg, Stuart. *The Heartland*. New York: Dorset, 1990. An epic history of the grassland empires of inner Asia.

Remnick, David. *Lenin's Tomb*. New York: Vintage Books, 1994. Pulitzer prize-winning account of the Gorbachev era.

Rzhevsky, Nicholas (Ed.). *An Anthology of Russian Literature from Earliest Writings to Modern Fiction: Introduction to a Culture*. New York: M. E. Sharpe, 1997. A useful anthology that includes information on related film, video, music, and art collections.

Solzhenitsyn, Alexandr. *The Gulag Archipelago* (3 volumes). New York: Harper and Row, 1974–78. An important and influential novel that describes the fate of political prisoners in the Soviet Union.

Stewart, John M. *The Nature of Russia*. London: Boxtree, 1992. A comprehensive volume on the natural history of Russia, with coverage of many environmental issues.

Taplin, Mark, *Open Lands: Travels Through Russia's Once Forbidden Places*. New York: Steerforth Press, 1998. An informed travelogue by one of the first writers to visit places and regions that were newly accessible to foreigners after the collapse of the Soviet Union.

Thubron, Colin. *In Siberia*. New York: HarperCollins, 1999. One of the first travel books to explore Siberia after it was opened to Western travelers.

Troiĕtìskiæi, Artemy. *Back in the USSR: The True Story of Rock in Russia*. New York: Omnibus Press, 1998. The definitive history of Russian rock music up to the late 1980s.

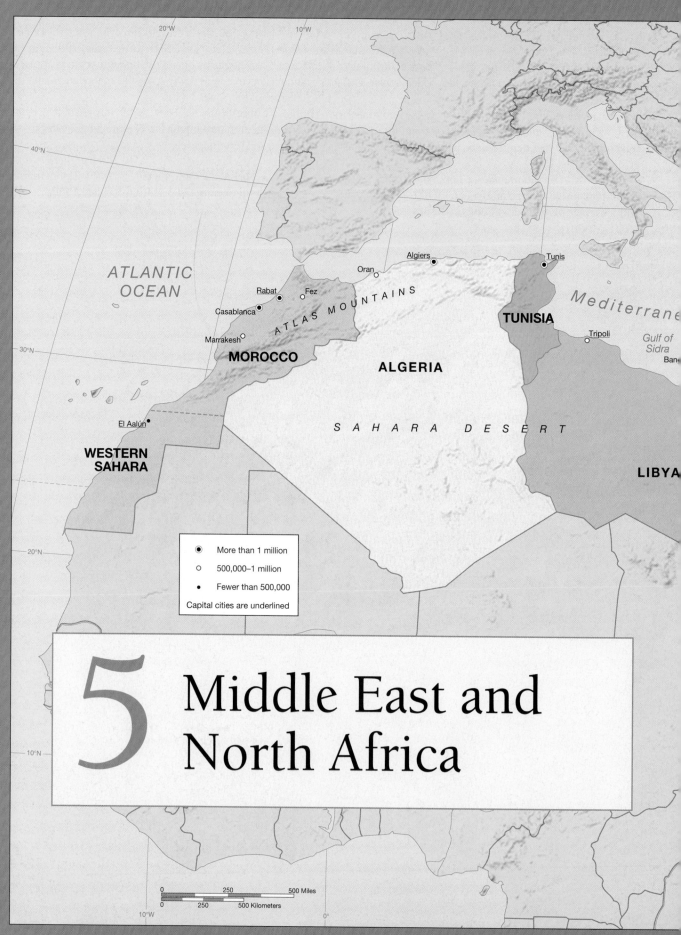

5 Middle East and North Africa

Figure 5.1

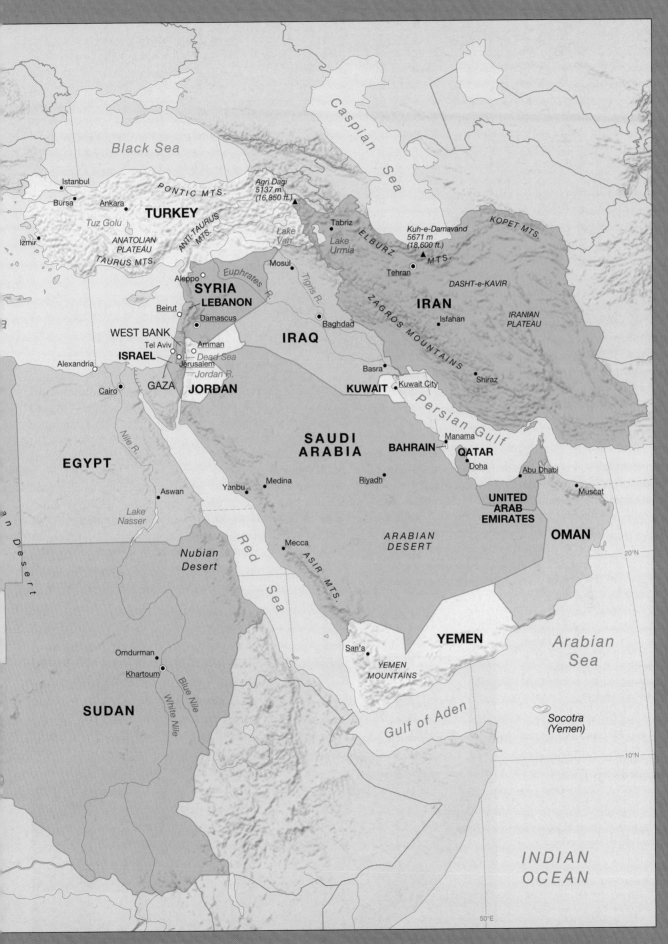

Black Sea

Istanbul
Bursa
Ankara
Izmir
TURKEY
Tuz Golu
PONTIC MTS.
ANTI-TAURUS MTS.
ANATOLIAN PLATEAU
TAURUS MTS.
Agri Dagi 5137 m (16,850 ft.)
Lake Van
Aleppo
SYRIA
LEBANON
Beirut
Damascus
Euphrates R.
Tigris R.
Mosul
Tabriz
Lake Urmia
ELBURZ
Kuh-e-Damavand 5671 m (18,600 ft.)
MTS.
KOPET MTS.
Tehran
ZAGROS MOUNTAINS
IRAN
DASHT-e-KAVIR
Isfahan
IRANIAN PLATEAU

Caspian Sea

WEST BANK
Tel Aviv
ISRAEL
Alexandria
Amman
Dead Sea
Jerusalem
Jordan R.
GAZA
JORDAN
IRAQ
Baghdad
Basra
KUWAIT
Kuwait City
Shiraz
Persian Gulf

Cairo
EGYPT
Nile R.
Aswan
Lake Nasser
Nubian Desert
Red Sea
Yanbu
Medina
ASIR MTS.
Mecca
SAUDI ARABIA
Riyadh
ARABIAN DESERT
BAHRAIN
Manama
QATAR
Doha
UNITED ARAB EMIRATES
Abu Dhabi
Muscat
OMAN
20°N

Omdurman
Khartoum
Blue Nile
White Nile
SUDAN
an Desert
San'a
YEMEN
YEMEN MOUNTAINS
Gulf of Aden
Arabian Sea
Socotra (Yemen)
INDIAN OCEAN
10°N
50°E

201

Located at the intersection of Europe, Asia, and Africa, the Middle East and North Africa together constitute a complex and exciting region linked by broad similarities of climate, religion, and culture. Physically, the Middle East and North Africa form something of a north-to-west arc. Iran and Turkey compose the northern tier. Southward are the Arab states of Lebanon, Jordan, Syria, and Iraq and the Jewish state of Israel. The southernmost boundary includes Saudi Arabia and the small Gulf states (Kuwait, Bahrain, Qatar, the United Arab Emirates), as well as Oman and Yemen. Moving westward from the Saudi Arabian Peninsula, the North African states of Egypt and Sudan, as well as Libya, Tunisia, Algeria, Morocco, and Western Sahara complete the region. As **Figure 5.1** shows, the region is also bordered by several crucially important water bodies—among them the Mediterranean, the Black, and the Red seas and the Persian Gulf. Also of significance are the Atlantic Ocean to the west and the Arabian Sea to the southeast. The region is often climatically known as the "dry world" due to its vast deserts—among them the Sahara, the Arabian, the Syrian, and the Nubian—as well as its overall low levels of precipitation. The Middle East and North Africa also contain impressive mountain ranges—from the Atlas Mountains in the west to the Zagros Mountains in the east—and critically important rivers including the Nile, the Tigris, the Euphrates, and the Jordan rivers (**Figure 5.2**).

The Middle Eastern and North African region has long been called a "cradle of civilization," the birthplace of the world's three great monotheistic religions (Islam, Christianity, and Judaism). The region is also the most important commercial crossroads of the ancient world and the base of several of the most sophisticated empires the world has ever known. Importantly, however, the region is not simply the site of ancient ruins and source areas. At the turn of the twenty-first century, it also possesses enormous economic and political significance as the site of most of the world's petroleum re-

serves and some of the world's most volatile and seemingly intractable political conflicts. As the Israeli-Palestinian conflict that reerupted in 2000 demonstrated, the fate of much of the core and large portions of the periphery is inextricably tied to what happens in the region. Even though the Middle East and North Africa loosely cohere around the broad similarities noted above, subregional divisions derive from very different histories, cultural systems, and political aspirations, each of which contribute to making the region an especially important and continuously evolving player on the world scene. The history and geography of the Middle East and North Africa indicate a high level of flexibility and a capacity for adaptation to a very challenging environment. These capacities persist in the region today.

Environment and Society in the Middle East and North Africa

The Middle Eastern and North African region is environmentally complex. The popular image of the region—fostered largely by the entertainment industry in commercial films such *Lawrence of Arabia* and *The English Patient*—is of vast, blazing hot deserts dotted with lush, but far-flung oases. The deserts of the region are certainly impressive. The Sahara, the largest desert in the world, has an average annual rainfall of less than 25 millimeters (1 inch). Incorporated within the larger framework of the Sahara are the Libyan and Nubian deserts of Egypt and northern Sudan. The other important desert of the region is known by two names: the Eastern Desert and the Arabian Desert. Although this desert lies between the Nile River and the Red Sea, the label *Arabian Desert* is also applied to the Rub al-

Figure 5.2   Middle East and North Africa from space This satellite photo highlights the extreme aridity of the Middle East and North Africa region. The deserts are shown as vast areas that dominate the southern part of the map. Along the coasts and in the mountains, high plateaus, and steppes, greater moisture availability means that more plants (and humans) can survive and thrive there. Few rivers and lakes exist in the region but the ones that do are crucial to human, plant, and animal life. (*Inset source:* Map projection, Buckminster Fuller Institute and Dymaxion Map Design, Santa Barbara, CA. The word *Dymaxion* and the Fuller Projection Dymaxion™ Map design are trademarks of the Buckminster Fuller Institute, Santa Barbara, California, © 1938, 1967 & 1992. All rights reserved.)

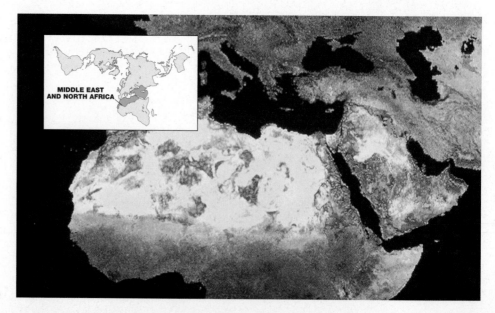

MIDDLE EAST AND NORTH AFRICA

Khali, or the Empty Quarter, a seemingly endless expanse of barren sand dunes, which occupies a substantial part of the Arabian Peninsula. Clearly, while deserts and oases like those portrayed in film do exist, surprisingly, they make up only a small percentage of the total land area of the Middle East and North Africa. The predominant landscape of the region is, in fact, vast grass plains, which receive substantially more precipitation than the deserts. Towering mountain ranges and extensive, treeless plateaus are also common landscapes. Trapped in among these more predominant landscapes are the isolated deserts, some of which are really vast seas of sand. Like other arid lands throughout the world, summers in the lowland areas of the Middle East and North Africa are extremely hot and dry, with daily high temperatures often at 38°C (100°F). The highland areas, such as the Atlas Mountains; the Iranian and Anatolian plateaus; and coastal areas of the Atlantic Ocean and the Mediterranean, Caspian, Arabian, Red, and Black seas experience more moderate daily summer temperatures and a predictable influx of visitors escaping the searing heat elsewhere. Winter temperatures, as would be expected, are more moderate in the lowlands and colder in the highlands.

The single most important climatic variable unifying the region is **aridity,** in that the climate lacks sufficient moisture to support trees or woody plants. This environmental characteristic is so pervasive that early geography textbooks labeled the Middle East and North Africa "the dry world." The result of aridity is landscapes with great stretches of little or no vegetation (**Figure 5.3**). Such landscapes are largely uninhabitable by humans unless substantial and expensive technologies are available that can introduce a dependable water supply. Yet to fully understand the environmental complexity of the region, one must first accept the principle that dryness is not synonymous with dullness or absence of diversity. Although the Middle East and North Africa share an important climatic variable, the countries and subregions contained within this world region exhibit a wide range of remarkable landscapes shaped by some small and some significant differences in the amount of surface water and annual precipitation. The absence or presence of water has had an important effect on the history of the interaction of peoples and environments of the Middle East and North Africa.

Landforms and Landscapes

The Middle East and North Africa contain a wide variety of physical landforms. As **Figure 5.4** shows, the region is interspersed with seas, and an ocean hems it on its western flank. There are substantial mountain ranges and two major river systems drain through its center. And while the region possesses the gamut of landscapes from seasides to mountains, the landscapes most heavily occupied by humans are the highland plateaus and the coastal lowlands as well as the floodplains of the major rivers where rain and surface water is most dependable. As is also clear from **Figure 5.5**, the region is located at the conjunction of three continental landmasses where active plates make the area highly prone to earthquakes.

The Arabian Peninsula, which is a tilted plateau that rises at its western flank on the Red Sea and slopes gradually to the Persian Gulf on its eastern flank, is part of the Arabian tectonic plate (see Figure 2.5). It is not hard to recognize that the Arabian Peninsula was once part of the African Plate tucked against the coastal areas of present-day Egypt, Sudan, and Eritrea. The separation of the Arabian Plate as it moved eastward from the African Plate millions of years ago resulted in the creation of the Red Sea, which initially formed as a rift valley and later filled with water. Both the African and the Arabian plates rub up against the Eurasian Plate along the Mediterranean Sea and at the mountains that separate the Arabian Peninsula from the Anatolian and Iranian plateaus. The mountain ranges of Turkey, Iran, and the Transcaucasus radiate out from this feature. Crustal plate contact also means that the subregion surrounding the contact zone is prone to severe earthquakes like the one that shook Turkey in August 1999, which killed close to 15,000 people and caused tens of billions of dollars of property damage.

Three mountain ranges dominate the region. While impressive in terms of their beauty and ruggedness, these ranges, more importantly, generate rainfall and are the sources of rivers and runoff for a region desperately short of water. The first set of ranges, which contains the most extensive and highest mountains, is the result of contact between the African, Arabian, and Eurasian plates at the center of the region (see Figure 5.5). These include the Taurus and Anti-Taurus mountains in Turkey and the Elburz and Zagros mountains in Iran. These ranges are higher than any in the continental United States and include Kuh-e-Damavand in Iran, which soars to 5671 meters (18,600 feet), and Agri Dagi, also known as Mount Ararat, which is only slightly less impressive at 5137 meters

Figure 5.3   Sahara Desert The Sahara is a vast desert that stretches from the Atlantic Ocean in the west, across the Nile Valley to the east, to the Atlas Mountains in the north, and south to central Sudan. Geological evidence suggests that between 50,000 to 100,000 years ago, the Sahara possessed a system of shallow lakes that sustained extensive areas of vegetation. Though most of these lakes had disappeared by the time the Romans arrived in the region, a few do survive in the form of oases.

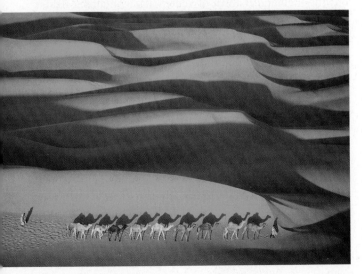

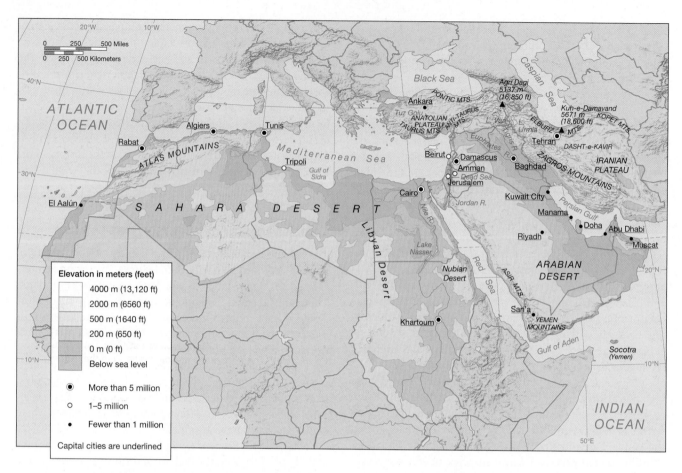

Figure 5.4 Physiographic features of the Middle East and North Africa Perhaps what is most consistent about the physiographic map of the region is the way that land and water features seem to alternate in a somewhat regular pattern. Also of significance is the scattering of plateaus and mountain ranges throughout the region that punctuate the vast lowland areas of desert and coastal plains.

(16,850 feet). The second important mountain range in the region is the Atlas Mountains of northwest Africa, which stretch along the southern edge of the Mediterranean from Morocco to Tunisia. The Atlas Mountains, a continuation of the European alpine range, are composed of a complex series of folded ridges separated by wide interior plateaus (**Figure 5.6**). The third set of mountains are those that border both sides of the Red Sea and are known as the Central Highlands or the High Yemen Mountains. These mountains are especially dramatic in Yemen, where the highest peaks are more than 3750 meters (12,300 feet) high. In addition to being critical sources of precious water, the mountains are also places of habitation for many people in the region. Though mountain environments can present substantial challenges to human habitation, the availability of moisture means that these environments can support agriculture over a somewhat shortened growing season. Historically, the mountains have also often provided safe havens for minority populations fleeing persecution and discrimination. The Druze in Syria and the Zayidis in Yemen are two such peoples who have sought mountain refuge from their oppressors.

While the highland areas are home to a small portion of the people of the Middle East and North Africa, many of the coastal areas, the floodplains, and the plateaus are the most densely populated landscapes of the region. The clustering of populations in these landscapes is hardly surprising given that they are the ones where water is most abundant and the environments are the least harsh. The Iranian and Anatolian plateaus are the most obvious examples of these landforms. But the highland plateau of Yemen also contains a sizable population. The coastal areas and floodplains of the region, excluding the coastal areas of the Persian Gulf, are equally attractive to human habitation and constitute some of the most remarkable, highly engineered, and scenic of the region's landscapes.

There are only two major river systems in the Middle East and North Africa. The Nile River system and the integrated Tigris and Euphrates rivers system are highly critical to the continued growth of the countries through which they flow and are, not suprisingly, also highly contested because of the precious resource they deliver. The Nile is one of the world's great rivers as well as being the longest. Its source is in the mountains of Ethiopia and East Africa. From the Ethiopian Plateau, the Blue Nile flows northward across the Sahara Desert, where it joins the White Nile at Khartoum, Sudan. It proceeds northward as the Nile River, finally emptying into

regions from Southeast Asia to Africa. Muslims represent more than 85 percent of the populations of Afghanistan, Algeria, Bangladesh, Egypt, Indonesia, Iran, Iraq, Jordan, Pakistan, Saudi Arabia, Senegal, Tunisia, Turkey, and most of the newly independent republics of Central Asia and the Transcaucasus (including Azerbaijan, Turkmenistan, Uzbekistan, and Tajikistan). In Albania, Chad, Ethiopia, and Nigeria, Muslims make up 50 to 85 percent of the population. In India, Burma (Myanmar), Cambodia, China, Greece, Slovenia, Thailand, and the Philippines, significant Muslim minorities also exist.

Plant Domestication   As mentioned in Chapter 2, the emergence of seed agriculture through the domestication of crops, such as wheat and barley, and animals, such as sheep and goats, replaced hunting and gathering as a way of living and sustaining human life. As cultural geographer Carl Sauer pointed out in his book *Agricultural Origins and Dispersals* (1952), these agricultural breakthroughs could take place only in certain geographic settings. These were settings where natural food supplies were plentiful; where the terrain was diversified (thus offering a variety of habitats and a variety of species); where soils were rich and relatively easy to till; and where there was no need for large-scale irrigation or drainage. In fact, the emergence of seed agriculture occurred during roughly the same period (between 9000 and 7000 B.C.) in several regions around the world, including the Middle East, parts of Latin America and the Caribbean (Chapter 8, p. 384), South Asia (Chapter 11, p. 531), and East Asia (Chapter 9, p. 429), as well as East Africa (Chapter 6, p. 268). In the Middle East, Mesopotamia was the source area. In North Africa, the Nile Valley in Egypt was a second important site of agricultural innovation.

The domestication of plants and animals represented a transition from hunting-and-gathering minisystems to agriculturally based minisystems that began in the early Stone Age, a period between 9000 and 7000 B.C. The transition was based on a series of technological preconditions: the use of fire to process food, the use of grindstones to mill grains, and the development of improved tools to prepare and store food. Before the first agricultural revolution, in prehistoric times, hunting-and-gathering minisystems were finely tuned to local physical environments. They were all very small in geographical extent and very vulnerable to environmental change. Because they did not have the ability (or the need) to sustain an extensive physical infrastructure, they were also limited in geographic scale. This transition to food-producing minisystems had several important implications for the long-term evolution of the world's geographies. First, it allowed much higher population densities and encouraged the proliferation of settled villages. Second, it brought about a change in social organization, from loose communal systems to systems that were more highly organized on the basis of kinship. Kinship groups provided a natural way of assigning rights over land and resources and of organizing patterns of land use. Third, it allowed some specialization in nonagricultural crafts, such as pottery, woven textiles, jewelry, and weaponry. This specialization led to a fourth development: the beginnings of barter and trade between communities, sometimes over substantial

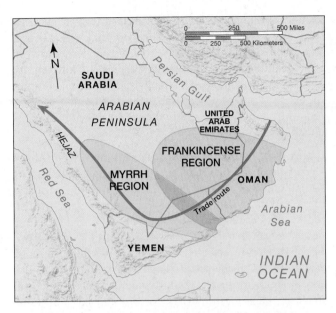

Figure 5.16 Frankincense trade routes in the ancient world For several hundred years before and after the birth of Jesus Christ, the southern rim of the Arabian Peninsula, within current-day Yemen and Oman, was an important transshipment area for the Middle East and North Africa. Goods that arrived from India, China, Ethiopia, and elsewhere were packed off by camel caravan to Egypt, Persia, Syria, and even Rome. The most prized shipments were two locally grown gum resin products: frankincense and myrrh. Myrrh was used in perfumes and cosmetics and as a medicine. The Egyptians used frankincense for embalming, as it was believed that burning it allowed the dead person to commune with the gods. Recently a team of American and British explorers discovered the ruins of two frankincense trading centers identified on a map of the region drawn by the ancient geographer Ptolemy in the second century A.D.

distances. **Figure 5.16** illustrates just one aspect of the widespread trade networks that criss-crossed the region. From precious metals, including silver, mercury, and other precious stones; to foodstuffs, including spices, fruits, rice, and dates; to textiles, carpets, brass, iron, and steelware; to more luxurious items, such as perfume, ebony, ivory, amber, and fur—a wide range of goods circulated throughout the Middle East and North Africa through the region's thousands of trading centers. We discuss the influence of trade on urbanization in more detail in the following section.

Irrigation was the key to the success of the many agriculturally based minisystems that emerged in the Middle East and North Africa when farmers there minimized their dependence on rainfall. Archaeologists and other scholars of the region argue that those settled minisystems that were able to control irrigated farming across large areas were able also to control weaker minisystems. By using food as a weapon of control, some minisystems were able to thrive, while others failed or were incorporated into stronger minisystems. An example is Babylon, which began as one of a number of regional minisystems in Mesopotamia but was able to turn itself into an enduring empire. Babylon's nearly 4000 years of dominance was achieved by systematically increasing control over regional agricultural production, through a buildup of its military strength (including a walled and fortified city center), the elaboration of a long-distance trade network (through extensive port facilities), and organizing extensive religious and symbolic political control. Evidence of a walled and fortified city, port facilities, and numerous temples and palaces demonstrate the extensive and well-established power and control of Babylon, which was, for a long period in its 4000-year history, the largest city in the world.

Cities and Trade   In the Middle East and North Africa, key urban centers, like Babylon, located at crucial points along natural and well-traveled human routes, were the organizational anchors of the region, shaping distinctive land-use patterns and serving as crucibles for significant cultural developments. In the Nile Valley, for instance, the urban-based Egyptian empire emerged, influenced by the cultures of Mesopotamia at the same time that it developed its own unique way of life. For instance, writing, which originated in Sumer in Mesopotamia for the purposes of recording inventories and trade transactions, was adapted in Egypt for religious and royal inscriptions. Like their neighbors, Egyptians were active city-builders, and they constructed monumental tombs, temples, and palaces whose engineering still baffles architectural historians (**Figure 5.17**). The culture was also premised on elaborate rituals and sophisticated body adornments that required significant quantities of gold, cedar, ebony, and turquoise. As a result, the Egyptians entered into active trade with settlements in the northern Red Sea, the upper Nile, and the eastern Mediterranean. These trade relations enabled the transfer of ideas and improvement of technologies that enriched all of the cultures of the Middle East and North Africa, as well as Greece and Rome.

It is widely held among scholars of ancient civilizations that the oldest cities on Earth were constructed along the valleys of the Tigris and Euphrates rivers as well as the Nile River (and possibly in the Indus River valley; see Chapter 11, p. 531) sometime during the fourth millennium B.C. The availability of the rivers for transportation and irrigation and the use of wheeled vehicles probably enabled the concentration of a surplus at a few regional centers. The need to protect inhabitants from flooding and to channel river water for irrigation suggests that populations concentrated to take advantage of these opportunities. Walled towns began to appear in Mesopotamia at least as early as 4500 B.C. These early cities probably contained between 7000 and 25,000 inhabitants, and the major producers were fishers and farmers who supported a nonproducing class of priests, administrators, and traders. Artisans also were clearly city dwellers. Houses for the producers, the lowest class of inhabitants, were likely to have been of mud construction. But the more elite urban dwellers probably lived in elaborate houses, many with courtyards and two stories. These houses were outfitted with systems for delivery of freshwater and removal of sewage. While some of the earliest cities were clearly planned, others were more randomly organized. What seems to be consistent across all of these early cities are three main elements: city walls; suburbs, including houses, fields, groves, pastures, and cattle folds; and a commercial district. As the historical evidence of urban commercial districts throughout the region suggests, trade was an essential part of life in these early cities and it not only helped to spread ideas but was also an important factor in encouraging and sustaining urban growth.

Many other empires of the region were also important in constructing cities and facilitating trade. The Egyptian empire had well-established trade relations with Crete, with cities along the Levant coast, with Anatolia in present-day Turkey, and with the people of the Sinai. As mentioned, Egyptian city building was also significant, and the ruins of temples, pyramids, and related monumental works attest to the sophisticated engineering skills of the ancient Egyptians. When the Egyptian empire declined after about 1090 B.C., other empires—among them the Assyrian, Persian, Roman, and the empire of Alexander the Great—all left their imprint on the region from Roman roads, aqueducts, and theaters to Assyrian palace complexes at Khorsabad and Nimrud. Of the more recent empires, the Islamic empire certainly had the most visible impact on contemporary urban patterns in the region. At its greatest extent, the Islamic empire reached westward as far as Tours in France and eastward into Turkey and the Iranian Plateau, throughout which local variations of the Islamic city can be found. From the tenth to the fourteenth centuries A.D., the Islamic empire's trade networks were so vast that they linked Mediterranean Europe to parts of the Transcaucasus and Pakistan. Such extensive trade networks are important evidence that parts of the world were highly integrated—politically, economically, and culturally—long before contemporary globalization occurred.

Figure 5.17   Pyramids of Giza, Egypt   The ancient Egyptians' belief in the continuity and stability of the cosmos was supported by a range of cultural activities from the preservation of wood, cloth, people, and animals to the construction of large-scale monuments. The pyramids are one example of this monumental architecture that represents Egyptians' view that life continues after death.

The Ottoman Empire, European Colonialism, and the Emergence of Modern States

The Ottoman empire based on the Anatolian Plateau in Turkey, which was in power for more than 600 years, was in decline in the Middle East and North Africa at the end of the nineteenth century. The Ottomans were Turkish Muslims who had replaced the Christian Greeks as the political power of the region after A.D. 1100. At its height, the Ottoman empire (named after the founder of the Ottoman dynasty, Osman) extended from the Danube River in southeastern Europe (including present-day Albania, Bosnia, and Kosovo) to North Africa and to the Arab lands of the eastern end of the Mediterranean. Within the region, only the Persian Empire based on the Iranian Plateau, the central Arabian Peninsula, and Morocco had been able to resist direct Ottoman control. By the early twentieth century, however, Ottoman rule was under seige from Europe both through the legacy of the French political revolution and the British industrial revolution. Already in the new century the edges of the empire were being nibbled away. Egypt had been lost to Britain and Algeria and Tunisia to France. By 1914 on the eve of the First World War, the Balkans and the remaining European possessions were lost, and the region was being carved up by various European powers, as well as Russia (**Figure 5.18**). At the same time, **nationalist movements**—organized groups of people, sharing common elements of culture such as language, religion, or history, who wish to determine their own political affairs—were erupting as the various subregions were exposed to European ideas about democracy. The nationalist movements were particularly problematic for the polyglot Ottoman empire, which had previously held itself together through an elaborate imperial legal and administrative structure that tended to allow for cultural differences united under Islam. When the Young Turk nationalist movement erupted onto the scene in 1908, a substantial wedge was driven between Turks and Arabs who had been long-term partners in the administration of the Ottoman empire. Already weakened by internal conflicts and external challenges, the Ottoman military, which had aligned with Germany, Austro-Hungary, and Bulgaria, was defeated by the Allied Powers during the First World War, resulting in the radical restructuring of the Ottoman empire.

The Arab provinces of the Ottoman empire, having negotiated their own separate treaties with the Allies, were carved up and were neither colonized nor allowed to be entirely

Figure 5.18 Europe in the Middle East and North Africa in 1914 The colonial presence of Europe in the Middle East and North Africa was short-lived—only about 50 years—but significant. The redrawing of boundaries in Israel/Palestine, Jordan, Lebanon, and Iraq has led to continued political conflict based on religious differences. Many of the most politically contentious areas of the world are in the Middle East and North Africa. (*Source:* Redrawn from *Hammond Times Concise Atlas of World History,* Maplewood, NJ: Hammond, 1994, pp. 100–1.)

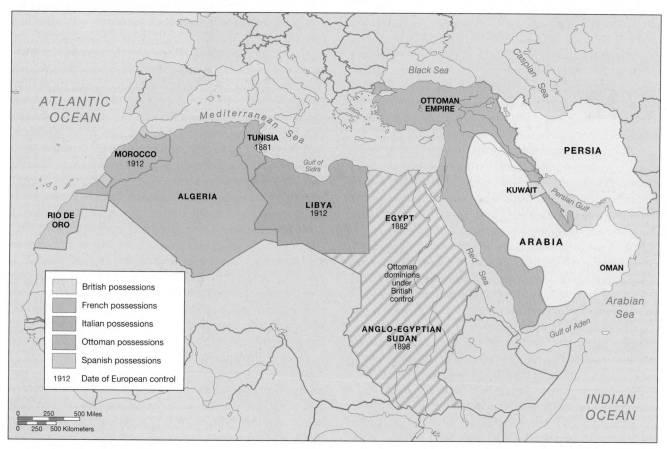

independent. Instead they were generally administered by a European power, with local people having limited access to political decision making. Syria and Lebanon were controlled by the French; Palestine and Trans-Jordan (present-day Jordan and Iraq) were relegated to the British. The idea of this sort of political arrangement—neither colony nor independent state—was heavily influenced by U.S. political ideas, which advocated self-determination and freedom over unmitigated colonization. The result was that a new form of external political control was created—the mandate—that legitimized French and British government dominance over their Middle Eastern and North African possessions. The mandate differs from outright colonial status because it requires the mandate holders to submit to internationally sanctioned guidelines that require that constitutional governments be established as the first step in preparing the new states for eventual independence. One of the most consequential mandates was the one determining the future of Palestine. As the mandate holder, Britain was obliged by treaty in 1917 to implement the provisions of the **Balfour Declaration,** named after British foreign secretary Arthur James Balfour, which committed Britain to the establishment of a Jewish national homeland in Palestine.

The political order that was imposed on the Middle East and North Africa following the First World War was seriously challenged throughout the region. Egypt, Iraq, Iran, Turkey, Syria, and Palestine, for instance, all revolted violently against the European presence in the region, though none were formally colonized. And while Turkey, Egypt, and Iran were able eventually to establish independent republics, by the mid-1920s, Britain and France exercised control—often somewhat tenuous control—over the rest of the Middle East and North Africa.

Due to the mandate system of external control, strictly speaking, there were only three true colonies—Aden (British), Libya (Italian), and Algeria (French)—in the Middle East and North Africa in the twentieth century. But a pattern of control did emerge such that the new states were heavily influenced, and in many cases, overly dominated, by their mandate holder.

The negative impacts of the mandate system helped to foment increasing regional dissatisfaction with outside dominance, and by the mid-twentieth century, aided by the crushing impacts that the Second World War dealt to Europe, all of the states of the Middle East and North Africa had gained their independence. It is one thing, however, to successfully win independence from colonizers who have effectively lost interest in their colony and quite another to obtain the allegiance of the diverse collection of new citizens. Many of the challenges that the new states of the Middle East and North Africa faced at mid-century continue to plague them in the new one.

The contemporary integration of the region into the global capitalist economy has brought increased wealth for some but also increased poverty or reduced living standards for many others, even in the wealthy oil-producing states. In the 1930s, entrepreneurial states such as Turkey and Egypt adopted **import-substitution** policies, which is the process by which domestic producers provide goods and services that were formerly bought from foreign producers. After the Second World

War, more comprehensive and aggressive approaches to state-led development were also undertaken, often in response to nationalist movements and anti-imperialist sentiments. Iran, Turkey, Egypt, Syria, Iraq, Tunisia, and Algeria were foremost among the Middle East and North African states who adopted the approach of the **nationalization** of economic development, which involves the conversion of key industries from private operation to governmental operation and control. The explicit goal of the nationalization of private enterprises was to improve the standard of living of working people, especially peasants in rural communities. The nationalization of the Anglo-Iranian Oil Company undertaken by the Iranian Prime Minister Muhammad Mossadegh illustrates how this trend often unfolded. Unfortunately, while the nationalization policies in the region did help to expand the public sector, they had an urban bias. As a result, the main beneficiaries were urban industrial, clerical, and service workers. While some middle-class peasants did benefit from land reforms that helped to expand their holdings, the condition of poor rural peasants either stayed the same or deteriorated.

Eventually, states in the Middle East and North Africa began to turn away from nationalization of industries as their economies began to stagnate, standards of living declined, and national debt skyrocketed. Pressured by the International Monetary Fund (IMF), the World Bank, and the U.S. Agency for International Development, states of the region were forced to initiate stabilization and structural-adjustment programs, also known as *neoliberal policies* (see Chapter 2, p. 87) in order to qualify for new loans and to reschedule old debts. The typical stabilization and structural adjustment programs raised the cost of food and other necessities, cut government spending on social programs, and generally reduced investments in the public sector. The impact of these programs was felt most directly and significantly by urban workers, government bureaucrats, and people on fixed incomes. Although the rural peasantry was supposed to benefit most from these neoliberal policies as consumer subsidies were dismantled and markets were privatized (allowing market-based prices for goods and the opportunity to market crops more freely), capitalist farmers have been the main beneficiaries of neoliberalism. Thus, the impact of neoliberal policies has been to put into motion a whole new set of forces in the Middle East and North Africa that have improved the lives of some but have mostly lowered the living standards of both urban workers and peasants. This is an especially troubling outcome; populations in the region are becoming increasingly urbanized as rural people move to the cities to find employment. Unfortunately, when they arrive they are confronted with decreased public services, not only in terms of schools and health care but also in terms of the most basic necessities, such as adequate housing and clean water. As a result, many people are forced to live in squatter settlements without sanitation. They are also often forced to eke out a living in the **informal economy**—that is, those economic activities that take place beyond official record and are not subject to formalized systems of regulation or remuneration, such as unregulated taxi driving and street vending.

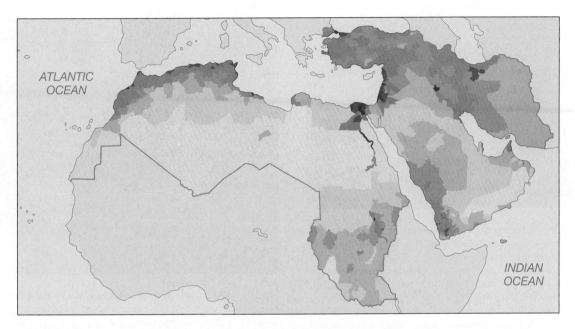

Figure 5.19   Population distribution in the Middle East and North Africa 1995   Population distribution is clearly heavily influenced by the availability of water. The intensity and unusual linear pattern of population concentration along the Nile River Valley is a perfect illustration of this point. Other concentrations, such as along the eastern end of the Mediterranean, the northern edge of Algeria, and the northern parts of Iran, Iraq, and Turkey, also reflect higher availability of freshwater. For a view of worldwide population density and a guide to reading population-density maps, see Figure 1.40.   (*Source:* Center for International Earth Science Information Network (CIESIN), Columbia University; International Food Policy Research Institute (IFPRI); and World Resources Institute (WRI), 2000. Gridded Population of the World (GPW), Version 2. Palisades, NY: CIESIN, Columbia University. Available at http://sedac.ciesin.org/plue/gpw/index.html?main.html&2)

Peoples of the Middle East and North Africa

The distinctive pattern of population distribution in the Middle East and North Africa shown in **Figure 5.19** reflects the influences of environment, history, and culture on the region. Environment is clearly a key factor in that population concentrations are found near rivers, streams, and oases or in areas of dependable precipitation. As a result, the coastal areas, the floodplains of the Tigris, Euphrates, and Nile rivers and smaller streams, and highland settings such as the Atlas Mountains are heavily populated. Other population clustering occurs around the region's cities, which have been well established for centuries but have grown especially rapidly since the independence period of the 1950s. Still, even though the Middle East and North Africa is more urbanized than is popularly assumed, many of the people of the region still live in rural villages.

The total population of the 21 countries that make up the Middle East and North Africa is well over 350 million, but given the inaccuracy, infrequency, and inconsistency of national censuses in the region, this number is only an approximation. Although most of the region's states have recently conducted censuses that are considered by population experts to be accurate and reliable, others, such as Turkey, tend to underestimate their minority populations for political reasons. Despite the fact that fertility rates have fallen over the last four decades, it needs to be pointed out that the region is currently experiencing rapid population growth. Both through natu-

ral increase as well as through immigration, a number of places in the Middle East and North Africa region, such as Libya, where traditional ways of life support the promotion of large families, and Iraq, where the government promotes high rates of fertility, are among the fastest growing in the world. For a number of the most populated countries in the region, including Egypt, Algeria, and Turkey, a large part of the population is younger than 15 years of age, meaning that the populations of these countries will continue to grow as these individuals reach their reproductive age.

Religion

The Middle East and North Africa is a region that is highly infused with especially active religious practices and belief systems. Unlike many parts of the core, where societies have become increasingly secular, religion in this complicated region is a central feature of everyday life for the vast majority of the inhabitants. Yet, the geography of religion is not uniform and the smaller the scale of analysis—as you move from the regional level to the neighborhood level, for example—the more likely there are to be significant differences in religious practice. For instance, although Iran is widely recognized as a state that is passionately committed to Islam, so much so that there is little separation between the state and the Islamic religion, in Tehran, the capital city of more than 7 million inhabitants, many upper- and middle-class households are likely to be secular and more aligned with Western values than with the teachings of Islam (**Figure 5.20**).

Figure 5.20 Satellite dish advertisement in Tehran
Although satellite dishes are technically illegal in Tehran, they are found throughout the city. Access to the Internet by middle-class families, who are likely to own their own satellite dishes, and poorer families, who are likely to share access to satellite television with a number of other families, means that residents of Tehran are well connected to global television culture. Geographer Farhang Rouhani believes that access to satellite television has been critical to the liberal cultural and political changes occurring over the last several years in Tehran. This photo shows a billboard advertising televisions, but a satellite dish is clearly visible.

Figure 5.21 Mecca, Saudi Arabia Mecca is the birthplace of Muhammad and the holiest city of Islam. It is a pilgrimage site for all devout believers of the faith. Every year, during the last month of the Islamic calendar, more than 1 million Muslims make a pilgrimage, or *hajj*, to Mecca. In addition to the required pilgrimage, Islamic traditions require Muslims around the world to face Mecca during their daily prayers. Pictured here is the Grand Mosque in the center of the city.

Despite the variation, however, about 96 percent of the population of the Middle East and North Africa is Islamic. After Islam, the religion with the second largest number of followers is Christianity (about 3 percent), with Judaism ranking third (less than 1 percent). Besides these three major religions, there are a number additional ones with much smaller numbers of adherents.

Islam   **Islam** is an Arabic term that means "submission," specifically, submission to God's will. A **Muslim** is a member of the community of believers whose duty is obedience and submission to the will of God. As a revealed religion, Islam recognizes the prophets of the Old and New Testaments of the Bible, but Muhammad is considered the last prophet and God's messenger on Earth. The Qur'an, the principal holy book of the Muslims, is considered to be the word of God as revealed to Muhammad by the Angel Gabriel beginning in about A.D. 610. There are two fundamental sources of Islamic doctrine and practice: the Qur'an and the Sunna. Muslims regard the Qur'an as directly spoken by God to Muhammad. The Sunna is not a written document but a set of practical guidelines to behavior. It is effectively the body of traditions that are derived from the words and actions of the prophet Muhammad. While Islam holds that God has four fundamental functions—creation, sustenance, guidance, and judgment—the purpose of people is to serve God by worshiping him alone and adhering to an ethical social order. The actions of the individual, moreover, should be to the ultimate benefit of humanity, not the immediate pleasures or ambitions of the self.

The emergence and spread of Islam is linked to the commercial history of the Middle East and North Africa. The geographical origin of Islam is Mecca, in present-day Saudi Arabia (**Figure 5.21**). When Islam first emerged, Mecca, where

Muhammad was born in A.D. 570, was an important node in the trade routes that at first connected Yemen and Syria and eventually linked the region to Europe and all of Asia. Today Mecca is the most important sacred city in the Islamic world. It also continues to be an important commercial center. Eventually Medina also became an important sacred city because it was the place to which Muhammad fled when he was driven out of Mecca by angry merchants who felt his religious beliefs were a threat to their commercial practices. There are five primary obligations, known as the five pillars of Islam, that a Muslim must fulfil: repeating the profession of the faith ("There is no god but God; Muhammad is the messenger of God"); praying five times a day facing Mecca; alms or charitable giving; fasting from sunup until sundown during the holy month of Ramadan; and making at least one pilgrimmage, or **hajj**, to Mecca if financially and physically able.

Disagreement over the line of succession from the prophet Muhammad occurred shortly after his death in 632 and resulted in the split of Islam into two main sects, the Sunni and the Shi'i. The Sunni faction, which argued that the clergy should succeed Muhammad, gained the upper hand and became dominant. The Shi'i had argued that Muhammad's son Ali should succeed his father, but Ali was killed, and the Sunnis have since been the mainstream branch of Islam. In specific countries, however, the pattern varies. The majority of Iran's 60 million people follow Shi'i, the official state religion of the Islamic Republic of Iran, founded in 1979. The majority of Iraq's population is also Shi'i, even though the government headed by Saddam Hussein is Sunni. It is also important to keep in mind that Islam is practiced differently in many different locales throughout the Middle East and North Africa and that Muslims who have migrated out of the region—to Europe and the United States, for instance—are shaped by and shape the practice of Islam in the Middle East and North Africa.

Perhaps one of the most widespread cultural counterforces to globalization has been the rise of **Islamism,** which is more popularly, although incorrectly, known as Islamic fundamentalism. Whereas *fundamentalism* is a general term that describes the desire to return to strict adherence to the fundamentals of a religious system, Islamism is an anticolonial, anti-imperial, and overall anticore political movement. In Muslim countries, Islamists resist the core, especially Western, forces of globalization—namely modernization and secularization. Not all Muslims are Islamists, although Islamism is the most militant movement within Islam today.

The basic intent of Islamism is to create a model of society that protects the purity and centrality of Islamic precepts through the return to a universal Islamic state—a state that would be religiously and politically unified. Islamists object to modernization because they believe the corrupting influences of the core place the rights of the individual over the common good. They view the popularity of Western ideas as a move away from religion to a more secular (nonreligious) society. Islamists desire to maintain religious precepts at the center of state actions, such as reintroducing principles from the sacred law of Islam into state constitutions.

Another important aspect of the Islamist movement is the concept of **jihad,** which is a sacred struggle. When this struggle is violently directed against the enemies of Islam, jihad is understood to be a holy war. But jihad can also be a more peaceful struggle to establish Islam as a universal religion through the conversion of nonbelievers. One example of jihad today is the struggle of Shi'ite Muslims for social, political, and economic rights within Sunni-dominated Islamic states.

As popular media reports make clear, no other movement emanating from the periphery is as widespread and has had more of an impact politically, militarily, economically, and culturally than Islamism. Yet Islamism—a radical and sometimes militant movement—should not be regarded as synonymous with the practices of Islam any more generally than Christian fundamentalism is with Christianity. Islam is not a monolithic religion, and even though all adherents accept the basic pillars, specific practices may vary according to the different histories of countries, nations, and tribes. Some expressions of Islam are moderate and allow for the existence and integration of Western styles of dress, food, music, and other aspects of culture, while others are extreme and call for the complete elimination of Western things and ideas.

Christianity, Judaism, and Other Middle Eastern and North African Religions

Although Islam is the most widespread and widely practiced religion in the Middle East and North Africa, it is by no means the only religion of political, cultural, or social significance. Christianity and Judaism are also important world religions of the region, and there are also other less widely practiced religions that are notable. There are more than a dozen Christian sects—among them Coptic Christians in Egypt, Maronites, the Chaldean Catholic Church,

and various orthodox affiliations, including Armenian, Greek, and Ethiopian, and even some Protestant faiths. Faiths not associated with any of the three world religions are largely concentrated in Iran and include Bahaism and Zoroastrianism. Most Jews in the Middle East and North Africa are either Orthodox or Hasidic, though there also exists an offshoot of Judaism, known as Samaritanism, the adherents of which are largely concentrated in the West Bank.

The three regionally predominant religions have been important forces in shaping the peoples and the landscape of the region. The most obvious and enduring landscape impacts have been places of worship and sacred spaces more generally. Nowhere is the enduring interrelationships of the three religions more apparent than in the ancient city of Jerusalem. As anthropologist Dale Eickelman describes it: "Jerusalem provides perhaps the most poignant symbol of all that Judaism, Christianity, and Islam share and contest in the Middle Eastern and North African context and beyond and how the significance given to religious symbols, space, and places ranges well beyond the local carriers of the three religious traditions."[1] The centrality of Jerusalem as an ancient religious space, as well as its contemporary significance as a place of pilgrimage for Jews and Christians, is very much tied up with Arab-Israeli conflicts (see Sense of Place: Jerusalem, p. 222). Modern constructions of the state that link territory with nationality are ill-equipped to deal with the religious significance of Jerusalem to Christians, Jews, and Muslims.

Culture and Society

The social organization of the Middle East and North Africa is as complex as that of any other region of the globe, with the social categories of gender, tribe, nationality, kinship, and family figuring prominently. Global media technologies such as satellite television and the Internet are increasingly penetrating the region, however, with the potential for new social forms to emerge and old ones to be reconfigured. Generally, the predominant forms of social organization in the region have persisted for hundreds of years. It would be incorrect, however, to assume that both subtle and dramatic changes within these forms have not already occurred. Whether the current impacts of globalization will be great or small, it is important to understand the contemporary social and cultural organization of the region in order to appreciate its past as well as its future.

Kinship and Family

In order to understand Middle Eastern and North African society, it is important to understand ideas of kinship, family, and other personal relationships. **Kinship** is normally thought of as a relationship based on blood, marriage, or adoption. However, this definition needs to be expanded to include a *shared notion of relationship* among members of a group. The point is that not all kinship relations are understood by social groups to be exclusively based on biological or marriage ties. While in the Middle East and North

---

[1]D. F. Eickelman, *The Middle East and Central Asia: An Anthropological Approach,* 3rd ed., Upper Saddle River, NJ: Prentice Hall, 1998, p. 250.

Sense of Place

Jerusalem

Although there are many cities in the world that have been the object of struggle and conflict over the centuries, none seems as endlessly beset by conflict as Jerusalem, the disputed capital of Israel. Visitors, writers, and residents believe Jerusalem to be the most beautiful city in the world. If there are other serious competitors to that coveted beauty title, Jerusalem certainly has few rivals for the title of the most sacred city in the world, possessing as it does an unmatched Christian, Jewish, and Islamic history. Jerusalem began as a small settlement on the slopes of Mount Moriah. In 997 B.C. it was captured by David, king of the Israelites, who made Jerusalem the capital of Israel. Solomon, David's son and successor, built the Great (First) Temple on Mount Moriah to commemorate the place where Abraham offered to sacrifice his son. Though the temple was destroyed centuries ago, the site is a central one to the Jewish faith.

The history of the city reflects the history of the various empires that dominated and were succeeded by yet new and more powerful empires (**Figure 1**). Nebuchadnezzar, king of Babylon, destroyed the Great Temple in 586 B.C. and banished the Jews. But the Babylonian control of Jerusalem eventually gave way to the Persians, under whose rule the Jews were allowed to return and rebuild their temple, known as the Second Temple. The Persian occupation of Jerusalem was swept aside by Alexander the Great (356–323 B.C.), the king of Macedonia and one of the world's greatest military leaders, who enabled the spread of the Greek Empire from the southern shores of the

Caspian Sea into central Asia. The Romans entered the scene around 63 B.C., and Herod the Great was eventually installed to command the Roman Kingdom of Judea from Jerusalem.

During the early Christian period in Jerusalem, the Jews revolted openly against the Roman occupiers. In A.D. 132 the Romans responded by destroying the Second Temple and banishing all Jews from Jerusalem and Palestine. As a result, the Jews scattered north into Babylon and later into Europe and North Africa. The popular myth is that this ancient Jewish diaspora remained in exile until 1948 when the state of Israel was created.

The major Christian influence upon Jerusalem, the "Holy City," began when Constantine I (A.D. 285?–337), the emperor of the Eastern Roman Empire, converted to Christianity in A.D. 313. This event led to the construction of churches and other buildings dedicated to celebrating the life of Jesus Christ. But Christian influence over the city ceased when Jerusalem eventually succumbed to Islam. In A.D. 638, Jerusalem was designated a holy city of Islam because it was believed that Muhammad had once visited heaven while in the city. Although for several centuries Jews, Christians, and Muslims were all allowed access to the city of Jerusalem, by the tenth century, the persecution of non-Muslims became common. Beginning in the eleventh century and continuing until the thirteenth, European Christians undertook military expeditions—called the Crusades—to the Holy Land in an attempt to wrest control of Jerusalem from the Muslims. In 1099 Crusaders captured the city but then lost it again to the Muslim

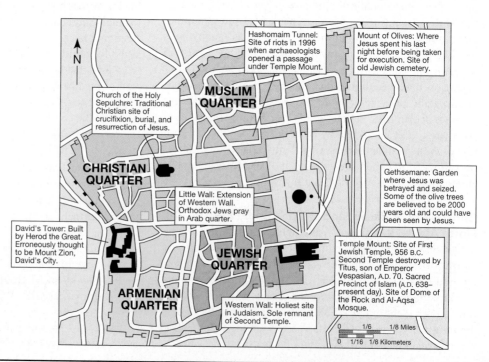

Figure 1   Jerusalem, the Old City This map of Jerusalem indicates the main sections of the city. Over the many years of the Israeli-Palestinian peace process, numerous proposals have been advanced about how to divide the city in order to satisfy the desire of both Palestinians and Israelis to possess it. The disposition of Jerusalem is one of the major sticking points in the ongoing peace process. (*Source:* Redrawn from *The Guardian*, October 14, 2000, p. 5.)

Hashomaim Tunnel: Site of riots in 1996 when archaeologists opened a passage under Temple Mount.

Mount of Olives: Where Jesus spent his last night before being taken for execution. Site of old Jewish cemetery.

Church of the Holy Sepulchre: Traditional Christian site of crucifixion, burial, and resurrection of Jesus.

MUSLIM QUARTER

CHRISTIAN QUARTER

Gethsemane: Garden where Jesus was betrayed and seized. Some of the olive trees are believed to be 2000 years old and could have been seen by Jesus.

Little Wall: Extension of Western Wall. Orthodox Jews pray in Arab quarter.

David's Tower: Built by Herod the Great. Erroneously thought to be Mount Zion, David's City.

JEWISH QUARTER

Temple Mount: Site of First Jewish Temple, 956 B.C. Second Temple destroyed by Titus, son of Emperor Vespasian, A.D. 70. Sacred Precinct of Islam (A.D. 638–present day). Site of Dome of the Rock and Al-Aqsa Mosque.

ARMENIAN QUARTER

Western Wall: Holiest site in Judaism. Sole remnant of Second Temple.

0   1/6   1/8 Miles

0   1/16   1/8 Kilometers

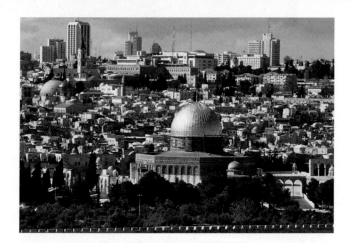

Figure 2   Dome of the Rock   Located in Haram Al-Shaif, or Temple Mount, the Dome of the Rock sits in part of Jerusalem that is technically neither Jewish nor Muslim but both. The Muslims want this entire holy site. The Israelis also want it because the Wailing Wall, sacred to Jews, also occupies part of the Temple Mount. Palestinian anger was unleashed—and the delicate peace process derailed—in the fall of 2000 when Ariel Sharon, a nationalist Israeli politician who several months later became Israel's president, visited the Temple Mount during the first days of the Muslim holy season of Ramadan.

military leader Saladin in 1187. In 1517 Jerusalem was absorbed into the Ottoman empire, and the city was ruled from Istanbul for more than 400 years. The Ottomans, however, had little interest in Jerusalem, and Jewish immigrants began returning to the city and Palestine starting in the mid-nineteenth century.

The contemporary history of the city derives from political and geographical implications of the Balfour Declaration, which stipulated that Jerusalem should be an international city with no one state claiming it as entirely its own. Today, Jerusalem is a highly contested city as Palestinians, Christians, Muslims, and Israeli Jews fight for control of it. An example of this contest for control is the continuing dispute over the Dome of the Rock, which was constructed between A.D. 688 and 691. Muslims claim the Dome as their most sacred site (**Figure 2**). Yet the Dome sits on a site sacred to the Jews, the Temple Mount, the site where the Great Temple and Second Temple were built and later destroyed. Indeed, the Dome is believed to enclose the sacred rock upon which Abraham prepared to sacrifice his son, according to Jewish tradition, and, according to Islamic tradition, the same rock from which the prophet Muhammad launched himself to visit heaven. Also located at the Temple Mount is the Al-Aqsa Mosque, a centrally important sacred site to Muslims (see Geography Matters: The Islamic City, Figure 2, p. 233).

While nationalist Israelis maintain that Jerusalem will be the "eternal and undivided capital" of Israel, Palestinians be-

lieve that Jerusalem is the future capital of the Palestinian state. In fact, the 1992 Oslo peace accords that lead to a Declaration of Principles between Israel and Palestine hint at the possibility of negotiating the future of Jerusalem. Further peace negotiations, including those at Wye River, Maryland, in 1998; at Sharm al-Shaykh in the Egyptian Sinai Desert in 1999; in Camp David, Maryland, in 2000; and again in Washington, D.C., in 2001 are cause for hope that a resolution to the future of Jerusalem that includes some control by the Palestinians can be achieved. At present, Jerusalem is entirely controlled by Israel. Yet in accordance with the Oslo peace agreements, in May 2000 Israel conceded three villages to the control of the Palestinian Authority. All three villages are part of greater Jerusalem in the West Bank, and all but one, Abu Dis, is actually a few meters beyond the Jerusalem municipal border (**Figure 3**). From Abu Dis, the Old City of Jerusalem is clearly visible, particularly the Dome of the Rock. In early 2001 it is still unclear how the contested control of Jerusalem might eventually be resolved. What is clear is that both sides are equally passionate about their claims.

Figure 3   Jerusalem and Abu Dis   This map shows the location of the village of Abu Dis within the city limits of Jerusalem. While an apparently minor victory for the Palestinians, the granting of this village to Palestinian control is a signal that politics is geographical in that control over space is at the heart of the disagreements and conflicts over the future of the city. (*Source:* Redrawn from *New York Times*, May 21, 2000, Section 4, p. 1.)

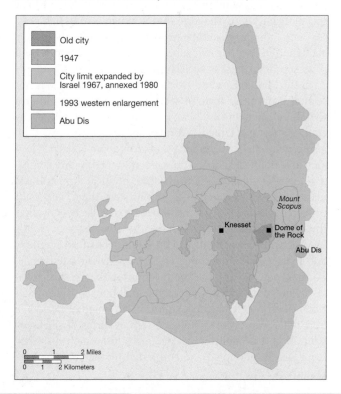

Old city

1947

City limit expanded by Israel 1967, annexed 1980

1993 western enlargement

Abu Dis

Mount Scopus

Knesset

Dome of the Rock

Abu Dis

0   1   2 Miles

0   1   2 Kilometers

Africa, biological ties, usually determined through the father, are important, they are not the only important ties that link individuals and families. In fact, though kinship is often expressed as a "blood" tie among social groups throughout the Middle East and North Africa, it is often the case that neighbors, friends, and even individuals with common economic or political interests are considered kin. Importantly, kinship is such a valued relationship for expressing solidarity and connection that it is often used to assert a feeling of group closeness and as a basis for identity even where no "natural" or "blood" ties are present.

For many Middle Easterners and North Africans, kinship helps to shape a whole range of social relationships from business to marriage to politics. This is true in ordinary households in urban and rural areas as well as in the monarchies of the region, such as those of Saudi Arabia, Oman, and Morocco. Kinship also figures largely in other regimes, such as in Iran, Egypt, Syria, Turkey, and Lebanon, where it is not unusual to find that the holders of many government offices are close relatives. The underlying assumption of such arrangements is that appointments based on kinship are not an abuse of political authority but a guarantee of loyalty.

Kinship is even an important factor in shaping the spatial relationships of the home as well as outside the home, determining who can interact with whom and under what circumstances. This is especially the case for the interaction of gender and kinship where women's and men's access to public and private space is sharply differentiated.

Social Order and Loyalty   The idea of the tribe is central to understanding the sociopolitical organization of the Middle East and North Africa. And while tribally organized populations do appear throughout the entire region, it is most important to understand that it is not a widely practiced form of social organization. The term *tribe* is a highly contested concept and one that should be treated carefully. For instance, it is often seen as a negative label applied by the colonizers to suggest primitiveness in social organization. Where it is adopted in the Middle East and North Africa, however, tribe is not seen as a primitive form of social organization but rather a valuable element in sustaining modern national identity. Generally speaking, a **tribe** is a form of social identity created by groups who share a common set of ideas about collective loyalty and political action. Tribes are grounded in any combination or single expression of social, political, and cultural identities that are created by those who share them. The result of shared tribal identity is the formation of collective loyalties that result in a primary allegiance to the tribe. External groups may recognize the existence of these self-defined tribal groups and may seek to undermine or encourage their persistence. For instance, in early twentieth-century Iran, the state ruthlessly and systematically attempted to eliminate tribal affiliations. In contrast, during the European colonial period in Sudan and Morocco, tribes were seen as forms of social organization that might inhibit nationalist movements, so they were largely promoted and supported by the colonial state (**Figure 5.22**).

Figure 5.22   Berber woman   The Berber have lived in North Africa since ancient times. Berber is the name applied to the language and people belonging to a number of tribes who currently inhabit large sections of North Africa. Islam is the main religion of Berbers, who constitute about 40 percent of the population of Morocco and about 30 percent of the population of Algeria. Berbers are largely rural and mountain-dwelling people. The traditional Berber occupations are raising sheep and cattle, but increasing numbers of Berbers raise crops.

One Middle Eastern and North African group that is frequently and proudly tribal is pastoralists. **Pastoralism** is a subsistence activity that involves the breeding and herding of animals to satisfy the human need for food, shelter, and clothing. Usually practiced in marginal areas where subsistence agriculture cannot be practiced, pastoralism can be either *sedentary* or *nomadic*. Sedentary pastoralists live in settlements and herd animals in nearby pastures, while nomadic pastoralists travel with their herds over long distances, never settling in any one place for very long. Most nomadic pastoralists practice **transhumance,** the movement of herds according to seasonal rhythms. Flocks are kept in warmer, lowland areas in the winter and in cooler, highland areas in the summer. Although the herds are occasionally slaughtered and used directly for food, for shelter, and for clothing, often they are bartered with farmers for grain and for other commodities. Female and younger members of pastoralist groups may also farm small plots. In such cases, mostly women and children split off from the larger group and plant crops at fixed locations in the spring. They may stay sedentary for the growing season, tending the crops, or they may rejoin the group and return to the fields when the crops are ready for harvesting.

Gender   Although gender differences play an important part in shaping social life for men and women in the Middle East and North Africa, as elsewhere around the globe, there is no single Islamic, Christian, or Jewish notion of gender that operates exclusively in the region. Many of us have formed stereotypes about the restricted lives of Middle Eastern and North African women because of the operation of rigid Islamic traditions. It is important to understand, however, that these are indeed stereotypes and do not capture the great variety in gender relations that exist in the Middle East and North Africa across lines of class, generation, level of education, and geography (urban versus rural origins), among other factors. What is pervasive throughout the Middle East and North Africa, as well as in many other societies throughout the world, is an ideological assumption that women should be subordinate to men. This view is largely held by both men and women in Middle Eastern and North African societies. Interestingly, it seems that men regard women's subordination as something natural, something that is effectively determined by biology. In contrast, most women in the Middle East and North Africa tend to regard their subordination as something social, something that is the product of the society in which they live and operate, and therefore something that can be negotiated and manipulated. The gender systems that operate in a wide variety of contexts in the Middle East and North Africa are derived in large part from some of the same notions about men and women that inform gender systems in Western societies. Although this view is being increasingly critiqued and dismantled in the West, it is still a powerful one there as well as in the Middle East and North Africa. Control over women in the Middle East and North Africa is frequently exercised by restricting their access to public space and secluding them within private space (**Figure 5.23**).

Sexuality and gender roles play an important part in how men and women see themselves and represent themselves publicly in the Middle East and North Africa. Some societies, such as in Yemen and Bahrain, exercise very strict control over women's public movements, and women are expected to cover themselves with veils and long, dark clothing when out on the streets. In some more generally secular of the region's societies, such as Turkey and Egypt, women's public movements are largely much less strictly regulated. The veil—from the all-encompassing full body garment, known as a **chador,** to a simpler head covering—has become the means by which women are able to effectively operate in public and yet remain in their personal space. **Figure 5.24** illustrates the variation that exists in national-level policies and practices with respect to women's access to, and behavior in, public space. While the map demonstrates national variation, it is important to keep in mind that subnational and local variations do exist, particularly differences in urban versus rural practices and even according to class differences within urban areas. Generally, urban women's public movements tend to be more restricted than those of rural women. This is largely because rural villages are usually composed of kin, and women can operate relatively freely among them, while urban women must move about in a world of both kin and strangers. But even within urban areas, highly restrictive practices often limit women's public movements, and in some cities middle- and upper-class women tend to have more constraints placed on their public social behaviors than do poorer women. Again, the strictures placed on women's movements vary throughout the region and even within particular countries and subregions within countries.

Figure 5.23   Gendered architecture Islamic architecture reflects gender differences within the culture, and in different places within the Middle East and North African region these differences can be either strictly or more loosely adhered to. A classic aspect of Islamic architecture is the screen placed across windows in the women's parts of the houses and in the interiors of some public buildings. The screens pictured here are from Tunisia. They allow women to observe activities outside their windows without being observed themselves.

Figure 5.24   Restrictions on women's movements Although Islamic law imposes restrictions on women's movements and dress in public space, there is a great deal of variation across the region with respect to adherence to these legal restrictions.

The most important aspect of gender systems in the Middle East and North Africa to remember is that social reality is not fixed and that cultural assumptions and practices around gender are subject to negotiation and change. Although the predominant gender theme in the Middle East and North Africa is that women are subordinate to men, women can and do exercise a great deal of household as well as political influence and independence across a range of societies in this region.

Migration and the Middle Eastern and North African Diaspora

In the Middle East and North Africa, populations have for thousands of years moved around within the region, at times as refugees, at other times voluntarily. They have also moved out of the region, settling all over the world. Generally speaking, migration into and out of the region since the end of the colonial period has largely been related to several very important factors that have pulled immigrants to the region and forced many others to leave (**Figure 5.25**). Internal regional and national migration has also been significant and is almost always related to the draw of urban economic opportunity as rural areas experience population increase or economic decline.

The strongest force drawing migrants to the Middle East and North Africa in the last 50 years has been the growth of the oil economy. This is not to say, however, that oil is the only attractive force. Other factors have also played a role in attracting migrants. For instance, the founding of the state of Israel at mid-century has drawn large numbers of European Jews to the region, particularly during and after the Second World War. More recently, Ethiopian and Russian Jews have also migrated, in the former case due to civil war and in the latter due to the end of the Cold War. Non-oil-related economic growth has also played a role in fostering migration to places like Beruit in Lebanon, Cairo in Egypt, and Istanbul in Turkey, all of which have increased their importance as core cities in the regional economy following the spread of political independence throughout the region.

Still, the most prominent attractive force for drawing in new migrants has been the job opportunities made possible through the wealth generated by the continued development of the petroleum industry (**Figure 5.26**). In the states of the Arabian Peninsula, several factors, including small populations, lack of skill, and a lack of interest in or possibly cultural resistance to the

Figure 5.25   Kurdish and Lebanese diaspora, 1990 The most significant diasporic populations of the Middle East and North Africa during modern times are the Palestinians, Lebanese, and Kurdish peoples. This map shows the scattering of the Lebanese and the Kurds. (The dispersal of the Palestinians is shown in Figure 5.29.) Over the last century, Lebanon has experienced various waves of diasporic migration due largely to the tensions that surround this multiethnic, multireligious society. The 1975–1990 civil war in Lebanon resulted in a particularly large number of emigrations. The failure of the Kurds to establish an autonomous region in the early twentieth century led to their being split among Iran, Iraq, and Turkey, with a small minority in Syria. Many Kurds have moved to different parts of the region or left it altogether during the twentieth century because of military aggression, persecution, and the failure of repeated attempts to establish a Kurdish state. Of course, Jews are also an important diasporic population but they have returned to the Middle East in very large numbers in recent times, rather than left it. (*Source:* Redrawn and modified from A. Segal, *An Atlas of International Migration.* London: Hans Zell, 1993, pp. 95 and 103.)

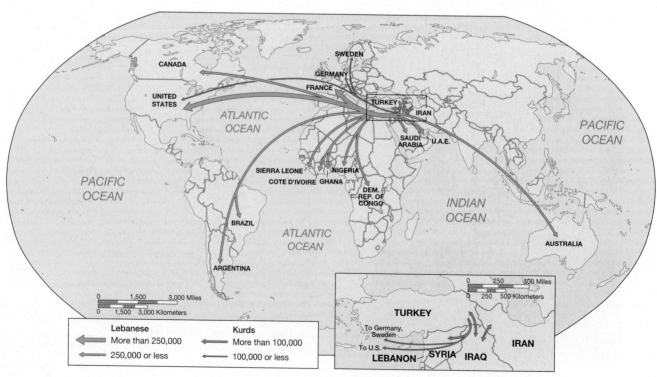

Figure 5.26   Labor migrants   Many immigrant workers are found in the oil-rich countries of the Middle East and North Africa. In some countries, these workers make up 80 or 90 percent of the workforce largely because there are so few local people to fill the jobs. Some states, such as Saudi Arabia and the United Arab Emirates, are becoming concerned about the large numbers of immigrants in their states and are passing laws to return many of them to their native countries and/or restrict the entry of others. This photo shows workers lining up for their lunch break.

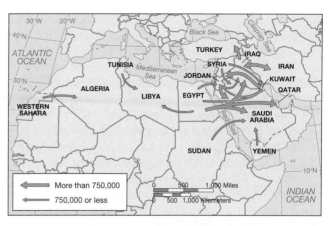

Figure 5.27   Internal migration in the Middle East and North Africa, 1990   Internal migration on a massive scale is a fairly recent phenomenon in the region. Until the decline of the Ottoman empire, most residents of the region lived and died close to where they were born. European colonialism and imperialism in the early twentieth century did result in some migration. But the most significant impetus for mass internal migration really began with the expansion of oil production in the Oil States in the 1950s. (*Source:* Redrawn and modified from A. Segal, *An Atlas of International Migration.* London: Hans Zell, 1993, pp. 45 and 77.)

kinds of jobs made available through the oil economy have meant that workers had to be imported. Nearly three-quarters of the labor force in the Gulf states are **guest workers**, brought in to work in all aspects of oil production from exploration and well development to drilling, refining, and shipping (**Figure 5.27**). And because oil revenues have been widely reinvested in economic development projects in the region, even more jobs have been created outside the petroleum industry, ranging from service sector positions to jobs in the building and construction industry. In order to lessen the potentially dislocating impact of foreign workers on local social and cultural systems, immigration policy among the oil-producing states of the Arabian Peninsula has favored Muslim applicants. Within the region, a large number of guest workers from Syria and Egypt, as well as Palestinian refugees, have come to participate in the Arabian Peninsula oil economy, filling both skilled and unskilled positions. Many additional guest workers have arrived from outside the region, especially from India, Indonesia, the Philippines, and Pakistan. While most of the migrants who have come from other Middle Eastern and North African countries have been male, a significant number of the labor migrants from Indonesia and the Philippines have been female. Although the region has experienced a great deal of in-migration from a variety of sources, much outmigration has also occurred and for very different political and economic reasons.

The most consistent forces pushing migrants out of the Middle East and North Africa have been the lack of economic opportunity, war, and civil unrest. The latter two have often either been fostered or exacerbated by the imposition of the core's political system of territorially bounded nation-states on populations previously organized around very different systems of political organization. The case of Lebanon is an illustration of this point. Lebanon began the twentieth century

following the breakup of the Ottoman empire as part of the Levant, a region that then combined both present-day Syria and Lebanon. At the beginning of the period of European imperialism in the region, Lebanon became a French mandate. During the Second World War, Lebanon was allowed to become fully independent. The way that independence was established was seen as especially problematic by the Arab majority of the region because the French gave a great deal of political power in Lebanon to the right-wing Maronite Christians who had been arbitrarily turned into a majority by the drawing of territorial boundaries. Since independence, Lebanon has been beset by cultural divisions, including rebellions, external attacks by Israel, and civil war between factions of Christian and Muslim militias and even within different Christian and Muslim groups. Since the twenty-first century dawned, Lebanon has been enjoying a period of relative civil tranquility, and the 2000 presidential election occurred in the absence of violence. The point of describing the Lebanese case in such detail, however, is to illustrate the difficult conditions that have compelled tens of thousands of Lebanese people—both Christians and Muslims—to flee the country to escape the violence. In 2001 more Lebanese lived outside Lebanon than in it.

British and U.S. political involvement in Iran also caused many Iranians to flee the country. With the beginning of the Cold War between Western countries and the Soviet Union following the Second World War, Iran's oil reserves were considered to be of great strategic importance to Britain and the United States. In the early 1950s, the United States and Britain feared that access to Iranian oil reserves was threatened by Iran's democratically elected but mildly anti-Western Prime Minister Mohammad Mossadegh because he had plans to nationalize the oil industry. Fearing a nationalized industry would cut off Britain, which had controlled Iran's oil through the Anglo-Iranian Oil Company, Britain appealed to the United States to help oust Prime Minister Mossadegh from office. The ouster was supported by President Eisenhower and carried out by the U.S. Central Intelligence Agency and British intelligence operators. The virulent anti-Western fervor that has circulated in Iran since the coup that ousted Mossadegh in 1953 and the revolution in 1979 that eventually deposed Shah Mohammad Reza Pahlavi is partly perceived to be the result of the interference of Britain and the United States in Iranian politics. Both events, but especially the fall of the shah, resulted in the exodus of tens of thousands of pro-Western Iranians to the United States, Britain, and Europe, as well as several Middle Eastern countries such as Egypt (**Figure 5.28**).

Other significant instances of migration created in the postcolonial Middle East and North Africa are the Algerians, Tunisians, and Moroccans, who have migrated mostly to Europe (see Geography Matters: Maghreb Workers in France, p. 230); Turks, who have followed a historical migration route to the Balkans and more recently to Germany; and Egyptians who have migrated to other Arab countries in the region as well as elsewhere in the world. The migrations of Turks and Egyptians, in large part, have occurred as educated and skilled workers have had to leave because economic growth is not able to keep up with population growth by providing high-paying jobs. Poorer Turkish and Egyptian migrants have also left. The artificial boundaries that were drawn around independent Iraq by the British in 1932 have also created fierce cultural tensions between the Shi'ite and Sunni Muslim sects that have contributed to nearly 50 years of enduring political instability in the country. One response to the impacts of this instability on daily life was emigration, which continued until 1979, when Saddam Hussein came to power and prohibited emigration entirely.

Figure 5.28 Iranian businesses in Los Angeles One of the most popular destinations for Iranian refugees during the Iranian revolution was Los Angeles, California. Geographer Ali Modarres found that while some of these immigrants (or political exiles) had substantial wealth—as did the exiled shah and his family—most were middle class or of lower economic status and came with limited financial resources. Many of them began small businesses that relied on other Iranian immigrants for their customer base.

The most dramatic instance of massive emigration in the Middle East and North Africa is that of the Palestinians, which began to occur when Israel became a state in 1948. Today Palestinians form the most widely scattered diasporic population in the world, with millions living outside the region altogether, while others live in various parts of the region, often in refugee camps (**Figure 5.29**). We discuss more of the background to the Palestinian diaspora and the current situation in Israel and the Occupied Territories later in this chapter (pp. 236–238).

The wide circulation of Middle Eastern and North African populations throughout the world has also resulted in the widespread distribution of their cultural practices, transforming the landscapes—in terms of buildings, tastes, sounds, and smells—of the places in which they have settled. The cuisine of the eastern Mediterranean, especially that of Lebanon and Syria, as well as of Turkey, Egypt, Iran, Morocco, Tunisia, and Algeria, is available in many large cities throughout most of the world's regions. In the United States and Europe, it is often available in smaller urban places as well. For example, many young Middle Eastern and North African men and women have gone off to study at universities in the United States and Europe where cafés offering strongly brewed coffee and regional cuisine have sprung up to serve them.

The types of food available from this region include what are known as *meze dishes,* which are predominantly subtly spiced appetizers or small dishes. The range of meze dishes broadly reflects the tastes and ingredients of a particular country or subregion within that country. Some of the most popular meze dishes include *baba ganoush,* a puree of toasted eggplant, sesame seeds, and garlic; *falafel,* a mixture of spicy chickpeas rolled into balls and deep fried; *fuul,* brown broad beans seasoned with olive oil, lemon juice, and garlic; and *tabouleh,* a salad of bulgur wheat, parsley, mint, tomato, and onion. The region also specializes in grilled meats, especially lamb and chicken, often served with rice.

A second cultural contribution of the Middle Eastern and North African region is music and dance. The most widespread of the region's dances is the traditional belly dance, a women's erotic solo dance done for entertainment. The dance is characterized by undulating movements of the abdomen and hips and by graceful arm movements (**Figure 5.30**). Belly dancing is believed to have originated in medieval Islamic culture, though some theories link it to prehistoric religious fertility rites. Middle Eastern and North African music has also become a staple of the contemporary world music scene. For instance, Googoosh, a celebrated female vocalist from Iran, has recently been allowed to tour outside the country and has been drawing record crowds. Googoosh was prohibited from performing in Iran for 20 years, beginning in 1979 when the Ayatollah Khomeini forbade all women from singing in public. Today Googoosh mania has hit Europe, the United States, Australia, and Japan, where her concerts have been sold out as Iranian emigrants, still suffering the pain of separation from Iran, weep as she sings a blend of traditional Persian and Western pop music.

Perhaps the most substantial evidence of the globalization of Middle Eastern and North African culture is the appearance

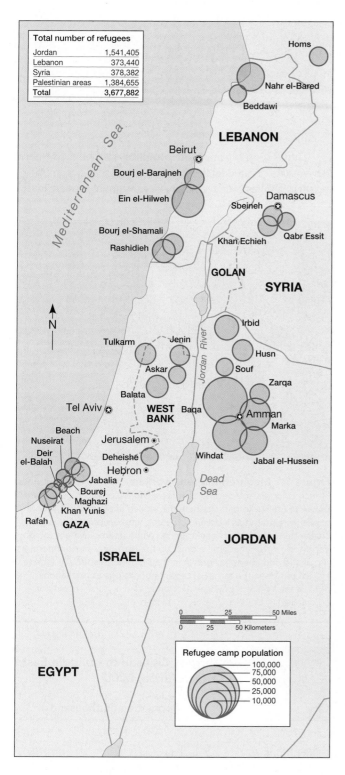

Total number of refugees	
Jordan	1,541,405
Lebanon	373,440
Syria	378,382
Palestinian areas	1,384,655
Total	3,677,882

Figure 5.29   Palestinian refugees in the Middle East This map shows the dispersion of Palestinian refugees—in camps and elsewhere—in the states around Israel. One of the biggest sticking points in the Israeli-Palestinian peace talks has been the question of refugee return and where Palestinians will be allowed to settle, given that most of their land has been occupied by Israeli settlers. More than 200,000 Jews have settled in formerly Palestinian communities. The settlers include some of the most radical Jewish nationalists in the region, and any attempt to remove them will inevitably lead to resistance and violence. (*Source:* Adapted from *The Guardian*, October 14, 2000, p. 5.)

Figure 5.30   Belly dancer in Cairo, Egypt The belly dancer's abdomen is exposed in order to emphasize the remarkably sinuous moves that the dancer performs.

of mosques throughout the world. Mosques serve as the main place of worship for Muslims, but they also serve many social and political needs, as forums for many public functions. Mosques also function as law courts, schools, and assembly halls. Adjoining chambers often house libraries, hospitals, or treasuries.

Cities and Human Settlement

The predominant pattern of settlement in the Middle East and North Africa is a relatively small number of very large cities, a substantial number of medium-sized cities, and a very great number of small rural settlements. Variation on this broad generalization among countries in the region is dramatic. Israel is mostly an urban country with 90 percent of its population living in cities. Sudan is largely a rural country with only about 30 percent of the population living in cities. The most relevant point about contemporary urbanization in the region is that cities are growing dramatically each year as more and more migrants come to live in them. Only about 50 years ago, most people in the region lived in small scattered rural settlements, but since then political independence and the economic development of the oil economy have been important underlying factors in the increasing urbanization of the population.

The Middle East and North Africa have a long and distinguished urban history. Beginning with the period of early empires, cities here have been important centers of religious authority, have played pivotal roles in trade networks and have reflected the complex culture that created them (see

Geography Matters

Maghreb Workers in France

The story of North African guest workers in the French automobile industry provides an excellent illustration of how international, national, and local conditions and processes interact to create global diasporas. The automobile industry in France actively recruited large numbers of North African guest workers in the 1950s and 1960s. Labor was needed because of the low birthrates that had occurred throughout France in the 1930s and 1940s. This was not the first time France had gone abroad for workers: Previous labor shortages had brought in Belgian, Spanish, Italian, and Polish workers. However, this was the first time that non-native-born Europeans had been intensively recruited. Some agents traveled to remote villages in Algeria, Morocco, Libya, and Tunisia (known as the Maghreb) and sought out illiterate males who were willing to leave their villages and families behind for work in French factories.

The workers constituted a cheap labor source: They worked for low wages and did not have to be given unemployment insurance, disability insurance, or healthcare benefits (**Figure 1**). They could be segregated from the French workforce, and through concentration in ghetto-like residential areas, they could be segregated from French society. Because they could not read and often did not speak French, there was little risk of their getting involved in union organizing. The North Africans were especially suited to auto industry needs because they were willing to withstand long hours of repetitive work on the assembly lines; could work night shifts; were easy to hire and fire; and rarely engaged in labor struggles.

Things soon became more complicated, though, for once the workers became established in France, they began bringing their families over, resulting in a substantial increase in the

Figure 1   Maghreb migrant workers As members of former French colonies, the people of Tunisia, Morocco, and Algeria were drawn to France to participate in the industrial economy. Although explicitly recruited as laborers, Maghreb workers present a challenge to the French government. When unemployment is high, imported workers become a drain on the economy. Furthermore, as nationalist sentiment has grown in the Maghreb, especially Algeria, France has been the target of terrorist bombings as retaliations against the Western political, economic, and cultural impacts of colonialism.

Geography Matters: The Islamic City, p. 232). In the early twenty-first century, cities in the region continue to play central administrative roles—though today their political significance is as important, if not more so, than their religious importance. Even though trade continues to be crucial to city building—especially oil-related trade—other economic sectors are also important stimuli to urbanization, including processing and manufacturing and services. What has been most remarkable about contemporary urbanization in the region is that its rapid pace has led to the emergence of one or two very large cities in each country that contain a large proportion of the country's population and disproportionately wield political and economic influence. The rapid pace and extreme degree of urbanization in the region can be traced to the migration of rural people in search of economic opportunity in the city as well as to natural increase among resident urban dwellers. **Table 5.1** lists the major cities of the Middle Eastern and North African region. Cities like Cairo,

Table 5.1	Ten Largest Cities in the Middle East and North Africa, 2000
City	**Population (in thousands)**
Cairo, Egypt	10,772
Istanbul, Turkey	9413
Tehran, Iran	7380
Baghdad, Iraq	4796
Algiers, Algeria	4447
Casablanca, Morocco	3635
Riyadh, Saudi Arabia	3329
Khartoum, Sudan	2748
Damascus, Syria	2335
Beirut, Lebanon	2058

country's foreign-born population between 1967 and 1974. In addition, by the mid-1970s, a global and national economic downturn, together with mechanization of many assembly functions, resulted in decreased employment opportunities. In response, the French government, which had previously been supportive of the recruitment effort, began restricting immigration. Automobile manufacturer Renault initiated a training program to teach its foreign workers French and technical skills. Nevertheless, massive unemployment occurred among the migrants, who were then replaced by French workers.

In 1977 the government began trying to repatriate the workers and their families to their home countries by offering them financial incentives to return. Some 45,000 people were repatriated, but many still remained, so the government applied further pressure in the form of housing restrictions. Massive layoffs and labor unrest persisted throughout the 1980s, however, and problems are still evident today, particularly as reflected in the xenophobic utterances from the far right of France's political spectrum.

Why do the workers remain, even in such hostile circumstances? Because going home may not be any better. For guest workers and their families, the prospect of return can be daunting: They may no longer be accepted in their local villages, and if their home countries have been experiencing severe economic and political problems, the prospects of finding any kind of job there become even dimmer. By now many second- and third-generation North Africans live in France (**Figure 2**). Having grown up in France, their attachment to their "homeland" in North Africa may be fairly weak, and they may decide to take advantage of French laws that allow them to become French citizens upon reaching adulthood.

Figure 2   Cheb Khaled, North African-French vocalist
Pictured here, center right, is Algerian singer Cheb Khaled, born in the Jewish-Spanish district of the oil-refining city of Oran, the most European of Algeria's cities. Oran is famous for its music and its seaside nightclubs. Khaled performs a modern version of Arab popular music known as *rai*, a style influenced by Bedouin chants and originally accompanied by a flute and a drum. The modern version of rai that Khaled performs fuses rai (accompanied by electric guitars, brass, drums, synthesizer, tambourine, and accordian) with jazz, funk, and, increasingly, rap. While extremely popular in the Maghreb, Khaled, known as the "King of Rai," has moved to Paris to escape the Islamist conservatism that has banned him from performing in his homeland. Khaled currently performs in Europe, the Middle East, and North Africa, as well as in North America.

with nearly 11 million inhabitants, and Istanbul, with just over 9 million, are in the top ten of the world's largest cities.

As in other parts of the periphery, rapid urbanization has severe consequences. One of the most widespread problems of rapid urbanization in the periphery is the inability of governments to meet the service and housing needs of growing urban populations. As a result, there is inadequate and often poor-quality water supplies, electricity, sewer systems, clinics, and schools, as well as air pollution and severe traffic congestion. Most critically, governments seem unable to provide housing for all who need it and squatter settlements have been assembled on unclaimed or unoccupied urban land (**Figure 5.31**). Unfortunately, the very largest cities in the region continue to draw in even more migrants who see the most well-known places as possessing the best opportunities for a better life. These primate cities continue to grow in disproportion to other urban places, compounding the severity of their problems. On the other hand, rapid urban growth in

a few of the very wealthy oil-producing countries such as Jubail (Saudi Arabia) and Doha (Qatar) has resulted in impressively modern cities with few of the urban problems of their neighbors. Their enormous wealth coupled with their very low populations has made the growth of some of their urban places a relatively uncomplicated process. Many of the other cities of the oil-producing region, including Jidda (Saudi Arabia) and Basra (Iraq), however, have not escaped the erection of shantytowns and the difficult social problems that accompany this type of urban change.

The highly diverse peoples of the Middle East and North Africa occupy an ancient region with a complex history and environment. As mentioned earlier, water is a critical variable in shaping where people live. Religion is a central force in shaping social interactions; kinship and tribe, as well as gender, play important roles. There are many important cities and several very large ones that connect the region to the world system. Just under half the population still lives in rural settlements, but

Geography Matters

The Islamic City

Islamic cities provide good examples of how social and cultural values and people's responses to their environment are translated into spatial terms through urban form and the design of the built environment. It is because of similarities in cityscapes, layout, and design that geographers are able to talk about the Islamic city as a meaningful category. It is a category that includes thousands of towns and cities, not only in the Middle East—the heart of the Islamic empire under the prophet Muhammad (A.D. 570–632)—but also in regions into which Islam spread later: North Africa, coastal East Africa, south-central Asia, and Indonesia. Most of the cities in North Africa and south-central Asia are Islamic, while many of the elements of the classic Islamic city can be found in towns and cities as far away as Seville, Granada, and Córdoba in southern Spain (the western extent of Islam), Kano in northern Nigeria and Dar-es-Salaam in Tanzania (the southern extent), and Davao in the Philippines (the eastern extent). Although urban growth does not have to conform to any overall master plan or layout, certain basic regulations and principles are intended to ensure Islam's emphasis on personal privacy and virtue, on communal well-being, and on the inner essence of things rather than their outward appearance (**Figure 1a**). The traditional Islamic city was walled for defense, with several lookout towers and a Kasbah, or citadel (fortress), containing palace buildings, baths, barracks, and its own small mosque and shops (**Figure 1b**).

(a)

(b)

Figure 1   The Islamic city (a) Seen from above, the traditional Islamic city is a compact mass of residences with walled courtyards—a cellular urban structure within which it is possible to maintain a high degree of privacy. (b) The traditional Islamic city was walled for defense, with several lookout towers and a Kasbah, or citadel (fortress), containing palace buildings, baths, and barracks. This photograph is of Yazd, Iran.

Figure 5.31   Squatter settlements in Istanbul, Turkey In Istanbul, squatter settlements are known as *gecekondu*, a Turkish word meaning that the settlements were built after dusk and before dawn. Geographer Paul Kaldjian's research in Turkey has shown that many of the residents of these settlements actually own land in the countryside upon which they grow foodstuffs for their urban livelihood. With no employment opportunities there, these individuals have been forced to migrate to the city in search of work. Urban unemployment is high, and those who can find work often do so for very low wages. Because land is so scarce in Istanbul, the *gecekondu* provide shelter but little opportunity to procure the other necessity for life: food. Relatives of the urban migrants often stay behind and maintain the family gardens during the growing season. At harvest time, the urban residents return to the countryside to help gather and divide up the crops.

Figure 2   The main mosque The dominant feature of traditional Islamic cities is the Jami, or main mosque. Pictured here is The Blue Mosque with its six minarets and gardens, in Istanbul, Turkey.

The single most dominant feature of the traditional Islamic city is the *Jami*—the city's principal mosque (**Figure 2**). Located centrally, the mosque complex is not only a center of worship but also a center of education and the hub of a broad range of welfare functions. As cities grow, new, smaller mosques are built toward the edge of the city, each out of earshot from the call to prayer from the *Jami* and from one another.

Traditionally, gates controlled access to the city, allowing careful scrutiny of strangers and permitting the imposition of taxes on merchants. The major streets led from these gates to the main covered bazaars or street markets called *suqs* (see Figure 5.41). The *suqs* nearest the *Jami* typically specialize in the cleanest and most prestigious goods, such as books, perfumes, prayer mats, and modern consumer goods. Those near-

er the gates typically specialize in bulkier and less valuable goods, such as basic foodstuffs, building materials, textiles, leather goods, and pots and pans. Within the *suqs,* every profession and line of business had its own specialized alley, and the residential districts around the *suqs* were organized into distinctive quarters, or *ahya',* according to occupation (or, sometimes, ethnicity, tribal affiliation, or religious sect).

Privacy is central to the construction of the Islamic city. Above all, women must be protected, according to Islamic values, from the gaze of men. Traditionally, doors must not face each other across a minor street, and windows must be small, narrow, and above normal eye level. Cul-de-sacs (dead-end streets) are used where possible in order to restrict the number of persons needing to approach the home, and angled entrances are used to prevent intrusive glances. Larger homes are built around courtyards, which provide an interior and private focus for domestic life (see Figure 5.23).

Because most Islamic cities in the Middle East and North Africa are located in hot, dry climates, these basic principles of urban design have evolved in conjunction with certain practical solutions to intense heat and sunlight. Twisting streets, as narrow as permissible, help to maximize shade, as does lattice work on windows and the cellular, courtyard design of residential areas. In some regions, local architectural styles include air ducts and roof funnels with adjustable shutters that can be used to create dust-free drafts.

All these features are still characteristic of Middle Eastern and North African Islamic cities, though they are especially clear in their old cores, or *medinas*. Like cities everywhere, however, Islamic cities also bear the imprint of globalization. Although Islamic culture is self-consciously resistant to many aspects of globalization, it has been unable to resist altogether the penetration of the world economy and the infusion of the Western-based culture of global metropolitanism. The result can be seen in international hotels, skyscrapers and office blocks, modern factories, highways and airports, and stores.

even with this large proportion of rural dwellers, the Middle East and North Africa are more urbanized than some other world regions, including South Asia and Sub-Saharan Africa. With a large population that derives its livelihood mostly from agriculture, the region, generally speaking, is most heavily involved in international trade around primary products such as minerals and agricultural goods (especially cereals and grains, cotton, and fruits and nuts). The most important of these products, as we shall see in the following section, is oil, and, in fact, the Middle East and North Africa possess and trade more oil on the world market than does any other region of the world. As we shall also see in the next section, despite its fantastic oil

wealth, which is by no means widely distributed across the region, the Middle East and North Africa is still predominantly a peripheral region in the world economy.

Regional Change and Interdependence

During the twentieth century, the countries of the Middle East and North Africa emerged from their colonial and dependent status with a range of economic and political problems. Many

Middle East and North Africa experts believe that most of the political and economic problems of the region are a direct result of the drawing of artificial political boundaries that have united peoples who were previously antagonistic or divided peoples who were once unified. In fact, while the region has experienced wars and conflict for hundreds, if not thousands of years, and certainly well before the Europeans arrived, it is generally agreed that most of its present conflicts stem directly from either one of two things. The first involves the boundaries and borders created by the colonial powers; the second is the strategic importance of the Middle East and North Africa to the political and economic interests of the core countries of the world-economy (**Figure 5.32**).

Remarkably, at the same time that the region has been the site for bitter and, in some instances, seemingly irresolvable conflicts, it has also been the site for a great deal of broad and sustained cooperation. The most significant unifying forces have been the religion of Islam and the Arabic language that have helped the vast majority of the peoples of the region to appreciate and nurture their common cultural heritage. Moreover, these unifying forces have not only been important culturally, they have also helped many of the peoples of the region to recognize that they have common political and economic goals as well.

New Political Geographies and Regional Conflicts

As mentioned, the postcolonial political geography of the region is seen as one of the most serious challenges to stability and peace there. Extreme stereotypes suggest that Arabs in general are naturally bellicose people. It is important to recognize, however, that while terrorist organizations do exist in the region (as they do in all regions of the world) and that armed conflict has been a sad reality of life in many parts of the region, such characterizations are fundamentally false at the same time that they grossly simplify the region's political, economic, and social history. Both conservative and more radical observers of the region agree that the boundaries drawn by Britain, France, Spain, and Italy in the Middle East and North Africa have been the single most important source of contemporary conflicts in the region, many of which are decades old. It is very important, therefore, to appreciate that our treatment of these conflicts in this text is meant to expose their structural sources so that they might be better understood not as stemming from the personal characteristics of the people who inhabit the region but from the difficult political situations they have inherited.

The most well known of these regional conflicts either exacerbated or created by the colonial powers during their occupation of the region include continuing tensions between Iran and Iraq and Iraq and Kuwait, the Arab-Israeli conflict, and the Palestinian self-rule movement. This is by no means an exhaustive list but one meant to represent the most highly publicized in U.S. newspapers and the ones seen to be most threatening to international security. Other conflicts that are a

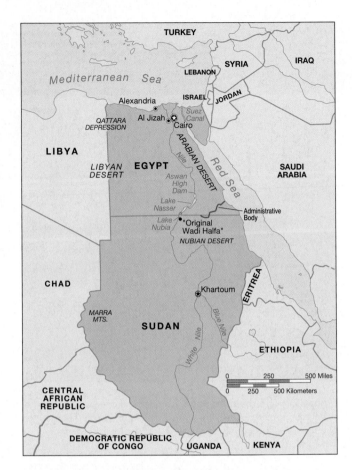

Figure 5.32   Border issues, Sudan The imposition of national boundaries in Sudan and Egypt by the departing British has resulted in conflicts among various groups within the Sudan as well as between Sudan and Egypt. Until the British presence in the region, no formal political boundaries existed. When the British left the region, however, the triangular area along the Red Sea on the eastern border of Sudan and Egypt that once belonged to Egypt was given to Sudan by Egypt in an attempt to induce it to join the now-disbanded United Arab Republic (UAR). Sudan never joined the UAR but was allowed to retain the land, which is a point of recurring controversy and sometimes outright conflict between the two countries.

legacy of the colonial past include Morocco's absorption of Western Sahara, the civil wars in Lebanon and Sudan, Turkey's persecution of its Kurdish minority, the Algerian nationalist movement, and Egyptian Islamists' persecution of the Coptic Christian minority.

Tensions Between Iran and Iraq and Iraq and Kuwait

The tensions that exist and the conflicts that have erupted between Iran and Iraq over the last quarter century are the result of a number of factors. One reason for the tension between the two countries is the cultural differences between Persians (Iranians) and Arabs (Iraqis). Though the majority of Persians and Arabs are Muslim, their ethnic origins, language, geography, and history are distinctly different. Furthermore, the Persians were unceremoniously conquered in the seventh century by the Arabs and converted to Islam beginning in the ninth century. Since then both countries have a very long history of

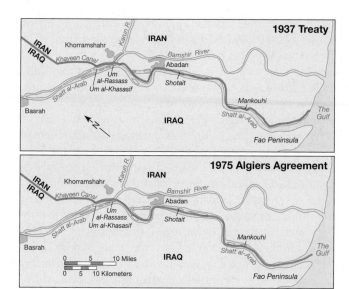

1937 Treaty

1975 Algiers Agreement

Figure 5.33   Shatt-al-Arab, Iran and Iraq   This map shows one of the areas of dispute that fuel the territorial conflict between Iran and Iraq. A shift in the boundary along the Shatt-al-Arab from the east bank to the deepest part of the channel in 1975 was one reason war erupted between the two states. While the enmity between the two countries goes back centuries, and certainly before they became modern states, the territories in question provide contemporary opportunities to ignite the ages-old antagonism. (*Source: Redrawn from T. Y. Ismael, Iraq and Iran: Roots of Conflict. Syracuse: Syracuse University Press, 1983, p. 23.*)

Figure 5.34   Persian Gulf War   Following the end of the Persian Gulf War in the spring of 1991, 732 oil wells were set ablaze in Kuwait by the retreating Iraqi Army. By early summer, some 550 wells were still burning. The smoke from the fires blanketed the entire Gulf region. In addition to setting the wells ablaze, Iraqi troops released oil into the Persian Gulf. Many environmental observers view the Persian Gulf War as one of the worst ecological disasters in history.

animosity that has more recently been complicated by their different dependent relationships with Britain and, later, the United States. The conflicts between Iran and Iraq remain unresolved today and are a source of continuing concern for the international community because both countries are important in global petroleum production.

Three small islands at the mouth of the Persian Gulf that had historically been part of Iran and later became part of the British protectorate over eastern Arabia were transferred to the United Arab Emirates by the British after the UAE became independent in 1971. The islands had historically been part of Iran, however, and in 1971 Iran seized them. Iraq pronounced this seizure an act of aggression and a violation of Arab sovereignty and unsuccessfully attempted to agitate the Arab population of Khuzestan Province to revolt against Iran. Less than a decade later, in 1980, war broke out between the two countries when Iraqi forces, in retaliation for Iranian artillery attacks and propaganda broadcasts against Iraq, occupied most of Iran's Khuzestan Province. Other points of tension occurred around Iran's support of the Kurdish guerrillas who had been fighting the Iraqi government for autonomous control over their mountainous region. And Iraq angered Iran when it gave asylum to the anti-shah Ayatollah Khomeini in 1963.

Yet another source of conflict has been a territorial dispute over the Shatt-al-Arab, a 204-kilometer (127-mile) stretch of water that connects the junction of the Tigris and Euphrates rivers to the Persian Gulf (**Figure 5.33**). Iraq controlled that part of the waterway, but Iran claimed it. Iraq eventually ceded it to Iran in a treaty that forced Iran to cease supporting the

Iraqi Kurds. Not long after the ceding of the Shatt-al-Arab by Iraq to Iran, however, President Saddam Hussein of Iraq came to regard it as an incident of humiliation, and he vowed to force Iran to return it. Years of attacks and counterattacks have ensued over the Shatt-al-Arab and other issues. For instance, the 1991 Persian Gulf War is seen as the latest continuation of the animosity between the two countries. During this brief encounter, U.S. forces—along with troops from other countries, including some Arab states—retaliated against Iraq's invasion of Kuwait (which Iraq has argued was historically part of Iraq). Even following the war, the international community has continued to be concerned over the conflictual relationship between Iraq and Iran and Iraq and Kuwait because together these three states possess one of the world's richest oil reserves (**Figure 5.34**).

The Arab-Israeli Conflict and the Palestinian Self-Rule Movement   The history of the Arab-Israeli conflict and the Palestinians' passionate desire for self-rule is a complex and highly volatile one despite persistent local and international efforts to bring peace to the region. The violence that erupted in fall of 2000, just as the peace process seemed to be most promising, underscores the complexity of the problem and the difficulty of resolution. As with the Iran/Iraq/Kuwait case, the chief factors that have inflamed this seemingly intractable political problem were exacerbated by British partitioning of the region.

The official Jewish state of Israel is a mid-twentieth-century construction that has its roots in the emergence of **zionism**, a late-nineteenth-century movement in Europe. Zionism's chief objective has been the establishment of a legally recognized home in Palestine for the Jewish people. Thousands of European Jews, inspired by the early Zionist movement, began migrating to Palestine at the turn of the nineteenth century. When the Ottoman empire was defeated in 1917, the British gained control over Palestine and the Transjordan area and issued the Balfour Declaration. The Balfour Declaration was highly problematic however, because indigenous peoples, the Palestinians, already occupied the area and viewed the arrival of increasing numbers of Jews and European sympathy for the establishment of a Jewish homeland as an incursion into the sacred lands of Islam. In response to increasing Arab-Jewish tensions in the area, the British decided to limit Jewish immigration to Palestine in the late 1930s through the end of the Second World War. In 1947, with conflict continuing between the two groups, Britain announced that it had despaired of ever resolving the problems and would withdraw from Palestine in 1948, turning it over to the United Nations at that time. The United Nations, under heavy pressure from the United States, responded by voting to partition Palestine into Arab and Jewish states and designated Jerusalem as an international city, preventing either group from having exclusive control. The Jewish state was to have 56 percent of mandate Palestine; an Arab state was to have 43 percent; and Jerusalem, a city sacred to Jews, Muslims, and Christians, was to be administered by the United Nations. The proposed United Nations plan was accepted by the Jews and angrily rejected by the Arabs, who argued that a mandate territory could not legally be taken from an indigenous population.

When Britain withdrew in 1948, war broke out. In an attempt to aid the militarily weaker Palestinians, combined forces from Egypt, Jordan, and Lebanon, as well as smaller units from Syria, Iraq, and Saudi Arabia, confronted the Israelis. Their goal was not only to prevent the Jewish forces from gaining control over additional Palestinian territory but also to wipe out the newly formed Jewish state altogether. This war, which came to be known as the first Arab-Israeli War, resulted in the defeat of the Arab forces in 1949, and later armistice agreements enabled Israel to expand beyond the United Nations plan by gaining the western sector of Jerusalem, including the Old City. In 1950 Israel declared Jerusalem its national capital, though very few countries have recognized this.

Israel maintained the new borders gained during the first Arab-Israeli War for another 18 years until the Six-Day War in 1967, which resulted in further gains for Israel, including the Sinai Peninsula and the Golan Heights, the southwestern corner of Syria. The eastern sector of Jerusalem and the West Bank, previously held by Jordan, were also annexed during the Six-Day War. As **Figure 5.35** shows, a long period of relatively little territorial change occurred until the 1970s and 1980s, when Israel moved toward reconciliation with Egypt through a series of withdrawals that eventually returned all of the Sinai to Egyptian control by 1988.

The territorial expansion of Israel has meant that hundreds of thousands of Palestinians have been driven from their homelands, and the landscape of Palestine has been dramatically transformed. Today Palestinians live as refugees either in other Arab countries in the region, abroad, or under Israeli occupation in the West Bank, the Golan Heights, and the Gaza Strip (also known as the "Occupied Territories"). The Arabs of the Middle East and North Africa and many other international observers are convinced that Israel has no intention of allowing the diasporic Palestinian population to return to their homelands. By the late 1980s, in fact, Palestinians who had remained in their homeland had become so angered by Israeli territorial aggression that they rose up in rebellion. This rebellion, known as the **intifada** ("uprising"), has involved frequent clashes between fully armed Israeli soldiers and rock-throwing Palestinian young men (**Figure 5.36**). The intifada, more than anything, is a reaction against 32 years of Israeli occupation of the Palestinian homeland and increasing Israeli settlement, particularly in the West Bank and the Gaza Strip. In addition to the intifada, other Palestinian groups have coalesced in opposition to the Israeli occupation. The Palestinian Liberation Organization (PLO) was formed in 1964 as an organization devoted to returning Palestine to the Palestinians. Since its official recognition, the PLO has become the Palestinian Authority. The Palestinian Authority, the chairman of which is Yasser Arafat, is seen as the only legitimate representative of the Palestinian people. Besides the Palestinian Authority, however, other, more extreme groups claim to represent the Palestinian cause. One of the most well-known is Hamas (Harakat al-Muqawama al-Islmiyya, or the Islamic Resistance Movement), whose activities are largely centered in the West Bank and Gaza Strip.

Since the mid-1990s, hopes for peace in the region have risen, fallen, risen, and most recently very decisively fallen once again. In October 2000, after weeks of very difficult, but promising, U.S.-sponsored peace negotiations between Yasar Arafat and Ehud Barak, then Israeli prime minister, violence broke out again in the West Bank. This new violence left little hope in Israel, the Occupied Territories, or elsewhere that the Arab-Israeli conflict will be resolved any time in the near future. Moreover, this conflict, as well as the conflict between Iraq and Iran, the civil war in Lebanon, and political unrest in other parts of the region, makes it very difficult for the countries involved to direct attention to the basic needs of its citizens, let alone the larger challenges of national economic development.

Regional Alliances

It is critically important to recognize that although the Middle East and North Africa is a region with more than its share of conflict, it is also one where a great deal of cooperation, coordination, and joint action exists. Many political, economic, and cultural cooperative organizations are operating in the region.

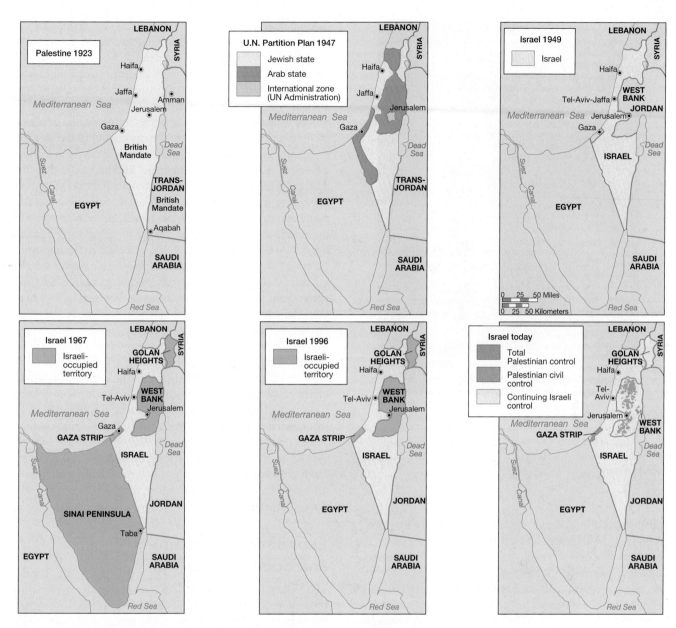

Figure 5.35   Changing geography of Israel and Palestine, 1923–2000 Since the creation of Israel out of part of the former Palestine in 1947, the regional political geography has undergone significant modifications. A series of wars between Israelis and neighboring Arab states and a number of political decisions regarding how to cope with both resident Palestinians and large numbers of Jewish people immigrating to Israel from around the world have produced the changing map we see here. (*Source:* J. M. Rubenstein, *The Cultural Landscape: An Introduction to Human Geography*, 5th ed. Upper Saddle River, NJ: Prentice Hall, 1996, p. 233.)

Organized in 1945, the Arab League is a voluntary association of Arab states whose peoples are mainly Arabic speaking. Formally known as the League of Arab States, it is the most unifying of all the Middle Eastern and North African regional organizations. The stated purposes of the Arab League are to strengthen ties among member states, coordinate their policies, and promote their common interests. The league is involved in various economic, cultural, and social programs, including literacy campaigns and programs dealing with labor issues. It is also a high-profile political organization that acts as an important sounding board on conflicts in the region, such as the Persian Gulf War and the Arab-Israeli conflict.

The league was founded by Egypt, Iraq, Lebanon, Saudi Arabia, Syria, Transjordan (Jordan, as of 1950), and Yemen. Other countries of the Middle East and North Africa that later joined the Arab League are Algeria (1962), Bahrain (1971), Kuwait (1961), Morocco (1958), Oman (1971), Qatar (1971), Sudan (1956), Tunisia (1958), and the United Arab Emirates (1971). The PLO was admitted in 1976. In 1979 Egypt's membership was suspended after it signed a peace treaty with Israel, but it was readmitted 10 years later.

Another central and widely known regional organization, this one based on economic interests, is the Organization of Petroleum Exporting Countries (OPEC). While organizations like

Figure 5.36 The intifada The intifada is an informal and mostly spontaneous expression of dissatisfaction with the Israeli occupation of Palestine, and usually takes the form of violent encounters between bands of rock-throwing Palestinian youths and Israeli military forces.

the Council of Arab Economic Unity deal with every aspect of economic development and change, OPEC, as its name suggests, is a specialist economic organization. OPEC's central purpose is to coordinate the crude-oil policies of its member states. Founded in 1960, OPEC has 12 members—Algeria, Gabon, Indonesia, Iran, Iraq, Kuwait, Libya, Nigeria, Qatar, Saudi Arabia, United Arab Emirates, and Venezuela—four of which (Gabon, Indonesia, Nigeria, Venezuela) are not part of the Middle Eastern and North African region. As is clear from the list, however, Middle Eastern Arab states dominate the membership. OPEC originally was formed in response to the dropping price of oil in the 1950s when supply greatly outstripped demand. In the 1970s, as oil supplies in non-OPEC countries were reduced, the organization lowered production, which had the effect of raising the price of oil. OPEC also sets production ceilings that specify how much oil may be produced by each member state. This practice ensures that the price per barrel does not fluctuate dramatically due to market gluts or scarcity.

Complementing as well as contrasting with the goals and objectives of OPEC is the Gulf Cooperation Council (GCC). The GCC coordinates political, economic, and cultural issues of concern to its six member states—Saudi Arabia, Kuwait, Bahrain, Qatar, the United Arab Emirates, and Oman (see Sense of Place: United Arab Emirates, p. 240). The members of the GCC have come together in order to exert a coordinated effort over the management of their substantial income from

their oil reserves, problems of economic development, and social problems, trade, and security issues. All six of the states in the GCC are politically conservative monarchies wary of the revolutionary republican urges that have swept the region and transformed the previous monarchies of Egypt, Iran, and Iraq, for instance. The GCC has made very large sums of money available to all Arab countries for economic development as well as military protection during political crises.

Many other important regional and international organizations have been established in the Middle East and North Africa. Additionally, some individual states have also begun to attempt to make connections with organizations beyond the region, tying the Middle East and North Africa more securely to the rest of the globe. For instance, Turkey, already a member of NATO, has applied for full membership in the European Union, though there are significant political barriers to achieving this. Morocco, Tunisia, Jordan, and Israel have also signed agreements, so-called "Euro-Med" agreements, with the European Union (an organization of European states dedicated to increasing economic integration and cooperation in Europe—see Chapter 2, p. 90) that are leading to increased transnational integration beyond the region. While some critics argue that alliances with organizations outside the region may erode unique regional identities or diminish local control over local processes, others argue that transnational integration beyond the region may increase political stability and decrease conflict. At this point it is unclear whether either or both might be true. What is true is that the cooperation brought on by the vast oil wealth of some states as well as the conflict brought on by cultural differences has helped to widen the gap between rich and poor in the region.

Economic Development and Social Inequality

The Middle East and North Africa is a region of extreme contrasts of wealth and poverty. For example, in 2000 Kuwait had the highest per capita income in the region ($22,700), while Yemen had the lowest ($740). Not surprisingly, however, national figures like per capita income hide all sorts of variation within different regions—oftentimes between the city and the rural areas—but also even within the same area where dramatic variation can occur between one urban neighborhood and the next. While Israel is the exception, most of the extreme wealth of the region comes from oil-based revenues to the states of Saudi Arabia, Kuwait, Iran, Iraq, Oman, Qatar, Libya, and the United Arab Emirates (**Table 5.2**). States with the largest populations tend to have the lowest levels of wealth. Despite the phenomenal wealth generated from oil production for some parts of the region, most of the Middle East and North Africa remains poor and highly dependent on an increasingly marginalized agricultural sector.

In addition to the impact that neoliberalism has had on increasing the levels of inequality in the region, other forces are limiting the life chances and standards of living of the region's population. For example, refugees and many migrant workers face very difficult economic circumstances. We have men-

Table 5.2	Richest and Poorest States (in 2000 $)	
State	**Per Capita Income**	**Human Development Index[a] Ranking**
Richest		
Kuwait	$22,700	54 (high)
Israel	$18,000	22 (high)
UAE	$17,400	48 (high)
Qatar	$17,100	57 (high)
Bahrain	$13,100	43 (high)
Poorest		
Yemen	$740	151 (low)
Sudan	$930	157 (low)
Iraq	$2400	127 (medium)
Syria	$2500	81 (medium)
Egypt	$2850	112 (medium)

[a]The human development index is based on measures of life expectancy, educational attainment, and personal income. The table shows the ranking of the richest and poorest countries in the Middle East and North Africa on this measure compared to countries in other world regions.

(a)

(b)

Figure 5.37　Beirut, Lebanon (a) The 16-year civil war that gripped Lebanon destroyed large parts of one of the most prosperous and Europeanized cities of the region. Shown is a war-damaged building on the former "Green Line" that separated Muslim West Beirut from Christian East Beirut. (b) In the last decade, redevelopment in Beirut has been steadily increasing, and a new city has arisen from the rubble of the old.

tioned the difficult situation of Palestinian refugees living in temporary camps throughout the region. Since the first refugees left Palestine in 1948, as many as 2 million Palestinians have been displaced to refugee camps in Lebanon, Syria, Jordan, the West Bank, and the Gaza Strip. Many persons are born and grow up in these camps, originally intended as temporary settlements, where basic provisions are poor. Other large refugee populations live in camps in Iran, which shelters 1.5 million Afghan refugees and 600,000 Iraqis, and Sudan, where 4 million displaced Sudanese people live outside their home territories. Migrant workers imported into the oil-rich states of the Persian Gulf as guest workers, though receiving much higher wages than they would in their home countries, have substantially lower standards of living than the resident Arab populations.

Finally, the many conflicts that have occurred and are occurring throughout the region can also economically dislocate resident populations and increase the plight of those who are socially or politically weak. For example, the civil war in Lebanon, which lasted for more than 16 years and ended in 1991, is a sad illustration of the toll that conflict exacts on ordinary people who are involuntarily caught up in the violence. Lebanon before its civil war was one of the most economically developed countries in the region, with a highly literate, entrepreneurial society and modernized economy (**Figure 5.37**). But Lebanon is also a diverse multiethnic society. Hostile minority groups, forced to live and work within the same state boundaries by the colonial decree, contributed to ethnic conflict that helped to drive the state into protracted political and economic chaos. The result was a dramatic decrease in

the standard of living of a large proportion of the Lebanese population. It has now been 10 years since the end of the civil war, and the country has begun to recover. Unemployment, however, is still extremely high (18 percent), though the agricultural sector, the mainstay of the economy, is nearly revived. While much of the Lebanese middle class was able to survive the civil war through investing their savings in foreign banks, poorer farm laborers were particularly hard hit.

Core Regions and Key Cities of the Middle East and North Africa

As we have already discussed, the Middle East and North Africa is a region of dramatic contrasts, with high-technology industrial production existing side by side with ancient agricultural techniques. The core regions within the Middle East

Sense of Place

United Arab Emirates

Travel agencies in Europe promote the United Arab Emirates (UAE) as a land of contrasts. With mountains and beaches, deserts and oases, camel racing, Bedouin markets, and the legendary duty-free shopping in Dubai, the promotion is an easy one to sell. Known as a destination for upscale travelers, the UAE is an interesting and exciting corner of the Arabian Peninsula. Once a region whose economic base was derived from breeding camels, some farming, and pearl fishing, the UAE has transformed itself from a set of sleepy, isolated, and poor shaykhdoms (sheikdoms) to dramatic wealth and development based entirely on oil. The UAE is relatively small at about 83,000 square kilometers (32,000 square miles), roughly the size of the state of Maine. The climate is hot and dry, though somewhat cooler in the eastern mountain areas of the region.

Politically, the UAE is a federation of seven independent states with a central governing council. The states are actually seven sovereign shaykhdoms, known as *emirates,* from the title of their rulers. They include Abu Dhabi, Ajman, Dubai, Fujairah, Ras al-Khaimah, Sharjah, and Umm al-Qaiwain. The history of human occupation of the area dates back to the Bronze Age (about 4500 B.C.). The contemporary history that has been most influential in shaping its current situation was early contact with Portuguese traders in the sixteenth century and contact with British naval operations in the eighteenth century. Through truces (as well as agreements and treaties) organized by the British to rationalize maritime relations, the seven shaykhdoms eventually came under British protection in the mid-nineteenth century and were given the name the Trucial States, or Shaykhdoms. For more than 100 years the British played a dominant role in the administration of the states, until 1971 when they were granted independence.

Since independence, and due to the discovery of oil in Abu Dubai in the late 1950s (and soon after, other parts of

Figure 1   Abu Dhabi Like cities throughout the world, Abu Dhabi also reflects the impact of economic globalization in international hotels and the offices of transnational corporations.

the UAE), the states have been dramatically transformed (**Figure 1**). The development of the oil reserves has stimulated population growth from about 75,000 in 1950 to about 223,000 in 1970 and to a twelvefold increase in 1999 to more than 2.7 million. A large proportion of the population is concentrated in the provisional capital of the UAE, Abu Dhabi—one of the world's most modern cities—but there are significant numbers in Dubai, certainly the most vibrant city on the Arabian Peninsula. Because of its relatively small domestic population, the UAE has had to depend to a significant extent on foreign workers to help exploit its oil wealth. The result is that

and North Africa are those whose economies are the most dynamic and productive and effectively set the pace economically, and to some extent politically, for the rest of the Middle East and North Africa. The two core regions of the Middle East and North Africa are the Oil States and the Eastern Mediterranean Crescent (**Figure 5.38**).

The Oil States

The Oil States subregion of the Middle East and North Africa includes Bahrain, Iran, Iraq, Kuwait, Oman, Qatar, Saudia Arabia, and the United Arab Emirates. Although Yemen, Al-

geria, and Libya are also regional oil-producing states, their situations are substantially different from the other eight states discussed here. Yemen's oil reserves are only newly discovered and it has yet to be able to exploit them to the level that its other Arabian Peninsula neighbors have. Algeria has substantial oil and gas reserves but a far more mixed economy than the states along the Persian Gulf. Its history as a French colony also sharply differentiates it from the others. Libya, like Algeria, has an economy that is not solely oil-based, and its colonial and modern political history make it an exceptional case. Whereas the once active agricultural economies of the Oil States have been largely eroded by an emphasis on

approximately 80 percent of its labor force is expatriate, many of whom have come from Pakistan, India, and other Arab countries. Upper- and middle-class expatriates who work in the UAE, many of whom are from Europe and the United States, find life in the UAE something of a fantasy world. There are no taxes on retail goods, and often these expatriates live in the UAE rent-free and are given free airfare home once or twice a year. A typical existence for a Western expatriate is one of nearly total disposable income, continuously sunny weather, weekend camping or beach trips, bars, and restaurants. Unlike Saudi Arabia and some of the other states of the Arabian Peninsula, most of the emirates of the UAE have a relatively relaxed atmosphere. But for the other expatriates in the UAE, Pakistanis, Indians, Sri Lankans, Bangladeshis, and some Chinese and Filipinos, who are employed at the lower end of the economic spectrum, life is not so glamorous and carefree. Many work on building sites, outdoors, where the temperatures in summer are searing. Most do not come with their families but work in the UAE in order to support their families back home. While working in the UAE offers an improved living situation to what they might expect in their home countries, many poor expatriate workers seldom return home but spend years sending what they earn to those depending on it back home.

As mentioned, the liveliest city-state in the emirates is Dubai. Dubai is a bastion of freewheeling capitalism akin to similar duty-free ports like Hong Kong. Cheap electronics goods, gold, and Persian carpets are the specialty items that can be purchased there. Dubai is actually two towns, however. There is Deira to the east and Dubai to the west separated by the *al-khor* (the Creek), an inlet in the Gulf (**Figure 2**). The Creek opens up into a picturesque harbor for *dhows* (small sailing craft) engaged in fishing, pearling, smuggling, and other maritime pursuits. Dubai also contains the enormous and extremely busy Jabal Ali free port and industrial park where its highly developed reexport system is in operation. European companies shipping their goods to Asian destinations send

Figure 2 Bastakia Quarter, Dubai The Bastakia Quarter sits in the old city of Dubai along the waterfront. It contains some of the oldest windtower houses built at the turn of the nineteenth century by wealthy Dubai merchants. The towers are designed to catch even slight breezes and to funnel the air down into the houses. The city of Dubai has created a program to conserve and restore the windtower houses.

them first to Dubai where they are offloaded, broken down into smaller shipments, and reloaded as reexports for more specific destinations. Alternatively, sea cargoes coming from Asia to Europe are also reexported, though the new shipments are routinely sent by air to Europe.

In 1971 the odds were poor that the UAE would be able to maintain its independence, let alone thrive. Yet in 2001, the UAE celebrated 30 years of independence. The changes that have taken place throughout the seven shaykhdoms have been phenomenal and have transformed it dramatically. With shopping malls, elegant hotels, Starbucks Coffee, and a whole range of Western fast-food franchises, the UAE is perhaps the most globalized of all the states on the Arabian Peninsula.

petroleum production, Algeria and Libya continue to possess productive agricultural sectors in addition to profiting from their oil reserves.

Generally speaking, the Oil States tend to be among the most culturally conservative in the region. Of all the countries of the Middle Eastern and North African region, the Oil States are the most committed to a strict interpretation of Islamic law. As a result, women's public movements are highly restricted. Socializing among middle- and upper-class women tends to take place in private spaces—in homes and sequestered spaces in restaurants. Religious police—charged with enforcing Islamic orthodoxy—can be found on the city streets of Saudi Arabia and Iran, for instance, ensuring that public telephones are not used during prayer time, and women, especially tourists, are properly dressed (for instance, no skirts above the knee, no sleeveless tops). But broad generalizations such as these gloss over the rich cultural and social life that exists among the eight states of this subregion. So, for example, while Saudi Arabia and Iran are especially conservative Islamic states, other states like Kuwait and Bahrain are more open and possess a relatively more lively public culture of cafés, horse racing, camel racing, and soccer where both women and men are spectators. Overall, poor people in the region, especially the guest workers who have come to work in the oil economies, have a very

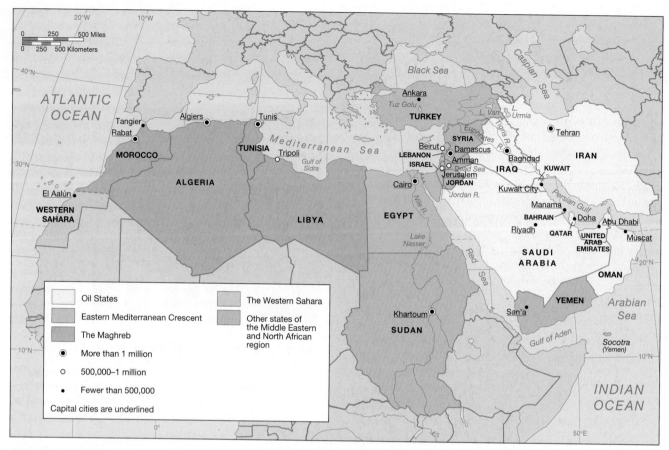

Figure 5.38   Middle Eastern and North African subregions This map shows the several subregions within the larger world region of the Middle East and North Africa. As with world regions, the boundaries of these subregions are somewhat arbitrary and are certainly subject to the possibility of future alterations. For instance, the Oil States may lose their core status when the oil reserves are depleted sometime in the early twenty-first century.

different social experience from the other classes. For instance, in the United Arab Emirates, the range of restaurants for foreigners includes small and inexpensive Indian, Pakistani, Lebanese, Chinese, and Filipino eateries, as well as American fast-food restaurants.

Although important social and cultural differences characterize the region, the one thing that undeniably unites it is oil and oil wealth. A visit to most of these countries will provide opportunities for partaking in activities that cater especially to the wealthy, including shopping in flashy new malls for luxury goods, attending sporting events, or participating in sporting activities catering to the rich (yacht clubs, golf courses, and the Dubai Cup, the richest horse race in the world). At some luxury hotels, spacious suites that are really more like lavishly appointed condominiums can cost up to $10,000 a night.

The Arabian Peninsula, which includes all but two of the Oil States (Iran and Iraq), is an area that is heavily dominated by the presence of huge reserves of oil—so huge, it is estimated, that Saudi Arabia alone may possess one-quarter of the world's known reserves. Saudi Arabia is the controlling political and economic force on the peninsula that was, until the early twentieth century, occupied mostly by rural villagers and nomadic peoples. After Saudi Arabia, Iraq possesses about 10 percent of the world's known reserves, Kuwait about 9 per-

cent, the United Arab Emirates 7 percent, and Iran about 6 percent. Oil production is the mainstay of each of the Oil States' subregional economies and contributes substantial revenues to each of the national economies. In some of the individual Oil States, revenues from oil production are so substantial that their respective national governments have a difficult time determining where to spend them. An embarrassment of oil riches hardly seems a problem anyone would seriously mind having. Yet it is a problem, and one that has a number of different facets to it. One facet is that a nearly exclusive dependence on one economic sector, and only one product within that sector, leaves the economy of such a state highly vulnerable to fluctuations in the demand for oil. For example, in the 1980s, a period of sustained expansion in the oil economy came to an end. As more fuel-efficient vehicles were developed, energy-conservation measures were taken, and cheaper fuels were substituted for oil. At the same time the world entered a period of economic recession, partly in response to high oil prices. As a result, oil profits fell dramatically, with implications for the Oil States moving like a wave through their economies.

When prices fall in an exclusively oil-based economy, there is nothing to fall back on to bring new revenues into the economy to supplement the decreasing profits from oil. In fact, for much of the early history of oil production in the region, the Oil States were content to use their wealth to benefit the kin of the ruling elite. Instead of investing in national infrastructure or developing human resources, these groups invested their wealth abroad or spent their money on luxurious living. As a result, until fairly recently, the economies of the subregion have been exceedingly vulnerable to falling oil prices. Recognizing this, all of the Oil States on the Arabian Peninsula have begun to diversify their economies (**Figure 5.39**). Some are attempting to introduce new industries, such as textile production and food processing plants. Others are developing port facilities. Still others are resuscitating or introducing agriculture, though the scarcity of water makes this an enormous technological challenge for all but Iraq and Iran. Yet, while large-scale, irrigated agriculture is a costly undertaking for any of the Arabian Peninsula Oil States, the feeling among them is that food is as much of a security issue as oil is.

Another less obvious problem that the abundance of oil wealth has generated in the Oil States is the dislocation of traditional social and economic systems as Western ideas have penetrated the region through the introduction of television and other consumer goods. The importation of migrant labor from other parts of the Middle East and North Africa to work in the oil industry or in the related jobs enabled by it has also tended to disrupt the local traditions as new migrants have brought their rather different cultures along with them. As participation in the Oil States' increased global economy has brought the region into closer contact with the West, the stress exerted on the local culture has sometimes intensified the already existing conservative interpretations of Islam among the Oil States.

The enormous wealth of the Oil States has made them a global economic player to be reckoned with. Understanding that power, the Oil States have spent at least two decades augmenting it with more military strength by improving their armies and purchasing sophisticated weaponry on the world arms market. Israel is perhaps the most heavily armed state in the region, and when its military might is included, it becomes clear that the Middle East and North Africa in general and the Oil States in particular are major players in the world military theater. Thus far, however, the new sophisticated weaponry has tended to be used largely within the region, against old rivals, rather than beyond the region.

Even without weaponry, the Oil States remain important players on the global stage for at least two additional reasons. First, many of the world states are highly dependent upon them for Gulf oil. Without petroleum from the Oil States, the ability of a large portion of the rest of the world to maintain productive economies would be severely hampered. Second, the impact of **petrodollars**—revenues generated by the sale of oil—are especially significant for the core economies of the world system where they are spent, invested, and banked. The Oil States of the Middle East are likely to continue to occupy a central role in the affairs of the region as well as the affairs of the core countries where guaranteeing a secure supply of oil is absolutely central to the smooth functioning of the global economy (see Geographies of Indulgence, Desire, and Addiction: Petroleum, p. 244).

Figure 5.39   Economic diversification in the Oil States
Mindful that petroleum is a nonrenewable resource and aware that the volatility of petroleum prices on the world market leaves them vulnerable, the Oil States have made efforts to diversify their economies. The development of tourism, agriculture, and industry is receiving special attention. In Kuwait, pictured here, as well as elsewhere in the region, there have been attempts to address the food security issue through the development of desert agriculture.

Riyadh   Riyadh is the capital city of Saudi Arabia, a country that is one of the world's top three oil producers. A modern city—built on the ruins of an old walled city—Riyadh has been the capital of the Kingdom of Saudi Arabia since 1932 (**Figure 5.40**). Riyadh is situated on a high plateau in the Najd region of the central Gulf Peninsula. Although designated as the capital of the Sa'ud dynasty in 1824, the city lost this status in 1881 when the Rashid family extended its control over the Najd. Ibn Sa'ud was able to regain control of Riyadh in 1902, and he used the city as the command center for his eventual conquest of all of Arabia, which he completed in 1930. The unified Kingdom of Saudi Arabia was proclaimed in 1932, and Riyad became its capital. Although it has officially been the capital for nearly 70 years, only in the last generation has Riyadh truly functioned as the capital, having been

Figure 5.40   Riyadh, Saudi Arabia   The capital of Saudi Arabia, Riyadh is a modern city serving as the political and economic center of the oil industry in the kingdom.

Geographies of Indulgence, Desire, and Addiction

Petroleum

Petroleum, also more commonly known as oil, is a naturally occurring oily liquid composed of various organic chemicals. *Petroleum* usually refers just to crude oil, but the term can also apply to natural gas and shale oil. Petroleum is most widely used as an energy source for industrial, commercial, government/military, and residential uses. Without petroleum, life as we know it in the core of the world system would collapse and life in the semiperiphery and periphery would be deeply disabled as well. In fact, semiperipheral countries base their development goals on petroleum availability.

While contemporary society is utterly dependent upon petroleum, ancient peoples also knew about petroleum and made use of surface deposits for waterproofing and as fuel for torches. Later, they learned to distill it and came to use oil as a lubricant and for medicinal purposes. In the nineteenth century, kerosene was extracted from petroleum and used as a cheap fuel for lamps and lanterns. Recognizing that the crude oil itself might have value, in the mid-nineteenth century drilling began in the state of Pennsylvania, marking the beginning of the modern petroleum industry. The invention of the internal combustion engine and the First World War helped to establish the petroleum industry as a foundation of industrial society.

The organization of the petroleum industry took several decades to consolidate and was more or less complete by the early decades of the twentieth century. The U.S. petroleum industry, through its five leading companies, along with two companies in Britain, dominated the petroleum industry worldwide throughout much of the twentieth century (**Figure 1**). In its early stages, the world petroleum market was organized by these seven companies, which produced and distributed abundant cheap oil first drilled around 1907 in Iran, by the 1920s in Iraq, and not until the 1930s in Bahrain and Saudi Arabia. In the United States, Standard Oil was the most prominent oil company, which by mid-century controlled 95 percent of the U.S. petroleum industry. In Britain, Shell Oil dominated.

In the middle decades of the twentieth century, the rumblings of resistance to foreign companies dominating their oil supply began to occur in the Middle East. Iran was the first to resist, and by the 1960s, enraged by the unilateral cuts in oil prices made by the seven big oil companies, the major oil-exporting countries formed the Organization of Petroleum Exporting Countries (OPEC) with the goal of controlling the price of oil. The so-called global "Oil Crisis" occurred in 1973 with another shock delivered in 1978 when the Middle East oil supply was disrupted during the Arab-Israeli War and the Iranian revolution, the latter being directly connected to the Western thirst for oil. The Oil Crisis dealt a major shock to the global economy when panic over supply led to wild speculation, and OPEC cut back on production and thus raised the barrel price in order to cash in on the panic (**Figure 2**). A worldwide recession resulted, and the oil market suffered greatly due to decreased demand. The conflicts and economic problems that emerged around the global oil supply in the 1970s made it very clear to the core that oil was a political issue that had to be monitored closely and carefully. The Persian Gulf War in 1990–1991 is an illustration of the strategic importance of oil to international relations and its central role in foreign policy.

Every day a globally coordinated system moves more than 60 million barrels of oil from producers to consumers. Al-

Figure 1   Oil tankers in Iran Oil is a strategic international resource, and the United States and other core countries are committed to protecting it from the political instability that sometimes overwhelms the Middle East region. Pictured here are oil tankers fueling at a station in Iran.

eclipsed by Jeddah, formerly the premier Saudi city. In the last three decades, however, the headquarters of the Saudi government have been moved to Riyadh, helping to confirm it as the practical as well as the symbolic center of the largest country in the Persian Gulf.

The discovery of vast petroleum deposits in Saudi Arabia in the 1930s helped to generate the wealth that transformed Riyadh from an old provincial town into a modern urban place. Once a cluster of mud-brick dwellings in a desert oasis, Riyadh is now home to more than 3.5 million people occupying a

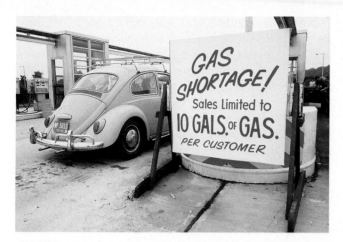

Figure 2   Oil crisis in the United States, 1973 Having developed a huge appetite for gas-guzzling automobiles, in the early 1970s Americans panicked when the global petroleum supply was threatened. Long lines at gas stations occurred throughout the United States, and energy-saving measures enacted through the government of President Jimmy Carter ensued. The interconnectedness of producers and consumers around oil is one illustration of how globalized resource use has become and how different regions play different roles in the world economy.

Figure 3   Oil slick off the coast of Qatar The most substantial ocean pollution from oil production occurs, not surprisingly, in the Middle East. Pictured here is a satellite image of an oil slick off the coast of Qatar indicated by the darker blue area at the top of the image. In addition to routine dumping, serious accidents have occurred as well. One of the most environmentally damaging occurred off Kharg Island in the Persian Gulf where an oil tanker disaster in 1982 resulted in a near continuous flow of oil into the sea for nearly a decade. As a result, an entire fishing industry has been destroyed, complete populations of some fish species are now extinct in that habitat, and desalination plants have become inoperative. It is unlikely that the Persian Gulf waters will return to normal for decades more.

though petroleum allows the core to enjoy a very high standard of living, there are serious implications to the widespread and increasing use of petroleum worldwide. The most critical problem is environmental pollution of both the air and the water. Air pollution generated from industries and vehicles burning petroleum or any of its products is significant but varies from region to region, depending upon any number of factors from how much petroleum-based energy is being consumed to what sorts of pollution-control systems are in place. The impact of oil pollution on Earth's oceans and rivers is also substantial. The amount of petroleum products ending up in the ocean is estimated at 25 percent of world oil production. About 6 million tons per year alone is discharged by tankers, passenger ships, and freighters (**Figure 3**). But the greatest volume of petroleum products dumped into the ocean is carried there by rivers. Through the discharge of industry, storage installations, refineries, and local gasoline stations, more than three times the quantity coming from all tankers and other ships travels to the ocean by way of the world's rivers.

The world's reserves of crude oil add up to about 700 billion barrels, more than half of which are in the Middle East and North Africa. Many energy policy experts believe that the utilization of oil as a major source of energy will end up being a very brief affair of little more than a century. Present estimates indicate that the supply of crude oil will probably be entirely exhausted sometime toward the middle of the twenty-first century, although new discoveries of oil continue to extend this deadline.

sprawling city of flashy contemporary buildings, wide boulevards, modern hospitals and schools, huge shopping centers that blend American-style malls and Middle Eastern and North African *suqs* (traditional markets; **Figure 5.41**). Because Riyadh is at the epicenter of the Saudi oil economy, its population, once

exclusively Najdi, is now quite cosmopolitan and includes a large number of other Arab nationals, Europeans, Americans, and Afro-Asians. In fact, although the most prestigious jobs are held by Saudis, most of the remainder of the private sector and government jobs are actually held by foreigners. In addition to

Figure 5.41   A traditional *suq* As the traditional centers of commercial life in the Middle East and North Africa, *suqs*—which are linear bazaars—can be found throughout the region. Pictured here are a traditional *suq* specializing in local, handicraft goods, and a modern one. The modern *suq* is something like an American-style mall and carries the usual traditional goods as well as more commercial products like jeans, electronics, and rock music.

its administrative role, Riyadh is also the kingdom's commercial and higher educational center, with an industrial base devoted almost entirely to oil production. As an important commercial hub in the region, Riyadh also possesses one of its most important markets, the camel market located about 30 kilometers (about 19 miles) from the city center. As the largest camel market in the Middle East and North African region, this is a key commercial site where the tenor of peak trading in the late afternoon resembles that in the pit of the Chicago Stock Exchange right before the closing bell. While Riyadh is a classic example of the impact of oil wealth on urbanization in the region, other cities also demonstrating this pattern include Abu Dhabi (UAE), Doha (Qatar), and Kuwait City (Kuwait).

Tehran   Tehran is the capital, the largest city, and the political, cultural, and economic center of Iran. Its rapid growth, especially over the last 50 years, has been the result of its role in administrating and centrally investing the massive inflows of capital from the sale of oil, agricultural products, and foreign trade. Tehran is located in the northern part of the country, on a high, sandy plateau surrounded on the north by the majestic, snow-covered Elburz Mountains. To the south, east, and west are deserts. The city gradually rises in elevation to meet the mountains from roughly 1160 meters (3480 feet) in the south to 1800 meters (5400 feet) in the north. Its climate is arid but temperate, with summer temperatures averaging around 22.6°C (around 73°F) and winter temperatures averaging around 11.5°C (around 52°F). In terms of land area, the city occupies about 600 square kilometers (232 square miles).

Tehran first became an important place in 1220 when it survived the sacking of the ancient city of Rayy by the Mon-

gols. At the time, Tehran was a small suburb of Rayy, probably founded around A.D. 300. Over the centuries Tehran grew slowly but steadily. In 1788 Agha Mohammad Khan, founder of the Kajar dynasty, made Tehran the capital of Persia. Beginning in 1925, when the Pahlavi dynasty came to power in Iran, Tehran became modernized, industrialized, and considerably rebuilt (**Figure 5.42a**). Although a modern city, there are parts of the city in the central and southern sections where traditional inward-looking courtyarded structures are more typical than the high-rise apartment structures found in the north. Moreover, at the center of Tehran are a number of beautiful old squares, among them Meydan-e Sepah (Sepah Square). This new and completely foreign urban form, with Renaissance proportions, replaced what used to be a farmers' market in the nineteenth century. Located nearby are important government buildings and several mosques and palaces (**Figure 5.42b**). Unfortunately, conservation of old buildings in the city core has not been a priority, and a more systematic plan of conservation and restoration is needed to guarantee that areas such as these beautiful squares will be maintained.

While the northern part of the city is mainly residential, the southern part contains a mix of residential, industrial, and commercial buildings. The southern and northern sections also reflect rather extreme social differences, with the north containing newer and mostly spacious middle- and upper-class neighborhoods, while the south contains densely occupied and underserviced poor and working-class neighborhoods (see A Day in the Life: Gharchak, p. 248).

Today, the population of metropolitan Tehran exceeds 7 million. In the last two or three decades it has experienced rapid growth, resulting in a number of serious problems, including traffic congestion, air pollution, water shortages, and inadequate housing. Transportation may be the biggest problem, as lack of planning has resulted in too many vehicles on too few thoroughfares. The large number of vehicles, many of them older and with no emission controls, has also created serious pollution problems. The levels of CO_2 found in the air are often four times that of safe levels, according to World Health Organization reports. On some particularly bad smog days, schools have to be cancelled, and people with respiratory problems are warned to stay indoors. The pollution is worst in the central and southern parts of the city, while many higher areas in the northern part are actually above the smog level or receive the benefit of the north–south breezes that drive the smog farther south. In order to address the private-vehicle pollution problem, the city heavily subsidizes its bus services and introduced a metro system in 1999.

Water shortages are also a problem in this arid region where periodic droughts occur. Although three large dams were built in the 1960s to help provide an adequate and more dependable water supply, population growth over the last three decades has begun to put new pressures on the water supply. There are also problems with water quality, particularly in the southern part of the city, where industrial pollutants have seeped into the water table.

While many cities in the periphery experience extreme problems around housing supply and quality, Tehran's prob-

(a)

(b)

Figure 5.42   Tehran   (a) The predominance of modern structures in Tehran is evidence of efforts by the Pahlavi dynasty to enable the city to become an important player on the world stage. The Iranian revolution in the 1970s slowed this process, but recent political changes have resulted in new development as well as redevelopment throughout the urban area. (b) The older parts of Tehran contain some very beautiful architecture dating back to the eighteenth century. Sepah Square is in the oldest part of Tehran and is the most important plaza in the city, acting as an artificial dividing line between the wealthier northern and the poorer southern parts of the city. The city's bazaar, the palace, old alleyways, and some of the city's oldest buildings are close to or on the southern side of the square. The northern side of the square is the site of banks, embassies, and other more contemporary buildings.

lems in this area are much less dramatic. The government has assumed an active role in attempting to provide housing for poor Tehranis. Still, the rapid growth in population has made it difficult for the government to keep pace, and squatter settlements and shanty towns have grown up in the southern section of the capital. On a more positive note, since 1990 the city government has been actively committed to creating and maintaining parks and green space throughout the city and has been remarkably successful in achieving this goal.

The Eastern Mediterranean Crescent

The Eastern Mediterranean Crescent, made up of Egypt, Turkey, Lebanon, and Israel, constitutes something of an awkward but compelling clustering of states (see Figure 5.25). It is awkward largely because none of the states view themselves as regionally coherent with the others. It is compelling because among them, the four states have the potential to be—or already have become—regional economic success stories. The greatest differences exist between Egypt, Turkey, and Lebanon on the one hand, and Israel on the other. While Egypt, Turkey, and Lebanon have had strong agricultural bases for hundreds, if not thousands of years, Israel's commitment to agriculture is more recent (**Figure 5.43**). Furthermore, while the former three employ agriculture as a major source of export revenues, for Israel, agricultural production is more about

achieving national food security, though foodstuffs are also exported. Finally, while Egypt, Turkey, and Lebanon are just beginning to encourage more industrial development—some more successfully than others—Israel already possesses a strong industrial base that is fairly diverse but receives a large share of its income from high technology production.

So why include Israel if it is really more different from than similar to the other three states? The problem is that Israel doesn't really fit anywhere very comfortably. As mentioned, Israel is a Jewish state in a predominantly Arab region (though neither Turkey nor Iran is Arabic). Its economy generates the highest GNP in the region, higher than any of the Oil States. It possesses a military that is as highly trained and technologically equipped as any European state. Yet, while Israel is really not like any of its neighbors, it is too important politically, culturally, or economically to be ignored. One reason for including Israel in the Eastern Mediterranean Crescent is that it functions as a kind of pacesetter for the other three, who are anxious to be more active players in the contemporary global economy. Furthermore, there are important similarities among the four states as well. For instance, all four possess more of a European orientation than many of their neighbors, certainly more so than the Oil States. All four have a sizable middle class that has been important to their political stability. And all four have the potential, because of their resource endowments, to continue to build a diverse economic base. While Israel,

Gharchak

In *Tehran: The Making of a Modern Metropolis*, author Ali Madanipour attempts to reveal the complexity of modern Tehran. One of the ways in which he approaches this task is to interview a whole range of Tehranis to get a sense of what it is like for different people to live in the city. This "Day in the Life" is based on material in Mandanipour's book.[*] No names were included in the original source.

One of Madanipour's interviewees is a middle-aged woman who lives in Gharchak, a neighborhood in the southern part of Tehran. Gharchak is a very poor district that serves as something of a working-class bedroom community for wage earners who commute to the central business district or other parts of Tehran for work. Gharchak and another southern Tehran neighborhood, Veramin, have been experiencing serious problems with polluted drinking water. The pollution has been caused by a rising water table that is bringing with it toxic chemicals left behind in the soil from decades of industrial-related processes.

This Gharchak resident works as a domestic helper and has done this kind of work since she moved to Tehran as a young married woman. She has worked for a number of different households over the years and even today she works for several employers. She travels by bus all around the city to get to these various households and therefore has seen lots of changes in the city over the many years she has lived there. She says of the city: "God knows how I got used to this huge city, which has no beginning and no end, after coming from our little village" (p. 151).

The small village where she was born is in Khyrasan, the northeasternmost province of Iran. The capital of Khyrasan is Mashhad, a significant center of religious pilgrimage among Shi'ite Muslims. Although she does not mention this in her interview, it is possible that this domestic helper left the province of Kyrasan because of the tremendous growth in population it has been experiencing over the last two or three decades. This growth is the result of an influx of Afghan refugees who have migrated there and transformed the labor market in both cities and rural areas (**Figure 1**). The increased number of refugees, coupled with the failure of the government's agricultural improvement policies, has resulted in too many people and too few jobs.

While living in Khyrasan, this domestic helper and her husband owned no land of their own and worked on other people's farms. As part of the post-revolutionary rural-to-urban migration in Iran, she and her husband moved to Tehran many years ago in order to improve their economic situation. Her husband eventually learned plastering skills and was employed on construction sites. He was often violent and abusive to her and her children, however, and became addicted to drugs in Tehran. Soon after his drug addiction began, he was killed in an accident on a construction site, and his wife was left to raise their eight children alone. While she still continues to work as a domestic helper, her children are now grown and are helping her. One of her sons bought her a house, where she currently lives with some of her grown children.

Interestingly, this woman's children appear to have exceeded the economic status that she was able to achieve through her move to Tehran. Several have moved away from the city to other parts of Iran and one of her sons was able to accumulate enough money to buy her the house. Although she grew up in a rural environment, she feels little ties to her home village. It has been too many years since she left, and many of her friends and neighbors who once lived there have left as well. She says of that village, "There is nothing left there for me" (p. 151).

Although this is a story specific to an Iranian woman living in Tehran, it is emblematic of the many stories of rural-to-urban migration that have been occurring throughout the Middle East and North Africa for the last several decades. As agriculture has become increasingly mechanized throughout the globe, and fewer laborers are needed to work in the fields, millions have moved to the cities to find work and, hopefully, improved economic circumstances. For some, hard work has paid off, and they have been able to improve their circumstances and move out of the poorest parts of the city. For others, lack of education, the high levels of urban unemployment and underemployment, and the low level or complete absence of social services in the cities of the region have meant that many more have been frustrated in their goals for a better life.

Figure 1   Afghan refugees in Khyrasan, Iran Iran has been the destination for an estimated two million refugees fleeing the Taliban regime in Afghanistan over the last several years. Tent cities are typical of the camps these refugees occupy. Since the events of September 11, 2001, and the commencement of the bombing of Afghanistan by U.S. and U.K. forces, Iran, as well as other neighboring countries, have sealed their borders with Afghanistan to prevent additional refugees from entering.

[*]*Source:* A. Madanipour, *Tehran: The Making of a Modern Metropolis.* Chichester: John Wiley & Sons, 1998, pp. 147–160.

Figure 5.43   Agriculture in Turkey   Much of Turkey's climate is conducive to agriculture, and the country contains numerous farming regions. Cotton is a major export crop, and Turkey is also the world's largest exporter of sultana raisins and hazelnuts. Other important crops are tobacco, wheat, sunflower seeds, sesame and linseed oils, and cotton-oil seeds. Opium was once a major crop, but its exportation was banned by the government in 1972. The ban was lifted two years later as poppy farmers were unable to adapt their land to other crops. The government now controls the production and sale of opium.

coupled with any of the other three states, would make strange bedfellows, and strong political and cultural differences have tended to prevent all four of these countries from acting in concert, there is much to suggest that cooperation would be mutually beneficial. The main reason for clustering these four countries is that they appear to possess the necessary ingredients to be important players in the world economy because of their histories, their economies, and their important roles in regional politics.

Turkey and Egypt are two states in the Eastern Mediterranean Crescent that once controlled long-lasting, influential, and extensive world empires. Although neither was colonized by Europe, both labored under the conditions of a foreign bureaucracy—Egypt longer than Turkey. Presently, both states have similarly sized populations—Egypt with around 67 million and Turkey with around 65.5 million—and both are burdened with the problems that large national populations present to economic development. While Turkey possesses a fairly diversified economy with a strong agricultural sector and substantial mineral wealth, national agriculture in Egypt is built upon a fairly narrow base, largely due to the environmental constraints of the desert. Both have significant manufacturing capacity across a range of products from food processing to heavy machinery.

The final important player in the Eastern Mediterranean Crescent is Lebanon. Lebanon possesses a strong agricultural base and was, for many years before its civil war, an important banking and financial center for the Middle East and North Africa, connecting it to the core of Europe and North America. It was not until the early 1990s, however, that Lebanon began to recover from the serious political problems that civil

war unleashed. Given its previous history as "the Paris of the Middle East," there is reason to be optimistic that Lebanon, under its recently elected president, can regain some of its former economic and political power and become an important force in the region once again.

Finally, the four states of the Eastern Mediterranean Crescent possess the most important cities of the Middle East and North Africa. Cairo in Egypt, Istanbul in Turkey, Beirut in Lebanon, and Tel-Aviv in Israel are all critically important nodes in the urban system of the world economy, enabling capital, ideas, and people to come together. Because capital is so highly urbanized under the current conditions of globalization, these four cities serve to integrate the Middle East and North African region more directly and securely into the global economy.

Cairo   Cairo is Egypt: So say all the travel guides. This so-called "Mother of the World" is the capital of Egypt and the largest city in North Africa and the Middle East. Located on both sides of the Nile River near its delta in northern Egypt, Cairo is a city that visitors either love or hate, but few are ever indifferent.

The history of Cairo reflects the history of the region, which is one of conquest and change. The origins of Cairo go back to early fourth millennium B.C. when Memphis, located on the site of present-day Cairo, was the capital of Egypt. As the city of Memphis grew, it spread along the east bank of the Nile. Later, the Roman conquerers constructed the city of Babylon upon the Memphis site. When the Fatimids, a dissident branch of Muslims, conquered Egypt in A.D. 969, they too established their headquarters at Babylon, but changed its name to the Arabic Al Qahira (Cairo), "the city victorious." When the Mamelukes rose to power in the region in the thirteenth century, they too established Cairo as their capital, and the city became renowned throughout Africa, Europe, and Asia. In 1517 the Ottomans conquered Cairo and ruled Egypt from there until 1798, when Napoleon I took possession of it. The Ottomans were able to regain control of Cairo in 1801, but by the late nineteenth century, Egypt's sizable foreign debt and the increasing dissolution of the Ottoman empire paved the way for British influence in the region, which continued until the end of the First World War in 1918. The postwar period marked the beginning of Cairo's rapid population growth, which continues today.

Metropolitan Cairo is home to 16 million people, overburdening the urban region with the world's highest density of people per square kilometer. In central Cairo, traffic crawls along the city's narrow streets; building pediments hug the street's edge, leaving little room for pedestrians to pass; and in the narrow walkways between buildings, pedestrians jostle for space. In addition to being the administrative capital of Egypt, Cairo is also a port and the chief commercial and industrial center of a country that produces cotton textiles, food products, construction supplies, motor vehicles, aircraft, and chemical fertilizers. It is also the cultural center of Egypt.

At the center of the city is the river island of Zamalik, where three bridges link the island to the mainland on both sides of the river's bank. Scores of novels about Cairo have used this backdrop as the setting for passionate scenes—both political and personal. Though located within close range of the pyramids, Cairo was built not by the pharaohs but by the Fatimid dynasty, beginning just over 1000 years ago. Much of the city that the Fatimids built has survived, including the Fatimid mosque, Al-Azhar University (the oldest in the Islamic world), and the three great city gates of Bab An-Nasr, Bab al-Futuh, and Bab Zuweila.

The neighborhoods of Islamic Cairo are located on the east bank of the Nile, where the streets are especially narrow and the bazaars especially crowded. Hundreds of mosques populate the neighborhoods of Islamic Cairo, which is perhaps why the area is known as such, since it is no more or less Islamic than the rest of the city. To the south of Islamic Cairo is Old Cairo, home to many of the oldest architectural monuments in the city. Cairo's Coptic Christian community also occupies much of Old Cairo where mosques are scarce and Coptic churches dominate the landscape (**Figure 5.44**).

Although an important world city for political and economic reasons, Cairo faces very serious social problems. The arrival of thousands of new migrants each month from rural areas as well as other from other parts of the region has been unanswered by any sort of additional housing provision. The city and the country have been unable to meet the land and housing needs of newly arriving migrants, as well as the need for schools, hospitals, and other social services. As a result, new arrivals to the city must find alternatives to traditional housing. Some sleep on the streets. Others take over abandoned buildings or land that is underutilized. The most famous example of alternative uses is the thousands of people who live among the tombs of one of Cairo's oldest cemeteries. Known as the City of the Dead, this cemetery, intermingled with houses and apartment buildings, is now occupied by nearly 1 million living residents. Cairo's urban planners are well aware of the many problems such a large city faces and are implementing a whole range of plans. The subway system that was expanded during the 1990s has helped to ease some of the extreme traffic prob-

(a)

(c)

(b)

Figure 5.44 Cairo, Egypt (a) Coptic Cairo. The architecture generally associated with the Copts, or Egyptian Christians, dates from about the third to the twelfth century. Monasteries and churches, scattered throughout Egypt, were built of unbaked brick on the basilica plan inherited from the Greco-Roman world. Characteristic features include heavy walls and columns and vaulted roofs. (b) Islamic Cairo. In Islamic architecture almost all mosques repeat the plan of the house of the prophet Muhammad, founder of Islam, and are composed essentially of an enclosed courtyard, a building at one end for prayer, and arcades on the sides. (c) Modern Cairo. Modern Cairo could easily be mistaken for any number of core cities with skyscrapers; broad streets packed with automobiles, trucks, and buses; and busy sidewalks loaded with pedestrians.

lems in the city, and new urban centers are being built to house some of the population away from central Cairo.

Today, Cairo continues to maintain its regional significance as the largest and most important city in the Islamic world. As such, its influence also projects outward, connecting the rest of the world to the Middle East and North Africa.

Distinctive Regions and Landscapes of the Middle East and North Africa

The Maghreb

The Maghreb is the region of northwest Africa that contains the coastlands and Atlas Mountains of Morocco, Algeria, and Tunisia, and the mostly desert state of Libya. Its people, history, and geography make it a distinctive region. As Figure 5.38 shows, part of the Maghreb includes a coastal plain along the Mediterranean Sea. Within this narrow band, the region enjoys a moderate, Mediterranean climate, with cool, wet winters and hot, dry summers. Agriculture thrives here, as well as tourism. Within this region are located the famous cities of Tripoli, Casablanca, Tangier, and the Barbary Coast—places of heroics and legend and Hollywood-style glamour that conjure in the imagination a landscape of mythic and romantic appeal (**Figure 5.45**).

Figure 5.45   Tangier port and city, Morocco   Tangier is a seaport on a small bay of the Strait of Gibraltar in northern Morocco. The city is a shipping center and has few other industries. Tangier was taken in 1471 from the Arabs by the Portuguese and given to Charles II of England as part of the dowry from Catherine of Braganza. The English abandoned the city to the Moors in 1684 when it became a pirate haunt. Together with a surrounding zone of about 360 square kilometers (about 140 square miles), Tangier was temporarily internationalized (1911–1912) and proceeded to be governed by a number of states, mostly by Spain, until Moroccan independence in 1956.

The Maghreb has a unique and perhaps even a legendary history that differentiates it, not only from the rest of Africa but also from the Middle East. In ancient times, the region was influenced by the Phoenicians, Carthaginians, Romans, Christians, Vandals, and Byzantines and then finally by the Arabs, who in the late ninth century A.D. converted the populace to Islam. While much of the material remains of that period have been destroyed or built over, important remnants have survived. For instance, in Tunisia, Phoenician merchants founded a number of trading posts several thousand years ago. The most important one was Carthage, founded in 814 B.C. When the Romans defeated the Phoenicians, this area was incorporated into the Roman Empire, providing wheat and other commodities to the population in Rome. The ruins of Carthage lie in a suburb of present-day Tunis, the capital of Tunisia (**Figure 5.46**).

During the most recent period of imperialism, from the mid-nineteenth to the mid-twentieth century, millions of Europeans, primarily French but also Spanish and Italian, flocked to the Maghreb and influenced its government, architecture, and language, especially in the city of Algiers, which is where most of the colonial Europeans lived. Significant numbers of Europeans also inhabited Casablanca and Tunis and formed a professional class that introduced many of the local elites to European cultural practices. In the early to mid-twentieth century, thousands of young people from the Maghreb went off to obtain university educations on the continent and then returned to the Maghreb with the seeds of nationalist ideology planted in their hearts and minds. Many of these individuals played important roles in the independence movements that occurred throughout the Maghreb.

In addition to possessing important material remnants of the ancient world, the region is also home to the Berber, a people of ancient origin who preceded the Carthaginians and the Romans and who still live in the region today (see Figure 5.22). The Berber appear to have been indigenous to the Maghreb region, though in more recent times, those who are attempting

Figure 5.46   Carthaginian ruins in Tunis   Carthage was one of the great cities of the ancient world. Artifacts unearthed by archaeologists suggest the city was probably established as a trading post by Phoenicians toward the end of the ninth century B.C. Pictured here are the Roman Baths of Carthage.

to maintain their traditional ways of life have tended to live in the mountains and deserts, away from the increasingly populated coastal area. In Algeria, the Tuareg, once a largely nomadic Berber people, live. For centuries the Tuareg were known as the "lords of the desert" because they patroled the caravan routes on their camels and acted as guides for caravaneers. The word *Tuareg* literally means "blue men," and the group was given the name because their indigo robes darken their skin blue. Like many other of the Berber peoples, though, the Tuareg are for the most part no longer nomadic and are more likely to be oil or service workers. Another aspect of Berber culture can be found in a number of place names in the Maghreb. The Barbary Coast, for instance, is a name derived from the word *Berber*. It became famous as a base of Arab and Berber pirates who launched attacks on Spanish and other European fleets as well as, at one point, on vessels of the United States.

The Maghreb region has a relatively strong economy based largely on oil and mineral exploitation, agriculture, and tourism. Algeria's oil industry provides nearly 90 percent of its export revenues. Libya too has substantial, high-quality petroleum reserves. Both Tunisia and Morocco are significant globally for their phosphate industries. All of the Maghreb countries are also agricultural producers, though none is self-sufficient. The most important agricultural products of the region include wheat, barley, olives, dates, citrus fruits, almonds, peanuts, beef and poultry, and vegetables.

Another significant aspect of the economy of the Maghreb is tourism. Hugging the southern coastline of the Mediterranean with rugged mountains rising up from the coastal plains and then trailing off to the desert, the Maghreb is a spectacularly beautiful setting for tourists. From lying on the beautiful beaches to trekking into the Atlas Mountains or the Sahara Desert to visiting ancient archaeological ruins, the Maghreb offers a range of tourist experiences, in both luxury and economy style. Europeans are frequent visitors to the Maghreb because after a short flight, they can enjoy warm temperatures, exotic landscapes, and inexpensive and sumptuous food.

Algeria and Morocco are two of the fastest-growing economies in the North African region, with Libya and Tunisia making substantial strides as well. Links across the Mediterranean with the European Union through the Euro-Med agreements discussed earlier are likely to boost all sectors of the Maghreb economy from resources to tourism. In the next 10 years or so, it is certainly possible that despite its strong Islamic history and resultant ties to the Middle East, the Maghreb may once again become especially close to Europe. This transformation is something of a tall order, however, as it requires the different states of the Maghreb to overcome their anxieties and animosities as former colonies or occupied territories and learn to wield power alongside their former colonizers and occupiers: Spain, Italy, and France.

The Western Sahara

What makes the Western Sahara unique is that this former Spanish colony did not gain independence as did most of the other former colonies or mandate territories of the region in the mid-twentieth century but instead was handed over to Morocco. It was incorporated as a province by the Moroccan state partially in the 1976 and completely in 1987. The population of this coastal area is just 200,000. The landscape is one of a wind-swept, flat, and monotonous desert where temperatures routinely reach 45°C (120°F) in the summer. Before the twentieth century, the area of Western Sahara was outside the control of any central political or military authority and was considered a geographical backwater. Other than some important phosphate deposits and offshore fishing rights, the Western Sahara would seem to be a generally resource-poor area and not obviously desirable as a territorial acquisition. Yet the Moroccans are aggressively determined to keep possession of Western Sahara, just as the Saharawis, as the people of Western Sahara are known, are determined to become independent. As in the Maghreb, a few Saharawis went to Spain in the early twentieth century, received a modern university education, and came back to the region with ideas of national independence. At present, the Western Saharan resistance movement operates under the aegis of the Polisario Front (an acronym for the Popular Front for the Liberation of Saguia al-Hara and Rio de Oro, the two divisions of the former Spanish colony).

In 1976, Polisario declared the formation of the Saharawi Arab Democratic Republic (SADR), "a free, independent, sovereign state ruled by an Arab democratic system of progressive unionist orientation and of Islamic religion" (**Figure 5.47**). While many of the African states have officially recognized the Sahrawi Republic, Morocco has refused to acknowledge it or to renounce control over it. After appeals to other states of the Maghreb for support and for the United Nations to intercede to enforce a cease-fire, a United Nations-sponsored referendum was to be held so that residents of Western Sahara could determine their fate through the ballot box. U.S. Secretary of State James Baker also attempted to work out a plan for full sovereignty over the area by the Saharawis.

The Western Sahara has been a valued acquisition for Morocco because it is known to contain phosphate reserves.

Figure 5.47   Saharawi refugees in Algeria   Political instability in Western Sahara caused the flight of tens of thousands of individuals to other parts of North Africa in the late 1970s and early 1980s. Pictured here are Saharawan women settled in a refugee camp in Algeria.

The rock phosphates found in the Western Sahara (and in other parts of the Maghreb) are converted to phosphoric acid and used in rustproofing metals and as an ingredient in soft drinks and dental cements. A variant, trisodium phosphate, is used as a detergent and water softener. In short, phosphates are globally important in both industrial production and household use, and Morocco appreciates the economic value of the Western Sahara to its development goals. Although Morocco is anxious to develop the phosphate reserves of the Western Sahara, the political situation has prevented even a full-scale exploration to proceed in the face of Polisario's determination to free itself from Moroccan control. Because the international community has recognized Polisario's claims, a United Nations referendum on sovereignty was to take place in March 2000. Unfortunately, the referendum was postponed because thousands of Moroccans migrated to the region to take advantage of free land there and have complicated the voter-registration process. At present no resolution to the dispute is in sight and the fate of the Saharawi fight for true independence remains uncertain. Moreover, it appears as if it may continue to be unresolved for a long time as other political conflict in the region, especially the Arab-Israeli conflict, diverts international attention from the Western Sahara.

Summary and Conclusions

In this chapter we examined the complex Middle Eastern and North African region. We discussed the region's unique physical geography and environmental history, the ancient origins of the region, the power of Islam and other religions, the particular cultural and gender systems that have emerged there, and the special political tensions that exist there.

The significance of the Middle East and North Africa to the rest of the world is substantial. This region is *a* primary, if not *the* primary, site of the origins of Western civilization, and its influence on world culture, politics, and technology has been phenomenal. As the possessor of the largest share of the world's oil reserves, the region is also strategically critical to the continued function of the global economy. While the region appears to be one in a state of economic transition, it is hard to predict what role political conflict might play in influencing its progress. While perhaps one of the least well understood and most complex of the world's regions, the Middle East and North Africa remains one of the world's most politically, culturally, and economically significant regions as well.

Key Terms

afforestation (p. 210)
aridity (p. 203)
Balfour Declaration
   (p. 218)
chador (p. 225)
desertification (p. 210)
dry farming (p. 206)

guest worker (p. 227)
hajj (p. 220)
import substitution
   (p. 218)
informal economy (p. 218)
intifada (p. 236)
Islam (p. 220)

Islamism (p. 221)
jihad (p. 221)
kinship (p. 221)
Muslim (p. 220)
nationalist movement
   (p. 217)
nationalization (p. 218)

oasis (p. 205)
pastoralism (p. 224)
petrodollars (p. 243)
transhumance (p. 224)
tribe (p. 224)
world religion (p. 212)
zionism (p. 236)

Review Questions

Testing Your Understanding

1. How has the scarcity of water affected aspects of the cultural, economic, or political history of this region?
2. How do the people of this region show adaptation to the generally hot, dry environment through their housing and clothing preferences?
3. How are Judaism, Christianity, and Islam linked? Why is Judaism so much smaller numerically than the other two religions?
4. What are the five pillars of Islam? What are the Qu'ran and the Sunna?
5. What is pastoralism? What is transhumance?

6. What factors bring migrants into this region? What factors motivate migration out?
7. In a few sentences each, discuss the Balfour Declaration (1917), British Mandate (1922), U.N. Partition of Palestine (1947), the Camp David Accords (1978), and intifada (1980s–present).
8. How does the Palestinean Authority (PA) differ from the Palestinean Liberation Organization (PLO) and from Hamas?
9. What are the Arab League and OPEC? In which ways are the goals of these groups the same or different from each other?

10. How have the neo-liberal policies of the IMF, World Bank, and U.S. Agency for International Development (USAID) affected the economic development of poor areas in the Middle East and North Africa?

Thinking Geographically

1. Using the Jordan River as an example, how well do countries in this region manage and share their water resources?
2. How is kinship viewed in this region? How do these views affect local government and society?
3. How do views about gender affect the use of public and private space? Are rules concerning the veiling of women uniform throughout the region? If not, then how do they vary geographically?
4. Guest workers both migrate into this region and migrate out. Why do so many guest workers in the Oil States come from predominantly Muslim countries worldwide? Why do guest workers from Turkey and North Africa primarily seek work in Europe?
5. What is the geographic distribution of Palestinians within the Middle East? How have colonial and post-colonial policies in the Middle East left Palestinians with no state of their own?
6. The Middle East and North Africa are becoming highly urbanized. What is driving urban growth in this region? How are cities with considerable oil wealth handling their rapid urbanization?
7. Why is the Maghreb a geographically distinct area within this region? What sort of tensions exist between indigenous populations who are trying to maintain their traditions and European-influenced populations who are trying to establish their national independence? Why is the Western Sahara a flashpoint for these tensions?

Further Reading

Amirahmadi, H., and el-Shakhs, S.S. (eds.), *Urban Development in the Muslim World*. New Brunswick, NJ: Center for Urban Policy Research, 1993.

Armbrust, W. (ed.), *Mass Mediations: New Approaches to Popular Culture in the Middle East*. Berkeley: University of California Press, 2000.

Beaumont, P., Blake, G. H., and Wagstaff, J. M., *The Middle East: A Geographical Study*. 2nd edition. New York: Halstead Press, 1988.

Blake, G., *The Cambridge Atlas of the Middle East and North Africa*. Cambridge: Cambridge University Press, 1988.

Bonine, M. (ed.), *Population, Poverty, and Politics in Middle East Cities*. Gainesville, FL: University Press of Florida, 1997.

Drysdale, A., and Blake, G. H., *The Middle East and North Africa: A Political Geography*. New York: Oxford University Press, 1985.

Esposito, J. (ed.), *The Oxford Encyclopedia of the Modern Islamic World*, 4 vols. New York: Oxford University Press, 1995.

Gilsenan, M., *Recognizing Islam: Religion and Society in the Modern Middle East*. London: I. B. Tauris, 1996.

Goldschmidt, A., Jr., *A Concise History of the Middle East*, 6th ed. Boulder: Westview Press, 1998.

Held, C. C., *Middle East Patterns; Places, Peoples, and Politics*, 3rd ed. Boulder: Westview Press, 2000.

Hillel, D., *Rivers of Eden: The Struggle for Water and the Quest for Peace in the Middle East*. Oxford: Oxford University Press, 1994.

Hourani, A., *A History of the Arab Peoples*. Cambridge, MA: Belknap Press, 1991.

Humphreys, R. S. *Between Memory and Desire: The Middle East in a Troubled Age*. Berkeley, CA: University of California Press, 1999.

Kamil, L., *Fueling the Fire: U.S. Policy and the Western Sahara Conflict*. Lawrenceville, NJ: Red Sea Press, 1996.

Mandanipour, A., *Tehran: The Making of a Metropolis*. Chichester: John Wiley & Sons, 1988.

Mostyn, T. (ed.), *The Cambridge Encyclopedia of the Middle East and North Africa*. New York: Cambridge University Press, 1988.

Owen, R., *Power and Politics in the Making of the Modern Middle East*. London: Routledge, 1992.

Peters, R. (ed.), *Jihad in Classical and Modern Islam: A Reader*. Princeton, NJ: Markus Weiner, 1996.

Rodenbeck, M., *Cairo: The City Victorious*. New York: Random House, 1999.

Roy, O., *The Failure of Political Islam*. Trans. by C. Volk. Cambridge, MA: Harvard University Press, 1996.

Spenser, W., *Global Studies: The Middle East*, 8th ed. Guildford, CT: Dushkin/McGraw-Hill, 2000.

Swearingen, W., and Bencherifa, A. (eds.), *The North African Environment at Risk*. Boulder: Westview Press, 1996.

Tessler, M., *A History of the Israeli-Palestinan Conflict*. Bloomington, IN: Indiana University Press, 1994.

Ventner, A. J. "The Oldest Threat: Water in the Middle East." *Middle East Policy* 6 (1998), 126–36.

Yergin, D., *The Prize: The Epic Quest for Oil, Money, and Power*. New York: Simon & Schuster, 1991.

Film, Music, and Popular Literature

Film

Bab El-Oued City. Directed by Merzak Allouache, 1994. Story of a district of Algiers as it recovered from the riots in 1989.

Battle of Algiers. Directed by Antonio Musu, 1993. Dramatization of the conflict between Algerian nationalists and French colonists that culminated in independence for Algeria in 1963.

Blackboards. Directed by Mezssam Makhamlbaf, 2000. Itinerant teachers wander the Iran-Iraq border searching for a village of children to teach.

Building the Impossible. Directed by Clive Maltby, 2000. Story of the human and technological requirements that enabled the building of the seven wonders of the ancient world.

Children of Shatila. Directed by Mai Masri, 1992. Documentary about Palestinian children in exile in Lebanon.

Coup. Directed by Elif Savas, 1999. Documentary about the 1960, 1971, 1980, and 1997 military interventions and coups d'état in Turkey.

Days of Democracy. Directed by Ateyyat El Abnoudy, 1996. Interviews with women running for Egypt's parliament.

The English Patient. Directed by Anthony Minghella, 1996. The story of Count Almasy, a Hungarian mapmaker employed by the Royal Geographical Society to chart the vast expanses of the Sahara Desert.

Grand Theatre: A Tale of Beirut. Directed by Omar Naim, 1999. A documentary viewing the Lebanese Civil War, as well as its roots and aftermath, through the eyes of an old theater.

Gunese Yolculuk. Directed by Yesim Ustaoglu, 1999. Depicts the discrimination and violence directed against the Kurdish minority in Turkey.

Lawrence of Arabia. Directed by David Lean, 1962. Story of T. E. Lawrence and his influence on the political history of Saudi Arabia.

The Message. Directed by Moustaffa Akad, 1976. The story of Islam.

Three Kings. Directed by David O. Russell, 1999. In the aftermath of the Gulf War, four soldiers set out to steal gold that was stolen from Kuwait, but they discover people who need their help.

Music

The Gnoua Brotherhood of Marrakech/The Master Musicians of Joujouka. *SUFI: Moroccan Trance Music II.* Sub Rosa, 1996.

Khaled, Cheb. *Aiyesha.* Movie Play Gold, 2000.

The Master Musicians of Jajouka, featuring Bachir al-Attar. *Apocalypse Across the Sky.* Axiom/Island, 1992.

Ocora. *Egypt: Les Musiciens du Nil.* Radio France, 1987.

Soleimani, Haj-Ghorban. *Music of the Bards from Iran.* Kereshmeh, 1995.

Sultan, Sa'ida. *Danna International.* IMP Dance, 1993.

Various Artists. *Best of Bellydance from Egypt and Lebanon.* Arc Records, 1996.

Various Artists. *Legends of Arabic Music.* Arc Records, 1998.

Various Artists. *The Music of Islam, Volume 11: Music of Yemen.* Celestial Harmonies, 1998.

Various Artists. *Samar: Music from Yemen Arabia.* Rounder Select, 1999.

Various Artists. *Traditional Music from Turkey.* Arc Records, 2000.

Popular Literature

Al-Shaykah, Hanan. *Beirut Blues: A Novel.* New York: Doubleday, 1995. Story about a well-to-do Lebanese woman who must decide whether to abandon war-torn Beirut, the city she loves, or stay and suffer the demeaning and difficult consequences of the civil war.

Al-Shaykah, Hanan. *Women of Sand and Myrrh.* New York: Doubleday, 1992. A story of four contemporary women living in an unnamed country in the Middle East who are confronting the powerful gender norms that created painful personal and social obstacles in their lives.

Anderson, Terry. *Dens of Lions: Memoirs of Seven Years.* New York: Del Rey, 1994. Moving account by the Associated Press's former chief Middle East correspondent of his 2454 days as a hostage of the Islamic terrorist organization Hezbollah.

Bahrampour, Tara. *To See and See Again: A Life in Iran and America.* New York: Farrar, Straus & Giroux, 1999. Memoir of the daughter of an American singer and an Iranian architect that does justice to both sides of her complex heritage.

Bowles, Paul. *The Sheltering Sky.* New York: Signet, 1955. The desert is itself a character in this book, which is about three young Americans of the postwar generation who go on a walkabout into Northern Africa's own arid heart of darkness.

Daneshvar, Simin. *Savushun: A Novel About Modern Iran.* Washington, DC: Mage, 1991. Set in the Iranian town of Shiraz, this novel chronicles the life of a Persian family during the Allied occupation in the Second World War.

Hodgson, Barbara. *The Tattooed Map.* San Francisco: Chronicle Books, 1995. An intriguing story of a traveller who wakes up somewhere in North Africa with a map tattooed to her hand and then follows the map to solve a mystery.

Kalifeh, Sahar. *Wild Thorns.* New York: Interlink, 1999. A portrait of everyday Arab life in the West Bank and Gaza Strip that describes the difficulties of survival under Israeli oppression.

Mafouz, Naguib. *The Cairo Trilogy: Palace Walk; Palace of Desire; Sugar Street.* New York: Doubleday, 1988, 1991, 1992. (Winner of the Nobel Prize, 1988.) The engrossing saga of a Muslim family in Cairo during Egypt's occupation by British forces in the early 1900s.

Manning, Olivia. *The Levant Trilogy.* New York: Penguin, 1982. Powerful political and romantic story of the role that the Levant played in the Second World War as Italy, Britain, and Germany used the region to launch various campaigns.

Munif, Abdelrahman. *Cities of Salt: A Novel.* New York: Vintage Books, 1993. A tale of indigenous people's exploitation and oppression by corporations and colonialists set against the background of the range of identity issues affecting the peoples of modern Saudi Arabia.

Said, Edward. *End of the Peace Process: Oslo and After.* New York: Pantheon, 2000. A collection of 50 impassioned, damning essays on the consequences of the Middle East peace process.

Soueif, Ahdaf. *Map of Love.* London: Bloomsbury, 1999. A massive family saga that draws its readers into two moments in the complex, troubled history of modern Egypt: the late nineteenth and the late twentieth centuries.

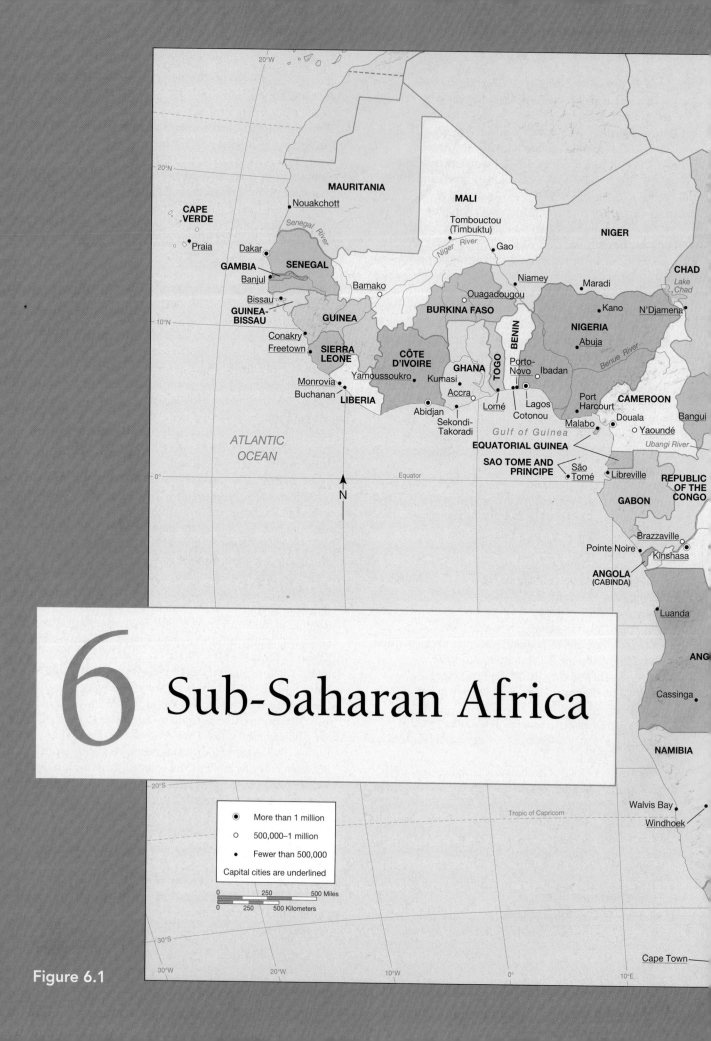

6 Sub-Saharan Africa

Figure 6.1

Nile River

Red Sea

Blue Nile

White Nile

ERITREA
Massawa
Asmara

DJIBOUTI
Djibouti

Gulf of Aden

Berbera

Addis
Ababa

ETHIOPIA

SOMALIA

**CENTRAL
AFRICAN
REPUBLIC**

Congo River

Kisangani

UGANDA

Lake
Rudolf
(Turkana)

KENYA

Kampala
Entebbe

Lake
Victoria

RWANDA

Kigali

Nairobi

Mogadishu

BURUNDI Bujumbura

Kigoma

Arusha

Mombasa

**DEMOCRATIC
REPUBLIC
OF THE
CONGO**

Mpanda

TANZANIA

Zanzibar

Dar es Salaam

SEYCHELLES

Victoria

Lake
Tanganyika

Lake
Nyasa

Moroni

COMOROS

Lubumbashi

MALAWI

ZAMBIA

Lilongwe

Nacala

**MAYOTTE
(FRANCE)**

Lusaka

Zambezi River

Blantyre

Lake
Kariba

Mozambique Channel

Harare

Antananarivo

Toamasina

Caprivi
Strip

ZIMBABWE

MOZAMBIQUE

MAURITIUS

Maun

Bulawayo

MADAGASCAR

Port Louis

BOTSWANA

Limpopo River

**RÉUNION
(FRANCE)**

Gaborone

Pretoria

Maputo

**INDIAN
OCEAN**

Kimberly

Johannesburg

Mbabane

Bloemfontein

Maseru

SWAZILAND

Orange
River

Durban

**SOUTH
AFRICA**

LESOTHO

Port Elizabeth

30°E

40°E

50°E

60°E

20°N

10°N

0°

10°S

20°S

30°S

257

Africa is a large, complex, and often misunderstood continent. Common perceptions range from a fertile tropical forest rife with exotic diseases to an idyllic game reserve, or from a harsh landscape devastated by war and drought to a place where rich cultural traditions reach back to the dawn of humanity. Africa has considerable mineral wealth and agricultural potential but is ranked lowest among world regions on almost all indicators of economic development and social and health conditions.

The continental landmass called Africa straddles the equator, stretching 8000 kilometers (5000 miles) miles from the Mediterranean Sea in the north to the southern tip in South Africa at about 35 degrees south latitude. At its widest, Africa spans 7400 kilometers (4600 miles) from Senegal on the Atlantic coast to Somalia on the Indian Ocean. The total area is about 30.4 million square kilometers (11.7 million square miles).

Geographers have argued that the countries of North Africa that border the Mediterranean Sea—Morocco, Algeria, Tunisia, Libya, and Egypt—have more in common with the Middle East than with the other countries of Africa that lie south of the Sahara Desert. North Africa is considered to share characteristics with the Middle East, including similar physical environments of dry climates and human geographies that reflect a dominant Arabic language and ethnicity and Islamic religion.

Sub-Saharan Africa has been defined and divided from North Africa based on common historical, physical, and social characteristics that include a legacy of European colonialism and slavery, a mostly tropical climate, and the darker skin of many inhabitants. The race-based definition of Sub-Saharan Africa is very controversial but has been used by both Africans and non-Africans to identify the region as "Black Africa." Sub-Saharan Africa includes 42 mainland countries, 6 island nations, and the French territories of Réunion and Mayotte (**Figure 6.1**). The region includes large populations growing their own food and living in small rural villages with less than 30 percent living in urban areas. Sub-Saharan Africa has an area of 22 million square kilometers (8.5 million square miles) and a population of 600 million. In this text, we discuss North Africa with the Middle East because of shared characteristics and because one chapter on the whole of Africa would be very large. However, the physical and human links across the continent of Africa are such that several sections of this chapter, including the section on humans and the environment that follows, discuss broader patterns across the whole continent and refer to Africa as a whole rather than Sub-Saharan Africa specifically.

We recognize that the geographical, racial, ethnic, and religious basis for dividing Africa into these two world regions is somewhat artificial, oversimplifying both the cultural and historical distinctiveness of the two regions, the overlaps between them, and the great variety that they contain. For example, the Sahara Desert is a large area, rather than a clear dividing line, and includes territory from both North and Sub-Saharan African

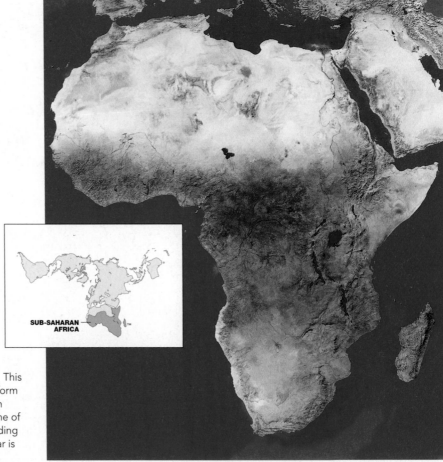

Figure 6.2 Satellite image of Africa This satellite image clearly shows the major landform regions of the Sahara Desert, the Congo rain forests, the highlands of Ethiopia, and the line of lakes along the East African Rift Valley, including Lake Victoria. The large island of Madagascar is also clearly visible.

SUB-SAHARAN AFRICA

countries (**Figure 6.2**). The Nile River links the North African countries of Egypt and Sudan through a long fertile corridor to the Sub-Saharan countries of Ethiopia, Uganda, Kenya, Tanzania, Rwanda, Burundi, the Central African Republic, and the Democratic Republic of the Congo (formerly Zaire). There are considerable populations of Arab and European ethnic groups and of Muslims in Sub-Saharan Africa and significant numbers of black Africans and non-Muslims in North Africa. Traders have linked the economies of North and Sub-Saharan Africa for centuries, and the Organization of African States includes members from throughout the African continent. Sudan exemplifies the challenges of treating Africa as two distinct world regions because the north part of the country is dominated by an Islamic and Arabic culture whereas the south hosts a predominantly black and Christian population. We discuss Sudan, the largest country on the African continent, in Chapter 5.

The world region of Sub-Saharan Africa is frequently divided into subregional clusters of countries that share common geographical characteristics and have some distinctive landscapes that include parts of several countries. Commonly discussed subregions include West Africa, East Africa, and southern Africa as well as the distinctive landscapes of Equatorial Africa, the Horn of Africa, the Indian Ocean islands, and the Sahel, a semiarid zone across the southern edge of the Sahara Desert (**Figure 6.3**).

Environment and Society in Sub-Saharan Africa

The continent of Africa is the heart of the ancient supercontinent called Pangaea, the southern part of which broke off to form Gondwanaland about 200 million years ago (see Chapter 2, p. 53). The theory of plate tectonics explains that when the regions that we now know as Latin America and Asia broke away from Gondwanaland, the high plateau that remained became the continent of Africa.

Half of the African continent is composed of very old crystalline rocks of volcanic origin that hold the key to Africa's mineral wealth. Ancient tropical swamps formed sedimentary rocks containing oil and other fossil fuels. These include coal in southern Africa and Nigeria, and oil and gas in West Africa, particularly Nigeria and Gabon. Iron and manganese are found in western and southern Africa, and most of the world's known chromium is found in Africa, especially in Zimbabwe and South Africa. Vast copper reserves are located in the copper belt of Zambia, where cobalt is found along with the copper, and in the southern Congo; bauxite, which is used in making aluminum, is found in a belt across West Africa, and uranium is found in Niger. These minerals are critical to industrial production elsewhere in the world. Gold is found in

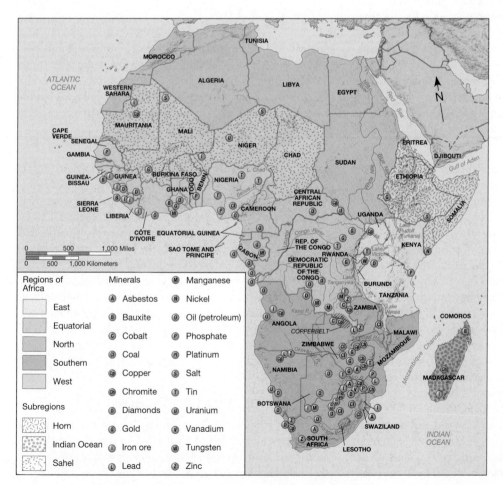

Figure 6.3   Major subregions of Africa Africa is commonly divided into several major regions, including southern, East, West, Equatorial, and North Africa as well as the distinctve landscapes of the Sahel, Horn of Africa, and Indian Ocean islands. The map also shows the location of the most important regions of mineral development in Africa, including oil, gold, and diamonds. South Africa, Zambia, and Sierra Leone are particularly rich in minerals, and Nigeria is a major oil producer.

Diamonds

For many consumers around the world, diamonds are associated with love and luxury, the symbol of engagement to marry and of wealth and sophistication. Larger diamonds, after grading, cutting, polishing, and setting in gold or other metals, are sold for high prices in jewelry stores, with about half of all purchases made in the United States. These glittering stones are also of considerable industrial value because of their hardness, forming a strong, sharp cutting edge. Africa is a major source of the world's diamonds, with about two-thirds of the trade controlled by a South African conglomerate, De Beers. The virtual monopoly held by De Beers permits careful control of diamond markets to ensure that prices remain high and supply stable, and the company has worked hard through advertising to maintain the romantic image of diamonds. Eighty percent of all diamonds are traded through the Diamond Center in Antwerp, Belgium.

Most diamond production takes place in South Africa, Botswana, and Namibia and contributes significantly to export revenue and local employment (**Figure 1**). In Angola, the Democratic Republic of the Congo, and Sierra Leone, diamonds are also mined or smuggled across borders and have become associated with corruption, violence, and warfare. Easy to transport, diamonds are increasingly used to purchase weapons, and some analysts suggest that these "blood diamonds" may now make up as much as 10 percent of all global trade in diamonds.

Diamonds are mined in large commercial mines such as those in South Africa but are also dug from mud and streams by hundreds of individuals dreaming of finding a large gem

Figure 1　Map of diamonds in Africa Diamonds are found in several regions of Africa including the famous mines at Kimberly in South Africa, in Botswana, the Congos, and Sierra Leone.

that will make their fortunes (**Figure 2**). Mines and miners in Angola and Democratic Republic of the Congo are often under the protection of armed guards or military forces. For example, in Angola, the Catoca mine, which produces $8 million worth of diamonds each month, pays $500,000 a month to the

several regions of Africa, including Ghana and Zimbabwe, and in South Africa, where as much as half of the world's gold reserves are found in the region around Johannesburg. South Africa is famous for diamonds, which are also found in Botswana and Namibia in southern Africa, at the edges of the Congo basin, and in Sierra Leone in West Africa.

These mineral resources have played important roles in African history. Gold was important in West Africa from early times and was worn and traded by kings and leaders; Mansu Musa, the emperor of Mali, carried and traded so much gold on a pilgrimage to Mecca in 1324 that his actions depressed gold prices worldwide. Salt also was a key commodity in trans-Saharan trade. Gold and diamonds spurred European colonial grabs for Africa and conflicts between colonial powers and with indigenous groups after the discovery of diamonds in 1867 at Kimberly and gold in 1886 on the Rand, a range of hills to the west of Johannesburg, in South Africa. Gold and diamonds, together with oil, continue to incite conflict within contemporary Africa and to amplify interest in African economies on the part of other states and multinational corporations (see Geo-

graphies of Indulgence, Desire, and Addiction: Diamonds, p. 260). Although these resources bring billions of dollars into Africa, they also make national and regional economies vulnerable to fluctuations in world market prices, especially where minerals dominate exports. In Africa as a whole, exports of mining products were valued at $44.7 billion in 1998, about 42 percent of total exports. The distribution of mineral wealth is uneven, with South Africa (gold and diamonds) and Nigeria (oil) accounting for more than half of total value. Sub-Saharan countries with mineral exports accounting for more than one-half of total earnings in 1998 include Angola (90 percent oil), Botswana (72 percent diamonds), Gabon (75 percent oil), Niger (65 percent uranium), and Nigeria (95 percent oil).

Landforms and Landscapes

Where the continental plates tore away from Africa during the breakup of Gondwanaland, they left steep slopes (called *escarpments*) that fell from the high plateau to the new oceans. Most of the rivers that had previously drained into the inland lakes of

Figure 2 Diamond diggers Hundreds of people have come to work in the diamond regions of the Democratic Republic of the Congo where they work deep in the mud of stream beds to sift sediment with the dream of finding especially valuable diamonds. However, the diggers receive only a small portion of the eventual value of the diamonds on the world market and are sometimes harassed by the military or others who use violence to gain access to the diamonds.

Angolan army for security against the rebel group UNITA who used to control the region. Money associated with diamond mines has funded an increase in number and magnitude of arms on every side of conflict in Angola, including the purchase of land mines that have maimed thousands of civilians.

The Democratic Republic of the Congo has also become a pawn in the struggle for diamonds, with Angola, Namibia, and Zimbabwe sending troops to protect the government, and Burundi, Rwanda, and Uganda assisting rebels. The area around the diamond zone near Kisangani in the eastern Democratic Republic of the Congo has been abandoned to fighting, and Zimbabwe and Rwanda are on opposing sides in attempts to obtain access to diamond deposits in the southern Democratic Republic of the Congo. Diamonds have also funded a brutal civil war in Sierra Leone, where rebels have even chopped off people's limbs with machetes to intimidate residents into leaving the diamond zones of eastern Sierra Leone. Pressure from human rights groups has led De Beers to refrain from purchasing diamonds that originate in conflict zones and to support research to develop a system to fingerprint diamonds based on chemical signatures from their area of origin.

Botswana, in contrast, is producing diamonds under peaceful conditions and with the guidance of traditional leaders. The mines employ more than 25 percent of the population and are responsible for a gross national product per capita and standard of living that is much higher than the average for Sub-Saharan Africa.

Source: Adapted from B. Harden, "Diamond Wars: A Special Report. Africa's Gems: Warfare's Best Friend," *New York Times,* April 6, 2000.

the supercontinent eventually found outlets to the sea. Because the continent was tearing apart, the tensions created trenches and volcanic activity, in contrast to other regions where colliding plates caused uplift and folding into high mountain ranges, such as the Andes in Latin America and Himalayas in South Asia.

Africa is still mainly a plateau continent, with elevations ranging from about 300 meters (approximately 1000 feet) in the west, tilting up to more than 1500 meters (approximately 5000 feet) in the eastern part of the continent (**Figure 6.4**). There are some significant mountain ranges in western Africa, including the Cameroon mountains and Fouta Diallon highlands, important for the rivers that flow from the uplands. Steep slopes, especially on the western edge of the plateau, fall to narrow coastal plains. Where rivers descend to the coast, they often cut deep valleys back into the plateaus and drop over rapids and waterfalls, such as Victoria Falls on the Zambezi River in southern Africa (**Figure 6.5c**). This poses a serious problem for navigation by boat into the continent but also offers the potential for hydroelectric development. This potential has been realized through major dams on many rivers in Africa, including

the Kariba on the Zambezi, Akosombo on the Volta, and Aswan on the Nile (see Geography Matters: Dams in Africa, p. 264).

The routes of Africa's major rivers reflect the legacies of inland drainage on the supercontinent, because many of them flow away from the coast and into inland wetlands and deltas before shifting back toward the ocean. For example, the immense Congo River, second only to the Amazon in terms of overall discharge, flows north before turning west toward the rapids that bring it down to the Atlantic. The Niger River flows north toward the Sahara into a large inland delta, before turning south toward its exit to the Atlantic in Nigeria. The Nile, discussed in more detail in Chapter 5 (p. 204), flows into the vast wetland known as the *Sudd.* Several river systems still drain to inland basins, including the Okavango River of southern Africa and the Chari-Logone river system, which drains into Lake Chad in the Sahel. These inland deltas create some of the richest ecosystems in the region, providing habitat for wildlife, fisheries, and grazing and irrigated land for human activities.

The higher areas of the plateau that are important to humans for their cooler temperatures and higher rainfall include the

Figure 6.4   Map of major landforms in Africa Africa is a plateau continent surrounded by steep escarpments, with rivers that often flow through inland deltas on the plateau or drop over waterfalls at the edge of the escarpment. One of the most significant features is the East African Rift Valley, filled with elongated lakes and several active volcanoes. (*Source:* Adapted from S. Aryeetey-Attoh (ed.), *The Geography of Sub-Saharan Africa.* Upper Saddle River, NJ: Prentice Hall, 1997, p. 5.)

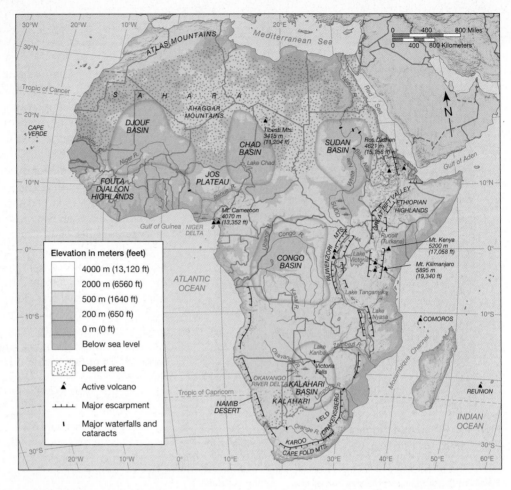

High Veld of southern Africa, the highlands of Kenya and Ethiopia, and the Jos plateau of West Africa. Volcanic peaks such as Kilimanjaro (5895 meters, 19,340 feet), Kenya/Kirinyaga (5200 meters, 17,058 feet), and the Virungas (4507 meters, 14,787 feet) rise from the eastern plateau, which is also split by a deep trough where tectonic processes continue to pull the eastern edge of Africa away from the rest of the continent (**Figure 6.5b**). This trough, which runs more than 9600 kilometers (6000 miles) from Jordan and the Red Sea in the north to Mozambique in the south, is called the *Rift Valley* and ranges from 50 to 100 kilometers (30 to 60 miles) wide. It has two major branches and is filled with deep elongated lakes, including Lake Tanganyika at 1473 meters deep (4832 feet). Lake Victoria, the third largest lake in the world, lies between the two branches of the rift valley. The age, size, and depth of these lakes make them diverse freshwater ecosystems with important fisheries.

African soils tend to be of low fertility because of the great age of the underlying geology and because of high rainfall that leaches (washes out) nutrients from exposed soils. Soil fertility tends to be higher in regions of recent volcanic activity, such as the East African highlands, and in wider river valleys where sediments settle and create alluvial (river) soils. The tropical soils of wetter zones, such as central Africa, lose their fertility rapidly once the forest is cleared and the soil is exposed to the elements. Between the dry and wet zones, such as between the coastal and Sahel regions of West Africa, soils have more organic material and are important for crops and pasture. Desert regions can have saline or alkaline soils that are toxic to crops. High iron and aluminum content is also poisonous to plants and crops in some regions.

Climate

Most of Sub-Saharan Africa lies between the tropics of Cancer and Capricorn and has a tropical climate with warm temperatures (higher than 20°C, 70°F) and little frost except in highland areas (**Figure 6.6**). The climate is dominated by two major features of the atmospheric circulation—the intertropical convergence zone (ITCZ) and the subtropical high (see Chapter 2, p. 55). As in Latin America and Asia, the ITCZ is a zone of rising air where moisture condenses and produces heavy rainfall. In Africa the ITCZ produces intense rainfall of more than 1500 millimeters (60 inches) a year over the Congo basin. The subtropical high is a zone of descending air resulting in dry and stable air that causes desert conditions over the Sahara and Kalahari. The regions between these two features experience seasonal rainfall as the ITCZ and subtropical high shift northward in April through September and southward in October through March. During December the southward shift of the ITCZ low pressure brings strong dry winds out of inland Africa. These winds, called the **harmattan,** carry large amounts of dust, which can be carried far out over the Atlantic and create stressful hot,

(a)

(b)

(c)

Figure 6.5 African landscapes (a) Elongated lakes line the bottom of the Great Rift Valley. Lake Bogoria, Kenya, shown here, has several geothermal hot springs as a result of the tectonic activity and a large population of flamingoes colored pink from the organisms that live in the lake. (b) Mount Kilimanjaro sits on the border between Kenya and Tanzania and at 5895 meters (19,340 feet) is the highest mountain in Africa. Each year thousands of tourists trek to the top of the mountain or visit the nearby game parks to see animals such as this giraffe in Amboseli National Park in Kenya. (c) Victoria Falls amazed explorer David Livingstone when he saw them on his travel along the Zambezi River in the 1855. Twice as wide and tall as Niagara Falls in North America, the gorge into which the 1700-meter-wide (5500 feet) falls drop has been cut back from the escarpment into the plateau by the force of the river.

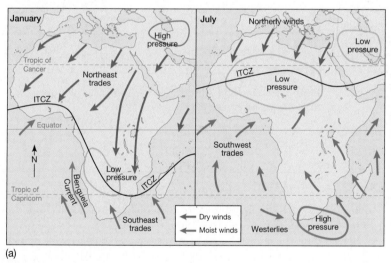
(a)

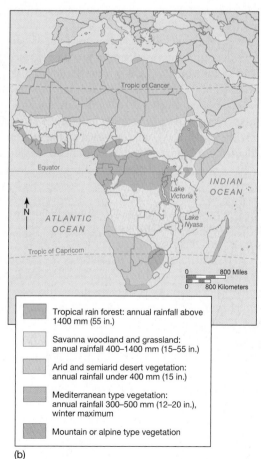
(b)

Figure 6.6 Climate and vegetation zones of Africa (a) The major wind and pressure patterns over Africa in January and July are shown here. Atmospheric circulations include the West African monsoon in July and the dry harmattan winds in December. Precipitation occurs generally to the south of the intertropical convergence zone (ITCZ); seasonal movement is shown on the map. (b) The tropical forest regions near the equator receive most of the continent's rainfall, and deserts occur where dry, descending air and cold, offshore climates inhibit rainfall. Southern Africa receives rainfall from storms in the westerly winds from April to August and has a winter rainfall maximum typical of a Mediterranean climate and vegetation. Climate and vegetation zones overlap closely in Africa and depend mostly on the amount and seasonality of precipitation. Large areas of Africa are covered with savanna vegetation and by deserts. (*Source:* Adapted from W. M. Adams, A.S. Goudie, and A. R. Orme, *The Physical Geography of Africa*. London: Oxford University Press, 1999, Fig. 10.5.)

Geography Matters

Dams in Africa

The rivers of Africa, especially where they descend over the coastal escarpment, provide considerable potential for hydroelectric development. Several large projects were initiated around the time of transition to independence to harness the energy of the rivers, to provide electricity to industry and cities, and to irrigate agricultural fields (**Figure 1**).

Proponents of large dams point to the social benefits and enhanced economic development that dams make possible, such as providing electric power, irrigation for agriculture, and water supply to growing towns and cities. Critics argue that project funding, both public and private, systematically downplays the adverse environmental, social, and economic impacts of dams.

The Federation of Rhodesia and Nyasaland (now the countries of Zimbabwe and Zambia) completed the Kariba Dam on the Zambezi at Kariba gorge in 1959 (**Figure 2**). This dam, which produces inexpensive electric power for this region of southern Africa, created Lake Kariba and required the resettlement of 57,000 people and the evacuation of thousands of wild animals isolated as the waters rose behind the dam through "Operation Noah." The areas to which many people were moved were infested with the tsetse fly, and many were exposed to sleeping sickness. Because of poor planning, some people ended up in places resembling refugee camps. The hygiene in these camps was very poor, and epidemics flourished. Some of the unanticipated benefits include development of a tourist industry and lush animal habitat around the new lake and a productive fishery. The electricity the dam provides is important to the copper-mining industry in Zambia.

The Akosombo Dam, completed on the Volta River in 1965 (**Figure 3**), was funded by the governments of Ghana, the Unit-

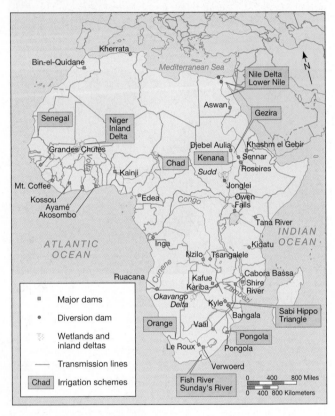

Figure 1   Dams in Africa This map shows the location of major dams and irrigation schemes in Africa.

dry conditions. In July, the northward shift allows the southwestern trade winds to blow onto the coast of West Africa, bringing seasonal rains to inland countries such as Mali.

This general pattern is modified by the regional effects of mountains, lakes, and ocean currents. The cold Benguela current creates cool, dry conditions along the coasts of Angola and Namibia and promotes the desert conditions of the Kalahari. Cold water and stable winds that blow along, rather than across, the coast also promote dry conditions in countries such as Somalia in the Horn of Africa. High-altitude regions, such as the East African highlands, have higher rainfall and more moderate temperatures, which are favorable to agriculture and human settlement. South Africa is located in more temperate latitudes and experiences a mild Mediterranean climate with

dry conditions from October to March and rainfall from the westerly wind belt from April to August.

The semiarid regions of Africa have highly variable rainfall and frequent droughts, which pose great challenges to agriculture and water resources management. Some of this variation is connected to the changes in Pacific Ocean temperatures known as El Niño (see Chapter 8, p. 380). El Niño is associated with severe droughts in southern Africa. Geographers have reconstructed African climate variability over several centuries using instrumental records as well as diaries and other observations and find extensive periods of drought in many regions. Over longer time spans Africa has experienced significant changes in climate, including a period about 6000 to 10,000 years ago when the Sahel was much wetter than it is now.

Figure 2   Kariba Dam   The Kariba Dam on the Zambezi River supplies electricity to Zimbabwe and Zambia and has become a destination for tourists who come to view game in the parks that line the shores of Lake Kariba.

Figure 3   Akosombo Dam   The Akosombo Dam on the Volta River in Ghana created the largest human-made lake in Africa and provides power to Ghana and its neighbors for domestic use and for aluminum production.

ed Kingdom, the United States, and the World Bank. The electricity was targeted for a large aluminum smelter on the coast at Tema. This is now U.S.-owned and consumes about 45 percent of the electricity the dam generates; the smelter pays very low taxes and and is thereby able to keep aluminum prices low. The remainder of the electric power either goes to domestic consumption or is exported to Togo and Benin.

The construction of the dam had a number of social and environmental impacts. The huge reservoir behind the dam, Lake Volta, reaches 400 kilometers (250 miles) northward and is the largest human-made lake in Africa. The area behind the dam now underwater, included 15,000 homes and more than 700 villages, and 78,000 people had to be resettled to make way for the lake. Many were unhappy with the quality of the new houses and land and have returned to live near the lake.

Prior to the dam, floodplains downstream had benefited from the annual renewal of sediment, and cattle had grazed on the lush grasses along the river. When the river flow declined and releases from the dam became more sporadic, agriculture and livestock production declined in the area below the dam. Sediment is now building up behind the dam, reducing storage, electrical potential, and the potential lifetime of the dam. Both Kariba and Akosombo have been affected by drought in recent years and have cut back on the amount of electricity they supply as a result. Another problem, common to most dams in Africa, is that the slower flow of the river and the stagnant water behind the dam has increased the incidence of several diseases, including schistosomiasis and malaria.

Environmental History

Ecosystems   African ecosystems, as in the rest of the world, are closely tied to climate conditions but also reflect a complex evolutionary history and physical geography that has produced great diversity, unique plants, and perhaps the most charismatic community of animal species. The major ecosystems are the forests (20 percent of the land area), the savanna woodlands and grasslands (40 percent), the deserts (38 percent), and the Mediterranean and montane (mountain) zones (2 percent).

The Congo basin hosts the world's second largest area of rain forest (after the Amazon), covering almost 2.6 million square kilometers (1 million square miles). Other important forests are found along the West African coasts, the coasts of Kenya, and on the island of Madagascar. These forests have great biodiversity, including monkeys and apes such as chimpanzees and gorillas, and tropical hardwoods of significant economic value. The forest and their wildlife are threatened by demands for timber and firewood, by poaching and foraging, and by conversion of natural habitats to cropland (see p. 298).

Drier regions have mixed wood and grasslands with open stands of trees interspersed with shrubs and grasses. The baobab tree is an important symbol of this landscape, which is found in West Africa centered on Guinea and in southern Africa near the Zambezi (**Figure 6.7**). The **savanna** grasslands are found where there are long dry seasons and provide expansive grazing areas for both wildlife and livestock. Grassy plains such as the Serengeti of Tanzania have some of

(a) (b) (c)

Figure 6.7 Ecosystems (a) Mountain gorilla in the forests of the Virunga volcanoes on the border of Uganda and Rwanda. (b) Baobab trees such as these are scattered across savanna grasslands in Tanzania. (c) The *fynbos* vegetation of the Mediterranean climate of South Africa, including the unusual protea flowers that have become popular for flower arrangements around the world.

the densest concentrations of wild, hoofed, grazing mammals (called *ungulates*) in the world, together with their predators such as the big cats. The larger herbivores include elephants, giraffes, zebras, and rhinoceroses.

Desert regions have very sparse and seasonal vegetation for the most part, with drought-resistant vegetation such as acacias and woody scrub. The Mediterranean climates of South Africa have produced a unique ecosystem dominated by *fynbos* shrubland, the vegetation of which is characterized by waxy or needlelike leaves and long roots that help plant life to survive long dry periods (**Figure 6.7c**). Finally, the highland, or *montane,* vegetation is found on mountain ranges such as the volcanoes of East Africa or the Drakensberg highlands of southern Africa. As in other world regions, African ecosystems show clear *vertical zonation* on mountain slopes because of higher rainfall and cooler temperatures at higher elevations. A hike up the slopes of Mount Kenya would reveal how vegetation and human land use change along this environmental gradient. The base of the mountain is nested in grasslands of the savanna, while its rocky peaks are covered with snow and ice. In between, a hiker would pass through zones of dry forest, bamboo forest, heathland, and alpine moorland.

Diseases and Insect Pests Africa's ecologies are notable for several pests and diseases that can have devastating impacts on human populations. Several of these diseases have *reservoirs* in certain wild species and are then transferred to humans or their domesticated animals by *vector* (transmission) organisms such as mosquitoes, flies, and snails. Some diseases are believed to have originated in Africa and to have spread globally. Malaria, a disease transmitted to humans by mosquitoes, causes fever, anemia, and often fatal complications, and it affects about 400 million people in Sub-Saharan Africa each year, killing more than 750,000 (**Figure 6.8**). Early European explorers and settlers were very vulnerable to malaria and suffered as much as 75 percent mortality in some regions. The discovery in 1820 that *quinine*—an extract from the cinchona tree, thought to have been brought to Europe from Peru by Jesuit priests—could partly control malaria fa-

cilitated colonialism and also allowed treatment of some local residents. But it did not cure the disease, and after the Second World War several new synthetic drugs such as chloroquine became popular cures. Unfortunately, several strains of the malaria parasite have developed resistance to these drugs and there is still no certain and cheap cure for the disease.

Yellow fever also has a mosquito vector and a reservoir in monkey populations. Mortality in Africa has been reduced through immunization, but many who are poor or live in remote regions still do not have access to vaccines, and as many as 20,000 people died in an outbreak of yellow fever in Senegal in the 1960s.

HIV/AIDS Today, the most serious disease affecting Sub-Saharan Africa—one that is thought to have emerged in central Africa—is HIV, the *human immunodeficiency virus* that causes fatal *acquired immunodeficiency syndrome,* or AIDS. Sub-Saharan Africa is more severely affected by AIDS than is any other part of the world, with the United Nations reporting more than 23.3 million people infected with the HIV virus—70 percent of the worldwide total. The infection rate is estimated at 8 percent of all adults compared to the 1 percent world rate, and more than 13 million Africans have lost their lives to AIDS since it was identified in 1981. It has become the main cause of death in Africa, killing more people than malaria and warfare.

The geography of AIDS in Africa varies by country, by regions within countries, and by social groups. The highest rates of infection were in eastern Africa in the early 1980s but have now shifted to southern Africa, especially Botswana, Zambia, and Zimbabwe, where more than 20 percent of adults are infected (**Figure 6.9**). Urban dwellers who have multiple sex partners, including young office workers and migrant workers, have a higher infection rate, as do women who work in the commercial sex trade and the wives and children of migrant workers. The incidence is lower in rural areas except along major truck routes and in areas where there are a lot of soldiers. The road from Malawi to Durban, where 92 percent of truck drivers are infected, has been called "the highway of death." Migrant workers have taken the disease back to their homes.

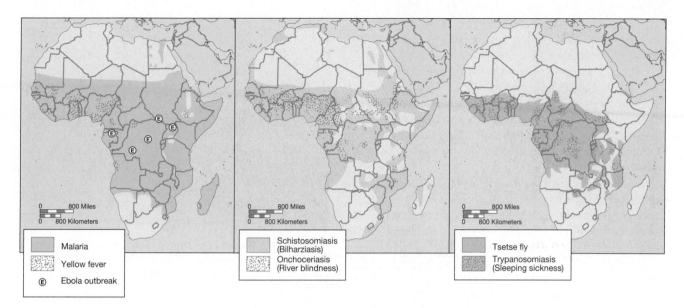

Figure 6.8   Maps of tropical infectious diseases and pests These maps show the distribution of some of the more serious tropical diseases in Africa, including malaria and disease carried by the tsetse fly. (*Source:* Modified from I. L. L. Griffiths, *An Atlas of African Affairs*, New York: Methuen, 1985, pp. 20–21.)

Unlike other regions, more women have AIDS than men in Sub-Saharan Africa, and mothers often transmit the disease to their children; frequently, married couples are both infected with AIDS and die from it, and this has created as many as 8 million orphans.

The death of skilled farm laborers has resulted in a decline in agricultural output, and many young professionals critical to the region's future have left their jobs because they have AIDS. Major industries and companies in southern Africa, such as diamond mines and banks, estimate that absence and loss of employees to AIDS is costing them more than 5 percent of their profits. AIDS is having serious impacts on life expectancy and population size and composition in Africa. Countries such as South Africa and Zimbabwe have adjusted projections of population and economic growth down to take account of the serious negative impact of AIDS on mortality and economic productivity.

Poverty exacerbates the AIDS problem in Sub-Saharan Africa because most Africans cannot afford prevention (for example, through the use of condoms), testing, or medicines that prolong the lives of those with HIV/AIDS. Governments have low healthcare budgets, and few people have health insurance. The average time between initial HIV infection and identification of AIDS is about three years later than in other regions because of the lack of diagnostic testing, and people often die earlier because of delayed treatment and the interactions between AIDS and other diseases that affect people with weak immune systems, such as tuberculosis. There has also been some reluctance on the part of African governments to admit to the extent of the AIDS epidemic.

Some countries have had success in combating AIDS. Uganda and Senegal have promoted aggressive and successful AIDS education and prevention campaigns and have cut infection rates in half. Unprecedented international agreements with

Figure 6.9   AIDS in Sub-Saharan Africa These maps of the progression of the HIV epidemic in Africa from 1982 to 1997 show both the increasing percentage of adults infected and the shift from West and East Africa to southern Africa. (*Source:* From the World Bank Group, "Intensifying Action Against HIV/AIDS in Africa: Responding to a Development Crisis," 1998, available at http://www.worldbank.org/html/extdr/offrep/afr/aidstrat.pdf)

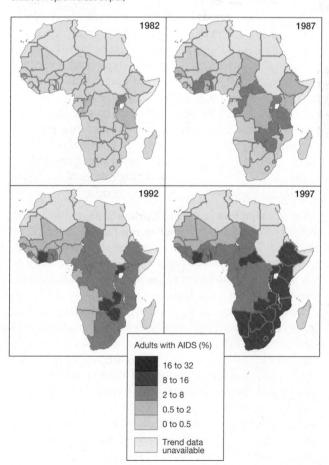

drug companies in combination with new assistance programs from the World Bank, charities, and donor countries are helping to bring down the cost of drugs.

There is also great concern about the potential impacts of some other emerging viruses in central Africa, specifically *Ebola fever*, which causes severe bleeding and kills more than 50 percent of its victims. So far, outbreaks such as the ones that occurred in the Congo in 1995 and in Uganda in 2000 have been contained, but only after killing more than 200 people in each case.

Another problem disease is *schistosomiasis* (also called *bilharzia*), which is associated with a parasite that causes gastrointestinal diseases and liver damage. Passed to humans who are exposed to a snail vector by working or bathing in slow-moving water, the disease is not fatal but reduces general health and energy levels. One of the unintended consequences of the construction of dams and irrigation canals in Africa has been the spread of schistosomiasis in the stagnant or slow-moving water systems (see Geography Matters: Dams in Africa, p. 264). The United Nations World Health Organization (WHO) estimates that 160 million people are infected in Sub-Saharan Africa. Another disease associated with African environments is *river blindness* (onchocerciasis) transmitted by the bite of the black fly that passes on small worms whose larvae disintegrate in the human eye and cause blindness. The eradication of river blindness has been relatively successful in West Africa by controlling fly populations with pesticides and treating victims with drugs. WHO estimates that 18 million people are infected, and 250,000 are blind as a result of this disease in Sub-Saharan Africa.

The tsetse fly, which lives in African woodland and scrub regions, is a vector for a virus with a reservoir in wild animals. It is associated with both human and livestock diseases. In humans, the bite of the tsetse fly causes sleeping sickness, or *trypanosomiasis*, with fever and infection of the brain that causes extreme lethargy and may end with the death of the victim. Half a million people are infected in Sub-Saharan Africa. Sleeping sickness can be treated in early stages and can be prevented by a variety of pest-control measures, including burning of brush, spraying with pesticides, and removing wild animals (including rodents) that serve as reservoirs for the disease. In domestic animals such as cattle and horses, the tsetse fly causes a disease called *nagana*, which is similar to sleeping sickness and causes fever and paralysis. This disease prevented the introduction of livestock into many parts of Africa and as a result preserved habitats for wild species. Areas with an elevation above 150 meters (480 feet), those with a long dry season, and those with sparse or no woodland are free from tsetse flies. Many of these debilitating diseases are associated with the tropical climate and diverse ecologies of Africa and their spread may have been facilitated by the expansion of human populations and the transformation of natural environments through deforestation and irrigation. Reducing the human toll from these diseases is a major challenge for scientific research, African governments, charitable organizations, and the World Health Organization, which has targeted Africa for extra funds and programs.

Figure 6.10 Termites in Africa Termite mound in the Okavango Delta of Botswana. Termites consume wood, cloth, and paper, disrupting crop and pasture land but contributing to soil formation.

One of Africa's most important and numerous organisms is the termite, an insect that lives in wood or the soil and often builds towering mounds (**Figure 6.10**). By consuming dead plant and animal remains, termites perform valuable work in the ecosystem in recycling organic matter and creating soil, rather like the earthworm in temperate regions. However, termites also destroy wooden buildings, cloth, and paper, contribute to global warming through methane emissions, and create holes in fields that are hazards for farm implements and can cause injuries to livestock. Termites are so numerous that they are extremely difficult to control and require applications of costly and sometimes risky pesticides.

Agriculture and Environmental History   The United Nations Food and Agricultural Organization (FAO) has estimated that less than 30 percent of the soils in Sub-Saharan Africa are suitable for agriculture. In addition, agriculture is hindered by an unsuitable climate and an environment that is prone to pests and diseases. However, Africa, as the birthplace of the human species, is also the region where humans first adapted to the constraints of the physical environment, finding sustenance through hunting wild animals, fishing, gathering plants, and domesticating a number of important crop and livestock species.

There is some disagreement about whether cattle were domesticated in Africa or introduced from the Middle East and Asia about 8000 years ago. Archaeological sites from this period have provided evidence of livestock living along with humans and also of domesticated and cultivated grains, including sorghum. The highlands of Ethiopia are considered one of the centers of **domestication,** producing coffee, millet, and an important local cereal called *teff*. Other important crops domesticated in Africa include yams, oil palm, cow pea, and African rice.

Traditional peoples developed several strategies for adapting to low soil fertility, including **shifting cultivation,** which involves moving crops from one plot to another to preserve soil fertility. As in other regions of the tropical world, one form of shifting cultivation is **slash and burn** agriculture, used to clear patches of forest, shrubs, or grassland through burning and then take advantage of the ash to fertilize their crops. When, after a few years, the nutrients are exhausted, farmers move on to a new area and leave the previous plot to return to forest or other vegetation. After a long fallow (rest) period, they return and clear and burn the land again. **Bush fallow** is a modification of shifting cultivation by which crops are planted around a village and plots are left fallow for shorter periods than in the slash-and-burn system. Soil fertility is often maintained through fallow or by applying household waste to the fields. Where household compost is used to grow crops within the village, the technique is called "compound farming" and is popular in forest environments as well as in some urban areas.

Intercropping—the planting of several crops together—is a technique for keeping the soil covered to reduce erosion, evaporation, and the leaching of nutrients from the soil. Where one of the crops, such as beans, can capture nutrients such as nitrogen, intercropping also improves soil fertility. *Floodplain farming* is also important in regions such as the inland delta of the Niger River and the Sudd wetlands along the Nile River in the Sudan (**Figure 6.11a**). More than 500,000 people make a living from growing crops, grazing livestock, and fishing along the Niger, adjusting their activities as the flooded area expands from 5000 to 25,000 square kilometers (3100 to 15,500 square miles) during and following the wet season.

Pastoralism—a way of life that relies on livestock raising—is the human activity best adapted to drier regions of Africa. Nomads, such as the Bedouin, migrate with their animals in search of pastures in the arid landscapes of the Sahel and North Africa. Other groups, such as the Fulani of West Africa (**Figure 6.11b**), practice a system of seasonal herd movements called **transhumance.** They move their herds to wells and rivers in the dry season and drive them northward to take advantage of new pastures in the wet season. In some regions, farmers let pastoralists graze their herds on harvested fields in the dry season, thereby fertilizing the land with animal manure in a mutually beneficial (*symbiotic*) relationship with the pastoralists, and in other regions pastoralists are also farmers.

Sub-Saharan Africa in the World

Sub-Saharan Africa's role in the world begins with evidence of human origins on the continent more than 2 million years ago and continues with the development of major trading societies from about 5000 years ago and the incorporation into a European-dominated colonial system about 500 years ago. Colonialism included the worldwide trade in African slaves, resulting in a diaspora of African peoples that has continued to influence the culture and societies of other world regions. It also resulted in political boundaries that split ethnic groups across territories or clustered enemies within one territory. Peoples from other world regions, including Europe and Asia, came to Africa under colonial rule and created hierarchies of power and politics. These hierarchies included the racial discrimination associated with apartheid in South Africa and tensions over land distribution in southern and eastern Africa, where countries gained independence and white farmers continued to hold the best land. Most of Sub-Saharan Africa was under European colonial domination by 1900 and did not become independent until after 1950. Independence also coincided with the height of Cold War tensions between the United States and the former Soviet Union and the consequent interventions of the superpowers in African political struggles and civil wars.

At the end of the twentieth century, much of Sub-Saharan Africa was still struggling with the transition to independent and democratic government and with economies that rely on a narrow set of exports to other world regions. A series of natural disasters, development failures, and wars has seriously hindered the ability of agricultural production to meet the food needs of growing populations, with large numbers of poor people unable to grow or purchase food or to find alternative employment. Yet, as we will see, some countries and some sectors within those countries have been able make considerable progress in improving economic and social conditions and are

Figure 6.11 Traditional agriculture in West Africa (a) The shores of the Niger River and its tributaries are important agricultural regions in West Africa. These traditional fields near Gao in Mali are shown in the wet season. (b) Nomadic Fulani herders milk a cow near Agadez, Niger.

(a)

(b)

actively debating the most appropriate way to participate in the global economy.

Sub-Saharan Africa is a major world region with a very low level of economic development compared with other regions. The region has a total GNP of $322 billion, only about 1 percent of total global GNP, and an average per capita GNP of $510, compared to world average per capita GNP of $4890 (see World Data Appendix, p. 639). Africa is also singled out for attention by many international agencies and receives the highest amount of development assistance per capita of any world region. The World Bank, for example, identifies Sub-Saharan Africa as "the most important development challenge of the twenty-first century."

Human Origins and Early African History

Africa is often called the "cradle of mankind" because archaeologists have shown that the earliest evidence of the human species (*Homo sapiens*) is found in Africa. Fossilized footprints of an earlier ancestor, the hominid (humanlike) *australopithecus,* were found by Mary Leakey at Laetoli in Tanzania and dated to 3.7 million years ago. Two-million-year-old stone

tools have been found at several sites in Ethiopia and East Africa, including the famous site at Olduvai Gorge in Tanzania (**Figure 6.12**). Anatomically modern humans, who walked upright and had larger brains, have been dated to at least 100,000 years ago from sites in southern Africa and along the Rift Valley, and many scholars now believe that these humans are the genetic ancestors of all modern humans and thus the most basic link between Africa and the world.

For most of human history in Africa, the only record of history is from scattered archaeological sites. More detailed written accounts begin with the development of complex societies in the Nile Valley about 5000 years ago with sophisticated irrigation systems, hieroglyphic writing, and the hierarchical social organization of the Egyptians under their king or pharaoh (see Chapter 5).

From this time onward, explorations, military campaigns, and European trading begins with Sub-Saharan Africa from bases in the Nile Valley and North African coast, such as the Phoenician city of Carthage (in today's Tunisia). About 2500 years ago the famous Greek geographer Herodotus described accounts of Saharan trade in salt and of kingdoms to the south of Egypt, and by 2000 years ago the Roman Empire had extended to most of North Africa. By A.D. 500 some Indonesians

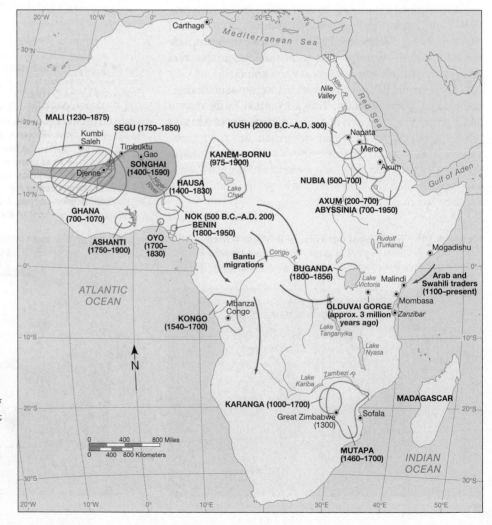

Figure 6.12 Map of African history Sub-Saharan Africa had a rich historical heritage prior to European arrival in the region. It is the place where the oldest human remains were found at Olduvai Gorge in present-day Tanzania. This map shows some of the locations and dates of the great kingdoms as well as the early migrations of Bantu people and Arab traders. (*Source:* Adapted from I. L. L. Griffiths, *An Atlas of African Affairs.* New York: Methuen, 1985; D. L. Clawson and J. S. Fisher (eds.), *World Regional Geography: A Developmental Approach.* Upper Saddle River, NJ: Prentice Hall, 1998, Fig. 23.2; and C. McEvedy, *The Penguin Atlas of African History.* New York: Penguin Books, 1995.)

had settled on the island now known as Madagascar, introducing yams and bananas to mainland Africa, and a strong kingdom had emerged at Aksum in Ethiopia and had adopted Christianity. Over many centuries the Bantu people had been migrating out from West Africa into most of Sub-Saharan Africa, bringing with them technologies such as iron smelting. Between A.D. 500 and 1000, several other important power centers with links to Roman and Arabic empires developed, including the kingdom of Ghana (centered in present-day Mali), where gold was mined and traded with Berber merchants from the Sahara and North Africa in exchange for salt. Trade also linked the Mediterranean coast with the kingdoms of Gao, Songhai (located in Niger), Kanem (near Lake Chad), and Mali, and this trade led to the conversion of many in these empires to the Islamic faith. The east coast of Africa was brought into the Arab system around A.D. 1100 through a series of trading posts that included Mogadishu, Malindi, Mombasa, and Zanzibar, cities that still exhibit the legacy of Arab and Islamic culture in their architecture, language, and religious traditions (see Sense of Place: Zanzibar, p. 272).

Great Zimbabwe is another example of a highly complex African society that existed prior to contact with Europe (**Figure 6.13**). In the fourteenth century, this city, constructed of massive stonework, housed up to 20,000 members of the Shona population. It was a center of metalworking, pottery production, and religion and traded gold with coastal ports. The ruins of this city are now a major tourist attraction and source of pride in southern Africa. Other early cities include several in the Sahel, such as Timbuktu and Djenné, that were centers of trade, religion, and scholarly learning.

The European and colonial tendency to discount indigenous achievements has been especially unfair in the case of Africa. This stems partly from explorers' accounts of Africa in which they constructed a vision of Africa as the "dark" continent by using words such as *uncivilized, savage,* and *primitive* to describe the landscapes and societies of Africa.

The Colonial Period in Africa

With the development of better ships in the fifteenth century, contacts with Spain, Portugal, and China were added to the regular interaction between the Middle East and Africa. The Portuguese traded for gold from coastal settlements in West Africa, and in 1497 the Portuguese explorer Vasco da Gama rounded the Cape of Good Hope at the southern tip of the African continent en route from Portugal to India. During the 1500s Portugal traded along both the west and southeast coasts of Africa and had made contact with empires in West Africa, Congo, and Zimbabwe. By 1600 most parts of Sub-Saharan Africa had some contact with the rest of the world through trading networks set up by the Arabs in the Sahel and Horn of Africa and by the Portuguese along the Atlantic coast and in southern Africa. In return for salt, horses, cloth, and glass, Sub-Saharan Africa provided gold, ivory, and slaves. Cities such as Timbuktu (in present-day Mali) were centers for long-distance caravans, and they had large populations, including many literate intellectuals, a university, and traders. For several centuries African slaves

had been in demand among the Arabs, who used the slaves as servants, soldiers, courtiers, and concubines.

European colonialism took some time to establish control in Africa, and for many years only the coastal ports and trading posts were under European command. One of the main reasons for European reluctance to move inland was the reputation of Africa as the "White Man's Grave" because so many Europeans were rapidly killed by malaria, yellow fever, and sleeping sickness, diseases against which they had no natural immunity. In addition, African armies attacked ports and resisted European attempts to move inland. Some of the most successful early traders were Lebanese who learned local languages.

Slavery   Even in the face of native resistance and the ravages of disease, the coastal regions generated enormous profits for European traders. The Portuguese started to take slaves for their own use on new sugarcane plantations on the Atlantic islands of Madeira and Cape Verde, and in 1530 the first slaves were shipped to the Americas to work on plantations in Brazil. By 1700, 50,000 slaves were being shipped each year to the Americas to provide labor on colonized lands and new plantations whose potential indigenous labor supply had been decimated by European diseases (see Chapter 8, p. 396). Slavery was an important income source for some African coastal kingdoms, such as Dahomey and Benin, who captured their enemies or residents of inland villages and sold them to the slave traders. It is estimated that 11 million slaves were shipped to the Americas from

Figure 6.13   Pre-European architecture (a) Great Zimbabwe is the most famous of a large group of stone-walled enclosures on the Zimbabwean plateau. The modern Zimbabwe nation took its name from this major cultural monument. According to scholars, the structure was erected by the Shona people over the course of about 400 years, beginning in the early eleventh century. At its height in the fourteenth century, Great Zimbabwe's capital was home to as many as 20,000 people. (b) Djenné was founded in the thirteenth century near the Niger River in southern Mali and became a center for scholarly and religious learning. A woman can be seen carrying a stack of firewood, a resource for everyday survival, in front of the Djenné mosque.

(a)

(b)

Sense of Place

Zanzibar

The island of Zanzibar, located just off the coast of Tanzania, has a distinct history and culture as a key port on the Indian Ocean trading with the Middle East and Africa (**Figure 1**). The oldest community on the island, Stone Town, has winding alleys, bustling bazaars, mosques, and grand Arab houses with ornate carved doors (**Figure 2**). The scent of spices wafts through the air.

Historians believe that Zanzibar was first inhabited by fisher people who traveled to the island from mainland Africa around 4000 B.C. By 1000 B.C., Zanzibar and the islands off the coast of East Africa were familiar to the Egyptians, Phoenicians, Greeks, and Romans. As these Mediterranean empires extended their trade routes to the south and east, Zanzibar became one of several major commercial ports along the East African coast. Around the third century A.D., the trade in goods attracted the attention of merchants from southwestern Arabia, who also began trading with the island residents, bringing weapons, wine, and wheat to barter for ivory and other luxury goods. The Arabs brought the religious traditions of Islam, and the Kizimkazi mosque dates from 1107. The local language, Swahili, has many Arabic words and became the trade language for eastern Africa. For a period in the eighteenth and nineteenth centuries, the island was controlled by the Sultan of Oman and was central to the slave trade. The Arab Fort was built between 1698 and 1701 by the Omani Arabs.

Figure 1   View of Zanzibar   A *dhow* sails into the harbor of Zanzibar. The *dhow* is the traditional boat used for trading along the coast of Africa and is still used for fishing and transport.

Zanzibar was transformed when in 1818 cloves were introduced to the island, and production grew rapidly to dominate world trade. Other spices, such as cinnamon, ginger, pepper, and cardamom, were introduced from Southeast Asia (**Figure 3**). Explorers, including David Livingstone, used Zanz-

Africa between 1500 and 1870, with at least 1.5 million slaves dying during the journey. The majority of slaves were male, and the conditions of capture and transport were inhuman; hundreds of slaves were packed into the holds of ships with little food and water and brutally abused by traders. Movements to abolish slavery were led by Quakers in Britain and the United States beginning at the end of the eighteenth century. The British abolished the slave trade with their colonies in 1807 and emancipated (freed) slaves in the Caribbean in 1834. Slavery in the Americas was abolished in most countries by the 1850s, and the slavery issue was key in the U.S. Civil War and was abolished by the 1863 U.S. Emancipation Proclamation (see Chapter 7, p. 326). Although slavery has been banned in most countries since about 1870, there are reports that people are still sold into servitude in countries such as Mauritania and Sudan. Slavery had severe impacts within Sub-Saharan Africa, especially through the loss of labor. In addition, some communities retreated into the interior.

The European names for coastal regions along the west coast of Africa clearly indicate the commodities that they provided from the Ivory Coast in the west to the Gold Coast (now Ghana) and Slave Coast (Nigeria and Benin) to the east. The exchange has been called the "triangular trade," with guns, al-

cohol, and manufactured goods being shipped to Africa in exchange for slaves who were then sent to the Americas. There the slaves were exchanged for gold, silver, tobacco, sugar, and rum, which were shipped to Europe (**Figure 6.14**).

The banning of slavery in Britain in 1772 resulted in the freeing of slaves in several regions under British control, and a large group of liberated slaves was shipped back to Africa to settle in Freetown, the present capital of Sierra Leone. The American Colonization Society subsequently settled 12,000 freed American slaves in 1822 at Monrovia, in Liberia. The British continued to intercept slave ships in the nineteenth century and settled the slaves in Sierra Leone. The descendants of these settlers, who were from many different regions of Africa, retained a separate identity from local cultural groups, and tensions between them and the locals still emerge in civil wars in Liberia and Sierra Leone.

European Settlement   European settlement was encouraged in southern Africa by the more temperate climate and the strategic significance of the trading routes around the southern tip of the continent. The Dutch established a community at Cape Town in 1652, which became surrounded with small farms grow-

Figure 2 **Stone Town, Zanzibar** A woman sits in one of the ornately carved doors in the old Stone Town district of Zanzibar.

Figure 3 **Fruits and spices of Zanzibar** A collection of fruits and spices grown on the island of Zanzibar.

ibar as a base to prepare for their expeditions. It became a British protectorate in 1890 and gained independence in 1963 prior to union with Tanganyika to create Tanzania in 1964. The current population is about 800,000, mostly living in Zanzibar City. Several other communities along the East African coast, including the Kenyan port of Lamu, have an architectural and cultural heritage similar to Zanzibar, and all are now popular with tourists.

ing wheat and raising cattle for supplying ships and the "Cape" communities, as the region around the Cape of Good Hope was called. The settlers evolved Afrikaans, a modified version of Dutch, as their language; belonged to the strict puritan Christian Calvinist religion; saw themselves as superior to black Africans; and became known as the *Boers* (Dutch for "farmer"). As their military and trading power grew in the 1800s, the British took control of the Cape trading route, and British immigrants were encouraged to settle in the Cape region from about 1820, mainly in Cape Town and Durban. When the British imposed laws on the Boers, including the banning of slavery in 1834, the Boers moved north of the Orange River in a great trek, settling on the high pastures called the *veld* in what is now known as the Free State. Some Boers also migrated eastward into the Natal region, where they came into conflict with the powerful Zulus. As we will see, the geography of this colonial settlement has framed the twentieth-century politics of South Africa.

The Scramble for Africa International interest in Africa increased dramatically after 1850, with growing competition between European powers for colonial control and the discovery that quinine could suppress malaria. Explorers, traders,

and missionaries moved into the interior of the continent seeking territory, the source of the Nile, commodities, and souls to convert (see Geography Matters: The Royal Geographical Society and Exploration, p. 278).

By 1880, new knowledge of African resources, including gold and diamonds, competition between European powers to dominate global empires and markets, and reduced risk of African diseases further increased interest in Africa. In 1882 the British claimed Egypt, prompting the French to exert their dominion over West Africa in Senegal and Gabon (**Figure 6.15**). The Portuguese made efforts to consolidate their holdings in Mozambique and Angola, and the Spanish did so in Equatorial Guinea. Inspired by the reports of explorer Henry Stanley, the personal crusade of King Leopold II of Belgium to establish colonies in Africa focused attention on the Congo basin; Germany sought colonies where they had missionaries in what is now Togo, Cameroon, Namibia, and Tanzania. Pressure from commercial companies and even missionaries drove these imperial ambitions and incorporated Africa into the emerging global capitalist economy.

The role of private companies in the colonization of Africa was important because European governments granted

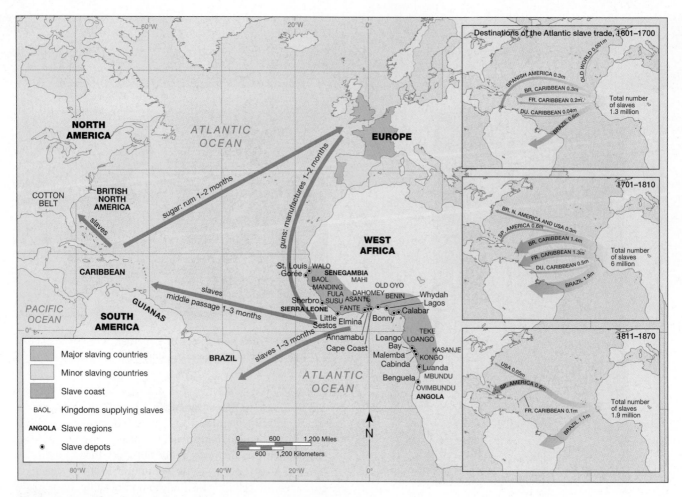

Figure 6.14 The slave trade Millions of slaves were exported from Africa between about 1600 and 1870, mainly from the west African coast. Some local leaders acted as suppliers in return for guns and manufactured goods. Slaves were sent to work in plantations in the Americas, which then sent sugar, rum, and other products back to Europe in a triangular trade. (*Source:* Adapted from A. Thomas and B. Crow (eds.), *Third World Atlas*. Buckingham, UK: Open University Press, 1994, p. 28; and J. F. Ade, A. Crowder, M. Crowder, P. Richards, E. Dunstan, and A. Newman (eds.), *Historical Atlas of Africa*. Harlow, Essex, UK: Longman, 1985, p. 67.)

exclusive concessions for trade and resource exploitation in Africa. These companies, which were often given the right to police, to conscript, and to tax local populations, included the Royal Niger Company and British South Africa Company, both of which received royal charters in the late 1800s.

This hasty "scramble for Africa" culminated in the **Berlin Conference** of 1884–1885, which was convened by German chancellor Otto von Bismark to negotiate European territorial claims in Africa. The 13 countries represented at this conference did not include one African representative from any state in Sub-Saharan Africa, even though more than 80 percent of Africa was at that time under African rule. The Berlin Conference allocated African territory among the colonial powers, according to prior claims and laying down a set of rather arbitrary boundaries that paid little respect to existing cultural, ethnic, political, religious, or linguistic regions (**Figure 6.16**). The British claimed what are now known as Gambia, Ghana, Nigeria, and Sierra Leone in West Africa; Kenya, Uganda, Sudan, and part of Somalia in East Africa; and southern Africa, except for German South West Africa (Namibia), Portuguese

Mozambique, the Cape Verde Islands, and Angola, and the independent Boer region of South Africa. Germany also claimed German East Africa (Tanzania) and Cameroon, Portugal claimed the Cape Verde Islands, and France and Belgium split the Congo. Italy took Somalia, Djibouti, and Eritrea and coastal regions of Libya, with ambitions for Abyssinia (now Ethiopia). The Spanish obtained a small coastal region of northwest Africa and Equatorial Guinea. Most of the remaining territory of West and North Africa allocated to or taken by the French.

The next 20 years saw some rearrangement and consolidation of the European colonies. The British created protectorates in what is now Botswana, Zambia, Zimbabwe, and Malawi and expanded their control over the Sudan. The French took control of many regions along the Niger, and the Italians unsuccessfully invaded Abyssinia. In southern Africa, British entrepreneurs, including the ambitious Cecil Rhodes, responded to the discovery of gold and diamonds between 1867 and 1886 by acquiring the mines at Kimberly and the Rand and sparking a gold and diamond rush. Growing tensions between the British and Afrikaners resulted in the Boer

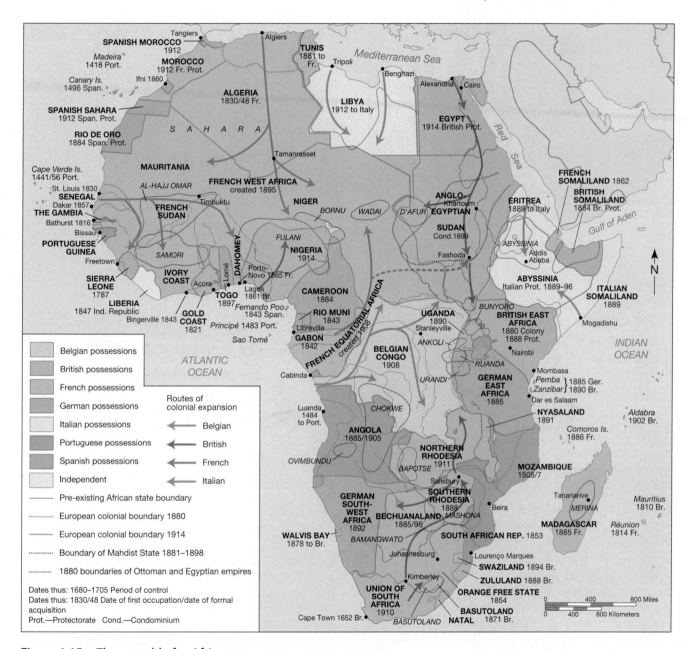

Figure 6.15   The scramble for Africa   Between about 1880 and 1914, European powers aggressively moved to colonize Africa, especially the British, French, and Belgian governments. This map shows the routes and dates of the takeover of Africa. (*Source:* Adapted from A. Thomas and B. Crow (eds.), *Third World Atlas.* Buckingham, UK: Open University Press, 1994, p. 35.)

War (1899–1902), which gave control of much of southern Africa to the British.

By 1914 almost all of Africa was under European colonial control except for Abyssinia, Liberia, and some interior regions of the Sahara Desert. A number of battles were fought in Africa during the First World War, but Germany's eventual loss redistributed the German colonies to Britain, France, and Belgium. Tanzania and South West Africa were assigned to the British, Rwanda and Burundi to Belgium. Togo and Cameroon were each split between Britain and France. Italy's long-standing imperial ambitions in Africa were temporarily achieved with Mussolini's conquest of Abyssinia (Ethiopia) in 1936, but the British soon moved to evict Italy from Africa, returning Ethiopia and Libya to independent rule by monarchy and taking over the Italian portion of Somalia.

The Impacts and Legacy of Colonialism   All of this reshuffling of African territory among European states overshadows the considerable and everyday impacts of colonial rule on African landscapes and peoples. The most general and enduring impacts of colonialism include the establishment of political boundaries; the reorientation of economies, transport routes, and land use toward the export of commodities; improved medical care; and the introduction of European languages, land tenure, taxation, education, and governance. As

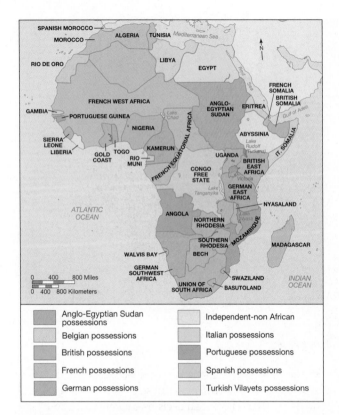

Figure 6.16 Map of colonial Africa 1914 This map shows colonial Africa in 1914 at the beginning of the First World War. Between 1884 and 1914, the colonial powers consolidated their territory and drew boundaries. (*Source:* A. T. Grove, *The Changing Geography of Africa.* Oxford: Oxford University Press, 1994, Fig. 7.1.)

Figure 6.17 African rail network in 1970s The railway network of Africa still shows the legacy of the colonial period, when railways were constructed from the interior to the coasts, originating especially from mining and cash-crops areas.

noted earlier, many of the new colonial boundaries divided indigenous cultural groups and in some cases placed traditional enemies within the same country. For example, the Yoruba were divided between Nigeria and Benin, and Nigeria itself comprised several competitive groups, including the Yoruba in the southwest, the Ibo in the southeast, and the Hausa in the north.

Mining activities were expanded in many regions, especially in southern and central Africa, with large amounts of gold, diamonds, and copper extracted and exported by European companies. New roads and railways were constructed from inland to the coasts to speed the export of crops and minerals, but few efforts were made to link regions within Africa. For example, a rail line from St. Louis to Dakar facilitated the export of peanuts from Senegal, and another from Kumasi to the coast speeded the export of labor, gold, and cocoa from the Gold Coast (Ghana). The line from Mombasa via Nairobi to Kisumu on Lake Victoria in Kenya linked the inland to the coast. The resulting infrastructure facilitated trade beyond but not within Africa (**Figure 6.17**).

Plantations were established to produce crops such as rubber, and a variety of means, including taxation and intimidation, were used to persuade peasant farmers to produce peanuts, coffee, cocoa, or cotton for global markets. In the temperate climates of the East African highlands and southern Africa, areas that were more attractive to European immigrants, the best land

was taken by white settlers for tea and tobacco plantations, livestock ranches, and other farming activities. By 1950, the geography of African agriculture illustrated this export orientation with vast rubber plantations owned by the Firestone Corporation in Liberia, cocoa dominating the cropland of Ghana and the Ivory Coast, cotton in Sudan, peanuts in French West Africa, and tea and coffee in East Africa (**Figure 6.18**). The British established "marketing boards" for several commodities with the goal of improving quality and smoothing out price fluctuations. Traditional African land-tenure systems of communal land and flexible boundaries were forced into privately owned and bounded plots, and traditional decision-making and legal systems were often replaced with European managers and courts.

Some African groups actively resisted European colonial expansion and policies. For example, the Ashanti kingdom in central Ghana and the Zulu in southern Africa fought British expansion, and local people who were forced to work on cotton plantations in East Africa resisted through work stoppages. More passive forms of resistance were widespread, including pilfering and poaching as well as "go slow" work habits and false compliance with colonial rules. Anthropologist James Scott has shown how these *everyday forms of resistance* are still used by those who are trying to show their opposition to exploitation and repression in Sub-Saharan Africa and elsewhere.

The impacts and process of colonial rule varied between European powers. The British chose a paternalistic indirect rule for most of their African colonies, making preexisting power structures and leaders responsible to the British Crown and colonial administrators in a decentralized and flexible administrative structure. For example, local leaders were required to

(a)

(b)

Figure 6.18   Export crops in Africa (a) Sisal is grown as a cash and export crop over large areas of Kenya and Tanzania and is used as a fiber for mats, sacks, and baskets. The sacks were used as containers for other export crops such as coffee and tea. (b) A small child carries a peanut plant near Djiffer, Senegal. Peanuts were introduced as a cash crop in French West Africa during the colonial period.

collect taxes—sometimes a *hut tax* based on the number of dwellings in a community, sometimes a *poll tax* based on the number of residents. In order to obtain money to pay taxes, people had to produce crops for sale to the Europeans, an indirect way of transforming economies and land use to commodity production. Foreign ownership of land was prohibited in some cases, and traditional legal systems were used to resolve local conflicts. The British, preceded by missionaries, also introduced some European-style schools, and by the 1940s a select group of Africans were attending overseas universities and given posts in government administration. The French colonial policy was one of assimilation, encouraging elites to evolve into French provincial citizens with allegiance to France, but with agriculture and mining under close supervision from the French capital in Paris. By 1946 there were about 20 Africans, elected from West Africa, in the French parliament. The Belgian and Portuguese modes of colonialism are described as much harsher, with direct rule and often ruthless control of land and labor. In the Congo local people were forced to gather rubber, kill elephants for ivory, and build public works under threat of death or severe punishment. These authoritarian forms of control, with little political participation, dominating official ideology, and frequent use of armed force, provided an unfortunate model for leadership in independent Africa.

Given the dramatic impact of the colonial period on contemporary Africa, it is important to note that in most of Africa, formal colonialism only lasted 80 years, from about 1880 to 1960. The legacies of the colonial period in specific regions and sectors will be discussed in more detail later in this chapter.

Independence

The period of decolonization in Sub-Saharan Africa was rapid and ranged from relatively peaceful handovers of leadership to well-prepared African leaders to more violent transitions of

power to divided or unprepared local leadership. South Africa was consolidated as an independent state—the Union of South Africa—in 1910. The British decision to grant India and Pakistan independence in 1947 gave hope to Africans, led by several foreign-educated activists such as Kwame Nkrumah of Ghana and Jomo Kenyatta of Kenya (**Figure 6.19**). A *Pan-African* movement, led by black activists in the United States—including W. E. B. Du Bois and Marcus Garvey—and others in the West Indies also promoted independence. The sixth Pan African Congress of 1945, held in Manchester, England, brought together leading African nationalists, including Nkrumah and Kenyatta, to discuss independence, supported by many Africans who had fought for the allies in the Second World War and demanded more equal treatment. During the war, which was fought partly in East and North Africa, 80,000 Africans fought for France and more than 400,000 for Britain. Another element in the transition to independence was the formation of organized nationalist groups or African Unions within such key African countries as Tanganyika, Zimbabwe, and Kenya. These groups provided the basis for political parties. In 1957 Ghana became the first country in Sub-Saharan Africa to have power handed over to local populations; 3 years later, Nigeria gained its independence.

Although most of the British handovers were relatively peaceful, countries with significant white settler populations endured more violent transitions. In Kenya, about 3000 white settlers controlled more than 2.6 million hectares (6.4 million acres) of the best land, especially in the highlands, adjacent to overpopulated Kikuyu farms. Whites also dominated the government and set policy in the interests of the 60,000 white residents. The *Mau-Mau* rebellion between 1952 and 1956 resulted in the deaths of 100 whites and more than 10,000 black Africans prior to independence in 1963. In Southern Rhodesia (now Zimbabwe) the population of about 250,000 white settlers, led by Ian Smith, made a Unilateral Declaration of Independence

Geography Matters

The Royal Geographical Society and Exploration

Some of the most famous explorers were associated with the British Royal Geographical Society (RGS), which was founded in 1830 for the "advancement of geographical science." The RGS supported and awarded their medal of honor to many explorers of Africa, including David Livingstone, Richard Burton, and John Speke.

David Livingstone, a Scottish missionary, arrived in South Africa in 1841, and for the next 15 years he traveled throughout the interior (**Figure 1**). He is best known for his explorations of the Zambezi and his encounter with the magnificent waterfalls that he named after Queen Victoria. His book on his travels in South Africa sold more than 70,000 copies. His last expedition from 1866 to 1873 attempted to reach Lake Tanganyika from southern Africa, and he traveled farther into west central Africa than had any previous European expedition. He was out of touch for so long that an American journalist, Henry Stanley of the *New York Herald* (**Figure 2**), was sent to find him and wrote popular reports that added to the fame of Livingstone, who died soon after Stanley found him in 1873.

Richard Burton spoke 27 languages and traveled widely through Asia, the Middle East, and Africa, writing dozens of reports about his encounters. His goal in Africa was to discover the source of the Nile. On his first expedition in 1855 he was wounded in Somalia; on his second in 1857 he reached Lake Tanganyika, but he was so ill from malaria that he decided to return to London. His companion on this expedition, John Speke, continued on to Lake Victoria and claimed it as the

Figure 1   David Livingstone Scottish missionary and explorer David Livingstone rests his hand on a globe.

Figure 2   Henry Stanley This photo of Henry Stanley contributes to European images of Africa as a hostile continent where Africans served the interests of Europeans.

(UDI) in 1965 rather than consider the possibility of rule by the 6 million black Africans. Only after 15 years of conflict and international trade embargoes did an independent Zimbabwe finally emerge in 1980, with a mostly black government.

The transition in French West Africa and French Equatorial Africa occurred dramatically in 1960, with France recognizing the independent countries of Mauritania, Mali, Niger, Senegal, Upper Volta (now Burkina Faso), Ivory Coast (now Côte d'Ivoire), Togo, Dahomey, Chad, the Central African Republic, Cameroon, Gabon, and the Congo. Guinea obtained independence slightly earlier in 1958. In most cases strong economic and cultural ties were maintained with France, the franc remained the currency, and French troops were stationed in most countries. Britain also continued to invest in its former colonies, especially in larger development projects, some of which were initiated before independence, such as the project to build a hydroelectric dam on the

source of the Nile, to great acclaim, despite opposition from Burton. Speke returned to travel the entire length of the Nile from Lake Victoria to the Mediterranean in 1862. Burton went on to work for the British Foreign Office in West Africa and published several books describing local customs.

Mary Kingsley (**Figure 3**) traveled through the Congo in 1893 and 1894 and was the first European to visit Gabon. She collected specimens of beetles and freshwater fishes for the British and lectured and wrote about her travels when she returned.

These Victorian explorers added greatly to geographic knowledge of Africa, and their reports fueled colonial interest in the continent's resources and peoples. Their lectures at the Royal Geographical Society and elsewhere increased interest in the discipline of geography and its role in Britain's colonial enterprise. However, their books and those of other explorers contained many Victorian prejudices and paternalistic attitudes that fostered the popular imagination of Africa as a barbarous and exotic continent in need of civilization and colonial supervision.

Figure 3 **Mary Kingsley** Mary Kingsley explored the Congo basin.

Volta River in Ghana. One of the groups to benefit most from decolonization were the transnational corporations that preferred to deal directly with African economies rather than through the mediation of colonial powers and their monopoly companies and marketing boards.

Belgium left the Belgian Congo suddenly in 1960, with chaos following as the army mutinied; separatist groups tried to form governments in the wealthier provinces; and a U.S.-sponsored army officer, Joseph Mobutu, won a military coup to depose and assassinate independence leader Patrice Lumumba. In Rwanda and Burundi, independence from Belgium in 1962 left a legacy of tension between ethnic groups because the Belgians had favored the Tutsi minority over the Hutu majority. Portugal hung onto its colonies of Angola and Mozambique until 1974,

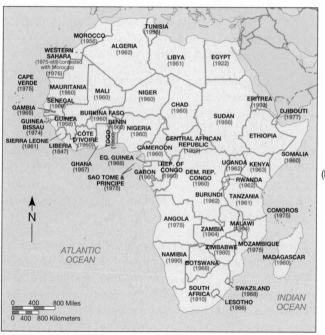

(b)

(c)

Figure 6.19 **African independence** (a) Dates of independence in Africa. (b) Kwame Nkrumah of Ghana. (c) Attending a "Big Four" meeting in Nairobi on October 16, 1964, are the African Leaders (left to right): Dr. Kenneth Kaunda, premier of Northern Rhodesia; Julius Nyerere, president of the United Republic of Tanganyika and Zanzibar; Jomo Kenyatta, premier of Kenya; and Dr. Milton Obote, premier of Uganda. [*Source:* (a) Adapted from I. L. L. Griffiths, *An Atlas of African Affairs.* New York: Methuen, 1985.]

by which time some groups demanding independence had come under the influence of the Soviet Union and Cuba. Independence groups in Angola, Mozambique, and Guinea also sought help from the United Nations in freeing themselves from colonial rule, while Namibia sought freedom from the control of South Africa. The best known of these underground independence movements were the MPLA (People's Liberation Movement of Angola), FRELIMO (Mozambique Liberation Front), and SWAPO (the South West African People's Organization).

South African History and Apartheid

The case of South Africa merits particular attention because of the policies of **apartheid,** a system of control of the movement, employment, and residences of blacks with the goal of separate development of the races within South Africa. South Africa is also significant for its eventual transition to black majority rule.

The history of racial segregation in South Africa is a long one, dating back to the establishment of a supply station by the Dutch East India Company in Cape Town in 1652. The Dutch, whose settlement developed slowly at first, were segregationists and attempted to prevent contact between whites and native peoples, although they did hold Africans as slaves. The Dutch grew wheat, planted vines, and introduced livestock, moving northward into southern Africa, displacing African native communities as the frontier expanded in the search for more pasture land. In 1806 Britain seized political control over the Cape in order to control the route to its empire in India. Like the Dutch, the British set about expropriating land and setting up defendable boundary lines between the European immigrant settlements and the largely Bantu-speaking Nguni and Sotho people. The Boer policies of strict racial segregation between blacks and Afrikaners (as the Dutch farm settlers were called) included the establishment of native reserves and the mandate that blacks needed permission to enter or live in white areas, restrictions known as the *pass laws*. Native peoples were incorporated into the economy as servants, squatter tenants, or semifeudal serfs. Ultimately, the "Fundamental Law," established in 1852, legally enshrined the inequality of blacks and whites.

British policies promoted more mixing between whites and blacks, who were intentionally exposed to white value systems and institutions, but still the British sought a cheap and docile workforce for farms and mines. The first half of the twentieth century witnessed the strengthening and extension of the Boer principles of racial segregation through territorial segregation. Black ownership of land was restricted, as was black settlement activity. In addition, the permanent residence of blacks in white urban areas was prohibited. The Natives (Urban Areas) Act codified this latter restriction, defining blacks as temporary urban residents who were to be repatriated to the tribal reserves if not employed. The act also required that blacks, while within urban areas, were to be physically, socially, and economically separated from the white population.

The goal was to have the tribal reserves operate as independent economies that supported the black population and were separate from the operations of the white economy. Unfortunately, low wages for black laborers as well as high rates of landlessness among blacks living on the reserves undermined the viability of an independent subsistence economy. The reserves were unable to support the black migrant labor system so necessary to the success of the white economy. In addition, blacks increasingly flowed into the urban areas for work, creating a growing and permanent black population in the white cities.

By 1946, blacks were the largest racial group in the urban areas, a direct result of the demand for black labor in the growing urban manufacturing sector. Clearly, territorial segregation was becoming increasingly ineffective as a method of separating the races. By mid-century, the policies of segregation were abandoned, and new policies of apartheid were introduced. By the mid-1940s, the separation and unequal treatment of races—white, colored (of mixed race), and black (or African)—were ubiquitous practices with a loose set of laws, practices, and procedures to uphold them. When the Anglo South African political groups lost control of national political power to Afrikaner parties in 1946, segregationist practices became more solidly codified. In the wake of their victory, the Afrikaners imposed strict racial separation policies transforming apartheid from practice to rule. Laws included the Group Areas Act of 1950, which established residential and business sections in urban areas for each race, and the Land Acts of 1954 and 1955, which effectively set aside more than 80 percent of South Africa's land for the white minority. In addition, the pass laws that required nonwhites to carry permits when in white areas were reinforced. The 1950 Population Registration Act classified all South Africans as either Bantu (black), colored (mixed race), or white, with Asian added as a later category.

Segregation was enforced through regulations to prevent social contact and marriage between races, the establishment of separate education standards and job categories, and the provision for separate entrances to public facilities. Some of the important acts and regulations enacted during this period to enforce social apartheid include the following:

- Prohibition of Mixed Marriages Act (1949)—barred interracial marriage
- Immorality Act (1950)—prohibited sexual relations between different races
- Reservation of Separate Amenities Act (1953)—set up segregation of train stations, buses, movie theaters, hotels, and many other public facilities
- Bantu Education Act (1953)—set up a separate and unequal education system for blacks

Large-scale segregation was established in 1959 through the creation of ten **homelands,** a new version of tribal reserves (**Figure 6.20**). The homelands were areas set aside for black residents as tribal territories where residents were given limited self-government but no vote and limited rights in the general politics of South Africa. Comparisons have been drawn between the homelands policy and the establishment of Indi-

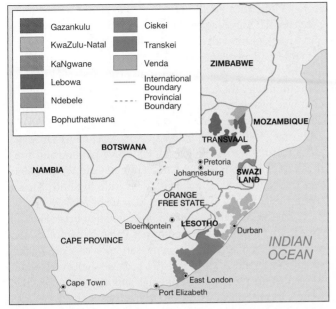

(a)

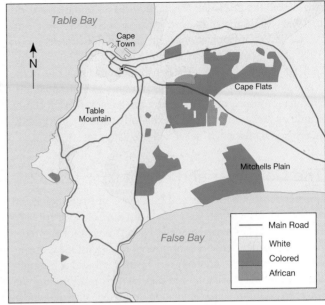

(b)

(c)

(d)

Figure 6.20 The different scales of apartheid (a) Map of the system of South African homelands. (b) Map of residential areas in Cape Town designated for different race groups prior to 1994. (c) Black woman sits outside a "whites only" waiting room in a train station. (d) Nelson Mandela in 1994. [*Source:* (a, b) Redrawn from D. M. Smith (ed.), *Living Under Apartheid*. London: Allen & Unwin, 1982; Figs. 2.1, 2.3, and 2.5.]

an reservations in North America, especially where the land provided was also of poor quality. In 1970 each black South African was made a citizen of one of these homelands, which were given limited independence in lieu of giving them general voting rights. Three million black South Africans were forced to resettle in their homelands.

For nearly 40 years, apartheid was the method of control of a white minority over a black majority through processes that were fundamentally geographical in the spatial separation of races by marriage, within buildings, within cities, in employment, and within the countryside. Through containment of urban blacks, regional decentralization of employment, and the suppression of dissent, Afrikaners attempted to maintain white supremacy while they continued to exploit black labor to fuel a burgeoning economy. Legislation to remove blacks from urban areas was also enacted. Industrial decentralization, though encouraged, was not a successful strategy and instead fostered the settlement of blacks in homeland townships close to white urban areas, such as Soweto near Johannesburg. This enabled the

white-controlled cities to have cheap (black) labor close by, without having to pay for housing, infrastructure, and services for the black population. Protests against apartheid were also quickly and ruthlessly repressed, with African National Congress leaders like Nelson Mandela being jailed and activists such as Steven Biko killed. Enforcement of the requirement that black students use the Afrikaans language led to the 1976 riots in Soweto. International pressure that contributed to the end of apartheid included sports boycotts that threatened white South Africa's passionate devotion to rugby and cricket, a forced withdrawal from the British Commonwealth, and economic and trade sanctions as well as voluntary investment bans by some major international corporations. A number of white South Africans were also vocal in their opposition to the system.

The 1990s saw the end of apartheid in South Africa—Nelson Mandela was freed from jail, and President F. W. de Klerk agreed to the sharing of political power between blacks and whites. In 1994 South Africa held the first election in its history in which blacks were allowed to vote, and Nelson Mandela was elected the first black president. With the end of apartheid in 1994, the homelands were incorporated back into South Africa, and their populations were granted full citizenship. In 1996 a new South African constitution was signed into law and took effect in early 1997. The constitution includes one of the world's most comprehensive bills of rights and prohibits discrimination based on race, gender, pregnancy, marital status, ethnic or social origin, color, sexual orientation, age, disability, religion, conscience, belief, culture, language, or birth.

The Cold War and Africa

Independence movements and transitions in Africa coincided with the global tensions associated with the Cold War between the United States and the Soviet Union. Several of the independence leaders had been introduced to socialist ideas during education overseas, and many other Africans found communist and socialist ideas of equity and state ownership appealing after the repression, foreign domination, and inequality of colonial rule. In countries with relatively peaceful transitions to independence, such as Ghana, socialism took a populist form, focusing on state control of the economy and nonalignment in foreign policy. In Tanzania, President Julius Nyerere developed the concept of an *African socialism* based on the traditional values of communal ownership and kinship ties to extended family expressed as *ujamaa* (familyhood). He believed that a socialist system of cooperative production would be more compatible with African traditions than would individualistic capitalism. He focused on rural development and self-reliance in agriculture, moving rural residents into village collectives, instigating campaigns for free and universal education and literacy.

In Angola and Mozambique, where the Portuguese fiercely repressed independence movements, revolutionary movements espousing leftist ideals attracted the interest of the Soviet Union, China, and Cuba, which provided military and economic assistance and trained young Africans in their universities. In Angola, early ties between the MPLA (People's Liberation Movement of Angola) and the Portuguese communist party led rebels to seek arms from the Soviet Union and China and to accept an offer of military training from Cuba. After the Portuguese colonial government fell in a coup in 1974, and a three-party governing coalition of independence movements collapsed, Angola became a focus for a Cold War power struggle. The United States provided funds to the pro-Western groups, and the Soviet Union and Cuba continued to support the MPLA government (**Figure 6.21**).

Tensions escalated after South Africa responded to Angolan support for rebels in Namibia (at that time under South African control) by sending troops toward Angola and supporting antigovernment and anticommunist rebels in the mid-1970s. Angola reacted by moving further into the Soviet sphere, signing a treaty that allowed the Soviets to use Angolan ports and airports and placing abandoned land and mines under state control. Other African countries supported the Marxist-Leninist government of Angola in opposition to South Africa. By the 1980s, there were 50,000 Cuban advisors and troops and millions of dollars' worth of Soviet military aid flowing into the

region. Only when South Africa agreed to grant Namibia independence did Angola agree to reduce Cuban and Soviet presence and to seek peace with the pro-Western rebel groups.

After independence in 1975, Mozambique was similarly caught in Cold War competition. The leftist FRELIMO (Mozambique Liberation Movement) government fought a civil war against groups supported by anticommunist white governments in Rhodesia (now Zimbabwe) and South Africa. Over time, FRELIMO shifted away from dependence on Soviet aid and strict Marxist policies of collective farming and suppression of religion. With the reduction in Cold War tensions and the beginnings of majority black rule in South Africa, peace was achieved in 1992 after 30 years of war.

The Cold War had a different manifestation in the Horn of Africa, where independent countries came into conflict and sought arms and assistance from the superpowers. Somalia had sought Soviet aid as early as 1962 because the United States was at that time supporting Ethiopian and Kenyan regimes that were resisting Somali expansion into adjacent territories with large Somali populations (a process called *irredentism*). In 1969 the regime of Siad Barre officially espoused *scientific socialism* with a mass-literacy campaign and attempts to break Somalis' traditional allegiances to their clans. In Ethiopia, Emperor Haile Selassie had ruled for more than 40 years over an economy organized on semifeudal lines in which the concentration of land and wealth contributed to periodic famines. Growing demands for reform grew during the 1960s and culminated in a takeover by a council of junior military officers (the "Derg") in 1974 led by Major Mengistu Haile Marian. Mengistu promoted an Ethiopian socialism of self-reliance and widespread land reform in support of peasants and workers and had some initial U.S. support. Only when Somalia invaded the Ogaden region of Ethiopia (where many ethnic Somalis lived) did Mengistu turn to the Soviets and Cubans for assistance, hosting an estimated 17,000 Cuban advisers and troops and receiving more than $13 billion in Soviet aid between 1977 and 1990. The United States then shifted its bases from Ethiopia to Somalia.

Africa provided fertile soil for Cold War rivalry as newly independent nations searched for political ideals, dealt with civil wars and incursions from their neighbors, and sought assistance to develop their economies. In many countries, millions of dollars were expended on arms and other military assistance, thousands were killed, and rural areas were abandoned. One of the most tragic legacies of the civil war in Mozambique was the more than half million land mines in the countryside that continued to kill, maim, and inhibit agriculture. On a more positive note, the literacy campaigns and land reforms of countries such as Tanzania improved conditions for many poor people, and nonmilitary aid from both the United States and Soviet Union mitigated some famines and assisted in some development projects.

Development, Debt, and Foreign Aid

Development theorists agonized over prescriptions for African development in the late twentieth century. *Modernization theory* was seen as a solution in the 1950s, explaining

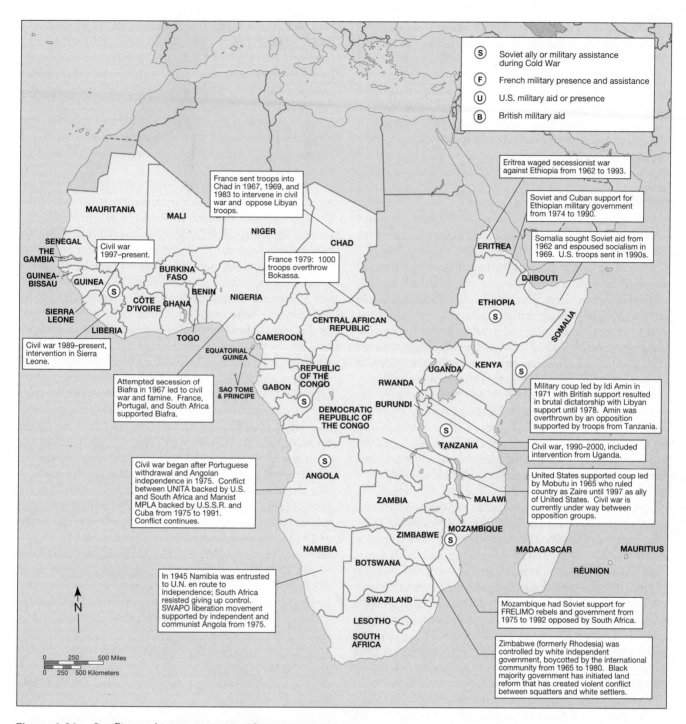

Figure 6.21   Conflict and intervention in Africa Africa has experienced many wars and conflicts since independence, some of them fueled by foreign intervention, including Cold War politics involving the Soviet Union, Cuba, and the United States. France and Britain have continued to provide some military assistance in former colonies, and countries such as South Africa, Liberia, Libya, Uganda, and Angola have provided military support for groups in neighboring countries.

African underdevelopment in terms of the lack of industrialization and proposing solutions of technology transfer, training, and large projects for power generation, often supported by international assistance. Modernization projects included road construction, the Volta and Aswan dams, and harbors such as Tema in Ghana (See Geography Matters: Dams in Africa, p. 264).

The *dependency theory* perspective that argued that African resources had been exploited by the colonial powers was represented by Walter Rodney's classic 1972 text *How Europe Underdeveloped Africa* and by the analysis of famous Egyptian economist Samir Amin, including his 1971 book about neocolonialism in West Africa. These texts argued that the dominant capitalist powers of England and other European colonial powers transformed the political and economic structures of Africa to serve their interests in obtaining cheap raw materials and in doing so undermined local agriculture and social development. Many countries emerged from colonialism with their economies and trade dependent on the production of just a few products.

The *terms of trade* for products produced by African countries, including minerals and export crops such as cocoa and coffee, deteriorated over the last part of the twentieth century, meaning that their value on the world market decreased in comparison with manufactured items and other goods that were being imported by African countries. This meant, for example, that in 1990 a farmer had to produce twice as much coffee as in 1960 to earn the money needed to purchase a bag of fertilizer.

For dependency theorists the remedy was to reduce the dependency on export revenue and avoid the high costs of imports by creating local capacity to produce goods that would replace those previously imported from other countries, especially manufactured industrial products (a policy called *import substitution*). Several African nations adopted strict import-substitution policies that included subsidization of local industry and protection against foreign imports through tariffs and other mechanisms. As in Latin America, import substitution was a mixed success and faced greater challenges because of Africa's generally low level of infrastructure, skills, domestic markets, and investment capital. While it fostered the development of some industries, particularly small-scale manufacturing in African capitals, it also led to inefficiency and poor quality.

Both modernization and import substitution required capital funds that were not easily available in Africa and so many countries looked outside the region to borrow money. Because many African countries had poor commercial credit ratings, most loans were made through country-to-country arrangements with governments such as the United States or through international agencies such as the World Bank. Although some funds were invested in infrastructural, industrial, and agricultural development, considerable sums were used to purchase arms or were diverted by ruling elites to increase their own personal fortunes as overall debt increased.

The total debt of Sub-Saharan Africa, at $230 billion in 1998 according to the World Bank, is only one-third of that of Latin America and Asia, but it is huge as a percentage of gross national product (GNP) in most African countries. In Angola, both Congos, Côte d'Ivoire, Ethiopia, Guinea-Bissau, Mauritania, Sierra Leone, and Zambia, the total debt exceeded GNP in 1998 (see World Data Appendix, p. 639). Several countries were paying more than five times the value of their exports to service their debt each year (**Figure 6.22**).

As in other regions, the multilateral agencies responded to the debt crisis by first stabilizing the loans through extending payment periods and adjusting interest rates and then by demanding *structural adjustment programs*. These programs required devaluation of currencies (to promote exports), liberalization of trade (by removing tariffs), reduced public spending, and the privatization of government-held compa-

nies. The first country to accept structural adjustment was Ghana in 1983, but by 1990, 30 countries had implemented programs with the insistence of the International Monetary Fund (IMF). In many of these countries the impacts of structural adjustment were severe, sending food prices spiraling and increasing unemployment as governments cut public sector jobs. Kenya and Nigeria both fired more than 150,000 government employees in response to IMF policies. Some critics suggested that IMF should be renamed "Imposing Misery and Famine."

The structural adjustment programs fit within the broader program of neoliberalism promoted in particular by the United States that includes reduction in government subsidies for social and agricultural programs, removal of barriers to trade, and privatization of publicly owned land and corporations (see the discussion in Chapter 8 of similar policies in Latin America). But neoliberal policies of free trade clashed with the special concessions that had been granted to Africa by the European Economic Community (EEC). The *Lomé convention,* named after the capital of Togo where it was first signed in 1975, is an agreement between the EEC and 66 countries, 43 of which are in Sub-Saharan Africa, by which the EEC provides economic assistance and trade concessions to promote exports of certain key commodities. The convention includes access to European markets, stabilization of export earnings on selected commodities, industrial technology transfer, project financing, and development aid. However, such preferential treatment has been criticized by the United States and nations in other regions who seek to compete for European markets.

The destitution created by economic crises, structural adjustment, and war in Africa prompted the international agencies and others to try to cushion the impacts of restructuring in Africa by providing programs for alleviating poverty. This assistance is built on several decades of humanitarian and economic assistance to those regions of Africa suffering from disasters and war. Africa is the largest recipient of what is called *foreign aid,* receiving one-third of the global total, equivalent to about 10 percent of the region's total gross national product

Figure 6.22 The debt crisis During the 2000 AIDS conference, southern Africans protest the burden of debt in Africa and its impact on health care.

and averaging about $34 per capita (that is, for each member of the population). In some countries, foreign aid, also called *official development assistance* (ODA), is more than 20 percent of the GNP and reaches $50 per capita (see World Data Appendix, p. 639). For example, Mozambique received a total of $1.03 billion in 1998, approximately 28 percent of GNP and amounting to $61 per capita.

As the twentieth century drew to a close, an international campaign was organized to pressure for debt relief, especially for the poorest countries in Africa. Official recognition that many of the countries had debt burdens that would permanently cripple development led to several debt relief programs. In 1996 the World Bank and International Monetary Fund (IMF) introduced the *Heavily Indebted Poor Countries* (*HIPC*) initiative to restructure and forgive part of the debt of poor countries that, over a five-year period, showed a willingness to pursue neoliberal economic policies of reduced government spending and free trade. Mozambique was one of the first of about a dozen African countries that qualified for this form of debt relief. The HIPC program was severely criticized because it was very specific about changes in policy and told countries how to run their economies and because the debt reductions were too small. An international protest movement, called the *Jubilee initiative,* petitioned to cancel the majority of debts owed by poorer countries by the year 2000. In 2000 several European countries and the United States did move to cancel bilateral debts owed by many countries in Africa, and the IMF increased the amount of debt relief under HIPC.

Peoples of Sub-Saharan Africa

The population of Sub-Saharan Africa was estimated at about 627 million (including Sudan) in 2000 and had a higher average birthrate, at more than 41 births per 1000 population per year, than any other continent. The overall population growth rate was about 2.5 percent a year, with a doubling time of about 27 years. The Population Reference Bureau has projected the 2025 population at just over 1 billion.

The total fertility rate (the average number of children born to a woman during her lifetime) is high in most of Sub-Saharan Africa, averaging 5.3 and reaching more than six children per woman in many regions (see World Data Appendix, p. 639). As a result, between 40 and 50 percent of the population of most African countries is under age 15. This has major implications for future population growth and its impacts as this group starts to have children and makes demands on education and employment systems. Fertility rates are lower in southern Africa.

What are the reasons for high fertility and birthrates in Africa and what are the prospects for slowing population growth? Population geographers and other researchers have focused on the study of African demography in trying to understand these questions. They have found that although religious prohibitions of contraception and lack of access to contraceptive devices may play a small role, other factors are much more important. Children are valued in Africa for many

logical reasons, including their ability to work in agricultural fields and as herders, to help with household work, and to care for younger siblings; children are also the possible source of financial or other gain when they marry. Children are the main source of security for elderly people in countries where there are few pensions or public services for the aged, and it is traditional for younger generations to respect and care for their elders. Large families are also often perceived as prestigious, a spiritual approach to linking past and present, and a way of ensuring family lineage; and in regions of ethnic strife, children represent a way of securing votes, warriors, or political power. Even though infant mortality rates have improved with better nutrition and health care in much of Africa, many African families have internalized the need to have many children in order to ensure that some survive to adulthood.

Many studies have also shown that conditions for women have a strong influence on fertility rates, with lower age of marriage, minimal female education and literacy, and low rates of female employment all contributing to higher fertility rates (**Figure 6.23**). Fertility rates tend to be lower in urban areas with high rates of female education, employment, and later ages for marriage. In Kenya, for example, women with secondary or higher

Figure 6.23   Status of women and fertility in Africa
Women with more education in Nigeria and Botswana tend to have fewer children and are more likely to use contraceptives. Contraceptive use is generally lower in Nigeria, partly as a result of larger Islamic populations and the Islamic prohibitions on contraceptive use. (*Source:* Population Reference Bureau. Available at www.prb.org)

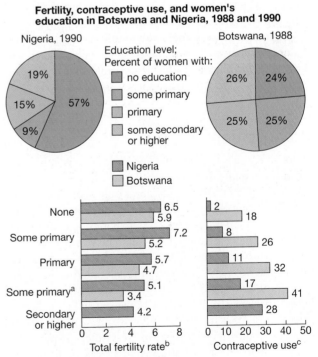

education have a total fertility rate of 4.9 compared to a rate of more than 7.0 for women with little or no schooling, and in Nigeria the fertility rate is 4.2 for better-educated women compared to 7.2 for women with little education.

Population projections for several African countries have been revised downward because of the high mortality and infection rates associated with the AIDS epidemic. In Zimbabwe and Botswana, where infection rates are as high as 25 percent, estimates of average life expectancy have been revised down by 20 years, and population growth rates have been reduced or even reversed. Overall population projections for Africa for 2025 have been adjusted down by 200 million people.

Nigeria is the most populous country in Africa, with a 1998 population estimated at 120 million, followed by Ethiopia with 61 million, the Democratic Republic of the Congo with 48 million, South Africa with 41 million, Tanzania with 32 million, and Kenya with 29 million (see World Data Appendix, p. 639). Some populations, such as Nigeria, have not been reliably censused for many years and are approximate estimates. Overall population density is relatively low at 27 people per square kilometer (74 people per square mile), about half the global average density. The only countries in the region with population densities higher than the global average are the Indian Ocean islands and the Central African countries of Rwanda and Burundi. The density of population per unit of arable land is sometimes considered a better indicator of population pressure on land resources because it measures the ability of land to support its population, or its *carrying capacity*. By this measure Sub-Saharan Africa appears much more densely populated, with an average of 400 people per hectare (162 per acre) of arable land, which is just about equal to the global average.

Figure 6.24 shows Africa with mostly scattered rural population. Population concentrations tend to be associated with better soils and climate, with colonial centers for mining and export crops, and with coastal ports.

Urbanization

Although Africa is the most rural of world regions, it has been urbanizing rapidly over the last 40 years. In 1960 the urban population of Sub-Saharan Africa was only 17 million people, about 20 percent of the overall regional population. By 1998, the urban population had reached 209 million people, 33 percent of the total, and was growing at about 5 percent per year. This is more urban than most of Asia, but much less urban than Europe or the Americas. The level of urbanization varies greatly by country, from South Africa with 53 percent of its people living in urban areas, to Ethiopia at 17 percent and Uganda at 14 percent. East Africa has lower level of urbanization (19 percent) than does southern (46 percent) or West Africa (33 percent).

About 30 percent of Sub-Saharan Africa's population live in the largest city in their country. Lagos and Johannesburg are the two largest cities with populations of about 11 million and 2.5 million, respectively. Geographer David Simon has used several indicators to identify those African cities of great-

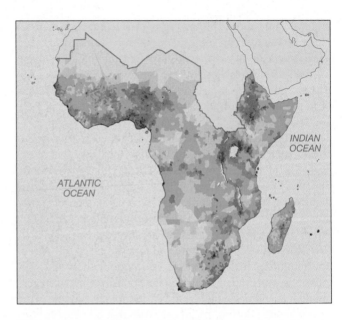

Figure 6.24   Africa population density Africa population is mostly scattered rural population, with the greatest density along the West African coast, the southeast coast of South Africa, the highlands or East Africa, and along major rivers. For a view of worldwide population density and a guide to reading population-density maps, see Figure 1.40. (*Source:* Center for International Earth Science Information Network (CIESIN), Columbia University; International Food Policy Research Institute (IFPRI); and World Resources Institute (WRI), 2000. Gridded Population of the World (GPW), Version 2. Palisades, NY: CIESIN, Columbia University. Available at http://sedac.ciesin.org/plue/gpw/index.html?main.html&2)

est regional importance and links to the global system, including the presence of stock markets, large numbers of embassies, air traffic, and headquarters of international or regional organizations and corporations. According to these criteria, Johannesburg, South Africa; Nairobi, Kenya; and Lagos, Nigeria, lead with more than 70 embassies and 40 regional or international headquarters in each city.

Although urban growth is partly driven by overall population growth, cities have been growing twice as fast as overall populations, and fertility rates tend to be lower in urban areas. As in many other regions, the major driver of urban growth is migration from rural areas to the cities, and the factors pushing people from rural areas and pulling them to the cities are somewhat similar. People are leaving rural areas because of poverty, lack of services or support for agriculture, scarcity of land, natural disasters, and civil wars. Urban areas are more attractive because they offer jobs, higher wages, better services (including education and electricity), and entertainment. Urban areas have benefited from the urban bias of both colonial and independent governments in Africa that tended to invest disproportionately in capital cities that housed centralized administrative functions; in addition, food prices were kept down in the cities to reduce wage demands and to decrease the risk of civil unrest.

There has been some emigration from African cities as a result of civil unrest and economic crises. For example, people left Mogadishu in Somalia and Kigali in Rwanda when fighting reached the cities. Another downside of city life is the problem of servicing the rapidly growing urban areas. For example, Accra, Ghana, often considered one of the more livable cities in Sub-Saharan Africa, has a population of 1.7 million, of whom only 12 percent have homes with connections to the sewer system. More than 15 percent of the population do not have access to safe drinking water.

Sub-Saharan African Diaspora

Migration into Africa from other regions is overwhelmed by the immense African diaspora and emigration from Africa to other regions. Millions of black Africans were captured and sent as slaves, initially to the Middle East, but more significantly to the Americas, where their descendants represent a high percentage of the populations in Brazil, the Caribbean and the United States (see Figure 6.14). A second wave of emigration was associated with the aftermath of colonialism, a time when many Africans retained British Commonwealth passports or French citizenship and moved to Britain and France (or other Commonwealth countries such as Canada) in search of work (see Chapter 3, p. 115). This included many people living in the Caribbean, often the descendants of slaves, who then were part of a secondary migration back to England or to Canada and the United States. Another secondary migration, of Asians who had settled in East Africa, was associated with Ugandan dictator Idi Amin's decision in 1972 to evict all Asians from Uganda because of their dominant role in commerce.

Other recent emigrations from Africa include movements of white populations from South Africa and other countries to Europe, North America, and Australia, and a "brain drain" of 20,000 African intellectuals per year to universities and companies in the core regions of Europe and the Americas. There are also African refugee populations in several world regions, including large numbers of Ethiopians and Nigerians in the United States.

Contemporary migration within Africa is mainly associated with movements in search of work and with refugees fleeing famine, floods, and violent conflict (**Figure 6.25**). Labor migrations emerged during the colonial period when loss of traditional land to colonists and high taxes forced people to look for other work, and employment became available in mines and on plantations. For example, Sahelian residents migrated to work in peanut-, cotton-, and cocoa-producing areas in Senegal, Côte D'Ivoire, Nigeria, and Ghana, and in East Africa, Hutu, and Kikuyu to work on Kenyan and Ugandan coffee and tea plantations and European farms. The most significant labor migration of the last 100 years is from southern Africa, especially Botswana and Zimbabwe, into South Africa to work in the mining industry. In 1960 more than 350,000 foreign workers were employed in South Africa. These migrations have disrupted family life, but the remittances that are sent back by workers have become an important contribution to local and, in the case of countries such as Botswana and Lesotho, national economies. Labor migration also continues from inland West Africa to coastal cities such as Abidjan.

Geographer David Rain has also documented more traditional and long-standing **circular migrations** that respond

to seasonal availability of pasture, droughts, and wage employment. In Niger, many people move to regional centers such as Maradi in the dry season and return to their villages to plant crops when the rains begin (See A Day in the Life: Balkissa Ourmarou, Niger, p. 290). Pastoralists move their herds south in the dry season seeking water, pasture, and the possibility of grazing on harvested cropland or on wetlands. These migrations are a rational response to the spatial and seasonal variations in environmental conditions.

The United Nations High Commission on Refugees (UNHCR) estimated that as of January 1999 there were more than 6 million refugees in Sub-Saharan Africa, more than one-third of the world total and the largest number of any world region. UNHCR reports that 3.3 million were official international refugees—involuntary migrants who have crossed a national frontier—with the remainder displaced within their countries. War in Liberia and Sierra Leone drove more than 685,000 people into neighboring countries, and civil war and famine have forced more than 400,000 Somalis and 340,000 Eritreans from their homes (**Figure 6.26**). Other refugees include those from conflicts in Rwanda and Burundi. The refugee populations place serious burdens on neighboring countries that lack the re-

Figure 6.25   Interregional migration in Africa Africans are moving within and beyond Africa in response to employment opportunities. Major flows include those into the mining zones of southern Africa, the oil regions, and the export crop production zones of West and East Africa. Some of these are circular migrations for seasonal temporary work. Africans are moving from Africa to Europe, the United States, and the Middle East. (*Source:* Redrawn from S. Aryeetey-Attoh (ed.), *The Geography of Sub-Saharan Africa*. Upper Saddle River, NJ: Prentice Hall, 1997, p. 136.)

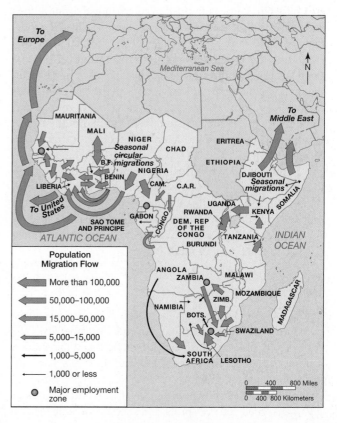

(a)

(b)

Figure 6.26   Refugees in Africa (a) Millions of Rwandans fled the country in the 1990s to camps such as this in Katale, Democratic Republic of the Congo. (b) The Horn of Africa has produced many internal and international refugees in the twentieth century. This family shelters in a camp at Las Dhure, Somalia, in 1981.

sources to feed and resettle the impoverished and starving arrivals. Guinea, for example, absorbed almost half a million refugees from Liberia and Sierra Leone, and Tanzania took in a similar number from Burundi. Most international refugees are housed in camps, supported by international organizations and charities. Disease spreads rapidly in the crowded conditions of the camps, and food supplies are sometimes interrupted or diverted by military groups and governments. Long-standing conflicts and loss of livelihoods at home mean that many refugees spend long periods in the camps with little hope of return. However, more peaceful conditions and carefully monitored repatriation have resulted in the return of refugee populations to countries such as Rwanda and Mozambique.

People forced to move within their own country, because they do not fall under international definitions or assistance for refugees, are some of the most desperate migrants. UNHCR estimates that there are 700,000 internally displaced people in the Democratic Republic of the Congo, most of them inaccessible to relief organizations.

Religion

Traditional African religions have been described as *animist* (worship of nature and spirits), but this overgeneralizes the wide variety of local religious beliefs in Africa. While natural symbols, sacred groves of trees, and landforms may have important religious significance, many African religions also feature a belief in a supreme being, several secondary gods or guardians, good and evil spirits, and ancestor worship. Ances-

tors, priests, or witch doctors mediate and interpret the wishes of the gods and spirits, and rituals ensure the stability of society and relations with the natural world. More than 70 million people (about 10 percent of the total population) are reported to practice traditional religions in the continent of Africa.

Christianity spread via North Africa and Ethiopia from about A.D. 300, but the pace of conversion accelerated rapidly under European colonial rule and European missionaries. Dutch Calvinists in southern Africa, Catholicism in French colonies, and Anglican beliefs in the British colonies all had strong influences. Of the 360 million estimated Christians in all of Africa, there are about 125 million Roman Catholics and 114 million Protestants. Islam is another major religion in Africa as a whole with 308 million adherents, and it is predominant in the Sahel, North Africa, and some East African coastal communities. It was spread by traders and drew some fierce defenders among West African groups, such as the Fulani, who went to war to eradicate animistic beliefs. As in Latin America, traditional religion has blended with Christianity and Islam to create forms in which local traditional rituals are incorporated into Christian services. Another parallel to Latin America is the recent rapid spread of evangelical Christianity in many regions of Africa.

Religious differences have fueled political conflict in some regions of Africa, most notably where Muslims and Christians were forced into the new colonial national boundaries. In West African countries bordering the Sahel, such as Nigeria, tensions exist between northern Islamic groups, such as the Hausa, and southern Christians.

Language and Ethnicity

The geography of languages in Africa is incredibly complex, with more than 800 living languages, 40 of them spoken by more than 1 million people (**Figure 6.27**). The dominant indigenous languages, spoken by 10 million or more, are Hausa

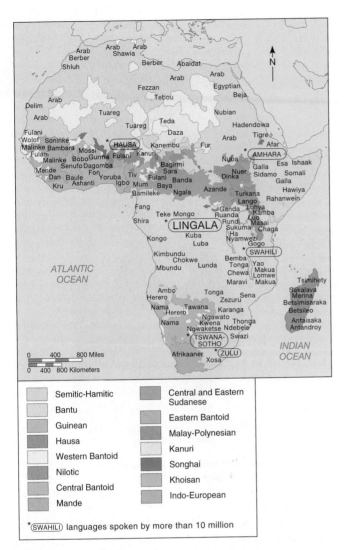

Figure 6.27   Language map of Africa The cultural complexity of Africa is clearly demonstrated in the variety of indigenous languages shown on this map. Swahili and Hausa are spoken by millions as the trade languages of East and West Africa, respectively. (*Source:* Redrawn from S. Aryeetey-Attoh (ed.), *The Geography of Sub-Saharan Africa.* Upper Saddle River, NJ: Prentice Hall, 1997, Fig. 4.6.)

in East Africa; and the largest Niger-Congo group that includes Hausa, Yoruba, Zulu, Swahili, and Kikuyu. A small family are the Khoisan languages spoken by the bushmen of southern Africa, which have a distinctive "click" vocalization.

The multiplicity of languages and dialects reflects the large number of distinct cultural or ethnic groups in Africa. Some writers use the term *tribe* to define these groupings and describe Africa as a *tribalist* society. The term *tribe* is used to describe group affiliation based on shared kinship, language, and territory, and while it is used by many groups to identify themselves, other groups see the term as negative (related to colonial perceptions of savagery) and now prefer to use the term *ethnic group*.

The largest ethnic groups in Africa are associated with the dominance of certain languages, such as Hausa, Yoruba, and Zulu, but almost all groups, however large or small, were either split geographically by colonial national boundaries or grouped together with their neighbors, enemies, or others with whom they shared no affinity. Attempts to consolidate ethnic groups across boundaries and struggles for power between groups within countries are a major cause of conflict in contemporary Africa.

For example, tensions between the Ibo and Yoruba in Nigeria led to civil war when the Ibo declared the independence of eastern Nigeria as Biafra in 1967. The conflict, which drew international attention and intervention because of starvation in Biafra and the presence of oil in the region, resulted in as many as a million deaths (mainly from hunger and disease) and lingering ethnic resentments after Nigeria was reunited. Another evident tension is between the ruling Kalenjin and other ethnic groups in Kenya, where opposition political parties have organized around ethnicity and threaten violence in the face of perceived election corruption and bias toward Kikuyu regions and individuals. Ethnic rivalries also play a role in several other regional conflicts discussed later in this chapter.

Culture

It is hard to draw cultural generalities from a continent as large and diverse as Africa. Those who do generalize highlight the importance of the extended family, ties to the land, oral tradition, village life, and music in traditional African culture. The importance of the extended family is linked to the supremacy of kinship ties in social relations and obligations and to a widespread respect for elders as sources of wisdom. Kinship ties going back multiple generations in the same region may define "clan" allegiances and may drive primary loyalties, as in contemporary Somalia, where interclan conflict has dominated recent political events. Different extended families are often linked through intermarriage, with a transfer of wealth, sometimes in the form of cattle, from the husband to the wife's family as a mark of respect and value of the woman's labor and companionship.

The tie to the land is connected to traditional forms of land tenure in Africa, where in many regions land was viewed as given by the spirits or held in trust for ancestors and future generations. The Elesi of Odogbolu (a traditional leader) in

(the Sahel), Yoruba (Nigeria), Ibo (Nigeria), Swahili (East Africa), and Zulu (southern Africa). Hausa and Swahili are known as trade languages (or *lingua franca*), spoken as second languages by many groups to facilitate trade. English, French, Portuguese, and Afrikaans are also spoken in regions of recent colonial control and education systems or white settlement, and Arabic is common toward northern Africa. Arabic has strongly influenced Swahili along the east coast of Africa. Because most countries have no dominant indigenous African language, they have often chosen a European language for official business and school systems. The countries with the most coherent overlap between their territory and a dominant African language are Somalia (Somali), Botswana (Tswana) and Ethiopia (Amharic).

The indigenous languages of Africa can be grouped into larger language families, including the Afro-Asiatic languages of North Africa, including Somali, Amharic, and Tuareg; the Nilo-Saharan languages that include Dinka, Turkana, and Nuer

A Day in the Life

Balkissa Oumarou, Niger

Balkissa Oumarou (**Figure 1**) is a schemer and one who truly "eats" the dry season. She is able to do so because she has an alternative source of survival income waiting in reserve. Balkissa does not have small children, and this enables her to move much more freely. Balkissa migrates to the West African coast, selling teething medicine for babies and dressing hair. This is her sixth seasonal migration to the coast. She traveled with a group of about twenty-five Bororo women from Maradi to Cotonou, Benin, by way of Dosso in southwestern Niger (**Figure 2**). In her words, "Getting to Cotonou took about ten days, but once we got there we stayed two months. Lomé was a bust— we only spent a day there and it was difficult—so we left. We took a bus to Accra [Ghana] and spent the next two months in the same compound." Like her husband, Balkissa grew up about twenty-five kilometers north of Dakoro. She is not formally educated. Her parents were always in motion, taking their large cattle herds north when the rainy season began and back down south in the dry season. Asked if her parents ever traveled the way she does now, to Cotonou and Accra, she replies that they did and that as kids she and her brothers and sisters went along, leaving the animals with other family members up north near the water points. I ask if much has changed in the environment of Dakoro since her childhood. "There is not as much land for animals, and the wells need to be dredged," she says. During the droughts, even the rich become poor." Her migrations afford Balkissa and the other Bororo women a chance to escape the harshness of the herding life. "As soon as we arrive in a new city, we all look for a place to stay, where we can leave our

Figure 1   Balkissa Oumarou Balkissa Oumarou sits with some of her belongings in Maradi, Niger.

things." It should be pointed out that they do not take very many things. She makes some money on these migrations. "Well, we eat and we save a little." The group left around late January and returned in May. Many Bororo women go on seasonal migrations like this every year now. Balkissa feels that the

---

Nigeria expressed this view in these words: "The land belongs to a vast family of which many are dead, few are living, and countless members are still unborn." This gives land a communal nature such that it could not be bought or sold by individuals. In some cases, land rights are held by the extended family or the community rather than by the individual, and in other societies the chief or king controls land. For example, in Ghana, some land rights are vested in the "stool," the seat of the Ashante leader. Because many of these land-tenure traditions were set aside during colonialism—and the land was taken over by colonial settlers or reallocated by administrators—after independence, the governments of some countries recovered this land and kept it as state land.

Traditions of reciprocity, where a gift is given in order to obtain a favor, and of helping family members are a major source of cultural confusion, according to African historian Ali Mazrui. He suggests that these traditions provide an ex-

planation for the way in which some leaders have favored family members with jobs in their administrations and for the use of bribes when making requests from government officials. He also notes that under colonialism, local residents viewed stealing from the government as a legitimate form of resistance because they felt that foreigners were robbing Africa of resources and funds through taxation.

Africa is associated with a rich tradition of music and the arts that has increasingly influenced other regions of the world. Slavery was one way in which African musical, artistic, and food customs spread around the world, especially to the Americas. Traditional music of the West African Sahel may have influenced the development of American blues, and West African coastal traditions influenced Afro-Caribbean music styles such as the Cuban rumba (**Figure 6.28**).

Africa is often associated with music from percussion instruments, especially drums. Other instruments with metal

Bororo women get business because they look unusual, and people are attracted to them and buy what they are selling. She believes that her husband Boubé could see the advantages of the migration as well as she could. Balkissa has an infectious sense of the possibilities that migration can afford. She brings back intangibles, like knowledge and news. She has seen the ocean, and she has found a place where she could enjoy herself and recover her health. Although she did not know anyone in Accra prior to going there, the next time she returns she will have many contacts. She learns a little more each time. Next year she plans to travel to Cameroon, and she says there is now a bus that goes directly from Kano to Douala. Of the money Balkissa earns on her migration, she plans to give about half to Boubé and a small amount to her father. She will save some for emergencies and keep the rest for herself to spend on clothes, condiments, and food.

She says, "If people migrate, they have the chance of making some money—and they can help somebody." Balkissa's migratory habits are significant on several levels. As a woman with the freedom to go where she likes and do what she pleases, she is unusual in the region. Balkissa has turned her own ingenuity into a valuable resource. Though not formally educated, Balkissa is what is called in Hausa a *matan zamani*, or "modern woman." In demographic terms, she has fewer children than the regional norm, and she is likely to have exercised more choice over her reproduction than many other women. In political-economic terms, although Balkissa as a female migrant is invisible to development efforts, she is critical to them because she has both escaped the bonds of the rural household and helped stretch norms. In environmental terms, she has

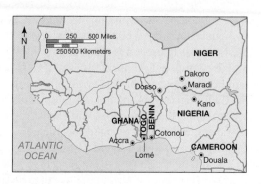

Figure 2   Circular migration   Map of Balkissa's migrations.

employed flexibility and knowledge to adapt to changes both in the grazing lands her family occupies and in the network of West African cities. In social terms, Balkissa has skirted oppression by moving out of the vise that rural life can be, and she has used mobility to gain access to the world and to satisfy her needs. For demographic, political-economic, environmental, and social reasons, Balkissa is a model for future Nigérien women. Women have shouldered much of the burden of the changes occurring in the rural sector, where population growth compounded by environmental change and political-economic change have presented a profound challenge for the future of the region. Women desperately need a model, and they need to move into spaces where they can make decisions governing their lives.

Source: D. Rain, *Eaters of the Dry Season: Circular Labor Migration in the West African Sahel.* Boulder: Westview Press, 1999, pp. 183–185.

(a)

(b)

(c)

Figure 6.28   African instruments and musicians   (a) San women playing traditional metal-keyed instruments called *lamellaphones*. (b) South African singer Miriam Makeba, who lived in exile during the apartheid era. (c) Nigerian musician King Sunny Adé.

keys that are plucked or tapped, as well as flutes and harplike stringed instruments, are also commonly used in traditional and popular music. Such musical traditions vary widely across the continent and include the complex rhythms of women drummers in Tanzania, of xylophone players in Uganda, and the chanting of the Zulu of southern Africa. In West Africa, oral traditions are associated with the singers and storytellers, some of whom receive the respected name of *griots*.

African popular music mixes indigenous influences with those of the West, especially those of the United States and the Caribbean. For example, the *highlife* music of West Africa derived from Caribbean calypso and military brass bands, adding stronger percussion, soul influences, and exchange between lead and background singers to emerge as the now internationally popular *juju* or Afrobeat sounds of Nigeria's King Sunny Adé and Fela Kuti. In French West Africa, singers such as Youssou N'Dour blend traditional African beats with powerful vocals, and in South Africa, singers such as Miriam Makeba received international recognition in the 1950s, presaging the popularity of a cappella South African black musicians such as Ladysmith Black Mambazo. Popular musicians have often expressed political opinions against apartheid or corruption.

African art is also incredibly varied and includes painting, metalwork, and sculpture (**Figure 6.29**). Artists were seen as important specialists in traditional Africa, often under the patronage of kings and producing works of spiritual value. The masks and wood sculptures of the Dogon and Bambara peoples of West Africa are now collected around the world while maintaining cultural significance within the region. *Kente cloth* designs from northern Ghana have become meaningful in African-American identity and clothing. As interest in travel and world culture has grown, artists and others have started to produce for sale to tourists and to international distributors, including some organizations that try to ensure fair trading principles of returning as much value as possible to local people.

Figure 6.29 **African arts and crafts** Traditional Kente cloth from northern Ghana has become popular worldwide, especially in North America.

Regional Change and Interdependence

Contemporary Africa faces challenges as a world region and as a collection of diverse countries and communities. The most important issues include issues of political stability, economic and social inequality, agricultural capacity, environmental conservation and degradation, and health (especially AIDS). Intertwined among these issues are interregional questions about economic integration, refugee flows, and development strategies.

Politics and Peace

The search for peace in Africa has been frustrated by the legacies of colonial frontiers and the Cold War, ethnic rivalries, and the special interests of powerful individuals and sectors. While some countries were able to create (or re-create) a sense of national identity following independence, others are still coping with internal struggles, contested nationalisms, and claims on land beyond their current borders. Although the end of the Cold War and the end of apartheid in South Africa opened new prospects for peace and cooperation, there are continuing wars and precarious coalitions in several regions of Africa. Geographers Ian Yeboah and Samuel Aryeetey-Attoh identify multiple causes for continuing political instability in Africa, including ethnic conflict, poor leadership, outside interference, and the legacies of recent independence struggles and racist government. They note the frequency of military coups with Ghana, Nigeria, and Uganda all experiencing at least five coups since independence just under 50 years ago. They are also concerned about the number of elected leaders who eventually drifted toward one-party states and dictatorships, with accompanying repression and restrictions on freedom of speech. Yeboah and Aryeetey-Attoh do see signs of optimism emerging in the political geography of Africa, as many countries moved toward democratic elections in the 1990s and as political and ethnic tensions were reduced with the end of the Cold War and of apartheid in South Africa. They and others point to the relative success of Botswana, where after independence, government structure included a strong role for traditional chiefs and public input into government. Economic growth, especially in the diamond and beef industries, the lack of civil unrest, and relatively democratic decision making have all contributed to improved social conditions, with literacy at 70 percent and per capita incomes averaging $3000.

However, the scars of conflict and genocide are still raw in Rwanda and Burundi, where tensions between the majority Hutu and powerful Tutsi peoples erupted into civil war in 1994. The ethnic divisions between these two groups were created or exacerbated by the Belgian colonists, who gave the Tutsis control over the Hutus, who were mostly peasant farmers. The Tutsis received education, training, and other benefits, while the Hutus were taxed heavily and given few privileges. The rapid withdrawal of Belgium resulted in a Hutu majority government in Rwanda and a population who harbored resentment against the Tutsi minority (about 20 percent of the population of the two countries). In Burundi, the Tutsis maintained power until the late 1980s, but with several internal military coups and severe repression of Hutu uprisings. Both countries have predominantly agricultural economies that are dependent on exports of coffee and tea. Prior to the outbreak of civil war in 1994, Rwanda was beginning to benefit from environmental tourism associated with endangered populations of mountain gorillas in the forests, and this was reducing the threat of deforestation as local people began to perceive benefits from conservation. Tensions increased after 1989, when international market prices for coffee dropped precipitously, and when Tutsi rebels, with the aid of Uganda, invaded Rwanda in 1990.

Burundi finally moved toward multiparty and multiethnic government in 1993, but the new president, Melchior Ndadaye, was killed in a coup, and subsequent ethnic violence killed more than 200,000 and sent 800,000 refugees into neighboring countries. When in 1994 a plane crash killed the presidents of Burundi and Rwanda (Cyprien Ntaryamira and Juvenal Habyarimana), some Hutus blamed Tutsis and moderate Hutus and initiated a massacre in which more than 500,000 died in Rwanda and many Tutsis fled to neighboring countries where rebel forces were organized. When these Tutsi rebels won control of Rwanda and Burundi, thousands of Hutus fled to avoid retribution. Two million refugees ended up in Zaire (now the Democratic Republic of the Congo) and thousands in Uganda, Kenya, and Tanzania. The challenge of reconciliation in these two countries is enormous because the memories of violence are so fresh and the divisions so deep. The international community has been accused of ignoring signs of imminent massacres and then delaying their response to the misery of refugees. As of 2000, considerable energy was being focused on reconciliation and peace, with attempts to form multiparty and multiethnic governments, thanks to the mediation of respected leaders such as Nelson Mandela of South Africa and Julius Nyerere of Tanzania.

Other conflicts seem even harder to resolve. In Liberia and Sierra Leone civil wars produced conditions of anarchy in the late 1990s. The economies of these two countries depend on a very narrow range of exports with volatile prices—rubber in Liberia and cocoa in Sierra Leone. The resettlement of freed slaves from other regions who saw themselves as an elite has also caused conflict with the indigenous residents, and there is also considerable resentment of urban wealth by rural residents. In the 1990s, struggles for power between opposition power groups within the countries were fueled by arms and capital obtained through the sale of diamonds, with many ordinary people fleeing as refugees (see Geographies of Indulgence, Desire, and Addiction: Diamonds, p. 260).

The cost of wars in Sub-Saharan Africa has been a hindrance to investments in other forms of development. The Stockholm International Peace Research Institute reports that military expenditures by governments in Sub-Saharan Africa ranged between $6.6 and $9.5 billion in the 1990s and that many governments were spending more on arms and the military than on education or health.

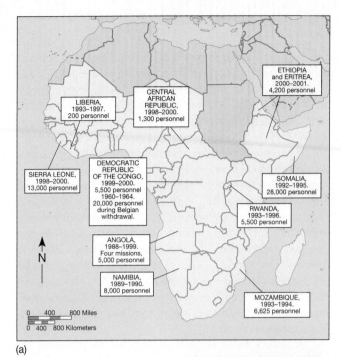

(a)

(b)

(c)

Figure 6.30   U.N. peacekeeping in Africa (a) This map shows the dates and locations of U.N. peacekeeping missions in Sub-Saharan Africa. (b) Photo of the U.N. peacekeeping mission in Sierra Leone, 2000. (c) Kofi Annan of Ghana, U.N. secretary-general. In October 2001, Kofi Annan and the United Nations were awarded the 2001 Nobel Peace Prize.

These conflicts are occurring within the context of a new post-Cold War international political geography in which the United Nations, African regional security forces, and the United States are all playing new roles. The United Nations Peacekeeping Forces operate under the authority of the United Nations Security Council to help establish and maintain peace in areas of armed conflict with the permission of disputing parties. In Africa, U.N. forces, with their distinctive pale blue helmets, have been deployed in Angola, the Democratic Republic of the Congo, Sudan, Ethiopia, Eritrea, Rwanda, Namibia, Somalia, Sierra Leone, Liberia, and Mozambique (**Figure 6.30**). The success of these missions has been mixed after initial success in monitoring transitions to peace in Mozambique and Namibia. U.N. forces failed to prevent massacres in Rwanda and Sierra Leone, were unable to establish peace in Somalia and the Democratic Republic of the Congo, and had inadequate human or financial resources to sustain several operations. The United States has been reluctant to send troops to Africa after negative publicity about its involvement in Somalia, and U.N. forces are increasingly composed of soldiers from poorer countries. The United States has been widely criticized for allowing foreign aid to be used for arms sales and for refusing to ban the sale of military hardware, including land mines. African leadership—especially from South Africa and Nigeria—in promoting peace within the region is of growing importance, including negotiations led by former South African president Nelson Mandela, and a West African peacekeeping force and monitoring group, ECOMOG, led by Nigeria under the auspices of the Economic Community of West African States (ECOWAS). The current head of the United Nations, Kofi Annan, is from Ghana and is particularly concerned with improving conditions in Africa.

ECOWAS is one example of programs for economic integration and political cooperation in Sub-Saharan Africa, established in 1975 to promote trade and cooperation with West Africa. ECOWAS includes the countries of Benin, Burkina Faso, Cape Verde, Côte d'Ivoire, Gambia, Ghana, Guinea, Guinea-Bissau, Liberia, Mali, Mauritania, Niger, Nigeria, Senegal, Sierra Leone, and Togo. African integration was the dream of several independence leaders, most notably Kwame Nkrumah of Ghana, who called for a Union of African States in his famous 1961 speech, "I Speak of Freedom." Nkrumah believed that only by joining together could independent Africa reach its full potential. Unable to convince others that complete unity was desirable, Nkrumah was able to lead the establishment of the Organization of African Unity (OAU) in 1963. The OAU, based in Addis Ababa, Ethiopia, promotes solidarity among African states, the elimination of colonialism, and cooperative development efforts. It was successful in mediating boundary disputes between Ethiopia and Somalia in the 1960s and in pressuring for the end of the apartheid regime in South Africa.

In southern Africa, economic integration has been promoted since 1979 through the Southern African Development Community (SADC), which promotes trade and development coordination, especially the improvement of transport links. Members include Angola, Botswana, Lesotho, Malawi, Mauritius, Mozambique, Namibia, Swaziland, Tanzania, Zambia, and Zimbabwe. South Africa finally joined in 1994 with the advent of black majority rule in South Africa. Similar regional programs have included the East African Cooperation (EAC) and East African Economic Union between Kenya, Uganda, and Tanzania, and the larger Common Market for East and Southern Africa (COMESA).

Social and Economic Inequality

The roots of political unrest in Africa also lie in the large regional and social inequalities in many countries. Sub-Saharan Africa ranks low on many measures compared to other world regions (see World Data Appendix, p. 639). Life expectancy averages 20 years below the world average of 70 years and is lower than any other world region, and GDP per capita is about $1600 compared to a $6500 world average. The United Nations estimates that 42 percent of Sub-Saharan Africans are poor as measured by income and 40 percent according to the more general measure of human poverty that combines life expectancy, literacy, and access to basic services such as clean water. Conditions in Africa are also difficult for children, who have an infant mortality rate of 106 deaths per 1000 children born (double the world average) and with low levels of child nutrition and immunization.

Within Africa, these generally gloomy average statistics do hide some regions with much better conditions. For example, the United Nations ranks Mauritius, South Africa, Swaziland, Botswana, Gabon, Ghana, Zimbabwe, and Lesotho higher on the Human Development Index (life expectancy, literacy, education and GDP per capita) than other countries of Africa. Life expectancies in these countries are generally above 50 (except for Botswana and Zimbabwe, where AIDS has caused life expectancy to drop), and levels of literacy, education, and income are also higher than the regional averages. These countries also tend to have better provision of basic services. For example, in South Africa, Zimbabwe, and Botswana, 75 percent or more of the population have access to safe drinking water. In almost all countries life expectancy, incomes, and services are better in urban areas than in rural ones.

Generally, conditions have improved over the last 25 years. For example, GDP per capita has increased from $780 in 1970 to $1590 in 1998, life expectancy has increased from 45 to 49 years, and infant mortality has dropped from 138 to 106 deaths per 1000 children born. Literacy has shown dramatic changes, increasing from 38 percent in 1980 to 61 percent in 1999.

Improvements in conditions in specific countries are reflected in life expectancy and infant mortality changes. In Ghana, for example, life expectancy increased from 45 to 60 years and infant mortality decreased from 131 to 65 deaths per 1000 children. But war and AIDS in have also had an effect in Africa. For example, life expectancy in Botswana and Zambia started to drop in the 1990s as a result of AIDS and in Rwanda as a result of war and genocide. In 1998 AIDS was also mostly responsible for the increases in infant mortality in Botswana, Kenya, Zambia, and Zimbabwe.

The lowest 24 countries on the Human Development Index are all in Africa. Sierra Leone, for example, has a life expectancy of only 37 years and an annual GDP per capita averaging $458, mainly as a result of the loss of life and economic collapse associated with civil war. Low levels of life expectancy reflect some of the deficiencies in service provision in Africa, where an average of 46 percent of the population lack access to safe drinking water and 52 percent lack sanitation. Two-thirds of the population lack safe drinking water in Ethiopia, Angola, Democratic Republic of the Congo, and Sierra Leone.

Economic and social conditions show great geographical and social variation within Africa countries. The core locations—urban areas, the formal sector, and producers of cash crops—generally have longer life expectancies, better service access, and higher incomes than the periphery, which includes rural areas and the informal and subsistence agricultural sectors.

Income concentration is high in many parts of Sub-Saharan Africa, with the richest 20 percent of the population receiving more than 60 percent of overall income in most of southern Africa, the Central African Republic, and Sierra Leone. More than 50 percent of the population earn less than $1 a day in most of West Africa, except Côte d'Ivoire.

Gender Differences

Women in Africa also tend to have less education and lower incomes than men but live slightly longer in most countries. The average female annual income is about $1150 less than men, and female literacy rates are 52 percent for women compared to 68 percent for men. Gender differences in Africa demonstrate the process of the **"feminization of poverty,"** whereby more than two-thirds of all people who join the ranks of the poor are women. Women are more likely to be poor, malnourished, and otherwise disadvantaged because of inequalities within the household, the community, and the country. Women are less likely to receive an education, and overall pay rates are lower in the workplace. Patriarchy and cultural traditions mean that women may be required to eat less than men and only after the men in the family have eaten, and women bear disproportionate responsibility for heavy household work such as collecting fuel wood and water (**Figure 6.31**). The tradition of female circumcision has become a controversial struggle between those who see it as a human rights abuse involving brutal genital mutilation and health risks and others who see it as an important religious and symbolic experience. More than 100 million women in Africa are estimated to have undergone the ritual.

Women are also disadvantaged by many traditional and modern institutions that define property rights. Land may be passed on only to male children, and new land titles are often granted to male household heads. Development policies for agricultural training and technology to make work easier have been directed at men, and new projects for tree planting or cash crops often focus on men. Although some African governments and development agencies have recognized these disparities and established programs meant to improve conditions for women, poverty reduction among women has been patchy and partial.

There are places in Africa where significant numbers of women are seen as powerful, especially in urban areas where female entrepreneurs have been successful. The best-known examples are the women who control the markets in Ghana and the women who control the cloth trade in Togo—women who are called the Mama Benz because of the expensive cars that they own.

Land Tenure

The allocation of land is also very skewed in many African countries, with a few people owning large areas of better land. Some countries have implemented extensive land reforms to try to increase productivity and quell social unrest.

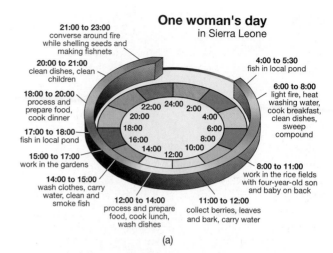

(a)

(b)

Figure 6.31 Women's work in Africa (a) This graphic from the United Nations Food and Agricultural Organization shows the typically large burden of work carried by many African women. (b) Woman drawing water from a well in Mopti, Mali. [*Source:* (a) "One Woman's Day in Sierra Leone," Food and Agricultural Organization of the United Nations, 1997. Available at http://www.fao.org/NEWS/FACTFILE/FF9719-E.HTM]

The problems of unequal distribution of land are dramatized in the case of Zimbabwe, where at independence in 1980 President Robert Mugabe promised to redistribute land owned by more than 5000 mainly white farmers (**Figure 6.32**). Twenty years later these large farms still occupy about 70 percent of the best land, and have large areas lying fallow or in ranches.

According to government figures, some 4400 whites own 32 percent of Zimbabwe's agricultural land—around 10 million hectares (almost 25 million acres)—while about 1 million black peasant families farm 16 million hectares (almost 40 million acres) in plots mostly less than one hectare. Although title to the lands was given to white settlers by the colonial

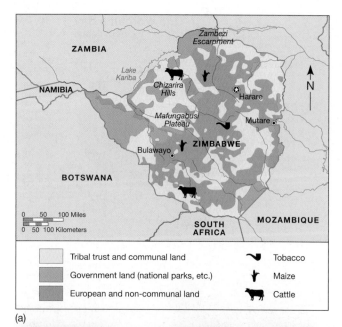

(a)

(b)

Figure 6.32   Land conflicts in Zimbabwe (a) In Zimbabwe, white farmers retained the better cropland after independence growing tobacco and other cash crops. Most of the crop area highlighted in yellow was in white ownership. (b) Expectations were high for land reform under black majority rule, and squatters have invaded white farms with the tacit support of the government in some cases. [*Source:* (a) Map from http://europe.cnn.com/interactive/world/0008/farm.map/land.html]

governments, many black Africans believe that these lands were seized unfairly and should be returned to the indigenous owners. In 2000, Mugabe's threat to confiscate the land was accompanied by invasions of squatters onto more than 500 farms, and by international and internal attempts to resolve the conflicts by compensation for land that is transferred. Critics of Mugabe argue that previous land redistribution failed because peasant farmers did not have the knowledge or resources to succeed in commercial and export-oriented farming and that many farm workers are at risk of losing their jobs. But resentment at colonial land expropriation and racially based land policies is widespread, and several other African governments supported Mugabe's policies. New post-apartheid

governments in South Africa have faced similar challenges in deciding how to provide better land to black peasant farmers without provoking serious conflicts with long-standing white landholders who control the land. More than 50,000 households have acquired land through government subsidies in South Africa, in some cases through buying shares in farms where they were formerly workers.

Development Debates

Africa provides an arena for struggles over development and the case studies for some of the best-known failures and successes of development. As discussed earlier, Sub-Saharan Africa was a focus of modernization in the pre- and post-independence periods, with many development projects that sought to transfer northern technology into Africa, including large dams and Green Revolution technologies. African countries such as Nigeria and Zambia then sought to reduce their dependency on expensive imports by seeking to develop national manufacturing industries and protectionist policies. In the most recent era of development thinking, Africa has shifted to the more neoliberal policies of free-trade-guided demands for economic restructuring from international agencies and financial and trading partners such as the United States.

Within these broad strategies, Africa provides examples of many less obvious shifts and debates in development policy, including those about the role of women, indigenous technology, credit, and social organizations. An early call for more attention to the role of women in Africa was that of economist Esther Boserup, whose 1970 book *Women's Role in Economic Development* documented the importance of women as farmers and resource managers in Africa. Geographers, among others, subsequently documented the work of African women in providing the basic needs of families through food preparation and collecting wood and water; as income generators in crafts and community work; and as agricultural producers in both subsistence and commercial sectors. They also showed distinctly gendered spaces in the African landscape, with parts of the home, the market, and certain trees and crops reserved primarily for women's lives and work. A 1996 study by the United Nations Food and Agriculture Organization (FAO) found that women's contribution to the production of food crops ranges from 30 percent in Sudan to 80 percent in the Republic of the Congo and that women are responsible for 70 percent of overall food production, 100 percent of food processing, 50 percent of animal husbandry, and 60 percent of agricultural marketing (see Figure 6.31).

Those who understood the importance of women's work in African communities and economies criticized development policies that ignored, undervalued, or displaced women. They also showed how women were often disproportionately affected by environmental degradation as deforestation and drought made the work of collecting wood, water, and food more difficult.

Development policies gradually began to incorporate these ideas. The "Women in Development" (WID) approach focused on women's productive roles with projects that provided technology, credit, and training to women. This approach was criticized for ignoring women's reproductive roles and the larger social processes such as discriminatory land-tenure policies that shape women's lives. In the 1980s the **gender and development (GAD)** approach was promoted as better linking women's productive and reproductive roles and trying to understand the gender-related differences and barriers to better lives of both men and women. Development agencies such as the World Bank incorporated elements of both approaches into programs that supported education, credit, and land-titling programs for women, women's organizations, and recognition of women's work.

Another trend in development thinking in Africa is the promotion of **microfinance programs,** which provide credit and savings to the self-employed poor, including those in the informal sector, who cannot borrow money from commercial banks. Based on the demonstrated success of the Grameen Bank program in Bangladesh (see Chapter 11, p. 556), which provided small loans to thousands, African microfinance projects offer loans and secure savings opportunities to people who want to start or expand their businesses. Examples include loans to purchase sewing machines, food-processing equipment, agricultural supplies, and shop inventories.

Governments and development agencies have also realized that development requires more than purely financial capital. Projects seem to be most successful in communities that have high levels of **social capital**—networks and relationships that encourage trust, reciprocity, and cooperation that share and expand on initial investments.

Agriculture and Development

Another shift in development thinking has been the increasing appreciation of *indigenous technology,* especially in agriculture. Researchers such as Paul Richards and Robert Chambers documented how colonial administrators and development institutions systematically devalued local knowledge, substituting imported European ideas about the appropriate management of soil and forests. Case studies showed how the substitution of new single-crop permanent systems in place of traditional shifting multiple cropping resulted in increases in soil erosion, pest damage, and nutrient losses. These criticisms also emerged in attempts to transfer the Green Revolution packages of improved seeds, irrigation systems, and agricultural chemicals to Sub-Saharan Africa. Many farmers could not afford the new technologies or wished to grow basic crops such as millet, yams, and sorghum for which improved varieties had not yet been developed. However, others, in Zimbabwe for example, were able to increase their yields of maize and to benefit from the new technologies. New approaches to agricultural development take more account of local expertise, asking farmers how they cope with climate variability and soil management. Local knowledge has also been valued in health programs where indigenous healers can make important contributions to the treatment of disease.

Sustainable development (see Chapter 1, p. 29) is a priority for Africa, because past development projects have caused serious environmental problems and because so many people depend directly on renewable resources, especially the productivity of agriculture, and on safe water supplies. The definition of sustainable development in Africa has ranged from narrow criteria of development that does not damage the environment for future generations or that is economically efficient to broader conceptions of development that is equitable and ecologically beneficial.

The challenges of agricultural development are an important theme across all of Sub-Saharan Africa, where most of the population still work in agriculture, thereby providing the foundation of food supplies and export earnings. Agriculture employs about 60 percent of the labor force in Africa and produces more than 40 percent of the regional GDP and up to 95 percent of the export revenue in some countries. Per capita agricultural production has fallen in the last 20 years because growth in agricultural production has not kept up with the growth in population and increased demand from urban populations. Of even greater concern is the fact that the benefits of agricultural progress have not raised the incomes or improved the nutrition of many of Africa's residents. As discussed earlier in this chapter, many regions have become dependent on food aid. Exports have declined relative to imports (**Table 6.1**).

Blame for agricultural problems in Sub-Saharan Africa has been attributed to environmental degradation, lack of infrastructure, government policy, and international market structures. While several of these factors are discussed later in this chapter in the context of particular regions, region-wide challenges, identified by geographer Godson Obia, include improving infrastructure for roads and storage and providing adequate incentives and rewards to local producers. Difficulties of getting crops to the market on Africa's dirt roads and tracks, especially in the rainy season, mean that farmers risk having to store grain and other products in granaries that are vulnerable to pests and molds.

Governments have controlled food prices to keep wages and unrest down in urban areas, and this has reduced the prices paid to farmers. In Zimbabwe, maize (corn) prices paid to black farmers were suppressed in order to benefit the mostly white commercial maize producers, and farmers were also banned from planting crops in wetland areas. When these restrictions were removed, maize production in Zimbabwe soared to levels that allowed export to neighboring Mozambique. The marketing boards established by colonial powers have smoothed out price fluctuations, but they have also kept the profits when world prices are good. International market volatility and a lack of information has made it difficult for farmers to move into new types of crops, and the general decline in the price of agricultural products in comparison to needed imports has also made it difficult to make a living in agriculture.

Parts of Kenya have also seen increases in agricultural production and reductions in soil erosion that contradict the often gloomy picture of African agriculture (see Geography Matters: More People, Less Erosion, in Machakos, Kenya, p. 300).

Geography Matters

More People, Less Erosion, in Machakos, Kenya

In the nineteenth century, when Thomas Malthus argued that population was growing faster than food supplies, he established a basis for what is now termed *Malthusian thinking* about the relationship between population growth and environment. Malthusians tend to argue that rapid population growth is associated with resource scarcity and environmental degradation, including soil erosion, deforestation, and food shortages. Overpopulation occurs when there are too many people for the carrying capacity of the environment and people are forced to overexploit resources or suffer hunger and destitution. This Malthusian specter has frequently been associated with Africa, and especially with famine and soil erosion in drier regions.

The case of Machakos, Kenya, has been used to argue against this negative view of population growth by researchers who found an improvement in environmental and economic conditions during a period of rapid population increase. The population grew fivefold from 240,000 in 1932 to 1,400,000 in 1990 (**Figure 1**). The Machakos district is about 50 kilometers (30 miles) southeast of Kenya's capital Nairobi and stretches another 300 kilometers (180 miles) halfway to the coast. Rainfall is variable, soils have low fertility, and the hilly topography creates significant risks of soil erosion. Under British colonial administration, the local Akamba people were contained in a small reserve to allow white farmers to use the region for grazing beginning about 1930. This resulted in more intense use of land by locals and the establishment of the Soil

Figure 1   Population changes in Machakos, Kenya   The location of the Machakos district, and two maps that show large increases in population density in the district from 1932 to 1979. (*Source*: Redrawn from M. Mortimore and M. Tiffen, "Population and Environment in Time Perspective: The Machakos Story," in T. Binns (ed.), *People and Environment in Africa* (pp. 69–89). New York: John Wiley & Sons, Fig. 7.3.)

Figure 2   Agricultural change in Machakos, Kenya   These maps show how sloped land was converted to terraces in the Masii area of the Machakos district between 1948 and 1978. (*Source*: Redrawn from M. Mortimore and M. Tiffen, "Population and Environment in Time Perspective: The Machakos Story," in T. Binns (ed.), *People and Environment in Africa* (pp. 69–89). New York: John Wiley & Sons, Fig. 7.11.)

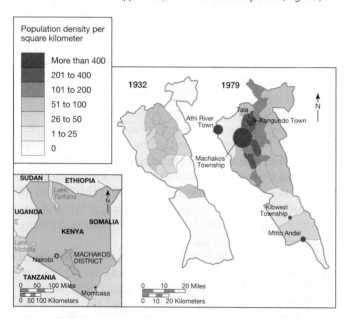

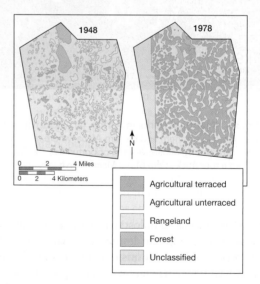

but only for South Africa, Zimbabwe, and Namibia and for carefully controlled sales to Japan.

The rhino is under much greater threats, especially from poachers who hope to sell rhino horn for dagger handles in the Middle East, especially in Yemen, and for highly valued medic-inal powders in Asia. Protecting the rhino from poachers who can make thousands of dollars from selling a horn is a full-time and costly enterprise. Many rhinos now have their own body-guards or have their horns regularly removed and replaced by bright plastic horns to reduce their attractiveness to poachers.

Conservation Service by the colonial government to promote conservation and prevent erosion. Compulsory labor, often by women, was required to build terraces and plant grass on hill slopes. Locally organized community work teams replaced compulsory labor in the 1950s, and the technique of terracing spread rapidly as local people observed the better soil and moisture and associated greater yields on the terraced fields. In some parts of the district, almost all of the farmland was terraced by 1980, encouraged by support from European governments and charities (**Figure 2**).

Production also benefited from the adoption of ox-drawn plows and from the fertilization by manure from livestock mostly fed fodder while kept in stalls rather than allowed to roam freely. People seem to have maintained and replanted trees while using dead wood and hedge cutting for fires.

Agriculture changed in the district from one based on subsistence cultivation of maize, beans, and pigeon peas to a more diverse crop mix that included coffee and, more recently, fresh vegetables for sale in Nairobi, to the canning industry, and even for export to Europe. In all but the driest years,

food production per capita was maintained. As **Figure 3** shows, overall output per square kilometer increased from about 10 tons in 1930 (mostly food crops) to 11 tons in 1987 (75 percent cash crops and horticulture).

Machakos provides a classic case of agricultural intensification, where according to the theories of economist Ester Boserup, population increases provide the labor and the incentive to apply more labor and capital to produce more on the same area of land (**Figure 4**). Population increase resulted in better environmental management and an increase in agricultural production, rather than soil erosion and impoverishment as predicted by a Malthusian perspective. Researchers Michael Mortimore, Mary Tiffen, and Frances Gichuki suggest that several factors help explain the relative success of Machakos. These include security of land titles, proximity to off-farm employment in Nairobi that provided capital for improvements, complementary efforts of husbands and wives, women's leadership in organizing community work teams, and improving educational levels.

Source: Adapted from M. Mortimore and M. Tiffen, "Population and Environment in Time Perspective: The Machakos Story," in T. Binns (ed.), *People and Environment in Africa* (pp. 69–89). New York: John Wiley & Sons, 1995.

Figure 3   Agricultural change in Machakos, Kenya From 1930 to 1987 agricultural production increased, with a shift to the production of cash crops and horticulture. (*Source:* M. Mortimore and M. Tiffen, "Population and Environment in Time Perspective: The Machakos Story," in T. Binns (ed.), *People and Environment in Africa* (pp. 69–89). New York: John Wiley & Sons, Fig. 7.6.)

Output per km² in constant 1957 maize prices

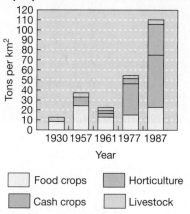

Figure 4   Agricultural terraces in Machakos

One of the more successful conservation programs that is said to benefit local communities is *CAMPFIRE (Community Areas Management Programme for Indigenous Resources)* launched in 1989 in Zimbabwe. Fees paid for park entry and hunting go directly to the community in which the use takes

place. Fees for carefully controlled hunting of "trophy" species such as elephants can reach $12,000 per animal; the money pays for schools, clinics, irrigation, and electricity. But this program too has been criticized for benefiting local governments and elites more than individual farmers.

Core Regions and Key Cities of Sub-Saharan Africa

Sub-Saharan Africa is seen as the periphery and does not contain industrial regions or world cities that drive the global economy. But within Sub-Saharan Africa there are important regions that historically produced commodities for the core and have developed regional industrial hubs with cities that act as centers of regional commerce, transportation, and culture. While cities such as Accra, Abidjan, Addis Ababa, Cape Town, and Dar es Salaam are important capitals and trading centers, their role on the continent is overshadowed in numbers of inhabitants by Johannesburg in southern Africa and Lagos in West Africa and by Nairobi as an international center in East Africa. These cities and their surrounding regions illustrate many of the challenges and characteristics of economic development and urbanization in Sub-Saharan Africa and the ways in which centers within Africa influence and manipulate their local peripheries. Although these are not world cities, they do act as the principal regional links to the world-system, and they are the principal agents in transmitting global flows of all sorts to and from the region.

Southern Africa and Johannesburg

By most measures South Africa is the economic powerhouse of Sub-Saharan Africa, producing more than one-third of Africa's manufactured goods and having the highest overall gross domestic product (GDP). The historical core of southern Africa's economy is the region around the city of Johannesburg on the high *veld* in the province of Gauteng (meaning "place of gold," formerly the southern Transvaal) that stretches from Pretoria southwest toward Kimberly (**Figure 6.36**). The high veld is a grassland ecosystem of short grasses and shrubs with an elevation of about 2000 meters (6000 feet). About 40 million people live in this zone, where frequent droughts bring dry and dusty conditions to the high plateau of the South African interior. Gauteng province has only 2.5

Figure 6.36 Core region of Gauteng and Johannesburg, South Africa (a) The core region of southern Africa is centered on the city of Johannesburg, South Africa, located on the high plateau called the *veld* adjacent to the Rand gold fields. (b) Johannesburg at dawn with the high-rise downtown headquarters for major mining corporations and the smoke drifting in from the townships on the outskirts, such as Soweto. (c) The seven members of the Qampie family are shown here with most of their possessions outside their rented house in Soweto. The parents work in downtown Johannesburg—Simon Qampie as a security guard in a department store and Poppy Qampie as an office assistant. Their aspirations include owning a home, a car, and a computer, and their major concern is the high level of crime in Soweto.

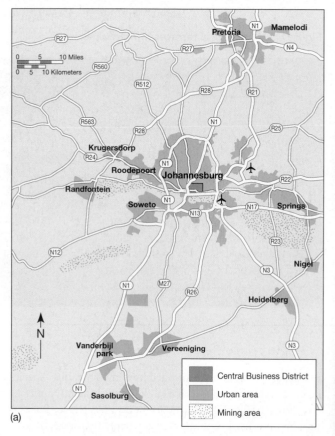
(a)

(b)

(c)

percent of the area of South Africa but is home to 25 percent of the population and 33 percent of the jobs, and it produces 45 percent of the GDP.

Johannesburg was founded in 1886, following the discovery of gold in the Witwatersrand, or Rand, a range of hills to the west of the present city that overlays the world's richest gold deposit. The region also contains large reserves of uranium, platinum, tin, and nickel, as well as diamonds and coal. The main diamond mines were developed around Kimberly, 325 kilometers (200 miles) south of Johannesburg. About 65 kilometers (40 miles) to the north of Johannesburg lies Pretoria, the administrative capital of South Africa, which is the location of dozens of government buildings.

The city of Johannesburg still bears the spatial mark of apartheid, although racial segregation is slowly being replaced by a residential pattern based more on economic status, and white residents have abandoned the city center. The majority (70 percent) of the population are black and live in "townships" such as Soweto, a community of 1.5 million people about 16 kilometers (10 miles) southwest of the city center; here most people live in small houses and commute into the city to work (Figure 6.36c). Soweto became famous for demonstrations and rent strikes during the struggle against apartheid and for the stark contrasts between the poverty of its residents and the wealth of white neighborhoods in the northern suburbs.

The city houses migrants from other parts of southern Africa who live in hostels and work in the mines and manufacturing plants. Some estimates place the number of illegal migrants in South Africa at as high as 8 million, many of them living in the Johannesburg region. Many families were divided by the pass laws, with men living in hostels and women working as domestic servants, leaving their spouses and families in rural areas. Manufacturing in the Gauteng region makes up half of South Africa's industry and includes iron and steel (using coal from the local area and iron from northeast and central South Africa) and textiles. In 1951, 18 percent of the population worked in mining and 55 percent in services, while in 1991 only 2.3 percent working in mining and 72 percent in services; this represents a substantial shift from the primary sector (mining) to the tertiary sector (services).

Today, many people work in the informal economy. Since the 1990s informal occupations of taxi driving, haircutting, small-scale manufacturing, urban farming, and street selling have boomed, and it has been estimated that more than 1.2 million people in Gauteng province are now active in the informal sector. The causes of this boom include unemployment, reduced regulation, and an increase in contracting work, such as sewing and child care, to women working in their homes.

The central area of Johannesburg has a grid pattern, with many high-rise office buildings in the central business district. Found here are the headquarters of mining corporations such as Rand Mines, the Johannesburg Stock Exchange, and evidence of foreign investment from both before and after the onset of black majority rule. The problems facing the new regional and city governments of the Johannesburg region include the provision of jobs, housing, and services to the poorer residents whose hopes were raised by the onset of black ma-

jority rule and controlling rapidly rising violent crime in the city. Many townships do not have adequate electricity and water supply, and residents must use wood or coal for heat during the chilly winter season, casting a pall of polluting smoke over the townships that contributes to health problems and high infant mortality rates.

The more optimistic city planners see signs that violence will decrease as living conditions improve in the surrounding communities and the city core becomes revitalized as a residential, shopping, and cultural center. Even the townships have become sites for tourists to visit local families, though most tourists to southern Africa still avoid Johannesburg and head straight for Cape Town or the game parks.

The West African Coast and the City of Lagos

The West African coast, with its warm, humid climate, lush forests, and abundant mineral resources, has been a center of traditional leadership, culture, trade, and population for centuries. The focus of the early slave trade and then of British and French colonialism, the West African coast is one of the most densely populated regions of Sub-Saharan Africa. The Europeans sited some of their trading ports and settlements near locations of earlier cities and took advantage of existing social and political hierarchies to obtain slaves and gold, and subsequently taxes and export crops, from the region.

The landscape of the West African coast was massively transformed by plantation agriculture, with thousands of hectares of land converted from indigenous cropland and forest into cocoa, oil-palm, and rubber trees for export. In the 1960s the Niger delta became the core of an oil-producing region that triggered the secessionist Biafra war by the Ibo against the mostly Yoruba Nigerian government and promised an economic boom for Nigeria. As Nigeria has become even more dependent on oil exports, which are more than 90 percent of all exports by value, political unrest has continued in the oil region. While some of the unrest still arises from ethnic conflict between Igbo and Yoruba and from resentment that oil wealth is not benefiting the residents of the oil region, the most recent tensions relate to environmental pollution of the lands and waters of the delta. The Nigerian government's hanging of Ken Saro-Iwa—a novelist and activist who protested oil spills in the area occupied by 500,000 Ogoni people—for his alleged role in political unrest in 1995, provoked international outrage and sanctions against the government.

Nigeria is the giant of the region, with a population estimated at 120 million and the largest GDP in West Africa. Lagos—the former colonial capital until it was moved to Abuja in the interior in 1992—has a metropolitan population of more than 12 million people, compared to only 1 million in 1960 (**Figure 6.37**). Sited on a natural harbor, Lagos was developed by the British as an important rail terminus and as a leading cargo port and industrial area and a center for the production of consumer goods beginning about 1880. About 80 percent of Nigeria's trade goes through Lagos, although some is shifting to the oil regions to the east. Lagos state has 53 percent of

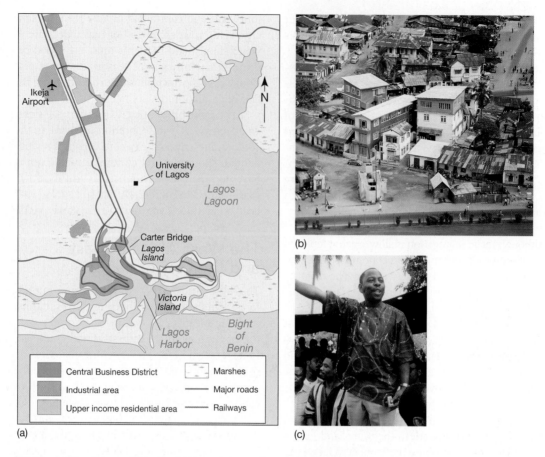

Figure 6.37 Lagos, Nigeria (a) Lagos is located on marshy land along the Bight of Benin in Nigeria, West Africa, with the central business district on an island that is linked by bridge to the mainland. (b) Mixed housing along a lagoon in Lagos. (c) Activist Ken Saro-Iwa, hanged for his opposition to oil development in eastern Nigeria.

Nigeria's manufacturing, 62 percent of the gross industrial output, and 22 industrial estates. Lagos is also a cultural center, original home of prize-winning writers Chinua Achebe, Ben Okri, and Wole Soyinka and of world-famous musicians Feli and Femi Kuti and King Sunny Adé.

The heart of the city is located on Lagos Island with major banks and commercial establishments. Wealthier residents live on Victoria Island, but millions of poor residents live in shantytowns to the north of the city and are unemployed or work in the informal sector. The location on islands and former mangrove swamps makes the city vulnerable to flooding and creates problems with drainage and construction. When structural adjustment caused food price increases and unemployment after 1989, rioting became a way to express the frustrations of city residents. Lagos is infamous for its traffic and crime problems. The average commute to work is more than 90 minutes in polluted air and tangled traffic jams made worse by inadequate bridges between islands. Electricity and other services are also insufficient; as a result of interruptions and lack of service, the National Electric Power Administration (NEPA) has been given the nickname of "Not Expecting Power Anytime."

The East African Highlands and Nairobi

Less than 20 minutes after leaving the international airport outside Nairobi, a visitor can be watching leopards hunting zebra across the grasslands in a national park while the high-rise buildings of downtown Nairobi glimmer on the horizon. Driving into the city, glimpses of former colonial mansions and luxury hotels surrounded by flowers shift rapidly to views across valleys crowded with slum housing and of a city center where business and government leaders drive Mercedes cars through streets crowded with traders, pickpockets, street children, and tourists (**Figure 6.38**).

Nairobi, Kenya, is the commercial and communications center of East Africa, a city of almost 2 million located at an altitude of 1800 meters (5900 feet) on the high plateau adjacent to Mount Kenya and the East African Rift Valley. Originally established at a railroad stop where there was a spring and low incidence of malaria, Nairobi became the center of British colonial rule and white settlement in East Africa. The temperate highlands and rich soils north of Nairobi were assigned to European farmers who raised wheat, vegetables,

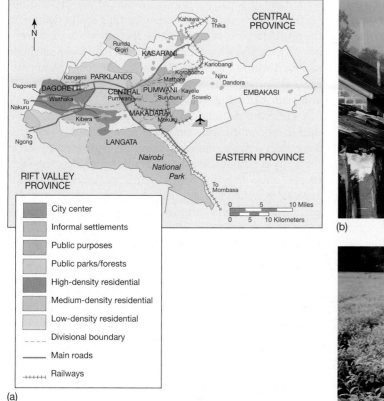

(a)

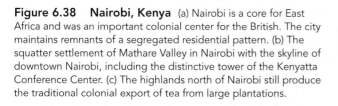

(b)

(c)

Figure 6.38   Nairobi, Kenya (a) Nairobi is a core for East Africa and was an important colonial center for the British. The city maintains remnants of a segregated residential pattern. (b) The squatter settlement of Mathare Valley in Nairobi with the skyline of downtown Nairobi, including the distinctive tower of the Kenyatta Conference Center. (c) The highlands north of Nairobi still produce the traditional colonial export of tea from large plantations.

and fruit and developed extensive plantations of tea and coffee for export. Cattle, sisal (a fiber), and food crops were also produced on white-owned estates in the Rift Valley to the west. While the lives of colonial settlers have been romanticized by films and books such as *Out of Africa,* their paternalistic treatment of black Africans and the eviction of people from their traditional lands eventually led to rebellions and then independence in 1963.

Like most colonial cities in Africa, Nairobi maintains elements of a segregated residential pattern that originated in race, with better-off residents living in the former white colonial neighborhoods to the west and north, while the rest live in cheaper high-density housing to the east. A population of Asians initially settled near the train station and has now moved into middle- and upper-class residential districts to the north, working mainly in trading and retail. The highest population densities are found in slum or squatter settlements such as Mathare Valley, where provision of basic services is inadequate and serious health problems are common.

Nairobi's population and manufacturing sector is smaller than that of other major African cities such as Lagos, but it is an important service center with headquarters of international companies, agencies, and nongovernmental organizations, including the United Nations Environment Programme (UNEP). Manufacturing includes an automobile plant, and textile, canning, and small metal goods factories, mainly to the east of the city. Most of the population works in manufacturing, construction, or services, and thousands of poorer residents work in the informal sector selling food, household goods, and crafts on the street or from small kiosks. The city lacks adequate public transportation and relies on the infamous private minibuses called *matutus* that swerve along the streets overloaded with passengers and contribute to Kenya's high motor vehicle accident rate. Nairobi is an important tourist center, serving as the hub for numerous tours to see the wildlife, culture, and landscapes of Kenya and adjacent countries such as Tanzania.

As in many other African cities, poorer communities in and around Nairobi have limited access to electricity, gas, or petroleum fuels for heating and cooking and rely on wood or charcoal as their major source of energy. The increasing wood and charcoal demands of the city have had a tremendous impact on forests in the region, with serious deforestation tied to Nairobi's energy needs occurring as far as 200 kilometers (124 miles) away. The difficulties of finding firewood and the increases in prices have disproportionately affected women who traditionally collect the wood and are responsible for cooking and heating the homes. Projects to reduce energy demands by using scrap metal to make more efficient stoves have complemented the efforts of female-led nongovernmental organizations to protect trees in and around Nairobi. The best-known social movement is the *Green Belt* movement, which counts 50,000 women as members. Led by environmental and political activist Wangari Maathai, Green Belt has planted thousands of trees around Nairobi and has been the model for similar groups elsewhere in Africa and the world.

The East African Highlands and the adjacent rift valley continue to be important agricultural production zones, serving urban demand in Nairobi and exports of coffee, tea, and sisal (see Geography Matters: More People, Less Erosion, in Machakos, Kenya, p. 300). The most rapidly growing export sector is fresh vegetables and cut flowers for export to Europe. Relying on refrigerated air transport out of Nairobi airport, Kenya now provides 40 percent of the European Union imports of fresh vegetables, sending 21,000 tons of mainly peas and green beans to Britain in 1997. The flower industry, centered on Lake Naivasha, sends more than 1 billion cut blooms to Europe each year, including carnations and roses, and is now the fourth largest flower producer in the world, after the Netherlands, Colombia, and Israel. While these new industries provide employment and higher wages than some other sectors, the strict quality standards, perishability, and need for air transport mean that small producers find it difficult to compete. There are also concerns about pesticide risks to workers in the growing and packing sectors.

Distinctive Regions and Landscapes of Sub-Saharan Africa

The news media and development agencies have traditionally grouped some of the more peripheral countries of Africa into regional groups that share distinctive characteristics and problems. For example, discussions of food security and environmental degradation often refer to the Sahel as a distinctive region in West Africa and to the Horn of Africa as a similar grouping in East Africa. As we discuss below, these large regions do have some coherence, but they include great regional and social variations. They also provide examples of some important processes, such as desertification and famine, that have been particularly associated with Sub-Saharan Africa.

Other distinctive landscapes in Sub-Saharan Africa include the islands of the Indian Ocean and the forests of Central Africa, illustrated in the discussion of Madagascar and of the Congo River basin.

The Sahel

The Sahel has both a physical and a political definition. Geographers define the physical Sahel as the semiarid zone across the southern edge of the Sahara Desert, where sea-

sonal rainfall usually brings a renewal of pasture and opportunities for crop production. According to this definition, the northern parts of several coastal countries such as Gambia, Nigeria, and Ghana and the Cape Verde Islands might be included within the Sahel. Politically, the zone encompasses six countries—Chad, Niger, Burkina Faso, Mali, Senegal, and Mauritania (**Figure 6.39**).

The Sahel is not a desert of wind and blowing sand, though when the harmattan winds blow in the dry season, it sometimes seems as though nothing could grow or survive in the dusty conditions. When good rains have fallen, the Sahel blooms into a grassland with scattered acacia bushes, and the expanding wetlands of inland deltas provide lush crop and pasture land as well as fishing opportunities. The cities of the Sahel, such as Kano and Timbuktu, were centers of trade, religion, and scholarship before European colonialism, mostly adopting the Islamic religion and customs.

The traditional livelihoods of the Sahel included nomadic pastoralism as an adaptation to rainfall that varied from year to year and place to place, farming of cereals and other crops, mining, small-scale manufacturing, and trade. Most of the Sahel came under French colonial rule at the end of the nineteenth century, with several regions encouraged to plant peanuts and cotton for export and trading actively with coastal regions. Colonialism also provided opportunities for seasonal work in mines and plantations, though it also brought the burden of taxes to many communities.

A limited network of meteorological stations provides data that show that rainfall is quite variable across the Sahel but that wetter conditions seem to have occurred in the 1950s and early 1960s. Rather than adjust their herds to average conditions, Sahelian pastoralists tend to be opportunistic, building up their herds in good years because their livestock are the most important way of accumulating wealth and investing capital.

Beginning in 1968, it appears that the rains failed in most parts of the Sahel for up to seven years. As herds began to die off, images of starving refugees from the drought began to appear in the international media, resulting in a relief effort and anguished debates among researchers and policy makers about what had gone wrong and what could be done to avoid future tragedy. As many as 3.5 million cattle died, and 15 million farmers lost more than half their harvests between 1968 and 1973. A quarter of a million people died from famine before food relief could reach them.

Although some researchers blamed nature and the irrational buildup of herds in the face of regular drought cycles in the Sahel, others argued that the roots of the crisis lay in changes in Sahelian political economy stemming from colonial structures and continued overreliance on export cropping. Geographer Michael Watts showed how in northern Nigeria people lost access to land and their drought-coping strategies because of the loss of traditional self-help institutions, colonial policies, and the marginalization of poorer farmers.

Others argued that decades of peanut and cotton production, particularly in Senegal and Mali, had exhausted the soil and left it unproductive. Climatologists suggested that de-

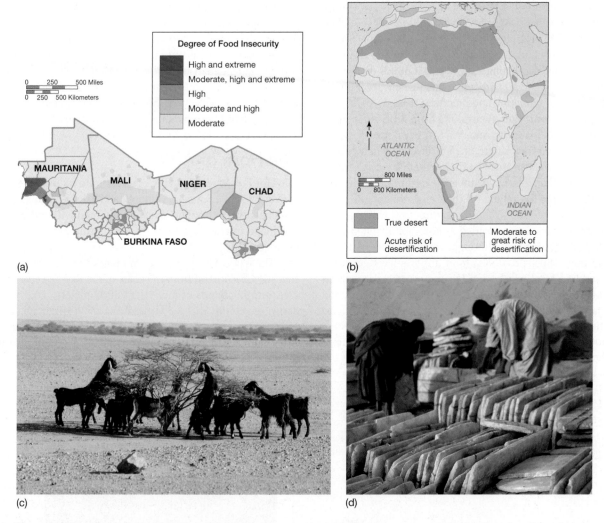

Figure 6.39   The Sahel (a) The Sahel region, showing the degree of food insecurity in 1999. (b) Spreading desertification threatens one-third of Africa. (c) Goats feeding on acacia in Sahelian desert landscape. (d) The salt market in Mopti, Mali.

forestation and overgrazing had increased the reflectivity of the land surface, which meant less warm air rising to form clouds, or that increased atmospheric dust was reducing the uplift of air, two processes that reduce rainfall.

International efforts to respond to the 1970s drought focused initially on food relief but then turned to longer-term technical efforts to reduce risk, including the drilling of deep wells for cattle herds. Unfortunately so many thirsty cattle gathered around the wells that all possible forage vegetation was consumed, and the herds starved. When food aid arrived in communities where some farmers still had crops to sell, food prices dropped, and farmers could not make a living.

In 1977 the United Nations held a conference on **desertification** to discuss the problems of the Sahel and other regions where the deserts appeared to be spreading into previously productive areas (see Figure 6.39). Desertification has a variety of meanings but is most generally viewed as the degradation of arid and semiarid lands to less productive conditions through drying, erosion, compaction, buildup of salts,

and loss of fertility and vegetation. Although the United Nations gave a high priority to the monitoring of desertification, differences in definition and measurement resulted in widely varying estimates of the area affected over the next few decades. While some sources reported that the desert had advanced hundreds of kilometers into the Sahel, others saw no long-term trend but only year-to-year variations in the vigor of vegetation. There were also disagreements about the relative role of different factors in causing desertification. The main culprits were seen as climate change, overgrazing, overcultivation (including cash cropping), deforestation for wood fuel, and unskilled irrigation that results in the buildup of salt in the soil (salinization).

Whether or not desertification is actually occurring on a large scale, a number of development projects have attempted to reduce the vulnerability of Sahelian people and countries to drought and famine. Reforestation projects in Mali have successfully created erosion barriers, forage for animals, and wood for fuel. Traditional rainwater-harvesting techniques using stone barriers to trap moisture have been diffused from one community to another. The United States has led in efforts to create a Famine Early Warning System (FEWS) for the Sahel. The FEWS combines multiple sources of social and environmental data to try and anticipate where and when drought, agricultural crisis, and famine may occur. It builds on the understanding of famine as a process of insufficient food intake resulting in acute starvation and increased death rates, distinguished from chronic hunger resulting from acute nutritional deprivation. Nobel laureate Amartya Sen developed a theory that famines occur as a failure of *entitlements*—when people lose their right to farm their land, sell their labor, or trade. The loss of entitlements can be invisible except through indicators that people are selling off their endowments of cattle and jewelry or that food prices are starting to slowly rise.

Recent studies have provided a more optimistic view of the Sahel, suggesting that although drought returned to the Sahel in the 1980s and 1990s, food imports have generally decreased and food production has increased, especially in wetland areas. Pastoralists have rebuilt their cattle herds, and some have switched to animals that are less vulnerable to drought, such as sheep and goats. It is argued that the solution to many problems in the region is to support the diversity,

flexibility, and adaptability that can be found in different combinations in different communities and households. Diversification strategies include farming and herding combinations of species in several microenvironments using various resource-management techniques, finding sources of off-farm income in the community or through seasonal migration, negotiating work within the household, and exchanging food, goods, or labor with neighbors or other towns.

The Horn of Africa

The Horn of Africa—usually considered to include the countries of Ethiopia, Somalia, Djibouti, and Eritrea—has also been associated with drought, which, combined with warfare, has produced images of famine and large numbers of refugees (**Figure 6.40**).

The landscape of this region ranges from the broad plain that stretches inland from the coast of the Red Sea and Indian Ocean to highlands of more than 3000 meters (10,000 feet) and receive rains that feed one of the main tributaries of the Nile. The countries of this region are fiercely independent. Ethiopia resisted colonization for most of its history, largely

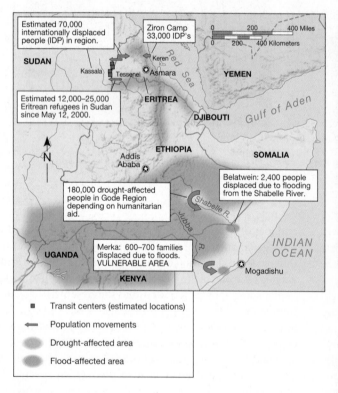

(b)

(c)

Figure 6.40   The Horn of Africa (a) The Horn of Africa includes the countries of Somalia, Ethiopia, Djibouti, and Eritrea and has experienced large-scale refugee flows as a result of conflict and weather disasters. (b) Agricultural terraces near Fasher, Ethiopia with erosion on the upper slopes. (c) Somalis watch a U.S. military plane land in Mogadishu.

had a common language (Amharic), and was ruled by a Christian monarchy (supposedly descended from Solomon and Sheba) for hundreds of years until Emperor Haile Selassie was deposed in 1974. Ethiopia is often thought of as a resource-rich country, with large areas suitable for wheat and other cereal production. Coffee was domesticated in the Ethiopian highlands and became an important export, contributing 66 percent of all export value in 1998. The traditional grain of Ethiopia is *teff*, which is used to produce a flat bread known as *injera*. Somalia coheres around the Somali language, clan systems, and Islamic traditions and, with little arable land, has focused on cattle for export to the Middle East and on fishing the productive coastal waters. Djibouti, formerly French Somaliland, became independent in 1977 and has used its position on the coast between the Red Sea and the Gulf of Aden to develop an economy based on trade and services. Eritrea, stretching 1000 kilometers (600 miles) along the Red Sea north of Djibouti, gained independence from Ethiopia in 1993 after many years or warfare and has an economy mainly based on livestock and subsistence agriculture.

Hunger in the Horn of Africa has some relation to failure of the rains, but it has been linked directly to continuing warfare and population displacements. A brief review of the complex political geography of the region is essential to understand the challenges of food production and development. Earlier we discussed how conflict in the Horn of Africa was fueled by Cold War tensions and ideology, especially the inequalities connected with the feudal agricultural system of Ethiopia and the incomplete agrarian reforms of the socialist government. The global strategic location of the Horn of Africa at the entrance to the Suez Canal and near the oilfields of the Middle East draws considerable attention from the United States and other major powers. U.S. involvement in Somalia was partly driven by the offer of military bases in Mogadishu.

Armed struggles between Ethiopia, Eritrea, and Somalia diverted resources from efforts to improve agriculture and reduce poverty for several decades. Peace is still not secure in the Horn of Africa because Somali *irredentism* (the attempt to create a unified cultural-linguistic nation) continues to drive interest in territory in Kenya and Ethiopia, and competition between clans within Somalia produced a state of civil war that was tentatively resolved in 2000 by a coalition government. Eritrea and Ethiopia (from which Eritrea seceded in 1993 after 30 years of rebellion) continue to struggle over the border between the two countries, driven by landlocked Ethiopia's concern over its access to ocean ports and Eritrea's bid for rapid economic development and independent currency.

In the 1984–1985 famine in Ethiopia, more than 1 million people died, and the images of dying children and starving refugees drew worldwide attention, including that of rock musicians who organized the Live Aid concert to raise funds for famine relief. The return of famine in 2000, threatening 8 million people especially in the eastern Ogaden region, was linked to drought and to continuing border conflicts with Eritrea. These conflicts focused resources on military activities rather than on humanitarian and agricultural programs and caused farmers to abandon their fields. Increases in the population since the mid-1980s and the associated divisions of farmland into smaller, overcultivated plots were also blamed for reduced harvests that were unable to meet the demand for food. The distribution of aid was insecure because Eritrea controlled the ports and wanted assurances that food would not benefit the Ethiopian army and because bandits within Ethiopia jeopardized transport and the safety of aid workers.

Madagascar

Rising to more than 3000 meters (10,000 feet), the island of Madagascar emerges from the Indian Ocean about 400 kilometers (250 miles) from the east coast of Africa (**Figure 6.41**). Like the other Indian Ocean islands included in the Sub-Saharan African world region (Réunion, Comoros, Mauritius, and the Seychelles), Madagascar's human geography has been influenced as much by Asia to the east as by Africa. Settled by migrants from Indonesia, the island was colonized by France and now has a population of 45 million people. From the outset, the colony's agricultural production was geared primarily for export, including coffee, rice, and beef. The Central Highlands became the primary irrigated rice-growing region for both subsistence and export. Cloves, vanilla, and sugarcane were cultivated in the north; cattle, rice, and maize were major crops in the west. Coffee, which remains the island's major export crop, was planted on the east coast, the region of the island with the largest remaining forest cover.

Madagascar is the fourth largest island in the world—after Greenland, New Guinea, and Borneo—with an area of 587,000 square kilometers (226,650 square miles). Madagascar is noted for the unique and diverse species of animals and plants that inhabit its forests, including 33 species of lemurs and 800 species of butterflies. It provides an important example of the theory of island biogeography that suggests that larger islands or forest areas will have many more species than smaller ones and that deforestation will thus reduce species diversity. The original forests developed on slopes that received heavy rainfall from the trade winds, but most of these slopes are now used for agriculture or ranching.

The most serious environmental problem in Madagascar is deforestation, which has a long history and is associated with clearing land for rice production, sugar plantations, and cattle ranches, and cutting trees for export of tropical hardwoods. Geographer Lucy Jarosz reports that up to 7 million hectares (17.3 million acres) of forest were cleared between 1900 and 1940. In 1909 the governor prohibited shifting cultivation (or *tavy* as it is called in Madagascar) and tried to move people into wage labor. Local people reacted strongly to the ban on their traditional practices and burned the forest in protest. The government also opened up the forest to logging concessions, which resulted in massive cutting of forests. Madagascar is one of the world's poorest nations, with a per capita income of approximately $240 per year. About 80 percent of the population are subsistence farmers. Only remnants of the original forests remain and are of concern to conservationists worldwide because of their unique ecology. Several important medicinal plants, including the

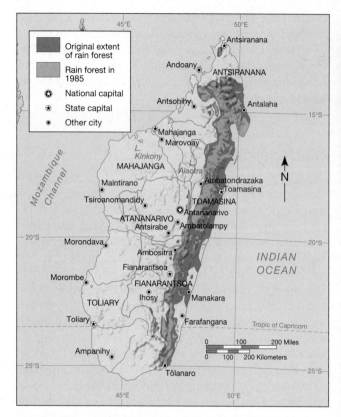

Figure 6.41 Map and images of Madagascar (a) Map showing severe loss of rain forest; original extent of rain forest is indicated by both shades of green, and 1985 extent by light green only. (b) Lemurs in the forest. (c) Erosion of hillsides near Ambatolampy.

rosy periwinkle, used to treat leukemia, were discovered within the rapidly disappearing forests. Deforestation on steep slopes has also placed human populations directly at risk from soil erosion, landslides, and floods that are much more severe on devegetated land. For example, more than 100 people were killed and 10,000 made homeless by floods associated with cyclones in 2000.

Recently Madagascar has established eight new protected areas totaling 6809 square kilometers (2630 square miles). The country's new National Association for Protected Area Management has taken over the management of several of the key national parks for ecotourism (Ranomafana, Isalo, and Montagne d'Ambre).

The economy remains dependent on agriculture, with coffee and vanilla as the dominant exports, and was seriously affected by growing debt during the 1980s. The socialist government, with close links to North Korea, then shifted from policies of nationalization of major sectors to a more free-market orientation, and France forgave a significant portion of the country's external debt.

Summary and Conclusions

The future of Sub-Saharan Africa is difficult to forecast because there are both positive and negative signs throughout the region. While some countries have improved living standards, established democratic governments, and increased food and economic production, others are mired in conflict over resources and political futures or face forbidding health challenges of malaria and AIDS. Social inequalities persist in many regions and contribute to migration, unrest, and famines. Although some Africans maintain rural subsistence lives, almost totally disconnected from the world economy, others are working in transnational corporations or are producing new exports for the global market.

The region is still adjusting to the enduring legacies of colonialism that include economies dependent on mineral or agricultural exports, unequal land distribution, and boundaries that divide or cluster groups with little attention to cultural values or political expediency. Many countries are also struggling with the transition to independence and the creation of representative governments, with high debts and over-reliance on foreign assistance. High fertility rates pose some challenges in terms of finding jobs, housing, food, and services for growing populations, especially where poverty, gender inequalities, and cultural factors promote large families, but also sustain the flow of labor and ideas for economic development.

Sub-Saharan Africa has many connections to other world regions and to global systems through trade, migration, and the diffusion of rich and varied cultures, yet it is often overlooked by the media and the core economies and portrayed only through negative images of poverty, war, and disease or as a vast nature reserve. Whatever the future may hold for Sub-Saharan Africa, it is a region that the rest of the world can ill afford to ignore.

Key Terms

apartheid (p. 280)
Berlin Conference (p. 274)
bush fallow (p. 269)
circular migration (p. 287)
desertification (p. 307)

domestication (p. 268)
feminization of poverty (p. 295)
gender and development (GAD) (p. 297)

harmattan (p. 263)
homelands (p. 280)
microfinance programs (p. 297)
pastoralism (p. 269)

savanna (p. 265)
shifting cultivation (p. 268)
slash and burn (p. 268)
social capital (p. 297)
transhumance (p. 269)

Review Questions

Testing Your Understanding

1. Why does Sub-Saharan Africa have such large regions of poor soils and hot, dry climates? Which regions have the more temperate climates and better conditions for agriculture and why?

2. What major crops or animals were domesticated in Sub-Saharan Africa and what are the traditional agricultural adaptations to low soil fertility and seasonal rainfall?

3. Where did modern humans originate and what were some of the major centers of African empires and trade within Africa prior to 1600?

4. Where did the Arabs, Portuguese, British, French, and Dutch begin their colonization and settlement of Sub-Saharan Africa? What was the role of slavery before and after European colonization? Who were some of the main European explorers and what image did they promote of the African interior?

5. How does Africa's dependency on mineral exports relate to the colonial period, the geographic pattern of roads and railways, and to contemporary conflicts?

6. What was apartheid and what were the ways in which it was implemented and enforced at different geographic scales? How and when was apartheid abolished and what challenges face post-apartheid South Africa?

7. How did socialist ideas and the Cold War influence politics and conflict in the countries of Tanzania, Angola, Mozambique, Ethiopia, and Somalia?

8. How do modernization theory, dependency theory, import substitution, and structural adjustment differ in their approaches to development and what have been their positive and negative impacts in Sub-Saharan Africa?

9. What hope do the following policies hold for the future in Sub-Saharan Africa: debt relief, gender and development programs, microfinance programs, sustainable development, social capital?

10. What factors may have lead to continued political instability, warfare and corruption in many Sub-Saharan African states? Which African countries are most seriously affected by conflict and to what extent does this explain their ranking on social indicators lists such as the Human Development Index? What role do organizations for regional cooperation and the United Nations play in the resolution or reduction of these conflicts?

11. Why has wildlife conservation become a priority in some regions of Sub-Saharan Africa and what strategies have been used to protect wildlife? Which countries opposed the ivory ban and why? What are the main causes of deforestation and associated wildlife depletion in Madagascar?

Thinking Geographically

1. What are the arguments for and against treating Sub-Saharan Africa as a major world region?

2. How does plate tectonics help explain the age, high elevation, and steep escarpments around much of Africa and the major feature of the East African Rift Valley? What are the implications for human activities including mining, settlement, river navigation, energy production, and freshwater fisheries?

3. What are the major pests and diseases in Sub-Saharan Africa and their geographical distributions? How have they limited human activity and what has been done to reduce the spread and incidence of these pests and diseases?

4. How did Europe divide up Africa after the Berlin Conference and what were the legacies of this geographical division on languages, boundaries, land tenure, and forms of rule? How did indigenous peoples respond to or resist European rule?

5. How do the following affect the geographical distribution of population in Sub-Saharan Africa and/or population growth rates: high infant mortality, high HIV/AIDS rates, education levels of women, coastal ports, highland areas, mineral deposits?

6. Using examples from at least two major urban areas in Africa, discuss what attracts people to these cities and some of the social and environmental problems of these cities.

7. What are the major patterns of migration between and within African countries and how do they relate to employment opportunities, natural disasters, and warfare? How does the case of the Horn of Africa illustrate the causes of refugee movements and the international responses to them?

8. How do the cases of the West African Sahel and Machakos Kenya illustrate the debates about the causes of environmental change in Africa? For example, what arguments have been made about the role of human activity in causing desertification or soil erosion?

Further Reading

Adams, W. M., Goudie, A. S., and Orme, A. R. (eds.), *The Physical Geography of Africa*. Oxford University Press, 1999.

Amin, S., *L'Afrique de l'Ouest Bloquée, l'Economie Politique de la Colonisation, 1880–1970*. Paris: Éditions de Minuit, 1971.

Anderson, D., and Grove, R. H. (eds.), *Conservation in Africa: Peoples, Policies and Practice*. Cambridge: Cambridge University Press, 1990.

Aryeetey-Attoh, S. (ed.), *The Geography of Sub-Saharan Africa*. Upper Saddle River, NJ: Prentice Hall, 1997.

Boserup, E., *Women's Role in Economic Development*. London: Allen & Unwin, 1970.

Binns, T. *Tropical Africa*. London: Routledge, 1994.

Carney, J. A., "Peasant Women and Economic Transformation in the Gambia," *Development and Change*, 23 (1992), 67.

Chambers, R., *Rural Development: Putting the Last First*. New York: Longman, 1983.

Christopher, A. J., *The Atlas of Changing South Africa*. London: Routledge, 2000.

Daniel, M. L. "The Demographic Impact of HIV/AIDS in Sub-Saharan Africa," *Geography*, 85 (1): 46–55.

Gourevitch, P., *We Wish to Inform You That Tomorrow We Will Be Killed with Our Families: Stories from Rwanda*. New York: Farrar, Straus, and Giroux, 1998.

Griffiths, I. L. L., *An Atlas of African Affairs*. New York: Methuen, 1985.

Grove A. T., *The Changing Geography of Africa*. Oxford: Oxford University Press, 1994.

Jarosz, L., Defining and Explaining Tropical Deforestation: Shifting Cultivation and Population Growth in Colonial Madagascar," *Economic Geography*, 64 (1993): 366–80.

Leach, M., and Mearns, R., *The Lie of the Land: Challenging Received Wisdom on the African Environment*. Portsmouth, NH: Heinemann, 1996.

Levin, R., and Wiener, D. (eds.), *"No More Tears . . .": Struggles for Land in Mpumalanga, South Africa*. Trenton, NJ: Africa World Press, 1997.

Lewis, L. A., and Berry, L., *African Environments and Resources*. Boston: Allen and Unwin, 1988.

Mazrui, A. A., *The Africans: A Triple Heritage*. London: BBC Publications, 1986.

Mortimore, M., and Adams W. M., *Working the Sahel: Environment and Society in Northern Nigeria*. London: Routledge, 1999.

Mortimore, M., and Tiffen, M., *Population and Environment in Time Perspective: The Machakos Story*, in T. Binns (ed.), *People and Environment in Africa* (pp. 69–89). New York: John Wiley & Sons, 1995.

Nicholson, S. E., "Environmental Change Within the Historical Period," in W. M. Adams, A. S. Goudie, and A. R. Orme (eds.), *The Physical Geography of Africa* (pp. 60–86). Oxford: Oxford University Press, 1996.

Obia, G. C., "Industry, Enterprises, and Entrepreneurship in the Development Process," in S. Aryeetey-Attoh (ed.), *The Geography of Sub-Saharan Africa* (pp. 286–324). Upper Saddle River, NJ: Prentice Hall, 1997.

Oliver, R., and Crowder, M. (eds.), *The Cambridge Encyclopedia of Africa*. Cambridge: Cambridge University Press, 1981.

Peil, M., *Lagos: The City Is the People*. Boston: G. K. Hall, 1991.

Rain, D. *Eaters of the Dry Season: Circular Labor Migration in the West African Sahel*. Boulder: Westview Press, 1999.

Rakodi, C. (ed.), *The Urban Challenge in Africa: Growth and Management of Its Large Cities*. Tokyo: United Nations University Press, 1997.

Reader, J., *Africa: A Biography of the Continent*. London: Penguin Books, 1997.

Richards, P., *Indigenous Agricultural Revolution: Ecology and Food Production in West Africa*. Boulder: Westview Press, 1985.

Rodney, W., *How Europe Underdeveloped Africa*. Washington, DC: Howard University Press, 1974.

Rogerson, C. M., "Urban Poverty and the Informal Economy in South Africa's Economic Heartland," *Environment and Urbanization* 6 (1996): 167–81.

Schroeder, R. A., *Shady Practices: Agroforestry and Gender Politics in the Gambia*. Berkeley: University of California Press, 1999.

Simon, D., *Cities, Capital, and Development: African Cities in the World Economy*. New York: Halstead Press, 1992.

Stock, R. F., *Africa South of the Sahara: A Geographical Interpretation*. New York: Guilford Press, 1998.

Udo, R., *The Human Geography of Tropical Africa*. Ibadan: Heinemann, 1982.

Watts, M., *Silent Violence: Food, Famine, and Peasantry in Northern Nigeria*. Berkeley: University of California Press, 1983.

World Bank, *Intensifying Action Against HIV/AIDS in Africa: Responding to a Development Crisis, Africa Region*, Washington, DC: The World Bank, 2000.

Film, Music, and Popular Literature

Film

Chinua Achebe: A World of Ideas. Directed by Bill Moyers, 1989. Bill Moyers interviews the Nigerian author about African literature.

Chocolat. Directed by Claire Denis, 1988. A young French woman returns to contemplate her childhood in Cameroon and the nature of the human damage exacted on both the colonized and colonizer.

Cry Freedom. Directed by Richard Attenborough, 1987. A white South African journalist investigates the death of black activist Steve Biko during the era of apartheid.

Everyone's Child. Directed by Tsitsi Dangarembga, 1996. The tragic story of one Zimbabwean family devastated by AIDS.

From Sun Up. Directed by Flora M'mbubu, 1987. A candid, authentic picture of the dawn-to-dusk, life-sustaining efforts of the women of black Africa to survive and prosper.

Mister Johnson. Directed by Bruce Beresford, 1991. An educated black man tries to succeed in colonial Nigeria.

Mountains of the Moon. Directed by Bob Rafaelson, 1990. Powerful epic of explorers John Hanning Speke and Sir Richard Francis Burton's quest to find the source of the River Nile during the mid-nineteenth century.

Out of Africa. Directed by Sydney Pollack, 1985. The life of Karen Blixen, who established a plantation in Kenya.

Yeelen. Directed by Souleymane Cissé, 1987. Set during the era of the powerful Mali Empire of the thirteenth century, this film adapts one the great oral epics of the Bambara people of West Africa.

Music

Adé, King Sunny. *Juju Music*. Uni/Mango, 1982.

Cabo Verde. *Cesaria Evora*. RCA Victor, 1999.

Kuti, Fela. *Best of Fela Kuti*. Uni/MCA, 2000.

Ladysmith Black Mambazo. *Best of Ladysmith Black Mambazo*. Shanachie, 1992.

Makeba, Miriam. *Homeland*. Putumayo, 2000.

N'Dour, Youssou. *Immigres*. Earthworks, 1998.

Sangare, Oumou. *Moussoulou*. Wea/Atlantic, 1999.

Various artists. *The Indestructible Beat of Soweto*. Shanachie, 1986.

Popular Literature

Achebe, Chinua. *Things Fall Apart*. Portsmouth, NH: Heinemann, 1996. Unsentimental rendering of Nigerian tribal life before and after the coming of colonialism.

Conde, Maryse. *Segu*. New York: Penguin USA, 1998. Depicts the Bambara of 1797, a kingdom in Africa flourishing before the coming of Islam and slavery.

Emecheta, Buchi. *The Slave Girl: A Novel*. London: Allison and Busby, 1977. An Ibo woman writes of tribal life and those of African women.

Farah, Nurridin. *Maps*. New York: Penguin USA, 2000. The story of Askar, a man coming of age in the turmoil of modern Africa.

Gordimer, Nadine. *Burgher's Daughter*. New York: Viking, 1980. A white woman comes to understand the impact of the South African political climate under apartheid.

Lessing, Doris. *Going Home*. New York: Harper, 1996. An account of her return to her childhood home in what is now Zimbabwe.

Mandela, Nelson. *The Long Walk to Freedom*. Boston: Little, Brown and Co., 1995. Biography of South Africa's first black president, who was imprisoned for 27 years by apartheid governments.

Matthiessen, Peter. *The Tree Where Man Was Born*. New York: Penguin USA, 1995. In this classic volume, Matthiessen exquisitely combines both nature and travel writing to bring East Africa to vivid life.

Ngugi, James. *Grain of Wheat*. London: Heinemann, 1967. A story about Kenya on the cusp of independence.

Paton, Alan. *Cry the Beloved Country*. New York: Milestone, 1959. Paton's deeply moving story of Zulu pastor Stephen Kumalo and his son Absalom, set against the backdrop of a land and people riven by racial inequality and injustice, remains one of the most famous and important novels in South Africa's history.

Soyinka, Wole. *Ake: The Years of Childhood*. London: Rex Collings, 1981. The autobiography of an important African writer, who describes his childhood in Nigeria.

7 North America

Figure 7.1

ARCTIC
OCEAN

GREENLAND
(DENMARK)

Baffin
Bay

U N A V U T

Iqaluit

Churchill

Hudson
Bay

MANITOBA

C A N A D A

ONTARIO

Winnipeg

NEWFOUNDLAND

QUÉBEC

Island of
Newfoundland

St. John's

St. Pierre and
Miquelon (Fr.)

Prince
Edward
Island

Cape Breton I.

NEW
BRUNSWICK

Charlottetown

NOVA
SCOTIA

St. Lawrence R.

Québec

St. John

Halifax

MAINE

NORTH
DAKOTA

MINN.

Minneapolis-
St. Paul

SOUTH
DAKOTA

WIS.

Lake Superior

MICH.

Milwaukee

Detroit

Lake Michigan

Lake Huron

Montréal

Ottawa

Toronto

VT.

N.Y.

N.H.

Albany

Rochester

Buffalo

Lake Erie

MA.

Boston

R.I.

Providence

CT.

Hartford

Lake Ontario

NEB.

IOWA

Chicago

IND.

Dayton

OHIO

Columbus

Cleveland

Pittsburgh

PENN.

New York

N.J.

Philadelphia

Indianapolis

ILL.

Kansas City

KANSAS

St. Louis

MO.

S T A T E S

KY.

Cincinnati

Louisville

W.VA.

Baltimore

MD.

DEL.

Washington, D.C.

Richmond

VA.

Norfolk

ATLANTIC
OCEAN

Bermuda
(Br.)

Oklahoma City

OKLAHOMA

ARK.

Memphis

TENN.

Nashville

N.C.

Charlotte

S.C.

Dallas-
Ft. Worth

TEXAS

MISS.

ALA.

Birmingham

Atlanta

GEORGIA

Houston

LA.

New Orleans

Jacksonville

Orlando

Tampa-
St. Petersburg

FLORIDA

Miami

Gulf of
Mexico

San
Antonio

Mississippi R.

Missouri R.

Ohio R.

Tropic of Cancer

70°N

60°N

Arctic Circle

50°N

10°W

20°W

30°W

40°N

40°N

30°N

20°N

40°W

50°W

60°W

70°W

90°W

N

The United States and Canada constitute the North American region (**Figure 7.1**). The United States covers about 9,666,861 square kilometers (3,732,397 square miles), including 48 contiguous states and two noncontiguous states: Alaska (northwest of Canada) and the volcanic islands of Hawai'i, 3220 kilometers (2000 miles) southwest of the North American continent in the Pacific Ocean. Canada covers about 9,970,610 square kilometers (3,849,652 square miles), including 10 provinces and 3 territories. Canada is the world's second largest country after the Russia Federation. Whereas residents are spread thinly across the country with significant population concentrations in large cities, most Canadians live in a 322-kilometer-wide (200-mile-wide) band along the 5635-kilometer (3500-mile) U.S.–Canada border (**Figure 7.2**).

Canada's political geography stretches from 49° N latitude to a few degrees beyond the Arctic Circle (66.5° N), making it mostly a high-latitude, or "northern," country. The continental landmass of the United States (excluding Hawai'i and Alaska) is predominantly a midlatitude country that occupies a swath of land area that extends southward from Canada to the tip of Florida, a few degrees above the tropic of Cancer (23.5° N). Both countries share many of the same physical features of rugged mountains in the west, older, more eroded ones in the east; vast plains, and extensive coastlines on three oceans (the Atlantic, Pacific, and Arctic). Both are established democracies modeled on European political traditions. Both have significant populations of native peoples. Both consolidated their leadership roles in the world economy early in the twentieth century. Currently, the United States and Canada are two of the world's most prosperous states. The North American region is certainly one of the most integral to, if not the hub of, the contemporary world system.

As with other New World lands, much of the recent history of these two countries is the result of European colonization. Because of this and because the current economic status of the United States and Canada is also comparable, academics as well as policy makers and government agencies treat the two countries as constituting a coherent region. But as has been discussed in Chapters 1 and 2, we are in a period of dramatic and rapid change as the world's political borders are being dismantled or rearranged around the new economic relationships wrought by the accelerating globalization of capitalism. It is impossible to know conclusively what these new relationships will mean for the future of today's world regions. For example, although the regionalization framework for this book locates Mexico with Latin America and the Caribbean, despite the fact that it too occupies the North American continental landmass, it is not inconceivable that in 20 years or so, a new regionalization will combine Mexico (and possibly other countries in Central America) with the United States and Canada because of immigration, internal demographic changes, or because of the impacts of economic alliances like the North American Free Trade Agreement (NAFTA). For the present, however, the United States and Canada persist in public imagination as well as in more formal political and economic frameworks that constitute a world region.

Environment and Society in North America

Perhaps the most significant aspect of North America's physical geography is that it contains an almost unimaginable bounty of resources from the whole range of minerals, to vast forests; fertile, highly productive land; extensive fisheries; varied and abun-

Figure 7.2   Population distribution of North America While the U.S. population is distributed thinly across the country, with the heaviest concentrations across the east and west coasts, the Canadian population has coastal clusters but only a thin band of settlement from east to west, hugging the U.S.–Canada border. For a view of worldwide population density and a guide to reading population-density maps, see Figure 1.40. (*Source:* Center for International Earth Science Information Network (CIESIN), Columbia University; International Food Policy Research Institute (IFPRI); and World Resources Institute (WRI). 2000. *Gridded Population of the World* (GPW), Version 2. Palisades, NY: CIESIN, Columbia University. Available at http://sedac.ciesin.org/plue/gpw)

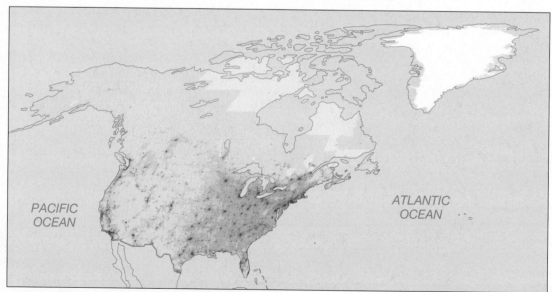

PACIFIC OCEAN

ATLANTIC OCEAN

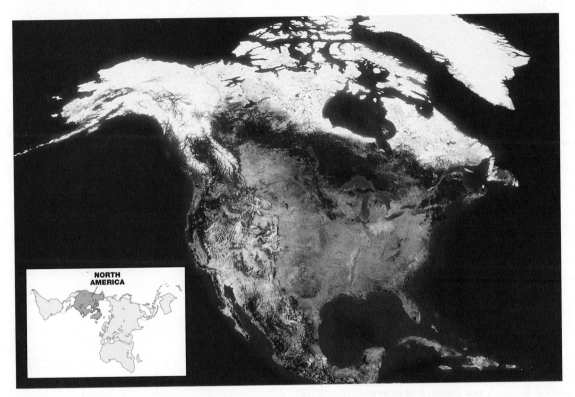

Figure 7.3 North America from space The image of North America from space shows a region surrounded by oceans, gulfs, and bays with vast interior plains and high mountains on the east and west coasts. The region also contains vast mineral wealth, extensive forests and supplies of clean water, and high agricultural productivity. [*Source:* (inset) Map projection, Buckminster Fuller Institute and Dymaxion Map Design, Santa Barbara, CA. The word *Dymaxion* and the Fuller Projection Dymaxion™ Map design are trademarks of the Buckminster Fuller Institute, Santa Barbara, California, © 1938, 1967 & 1992. All rights reserved.]

dant wildlife; and magnificent and unique physical beauty (**Figure 7.3**). Accompanying this extensive physical wealth has been the technological capability and the drive to exploit these resources to an extent that few other regions on Earth have been able to accomplish. One result of this combination of physical resources and human ingenuity is a region that experiences an extraordinarily high quality of life. A second result is a high level of consumption that produces a region with elevated levels of air and water pollution, numerous sites—often on the fringes of major urban settlements—of soil contamination, an ongoing problem of nuclear waste disposal, acid rain, extinct and endangered species, and increasingly frequent reports of insect and animal genetic mutations, just to name the most commonly known environmental problems.

The physical geography of the region has been an important context for, and an inescapable reminder of, the enormous benefits and the sometimes devasting costs of the "North American way of life." For instance, the Mississippi–Missouri river system is the most extensive and navigable river system in the world. This system, as well as the five Great Lakes (Superior, Michigan, Huron, Erie, and Ontario), undeniably enabled the early and formidable growth of the interior of the North American region in ways that no railway system could have. At the same time that the river system enabled growth and prosperity, however, the engineering that has been applied to the Mississippi–Missouri system (including its major tributaries) over a century or more has also irretrievably altered its ecosystem and exposed vast numbers of human settlements to extreme danger from flooding. De-

spite the awesome technological power that humans have developed, the physical world still acts as a centrally significant factor in North American regional growth and change.

Landforms and Landscapes

As **Figure 7.4** shows, the North American region is centered on a vast central lowland that includes the Canadian Shield, the Interior Lowlands, and the Great Plains. To the east of this central lowland are the Appalachian Mountains, which descend gradually to the Gulf-Atlantic Coastal Plain that becomes more broad the farther south one travels. To the west of the central lowland are three distinct topographical regions. Moving from west to east are the mountains and valleys of the Pacific coastal ranges; then an **intermontane** set of basins, plateaus, and smaller ranges; and finally the great Rocky Mountain range, which rises steeply and imposingly at the western edge of the central lowland region.

The western coastal formation sits along the fault line of two active crustal plates, the Pacific Plate and the Juan de Fuca Plate (see Figure 2.5, p. 55). Both plates are traveling northward rubbing against the more stationary North American Plate upon which the continental landmass sits. The coastal area from San Diego through British Columbia to Alaska is subject to frequent tremors. Extreme and devastating earthquakes have occurred in the past and are likely to occur again in the near future. The extraordinary views of the Pacific Ocean provided by the mountainous topography of the Pacific coastal

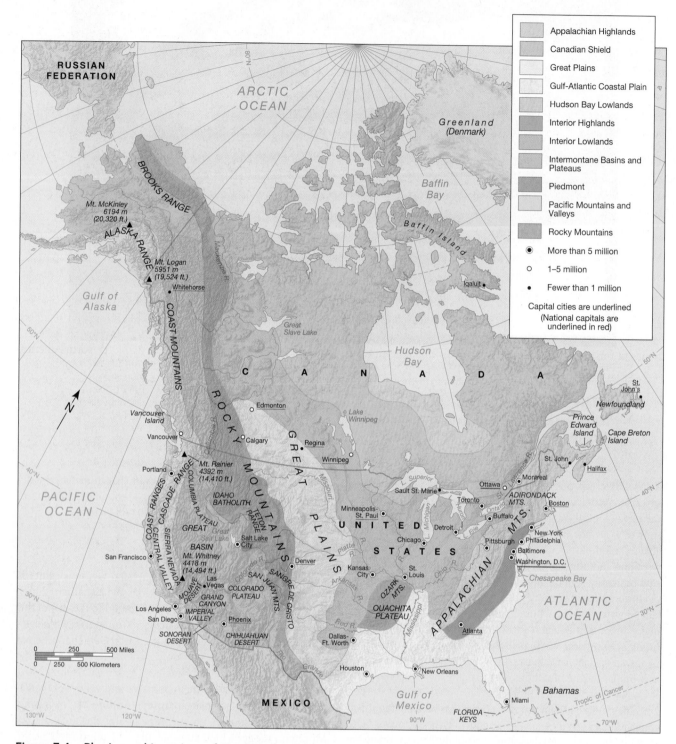

Figure 7.4   Physiographic regions of North America Various landforms and significant features define the North American region. The Interior Lowlands and Great Plains are of central importance for agricultural production. The mountain ranges in the west are significantly higher and steeper due to their more recent emergence, while the Appalachian Mountains in the east are lower and more sloping due to their older age and the effects of erosional forces. While the eastern coastal region slopes toward the sea, the West Coast is characterized by steep cliffs and fairly narrow beaches. Waves are also more dramatic in the west, as the steep land/sea interaction creates the conditions that push the advancing surf higher and steeper. The seashore is also more dangerous and prone to riptides and deadly undertows. Not surprisingly, because of the higher waves, the West Coast attracts a large number of surfers.

region have attracted a great deal of home development on mountain slopes, and past earthquakes have destroyed many of these homes. The extensive rainfall that marks this area has also wreaked havoc despite extraordinary feats of engineering that have placed multimillion-dollar homes on steep mountainsides. When the soil becomes saturated with rain water, liquefaction occurs, and the soil literally moves in one massive slump, carrying very large structures along with it.

In the western United States, the intermontane basin and plateau formation that lies between the Pacific coastal range and the Rockies includes four major physiographic provinces: the Columbia Plateau to the north, which begins at the headlands for the Columbia River; the Colorado Plateau in the south, which includes the erosional landscape of the Grand Canyon; in between, the Great Basin in Nevada and Utah, which includes extinct lakes as well as the Great Salt Lake; and the southwestern deserts (Mojave, Sonoran, and Chihuahuan). Ranching and mining have historically characterized the Great Basin region, but recreation and tourism is now the leading economic sector. Copper mining, as well as oil, gas, coal, and uranium mining are still important primary sector activities in this region (see Chapter 1, p. 43). In addition, hydroelectricity is generated for much of the western United States from the many dams that have been built in this region.

The Rocky Mountains, extending from Alaska to northern New Mexico, constitute the eastern edge of the intermontane basin and plateau region. A significant aspect of both western and eastern North American mountain ranges is that their north-south orientation does not form a barrier to the southward flow of polar air masses or the northward flow of tropical ones. For places like Arizona and New Mexico, northward-flowing air masses bring summer monsoonal precipitation patterns that support the region's unique desert vegetation. The summer and winter rains there allowed the ancestors of contemporary Native Americans to introduce the cultivation of beans, squash, and corn using sophisticated irrigation systems in a landscape that would otherwise seem inhospitable to subsistence agriculture. The absence of mountainous barriers to the north means that polar air sweeps down regularly in the winter months all the way into the midsection of the Interior Lowlands.

The extensive lowland region that extends from the mouth of the Mackenzie River in Alaska to the Gulf of Mexico can, as mentioned earlier, be further subdivided into three provinces. The first of these is the Canadian Shield, a geologically very old region rich in mineral wealth. The second is the Interior Lowlands, a glaciated landscape that contains fertile soils and an abundance of lakes and rivers. This region is devoted mostly to agriculture, with some industry present as well. The third subdivision is the Great Plains, an area that slopes gradually upward toward the Rocky Mountains. The Great Plains region is also delimited by the fact that it experiences more rainfall than the regions surrounding it. This is an area of extensive gently rolling and flat terrain with excellent soils and some of the world's most productive farms. Beef, pork, corn, soybeans, and wheat are produced here.

The major landform feature of eastern North America is the Appalachian mountain chain. Surrounding it in its eastern and southern flanks is a coastal plain that can be divided into two subregions, the Piedmont area of hills and easterly sloping land and the Gulf-Atlantic Coastal plain, a lowland area that extends from New York to Texas. The coastal plains are generally level, with soils that are sandy and relatively infertile. Where agriculture does occur, it is intensive and scattered. For instance, the area of the lower Mississippi is an area of good soils and intensive agriculture. The Piedmont is the historic region of plantation agriculture—cotton, tobacco, and corn—but much of the soil in this region is depleted or eroded from overfarming.

Climate

Because temperature, precipitation, and terrain patterns combine to influence vegetation, and to some extent soil, it is important to understand the role that climate plays in the subregions of North America. With respect to temperatures, North America ranges from the Arctic Circle in the north, where very cold temperatures persist for most of the year, to very close to the tropic of Cancer in the south, where warmer, tropical temperatures are the norm. In short, North America contains nearly every type of climate condition possible, with temperatures varying quite dramatically on any one day of the year from north to south. Of course, within-region differentials also occur due to the moderating influence of the oceans on three sides, the Gulf of Mexico, and the interior Great Lakes, as well as mountains and plateaus. Generally speaking, central North America experiences the most dramatic temperature ranges, while the coasts have far less of a range due to the influence of the oceans.

Variations in amounts of precipitation also have a profound effect on climate. While the east and west coastal areas tend to be mild and moderately wet, the interior is largely arid as the north-south mountain chains prevent moisture-bearing clouds from moving inland to drop their moisture. Because of this, a moisture gradient exists that declines slowly but continuously from east to west as far as the three significant mountain ranges—the Rockies, Sierra Nevadas, and Cascades (see Figure 7.4). Once beyond these mountains, the moisture gradient rises dramatically toward the Pacific. In the southeastern part of the United States, where no significant coastal ranges exist, moisture-bearing clouds are able to more readily condense into rain. In the Arctic north, annual precipitation approximates desert conditions due to the dominance of very stable air masses with low moisture content. It is also important to remember that the jet stream (see Chapter 2, p. 55) brings precipitation to most of the continent in the winter months. While the warmer parts of the region—in the southern United States—experience this precipitation as rain, the colder, more northerly parts experience snow. In areas around the Great Lakes, the warming effects of these large bodies of water add even more moisture to the mix, bringing especially heavy snowstorms (called "lake-effect" snow) to places like Buffalo, New York, on the northeastern tip of Lake Erie, and Sault St. Marie, Ontario, on the channel between Lake Superior and Lake Huron.

As described in Chapter 2, climate is directly linked to vegetation and soil and the conditions for agriculture and forest growth. Overall conditions are very good for agriculture as one moves eastward until the precipitation gradient drops to a very low level just beyond the Interior Lowland and Great Plains regions. Beyond that point, soil fertility is low and rainfall is limited and infrequent. Agriculture picks up again in the valleys along the Pacific Coast. As expected, most of the agricultural productivity of the North American region is concentrated in the Interior Lowland and Great Plains regions. Although natural conditions favor these subregions for the highest agricultural productivity, other parts of the North American region are also important agriculturally, largely because farmers there have overcome the natural barriers to production through fertilizers, irrigation, pesticides, and other technological applications. This is the case in the Pacific valley areas, for instance, where irrigation water drawn from the Colorado River enables agriculture to flourish, and in the Southwest, where intensive irrigation enables significant cotton and citrus production.

Environmental History

The Europeans who arrived on the Atlantic Coast of the United States and Canada beginning in the late fifteenth century encountered an environmentally diverse landscape thinly populated by native peoples. These were the peoples who first "discovered" North America, having traveled as Old World hunters and gatherers across the broad low-lying belt of tundra known as the Bering Strait land bridge when sea levels dropped and more land was exposed. It is estimated that more than 30,000 years ago these peoples began the process of populating the continent and altering and permanently changing the environments they encountered. Despite what written accounts of initial contact between European explorers and native peoples record, Europeans did not discover a "pristine" or "virgin" land but one that had already been transformed by tens of thousands of years of human settlement.

Most archaeologists, anthropologists, historians, and geographers believe that during the last great ice age the gradual retreat of the great ice sheet that had once covered much of North America enabled the first group of Arctic hunters to cross from Siberia into present-day Alaska, between 20,000 and 35,000 and possibly up to 60,000 years ago (**Figure 7.5**). They moved into North America by traveling southward along the western edge of the North American icecap. Because Canada was probably entirely covered by ice during this period, it is likely that people settled the southern part of North America before the northern part was free of ice and that the Inuit, who currently live in Canada's Arctic region, were the last of the aboriginal people to reach North America. These neolithic, or Stone Age, hunters probably originally came from northern China and Siberia. The descendants of these first hunters gradually moved farther and farther southward into the North American continent, advancing eventually into Mexico.

There is abundant evidence that the ancestors of the neolithic hunters eventually spread throughout North America (as well as Central and South America) and adapted their way of life to the particular conditions they encountered as they moved and settled. As they began to settle, different native groups introduced agriculture in different places. With game, fish, and wild and cultivated foodstuffs available, an economic system based on subsistence production and trade became established. The cultivation of maize spread to wherever it could be grown, and new wild foods, like potatoes and tomatoes, were eventually domesticated.

Significant technological accomplishments also characterized the cultures of early North Americans. For example, in the area around the Ohio and Mississippi valleys, the Hopewell culture arose around 500 B.C. Hopewell settlements were characterized by earthen mounds used for defense and later as burial sites (**Figure 7.6**).

In New England, prior to European contact, hunting, gathering and some shifting cultivation existed among the indigenous peoples. Hunter-gatherers were mobile, moving with the seasons to obtain fish, migrating birds, deer, and wild berries and plants. Shifting agriculture was organized around the planting and harvesting of corn, squash, beans, and tobacco. For both groups, a wide range of resources were identified and used. The economy was based on need. Need was attended to by planting or foraging or through barter (for example, trading corn for fish). Moreover, the prevailing ethos was to take only what was needed to survive. In addition to native peoples having no concept of private property or land ownership, there is also no evidence that a profit motive existed before contact with Europeans. Land and resources were shared in common. Still, while New England native peoples did appear to live in something of a balance with the natural resources they exploited, substantial vegetation change occurred as a result of their settlement and hunting activities (often using fire), which resulted in some species depletion.

Figure 7.5   Migration of neolithic hunters into the Americas It is believed that the first humans to enter the North American continent by way of the Bering Strait land bridge, which had been exposed through the gradual retreat of ice sheets, were hunters pursuing large mammals and mastodons. When these animals grew scarce, these small bands of hunters moved on, looking for new prey for sustenance. They also fed themselves with roots, plants, and berries and learned to fashion clothing, weave baskets, and construct fishing nets to take advantage of the available fish populations. Because of the limits of the environment, the bands tended to remain small, and as the population size increased, small groups would break off and move on to increase their chances of survival in more plentiful surroundings as yet unoccupied or hunted. Within several generations the multiplying bands of hunters became increasingly differentiated in language and custom as they interacted with the environment around them. (*Source:* Redrawn and adapted from E. Homberger, *The Historical Atlas of North America.* London: Penguin Books, 1995, p. 21.)

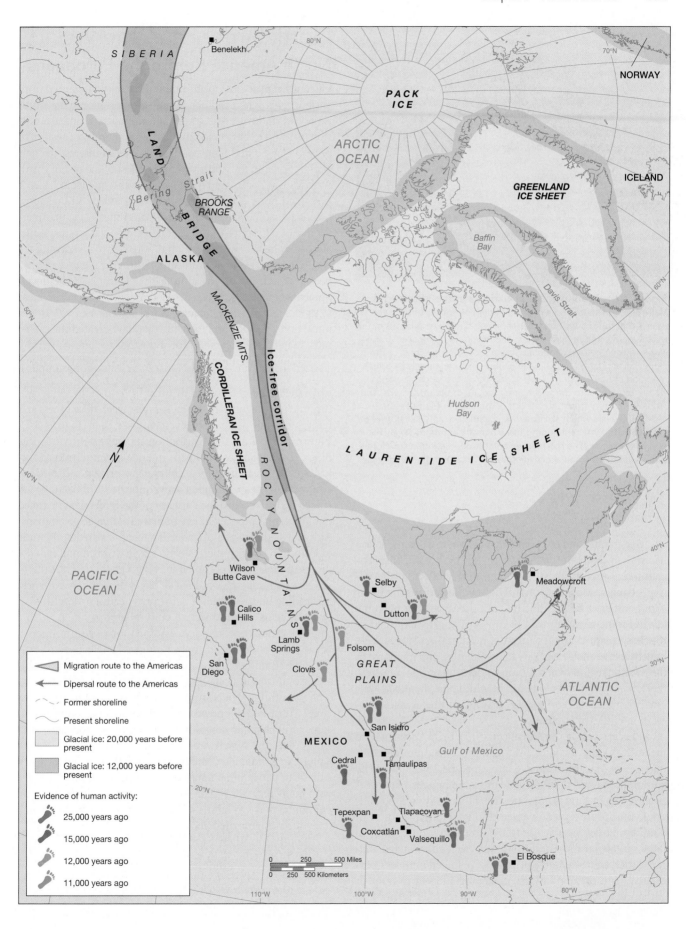

SIBERIA

Benelekh

80°N

70°N

NORWAY

PACK ICE

ARCTIC OCEAN

ICELAND

GREENLAND ICE SHEET

L A N D

Bering Strait

BROOKS RANGE

Baffin Bay

60°N

ALASKA

B R I D G E

MACKENZIE MTS.

Ice-free corridor

Davis Strait

50°N

CORDILLERAN ICE SHEET

R O C K Y   M O U N T A I N S

Hudson Bay

L A U R E N T I D E   I C E   S H E E T

N

40°N

PACIFIC OCEAN

Wilson Butte Cave

Selby

Meadowcroft

Calico Hills

Dutton

Lamb Springs

Folsom

San Diego

Clovis

GREAT PLAINS

30°N

ATLANTIC OCEAN

San Isidro

MEXICO

Gulf of Mexico

Cedral

Tamaulipas

20°N

Tepexpan

Tlapacoyan

Coxcatlán

Valsequillo

El Bosque

Legend	
◁	Migration route to the Americas
←	Dipersal route to the Americas
⌇	Former shoreline
∿	Present shoreline
▢	Glacial ice: 20,000 years before present
▨	Glacial ice: 12,000 years before present

Evidence of human activity:

25,000 years ago

15,000 years ago

12,000 years ago

11,000 years ago

0   250   500 Miles

0   250   500 Kilometers

110°W   100°W   90°W   80°W

Figure 7.6 Mound town of Cahokia in the Mississippi Valley In the south and along the fertile floodplains of the river valleys that ranged east from Oklahoma and north to Wisconsin, the Mississippian culture flourished as the inheritors of the ancestral Hopewell culture. As mentioned, mound building was part of the Hopewell and later Mississippian culture. Pictured here is an artist's rendition of how those mounds appeared on the landscape and were integrated into the spatial organization of the settlement. The Hopewell culture dominated much of the eastern Midwest for nearly 500 years where skilled artisans gathered materials from the eastern half of the continent to make elaborate ornaments and everyday objects. The mounds were flat-topped earthen structures that elevated ceremonial activities above the plane of everyday life.

Geographer and environmental historian William Cronon has shown that Europeans saw the natural world they encountered in North America much differently from the native peoples. Most importantly, they viewed resources as commodities to be accumulated not necessarily for personal use but to be sold for profit or export. The arrival of the Europeans meant that pressures upon natural resources were hugely accelerated, especially for wood, furs, and minerals. In the Atlantic region, where European settlement first occurred, there was extreme exploitation of white pine, hemlock, yellow birch, beaver, and whales that led ultimately to deforestation and extinctions.

The arrival of the Europeans in North America also meant a dramatic change in prevailing social understandings of the nature of land. Native perspectives about the communality and flexibility of land were replaced by European views of land as private and as having fixed boundaries. European settlers wanted to own and fence a plot of land, which led eventually to the concentration of land in large private farms, plantations, and haciendas. Increasingly, North American native peoples were forced onto less productive land or reservations, not being allowed to hunt or gather on private lands or to move with seasons as they had previously.

North America in the World
North America Before the Europeans

Fifteenth-century European views of the world inherited from ancient geographers did not include the existence of the North American landmass. And while sixteenth-century Spanish missionaries and explorers identified the southernmost section of present-day United States and Mexico as of interest to their exploration and missionizing efforts, most of the rest of North America was considered to be of little consequence because it presented none of the appearances of grandeur and resource potential that the Aztec, Maya, and Inca empires of Latin America did (see Chapter 8). It is therefore especially remarkable, as geographer Robert Mitchell has noted, that ". . . during the last 400 years this least promising part of the New World has been transformed, in the culture-bound terms of the development specialist, from one of the world's most backward areas to its most advanced: a dynamic world comprising two vast political units, Canada and the U.S., that have created the earth's most materially successful post-industrial societies. . . ."[1]

Although the Inuit appear to have had trading relations with Norse settlers at L'Anse aux Meadows on the northern tip of the Island of Newfoundland at the end of the tenth century, the most significant interactions between newcomers and natives did not occur until five centuries later (**Figure 7.7**). **Figure 7.8** portrays the distribution and subsistence practices of the native peoples of North America around 1600 and illustrates the great diversity of cultural groups that occupied the continent. When Columbus arrived in the Caribbean, he assumed he had reached the Far East, or the Indies, and he called the aboriginal peoples he encountered *los indios,* or indians. And just as thousands of native languages existed at the time of European contact, there was no single word common to the diverse peoples who occupied the continent to describe themselves. As a result, the word *Indian* has endured as an extremely misleading and sometimes derogatory term for describing a wide range of North American cultural groups.

The estimates of the native population of North America at the point of European contact range widely between 1 million and 18 million individuals. More than anything, the differing estimates indicate how little is known about the peoples who were thriving in North America before the Europeans arrived. What is well known is that there was no common culture, particularly no common language, among the many who first populated North America widely but sparsely. Native American scholar Jay Miller estimates that in 1492 North American native inhabitants spoke 2200 different languages, with many regional variations as well. Tribal culture and local environmental conditions were the frameworks within which daily life was governed and lived.

Europeans originally made contact with various individuals and tribes for help with resource extraction, including

[1]R. D. Mitchell, "The North American Past: Retrospect and Prospect," in R. D. Mitchell and P. A. Groves (eds.) *North America: The Historical Geography of a Changing Continent,* London: Rowman & Littlefield, 1987, p. 3.

Figure 7.7   Norse settlement in L'Anse aux Meadows, Newfoundland   Seaborne Norse adventurers pushed out of the Scandinavian peninsula to Britain, Ireland, and northern Europe at the beginning of the ninth century A.D. By the mid-ninth century, a number of Norse craft had reached Iceland, establishing a permanent settlement there. Near the end of the tenth century, the Norse reached Greenland and ventured to the coast of North America, establishing themselves at L'Anse aux Meadows on the northern tip of the Island of Newfoundland. This photo shows the remains of what are believed to be as many as three Norse settlements. According to available evidence, the Norse settlers and the Inuit at first fought each other, but then established a regular trading relationship. The Norse settlements were soon abandoned, probably as the Norse withdrew from Greenland.

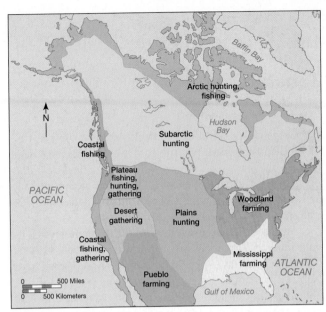

Figure 7.8   Subsistence practices of North American native peoples around 1600   This map, which shows the subsistence practices of indigenous peoples of North America at the point of European contact, testifies to the extensive spread of native population on the continent before the Europeans began to colonize. The geographic distribution of different tribes indicates their relationship to cultures and subsistence practices developed by the ancient people who preceded them. Compare this map to Figure 7.5 for an understanding of the long history of occupation of the North American continent before the arrival of Europeans and their current settlement distribution.

animal furs; naval stores, such as tar and turpentine for ship-building; fish; and other primary-sector products that would be exported back to the continent. The experience of the Dutch, French, and English in North America also differed substantially from that of the Spanish in Latin America (as described in Chapter 8). The Spanish conquistadors vanquished sophisticated civilizations in order to plunder their gold and silver treasuries and make themselves irresistibly rich in the process. While the Dutch, French, and English were also interested in improving their financial situation, they encountered a very different set of cultures with no centralized system of social control that they could exploit as the Spanish had done with the Aztecs in Mexico. Instead, the eastern tribes Europeans first encountered in North America were small, autonomous groups, possessing a lively sense of rivalry and competition with their neighbors. While the Spanish were interested in massive occupation and exploitation, the Dutch, French, and English considered exploration and colonization to be commercial ventures.

Colonization and Independence

No other landmass already occupied by a diverse range of complex and widely distributed societies, with the possible exception of Australia, has undergone such a dramatic transformation in such a short period of time. Moreover, lumping the settlers into one category of Europeans vastly simplifies the very complicated process of settlement that different peoples from different parts of Europe brought with them to North America. So, while the occupation by missionaries and settlers who came to North America in the sixteenth and seventeenth centuries is widely known as the period of **europeanization,** europeanization was actually a highly selective process. Only a few western European countries—France, Spain, the Netherlands, and Great Britain—dominated colonization in the United States. In Canada, the European colonizers were predominantly Britain and France. In both Canada and the United States, Great Britain was far and away the most influential of the four, though others did have substantial impacts, particularly the French in Canada and the Spanish in the United States. It is also important to point out that even within the four countries that dominated North American colonization, different groups from different regions settled in different parts of North America. Finally, the unusual individuals and groups who assumed great risks in coming to North America represented only a small sample of their national cultures. As a result, the europeanization process, as it unfolded along the Atlantic seaboard of the United States, was accomplished through the mixing of a very wide range of native and imported traditions that eventually created distinct colonial cultures and societies in different places.

The process of European settlement was a complex one that relied on physical as well as mental negotiations. At the

Geographies of Indulgence, Desire, and Addiction

Tobacco

Written by Jamey S. Essex*

The global spread of tobacco cultivation, manufacturing, and consumption over the past five centuries has made tobacco one of the world's most widespread agricultural products. Tobacco's development as a global commodity presents an excellent example of the diffusion and diversification of social practices and values.

Archaeologists believe Native Americans domesticated the wild tobacco plant more than 5000 years ago, based on evidence such as clay pipes found at several prehistoric sites. Native-American cultivators grew two main types of tobacco, *Nicotiana rusticum* in northeastern North America and *Nicotiana tabacum* in Central and South America. Practically all of the tobacco produced today is of the *tabacum* species. Native Americans ascribed to tobacco a number of economic, social, and cultural purposes, and by the time of sustained European contact in the fifteenth century, tobacco use was ubiquitous in the Americas.

Native Americans also developed all of the principal means of tobacco consumption—smoking, inhaling, and chewing—before European contact. Sailors and merchants carried these customs back to Europe during the sixteenth century, where tobacco found favor first in the ports and royal courts of Spain and Portugal. From the ports of these early colonial powers, tobacco use spread along trade routes to Africa, Asia, and the rest of Europe. By the end of the seventeenth century, the tobacco trade formed one of the most important parts of the colonial economy connecting the Old and New worlds.

With the constant spread of tobacco consumption came the extension of tobacco cultivation. Tobacco is a highly adaptable plant that can grow in a wide range of climatic and soil conditions, and has even been successfully cultivated as far north as Sweden and as far south as New Zealand (**Figure 1**). Despite its adaptability, tobacco is also quite sensitive to climatic and soil conditions. The result has been the development of hundreds of different tobacco types, each with its own regional complex of production and specific commodity uses based on the leaf's subjective qualities, such as taste and aroma. Tobacco types have changed, evolved, and even disappeared over the centuries of cultivation and use, with changes in consumer preference and market structure. The history of one particular tobacco commodity, cigarettes, illustrates the changing social and economic geography of tobacco production and consumption (**Figure 2**).

The modern cigarette did not enjoy widespread popularity until the middle of the twentieth century. Spanish and French consumers were the first to take up cigarette smoking in the early nineteenth century, and British soldiers brought Turkish cigarettes (probably developed from Spanish and

Figure 1 Tobacco plantation, Chiapas, Mexico The world's three largest multinational cigarette companies grow tobacco in scores of countries throughout the world. Mexico is just one of the countries that has a favorable climate for tobacco production.

French varieties) back with them from the Crimean War in the late 1850s. From the fashionable officers' clubs of London, cigarette smoking spread to New York's social elite in the 1860s. Manufacturers sprang up across Britain and the United States to feed and expand the growing demand for cigarettes, but several obstacles stood in the way of large-scale production and consumption. These early cigarettes required imported Turkish tobacco and expensive skilled labor to roll and package the finished product. In general, cigarettes were an upper-class urban luxury item, too expensive for the common consumer. In the United States, many considered cigarettes a passing fad, too expensive, European, and effeminate to be a viable long-term commodity.

The mechanization of cigarette production in the United States during the 1870s and 1880s made cigarettes available and affordable to most tobacco users. Large-scale mechanized production reduced labor and material costs and changed the structure of the entire tobacco industry, encouraging makers to expand production and create demand through intensive advertising campaigns, brand-name recognition (brand names often referred to British aristocracy or Middle Eastern luxury), and worldwide sales distribution. By the beginning of the twentieth century, American and British producers had pushed into markets from Shanghai to Cairo as cigarettes became an increasingly globalized commodity.

By 1910, the tobacco industry had developed into a "big business" dominated by the American Tobacco Company and its subsidiaries and partners around the world. The Supreme Court dissolved this monopoly in 1911 and divided the to-

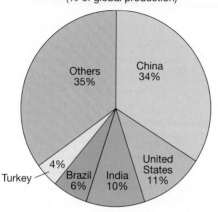

Top five tobacco–producing countries, 1998
(% of global production)

China 34%
Others 35%
United States 11%
India 10%
Brazil 6%
Turkey 4%

(a)

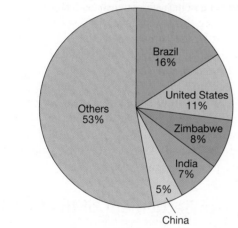

Top five tobacco–exporting countries, 1998
(% of global exports)

Brazil 16%
United States 11%
Zimbabwe 8%
India 7%
China 5%
Others 53%

(b)

Figure 2   Tobacco-producing and tobacco-exporting countries, 1998 (a) Global tobacco production serves 1.1 billion smokers. In China, which is the largest producer, 63 percent of males between the ages of 15 and 69 smoke. Only 3.8 percent of Chinese women smoke. (b) Despite the fact that cigarettes are a multibillion-dollar industry worldwide, in only four countries—Kyrgyzstan, Macedonia, Malawi, and Zimbabwe—do tobacco exports amount to more than 5 percent of total export earnings. (*Source:* The World Conference on Tobacco OR Health, 2000, Tobacco Fact Sheet, p. 2, http://tobaccofreekids.org/campaign/global/docs/facts.pdf)

Table 1	Number of Smokers, by Region, 1997*		
	Male	**Female**	**Total**
North America	27.80	25.50	53.30
Latin America and the Caribbean	64.06	37.24	101.30
Western Europe	62.11	43.81	105.92
Eastern Europe	76.38	46.45	122.83
Africa and the Middle East	90.68	44.16	134.84
Asia-Pacific	517.59	104.44	622.03
World Total	**838.62**	**301.60**	**1140.22**

*Numbers reported in millions.

Source: The World Conference on Tobacco OR Health, 2000, Tobacco Fact Sheet, p. 11, available at http://tobaccofreekids.org/campaign/global/docs/facts.pdf

produced cigarettes. The contours of the modern industry developed at this time, with several current manufacturers and their associated brands appearing in the aftermath of the Supreme Court's decision.

Through the twentieth century, the cigarette has become the most popular and widespread tobacco product, constituting more than half of British tobacco sales in 1920 and more than half of U.S. sales in 1941. Some of this market expansion came at the expense of other tobacco commodities but much also came during the 1920s when the gendered constructions of cigarette smoking changed and large numbers of urban women took up the habit. Manufacturers diversified their brands and advertising strategies accordingly, a process they have repeated to tap growing markets in the countries of the periphery. Today an estimated 46 million U.S. residents smoke cigarettes, while worldwide estimates place the number of adult smokers at more than 1 billion (**Table 1**).

Even as cigarette makers have opened new markets and realized astounding profits, they have had to answer to concerns about their products' social costs, particularly in relation to the impact on public health. Several state- and federal-level legal battles in the past decade have forced "Big Tobacco" to reassess its product and marketing strategies and pay out billions of dollars to cancer victims. These court decisions have set off further political debates about corporate responsibility, public health, and the allocation of settlement money. The trend of litigation against tobacco companies within the United States may set important precedents for makers in other countries as well and has forced a reassessment of tobacco's social value.

bacco industry among a handful of successor companies carved from American Tobacco. The dissolution sparked a new round of market competition and brand innovation, the most important of which was the development of the blended American cigarette. This new product added sweetened Kentucky burley tobacco to the Turkish tobacco and Virginia and North Carolina bright tobacco that had originally filled mass-

*Jamey S. Essex earned his M.A. in geography from Syracuse University.

level of ideas, the Europeans understood there to be a direct relationship between the appearance of the landscape and the capacities of its inhabitants. A wild and untamed landscape needed taming and transforming, and the apparent wildness of the native peoples suggested an inferior status for them in the grand scheme of building a "new world" on a new continent. As a result, interactions between native peoples and explorers and colonists routinely resulted in the overwhelming exploitation and abuse of the former by the latter. The history of colonial settlements, while at first peaceful, over time erupted into disputes over land claims that ended in violence and often outright massacres on both sides.

After a time, not only were there conflicts between tribal people and colonists, but direct conflict emerged between the various European groups that vied for control over land. Additionally, colonists fanned the flames of rivalry between opposed tribal groups by providing them with arms, thus elevating the level of technology and intensifying the degree of violence in armed conflict. Finally, as has been widely documented, exposure to the Old World diseases that the colonists brought with them had a devastating impact on native populations (see Chapter 8, p. 387). As native populations were decimated, defeated, or demoralized and pushed farther into the interior of North America, the various European groups increasingly came into direct conflict. Small-scale skirmishes and competition for land eventually led to the Seven Years' War (1756–1763), whose American phase is known as the French and Indian War (1754–1763). This war left the British more or less triumphant over the whole of the European-inhabited territory of North America. In the following two decades, residents of the original 13 colonies of the United States became disillusioned with their administrators in Britain—who were taxing them to help recoup the high cost of the Seven Years' War—and launched their own war, leading to the creation of a new, independent nation in the late eighteenth century. In Canada, a bloodless separation from Great Britain would not occur until well into the nineteenth century.

Even before the American Revolution, however, a process of **americanization** had begun as a generation of individuals of European parentage born in the U.S. colonies felt less loyalty and fewer cultural ties to the mother country. As a result, a new ethos of liberalism, individualism, capitalism, and Protestantism emerged, gained currency, and ultimately came to define a U.S. national character. The successful outcome of the Revolutionary War with Britain (1775–1783) left the continent with a robust new nation dominated by Anglo-American institutions and with the addition of slavery. Canada remained a colony under British control composed of both French- and English-speaking settlers.

The Legacy of Slavery in the United States

Although all along the Atlantic seaboard of the United States and Canada the impact of European colonization in the sixteenth and seventeenth centuries was felt, the subsequent development of the U.S. South following the end of the revolutionary period differed dramatically from that of its northern neighbors (**Figure 7.9**). Before the arrival of the Europeans, the area that now coincides with the southeastern United States was inhabited by a wide range of native tribes, among them the Cherokee, Choctaw, Chickasaw, Creek, and Seminole peoples. Early on, the region was occupied by military personnel living in scattered outposts like Jamestown in Virginia. However, by the mid-seventeenth century the military outposts had given way to tobacco farms (see Geographies of Indulgence, Desire, and Addiction: Tobacco, p. 324). At first, **indentured servants** from Britain were the primary source of labor on the tobacco and later indigo and cotton plantations. Increasingly, however, slaves from Africa replaced the servants who had earned their freedom, at the same time that disease, armed conflict, and demoralization reduced the native populations that would have been another source of laborers.

African slaves had been a well-established commercial staple of the Mediterranean well before the Spanish and Portuguese introduced them to their newly captured territories in Latin America and the Caribbean, thereby establishing the Atlantic slave trade (see Figure 6.14, p. 274). By the early fifteenth century, Dutch and English raiders attacking Spanish and Portuguese ships were able to wrest control of the slave trade so that by the early seventeenth century, England became the dominant slaving nation. As a result, slaves were a part of the social and economic system of the American colonies beginning, practically simultaneously, with their founding. As an institution of formal social and economic organization, slavery endured in the South for more than 250 years, ending officially in 1870 following the end of the U.S. Civil War (1861–1865). Its legacy, however, continues to shape the landscape and identity of the region.

European Settlement of the Continent

With the creation of new nations and the transformation of colonies into states, the steady but relentless process of European settlement of the North American continent accelerated, more so in the United States than in Canada, however. By the middle of the nineteenth century, settlement in the United States had pushed beyond the Appalachian Mountains into the interior lowland region, including the upper Ohio and Tennessee river valleys and the interior South. France's loss of control of Canada to Britain in 1763 had little impact on new settlement there. By the end of the century, however, southern Ontario, in and around present-day Toronto, became attractive to settlers (**Figure 7.10**).

Historian Frederick Jackson Turner in the United States and the Laurentian school in Canada (so called because of its association with the St. Lawrence region, the earliest settled part of Canada) have demonstrated the significance of the frontier to the North American psyche. Historians and social scientists have argued that frontier settlement involved a continual process of national and personal reappraisal as well as of increasing geographical divergence as new settlers encountered new landscapes. It was also a process of sustained mobility so much so that mobility has come to be seen as a

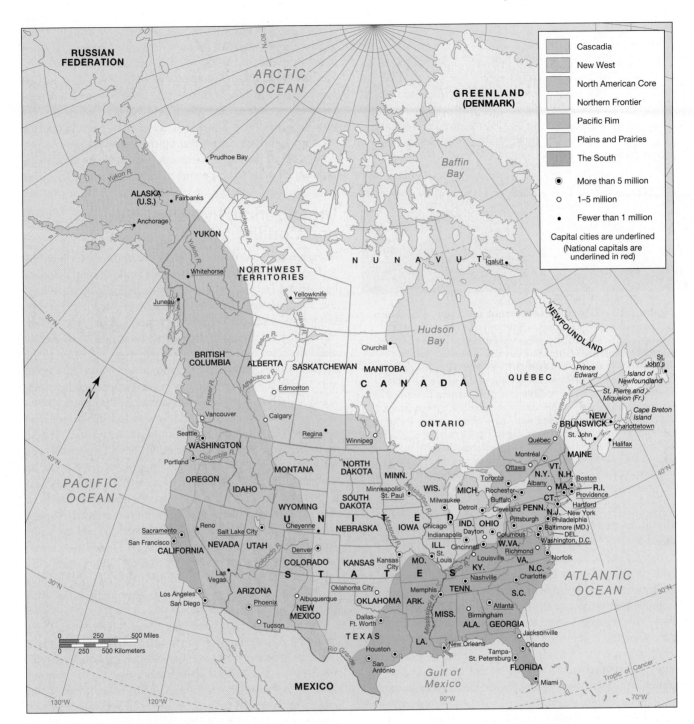

Figure 7.9   North American subregions North America consists of any number of cultural subregions, only some of which are identified on this map. The most central and most widely known subregions are shown, and we will have much more to say about these subregions in the last section of the chapter. It is a fairly constant phenomenon of North American journalism that every five or so years, another popular book on North America's new subregions is published. This phenomenon of serial publication has as much to do with the fact that capital and people are increasingly mobile, thereby creating new subregions as they participate in contemporary economic transformations, as it does with the hunger of the public for new ways to think about themselves.

North American characteristic. In the United States the movement of the frontier was continuous *and* mostly contiguous, at least until settlers reached the Great Plains in the middle of the nineteenth century and confronted significant mountain ranges at its western edge. In Canada, contiguous westward expansion was interrupted early on by the vast, generally infertile, though heavily forested Canadian Shield, which separated Ontario from the prairies. As Figure 7.2 shows, many Canadian settlers leapfrogged across to the northern midsection of the country in order to acquire suitable farmland.

Figure 7.10 Canadian settlement patterns In Québec the first settlers laid off long, narrow lots from the shores of the St. Lawrence River area into the interior, as shown in the above image. As settlement moved farther inland, roads were built parallel to the waterways with narrow lots extended on either side. The pattern is duplicated in the Red River valley of Manitoba, where the early settlers were also French. In Ontario and the eastern townships of Québec, land subdivision was made according to preconceived plans. Although the townships were more or less square, the grid became irregular because it was started from a number of different points, each of which used a differently oriented base. In the prairies the grid is much more regular, partly as a result of the topography, partly because a plan for the subdivision of the whole region was laid out in advance of settlement.

In the United States the pace of westward expansion was accelerated by the federal government's decision in the late 1780s to transfer public lands into private hands by selling land cheaply to citizens. By 1850, the development of the railroads reoriented the pace and direction of continental settlement—eastward from the Pacific Coast to the interior west rather than from the East Coast westward—at the same time that it diminished the previous isolation of pioneer settlements. By the close of the nineteenth century, the frontier process had resulted in a set of rural and agrarian regions and subregions that stretched across the continent. Each of these regions was defined by its own experiences of the history and particular conditions of settlement and by its distinctive regional economic development.

It should be recalled that the North American economy during this period was oriented to agro-mercantile activities. This means that the trading of agricultural crops and primary resources, such as fish, timber, and minerals, provided an economic base for the expanding population. Yet, in the United States by the mid-nineteenth century, a new economy based on manufacturing was rapidly gaining momentum along the northern Atlantic seaboard, especially in and around southern New England and New York. At the same time the remainder of the United States and Canada was being settled by European immigrants or their descendants, making their livelihoods largely by farming. By the early twentieth century, however, with the continent occupied from east to west mostly by

Europeans and Euro-Americans, the industrialization of the U.S. economy was well on its way to transforming the landscape from one of rural agricultural settlement to one of urbanization and industrialization. The 1920 census documented for the first time in U.S. history that there were as many people living in cities as there were in rural areas. From that point onward, the United States and, soon after, Canada became increasingly urbanized so that by the late twentieth century, 75 percent of the North American population lived in cities, up from 25 percent in the mid-nineteenth century.

Urbanization, Industrialization, and Conflict

As mentioned, North America had experienced city-building among native peoples long before the explorers and colonists had arrived. Yet the colonial era is the period when an extensive and intensive urbanization process began in earnest. The Europeans who colonized North America were part of commercial urban systems in their home countries. As they colonized the new lands they encountered, they responded to the need for central places for organizing commerce, defense, communication, and, later, administration and worship by building cities. The Spanish colonizers founded cities in Florida and in the southwestern United States because cities were symbols of political and military authority. St. Augustine, Florida; Santa Fe, New Mexico; San Antonio, Texas; and San Diego and Los Angeles, California, bear the visual imprint of the Spanish influence on American urbanization. The French explorers came not to settle but to reap commercial rewards, and they established urban centers in order to facilitate the exchange of goods. Québec, Montréal, Detroit, and St. Louis were early commercial ports (**Figure 7.11**). The Dutch also established urban settlements for trading centers, including for furs and slaves. Their most impressive accomplishment is New York City, once known as New Amsterdam, now North America's most important city.

With few exceptions, however, the British played the most important role in shaping North American urbanization and urban life. Sustained by trade based on an agrarian economy, the U.S. Atlantic coastal cities established by the English colonists were also oriented around a kind of corporate communalism tempered by notions of social and religious harmony. As geographer Alan Pred has shown, in addition to being administrative centers, Boston, Providence, Baltimore, Philadelphia, Charleston, and Savannah were also ports and key nodes in a globally expanding mercantile system. As such, they enabled the transfer of resources, goods, and people, not only from the interior hinterland into the cities but also outwardly to Europe. At the same time these cities received goods shipped from England for North American consumer markets. It is also important to point out that the colonists saw their burgeoning cities not only as commercial centers but also as places where new ideas could be hatched and nurtured and as hearths of "civilization" where Old World cultural practices confronted those of the New World, creating in the process uniquely North American urban places.

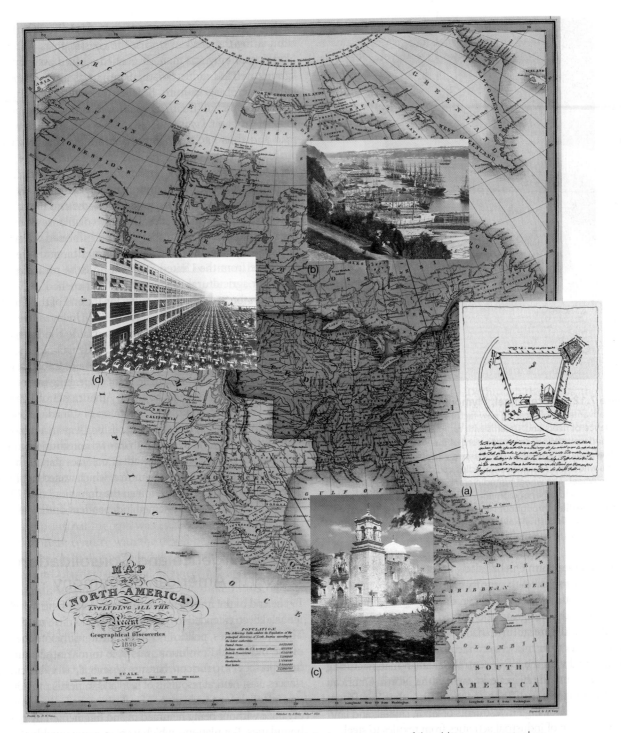

Figure 7.11   Early North American urbanization (a) St. Augustine is one of the oldest permanent urban settlements in North America, having been established by the Spanish in 1565. (b) French cities in Canada were originally important as centers for the fur trade. Today Québec City is a modern city with a center of historic architecture. (c) A significant site in the missionizing of native populations, San Antonio is now an important economic node in the American Southwest. (d) Like Québec City, Detroit began as an important fur trading center. In the twentieth century it became the most important site for automobile manufacturing in the United States. Today its economy is largely based on services.

As sites of innovation and cradles of culture, cities were also largely, though not exclusively, the places where, by the early nineteenth century, a new economy based on manufacturing was born and flourished. At first factories were located along the waterways of New England to take advantage of the power that could be harnessed from rushing rivers and streams that often dropped dramatically over steep falls (**Figure 7.12**). By the mid-nineteenth century, manufacturing sites were springing up along the Atlantic seaboard from north of the Chesapeake Bay to Maine as well as in the interior in southern

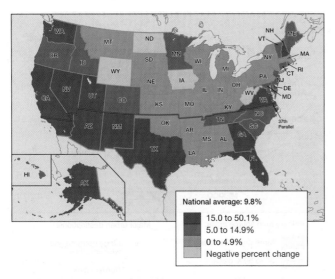

National average: 9.8%

15.0 to 50.1%
5.0 to 14.9%
0 to 4.9%
Negative percent change

Figure 7.20   Rustbelt and Sunbelt transformation, 1980–1990 The figure provides insight into the phenomenon of "people following jobs." As capital was disinvested from declining industries in the old Manufacturing Belt, it was being reinvested in new industries and services in the south and southwestern United States, especially the region south of the 37th parallel. Economic imperative—the need to have a job—impelled many to leave their families and hometowns for new lives in the relatively unsettled parts of the United States. While it is sometimes argued that people moved because the weather is nicer in the Sunbelt (hence the name), a new life in a warmer winter climate would not have been possible if jobs were not also migrating. (*Source: P. L. Knox and S. A. Marston, Places and Regions in Global Context,* 2nd ed. Upper Saddle River, NJ: Prentice Hall, 2000, p. 135.)

Figure 7.21   Bisbee, Arizona Bisbee is fairly typical of many of the mountain towns that have become attractive for settlement by those seeking a less frenetic lifestyle (so-called "lifestyle refugees"). Whereas places like Aspen, Colorado, have long since lost their original flavor as more and more wealthy residents have moved there, bringing their expensive lifestyles with them or demanding that they be created for them, Bisbee is still in the early stages of its new growth. It is too soon to tell whether Bisbee will undergo the same sorts of transformations that Aspen has, especially given that downhill skiing—and the amenities that accompany it—is not a sustainable economic option in the warmer Arizona climate.

The third wave of internal migration began shortly after the Second World War ended in 1945 and continued into the 1990s. Between the end of the war and the early 1980s and directly related to the impact of governmental defense policies and activities on the country's politics and economy, the region of the United States now commonly called the Sunbelt, and including most of the states of the U.S. South and Southwest, experienced a 97.9 percent increase in population. During the same period, the Midwest and Northeast, known as the Snowbelt or Rustbelt, together grew by only 33.3 percent. **Figure 7.20** shows a U.S. regional map illustrating the population changes, in terms of percentages, for the years 1980 to 1990. This map shows that the western, southwestern, and southeastern states experienced substantial population increases during that decade.

The first evidence of **suburbanization**—the growth of population along the fringes of large metropolitan areas—can be traced back to the late eighteenth century, when real estate developers looked beyond the city for investment opportunities and wealthy city-dwellers began seeking more scenic residential locations. Later, residents fled to the suburbs to get away from the new immigrants and their increasing hold over urban machine politics.

The process was rapidly accelerated, however, with the introduction of new transportation technologies—first horse-drawn streetcars, then commuter rail services, and, finally, automobiles. Each innovation in transportation allowed people to travel longer distances to and from work within the same or shorter time period. North Americans chose to move to the suburbs in massive numbers, not in the least because the suburbs were, arguably, considered by many to be more healthful places to raise a family. Suburbanization continues today in both the United States and Canada with a new wrinkle—a slight reversal of migration from urban to rural areas as retirees especially search out the good life on the far fringes of the metropolitan core in small towns like Bisbee, Arizona, and the Okanagan Valley in British Columbia (**Figure 7.21**).

The most compelling explanation for the large-scale population shift characteristic of the third migration wave is the pull of economic opportunity. Rather than reinvesting in upgrading the aged and obsolescent urban industrial areas of the Rustbelt, venture capital was invested in Sunbelt locations, where cheaper land and lower labor costs made the introduction of manufacturing and service-sector activity more profitable. The 2000 census shows a decrease in the rate of in-migration to the Sunbelt, but the changes in the geography of population at the beginning of the twenty-first century are dramatically different from the patterns of 150 years ago. This new population distribution illustrates the way in which political and economic transformations play an especially significant role in shaping individual choice and decision making.

North American Cultural Contributions

Despite the discrimination and bigotry that immigrants have experienced in becoming members of North American society, they have made significant and transformative contributions to enriching North American culture and influencing taste around the world. Music, art, literature, dance, architecture, film, photography, sports, fashion, journalism, and cuisine have all been shaped by the contributions of immigrants, not to mention the impact of immigrants on science, medicine, and technology. The influence of immigrants on music has been particularly impressive. Country, bluegrass, jazz, the blues, and rap are types of music that all originated in the United States but have deep roots in the Old World. From jazz to rap, African Americans have been responsible for musical innovations that have been widely accepted, applauded, and imitated throughout the world.

The early twentieth century origins and particularly U.S. expressions of jazz have been influential worldwide at the same time that they reveal a complex but clear lineage back to the African musical roots that slaves left behind in the wake of their terrible Atlantic passages. West African folk music forms one of the central foundations of jazz. But jazz was also influenced by European popular and light classical music of the eighteenth and nineteenth centuries. The earliest documented jazz style was Dixieland jazz, which emerged from New Orleans and was played by white musicians who recorded the new music form on phonograph records. The spread of these recordings helped jazz to become a sensation in the United States and Europe. Soon African-American jazz groups—the originators of the jazz style that was expressed through the related styles of ragtime, marches, hymns, spirituals, and the blues—were able to capitalize on the popularity of white Dixieland largely through the improvisational style of trumpeter Louis Armstrong. Armstrong migrated to Chicago in the 1920s, influencing local musicians and stimulating the evolution of the Chicago style. About the same time as jazz caught on in Chicago, Harlem was emerging as a center for jazz, organized around a highly technical, hard-driving piano style. Regional variations on the original Dixieland style emerged in the urban areas, where significant populations of African Americans had settled. Jazz continued to flourish from the 1930s through the 1950s with important regional flavors being developed. In the 1960s jazz began to lose popularity as audiences embraced mainstream rock and roll, which had itself been influenced by jazz and the blues. In the 1980s, jazz experienced a revival as a serious form of music, which it continues to enjoy today. Other distinctly North American musical and performance styles include rap, bluegrass, and musical theater, the latter having important roots in European opera.

The game of baseball is an another uniquely U.S. innovation, and it too has enjoyed widespread popularity beyond the national boundaries, especially in Caribbean countries like the Dominican Republic, Venezuela, and Cuba, but also in Europe and Japan, among other places. The composition of many U.S. and Canadian Major League Baseball teams (two—the Toronto Blue Jays and Montreal Expos—are Canadian) demonstrates just how popular this American sport has become worldwide. As a high-stakes commercial enterprise, baseball has traveled well. Consider the following: On opening day of the Major League Baseball season in 2001, nearly a quarter of all the players on the 30 team rosters were from countries other than the United States and Canada. These include Aruba, Australia, Colombia, Cuba, Curaçao, Dominican Republic, England, Jamaica, Japan, South Korea, Mexico, Nicaragua, Panama, Venezuela, and the Virgin Islands. The New York Yankees, Atlanta Braves, Florida Marlins, and Montreal Expos led all major league teams with 10 non-North American players each (**Figure 7.22**).

Figure 7.22   The globalization of North American major league baseball, 2001 The old saying that something is as "American as baseball or apple pie" may need revising given the dramatic transformations that have occurred over the last 10 years in the demographics of players on North America's Major League Baseball teams.

Canadian culture tends mostly to represent a mix of immigrant and British settler influences. Perhaps the most significant aspect of Canadian culture is the fact that it has had to battle the tremendous influence of commercialized U.S. culture, which because of its geographical proximity has been enormously difficult to resist. Canada has been very aggressive in its attempt to ward off the invasion of American cultural products and has developed an extensive and very public policy of cultural protection against the onslaught of music, television, magazines, films, and other art and media forms.

In early 1995, for example, the Canadian government levied an 80 percent excise tax against Time, Inc.'s Canadian version of *Sports Illustrated* because it was viewed as not being Canadian enough. The authorities complained that too many of the articles were directed at U.S. sports issues and not enough at Canadian ones. Other government bodies, such as the National Film Board of Canada and the Canadian Radio and Television Commission, are also active in monitoring the media for the incursion of culture. For example, 30 percent of the music on Canadian radio must be Canadian. Nashville-based Country Music TV was discontinued from Canada's cable system in the early 1990s and replaced with a Canadian-owned country music channel. Besides regulating how much and what type of culture can travel north across the border, the Canadian government also sponsors a sort of "affirmative action" grant program for its own culture industries (**Figure 7.23**).

Although they are heavily exposed to U.S. cultural products, Canadians have been able to produce distinctively Canadian film and literary pieces. Some of the most highly regarded authors among anglophone audiences are Canadians. These include Michael Ondaajte, *The English Patient;* Robertson Davies, *Deptford Trilogy;* Annie Proulx, *Shipping News;* and Margaret Atwood, *The Handmaid's Tale.* Prominent francophone writers in Canada include Michel Tremblay, *Les Chroniques du Plateau Mont-Royal;* Louis Hémon, *Maria Chapdelaine;* and Gabriel Roy, *Bonheur d'occasion.*

Certainly the most significant aspect of North American culture is the fact that it has been globalized and is widely imitated—and often embraced as much as it is vilified—around the world. Many scholars argue that "globalization" is really just a euphemism for "americanization," and it is difficult to argue against this perspective. It is also important to recognize, however, that just as U.S. culture is circulating intensively beyond its national borders, other cultures have come to influence U.S. culture in numerous and distinctive ways. The following examples help to illustrate the fact that the globalization of culture, though largely dominated by the United States, is not exclusively so. In 1999 other core countries began to exert a major cultural influence on the United States, for instance. It was the peak year of *Pokémon* trading cards and related commercial products manufactured in Japan. It was also a popular year for *Teletubbies,* a British children's television show that has become a serious rival to the U.S.-made *Sesame Street.* And Britain also delivered the Harry Potter children's fiction trilogy that occupied first, second, and third place for months and months on the *New York Times* fiction best-seller list. Clearly, globalization has made it possible not only for North American culture to circulate widely but for cultural products from other parts of the world to penetrate U.S. culture as well. It will be interesting to see what impacts the flows on different cultures into North America will have for existing traditions and practices.

Regional Change and Interdependence

Today the United States and Canada together produce more than one-quarter of the world's GNP. The United States has the world's largest economy; Canada, the ninth largest. Their resources are extensive and varied, and their ability to exploit them is high. As part of the recent restructuring of the global economy discussed in Chapter 2, the various regions of both the United States and Canada have experienced significant transformations in their economies, societies, political institutions, and even their physical environments. In this section we examine those changes and their implications, paying particular attention to the ways in which economy and politics have been reorganized to facilitate contemporary globalization and how social groups and the physical environment have shaped—and in turn have been shaped by—these changes.

The New Regional Economies

Political, economic, and social geographers would agree that the most important transformation of the last 25 years has been the rise of the service economy and the relative decline of manufacturing as the most significant employment sector in the United States and Canada. This phenomenon as it occurred in the United States was partially described in a previous section where the most recent wave of internal

Figure 7.23   Crash Test Dummies The internationally famous music group Crash Test Dummies produced its first album with the help of a $40,000 grant from the Canadian government. In short, the story surrounding cultural nationalism is that the effort to control cultural production is an intense one. This is especially true for Canada, which continues to struggle to establish an independent identity beyond the shadow of the United States.

migration, the movement of U.S. residents from the Rustbelt to the Sunbelt, was discussed. The rise of the U.S. Sunbelt is a classic case of how regional core-periphery patterns are modified to facilitate the accumulation of capital (as discussed in Chapter 2, p. 62). Between the 1960s and the 1970s, the historic core of North American industrialization, the Manufacturing Belt, began to experience economic problems in the form of high labor costs and aging infrastructure, mostly manifested in outdated technology systems. Once peripheral regions of the country, the South and Southwest began to become attractive to investors. The military had become an important investor in this region during the Second World War, when bases had been established and training exercises had been held. Following the war's end, the government continued to invest in this region as it built up its military capacity during the Cold War. Thus, by the 1960s, the South and the Southwest had substantial infrastructural development and a high level of technological sophistication organized around military applications—which have historically preceded the application of technology for civilian purposes.

As the computer age dawned, numerous places in the South and the Southwest were ripe for civilian investment opportunities, possessing abundant land and labor forces that were highly educated at the same time that they were unused to unions and high wage rates. The result was a shift, which has since been rebalanced, in the core-periphery patterns of the United States as the profitability of old, established industries in the Manufacturing Belt—or the Rustbelt, as it came to be called—declined compared to the profitability of new industries in the fast-growing new industrial districts of the Sunbelt. Once the profitability differentials between the two places became significant, disinvestment began to occur in the Rustbelt. Manufacturers there began to reduce their wage bill by cutting back on production; to reduce their fixed costs by closing down and selling off some of their factory space and equipment; to reduce their spending on research and development for new products. This disinvestment, in turn, led to deindustrialization in the formerly prosperous industrial core regions of the Midwest and Northeast. **Deindustrialization** involves a relative decline (and in extreme cases an absolute decline) in industrial employment in core regions as firms scale back their activities in response to lower levels of profitability (**Figure 7.24**). In effect, technological innovations in computerized production systems enabled new industrial applications, and investors and manufacturers began to look around for new places in which to invest and build. Innovations in transport and communications technology, combined with these production innovations, created *windows of locational opportunity* that resulted in the movement of capital investment in manufacturing away from the old industrial districts of the Manufacturing Belt and into small towns and cities in the Sunbelt region of the United States, to suburban fringe areas near some of the old industrial districts, and offshore to other countries that had lower-cost workforces.

Meanwhile, the capital made available from disinvestment in the Rustbelt became available for investment by entrepre-

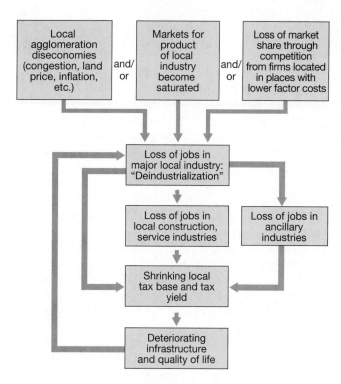

Figure 7.24   Spiral of deindustrialization When the locational advantages of manufacturing regions are undermined for one reason or another, profitability declines and manufacturing employment falls. This can lead to a downward spiral of economic decline, as experienced by the traditional manufacturing regions of North America during the 1970s and 1980s. (*Source:* P. L. Knox, *Urbanization.* Upper Saddle River, NJ: Prentice Hall, 1994, p. 295.)

neurs in new ventures based on innovative products and innovative production technologies. Old industries and a large proportion of an established industrial region were "dismantled" in order to help fund the creation of new centers of profitability and employment. This process is often referred to as **creative destruction**, something that is inherent to the dynamics of capitalism. Creative destruction provides us with a powerful image to understand the need to withdraw investments from activities (and regions) yielding low rates of profit and to reinvest in new activities (and, often, in new places).

The process does not stop there, however. If the deindustrialization of the old core regions is severe enough, the relative cost of their land, labor, and infrastructure may decline to the point where they once again become attractive to investors. As a result, a seesaw movement of investment capital occurs, which over the long term tends to move from developed to less-developed regions—then back again, once the formerly developed region has experienced a sufficient relative decline. "Has-been" regions can become redeveloped and revitalized, given a new lease on life by the infusion of new capital for new industries. This is what happened, for example, to the area in and around Pittsburgh in the 1980s, resulting in the creation of a postindustrial economy out of a depressed industrial setting. The USX Corporation, a worldwide producer of steel and oil and natural gas products, reduced its workforce in the Pittsburgh region from more than 20,000 to

The New Economy

In the last decade and a half, a new economy has emerged in North America that has fundamentally transformed industries and jobs through the revolutionary changes that have been brought about by information technologies (IT). These changes have been facilitated by a high degree of entrepreneurialism and competition, transforming all of the regions of North America as well as many other regions around the globe. But the new economy was born in North America, sired by the technological changes that emerged from Silicon Valley, California, nearly 50 years ago.

It is generally agreed that the previous economic order, the "old economy," lasted from 1938 to about 1974. The year 1974 was a critical year in economic history as it was a time when oil prices were skyrocketing, and the corporate rate of profit was falling in the core. The foundation of that economy was manufacturing geared toward standardized mass-market production and run by stable, hierarchically organized firms focused on the North American market. Massive political and economic restructuring was the response to the crises that rocked core regions, North America among them, which many regard as the transitional period (1975–1990) from the old manufacturing-based economy to the new technology, IT economy (**Table 1**).

The new economy, however, is about more than just new technology. It is also about the application of new technologies to the organization of work—from the impacts of biotechnology on farming to the impacts of IT on organizing management hierarchies in the insurance industry. In short, the new economy has applied IT to transform the organizational practices of firms and industries. Dynamism, innovation, and a high degree of risk are at the center of the new economy.

For instance, while in 1990 it took 6 years to produce an automobile—from conceptualization to final production—it now takes 2 years. Technological innovation has become remarkably rapid, and change is now measured in "web years" (which amounts to roughly one fiscal quarter) rather than in calendar years. Moreover, new jobs are being created largely by new firms that did not even exist 5 years ago. But the dynamism and innovation of the new economy is premised on a great deal of risk such that many ventures fail, and almost a third of all jobs are in flux every year (which means that these jobs have only recently been added or will soon be eliminated from the economy). In this high-risk economy, failure has become a badge of honor, signaling the willingness to jump headlong into this fast-break new economy where firms come and go in the click of a computer mouse.

Table 2 illustrates how well the 50 states of the United States are performing in the new economy. Performance was measured by exploring 17 indicators that were grouped into five categories that best capture the key components of the

Table 1	Keys to the Old and New Economies	
Issue	**Old Economy**	**New Economy**
Economy-wide Characteristics		
Markets	Stable	Dynamic
Scope of competition	National	Global
Organizational form	Hierarchical, bureaucratic	Networked, entrepreneurial
Potential geographic mobility of business	Low	High
Competition between regions	Low	High
Industry		
Organization of production	Mass production	Flexible production
Key factor of production	Capital/labor	Innovation/ knowledge
Key technology driver	Mechanization	Digitization
Source of competitive advantage	Lowering cost through economies of scale	Innovation, quality, time to market, and cost
Importance of research/innovation	Moderate	High
Relations with other firms	Go it alone	Alliances and collaboration
Workforce		
Principal policy goal	Full employment	Higher wages and incomes
Skills	Job-specific skills	Broad skills, cross-training
Requisite education	A skill	Lifelong learning
Labor-management relations	Adversarial	Collaborative
Nature of employment	Stable	Marked by risk and opportunity
Government		
Business-government relations	Impose requirements	Assist firms' innovation and growth
Regulation	Command and control	Market tools, flexibility

Source: R. D. Atkinson, R. H. Court, and J. M. Ward, *The State of the New Economy Index: Benchmarking Economic Transformations in the States.* Washington, DC: Progressive Policy Institute, 1999, p. 5.

Table 2	State Standings in the New Economy							
Rank	State	Score	Rank	State	Score	Rank	State	Score
1	Massachusetts	82.3	18	Vermont	51.9	35	Missouri	44.2
2	California	74.3	19	New Mexico	51.4	36	Nebraska	41.8
3	Colorado	72.3	20	Florida	50.8	37	Indiana	41.0
4	Washington	69.0	21	Nevada	49.0	38	South Carolina	39.7
5	Connecticut	64.9	22	Illinois	48.4	39	Kentucky	39.4
6	Utah	64.0	23	Idaho	47.9	40	Oklahoma	38.6
7	New Hampshire	62.5	24	Pennsylvania	46.7	41	Wyoming	34.5
8	New Jersey	60.9	25	Georgia	46.6	42	Iowa	33.5
9	Delaware	59.9	26	Hawaii	46.1	43	South Dakota	32.3
10	Arizona	59.2	27	Kansas	45.8	44	Alabama	32.3
11	Maryland	59.2	28	Maine	45.6	45	North Dakota	29.0
12	Virginia	58.8	29	Rhode Island	45.3	46	Montana	29.0
13	Alaska	57.7	30	North Carolina	45.2	47	Louisiana	28.2
14	Minnesota	56.5	31	Tennessee	45.1	48	West Virginia	26.8
15	Oregon	56.1	32	Wisconsin	44.9	49	Arkansas	26.2
16	New York	54.5	33	Ohio	44.8	50	Mississippi	22.6
17	Texas	52.3	34	Michigan	44.6		U.S. Average	48.1

Source: R. D. Atkinson, R. H. Court, and J. M. Ward, *The State of the New Ecomony Index: Benchmarking Economic Transformations in the States.* Washington, DC: Progressive Policy Institute, 1999, p. 5.

new economy. These include the number of *knowledge jobs* occupied; the level of *globalization* measured by the *export orientation of manufacturing* and *foreign direct investment; economic dynamism and competition* as measured by the number of jobs in fast-growing companies; the move to a *digital economy* as measured by the percentage of various groups online; and *technological innovation capacity* as measured by indicators of high-tech and knowledge-based jobs and venture capital activity.

The percentage of adults with Internet access in each state is a significant indicator of which regions of the United States are the most active participants in the digital economy (**Figure 1**). In 1997, 25 percent of households were online across the United States; by the end of 1998, the percentage was up to 33; by the year 2003, it is projected to be around 50 percent. And while the average income of Internet users is dropping, as is the average education level, there are still huge disparities in online access between blacks and whites in the United States. A late-1990s study published in the journal *Science* found that black Americans are far less likely to use the global computer network than are whites, even controlling for class. Lower-class whites were twice as likely to own a home computer than blacks. The *Science* study, which has been replicated by other less extensive surveys, raises the very serious concern that the recent exponential growth of the Internet will further exacerbate the country's social inequalities.

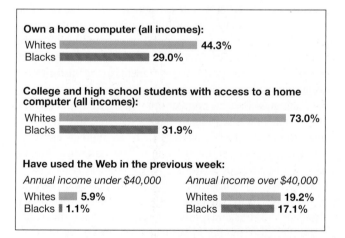

Figure 1 Computer use in black and white, 1998 As this figure indicates, race is an important variable in computer use. It is likely that race, combined with poverty, makes access to computers difficult for some. (*Source:* A. Harmon, "Blacks Found to Trail Whites in Cyberspace," *The New York Times*, April 17, 1998, p. A1.)

less than 5000 between 1975 and 1995. These losses have been more than made up, however, by new jobs generated in high-tech electronics, specialized engineering, and finance and business services.

The dramatic and often very painful changes that occurred around the decline of the Rustbelt and the rise of the Sunbelt have resulted in a major reorganization of the U.S. regional landscape. The new technology systems—using robotics, telematics, biotechnology, and other knowledge-based systems that have emerged and been refined and improved over the last 25 years—have helped to encourage the growth of new regions with very different economic bases than they possessed only 25 years ago. While the Rustbelt experienced extreme and crippling decline in the 1970s and early 1980s, by the mid-1990s it was again booming, having reorganized its economy and political institutions around service-related employment. The Sunbelt has maintained its strong economy and has continued to experience phenomenal population growth into the twenty-first century. Its economic base is a rich mix of different economic sectors from resource extraction to knowledge-based industries. The New West, or Intermountain West, an area known historically for ranching and other primary-sector activities, has made its presence felt by expanding its economy to include more service-based activities by mixing tourism and recreational activities with product-support service activities, as well as second-home and retirement residential developments. Some high-technology development is also part of the mix, especially in and around the places where universities are located, such as the Denver, Colorado, and Albuquerque, New Mexico, metropolitan regions. Cascadia is another region that has begun to boom due to the restructuring of the U.S. and Canadian economy over the last two decades. This region, which includes Alaska, parts of the Canadian Yukon and Northwest Territories, British Columbia, and Alberta, as well as parts of the states of Washington, Oregon, and California, is heavily involved with the full range of economic sectors from high-technology research, development, and manufacturing to farming, forestry, and fishing.

The new regional geographies of the United States have resulted in massive population redistribution so that more than 50 percent of the U.S. residents now live west of the Mississippi River whereas 100 years ago the reverse was true. In the last 15 years, as a result of the dramatic changes brought about by the emergence of a fifth technology system based on solar energy, robotics, microelectronics, biotechnology, advanced materials, and information technology (see Chapter 1, p. 43), a "new economy" has emerged in North America (see Geography Matters: The New Economy, p. 344). The new economy, though caused by the practical implementation of the elements of the fifth technology system, is also about the dramatic transformation of work and the labor force, markets, and the nature of competition between regions. In addition to reshaping regional economies, this new economy is also effecting great changes in local and regional politics as well as in the relationship between regional and national politics, particularly in federal elections.

Restructuring the State

The United States, like Canada, is a **federal state**, which is a form of government in which power is allocated to units of local government (province/state, county, and city/town government) within the country. Federalism leaves many political decisions to the local governments. During the first 100 years of the U.S. republic, the federal government spent most of its time directing attention to regulating commerce. But beginning in the late nineteenth century, urged by constituents across the country, the federal government began to take more and more of an active and direct role in regulating and supporting all aspects of American social and economic life, particularly with respect to providing for social welfare; developing infrastructure, such as dams and highways; and transferring large amounts of tax dollars to contractors for the buildup of U.S. defense systems, especially during the Cold War. Because the state was so heavily invested in all aspects of U.S. society, but especially in the economy, when the global economy experienced shock waves during the 1970s brought about by the oil crises and the falling corporate profits, the government was hit very hard. And as corporations and business in the United States and elsewhere searched for remedies to their economic problems, the government did likewise and imposed dramatic restructuring upon its own operations and programs.

The changes in core-periphery patterns of U.S. regions that have been brought about by the recent restructuring of the national and global economy were facilitated both by state intervention and retrenchment as well as by changes in the popular understanding of the role of government. As deindustrialization accelerated in the Rustbelt, government agencies in the Sunbelt helped to lure investment to their region by offering tax breaks, by creating needed infrastructure, and by providing subsidies for private investment. While the view that government's primary responsibility was as a guarantor of social welfare had dominated popular understanding since the recovery efforts launched during the New Deal in the 1930s, by the 1970s and 1980s, as local governments in the Rustbelt were declaring bankruptcy and the federal government was accumulating massive debt, popular opinion changed and the role of government was reconfigured. Since the late 1980s, it has become routine for local governments to act more as entrepreneurs than as managers of the social welfare. As a way of cutting back its mounting debt, the federal government also began to shed its responsibilities for social welfare, passing these responsibilities on to state governments. The federal government also began to shut down military bases throughout the country as the fall of the Berlin Wall signaled the end of the Cold War. With decreased responsibilities for social welfare and lower military spending, the federal government has reoriented its role away from these functions and toward a more active practice of facilitating the free flow of trade and the operations of transnational corporations abroad.

Since independence, Canada has fostered a government that has been far more inclined to guarantee social welfare than has the United States, though it should be pointed out that Canada also has a tradition of entrepreneurialism in government. Many scholars of Canadian history believe that federation of the former Canadian colonies into the Dominion of Canada was driven by capitalists interested in supporting the burgeoning industrialization of the country. More recently, in addition to continuing its tradition of providing social welfare, the state in Canada has accelerated its entrepreneurialism by directing support to expanding its tertiary sector (activities involving the sale and exchange of goods and services) and quaternary sector (activities involving the handling and processing of knowledge and information), particularly with respect to high-technology development.

Wealth and Inequality

Democratic politicians in the 2000 U.S. presidential campaign confidently used a recycled Republican slogan (from an earlier campaign of Ronald Reagan's) to garner votes by asking U.S. voters if they were better off since the Democrats had taken office 8 years previously. While globalization and the new economy has helped to improve the employment opportunities and level of wealth of many in the United States, it has also seriously set back many others (**Table 7.2**). From the impact of transnational agribusiness corporations on U.S. farmers to the impact of increasing wealth for some on the cost of housing for everyone, the structural transformations in the U.S. economy have left many people behind. In 1996, 36.5 million in the United States (or 13.7 percent of the population) lived in families that did not earn enough to rise above the official poverty threshold. And though many fewer individuals and families are on welfare than there were in 1992 when Bill Clinton was first elected President, many of those former welfare recipients are employed in low-paying jobs that provide insufficient income to support themselves and their families. In 1996 a family of two adults and two children with an income of $15,911 was not counted as officially poor because their income was too high. If a family of four with an income below this level is officially living in poverty, one in five children in the United States lives in poverty. Moreover, it is likely that

up to 40 percent of American children will experience poverty at one time or another, because many families move in and out of poverty over time. According to a United Nations survey, the United States has the highest child poverty rate of 18 industrialized nations. One reason is that the United States also has the lowest government benefits to families with poor children. Another reason is the stagnation of wages at the lower end of the wage spectrum. In short, the minimum hourly wage is inadequate for improving the lives of poor families.

Although the poor are found throughout the U.S. population, they can be roughly categorized into two geographic groups: the rural poor and the urban poor. Another form of diminishing economic well-being is declining household incomes that do not result in official poverty but can dramatically reduce the living standards of individuals and households.

Widespread instances of rural poverty and reduced income in the United States can be traced directly or indirectly to the transformation of family farms into transnational agricultural corporations. For instance, the number of family farms that have been in serious financial difficulty since the early 1980s continues to grow. The **farm crisis**—the financial failure and foreclosure of thousands of family farms across the U.S. Midwest—has not ceased, although it no longer gets the media attention it once did when Hollywood movie stars and rock stars championed the cause of the small farmer against the giant corporation. Fewer and fewer farms are the type of family businesses that were once believed to constitute the backbone of the U.S. economy. Farming is increasingly becoming a business run by transnational corporations that are able to use massive economies of scale to buy land, plant seed and fertilize it more efficiently, and then process and market the products of the harvest (see A Day in the Life: LaVerne Neal, p. 348). So while most of the American economy is buoyant and growing and food prices are lower thanks to transnational production practices, many rural areas in the United States are experiencing a painful restructuring. Large numbers of small farmers and ranchers are being pushed into poverty and large parts of rural America are being depopulated as land is sold off to corporations,

Household Groups	Share of All Income*		Average After-Tax Income (Estimated)		Change
	1977	1999	1977	1999	
One-fifth with lowest income	5.7%	4.2%	$10,000	$8800	▼ 12.0%
Next lowest one-fifth	11.5	9.7	22,100	20,000	▼ 9.5
Middle one-fifth	16.4	14.7	32,400	31,400	▼ 3.1
Next highest one-fifth	22.8	21.3	42,600	45,100	▲ 5.9
One-fifth with highest income	44.2	50.4	74,000	102,300	▲ 38.2
1 percent with highest income	7.3	12.9	234,700	515,600	▲ 119.7

Table 7.2 Growing Income Disparity in the United States

*Figures do not add to 100 due to rounding.

Source: *New York Times*, September 5, 1999, p. 14.

A Day in the Life

LaVerne Neal

LaVerne Neal is a 73-year-old cattle rancher in McPherson County, Nebraska (**Figure 1**), where the per capita income was $3961 in 1997 (the per capita income in Manhattan that same year was $68,686). All his life, Mr. Neal assumed that his family would always be ranchers. Yet while most of the rest of the North American economy is growing, the agricultural sector is in deep distress as it experiences a dramatic restructuring that is making rural people like Mr. Neal very pessimistic about the future of family farming and sending other farmers and ranchers into poverty. Mr. Neal has 240 cows and 3840 acres, but he now feels his acreage is too small to support future generations of ranch families. The primary problem for small farms and ranches is a very small return on investment. Despite the fact that most small farms and ranches have invested millions of dollars in machinery, land, inputs, and other aspects of production and harvesting, they are losing money.

Because of the poor prospects for success, many farmers and ranchers are simply abandoning their dreams and moving to cities for jobs with more security and fewer risks. As a result, agricultural restructuring is depopulating the rural landscapes of North America and changing the North American food system in significant ways. While transnational corporate farming is increasing efficiency and productivity, it is also undermining a way of life that has been a part of North America for hundreds of years. Interestingly, only about 2 percent of farms are run by corporations, but in 1997 corporations owned more than 14 percent of U.S. farmland (**Figure 2**). Corporations dominate both the input (such things as seeds,

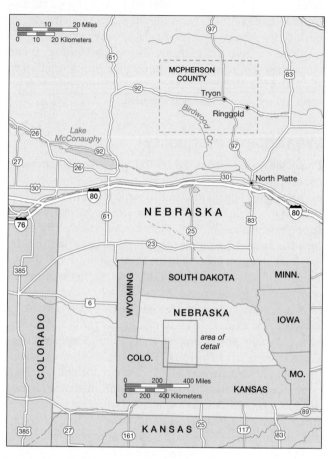

Figure 1   McPherson County, Nebraska McPherson County had about 540 residents in spring 2000. The number was 1692 in 1920.

forcing farmers and ranchers to move to into towns and cities to seek other ways of making a living. Canada has also experienced a precipitous drop in the number of farms, with the accompanying social distress (**Table 7.3**). A second aspect of the way in which the growth of transnational corporations has affected rural Americans is in the departure of rural food processing industries to other U.S. regions or parts of the globe where labor is cheaper and environmental regulations are less stringent.

Urban poverty is just as grinding and difficult as rural poverty, though its causes and outcomes may differ from those that characterize rural poverty. While rural poverty tends to affect poor white Americans, urban poverty tends to affect poor Americans of color. A particularly pressing component of contemporary urban poverty is the high cost of urban housing. In Silicon Valley, California, for example, many people with full-

time jobs are homeless because the cost of real estate is so extraordinarily high. Where the median income is $82,000, and in 1999 an average of 63 people a day became millionaires, people with non-technology-related jobs cannot compete in the housing market. Silicon Valley is a startling illustration of the gap between rich and poor in the United States that is widening every day. Consider the eye-opening contrasts between the following two related stories. A four-bedroom contemporary-style house in Palo Alto was offered on the local real estate market for $2.2 million in the early months of 2000. Because of the scarcity of housing in this, the most expensive housing market in the country, the house sold for $3.2 million, while a one-bedroom cottage listed for $495,000 sold for $750,000. Contrast the astronomically high cost of housing in Palo Alto with the fact that 34 percent of the estimated 20,000 homeless people in Santa Clara County (where Palo Alto is lo-

Figure 7.33   Boston, Massachusetts Early in U.S. history, Boston was the foremost port and largest and most important city of the colonies. The American Revolutionary War changed all that as New York seized a series of opportunities that enabled it to become the premier city of the United States, a place it has held since the late eighteenth century.

Figure 7.34   Chicago, Illinois Chicago, also known as the "second city," rose to second place in the U.S. urban hierarchy in the nineteenth century. An important distribution site for the Plains and Prairies region agricultural output, it is also a significant manufacturing and banking center.

biotechnology, and other knowledge-based systems—overwhelmed the ability of the region to respond, and disinvestment in the form of deindustrialization occurred (see Figure 7.24). For much of the 1970s and 1980s, the old Manufacturing Belt suffered population loss and capital flight as workers, entrepreneurs, and investors went to other parts of the North American region and the world to take advantage of more attractive employment and investment opportunities. But by the early 1990s, the old Manufacturing Belt was showing signs of recovery and by the middle of the decade it had regained its dominance. As Table 2 on page 345 shows, the key indicators of the new economy substantiate that a large proportion of the North American Core states are especially active participants in the new economy as they hold the largest share of the top places. Of the top ten ranked states, five are North American Core states (Massachusetts, Connecticut, New Hampshire, New Jersey, and Delaware), the other five are scattered among three fast-growing and relatively newer regions: Cascadia, the Pacific Rim, and the New West.

It is therefore no exaggeration to claim that the North American Core, with the metropolitan region of New York as its primary city, is also the primary core region of North America. While Cascadia and the Pacific Rim possess dynamic and complex economic structures with growing populations and lively cultures, they have not yet superseded the power—economic, political, and cultural—that the North American Core commands. While the North American Core has been built upon urban regions, all of which experienced significant and wrenching declines brought about by the economic crises of the mid-1970s, it has all also enjoyed remarkable renewal as these former heavy industrial urban areas replaced the outmoded components of the old economy (crumbling physical infrastructure, heavy dependence on unionized labor, inflexible production systems) with the more effective and leaner components of the new economy (high-technology and knowledge-based jobs, high levels of venture capital investment, a nonunionized workforce).

It is important to point out, however, that although the North American Core is once again prosperous, having remade itself as a key player in the new economy, not all areas of the region are enjoying that prosperity. Newark, New Jersey, a city that sits across the Hudson River in the long shadow of New York City, is still surrounded by pockets of extreme poverty, despite other pockets of progress and prosperity. Newark was the site of riots in the very hot summer of 1967 when other cities, also with large African-American populations, violently erupted to protest the institutionalized racism of U.S. society epitomized in the extreme poverty of too many of its black citizens. More than three decades later, Newark is hardly better off than it was in 1967. Although a $200-million performing arts center has been built along with hundreds of middle-class townhouses, and the University of Medicine and Dentistry of New Jersey and the New Jersey Institute of Technology are undertaking new construction, Newark continues to lose population as more and more manufacturing jobs disappear. Those who remain are more likely than ever to be poor.[5]

New York City New York City has been the premier urban center of North America since the early nineteenth century, eclipsing Boston at the end of the Revolutionary War. Over the last two centuries New York has attracted the largest population (21,199,865 in the metropolitan area in 2000)—and usually highly polyglot—and the most advanced and dynamic economy of any of its competitors both in North America and abroad. A series of forces enabled New York to catapult to the top of the national urban hierarchy and, by 1920, to the top of the global

[5]For more details see R. Smothers, "In Riot's Shadow, a City Stumbles On," *New York Times*, August 14, 1997, p. 1.

urban hierarchy. In the nineteenth century, New York proved especially hospitable to innovation and new commercial elites, and its port become the primary center for trade with Europe. The Erie Canal (1817–1825) expanded the city's hinterlands to the far western end of the Great Lakes. By 1867, the Canal carried nearly 70 percent of all commerce, double that of all the national and regional railroads combined, thereby turning New York into North America's entrepôt, a center for the storage and trans-shipment of goods. By 1850, 70 percent of imports and exports passed through the port of New York. With such an enormous volume of trade under its control, the urban region naturally also attracted bankers, financiers, stock, produce, and cotton exchanges, shippers, warehousing, law practices, insurance companies—in short, the full range of economic institutions.

The economic base of New York gradually shifted from commerce to light manufacturing over the course of the nineteenth century. Between 1850 and 1950, clothing and garment manufacturing employed nearly 50 percent of the population. The predominance of light manufacturing attracted cheap labor, most of whom were immigrants from Europe. By 1860, the foreign born constituted nearly 50 percent of the population. The explosive growth in population and in the economy resulted in dramatic physical development. Because the core of the New York metropolitan region, Manhattan, is a long narrow island—20.9 kilometers (13 miles) north to south; 3.2 kilometers (2 miles) east to west—building had to reach skyward to accommodate growth for both residences and economic activity. Tall buildings are emblematic of New York, so much so that at various times over the last century, the world's tallest building was in New York: The Park Row Building (1899), the Singer Tower (1908), the Metropolitan Life Tower (1909), the Woolworth Building (1931), and the Empire State Building (1931). In 1974 New York had as many 60-story buildings as all of the rest of the world combined and twice as much office space in its central business district as any other CBD on Earth (**Figure 7.35**). With building densities intensifying in Manhattan as the New York economy continued to prosper, by 1950, the resident-per-square-mile density for Manhattan was a staggering 86,730 as compared with Chicago at 16,165 and Los Angeles at 4391.

The 1950s was also the beginning of the end of manufacturing as the primary economic base for New York. Between 1950 and 1990, 70 percent of New York's manufacturing jobs were lost. Approximately 250,000 jobs in wholesale and retail were also eliminated. As industry declined and took related jobs with it, growth occurred in the service sector, a significant proportion of which was in finance, insurance, and real estate. But many banking jobs were also lost in the late 1980s and early 1990s as large corporations downsized their workforce or left the metropolitan region altogether. As a result of a declining tax base with the departure of jobs and people, New York City government experienced fiscal crises as increasing demands for services, expanding welfare rolls, and deteriorating infrastructure placed a heavy burden on a shaky budget. In 1975 the city nearly went bankrupt but was saved when the state of New York intervened. New York City's municipal debt was $22.5 billion in 1995.

Figure 7.35   World Trade Center The financial heart of New York City, the twin towers of the World Trade Center have also been taken to symbolize the heart of global capitalism. On September 11, 2001, a terrorist attack completely destroyed the towers, other parts of the complex, and surrounding buildings, killing thousands of people, paralyzing the city and the region, and profoundly disrupting the U.S. economy.

Since 1995, however, the economy of New York, as throughout North America, has improved dramatically. At the beginning of the twenty-first century, New York—and its economic institutions—is strong. It has a highly transnational population with more than 150 nations represented among its documented immigrants (**Figure 7.36**). For the first time in the city's history, no single ethnic group constitutes a majority of the population. A prosperous population of highly educated and skilled workers has increased the demand for housing and raised the value of real estate to astronomical levels in Manhattan. Geographer Neil Smith has shown that the housing market has responded by renovating areas of the city that were traditionally occupied by lower-income workers and immigrants. The result is **gentrification,** the process whereby the middle class, also know as the *gentry,* move into areas where real estate values have plummeted because of political and economic disinvestment. Gentrifiers transform the housing stock, which is usually architecturally interesting and structurally sound, making it difficult for the residents who first lived there to afford the higher rents or taxes that ensue from redevelopment.

Distinctive Regions and Landscapes of North America

Beyond North America's core regions and principal metropolitan areas lie numerous other landscapes and regions whose relationship to the environment, history, economic contribution, or political background makes them distinctive. The Prairie

Figure 7.37   Village green, New England Many New England towns and villages were settled around a commons. The commons, originally a space for grazing livestock, eventually became more of a recreational space. Central spaces like commons and squares (which were usually small commercial spaces) distinguish New England villages from western frontier towns, with their long parade of shops running down both sides of a fairly short street.

Figure 7.36   The Lower East Side, Manhattan In the nineteenth century the Lower East Side of the New York City borough of Manhattan was home to thousands of immigrants who worked in the manufacturing sector of the city's economy. More recently, the Lower East Side has experienced reinvestment and the migration of gentrifiers attracted to the undervalued and historic housing stock as well as the excellent location. A neighborhood once home to ethnic working-class people is increasingly being occupied by well-educated service workers employed in the city's finance, insurance, real estate, and government sectors.

provinces of Canada are critical to the world wheat market and to Canada's strong economy. Flat, rolling grasslands in a semi-arid climate characterize this landscape. New England and the Maritime Provinces are not only visually charming, as they reflect some of the earliest impacts of colonial architecture, but they are economically and politically significant for both historical and contemporary reasons (**Figure 7.37**). New England is the home of the North American industrial revolution and currently contains an important center of high-technology development clustered around some of the oldest and most prestigious educational institutions in the country. Its environment is one of woodlands and rolling terrain, with villages and towns regularly punctuating the natural landscape. Similarly, Atlantic Canada is also steeped in the colonial past. Like New England it possesses charming fishing villages and a ruggedly breathtaking coastline, but it is even more austere in some ways than its southern neighbor. Like northern New England, however, Atlantic Canada has been largely left behind by the most recent restructuring of the economy. As the Canadian population moved in greater numbers westward to the Pacific, Atlantic Canada has become more peripheral. This is especially so since the cod fishery has collapsed due to overexploitation. Atlantic Canada is attempting to make a comeback, however, through in-

vestments in the high-technology economy, and particularly through job development in telemarketing. While progress is encouraging in that direction, it is important to point out that Atlantic Canada as well as northern New England remain mostly peripheral to the larger North American economy.

The U.S. South is also distinctive from the other regions of North America in terms of environment, history, politics, and culture. Even within the region there is a great deal of variation, particularly with respect to coastal and inland areas as well as the Deep South (the states of Alabama, Georgia, Louisiana, Mississippi, and South Carolina), and the other Southern states (Kentucky, Florida, North Carolina, Tennessee, Virginia, West Virginia, and sometimes Texas). Whereas the Deep South is often called the Bible Belt, because of the attachment of much of the population to Protestant fundamentalism, states like Florida, for instance, hardly qualify as part of the Southern region as their populations are ethnically diverse and their orientation is more cosmopolitan than rural. For instance, two of Florida's key cities—Miami and Tampa—have strong economic, political, and cultural connections to Latin America. Miami is unarguably the business hub of all of Latin America. The South has also produced a distinctive literary voice from Carson McCullers (*A Member of the Wedding*) and William Faulkner (*The Sound and the Fury*) to the more contemporary Dorothy Allison (*Bastard out of Carolina*). Hawai'i, disconnected from the mainland United States far off in the Pacific, is also a unique region with the thick gloss of americanization overlying Polynesian cultural roots and a strong connection to East Asia, especially Japan.

Settlement patterns, agricultural patterns, physical landscape, dominant culture, history, economy, and society have all contributed to making the regions of North America distinct and interesting. The various groups who have lived and worked across the North American region have left telling legacies that have been embraced, ignored, sometimes destroyed, but often "remodeled" by the groups who have followed. The distinctive landscapes of North America, like elsewhere around the globe, can often be recognized at any number of scales, from a commercial airliner looking down upon the center-pivot irrigation schemes of the New West, to the front yards of Mexican-American residences in the Southwest.

Northern Frontier

The Northern Frontier, including the new territory of Nunavut (created in 1999 out of the eastern portion of the Northwest Territories and turned over to the control of the region's First Nations), sits at the apex of the North American continent. As Figure 7.9 shows, the Northern Frontier is a vast area, occupying about one-third of Canada (an area roughly the size of India) and home to only 60,000 people, nearly half of whom live in or around the settlement of Yellowknife. The distinctiveness of the Northern Frontier landscape lies in the fact that wildlife vastly outnumber people (and in the most northerly parts, human habitation is impossible); that much of the region is beyond tree line, while mountain chains in the east and west rise majestically from the slope and sedimentary plain of the interior of the region; that short growing seasons make it impossible to sustain agriculture; and that ice, snow, or **permafrost** are the norm. The Northern Frontier is a place of incredible, near-unimaginable beauty with relatively little impact from humans across its vast extent.

Nunavut, which is the largest territorial unit within Canada, is home to about 22,000 people, most of whom are Inuit, which makes the Northern Frontier unique. In response to claims by native peoples for land and political power, the Canadian government began to negotiate settlements beginning in the 1970s. While millions of acres of land and control over it have been ceded to native peoples in the last 25 years, Nunavut stands out as a monumental concession and an important indication of Canada's ability to make substantial reparations for its imperial past.

Until its transfer to native peoples, Nunavut was the central and eastern part of the Northwest Territories. Covering about 2 million square kilometers (about 772,000 square miles), Nunavut includes Baffin and Ellesmere islands and the surrounding region, stretching almost to the North Pole (**Figure 7.38**). For the most part, Nunavut is a flat **tundra** where average temperatures range from –32°C (–25°F) in January to 5°C (41°F) in July. Wildlife is abundant, including white fox, caribou, and seals. Geological surveys have shown that Nunavut is rich in copper, lead, silver, zinc, and iron. But the severity of the climate prohibits any kind of large-scale mining. The settlement pattern of the Inuit is along the coast of Hudson Bay and the Labrador Sea (**Figure 7.39**).

While the northern portion of the Northern Frontier is sufficiently inhospitable to deter human settlement, the southern part is inhabited and, moreover, is the destination for adventure tourists. For instance, the area in and around the settlement of Yellowknife, the largest city in the Northern Frontier, is relatively developed with transportation linkages and other aspects of urban infrastructure (see Sense of Place: Yellowknife, p. 362). The area around Yellowknife is also home to native peoples, the Dene Nation, who are attempting to fight off the encroachment of any additional development in the region in order to protect their culture.

Figure 7.39 Inuit boy The Inuit are one of the native groups who have occupied the Nunavut territory for thousands of years. In an attempt to protect their culture, the Inuit are moving toward controlling development in the territory, particularly development of tourist industries.

Figure 7.38 Coastal settlement, Nunavut Population densities are sparse in the Nunavut territory. Most people make their living from fishing and hunting.

The Plains and Prairies

At the geographic center of North America lies a region whose economy is primarily based upon agricultural production and related activities. The Prairie Provinces (Manitoba, Saskatchewan, and Alberta) in Canada and the Great Plains and Interior Lowlands in the United States (Ohio, Indiana, Illinois, Iowa, Kansas, Michigan, Minnesota, Nebraska, North Dakota, South Dakota, Wisconsin, and parts of Oklahoma, Texas, New Mexico, Colorado, Wyoming, and Montana; see Figure 7.9) taken together constitute the Plains and Prairies region. As with all regionalizations, this one is somewhat arbitrary, because although there are many similarities shared by these states and provinces, there are also some substantial differences. The similarities, which are the basis for this regionalization, consist of (1) location in the interior of the continent; (2) prairie/plains environment based on a range of grassland ecosystems; (3) high agricultural productivity; and (4) predominance of rural landscape settlement patterns. Most importantly, this region produces more agricultural output than any other place of comparable size on Earth. While the U.S. portion of the Plains and Prairies tends also to include large-scale industry, the Canadian side is involved in mineral and oil extraction. Moreover, various subregions within this larger agricultural region specialize in different crops or food products (**Figure 7.40**). And,

whereas in the Canadian portion of the region, the grasslands are bordered by extensive forests, in the U.S. portion there is little forested land of any significance.

The Plains and Prairies region was settled mostly during the late nineteenth and early twentieth century. Many of the settlers in the region were migrants from the eastern part of the United States and Canada, who were pushing westward looking for better opportunities beyond the rapidly urbanizing areas of the East. Europeans were also a significant part of the migrant stream to the Plains and Prairies, including large numbers of German, Scandinavian, and Ukrainian immigrants. Unlike the dense settlement patterns that characterize much of the North American Core, settlement in this region is widely spaced but fairly regular, with neat and prosperous-appearing homesteads.

While the Plains and Prairies cultural landscape is dominated by rural settlement, it is also home to many substantial and important urban regions. In Canada, Edmonton and Calgary in Alberta and Winnipeg in Manitoba are important regional centers and act as critical nodes in the Canadian national urban system, ultimately linking Canada to the rest of the globe. In the United States, the Plains and Prairies region is anchored by Chicago, Cleveland, Detroit, Cincinnati, Minneapolis-St. Paul, St. Louis, Kansas City, and Omaha (**Figure 7.41**). Chicago sits at the top of this regional urban system and acts as primary transportation hub in addition to offering a complete range of quarternary level services. The other cities are more subregional centers with each offering slightly different services for the particular needs of the economic specialties of its

Figure 7.40   Wheat fields   Different subregions within the larger Prairies and Plains region specialize in different agricultural products. This is caused in part by market demand, but it also has to do with an initial advantage leading to specialization. The initial advantage may exist for any number of reasons. For instance, it may be environmental—the availability of water and good soil, for instance—or it may be accidental—such that an innovator decided that proximity to a certain feed supply could enable the production of pigs. Wheat is one of the most widespread crops of the region, where environmental factors enable its production and a vast global market keeps demand high.

Figure 7.41   Mississippi River barge   The Mississippi River is a centrally important corridor for transporting the enormous agricultural bounty of the Plains and Prairies region. Huge barges ply the waters of the river loaded with wheat, corn, soy, and other such products destined for markets as far away as Asia and as close as Europe.

Sense of Place

Yellowknife

Yellowknife, Northwest Territories, is situated on the northern tip of the Great Slave Lake. The settlement began in the 1930s in response to the discovery of gold in the region. At first only a sledge trail created by three Caterpillar tractors connected Yellowknife to the Peace River area in northern Alberta. Slowly the trail was improved and, by 1957, a crude road known as the Mackenzie Highway was constructed. Since then new roads have been added and are pushing farther north into the Mackenzie Valley. Yellowknife is the territorial capital and contains most of the territorial population (**Figure 1**). It has traditionally been a Saturday night destination for trappers and gold miners seeking comfort from the challenges of Arctic weather and overall environmental conditions. Recently, however, Yellowknife has become a tourist destination for international travelers, most of them fairly well off, and all of them seeking a travel adventure (**Figure 2**). Tourism has become the number-one economic activity in Yellowknife. At one time places like Yellowknife provided rather rustic accommodations for those seeking their fortune in furs and gold, but today it provides luxury hotel accommodations and the whole range of fine dining. In a town with a resident population of 18,000 people, that is pretty impressive. The increasing wealth of many of the residents of core countries, and the increasing stress of their jobs, is pushing adventure travel more into the mainstream. Yellowknife has responded to this trend by upgrading its position as the jumping-off point for the Arctic, mostly through expansion of its infrastructure, like its airport, but also by expanding

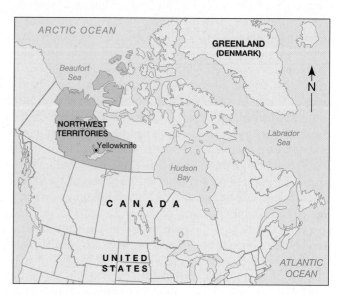

Figure 1   Yellowknife The last substantial settlement before the great northern frontier truly opens up, Yellowknife is connected to the more populous areas to the south by an all-weather road as well as frequent air service.

its commercial offerings such as supplies for prospectors and amenities for sports fishers and other adventure tourists. It also helps that the value of the Canadian dollar relative to the U.S. dollar is reversing the long-standing trend of Canadian tourists

economic hinterland. For instance, Minneapolis-St. Paul specialized in flour milling and the grain trade, acting as a collection point for the harvests of its hinterland. Its location as a rail hub enabled it to become a central place for linking the Plains and Prairies region to western North America. Today, Minneapolis has diversified its economic base by building high-level services upon agricultural processing activities.

Not surprisingly, although the Plains and Prairies region possesses a highly efficient and productive economic base, it is also a region experiencing a significant decline in rural employment as corporate farming increasingly becomes the norm. In an attempt to create a new employment base for its formerly farming (and sometimes mining) residents, many of the more rural counties have encouraged telemarketing and related companies to locate in their jurisdictions. North Dakota is just one state where telemarketing employment opportunities are booming.

While the Plains and Prairies region constitutes an important North American region linking the United States

and Canada around large-scale agricultural production, it should be pointed out that there are some strains in the relationship between the two countries. The recent controversy over genetically modified food has begun to drive a wedge between farmers on the two sides of the border. While U.S. farmers seem largely willing to continue to sow their fields with genetically engineered seeds, Canadian farmers are beginning to become wary, and in some cases even litigious, over the use of genetically modified seeds. Canadian farmers are now beginning to act more like Europeans than North Americans when it comes to the genetically engineered food controversy. One Saskatchewan farmer, for instance, is suing Monsanto, a biotechnology transnational corporation, claiming a genetically modified rapeseed (canola), which probably blew in from his neighbor's field, has established itself on his farmland and has now "contaminated" his crops.[6]

---

[6]"Food Fight," *The Economist*, December 23, 1999, p. 22.

(a)

(b)

Figure 2   Yellowknife, then and now   (a) Once a very sleepy outpost, Yellowknife has grown into a key urban place (b). Government services are located there, as well as a host of tourist services. Other locally oriented infrastructure has also developed, including schools, hospitals, and community recreation sites.

flocking in larger numbers to the United States. Overnight trips to Canada by U.S. residents increased 24 percent in the 1990s, and places like Yellowknife are responding to this increased demand. They are adding more hotel beds (5 years ago there were no bed and breakfasts in Yellowknife, today there are about 25), and packaging tours for the adventurous that would be hard to duplicate anywhere south of the Arctic Circle.

There is a downside to all this development and one that the native peoples of the region anticipated when the Mackenzie Highway was first contemplated: They see all the tourism and increased economic activity as a threat to their culture. This region is also the homeland of the Dene nation. Fur trapping and, later, mining brought large numbers of people into the area in the first half of the twentieth century. Since then, modern industrial society has been creeping ever northward into the Dene preserve. Tourism is the most recent threat to their culture and landscape and one that is hard to contain or resist.

New West

The New West, sometimes also called the Intermountain West, is a region that encompasses part or all of 12 states (all of Arizona, Utah, Nevada, and Idaho, as well as parts of Colorado, Wyoming, Montana, Washington, Oregon, California, New Mexico, and the southwestern corner of Texas), though the borders of the region are not entirely consistent with state borders, determined more by the edges of the Rocky Mountains in the east and the Sierra Nevadas in the west (see Figure 7.9). It is a landscape of many contrasts, with some of the most spectacular scenery—deep canyons, majestic mountains, and unique deserts—in the United States. The New West occupies an area of dry climates, basin and range topography (with north-south trending mountains), thin vegetation, and a general absence of many perennial surface streams. Wildfires, droughts, floods, and landslides are a continuing threat in the region. President Clinton declared the summer of 2000 the "worst fire season in memory," for example, with thousands of

wildfires burning in various parts of the New West. As geographer William Riebsame Travis has demonstrated cartographically, another important aspect of the New West is that fully one-half of it is federal land, one-quarter of which is national forest (**Figure 7.42**). The traditional economic base of this region was built upon mining, ranching, and logging, with ranches, small settlements, and cities spread seemingly haphazardly across the landscape, their location largely determined by the presence of a water source.

The traditional pursuits of this region, however, are being rapidly overtaken by a new economic orientation. Despite the dry climate and the sometimes biblical dimensions of the natural disasters, the New West is one of the fastest-growing regions in the United States. The previously dominant primary sector industries are now second to a rapidly growing service sector that includes everything from tourist-oriented services—such as food preparation and mountain bike rentals—to high-technology services—such as software development. Besides moving for regionally generated jobs, many migrants are

Sense of Place

Las Vegas

Las Vegas is an 805-square-kilometer (311-square-mile) area of desert surrounded by unvegetated, rusty-brown mountains. At the end of the twentieth century, well over 1 million people lived in Las Vegas, and projections suggest that the population will hit 2 million during the next decade. Las Vegas is a single-industry town: the industry is casino gambling, or the gaming industry, or "the Industry," as local casino elites like to call it. When most people think of Las Vegas—and a lot of people do, considering it is the number-one tourist destination in the United States—they think of the Strip and Downtown. The Strip, a 6.4-kilometer (4-mile) stretch of Las Vegas Boulevard, contains all the big new casinos and then some. Downtown, which really is not a downtown at all, is a secondary tourist district of 10 major and several minor casino hotels on Fremont Street. In 1995–1996, the Strip produced $3.7 billion in gambling revenues compared to $638 million in Downtown.

Las Vegas has been written about extensively over the last 10 years by everyone from cultural critics to urban planners. Some have seen Las Vegas as the capital of the twenty-first century, "a virtual capital of virtual capitalism," the embodiment of the post-industrial, new economy—a city with no past, a city without a sense of time. Las Vegas does have a past, and a present, and, very likely, a complicated future. Although it is a fantasyland of extraordinary dimensions, where the Eiffel Tower stands as a silent observer of the twice-hourly fountain shows on a faux Lake Como, it is also a real place, where real people live real lives, most of them directly or indirectly connected to the Industry.

Ironically, the first non-Native American settlers in Las Vegas were Mormon farmers who soon gave up the frustrations of agriculture in a desert landscape and began to offer services to train travelers. When the Hoover Dam, one of the

Figure 1 Hoover Dam It is no exaggeration to state that without water and electricity, Las Vegas could not have grown the way it has over the last decade. Hoover Dam is an important source of both, but it is also a phenomenal example of human engineering.

great engineering projects of the twentieth century was completed in 1936 in Henderson, Nevada, important supply lines of water and hydroelectricity for the future growth of Las Vegas were established (**Figure 1**). The most significant element of

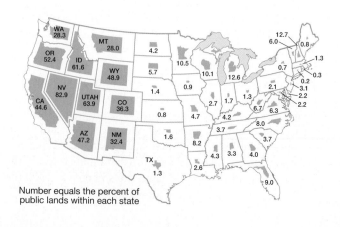

Figure 7.42 Public lands in the New West The largest landowner in the New West is clearly the federal government. Much of the land in this region has been stewarded by the government for public use. Other lands have been leased to ranchers for grazing cattle. These two uses often pit environmentalists against ranchers over the highest and best use of the land. (*Source:* W. Reibsame Travis, *Atlas of the New West.* New York: W. W. Norton, 1997, p. 58.)

Number equals the percent of public lands within each state

Figure 2   Paris hotel-casino, Las Vegas, Nevada   The city-themed hotel and casino has become ubiquitous in the fantasy landscape of Las Vegas. No need to book an expensive European vacation when you can get all you need of Paris, Venice, or Bellagio right in your own backyard—they speak English in Las Vegas.

the early history of Las Vegas was the arrival of members of the East Coast mafia, who transformed the dusty Mormon cross-roads from a site of somewhat disreputable gambling parlors to a site for the development of the gaming industry. In the late twentieth century, with its place- or theme-based 50-story hotel-casinos—Bellagio; New York, New York; Paris; Rio; San Remo; and Venice—Las Vegas has simply reinvented itself once again (**Figure 2**). This time, though, it has done so on the grandest and most outlandish scale imaginable.

The present incarnation of Las Vegas is branching into a very different aesthetic. While glitter and glitz are still absolutely central to the gaming experience, the new Las Vegas now also boasts itself as a more cultured city. *Wine Spectator* magazine featured Las Vegas as the "new mecca for fine dining in America." A *New York Times* opera critic raved that "O,"

a show performed by Cirque du Soleil in Las Vegas, was a spectacle to rival some of the best New York dramatic performances. The Bellagio Hotel houses a select fine art exhibit that includes the likes of Degas and Matisse. Entrepreneurs have understood the structure of feeling and taste of the new economy and have made it abundantly and even tastefully available in Las Vegas.

But perhaps the more interesting aspect of Las Vegas is the side that very few tourists see. This is the side that effectively enables the gaming industry to function. This is the part of the city where the people who work directly or indirectly for the gaming industry live. Surprisingly for some, Las Vegas was the fastest-growing city in the United States for most of the 1990s. The second-largest industry in Las Vegas, after gaming, is construction. In short, the gaming industry that has been created in Las Vegas requires an enormous workforce. Over the last decade, Las Vegas built new homes at a higher rate than any other city in the United States. The workforce for whom these houses are being built has been arriving in astonishing numbers and is helping to create a metropolitan area that, while growing in leaps and bounds, contains all the elements of most metropolitan areas: schools, government buildings, law offices, a major university, shopping centers, hospitals, and funeral homes. Along with this growth, metropolitan Las Vegas, like most U.S. cities, has its share of urban problems.

In 1995, Las Vegas had the highest total crime rate in the United States, according to F.B.I. statistics. In 1992, the high school dropout rate for Clark County (in which Las Vegas is located) was the highest in the country. It has been suggested that Las Vegans use more water per resident than do residents of any other city in the world. Bankruptcy declarations there are much higher than the national average. Air pollution is a serious environmental problem. Significant social problems include drug abuse, alcohol abuse, compulsive gambling, gang violence, and domestic violence.

moving to the New West as "lifestyle refugees"—individuals escaping the grind of high-stress urban lives who move their jobs with them and telecommute from their homes. In fact, some of the fastest-growing areas of the New West are its rural counties, where retirees and still economically active adults are seeking a quieter, gentler way of life, but a life that still has at least one of each of the amenities they left behind, such as Starbucks, upscale food markets and restaurants, and high-tech sporting goods shops. Colorado contained 10 of the fastest-growing counties in the country between 1990 and 1995. Idaho, Colorado, Montana, Arizona, and Utah grew faster than all the remaining states during 1995. The rural orientation of the new migrants is in sharp contrast to past set-

tlement geographies of the West, where migrants tended to settle in cities, more so even than they did in the East.

But urban growth is also being fed by the new migration to the West. And one of the fastest growing of the New West's urban places is Las Vegas, which during most of the 1990s was the fastest-growing city in the country (see Sense of Place: Las Vegas, p. 364). The phenomenal growth of New West urban areas like Las Vegas (as well as Salt Lake City, Denver, Albuquerque, Phoenix, and Tucson, for example) raises the nagging problem of adequate water supply and whether such unprecedented growth can continue in a region that has very little of it. Residents of the New West in no way appear to want to limit their water usage. In places like Phoenix or Palm

Springs—with their green lawns, ubiquitous swimming pools and fountains, and elaborate outdoor landscaping, driving water usage to 300 gallons a day per resident—residents appear not to know they live in an arid climate. Add to the inappropriate residential use the vast amounts of water used by agriculture and ranching, and it is undeniable that the extravagant use of water in the arid New West is causing serious environmental and human problems in the region.

In Las Vegas, the earth has subsided by 5 feet due to groundwater pumping. Subsidence is a problem in other urban areas throughout the New West as well. In places like Phoenix, excessive outdoor watering has created a humid microclimate, encouraging the growth of pollen and molds that cause serious health problems for the population.

A second significant problem of the New West is the clash of cultures between long-term residents and new migrants. Geographer Bryant Evans has written about the conflicts be-

tween newcomers and long-term residents in the old mining town of Bisbee, Arizona (see Figure 7.21). There the long-term residents are likely to have been associated with the copper mines. Newcomers have moved to Bisbee because they like the small-town atmosphere, relatively inexpensive housing, and the beautiful setting. Long-term residents and newcomers have come into conflict over the significance of the Lavender Pit, a giant open pit mine that is no longer active. Newcomers see the mined landscape—this enormous, graded concave depression in the ground—as an eyesore and a reminder of corporate irresponsibility toward the environment. Natives see it as a symbol of the difficult life of mining and the hard work they endured to help contribute to the growth and development of the nation. The conflict between these two groups is emblematic of the sorts of conflicts that occur between long-term residents and newcomers throughout the New West.

Summary and Conclusions

In this chapter we examined the North American region, looking specifically at the similarities and differences within and between Canada and the United States. While there is much parallelism in the history of the two countries, there is also much that is different. Differences in the type of early European settlement, political independence, and economic history and structure have helped to shape differences in political and economic structure, among other things. The same is true of subregional differences. And while the differences between the two countries are many and there have even been episodes of outright conflict, there is also a long history of cooperation and other forms of interaction between them. Still, many critics of U.S.–Canada relations would argue that "cooperation" and "interaction" are euphemistic terms for the heavy-handed dominance that the United States exerts over much of Canadian economic, cultural, and social life.

Canada and the United States continue to enjoy a strong relationship despite the friction that regularly emerges between them. The United States continues to be the largest importer of Canadian exports, and Canada continues to import a significant amount of U.S.-produced goods. The United States and Canada also have a reciprocal relationship when it comes to capital, with the United States investing large amounts in Canadian industry and new economy ventures and Canada investing in U.S. real estate, among other things. The trade alliances that created one market between the two countries continue to unite them.

The influence of the North American region on the world continues to be substantial. While U.S. economic, cultural, and political ideas are likely to dominate the processes of globalization, Canada will remain an important player in the development of the Pacific Rim and in the peaceful resolution of international conflict.

Key Terms

acid rain (p. 351)
americanization (p. 326)
assimilation (p. 337)
creative destruction (p. 343)
deindustrialization (p. 343)

europeanization (p. 323)
farm crisis (p. 347)
federal state (p. 346)
gentrification (p. 358)
ghetto (p. 337)
hate crime (p. 337)

indentured servants (p. 326)
intermontane (p. 317)
internal migration (p. 339)
Main Street (p. 356)
Megalopolis (p. 356)

multiculturalism (p. 338)
permafrost (p. 360)
staples economy (p. 331)
suburbanization (p. 340)
superfund site (p. 351)
tundra (p. 360)

Review Questions

Testing Your Understanding

1. Which mountains, rivers, and lakes contribute most to North America's prosperity? What role do the Great Plains play in North American economic development? How does the abundance of water moderate climate in the North American region?

2. Describe the process of human settlement in North America.

3. How did indigenous North Americans view concepts such as private property or land ownership? How culturally uniform were indigenous peoples at the time of European contact 500 years ago?

4. When and where was the first European settlement in North America located? Centuries later, while Spain colonized Florida, which other European powers followed to colonize North America? How did Europeanization differ from Americanization?

5. Why did European colonists arm indigenous peoples? How did disease and demoralization affect native populations?

6. How did the U.S. government encourage mobility and manifest destiny along a steadily moving frontier from the late 1700s to the mid-1800s?

7. What factors led to the establishment of the Confederate States of America in 1861 and the Dominion of Canada in 1867?

8. Why was the United States able to make the transition from a periphery to a core state by 1900? Why at that time was Canada unable to do the same?

9. How does Canada protect its own distinct national culture?

10. Describe the characteristics of the Rustbelt and the Sunbelt.

11. What factors influence rural and urban poverty in America today?

Thinking Geographically

1. Which areas of North America suffer most from acid rain, deforestation, chemical or nuclear toxic waste, and flooding? What can be done to reduce or remedy ecological damage?

2. What factors make Québec unique? Why do so many Québeçois want to secede from Canada? How might increasing globalization affect Québec?

3. Discuss the three main waves of imigration into the United States. When did they occur and where did the imigrants come from? Discuss the three main waves of internal migration within the United States. How were African Americans, many of whose ancestors were brought to the United States long before other immigrant groups arrived, generally affected by immigration and internal migration?

4. Which areas have been primarily Hispanic since their incorporation into the United States? Which areas have seen a large increase in Hispanic population over the past 30 years, and why? What roles do produce farms, maquiladoras, and NAFTA play in keeping laborers from Mexico in or out of the United States?

5. Contrast and compare the changing geography (since 1980) of the Pacific Rim, Cascadia, and the North American Core. What factors make Silicon Valley, Vancouver, and New York City distinctive?

6. What factors make some North American regions, such as the Prairie Provinces/States, New England States-Maritime Provinces, and U.S. South, peripheral to the core regions?

7. What concerns accompany rapid urbanization in the New West? What pull factors bring new migrants into the area?

Further Reading

Allen, L., *Capitalism*. Santa Barbara, California: ABC-Clio, 1998.

Ayers, E. L., and Mittendorf, B. C., *The Oxford Book of the American South: Testimony, Memory, and Fiction*. Oxford: Oxford University Press, 1997.

Birdsall, S., and Florin, J. *Regional Landscapes of the U.S. and Canada*. New York: John Wiley & Sons, 1992.

Bottles, S., *Los Angeles and the Automobile: The Making of the Modern City*. Berkeley: University of California Press, 1987.

Brodie, J., *The Political Economy of Canadian Regionalism*. Toronto: Harcourt Brace Jovanovich, 1990.

Cameron, D., and Watkins, M. *Canada Under Free Trade*. Toronto: James Lorimer, 1993.

Christian, S., "Hispanic Workers Revitalize a Town," *New York Times*, p. A1, January 29, 1998.

Chudacoff, H. P., and Smith, J.E., *The Evolution of American Urban Society*, 4th ed. Englewood Cliffs, NJ: Prentice Hall, 1994.

Clements, W., and Williams, G. (eds.), *The New Canadian Political Economy*. London: McGill-Queens's University Press, 1989.

Earle, R. L., and Wirth, J. D., *Identities in North America: The Search for Community*. Stanford, CA: Stanford University Press, 1995.

Harris, C., and Warkentin, J. *Canada Before Confederation*. New York: Oxford University Press, 1974.

Issel, W., "San Francisco, California," in L. Shumsky (ed.), *Encyclopedia of Urban America: The Cities and Suburbs*, Vol. 2 (pp. 688–91). Santa Barbara, CA: ABC-Clio, 1999.

Kerr, D., and Holdsworth, D. (eds.), *Historical Atlas of Canada. Volume 3: Addressing the Twentieth Century*. Toronto: University of Toronto Press, 1993.

Kotkin, J., *The New Geography: How the Digital Revolution Is Reshaping the American Landscape*. New York: Random House, 2000.

McCann, L. D. (ed.), *Heartland and Hinterland: A Geography of Canada*. Scarborough, Ontario: Prentice Hall, 1982.

Meinig, D. W., *The Shaping of America: A Geographical Perspective on 500 Years of History, Volume 1, Atlantic America, 1492–1800*. New Haven: Yale University Press, 1986.

Mitchell, R. D., and Groves, P. A. (eds.), *North America: The Historical Geography of a Changing Continent*. London: Hutchinson, 1987.

Mohl, R. (ed.), *The Making of Urban America*. Wilmington, DE: Scholarly Press, 1984.

Morrison, R. B., and Wilson, C.R. (eds.), *Native Peoples: The Canadian Experience*, 2nd ed. Toronto: McClelland and Stewart, 1986.

Reibsame, W., *Atlas of the New West: Portrait of a Changing Region*. New York: W. W. Norton, 1997.

Smith, N., *The New Urban Frontier: Gentrification and the Revanchist City*. New York: Routledge, 1996.

Stearns, P. N., and Hinshaw, J. H., *The Industrial Revolution*. Santa Barbara, CA: ABC-Clio, 1996.

Vance, J. E., Jr., *The North American Railroad. Its Origins, Evolution, and Geography*. Baltimore: Johns Hopkins University Press, 1995.

Warkentin, J., *Canada: A Regional Geography*. Scarborough, Ontario: Prentice Hall, 1997.

Zelinsky, W., *The Cultural Geography of the U.S.: A Revised Edition*. Englewood Cliffs, NJ: Prentice Hall, 1992.

Film, Music, and Popular Literature

Film

The Black Robe. Directed by Bruce Beresford, 1991. A tale of missionizing in eighteenth-century Canada.

Boyz N the Hood. Directed by John Singleton, 1991. Young men grow up in a Los Angeles ghetto.

Giant. Directed by George Stevens, 1956. Conflicts arise between ranching and oil exploration in Texas.

Lewis and Clark: The Journey of the Corps of Discovery. Directed by Ken Burns, 1993. Story of the famous expedition of Lewis and Clark across the United States.

Kanehsatake: 270 Years of Resistance. Directed by Alanis Obomsawin, 1993. This film documents the 1990 crisis when Native Americans of the Mohawk Nation blocked access to reserve land that was being appropriated against the nation's will by the white community of Oka, Québec, Canada.

My Father's Angel. Directed by Mark Bauche, 1999. A Muslim Serbian couple migrate to Vancouver.

Mi Vida Loca. Directed by Allison Anders, 1993. Shows what life is like for girl gang members in a Los Angeles barrio.

Out of Ireland. Directed by Paul Wagner, 1994. History of Irish immigration to the United States following the Great Famine in Ireland in the 1840s.

Roger and Me. Directed by Michael Moore, 1989. Explores the closure of the General Motors plant at Flint, Michigan, which resulted in the loss of 30,000 jobs.

Well-Founded Fear. Directed by Michael Camarini and Shari Robertson, 2000. A story about refugee asylums in the U.S. immigration service.

Music

Bragg, Billy, and Wilco. *Mermaid Avenue, Vols. I and II*. Electra, 1998.

Four Mountain Nation Singers. *Navajo Chants, Vol. 1: Pow Wow Songs*. Astro Music, 2000.

Johnson, Robert. *The Robert Johnson Songbook*. King of Spades Music, 1998.

Monroe, Bill. *What Would You Give in Exchange for Your Soul?* Rounder, 2000.

NPR Radio. "American Routes." Radio program featuring varieties of American music.

Original Dixieland Jazz Band. *The Complete Original Dixieland Jazz Band*. BMG/RCA, 1995.

The Riverside History of Classic Jazz. Fantasy/Riverside, 1994.

Sugar Hill Gang. *Rapper's Delight: The Best of Sugar Hill Gang*. Rhino, 1996.

Various Artists. *Alligator Stomp 1: Cajun and Zydeco Classics*. Wea/Atlantic/Rhino, 1990.

Various Artists. *Masters of Tejano Music*. Sony Discos, 2000.

Various Artists. *Weaving the Strands: Music by Contemporary Native American Women*. Red Feather, 1990.

Popular Literature

Berton, Pierre. *The Klondike Fever*. New York: Knopf, 1958. A true story of historical figures and ordinary people who persevered and often fell victim to their own ill-conceived dreams in the Klondike.

Brown, Dee. *Bury My Heart at Wonded Knee: An Indian History of the American West*. New York: Henry Holt, 1991. Beginning with the Long Walk of the Navajos in 1860 and ending 30 years later with the massacre of Sioux men, women, and children at Wounded Knee in South Dakota, this book tells how the American Indians lost their land and lives to a dynamically expanding white society.

Cather, Willa. *My Antonia*. Boston: Houghton Mifflin, 1954. Set in Nebraska in the late nineteenth century, this is the story of the immigrant settlers of the American plains told through the experiences of the spirited daughter of a Bohemian immigrant family planning to farm on the untamed land.

Churchill, Ward. *A Little Matter of Genocide: Holocaust and Denial in the Americas, 1492 to the Present*. San Francisco: City Lights, 1998. An extremely well-documented historical accounting of targeted racial/ethnic killings from 1492 to the present, resulting in the near-extermination or genocide of the once-populous native North American Indians.

Cronon, William. *Nature's Metropolis: Chicago and the Great West*. New York: W. W. Norton, 1992. History of nineteenth-century Chicago and the widespread effects it had on millions of square miles of ecological, cultural, and economic frontier.

Lee, Harper. *To Kill a Mockingbird*. Philadelphia: J. B. Lippincott, 1960. Story told through the eyes of an 8-year-old girl named Scout, living in Maycomb County, Alabama, in the 1930s, who is exposed to issues of class, justice, and race in the American South.

McMurtry, Larry. *Lonesome Dove*. New York: Simon & Schuster, 1985. Pulitzer Prize-winning novel that depicts the degeneration of the myth of the American West through the story of cowboys herding cattle on a great trail-drive set in the nineteenth century.

Morrison, Toni. *Beloved*. New York: Knopf, 1987. Pulitzer Prize-winning novel about the years following the Civil War, focusing on a murdered child who haunts the Ohio home of a former slave.

Proulx, Annie. *The Shipping News*. New York: Scribner's, 1992. This darkly comic, wonderfully inventive work, winner of the 1993 National Book Award, transforms the lore of Newfoundland—including shipwrecks, nautical knot-tying, horrid weather, and family legend—into brilliant literary art.

Smiley, Jane. *A Thousand Acres*. New York: Knopf, 1991. Set in Zebulon County, Iowa, this novel is a portrait of the American family farm at the end of the twentieth century.

Stegner, Wallace. *Angle of Repose*. New York: Doubleday, 1971. A novel of discovery—personal, historical, and geographical—that comes together in an enthralling portrait of four generations in the life of an American family.

Steinbeck, John. *The Grapes of Wrath*. New York: Viking, 1939. Pulitzer Prize-winning novel about a family of dispossessed Oklahoma farmers who migrate to California to begin their lives anew.

Twain, Mark. *The Adventures of Huckleberry Finn*. New York: Signet Classics, 1959. A classic novel that tells the story of a teenaged misfit who finds himself floating on a raft down the Mississippi River with an escaping slave.

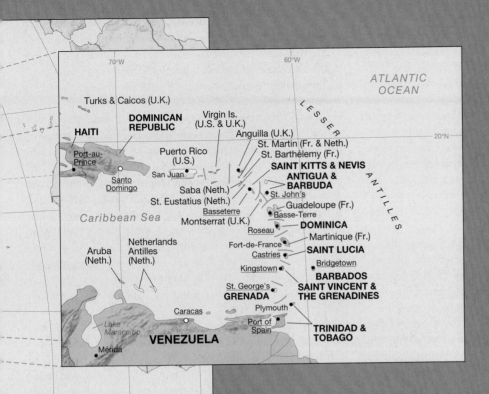

8 Latin America and the Caribbean

Figure 8.1

atin America is the southern part of the large landmass of the Americas that lies between the Pacific and Atlantic oceans (**Figure 8.1**). Traditionally, the Americas are divided into the two continents of North America (Canada, the United States, and Mexico) and South America (all the countries south of Mexico). Latin America includes all of the countries south of the United States from Mexico to the southern tip of South America in Chile and Argentina to form a world region of considerable physical and social coherence covering more than 20 million square kilometers (7.7 million square miles). The region includes 30 independent countries along with many islands and one mainland territory that are still under the political control of the United States, the United Kingdom, France, or the Netherlands.

The term *Latin America* was coined in the nineteenth century by the French, who sought to discourage British interests in the region and justify their own imperial ambitions there by asserting that the shared Romance languages of Spanish, French, and Portuguese—all of which were derived from Latin—were the defining characteristic of the region. In this book, we include the Caribbean islands in the Latin American region because of the physiographic links from the island chains to the South American mainland, because Spanish is spoken by many Caribbean inhabitants, and because of a shared legacy of European colonial domination.

Much of Latin America and the Caribbean can be characterized by a common experience of *colonialism* that included the dominance of the Spanish and Portuguese languages, religion (Roman Catholicism), legal and political institutions, and European control of resource extraction, trade links, and other economic activity (**Figure 8.2**). Most of the region became in-

dependent in the nineteenth century and was drawn into global trade relations, especially with Britain and the United States. In the twentieth century, the entire Latin American region experienced rapid integration into global markets and the transition from revolutionary and military governments to democratic ones.

Despite an apparent physical coherence and a shared historical experience of colonialism and economic development, like many other world regions, the regional definition of Latin America and the Caribbean is contested and unclear. The clearest physical breaks between North America and South America probably occur at two narrow isthmuses (narrow necks of land between major seas)—in Mexico at the Isthmus of Tehuantepec and in Panama where the Panama Canal now cuts the Isthmus of Panama. The islands of the Caribbean Sea are a chain of volcanic peaks and coral reefs that are physiographically linked to the U.S. state of Florida in the north and to Venezuela in the south, ringing the waters of the Caribbean and the Gulf of Mexico.

Several important subregions have been defined in Latin America, including Central America and the Southern Cone. Central America usually includes the countries of Belize, Costa Rica, El Salvador, Honduras, Guatemala, Nicaragua, and Panama. The Southern Cone encompasses Argentina, Chile, Paraguay, and Uruguay. The Caribbean region includes all of the islands in the Caribbean Sea, a suboceanic basin of the western Atlantic Ocean encircled by the northern coast of South America to the south, Central America and Mexico to the west, and the United States to the northwest. The large islands of Cuba, Hispaniola, Jamaica, and Puerto Rico are included as a northern boundary, and the eastern edge includes the Lesser Antillean chain of islands from the Virgin Islands in the northeast to Trinidad, off the Venezuelan coast, in the southeast.

Sometimes the Latin American region is defined to include only those countries where either Spanish or Portuguese is the official language, thus excluding those countries colonized by other European powers. Included in this latter group are the French-speaking islands of the Caribbean, such as Haiti, and the mainland country of French Guiana; the English-speaking islands of the Caribbean, such as Jamaica and Barbados, and the mainland countries Belize and Guyana; and the former Dutch mainland colony of Suriname and Caribbean islands still under Dutch political control, such as Bonaire and Curaçao. A legacy of Spanish colonialism and language could also bring the southwestern region of the United States into Latin America.

But the Latin American region is also characterized by diversity. Even within the Spanish- and Portuguese-speaking countries, many people speak only indigenous languages or languages other than Spanish or Portuguese, practice a religion other than Catholicism, and may have an Asian or African heritage. Furthermore, some Latin American countries are not well integrated into world economic markets and do not have democratic governments.

Thus, the definition of the Latin American world region arose from the period of global integration associated with European colonialism, but it oversimplifies the physical and social diversity within the region and its links with many re-

Figure 8.2   Overlapping cultural landscapes of Mexico City At the Plaza of the Three Cultures in Tlatelolco in Mexico City, the ancient Aztec ruins, a Spanish colonial church, and modern apartment buildings stand together, symbolizing the links and overlaps from the past to the present in Latin America. The common themes illustrated here include the legacies of highly developed indigenous societies, the enduring economic and cultural imprint of Spanish (or other European) colonialism, and recent integration into a world economy of cities of modern architecture and business districts and urban environmental problems.

gions of the world other than Europe and North America. In contemporary popular culture, the terms *Latin American* or *Caribbean* are more often associated with musical trends such as salsa and reggae, with certain foods and fashions, or with stereotypical images of tropical cultures and ethnicity (party loving, fiestas, black or brown skin, colorful fashions, eroticism—in effect, the image of "Carnival"). Although these stereotypes may be embraced for purposes of identity or tourism, some people feel that lumping this diverse region under a single definition and perception of Latin America is an oversimplification that contains unacceptable echoes of colonialism, racism, and environmental determinism.

Environment and Society in Latin America and the Caribbean

Latin America and the Caribbean's physical environments include vast areas of forests, grasslands, mountains, and deserts that, at first glance, seem minimally transformed by human activity. There are also areas of intensive human occupance where the physical limits of aridity and disease have been apparently overcome by modern technological innovations in irrigation, plant breeding, air conditioning, and medicine.

However, research has shown that many of the seemingly pristine forests were cleared centuries ago, the grasslands selectively burned or grazed, the mountains carved into terraces, and the infrequent waters of the deserts stored or diverted—thus, Latin American landscapes have long been transformed by humans. Similarly, studies show that human settlements and high-technology agriculture are extremely vulnerable to natural disasters and epidemics and therefore that the human geography of Latin America and the Caribbean continues to be guided by environmental conditions and events.

It seems that an understanding of the physical geography and the ways in which people have modified their environmental surroundings is integral to understanding the historical and contemporary human geography of the Latin American region and the challenges to its sustainable development.

Landforms and Landscapes

The physical landscape of Latin America and the Caribbean varies widely and includes striking mountain ranges, high plateaus, and enormous river networks (**Figure 8.3**). The two largest-scale physical features in Latin America and the Caribbean are the Andes Mountains and the Amazon basin or Amazonia, both easily seen from space (**Figure 8.4**). The Andes are an 8000-kilometer-long (5000-mile-long) chain of high-altitude peaks and valleys that for the most part run parallel to the west coast of South America (**Figure 8.5a**). The highest peak, Aconcagua at 6960 meters (22,834 feet), is a dramatic pinnacle revered and admired by indigenous peoples, tourists, and mountain climbers. The Amazon River tributaries flow downward and eastward from the Andes into an enormous

river network that covers a basin of more than 6 million square kilometers (2.3 million square miles), which includes the river itself and the surrounding landscape, about two-thirds in Brazil, but also including parts of Peru, Ecuador, Bolivia, Colombia, Venezuela, and the Guianas (Guyana [formerly British Guiana], Suriname [formerly Dutch Guiana], and French Guiana). The Amazon River and its tributaries carry 20 percent of the world's freshwater and provide water, sediment, and fish that support the agriculture and diets of the peoples of the basin while serving as a transport network (**Figure 8.5b**). This watershed also nourishes the Amazon rain forest, which is home to more than 100,000 different species and is often termed the "lungs" of the world because of its key role in recycling the oxygen, carbon, and water resources that are critical to life on Earth.

Other important physical features include the mountainous spines of Mexico and Central America and the high-altitude flatter areas, or plateaus, that lie between or adjacent to the mountain ridges. The Andean high plateau, the **altiplano,** and the Mexican plateau, or Mesa Central, are important areas of human occupance, because they provide flatter, cooler, and wetter environments for agriculture and settlement than do the adjacent steep-sloped mountains, dry lowland deserts, and humid lowlands. The Caribbean basin has large areas of limestone geology where water tends to flow underground and create large cave systems such as those in the Yucatan peninsula of Mexico and in Puerto Rico. Where the surface collapses into limestone caverns, deep, crystal-clear ponds, called *cenotes* in Mexico and *blue holes* in the Bahamas, are formed. Coral reefs, a key feature of the Caribbean landscape, are created when living coral organisms build colonies in warm, shallow oceans. These reefs, hosts to myriad other marine animals, are fragile ecosystems that are easily damaged by boats, divers, pollution, and environmental change.

The configuration of high-elevation and low-elevation land areas and the location of island chains is the result of a long history of tectonic activity in the region. Latin America is on and near several major tectonic plates (see Chapter 2, p. 52, and Figure 2.5) and Central America is one of the most tectonically active regions in the world. The region south of Panama sits on the South American plate, whose slow westward drift causes it to collide with the adjacent Nazca plate. This results in a folding and uplifting of the western edge of the South American plate to form the Andes Mountains, thus forcing the Nazca plate downward and under South America in the process of subduction. Similarly, Mexico sits on the North American continental plate, also drifting westward and causing the subduction of the Cocos plate and the uplift and folding of the Sierra Madre mountains. The Caribbean plate, in contrast, is moving eastward, pulling away from the Cocos, moving under the South and North American plates and producing geological tensions and cracks that produce earthquakes and volcanic activity in Central America as well as the formation of several volcanic islands in the Caribbean.

This geological activity and history of Latin America and the Caribbean has affected human history and activity in many important ways. The mineral wealth of Latin America is typically found on the old crystalline Guiana shield (or highlands) and

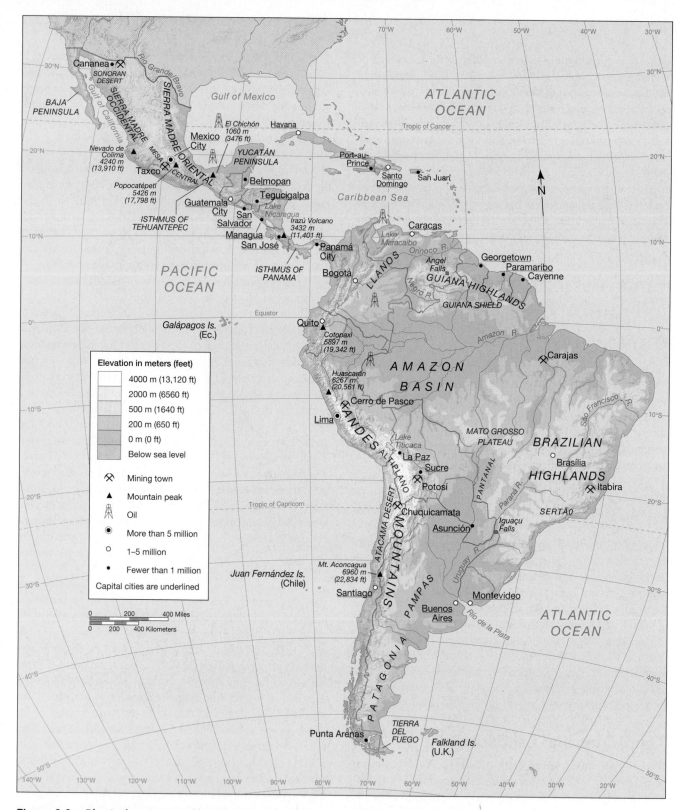

Figure 8.3 **Physical regions and landforms of Latin America and the Caribbean** The Amazon basin and the Andes mountain range are the two largest physical features in Latin America. Major South American rivers such as the Plata and the Orinoco provide transport routes into the interior as well as water resources for agriculture and hydroelectricity generation. The two major deserts—the Sonoran and Atacama—are located along the Pacific coasts of Mexico and Chile, and important grasslands are located in the pampas region of Argentina and the llanos of northern South America. Past and current human populations settled on highland plateaus, such as the altiplano of the Andes and the Mesa Central of Mexico. The region has many active volcano and earthquake zones associated with the movements of tectonic plates. Older geological formations are associated with important mineral deposits, such as the silver mines of the Andes and central Mexico, copper in central and northern Chile and in northern Mexico, and gold and iron along the edges of the Brazilian and Guiana highlands. World-class oil deposits occur in the Gulf of Mexico and in Venezuela, especially Lake Maracaibo, and in the western Amazon basin.

Figure 8.4   Satellite image of Latin America and the Caribbean   Certain physical features of Central and South America are clearly visible in this satellite image, including the verdant green of the Amazon basin and the mountain ridges of the Andes and Central America as well as Lake Maracaibo on the northern coast of Venezuela.

(a)

(b)

Figure 8.5   The varied landscapes of Latin America and the Caribbean   (a) The dramatic snow-capped peaks of the Andes soar above the city of Santiago, Chile, providing a dramatic contrast between cooler mountain climates and the warmer conditions of the central valley of Chile, which has a lush Mediterranean-type agriculture. (b) The Amazon River carries 20 percent of the world's freshwater and winds across an extensive basin covered by tropical rain forests. The floodplain of the Amazon is important for agriculture, with soils renewed by annual floods, and the river provides an important transport route and fishing ground. Here riverboats are moored on the Amazon at Iquitos, Peru.

(a)

(b)

Figure 8.6   Mineral resources of Latin America and the Caribbean (a) The famous Cerro Rico (rich hill) mountain in Potosí, Bolivia, which produced more than half of the world's silver in the sixteenth century and is still actively mined. (b) Oil wells in the Gulf of Mexico operate under the authority of the Mexican government's oil company, PEMEX.

in those areas where crustal folding brings older rocks near to the surface (see Figure 8.3). The most important precious-metal mining districts in the region include the Peruvian and Bolivian Andes, where mountains of silver were excavated in the colonial period at Cerro de Pasco and Potosí (**Figure 8.6a**), and lead, zinc, and tin are still important; the silver region of the Mexican Mesa Central; and the gold and iron mines at Carajas on the edge of the Brazilian plateau. World-class iron deposits are found on the southern edge of the Brazilian shield at Itabira, on the northern edge of the Guiana highlands at Cerro Bolivar, and in northern Mexico. Copper is the geological treasure of the southern Andes, especially northern Chile, and is also important in northern Mexico. The shores of the Caribbean, including the Guianas and Jamaica, have important bauxite deposits (a mineral used in the aluminum industry). These minerals, especially gold and silver, were foundations of the European colonial economies (see Figure 2.33) and now dominate the export economies of countries such as Chile and Bolivia. They are a focus of foreign interference and ownership and have often transformed local labor and environmental conditions.

The other critical resources associated with Latin America's geology are oil, gas, and coal. The earliest oil booms and later gas developments occurred on Mexico's Gulf Coast and in Venezuela around Lake Maracaibo. In all of Latin America's oil regions, environmental pollution has been a serious problem, leading to waterways contaminated with waste oil, widespread ecosystem damage, and serious health problems among local residents. Oil was discovered in Venezuela in 1917, and the country became one of the founding members of the Organization of Petroleum Exporting Countries (OPEC) in 1961. Oil became an important foundation of the national economy, and the oil industry was nationalized in 1976, but the economic benefits of oil production reached only about 20 percent of the population. Lake Maracaibo, the site of about 4 percent of world oil reserves, is now crowded with thousands of oil derricks that produce about two-thirds of Venezuela's oil output.

The Mexican oil deposits were first commercially exploited in the 1890s and were nationalized in the 1930s under the government oil company PEMEX (Petroleos Mexicanos). PEMEX developed petrochemical industries, but the debt crisis and possible corruption limited the investments and benefits from Mexico's oil resources (**Figure 8.6b**).

The most recent oil developments are in the Amazon, where oil was discovered in 1967. The Amazonian oil deposits are found mostly in remote forest areas where land rights of indigenous peoples are not secure, and this has resulted in conflicts between Peru and Ecuador and between governments, corporations, and indigenous groups. Mining in the Amazon has also created controversy in regions where migrants have gathered to work in the gold mines. The mining process uses mercury, a hazardous chemical element that is now polluting ecosystems and causing health problems among the miners.

Geology also affects soil fertility and agricultural potential through the influence of parent rock and volcanic activity. Some mineral deposits produce soils that are toxic and cannot be used for agriculture. The older shields often bear less fertile soil, whereas volcanic ash provides important nutrients.

Volcanic activity also poses threats to human activity when eruptions and ash destroy crops and lives. Throughout Mexico, Central America, the southern Caribbean, and the Andes, erupting volcanoes have frequently threatened settlements, and the archaeological and historical record documents many major disasters (**Figure 8.7**). The tensions associated with shifting tectonic plates have also produced devastating earthquakes that have ravaged the capital cities of some Latin American countries—for example, in Mexico City, Mexico; Managua, Nicaragua; Guatemala City, Guatemala; and Santiago, Chile (**Figure 8.8**). Such natural disasters cannot be blamed solely on geophysical conditions. The greatest damages occur when people are forced to live in unsafe houses or on unstable slopes because they lack the money or power to live in safer places, cannot afford insurance, or are unable to obtain warnings of impending natural disasters such as volcanic eruptions and hurricanes. (See Geography Matters: Hurricane Mitch in Honduras, p. 378.)

Figure 8.7 Volcanic eruption on Montserrat The people of the island of Montserrat in the Caribbean have lived with the threat of volcanic eruptions for centuries. Eruptions in the 1990s in the Soufriere Hills forced the evacuation of more than half of the residents. Most left their homes, farms, beach houses, and hotels, abandoning their entire lives at the foot of the volcano. The capital city, Plymouth, was burned and half-buried by waves of gas and ash.

Climate

The geological configurations also influence the climate and hydrology of Latin America and the Caribbean. The overall climates of the region are determined by global atmospheric circulation, including the positions of the equatorial high and tropical low pressure zones and the major global wind belts (see Figure 2.8). Because Latin America straddles the equator, reaching north of the tropic of Cancer in the North-

Figure 8.8 Mexico City earthquake, 1985 The 1985 earthquake in Mexico City killed as many as 10,000 people and devastated downtown buildings. The city is especially vulnerable because it is built on a former lake bed with sediments that act almost as liquids when shaken by earthquakes.

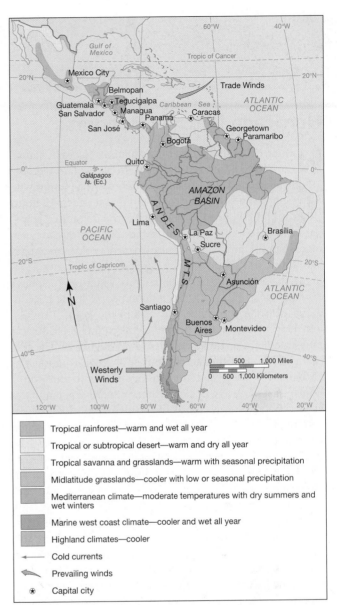

Figure 8.9 Climate regions of Latin America and the Caribbean Latin America's climate is influenced by major wind and pressure belts and the configuration of land and oceans. The average pattern shown in this map tends to shift northward in June and southward in December, bringing seasonal changes to many regions.

ern Hemisphere and south almost to the Antarctic, climatic patterns are relatively simple to understand and provide good general examples of how global circulation affects regional climate, vegetation, and human activity (**Figure 8.9**).

As we discussed in Chapter 2 (pp. 53–58), the general circulation of the atmosphere is driven by the differential heating and rotation of Earth, with warm air rising at the equator, then cooling as it rises, thereby producing high rainfall, then flowing poleward, and finally sinking over regions around the tropics of Capricorn and Cancer. The equatorial zones of high temperatures and rainfall provide conditions for the rapid growth of vegetation in the form of the rain forests of

Geography Matters

Hurricane Mitch in Honduras

Natural disasters are never just about the forces of nature; they have human causes and effects. In late October 1998, Hurricane Mitch dumped a year's worth of rain (about 1.2 meters, or nearly 4 feet) on Central America in 48 hours (**Figure 1**). Flash floods and mudslides on deforested slopes left nearly 10,000 people dead, almost 20,000 missing, and more than 2.5 million temporarily dependent on emergency aid. It was the Americas' worst hurricane in 200 years. Honduras, the second-poorest nation in the Western Hemisphere, was the hardest hit. Of the 6 million people living in Honduras, nearly 2 million were affected by the storm, 1 million lost their homes, and 70 percent of the country's productive infrastructure was damaged or destroyed (**Figure 2**). The government's initial estimate of the cost of reconstruction was $5 billion. What turned Mitch from a natural hazard into a human disaster was a chain reaction of social vulnerabilities created by long-term climate change, environmental degradation, poverty, social inequality, population pressure, rapid urbanization, and international debt. Geographers such as Piers Blaikie argue that the social, political, and economic environment is as much a cause of disasters as the natural environment. Nature creates hazards, says Blaikie, but it is humans who create *vulnerability*, through social inequality and unequal access to resources.[*] Poor people in poor states are the most vulnerable to natural hazards and the least able to cope. In the coastal plains north of Honduras' second largest city, San Pedro Sula, the flooding was 15 kilometers (9.3 miles) wide, devastating the vital banana-growing industry. To the south, the Choluteca River burst its banks downstream of the country's third-largest city, Choluteca City, reverted to its former delta area, and destroyed a coastal belt of export-oriented shrimp farms that employed 20,000 local people. In the aftermath of the flooding came the threat of

Figure 1   Satellite image of Mitch's devastation
Hurricane Mitch was one of the strongest hurricanes on record when it hit Central America in October 1998.

disease: dengue fever, diarrhea, typhoid fever, and cholera. More than 70 percent of the agricultural sector of Honduras was wiped out, creating immediate food shortages and decimating the vital export crops of bananas, coffee, and shrimp responsible for half the country's annual export revenue of $3 billion.

There is a connection between deforestation and environmental disasters. Forests play a critical role in stabilizing soil and storing water. When the forest is cleared, the ground is exposed, and this results in an increase in surface runoff when it rains and induces soil erosion. Since 1960, Honduras has lost 25 percent of its forest cover, which is now reduced to 36 per-

---

[*]P. M. Blaikie, T. Cannon, I. Davis, and B. Wisner, *At Risk: Natural Hazards, People's Vulnerability, and Disasters.* New York: Routledge, 1994.

---

the Amazon. Rainfall in the Amazon basin ranges from 1.5 to 2 meters (60 to 80 inches) a year. As the air sinks over the Tropics it becomes warmer and drier, holding so little moisture by the time it reaches ground level that these regions are characterized by the very low rainfall, sparse vegetation, and dry conditions of deserts. In Latin America and the Caribbean, the Sonoran and Chihuahuan deserts of Mexico and the Atacama Desert of Chile are partly associated with this type of large-scale atmospheric subsidence. Some air moves back toward the equator to rise again at the *intertropical convergence*

zone in the Hadley circulation (see Chapter 2, p. 55). The convergence zone of high rainfall and the dry zones of sinking air shift north in the June–August season and south in the November–February period.

Air flowing from the Tropics to the equator is dragged by the spinning Earth into an east–west flow called the *trade winds,* and air flowing poleward from the Tropics is dragged into a west to east flow called the *westerlies*. These wind belts were very important in the early exploration of the Americas because sailing ships used the trade winds to sail across the

Figure 2   Hurricane Mitch Hurricane Mitch inundated vast areas of Honduras, dropping about 1.2 meters (4 feet) of rainfall over a 48-hour period. Here the flood waters rage through the Honduran capital city, Tegucigalpa.

cent of its land area. Some 63 percent of the farmers of Honduras have access to only 6 percent of its arable land. Large-scale beef ranching and banana plantations have displaced poor peasants over decades, forcing them to live in isolated valleys, on riverbanks, and on steep hillside farms. Forced to live in marginal areas, the peasants' efforts to carve out subsistence farms have made surrounding hillsides and river banks even more unstable.

In urban areas, poorly developed shantytowns in marginal and risky areas is a common phenomenon; the development of urban infrastructure is outpaced by illegal, uncontrolled, and unplanned urbanization. Honduras has the most rapid urbanization rate in Central America, with the proportion of the population living in cities rising from 18 percent in 1950 to 44 percent in 1990 and projected to reach 60 percent by 2010. Tegucigalpa, the capital, has grown from 75,000 people in 1950 to more than 1.5 million today. The pine-clad hills surrounding the city have been carpeted by tens of thousands of wooden shacks as successive generations of squatters stripped the terrain bare for shelter and fuel, rendering the hills increasingly unstable.

The international community responded generously to Mitch with technical and financial assistance. Emergency food, shelter, and medical care were provided for the disaster-hit population in a matter of days. But weeks and months are required to rebuild infrastructure, community centers, schools, and rural health clinics. Hurricane Mitch washed the remaining topsoil from hillside marginal farms down onto the valley plantations of agribusiness. Where rivers burst their banks, they left sand 2 meters (6.5 feet) deep, destroying or degrading many small plots. Beans and other staples, ready to be picked, were destroyed and, with them, six months' income and seed for the next crop. Since Mitch, thousands of Hondurans have also lost their jobs, many in the export agricultural sector. The country lost 25 percent of its coffee plantations and 50 percent of its banana plantations—in all, 15 percent of exports. The two major banana-producing companies, Standard Fruit and Chiquita, laid off 25,000 workers for 12 months, claiming the banana crop would not recover for several years. A U.S. government study estimated that more than 400,000 Hondurans would leave their country for the United States as a result of Mitch.

Disasters create long-term effects and vulnerabilities, which like their causes are not immediately visible. The most pervasive effect is that the poor get poorer because they lack the money or the social capital to rebuild. Almost 80 percent of the Honduran population already lives below the poverty line. With malnutrition and illiteracy rates higher than those of many African countries, the next generation in Honduras will suffer the effects of Hurricane Mitch for many decades.

Source: Based on information produced by the Consultative Group for the Reconstruction and Transformation of Central America. http://www.iadb.org/regions/re2/consultative_group/groups/ecology_workshop_1.htm, and from _Natural Hazards Research Working Paper #103, Mitigation and the Consequences of International Aid in Postdisaster Reconstruction_ by Priya Ranganath, http://www.colorado.edu/hazards/wp/wp103/report.html.

---

tropical Atlantic from Europe to the Americas and then sailed up the coast to catch the westerlies across the northern Atlantic back to Europe. Ships sometimes got stuck for days or weeks in the zone of stable sinking dry air in the subtropics, where there are few horizontal winds, and were stuck for days. To conserve water, sailors on ships transporting horses sometimes threw some of them overboard, thus giving rise to the term _horse latitudes_ for this windless zone.

The trade winds flowing across the Atlantic frequently produce rain on the Caribbean islands and east coasts of Central America in the Northern Hemisphere. In southern Latin America and the Caribbean the trades bring rain to the east coast of Brazil but shift north and south because of the global circulation. The regions on the margins of the trades and at the edges of the equatorial rainfall zone have highly seasonal climates with a distinct rainy season.

The westerlies bring heavy rains to southern South America. In Chile, seasonal shifts in pressure and wind belts mean that the westerlies move southward in December, bringing rain to southernmost Chile, and northward in June, bringing rain to the central valley of Chile, resulting in distinct wet and dry seasons on the margins of the westerly circulation. When the global circulation shifts southward in December, storms spinning out of the Northern Hemisphere westerlies also bring rain to northern Mexico.

Latin America and the Caribbean's extensive grasslands occur where seasonal shifts in wind and pressure belts result in a distinct rainy season, especially on the margins where the rains are fairly moderate.

The coastal mountain chains of Latin America and the Caribbean clearly illustrate the role of topography in regional climate. First of all, ocean winds that encounter coastal mountains are forced to rise even higher, cooling to the point that they release most of their moisture in the form of rain and snow. This is a very clear feature of the Andean climate but also explains the abundant rainfall of highland Central America and Mexico. The high precipitation over the Andes feeds the rivers that water the lowlands east of the mountains, most notably the Amazon. However, mountains also create a *rain-shadow* effect because winds passing over mountains from the coast to the interior lose their moisture over the higher altitudes and then become warmer and drier as they descend to the interior, creating arid conditions to the leeward of mountain ranges (see Figure 2.9). Examples include the dry region of southern Argentina, known as Patagonia, in the lee of the Andes, and the drier regions of Chihuahua in northern Mexico in the lee of the Mexican Sierra Madre. Higher altitudes are also cooler, so despite the intensity of the sunlight, large regions of the Latin American Tropics have cooler temperatures more conducive to human activities and agricultural crops generally found in cooler climates, such as wheat, apples, and potatoes.

The **altitudinal zonation** of climate, vegetation, and human activity has led to a simple classification of Latin American mountain environments into the Tierra Caliente (up to 900 meters, 2950 feet), Tierra Templada (900–1800 meters, 2950–5900 feet), and Tierra Fría (1800–3600 meters, 5900–11,800 feet).

The very high altitudes (higher than 3600 meters) are called the Tierra Helada. Each of these zones is associated with characteristic vegetation types (e.g., rain forest in the Tierra Caliente and grasslands in the Tierra Fría) and with agricultural activities (tropical fruits in the Tierra Caliente, coffee in the Tierra Templada, and potatoes in the Tierra Fría) (**Figure 8.10**).

Latin America also provides a classic case of how the temperatures of the ocean can influence the climate of adjacent landmasses. Colder air holds less moisture and promotes less evaporation than does warm air. Winds flowing across the very cold ocean current that normally flows northward off the coast of Peru and Chile pick up very little moisture and exacerbate the already dry conditions promoted by descending air over the Tropics. The Atacama Desert is one of the driest spots on Earth.

In contrast, easterly winds moving across the warm Caribbean absorb a lot of moisture, especially during the fall when the sea surface is warmest. When storms start to circulate, the warm sea fuels both the moisture and energy of the storms, producing the hurricanes that regularly cross the Atlantic coast of Latin America and the Caribbean with some benefit to water resources but often threatening human settlements and lives with the power of their winds and the flooding from heavy rainfall.

One of the most significant features of Latin America and the Caribbean climate is that it does not remain constant from one year to another. One of the most important causes of this climate variability is the phenomenon known as **El Niño.** El Niño, which is now known to influence weather worldwide, occurs when the normally cold seas off Peru start to warm, and winds and weather patterns that characterize the southwestern Pacific shift eastward near the South American coast in what is called the *El Niño-Southern Oscillation (ENSO)*. The local effect of El Niño is to bring warmer and wetter winds to the coasts of Peru and Ecuador with high rainfall and flooding. But the sensitivity of the global atmospheric circulation is such that the links between Pacific Ocean temperatures and conditions elsewhere produce droughts in northeast Brazil,

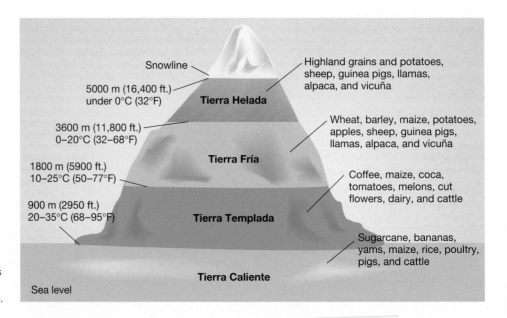

Figure 8.10 Altitudinal zonation The altitudinal zonation of climate and vegetation in mountainous regions such as the Andes creates vertical bands of ecosystems and provides a range of microenvironments for agricultural production.

Snowline

5000 m (16,400 ft.)
under 0°C (32°F)

Tierra Helada

Highland grains and potatoes, sheep, guinea pigs, llamas, alpaca, and vicuña

3600 m (11,800 ft.)
0–20°C (32–68°F)

Tierra Fría

Wheat, barley, maize, potatoes, apples, sheep, guinea pigs, llamas, alpaca, and vicuña

1800 m (5900 ft.)
10–25°C (50–77°F)

Coffee, maize, coca, tomatoes, melons, cut flowers, dairy, and cattle

900 m (2950 ft.)
20–35°C (68–95°F)

Tierra Templada

Sugarcane, bananas, yams, maize, rice, poultry, pigs, and cattle

Tierra Caliente

Sea level

floods in southern Brazil and northern Mexico, fewer hurricanes, as well as droughts in southern Africa, Australia, and Indonesia. In some years, the ocean off Peru gets colder than normal, and this produces a contrasting global pattern called **La Niña**, with floods in northeast Brazil, drought in northern Mexico, and more intense Pacific hurricanes.

One of the most exciting scientific developments in recent decades is an improved ability to monitor what is happening to sea surface temperatures in the Pacific and to forecast the onset of an El Niño up to a year in advance. These forecasts can help farmers and disaster relief agencies prepare for the droughts and floods associated with El Niño (or La Niña) and hopefully reduce losses, or even take advantage of the changed conditions by planting different crops.

The highest and coldest parts of the Andes store precipitation in mountain glaciers that act as important natural storage for water resources. One of the indicators that global climate may be becoming warmer is the clear evidence that the Andean glaciers are melting and shrinking.

The spatial and seasonal variation in climate is reflected in the hydrology and natural vegetation of Latin America. The region's major rivers originate in higher rainfall areas but their flow can vary seasonally and from year to year as a result of factors such as El Niño or hurricane intensity.

The three largest river basins in Latin America are the Amazon, the Plata, and the Orinoco, all flowing to the Atlantic Ocean (see Figure 8.1). The rivers in the Plata basin (including the Paraná, Paraguay, and Uruguay rivers) originate in the Andes and the Brazilian highlands. The Paraguay flows through the flooded wetlands of the Pantanal in Brazil and Paraguay and joins the Paraná, which flows across the northern end of the *pampas,* the arid grasslands of Argentina. The Paraná has become an important source of energy through large hydroelectric dams such as the Itaipu dam (**Figure 8.11a**). The Orinoco drains the *llanos* grasslands of Colombia and Venezuela. The Amazon drains a vast basin that includes parts of Brazil, the Guianas, Venezuela, Colombia, Ecuador,

Peru, and Bolivia. In other parts of the region, such as the western Andes, Mexico, and Central America, many smaller rivers drain from the mountains with highly seasonal flows. The rivers flowing from the Sierra Madre Oriental provide water resources to irrigated export agriculture in northwest Mexico; those rivers flowing eastward to the Gulf of Mexico such as the Grijalva have been extensively dammed and developed for hydropower. In cities such as Bogotá, Colombia, and Lima, Peru, increased water demands and climatic variations associated with El Niño have combined to threaten water and energy supplies with frequent droughts, water rationing, and electricity brownouts.

Latin America has several large freshwater lakes, among them Lake Nicaragua in Nicaragua and Lake Titicaca at the border of Bolivia and Peru (see Sense of Place: Lake Titicaca, p. 382). Major waterfalls such as Iguaçu Falls—where Brazil, Argentina, and Paraguay meet (**Figure 8.11b**)—and Angel Falls, Venezuela—the tallest waterfall in the world at a height of 985 meters (3230 feet), falling off the flat-topped mountain of Auyantepui—have become increasingly popular tourist destinations.

The diversity of Latin America's physical environments has produced a large variation in species, or *biodiversity.* Latin America's biodiversity is large because of the size of the continent, the range in climates from north to south, altitudinal variations within short distances, and a comparatively long history of fairly stable climates and isolation from other world regions. Many tourists are attracted to the colorful birds and verdant plants associated with the tropical regions of the Americas, also called the **neotropics.**

The wetter climates of Latin America are associated with magnificent forest ecosystems, including the tropical rain forests of the Amazon, Central America, and southern Mexico and the temperate rain forests of southern Chile. The Amazon forest ecosystems are notable for the sheer number of species found within small areas of forest (**Figure 8.12a**). The giant alerce trees of Chile resemble the redwoods of the west

Figure 8.11   Water resources of Latin America (a) The Itaipu dam on the Paraná River between Brazil and Paraguay generates more than 80 percent of Paraguay's electric power and 25 percent of Brazil's. (b) Iguaçu Falls are on the Paraná River, where Brazil, Argentina, and Paraguay meet, and have become a major tourist destination.

(a)

(b)

Sense of Place

Lake Titicaca

At 3820 meters (12,580 feet) above sea level, Lake Titicaca, with an area of 9064 square kilometers (3500 square miles) and a depth of up to 150 meters (500 feet), reflects the luminescent blues of the Andean sky, framed by the towering Andes mountains and the browns and greens of the grasses of the *altiplano,* upon which the herds of alpaca and llama graze. The lake, which now forms the border between Bolivia and Peru (**Figure 1**), is surrounded by ruins that archaeologists have linked to the Inca civilization. One legend tells that the first Inca emerged from the lake, and several of the 41 islands are still significant religious sites.

The indigenous Aymara residents live by growing crops such as potatoes, barley, and the high-protein Andean cereal called *quinoa* and by using the milk and wool from the al-paca and llama. Together with the Uros people, who live on Lake Titicaca on floating islands made from reeds (**Figure 2**), the Aymara increasingly derive an income from tourism. People fish on the deep lake from canoelike boats also made from reeds.

Lake Titicaca is now a key stop on the international tourist route that includes a visit to the dramatic ruins of Machu Picchu. For several years tourism was reduced by fears of unrest associated with the Sendero Luminoso guerrilla movement in Peru, but more peaceful conditions in the highlands are producing a rapid increase in tourist visits. Tourists are attracted by the dazzling landscape, Inca ruins, and the indigenous cultures with their colorful woven clothing, pipe music, and fiestas. Many tourists purchase alpaca sweaters to cope with the cold evenings, and some become unwell as a result of the high altitude. Although some local residents make an income from tourism, many are still very poor as a result of the social inequality that persists in both Peru and Bolivia. Migration to work in mines and cities continues, and some young people have become discontented with traditional lifestyles.

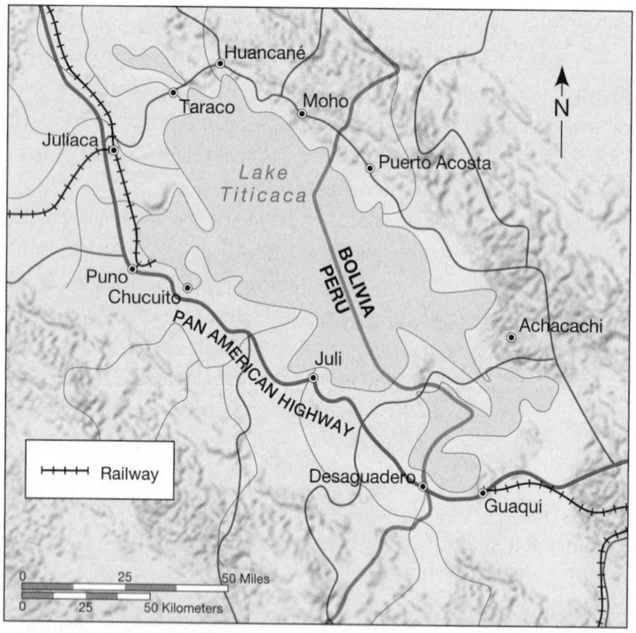

Figure 1   Lake Titicaca Lake Titicaca sits high on the Andean altiplano.

Figure 2   Local life Local people live on islands in the lake and fish from boats made from reeds.

---

coast of North America because both species rely on the heavy seasonal rains from westerly storms.

Desert ecosystems, such as in the Atacama Desert of northern Chile and Peru, are associated with drier climates where species have developed many interesting adaptations to water scarcity (**Figure 8.12b**). *Xerophytic* (dryland) plants such as cacti conserve water by having few leaves, waxy skins, and shallow roots. They store water in their stems and are covered with sharp spines to discourage thirsty animals. Other plants bloom only during the rains and are called *ephemerals.* Animals such as lizards and rodents are active only in the cool of the night and live in burrows during the day.

Between the moist forests and dry deserts lie ecosystems where alternating wet and dry seasons produce vegetation ranging from scattered woodlands to dry grasslands. Grasslands are also found at higher altitudes where there is not enough precipitation or temperatures are too low to support highland forests. In Argentina the *pampas* grasslands cover

(a)

(b)

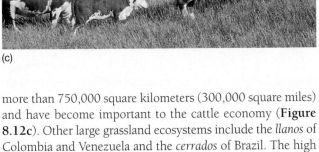

(c)

Figure 8.12   Latin American ecosystems   Latin American ecosystems range from forests and grasslands to deserts and coastal mangroves. (a) The Amazon rain forest is composed of several layers of diverse trees that house a great variety of mammals, birds, insects, and other species. During high flows, areas adjacent to the river are flooded for several weeks and fish are abundant. Nutrients cycle rapidly in the warm, wet climate. (b) The Atacama Desert of Peru and northern Chile is one of the driest locations on Earth. However, the area is economically important because of the copper and other minerals that lie beneath the surface and that support one of the world's most important mining areas. (c) The *pampas* of southern South America. The fertile soils of the extensive grasslands traditionally supported a livestock economy associated with the *gaucho* cowboy. It was the wheat and beef of the *pampas* that made Argentina one of the richest countries in the world a century ago.

more than 750,000 square kilometers (300,000 square miles) and have become important to the cattle economy (**Figure 8.12c**). Other large grassland ecosystems include the *llanos* of Colombia and Venezuela and the *cerrados* of Brazil. The high grasslands of the Andean altiplano provide habitat for grazing animals such as the llama, guanaco, and vicuña.

The long coasts of Latin America and the islands of the Caribbean include about 50,000 square kilometers (about 19,000 square miles) of mangrove ecosystems, or about 25 percent of the world's total. Mangrove ecosystems are important in protecting the coasts from storms and in providing breeding areas for fish and other marine animals.

Environmental History

The natural environments of Latin America and the Caribbean offer both constraints on and opportunities for human activity, especially agriculture. Some areas are much more productive than others, especially the flatter river valleys, with annual renewal of soils by sediment deposition and easy access to water, and regions where ash from volcanic activity provides nutrients to the soil.

A constraint on human development, particularly in the warm and wet climates of much of Latin America and the Caribbean, is the large diversity and prevalence of pests and diseases that weaken and kill plants, animals, and humans in the Tropics around the world. For example, malaria is endemic in much of the Amazon basin and lowland Central America.

As in Africa, Europeans found warm, wet tropical climates arduous in Latin America and the Caribbean because settlers were vulnerable to disease, and their crops and animals could not cope with local pests. In books, magazines, and newspapers, accounts of the settlement of Latin America in the nineteenth and early twentieth centuries portray the region as a lush but dangerous place. But as we shall see, European pests and diseases introduced into Latin America and the Caribbean wrought far greater damage on indigenous ecosystems and peoples.

Despite the constraints posed by the natural environment in Latin America and the Caribbean, any assertion of environmental determinism (see Chapter 1, p. 26) is unwarranted because the ability of humans to overcome many of these constraints is so clear throughout the region. Geographers such as Carl Sauer, William Denevan, and Billie Lee Turner III employed the approaches of *cultural ecology* to show how native Latin American populations used technology and social organization to adapt the harsh physical environments to their needs and take full advantage of more favorable environments.

Adaptations were taken by early hunter-gatherer populations who crossed the land bridge over the Bering Strait between Siberia and Alaska more than 15,000 years ago and spread into the Americas. In Mexico, archaeologists have found

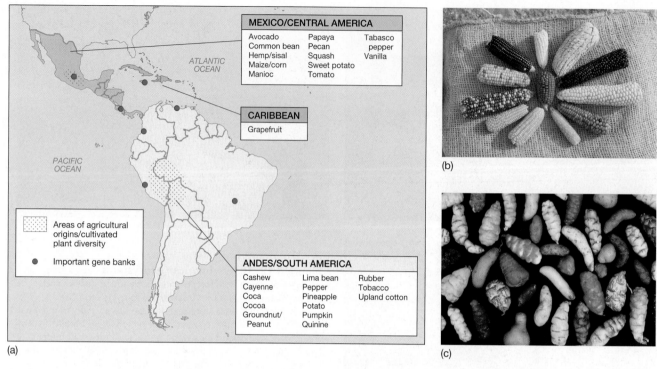

Figure 8.13 Domestication of food crops in Latin America (a) Latin America has two important centers of domestication in Mexico and the Andes. The red dots on the map denote research centers where attempts are being made to preserve crops' genetic diversity in "gene banks." (b) Maize (corn) was domesticated in Mexico from a wild grain called *teosinte*. Bred for a variety of microenvironments and tastes, traditional maize has many shapes and colors. (c) The many colors and shapes of potatoes grown in the Andes are indications of the diversity of varieties domesticated in these regions.

evidence of improvements in hunting technology such as the use of the more sophisticated double-sided Clovis hunting blade and the widespread use of fire. Fire was used to intimidate predators, to assist in the hunt by driving game animals, and to create deliberate burns of grasslands and forests to promote the regrowth of grasses attractive to game animals such as deer.

The most dramatic transformation of nature by early peoples in Latin America was the *domestication* of flora and fauna through control and selective breeding of wild plants and animals starting more than 10,000 years ago. Many of the world's major food crops were domesticated by native Latin Americans, including the staples of maize (corn), manioc, and potatoes, as well as vegetables and fruits such as tomatoes, peppers, squash, avocadoes, and pineapples (**Figure 8.13**). Tobacco, cacao (chocolate), vanilla, peanuts, and coca (cocaine) were also domesticated in Latin America. In dry areas, people tried to ensure water supplies by building small dams and channels to bring water to these crops.

Latin America has very few domesticated animals—the llama (and its relatives the alpaca and vicuña) was tamed and bred for wool, milk, meat, and transport, and dogs and guinea pigs were also used for pets and meat. As in other regions of the world, the increased yields from domesticated crops created a surplus that permitted the specialization of tasks, the growth of settlements, and ultimately the development of highly complex societies and cultures. In Latin America, the complex societies included the great Mayan, Incan, and Aztec empires (**Figure 8.14a**). These groups all modified their environments to increase agricultural production and to exploit water, wood, and minerals to support their cities, metal production, and trade.

These environmental transformations were widespread, and in some cases—most notoriously the Mayan civilization—people placed so much pressure on regional landscapes that environmental degradation may have precipitated social collapse. The Mayas occupied the Yucatán Peninsula as well as a considerable portion of Guatemala and Honduras, with a period of expansion beginning about 3000 B.C. and reaching a peak of control and social development from about 600 B.C. to A.D. 800. Faced with rapid declines in the fertility of soils after clearing the rain forest, the Mayas adapted by burning the forest to capture the nutrients in the trees through the ash and then by moving on to clear another patch of forest once the declining fertility of the previously cleared area resulted in reduced yields. It took up to 30 years for the forest to regrow on a plot that had been cleared, farmed, and then abandoned. This adaptation to rainforest environments mirrors those in other parts of the world and has been termed *slash and burn,* or *swidden,* agriculture. The Mayas also developed methods for growing crops in wetland areas by building *raised fields* that lifted plants above flooding but took advantage of the rich soils and reliable moisture of wetland environments (**Figure 8.14b**). There is evidence that the Mayas cleared vast areas of forest during this period.

(a)

(b)

(c)

(d)

Figure 8.14   Maya, Aztec and Inca adaptations to environment   (a) Map of the extent of Maya, Inca, and Aztec empires. (b) The Maya, Inca, and Aztec cultures adapted to environmental constraints in many resourceful ways. Around Mayan cities—such as Palenque, shown here with its pyramids in the Yucatán of Mexico—the forest was cleared using slash-and-burn agriculture, and elsewhere in the Yucatán flooded areas were farmed using raised fields. Palenque was abandoned sometime after A.D. 500 and is now a major tourist destination. (c) The Inca constructed terraces, such as these near Cuzco, Peru, so that they could create level surfaces for irrigating and growing crops and reduce frost risks by breaking the downslope flow of cold air. Constructing the terraces required considerable technical expertise and social organization, and the Inca rulers conscripted large numbers of laborers from local communities. Many terraces were abandoned as the native labor force was reduced due to the ravages of newly introduced European diseases and the need to shift laborers to the Spanish mines in the sixteenth century. (d) The Aztecs cultivated wetlands through the *chinampa* system of fields built from mud and vegetation and anchored to lake beds.

Between A.D. 500 and A.D. 1000 the great Mayan cities such as Copán, Palenque, and Tikal were abandoned, and overall population declined dramatically. Many scholars believe that one reason for the Maya collapse was their overuse of the soils. Population growth and tribute demands by the Maya elites required increases in agricultural production, often involving the clearing of sloped lands, and there was no time to allow plots to recover before planting them again. Large-scale forest clearing has also been linked to regional changes in climate, with increases in temperature and decreases in rainfall that would threaten agriculture. Soil erosion, droughts, and declining soil fertility would have contributed to a decline in the amount of food available to feed the large population and would cause some of the nutritional stresses that

archaeologists have detected in human skeletons from the period of collapse.

The Incas also responded to the difficulties of living in a mountain environment in a variety of ways, including through the construction of many miles of agricultural terraces on the steep hill slopes of the Andes during the height of their empire, which in A.D. 1400 stretched from northern Ecuador to central Chile (**Figure 8.14c**). These Andean terraces not only reduced soil erosion and provided a flat area for planting, but also decreased frost risks by breaking up downhill flows of cold air and allowed for irrigation canals to flow across the slopes in efficient ways. The construction and maintenance of these terraces and irrigation systems required large-scale social organization in which the Incan empire excelled. People were organized into small groups called *ayullu,* and labor for agriculture and mines was commanded through the *mita* labor system by which communities were required to provide a certain number of days of work to the central authorities.

The Aztecs, who settled in central Mexico in the 1300s, were experts in the control of water. They constructed an extensive network of dams, irrigation systems, and drainage canals in the basin of Mexico to cope with the highly seasonal and variable rainfall pattern that in some seasons and years produced droughts, and in other seasons and years produced large lakes and wetlands. They also developed the *chinampa* agricultural system, which permits agriculture in lake and wetland environments (**Figure 8.14d**). *Chinampas* were constructed by building up an island of soil and vegetation, usually attached to a lake bed or shore, planted with crops and trees that grew above the water level but benefited from abundant moisture and reduced frost risks. There is evidence that as they cleared forests in the basin of Mexico, the Aztecs may have contributed to a drop in the water table and to a resulting water crisis that led to the abandonment of some settlements.

These widespread modifications of the environment of Latin America are evidence that has been used to debunk what geographer William Denevan has called the **pristine myth** of an untouched continent, where native peoples lived in harmony with nature that was only degraded after the arrival of the first Europeans in 1492. The environmental adaptations and impacts of of the Mayas, Incas, and Aztecs still echo in the traditional technologies used in some regions of Latin America and in the continual efforts to use technology to benefit from the physical environment and avoid its hazards. Geographers are among those who have pointed to the ways in which overuse of their environment contributed to the collapse of the Mayan society and to the warning that this implies for current patterns of widespread deforestation, overuse of the land, and depletion of water resources.

Latin America and the Caribbean in the World

The Colonial Experience in Latin America and the Caribbean

The integration of Latin America and the Caribbean into a global system of political, economic, ecological, and social relationships began more than 500 years ago with the arrival of Spanish and Portuguese explorers at the end of the fifteenth century and their *colonial* activities that inexorably linked a Latin American periphery to a European core. As described in Chapter 3, the fifteenth and sixteenth centuries were a period of innovation in Europe with changes in manufacturing technology and the development of an economy of *merchant capitalism* (*mercantilism,* see Chapter 2, p. 66). Improvements in shipbuilding and navigation allowed Europe to explore—and

then expand trade with—other regions of the world, including Asia to the east and Africa to the south, and to sail to the west in search of new routes to Asia.

The most famous of these explorers was Christopher Columbus (known as Cristóbal Colón in Latin America), an Italian from Genoa who lived in Lisbon, Portugal, along with his brother Bartholomew, an expert chart maker. Under the sponsorship of Queen Isabella of Spain, Columbus was to search for new territory and trading opportunities on a western route to the Indies (as Asia was then known). Having set sail from southern Spain on August 3, 1492, with three small sailing ships—the *Santa María,* the *Pinta,* and the *Niña*—Columbus, commanding the *Santa María,* arrived in the Caribbean in October and landed on Watling Island, a small island in the Bahamas, to which he gave the name San Salvador (**Figure 8.15**). On his first voyage, Columbus also visited Cuba and another island that he called Hispaniola (now Haiti and the Dominican Republic). When the *Santa María* was wrecked on the north coast of Hispaniola, Columbus left behind 21 volunteers to found a colony and returned to Spain on the *Niña,* bringing with him six locals, several parrots, and some gold ornaments. His second voyage in 1493 was much larger, with 17 ships and 1500 men, because he intended to establish permanent settlements, but he was frustrated by divisions within his team and hostility from local residents. On his third and fourth voyages he explored the island of Trinidad and coasts of Venezuela and Central America. With the promise of new lands in the Western Hemisphere, the Spanish crown appealed to Pope Alexander VI for a ruling that would assign the new lands to Spain rather than to Portugal. The resulting **Treaty of Tordesillas** in 1494 drew a demarcation line 370 leagues (about 1800 kilometers, or 1100 miles) west of the Portuguese Cape Verde Islands (about 47 degrees west longitude), granting non-Christian lands west of this line to Spain and east of this line to Portugal, including the as-yet-undiscovered territory of Brazil that jutted out eastward into the Atlantic from the Americas, as well as parts of Africa.

Columbus was followed in subsequent decades by others seeking gold, territory, and other resources in Latin America and the Caribbean. The most notable explorers, or *conquistadors,* included Hernán Cortés, who landed in Veracruz, Mexico, in April 1519 and went on to conquer the Aztec empire and its capital of Tenochtitlán in the basin of Mexico; and Francisco Pizarro, who seized control of the Incan empire centered in Cuzco, Peru, in 1533. The Portuguese began their colonization of Brazil with the landing of Pedro Alvares Cabral in April 1500 at Porto Seguro in southeast Brazil.

The Spanish expanded and administered the new Latin American colonies through the two **viceroyalties** of New Spain (Mexico and Central America) based in Mexico City and of Peru (Andean and southern South America) based in Lima (**Figure 8.16**). The Peruvian viceroyalty was subsequently divided into New Granada, Peru, and the Río de la Plata as colonial power expanded into the interior of the continent and new cities were established. These viceroyalties were subdivided

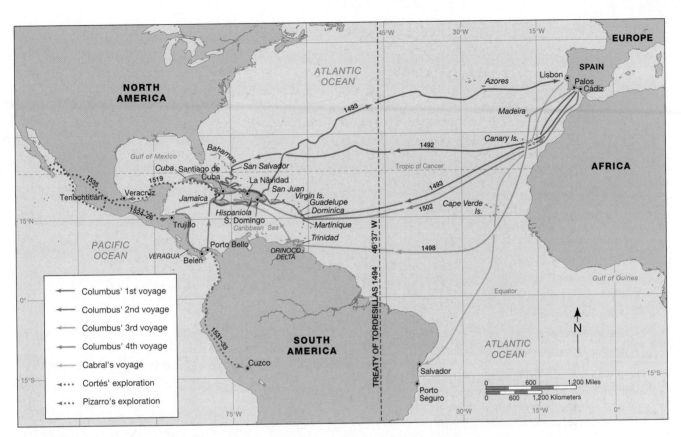

Figure 8.15 Colonial voyages and Treaty of Tordesillas The map shows the major voyages and missions of Columbus, Pizarro, Cabral, and Cortés and the division of Latin America between Spain and Portugal under the Treaty of Tordesillas in 1494. The initial line set by the pope in 1493 was contested by Portugal and shifted further west. (*Source:* Adapted from P. L. Knox and S. A. Marston, *Human Geography: Places and Regions in Global Context*, 2nd ed. Upper Saddle River, NJ: Prentice Hall, 2001.)

into judicial regions called *audiencias*, administered from regional centers in the territories of Guatemala and Panama and in six cities that are now the capitals of Latin American countries: Bogotá (Colombia), Caracas (Venezuela), Quito (Ecuador), La Paz (Bolivia), Santiago (Chile), and Buenos Aires (Argentina). The contemporary significance of this political organization is dramatic in that these centers, and the boundaries of the *audiencias* that they controlled, are even today important locations in Latin America. The Spanish charged the new administrators with obtaining gold and silver for the Spanish crown, converting the native people to the Catholic religion, and making the colonies as self-sufficient as possible through the use of local land and labor. The Spanish crown demanded 20 percent of all mine profits, the so-called *Quinto Real,* or royal fifth. These goals were based in concepts of nature as a commodity, with a secular identity to be governed by humans, and of land as a private property right (see Chapter 2, p. 69).

The search for local labor to work in the mines and fields of the Spanish colonizers was frustrated by one of the most immediate and significant impacts of the European arrival in Latin America and the Caribbean, the **demographic collapse** of indigenous populations as a result of diseases brought by the Europeans. Because of the long isolation of the Americas from other continents, native peoples lacked resistance and immunity to European diseases such as smallpox, influenza, and

measles. When they caught these diseases from Europeans and then from each other, mortality rates were very high. Researchers have estimated that up to 75 percent of the population of Latin America died in epidemics in the century or so after contact. This massive mortality demoralized local people, led to the abandonment of their settlements and fields, and meant that there was a scarcity of labor to work in the mines, missions, and agricultural activities with which the Spanish, for example, hoped to support their colonial enterprise.

The introduction of European diseases into the Americas is just one example of the interaction between the ecologies of the two continents that historian Alfred Crosby has called the **Columbian Exchange**. When the Spanish and other colonial powers arrived in new lands, they brought with them favorite plants and animals that they planned to introduce into the new colonies, but also, unintentionally, diseases, weeds, and pests such as rats that were stowaways on their ships. In return, the explorers and colonists collected species that they hoped could be sold or traded back to Europe and elsewhere.

In Latin America and the Caribbean, the Spanish introduced the crops and domesticated animals of their homeland—especially wheat and cattle, but also fruit and olive trees, horses, sheep, and pigs. Sugar, rice, citrus, coffee, cotton, and bananas, which had originally been brought to Spain from North Africa and the Middle East after the Moorish invasions

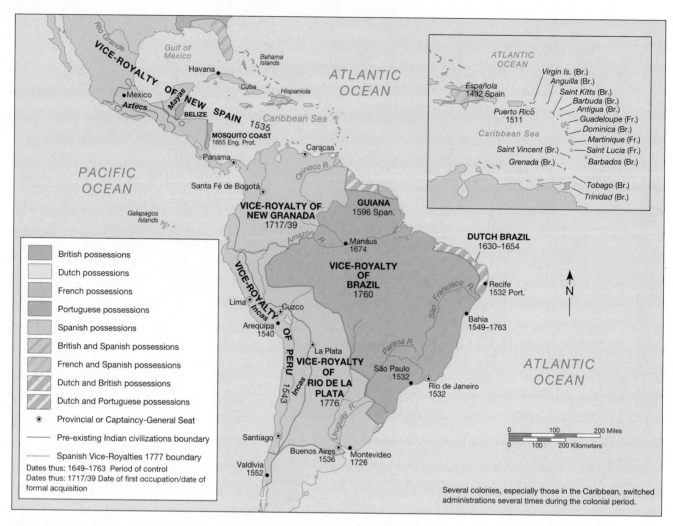

Figure 8.16   Colonial administration of Latin America and the Caribbean   The administration of colonial Latin America was organized around a system of viceroyalties and seats of provincial government called *audiencias*. This map shows the date and extent of European control from 1492 to the late 1700s. (*Source:* Adapted from A. Thomas and B. Crow, *Third World Atlas.* Buckingham, UK: Open University Press, 1994, p. 38.)

of the sixth century, were also transported to the Americas. The Spanish colonizers took back to Europe corn, potatoes, tomatoes, tobacco, and possibly the human disease syphilis.

Over longer periods these exchanges had other important effects. The clearing of land for European crops such as wheat and sugar and the overgrazing by cattle contributed to soil erosion and deforestation in Latin America and the Caribbean. Newly introduced rats, pigs, and cats ate the food that traditionally supported local species or consumed the species, especially ground-dwelling birds. One of many species of Andean potatoes carried back to Europe became the foundation of the Irish diet and one cause of a famine and migration to the Americas when disease destroyed potato harvests in Ireland in the mid-nineteenth century. Corn and manioc were introduced into Africa and became new staples, whereas peanuts and cacao were the basis of new African export economies. Cotton was introduced to India and grown for British textile mills; pineapple was distributed to the Pacific, including Hawaii; and tobacco became an addictive habit, eventually throughout the world.

In order to wrest profits and products from their new lands, the Spanish introduced several new forms of land tenure and labor relations into Latin America and the Caribbean that still influence contemporary landscapes. Where the colonizers wished directly to control the land, they granted land rights over large areas to Spanish colonists, often military leaders, and to the Catholic Church, ignoring traditional local uses and establishing fixed property boundaries. These latifundia typically occupied the best land, forcing other farmers onto small plots of land or minifundia (see Chapter 3, p. 146). Large estates, called **haciendas,** were established to raise cattle, wheat, olives, and fruit to support the mines and missions and for export in small amounts back to Europe (**Figure 8.17**). But the major export sector was the **plantation,** where single crops such as sugar or tobacco were grown for export, mainly in the wetter coastal areas. Labor for the haciendas, plantations, and mines was obtained initially through the institution of **encomienda,** in which local populations were brought under the "protection" of Spanish authorities and given small plots of land in return for tribute in the form of crops or other commodities and a certain number of days or men assigned to work in the mines and fields of the Europeans. Because some colonists abused the system, using the majority of those under their control as slaves, this form of labor control was replaced in 1542 by the slightly more flexible institution of *repartimiento,* in which indigenous communities received some autonomy and land and were only required to send a selection of their members to work for the Europeans.

These forms of labor control did not produce a large enough workforce, especially where the Europeans wanted to establish export plantations with high labor requirements and in the Tropics where the demographic collapse had devastated local populations that may have been small to begin with. In this case, colonial trading routes were used to import slaves, mainly from Africa, to the Caribbean, Central America, and Brazilian plantations to work in the production of sugar. Slave imports from Africa to Latin America and the Caribbean eventually totaled more than 5 million people, including 3.5 million to Brazil and 750,000 to Cuba (see Geographies of Indulgence, Desire, and Addiction: Sugar, p. 390).

The Spanish also maintained strict control over the ports through which goods were shipped (for example, Veracruz, Mexico) and over who could participate in maritime trade. They made profits not only from resource extraction from the Americas through agriculture, mining, and taxation, but also from goods that were imported exclusively from Spain into Latin America and the Caribbean for sale to both European settlers and local peoples.

It is important to recognize that the colonial effort in Latin America and the Caribbean was a process that took place over

Figure 8.17   Hacienda Many Spanish haciendas, such as this one in Mexico, still stand in Latin America. Hacienda Tenango, which was founded at the beginning of the seventeenth century near Cuautla in the state of Morelos, was the first sugar hacienda in Mexico.

at least two centuries, with some places incorporated earlier than others and some regions never really coming under complete colonial control because of their remoteness (the Amazon) and local resistance (parts of the Andes). Many accounts have portrayed the Spanish in very negative terms—what historians have termed the "black legend" of a Spanish conquest characterized by greed, cruelty, environmental destruction, and insensitivity to local peoples. But there were also Spanish settlers, such as the conquistador-turned-priest *Bartolomé de las Casas* and Jesuits in Brazil and Paraguay, who were concerned about the rights of local people. There were also indigenous local leaders and groups who joined the Europeans in exploiting labor and conquering their rivals, such as the Tlaxcalans who helped Cortés vanquish the Aztecs in Mexico. In some cases, the Spanish were able to capitalize on existing traditional hierarchies that demanded tribute and labor from local people, such as the Inca system of *mita* labor that required all communities in the empire to provide labor for mines and maintenance of terraces and other infrastructures.

As Europe consolidated colonial control of Latin America and the Caribbean, changes occurred in global and regional economies and political geographies that brought new colonial powers, trading patterns, and institutions to the region. During the sixteenth century, the Portuguese expanded their interests in Brazil from a few trading stations on the coast, exporting wood used to produce dyes, to the development of large coastal sugar plantations using local (indigenous) forced labor. When disease and retreat of local populations into the interior created a labor scarcity, the Portuguese expanded their role in the slave trade along the African coast and started importing thousands of African slaves to Brazil. The Portuguese crown controlled the licenses for shipping both sugar to Europe and slaves from Africa. Portuguese Brazil, in general, was more integrated into global markets at this time than were colonies in Spanish America, where the focus was more on domestic production and export of precious metals to Spain. In the Caribbean, the growing maritime and economic power of the British, French, and Dutch resulted in several early efforts to wrest control from the Spanish through piracy or state-supported military expeditions.

The most important export commodities in Spanish colonial America were silver, produced mainly from mines in Mexico and Bolivia; sugar, grown on plantations in Cuba and southern Mexico; tobacco from Cuba; gold from Colombia; cacao (for chocolate) from Venezuela and Guatemala; and indigo, a deep blue dye, from Central America. In the first phase of the developed colonial economy (1540–1620), Spain derived enormous wealth from the bonanza of the silver mines at Potosí (now within Bolivia) that produced half of the world's silver in the sixteenth century (see Figure 8.6a). This rapid influx of money led to inflation in Europe and Spanish industry suffered as upper classes in both Spain and the colonies chose to purchase luxuries from other parts of Europe.

By 1620, Spain was embroiled in expensive wars with England, France, and Germany, partly over control of trade with the Americas, and the demographic collapse and exhaustion of surface silver deposits was resulting in lower revenues from the

Sugar

No other food has had the historical and geographical impact of sugar, a sweet substance prevalent in most contemporary diets, yet unknown outside of some South Pacific islands and Asia for much of human history. Millions of people now consume sugar in almost addictive quantities—in soft and alcoholic drinks and in candy and as an additive to most processed foods. Too much sugar can cause health problems such as diabetes and tooth decay, and sugar is especially appealing to children because it provides a quick boost of energy when it is absorbed into the bloodstream. About 60 percent of sugar is produced from a tropical grass known as sugarcane, and the remainder comes from sugar beets, which are grown in many of the world's temperate regions. As demand for sugar, known as "white gold" because of its high value, grew in the seventeenth century, millions of Africans were transported to the Caribbean and Latin America and forced to work as slaves on sugarcane plantations. Sugarcane production was labor-intensive, especially the arduous process of burning the cane fields and then cutting the cane with machetes. The cane is then ground into pulp and boiled at the sugar mill to make sugar, using technologies pioneered by the Arabs. In fact it was Arabic-Moorish culture that introduced sugar to Spain and Portugal, which then disseminated it to their colonies in the Caribbean and Latin America.

The initial expansion of sugar production to the New World occurred in the 1500s when the Portuguese established plantations along the coast of northeast Brazil. Today the descendants of the African slaves who worked on the those plantations now make up a large segment of Brazil's large Afro-Brazilian population and dynamic culture. Geographer Jock Galloway describes how, by 1800, the Caribbean, under British and Dutch colonialism, had become the world's most important sugar production region, providing 80 percent of Europe's needs and the primary destination of slave arrivals from Africa (**Figure 1**). Sugar provided enormous profits to those capitalists who controlled its production and who often gained considerable political power in the core countries. George Washington and Thomas Jefferson owned sugar plantations on the British Caribbean island of Barbados, for example. Anthropologist Sidney Mintz, in his book *Sweetness and Power,* argues that there is a clear but complex link between slavery, sugar plantations in the New World, and the rise of industrial capitalism in Europe. It is sugar, and the greed that sought slave labor to produce it, that is mostly responsible for the suffering of millions of slaves in the Americas and for the high proportion of people of African heritage in countries such as Brazil, Cuba, Jamaica, and Haiti. After slavery ended, sugar prompted another diaspora of indentured contract workers from Asia who were lured to Caribbean islands such as Trinidad and Jamaica. Sugar, in the form of rum, was the drug that propelled the British and French navies in their battles and colonizing

Figure 1 Sugar production dominated the Caribbean in the nineteenth century. This print shows a plantation with the sugarcane fields, sugar mill (lower right), and housing for the workers (left). Labor needs for the sugar plantations drove the importation of millions of Africans as slaves to Brazil and the Caribbean until the mid-nineteenth century. The African diaspora to Latin America and the Caribbean is evident in the racial composition of contemporary populations of countries such as Jamaica, Cuba, Haiti, and Brazil.

expeditions into Asia and the Pacific. Sugar producers in Britain encouraged the consumption of sweetened tea and thus drove the development of tea plantations in Africa, Australia, and Southeast Asia, as well as on Pacific islands such as Hawaii and Fiji, and caused the shift from sugar as a luxury item to a virtual necessity in the diet of working-class Britain.

Slavery persisted longest in Cuba under Spanish control, and the sugar barons started to shift their operations there in the nineteenth century. The Cuban cane was of high quality, and before long the large island was covered with fields of sugarcane. By the mid-twentieth century, sugar dominated the economy, and many fields had foreign, and especially U.S., owners. After the Cuban Revolution in 1959, the sugar industry was nationalized by Fidel Castro's government, which found ready markets in the Soviet Union.

India is now the world's major sugarcane producer, followed by Brazil, Cuba, and Thailand. Sugar consumption is very high within Latin America and the Caribbean, especially in Brazil and Cuba. Sugar has had profound economic and social consequences on those parts of the world in which it became a major crop. It has created highly divided societies, with a wide gap between owners and workers, and has produced multiethnic societies with internal disparities and conflicts. Where sugar remains a main export crop, it preserves a legacy of dependency and vulnerability to world markets, exacerbated by surpluses and new diet trends that have reduced the demand for sugar and depressed the price of sugar relative to other crops.

Figure 8.18 Buenos Aires The Argentinian port city of Buenos Aires was one of the world's most important cities in the eighteenth and nineteenth centuries. Buenos Aires flourished from 1890 to 1940 when refrigerated shipping and European—especially British—investment produced a vibrant urban center called the "Paris of the South." Like the city of Chicago in the United States, it became the center of a productive agricultural region and received thousands of European immigrants in the nineteenth and early twentieth centuries. With a contemporary population of 11 million people, it is the capital of Argentina and an important economic and cultural center for southern South America. The view is from the Río de la Plata looking across the port toward the modern city skyline.

Figure 8.19 Simon Bolívar Statues of *The Liberator*, Simon Bolívar, throughout Latin America commemorate nineteenth-century independence movements throughout the region.

colonies. Merchants and landowners in the colonies were also starting to resent the strict control of trade and taxation by Spain and were using positions of power and smuggling to keep revenue for themselves, thereby contributing to the overall weakening of Spanish power and economy. The last phase of colonial political and economic control is associated with the *Bourbon Reforms* that occurred in 1713 when the French won the war to determine who should succeed the last Hapsburg king of Spain. These reforms were a significant step in Latin America's integration into the global economy because they expanded the number of ports that could be used for export, initiating, for example, the growth of Buenos Aires as a key city while increasing taxes and professionalizing the colonial administration and military to prevent corruption and to defend against other European interests (**Figure 8.18**). They produced economic growth, especially on the frontiers of the Spanish empire, but they also created growing resentment among those settlers and local people who had benefited from previous relaxed controls.

Independence Movements and the Export Boom

Independence movements arose in Latin America in 1808 when Napoleon conquered Spain and threatened to tighten trade controls. Revolutionaries, drawn from both Spanish-American and indigenous leaders, set out to liberate Latin America from Spain, partly inspired by the French and American revolutions. Between September 16, 1810, when the priest and peasant leader Miguel Hidalgo called for Mexican independence in the

famous *Grito* ("cry"), and 1824, when Simon Bolívar, known as *The Liberator* (**Figure 8.19**), finally led northern South America to independence, a series of regional revolts led to the formation of independent republics in Mexico, Argentina, Peru, Colombia, Chile, and Brazil. The boundaries and capital cities of most of these new countries originated in the colonial system of viceroyalties and *audiencias,* and this legacy was reinforced further, when, in the instability following independence, the republics of Venezuela and Ecuador split off from Colombia, Bolivia and Paraguay from Argentina, and the Central American countries from Mexico.

In the Caribbean, after several rebellions against France (inspired by the French Revolution), former slaves declared Haitian independence in 1804 and occupied the rest of the island of Hispaniola until the Dominican Republic gained independence in 1844.

In the first half of the nineteenth century, the loss of colonial trade routes and protections against competition, civil wars led by regional strongmen, foreign reluctance to invest capital in the new and unstable republics, and a brain drain of skilled Spaniards back to Europe all combined to produce an economic decline in Latin America. But around 1850, as the political situation stabilized and industrialization in Europe and North America created investment profits and new demands and consumers, capital became available for the Latin American economies in which liberal political thought supported free trade and foreign investment.

Foreign capital helped to develop export economies for nitrate (used to make fertilizer) and copper in Chile; livestock in Argentina; coffee in Brazil, Colombia, and Central America; bananas in Central America and Ecuador; tin in Bolivia; and silver and henequen (a fiber used in making sacks and matting) in Mexico. Foreign-owned companies, which were mostly British in the nineteenth century, ran many of the new export activities. The subsequent export boom led to some modernization of production methods, improved transportation, and some investment in local light industries, such as textiles and food processing. But many of the foreign companies made little effort to promote local markets and infrastructure, and the bulk of the profits were sent back to their home countries rather than reinvested locally. The basic mineral and agricultural exports did not command high prices in relation to the cost of manufactured imports (what is called poor *terms of trade*), and exports were very vulnerable to changes in world prices. Countries that relied on these exports developed economies highly dependent on, and closely linked to, a volatile world market—a condition still evident today in the vulnerabilities of Chile (35 percent of total export value in 1998 from copper), Dominican Republic (29 percent bananas), Ecuador (26 percent bananas), El Salvador (26 percent coffee), Honduras (44 percent coffee), and Venezuela (71 percent oil). This period also saw the emergence of a middle class as well as urban and rural working classes and related challenges to political systems dominated by elites, whose power stemmed from pre-independence colonial structures, their links to Spain, and large landholdings.

Independence came much later to most of the Caribbean, especially to countries like Jamaica and Trinidad, which were under British control. When slavery was abolished by the British in 1807, many emancipated slaves eventually became small farmers. But a decline in sugar prices produced poverty and unemployment that provoked rebellions in Jamaica beginning in 1865 that eventually resulted in a more representative government—but one that was still controlled by Britain as one of its crown colonies. Other island groups, such as the Leeward Islands (including Antigua and St. Kitts and Nevis), were also administered as crown colonies from the 1870s until the 1950s, with discontent with colonial rule dating from the 1930s.

In 1958 an attempt was made to establish the West Indies Federation as an independent unit within the British Commonwealth, but the federation collapsed in 1962 when Jamaica seceded and became fully independent. The first phase of independence from Britain also included Trinidad and Tobago in 1962, Barbados in 1966, and Dominica and Grenada in 1967. The second phase included the Bahamas in 1973, after a typical decolonization period of internal self-government with a British governor. Other British island colonies such as Antigua and Barbados, St. Lucia, and St. Kitts and Nevis did not become fully independent until the beginning of the 1980s.

Cuba remained under Spanish control until after the Spanish-American War in 1898, when limited independence was granted under U.S. influence and frequent intervention. Puerto Rico also shifted from Spanish control to a U.S. territory after 1899 and is still currently part of the United States, with some residents desiring independence and others full status as a U.S. state. Six islands remain as Dutch protectorates with full autonomy for internal affairs (Aruba, Bonaire, Curaçao, St. Martin, Saba, St. Eustatious), and Martinique and Guadeloupe are overseas departments of France. The British Virgin Islands, Turks and Caicos, Montserrat, Anguilla, and the Cayman islands are still colonies of Britain.

U.S. Dominance, Latin American Revolutions, and the Cold War

An important geopolitical step was taken in 1823 when U.S. president James Monroe issued his **Monroe Doctrine** declaring that any further European colonization or interference in the Western Hemisphere, including Latin America, would be considered a threat to the peace and security of the United States and a hostile act. It also stated that in return for European noninvolvement in the Western Hemisphere, the United States would not interfere in European affairs. This doctrine set the stage for subsequent U.S. involvement and intervention in Latin America and the growth of U.S. economic and political dominance in the region. In 1848 Mexico was defeated in its war with the United States and was forced to cede large portions of its territory to the state of Texas and to what would become the states of Arizona, California, Colorado, Nevada, and New Mexico, leaving in return an enduring Hispanic cultural legacy in these regions.

Latin America and the Caribbean was drawn more explicitly into the new U.S. political and economic sphere of influence with U.S. interventions (**Figure 8.20**) to maintain stability and economic access in Cuba (1896–1922), Haiti (1915–1934), Nicaragua (1909–1933), and Panama (from 1903 onward to control the canal). The First World War (1914–1918) was a positive stimulus to Latin America's industrial development because of an increased demand for raw materials and accelerated development of manufacturing industry to fill hemispheric demands.

Increasing concentration of land and wealth in the hands of the few and expanding foreign ownership and export orientation at the beginning of the twentieth century produced growing frustration among the poor, the landless, and opposition or regional factions in several countries. Internal tensions between elites and other groups, especially landless peasants, complicated relationships with the United States, especially as the Cold War between the democratic West and the Soviet Union intensified in the 1950s. A series of twentieth-century revolutions in Mexico, Guatemala, Cuba, Bolivia, and Nicaragua reverberated around the hemisphere and the world.

Urbanization and industrialization also created urban middle and working classes who wanted a role in governments that were dominated by *oligarchies*—small groups of powerful and wealthy families. In Central America these oligarchies included large coffee-producing landowners of El Salvador and the political elite that managed Guatemala and Honduras in the interests of the multinational fruit companies as so-called **banana republics**.

The Mexican revolution stemmed from a range of factors that included the demands of landless peasants for land and the desires of regional factions to overthrow the centralized dictatorships that were oriented toward the United States and depended heavily on foreign capital. These regional and social tensions that had arisen in independent Mexico resulted in the turmoil of the Mexican Revolution, which after 1920, finally produced a new post-revolutionary constitution and a government

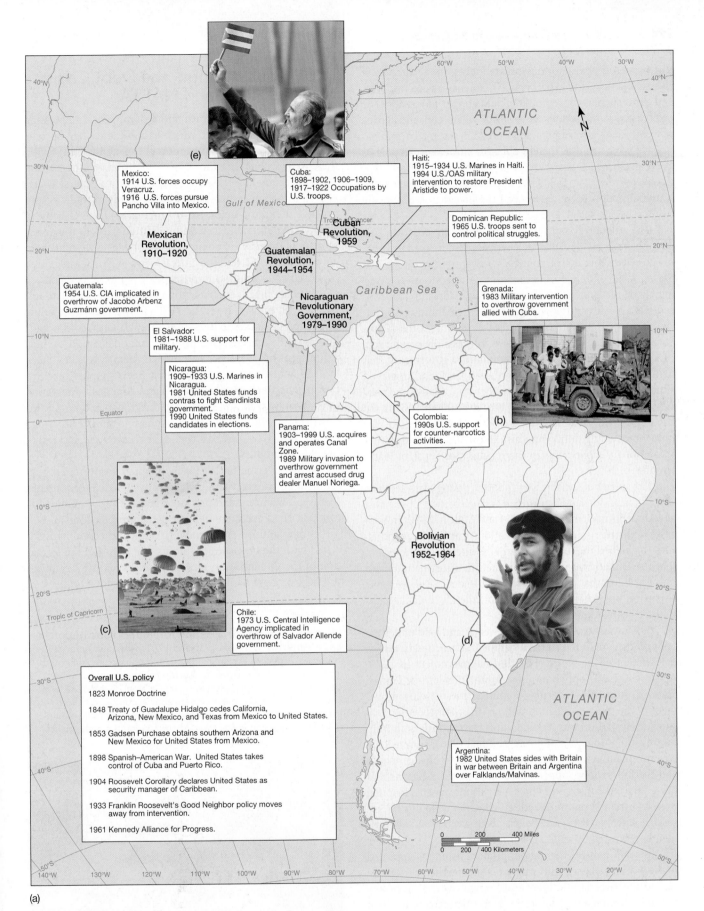

(e)

(a)

Mexico:
1914 U.S. forces occupy Veracruz.
1916 U.S. forces pursue Pancho Villa into Mexico.

Mexican Revolution, 1910–1920

Cuba:
1898–1902, 1906–1909, 1917–1922 Occupations by U.S. troops.

Cuban Revolution, 1959

Haiti:
1915–1934 U.S. Marines in Haiti.
1994 U.S./OAS military intervention to restore President Aristide to power.

Dominican Republic:
1965 U.S. troops sent to control political struggles.

Guatemala:
1954 U.S. CIA implicated in overthrow of Jacobo Arbenz Guzmánn government.

Guatemalan Revolution, 1944–1954

Nicaraguan Revolutionary Government, 1979–1990

El Salvador:
1981–1988 U.S. support for military.

Nicaragua:
1909–1933 U.S. Marines in Nicaragua.
1981 United States funds contras to fight Sandinista government.
1990 United States funds candidates in elections.

Grenada:
1983 Military intervention to overthrow government allied with Cuba.

(b)

Panama:
1903–1999 U.S. acquires and operates Canal Zone.
1989 Military invasion to overthrow government and arrest accused drug dealer Manuel Noriega.

Colombia:
1990s U.S. support for counter-narcotics activities.

Bolivian Revolution 1952–1964

(d)

(c)

Chile:
1973 U.S. Central Intelligence Agency implicated in overthrow of Salvador Allende government.

Overall U.S. policy

1823 Monroe Doctrine

1848 Treaty of Guadalupe Hidalgo cedes California, Arizona, New Mexico, and Texas from Mexico to United States.

1853 Gadsden Purchase obtains southern Arizona and New Mexico for United States from Mexico.

1898 Spanish–American War. United States takes control of Cuba and Puerto Rico.

1904 Roosevelt Corollary declares United States as security manager of Caribbean.

1933 Franklin Roosevelt's Good Neighbor policy moves away from intervention.

1961 Kennedy Alliance for Progress.

Argentina:
1982 United States sides with Britain in war between Britain and Argentina over Falklands/Malvinas.

Figure 8.20 **U.S. interventions in Latin America and the Caribbean, and Latin American and Caribbean revolutions** (a) This map shows the dates and locations of U.S. intervention in Latin America and those areas where there were major Latin American revolutions in the twentieth century. In most cases the United States intervened as a result of Cold War politics to prevent governments perceived as allied with the Soviet Union and in the interests of U.S. national security. Most revolutionary movements have been inspired by calls for land reform and socialist policies. (b) U.S. marines in Grenada in 1983. (c) U.S. paratroopers during the invasion of Panama, December 21, 1989. The invasion of Panama was ordered by U.S. president George Bush. The military were ordered to seize Manuel Noriega to face charges on drug trafficking in the United States. (d) Guerrilla leader Che Guevara, who fought in the Cuban and Bolivian revolutions. (e) Cuban revolutionary leader Fidel Castro.

that promised land reform, workers' rights, the separation of church and state, and expropriation of foreign-owned resources and firms such as those in the oil and copper industries. The *Institutional Revolutionary Party* (PRI), which consolidated its power through government-sponsored unions, media, and patronage programs, was thereby able to control Mexican politics at all levels for 70 years. The PRI lost its seven-decade monopoly over Mexican politics in 2000, when Vicente Fox of the National Action Party (PAN) won the presidency, promising many political, social, and economic reforms.

The Cuban revolution of 1959 had more complex origins, beginning with Cuba's independence from Spain in 1898, which resulted in active political intervention by the United States, a dependence on a sugar-export economy focused mainly on the United States, and political leadership that concentrated wealth and power in an elite segment of the population, most notably under the Batistas beginning in the 1930s. Socialist leaders, led by Fidel Castro, successfully organized a rural rebellion and established a communist government that nationalized corporations, redistributed land, and managed almost all aspects of everyday life under state control and subsidy. Partly in response to a failed U.S. invasion in 1961 at the Bay of Pigs, Cuba came under the Soviet sphere of Cold War influence, fueling anticommunist alarm and inspiring other revolutionary movements in the region. (See Sense of Place: Havana, p. 395.)

The spread of socialist ideas about working-class activism and the need for land reform led to the election of socialist governments in Guatemala in 1954 and Chile in 1970. In both cases, redistribution of land and nationalization of key industries threatened the local elite and U.S. interests to the extent that the United States was implicated in assassinations and military coups that overthrew socialist leaders Jacobo Arbenz Guzmánn in Guatemala and Salvador Allende in Chile within three years of their election.

In Nicaragua, concentration of wealth and land under the Somoza dictatorship fostered rebellion that resulted in the establishment of the socialist Sandinista government in 1979. Again, Cold War anticommunist sentiments led the United States to support—covertly—a counterrevolutionary movement of the *contras*. When funding to the *contras* by the Reagan administration was linked to illegal arms deals with Iran, the resulting domestic and international opposition to U.S. covert operations led the U.S. government to change tactics and instead support opposition candidates in elections that ousted the Sandinistas in 1989. Guerrilla movements inspired by socialist and communist ideas, which emerged in El Salvador, Colombia, Peru, and Bolivia, were severely and often violently repressed by ruling governments.

In other parts of the region, political aspirants appealed to the working classes with strong nationalist rhetoric that attacked the large landholders and foreign ownership of key economic resources. Juan Perón, who ruled Argentina from 1946 to 1955 and from 1973 to 1976, epitomized populism and the style of Latin American government that relied on the personal charisma of strong leaders (in this case of Perón, extending even to his spouse, Evita).

Import Substitution, the Debt Crisis, Neoliberalism, and NAFTA

The 1929 stock market crash and ensuing world depression demonstrated the extent to which Latin America had become integrated into the global economy: throughout the region there were declines in exports, restrictions on investment, and a general economic crisis. This, together with a general awareness that foreign ownership and poor terms of trade for unprocessed exports made Latin America and the Caribbean economies vulnerable to world conditions, led to the development of the new economic strategy of *import substitution industrialization* (ISI) and the critical views of global integration espoused by *dependency theorists* (see Chapter 2, p. 86).

Mexico, Brazil, and Argentina moved aggressively to implement ISI policies from the 1930s to the 1960s, including protection of domestic industries through tariffs and import quotas, and state investment in, or nationalization of, new manufacturing industries to produce chemicals, steel, automobiles, and electrical goods (**Figure 8.21**). Import-substitution policies temporarily slowed Latin America's integration into global markets and stimulated the growth of domestic industry and workforce in regions such as northeastern Mexico (steel), Mexico's Gulf Coast (petrochemicals), and São Paulo, Brazil (automobiles).

Growing criticisms of ISI highlighted an oversized government bureaucracy and high costs of subsidizing industries that were inefficient and produced goods of poor quality due to a lack of competition and government protectionism.

A new infusion of capital into the world economy, associated with the increased oil profits that followed the formation of the Organization of Petroleum Exporting Countries (OPEC), brought banks to Latin America seeking to invest in

Figure 8.21 Industrialization for import substitution in Latin America Between 1930 and 1970 many Latin American countries, especially Argentina, Brazil, and Mexico, implemented import-substitution policies to develop national manufacturing capacity. Governments invested heavily in steel, automobile, and chemical plants and protected them against competition from imports through tariffs. This photo shows a General Motors auto-manufacturing plant in São Paulo, Brazil, during the early import-substitution period.

Sense of Place

Havana

Havana, Cuba, has played a central role in Caribbean culture and commerce since the seventeenth century, when it was one of Spain's most important ports located on the northern shore of the Caribbean's largest island and the gate to the wealth of the Americas. As trade in slaves, sugar, and tobacco grew during the colonial period, imposing buildings were built around the main plaza and along the harbor front (**Figure 1**).

U.S. domination of Cuban trade and land use in the first half of the twentieth century combined with Havana's development as a center for U.S. tourism including gambling and prostitution during the Prohibition period in the United States (1920–1933), to create an American-style resort center only 200 kilometers (125 miles) from Florida.

The Cuban revolution brought a sudden end to U.S. tourism, however, and the revolutionary government's policy to keep people in the countryside meant that Havana's population grew slowly. A vibrant musical tradition includes salsa and *son,* increasingly popular around the world.

A recent resurgence in tourism, mainly from Europe and with modest investment from the Cuban government, has brought new prosperity to some sectors of Havana, which has seen construction of new hotels, the opening of stores accepting U.S. dollars, and renovation of older colonial architecture.

Figure 1   The Cuban economy Economic crises associated with the U.S. economic embargo on Cuba have contributed to the deterioration of many hotels and monuments dating from the colonial period and to the continuing presence of U.S. automobiles dating from the 1950s and still maintained by their owners.

---

what they viewed as stable and rapidly growing economies. Mexico, with the promise of its own oil bonanza and industrial expansion, as well as the more industrialized countries of Brazil and Chile, were offered the largest loans, but almost all Latin American governments took advantage of the initially low-interest loans to support development and other projects. When interest rates rose and debt payments soared in the early 1980s, Latin American governments were unwilling to cut back on popular subsidies and programs and instead borrowed more money, ran budget deficits, and overvalued their currencies. The resulting runaway inflation and debt reached unprecedented levels. By 1989 Brazil owed $111 billion, Mexico $104 billion, and Venezuela $33 billion, with annual payments reaching more than half of the annual gross national product (GNP). Mexico and Venezuela, whose foreign exchange earnings were heavily dependent on oil exports, suffered when oil prices declined in the 1980s and 1990s. Bolivia had an annual rate of inflation of 23,000 percent in 1985. The 1980s have been called Latin America's "lost decade" because of the slowdown in growth and deterioration in living standards that occurred during that decade.

The resulting decline in purchasing power and living standards, and the likelihood that suspension of debt payments and default (especially in Mexico) would destabilize the international financial system, prompted international financial institutions and the U.S. government to seek a solution to Latin America's debt crisis. The United States extended the repayment period for debts and lent more money, while the International Monetary Fund moved to restructure loans on condition that governments initiate stabilization and structural adjustment policies (see Chapter 2, p. 87). Mexico got a $48 billion bailout, paid mainly to the banking sector.

Stabilization policies set out to curb inflation by cutting public spending on government jobs and services, increasing interest rates, controlling wages, and devaluing currencies to increase exports. **Structural adjustment policies** required the removal of subsidies and trade barriers, the privatization of government-owned enterprises such as telephone and oil companies, reductions in the power of unions to demand higher wages, and an overall focus on export expansion. These policies, while reducing inflation and debt in a number of countries, had very negative effects on some people and sectors. Increased food prices and reduced health and education services due to the withdrawal of subsidies, as well as rising unemployment as government jobs were cut, hit the poor particularly hard with increases in malnutrition and destitution.

As a result of structural adjustment in Peru in 1990, gas prices went from 10 cents to $2 per gallon.

Free-trade policies were introduced in many countries as political power shifted to those with a belief in neoliberalism, echoing the views of the nineteenth-century liberals who believed in free trade and reduced government. Neoliberal governments were open to the possibility of expanding free trade through regional agreements that would take down barriers between trading partners.

The most dramatic step was taken by Mexico, which in 1994 joined the **North American Free Trade Agreement (NAFTA)** with the United States and Canada, creating a free-trade region of 400 million people with a combined GNP of more than $9 trillion (see Geography Matters: The Economic and Environmental Effects of the North American Free Trade Agreement, p. 397). Other initiatives include MERCOSUR (Spanish for "southern market"), initiated in 1991 and linking Chile, Argentina, Brazil, Paraguay, and Uruguay in a trade agreement, and CARICOM, formed in 1973 to create a trade zone in the Caribbean. The Andean Pact originally linked Peru, Ecuador, Colombia, Venezuela, Bolivia, and Chile in a 1969 agreement, but Chile has withdrawn and is actively negotiating with the United States for free-trade conditions similar to those under NAFTA.

The World Bank estimates the overall GNP of the Latin American and Caribbean region at $1.95 trillion, about 7 percent of the world's total, higher than that of Africa, South and Southeast Asia, the Middle East, East Asia, and the Pacific. Per capita GNP at $6280 is also higher than that of any other low- or middle-income region (all world regions except Europe, Australia, New Zealand, Japan, the United States, and Canada). Brazil is ranked the world's eighth largest economy and Mexico is ranked twelfth.

Total exports in 1997 were valued at $338 billion and imports at $377 billion (each about 5 percent of the global total). In that same year the Latin American region also received more foreign investment in private capital than any other low- or middle-income region but also had the largest total debt— more than $700 billion. The strongest export economies were those of Mexico (almost $130 billion in 1998) and Brazil ($59 billion), but these countries also had the highest debts at almost $160 billion in Mexico and $232 billion in Brazil.

These data suggest that the Latin American and Caribbean region is highly linked into the global economic system, with considerable flows of capital and goods to other world regions, especially North America. Although GNP ranks the region higher than many others on economic indicators, as we will see, these indicators hide tremendous variations in economic conditions within the region and living conditions that do not always reflect seemingly favorable economic statistics.

The highest average incomes are found in the Caribbean islands of the Bahamas, Barbados, and Puerto Rico (more than $12,000 per year), and Argentina also reached $12,000 per capita in 1998. The lowest per capita values in 1998 were reported for Nicaragua, Haiti, Honduras, and Bolivia at less than $2500 per capita. As we will see later, some of these countries also have a very uneven distribution of wealth, with a large percentage of very poor people.

The Peoples of Latin America and the Caribbean

History and Composition of the Peoples of Latin America and the Caribbean

Prior to the arrival of the Europeans around 1500, Latin America is estimated to have had a population of around 50 million people, including large concentrations within the empires of the Aztecs and Incas and many smaller groups of hunters, gatherers, and agricultural communities. The demographic collapse dramatically reduced indigenous populations, but significant Indian populations remained in Mexico and northern Central America and the Andes.

Colonialism changed the demographic profile of Latin America and the Caribbean through the intermixing of European and Indian peoples and the importation of slaves from Africa to the Americas. Few European women accompanied the early Spanish and Portuguese explorers and settlers, and many of the newcomers fathered children with Indian women through force, cohabitation, or marriage. The resulting mixed-race populations were called *castas,* or castes, and classified according to their racial mix. The most common category was that of **mestizo,** indicating someone of mixed Spanish and Indian heritage; others included **mulatto** (Spanish/African) and **zambo** (African/Indian). These racial categories reflected racist perceptions that permeated society and correlated strongly with social class and culture. Even the Spanish divided themselves between *peninsulares* (those born in Spain) and *criollos* (those born in the Americas), with the elite sending their pregnant wives to Spain so that their children would be born there and thus have the highest social status.

Because "whiteness" carried social and economic advantages, some mixed-race families tried to change their class by dressing, talking, and eating like those with whiter skin and higher class, wearing shoes rather than sandals, and eating wheat bread rather than corn tortillas, for example. The construction of race by styles of dress and diet continued into the twentieth century. The 1930 Mexican census includes wearing sandals and eating corn tortillas, together with indigenous language, as indicators of Indian race, and hence, lower class.

Slave imports to Latin America and the Caribbean from Africa totaled more than 5 million people during the colonial period, including 3.5 million to Brazil and 750,000 to Cuba. Many of the Caribbean islands, including Haiti and Trinidad, with very small indigenous and European populations, had a large number of African slaves working on plantations, and African populations also settled along the plantation coasts of Mexico, Central America, northern South America, and Ecuador. Although slavery was not abolished until the mid-1800s (1888 in Brazil), escaped and freed slaves formed communities as early as 1605, most famously the African community of Palmares, which was an autonomous republic from 1630 to 1694 in the Brazilian interior. These settlements,

The Economic and Environmental Effects of the North American Free Trade Agreement

The North American Free Trade Agreement (NAFTA) was implemented in January 1994 among the United States, Canada, and Mexico with the goal of creating a free market among the three countries by eliminating tariffs, quotas, and other trade protections. The signing of the agreement was preceded by a vigorous debate about the pros and cons of free trade, especially the possible impacts on labor and environmental conditions in Mexico.

Advocates of NAFTA argued that free trade would create thousands of jobs in Mexico with higher wages and that these opportunities would reduce migration to the United States. Mexican agriculture would shift to growing high-value fruit and vegetables where it had a comparative advantage during the winter and would be able to reduce food prices by importing low-cost grain from the United States and Canada. Free trade was also linked to financial stabilization and to promises of more democratic government in Mexico.

Critics of NAFTA felt that free trade would cause job losses in all three countries as industries abandoned the United States and Canada to take advantage of lower wages in Mexico and as thousands of Mexican farmers left the countryside because of the elimination of agricultural subsidies. There was concern that inadequate environmental and workplace standards would result in increased pollution, child labor, and worker exploitation in Mexico, or that standards in the United States and Canada would be lowered in order to remain competitive. Many Mexicans were also worried that NAFTA would mean even greater domination by the United States and a loss of Mexican sovereignty.

The coalition to oppose NAFTA brought together environmentalists and labor activists in all three countries working within trinational coalitions that successfully petitioned for two side agreements to be signed as part of NAFTA. The environmental side agreement established the Commission for Environmental Cooperation (CEC), and led to the creation of two related institutions, the Border Environment Cooperation Commission (BECC) and the North American Development Bank (NADB). The CEC would monitor environmental impacts and enforce regulations, the BECC would certify new water and sewage projects along the U.S.–Mexico border, and NADB would fund environmental infrastructure improvements certified by the BECC. The labor side agreement included commitments regarding minimum wages, child labor, and rights to unionize and was matched by a U.S. program to compensate U.S. workers who could prove they had lost their jobs as a result of NAFTA.

On January 1, 1994, the day that NAFTA came into force, rebels in the Mexican state of Chiapas announced that they had taken over several cities, including the tourist destination of San Cristóbal de las Casas, and had widespread support in rural areas of the Lacandón rain forest. The Zapatista Army of National Liberation (EZLN) demanded autonomy for indigenous communities, democracy, and economic policies that would benefit the poor. They opposed NAFTA because they felt that it would have a negative impact on the viability of growing corn and could result in foreign exploitation of forest and other resources. The subsequent conflict between the Zapatistas and the Mexican government, involving the progressive militarization of Chiapas and more than six years of peace negotiations fraught with social unrest, garnered worldwide attention, partly because Chiapas became a symbol for those who opposed economic globalization and its impacts on indigenous peoples and local ecologies. Geographer Karen O'Brien, who has worked with the NAFTA Commission for Environmental Cooperation, has pointed out that centuries of resource exploitation, unequal land tenure, and discrimination against indigenous groups in Chiapas have resulted in rebellion and forest loss in this region, which is the poorest in Mexico.[*]

Other studies do suggest that NAFTA has led to the creation of thousands of new jobs in Mexico and to increased wages in some industries. However, many of the hoped-for benefits of NAFTA were frustrated by the economic crises that followed currency devaluation in 1994 and by continuing inequality in both urban and rural areas, which means the benefits have not reached the poor. The new environmental agreements have created an important space for Mexicans to protest lack of environmental enforcement and to seek funding for water and sanitation projects, but this has not yet resulted in an overall improvement in environmental conditions in Mexico.

[*]K. L. O'Brien, *Sacrificing the Forest: Environmental and Social Struggles in Chiapas*, Boulder: Westview Press, 1998.

---

also called **maroon communities**, were created by escaped and liberated slaves in other regions such as Jamaica. Racial mixing occurred between European, Indian, and African populations, especially in Brazil, where by the twentieth century some scholars were promoting an image of Brazilian *racial democracy* and equality; skin color had merged to what was called "coffee"; and musical, religious, and dietary traditions had merged into a uniquely Brazilian culture.

This *myth of racial democracy* is contradicted by evidence of continuing racism in Brazil and other Latin American and the Caribbean countries. Studies show that race and class correlate strongly, with Afro-Brazilians being on the whole poorer, less healthy, less educated, and more discriminated against in employment and housing. In Mexico, the media have tended to promote lighter skin as more desirable through the choice of more European-looking actors in commercials and other programs, and job advertisements still ask for "good appearance," hinting at a preference for nonindigenous features.

There is a legacy of other diasporas in contemporary Latin America and the Caribbean populations. Asian immigration to the region began during the colonial period and picked up after the end of slavery, with Chinese, Indian, and Japanese workers brought to work on plantations and in construction as workers who had to pay off the cost of their travel and sustenance (*indentured* workers). Europeans other than the Spanish and Portuguese settled in the more temperate climates, especially in Argentina, where many families have German, Italian, or British names. Six million Italians and Spanish were recruited to Argentina as indentured laborers. Some regions, such as Patagonia, are associated with Welsh immigration and culture. French colonial links resulted in the 1977 resettlement of 2000 Hmong from Indochina to French Guiana, where they are reinvigorating agriculture through their market gardens adjacent to the European Space Agency's rocket facility staffed by French expatriates.

Recent population censuses have attempted to record race and ethnicity and show some general patterns that correlate with the population history as described above. Brazil, Cuba, and Haiti record large proportions of people of African heritage, and Argentina and Costa Rica report significant numbers of Europeans (**Table 8.1**). Peru, Ecuador, Bolivia, and Guatemala have a significant percentage of their population defined as Indian, and Colombia, Chile, Venezuela, and Mexico are more than half *mestizo*.

These numbers hide subtle differences in how different countries record, construct, and perceive race and ethnicity. For example, a tendency to identify with Europe may increase the proportion of those who report themselves as European in Argentina, whereas a national pride in *mestizo* heritage increases self-identification as being of mixed-race heritage in Mexico.

Population Growth and Urbanization

The overall population of contemporary Latin America and the Caribbean totals about 520 million people, and the distribution is clustered around the historical highland settlements of Central America and the Andes, and in the coastal colonial ports and cities (**Figure 8.22**). Geographers have compared the relatively denser *mainland* population of highland Mexico and Central America and the *rimland* populations around the coast and in the Andes of South America, with relatively unpopulated interiors, including the Amazon.

Population has grown rapidly since 1900, when the regional total was 100 million, mainly as a result of high birthrates and improvements in health care. Brazil (170 million) and Mexico (100 million) have the largest populations, and fertility rates reach as high as 4.7 children per woman in Haiti. (The fertility rate is the average number of children a woman in a particular population group is projected to have during the childbearing years—age 15–49.) Although many countries are still growing at more than 2 percent per year, and population-doubling times are at less than 35 years, placing pressure on food, water, housing, and infrastructure, fertility rates have declined through much of the region. However, because a large percentage of the population is under age 15, especially in Central America, populations are likely to continue to grow as this cohort enters its reproductive years. For example, Guatemala's population is predicted to almost triple to 32 million people by 2050. The highest population densities (more than 500 people per square mile, or about 200 people per square kilometer) are found on the Caribbean islands and in El Salvador.

Table 8.1	Ethnic and Racial Composition of Latin American Populations			
Country	Amerindian	Mestizo	Euroamerican	Afroamerican
Argentina	5	10	85	
Bolivia	55	31	14	
Brazil	1	12	54	33
Chile	7	91	2	
Colombia		59	22	19
Costa Rica	1	9	86	4
Cuba			66	34
Ecuador	40	40	15	5
Guatemala	46	46	6	2
Haiti			1	99
Mexico	30	61	9	
Peru	55	33	12	
Venezuela	2	69	20	9

Source: Cambridge Encyclopedia of Latin America, 1985.

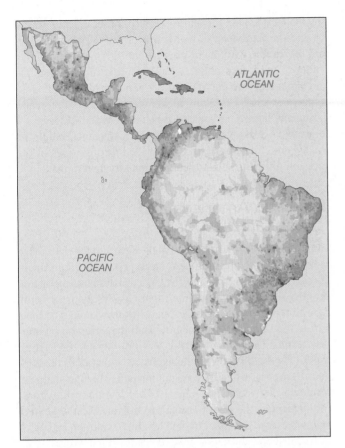

Figure 8.22 Population distribution of Latin America and the Caribbean 1995 The general distribution of population in Latin America and the Caribbean includes sparsely settled interiors and high population densities around the historical highland regions of the Andes, Mesoamerica, and the coastal regions, especially former colonial ports and cities and their hinterlands. For a view of worldwide population density and a guide to reading population-density maps, see Figure 1.40. (*Source:* Center for International Earth Science Information Network (CIESIN), Columbia University; International Food Policy Research Institute (IFPRI); and World Resources Institute (WRI). 2000. *Gridded Population of the World* (GPW), Version 2. Palisades, NY: CIESIN, Columbia University. Available at http://sedac.ciesin.org/plue/gpw)

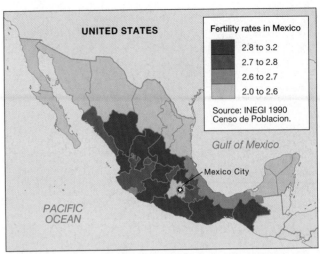

Figure 8.23 Map of fertility rates in Mexico in 1990 Fertility rates in Mexico are lower in urban areas such as Mexico City and areas of higher average incomes along the border, and they are higher in poorer areas such as southern Mexico. (*Source:* Data from Instituto Nacional de Estadistica, Geografía e Informática)

High fertility rates are characteristic of poorer, rural regions where infant mortality is high, children can contribute labor in the fields, and women do not have access to education, employment, or contraception. Fertility rates have tended to drop as people move into the cities, as health care improves, and as more women work and are formally educated. A map of Mexico illustrates this pattern with lower fertility rates in urban, industrial, and higher-income states near Mexico City and the U.S. border, and higher fertility rates in the poorer, more rural southern states (**Figure 8.23**). Attitudes toward family size in Latin America are also affected by the Catholic Church's position against contraception and the culture of machismo, which views high male fertility as a measure of status.

Most people in Latin America and the Caribbean now live in cities, and the levels of urbanization are among the highest in the world, ranging from about 50 percent in most of Central America to more than 80 percent in Argentina, Chile, Uruguay, and Venezuela, compared to a regional average of only 10 percent in 1900. The region also hosts three of the world's 10 largest cities including Mexico City at 21 million; São Paulo, with about 18 million in the metropolitan area; and Rio de Janeiro at 11 million. The major cause of urban growth is migration, although the redefinition of city boundaries (to include metropolitan regions) and internal population growth have also played a role. In many countries, one city dominates the country and this so-called **urban primacy** is characteristic of Argentina (Buenos Aires has 34 percent of the national population), Peru (Lima, 30 percent), Chile (Santiago, 30 percent), and Mexico (Mexico City, 19 percent). This concentration of population and development in one or two cities within a country can create problems when physical and human resources, political power, and pollution are all focused in one major settlement.

Migration

More than 150 million people are estimated to have moved from rural areas to cities in Latin America and the Caribbean in the twentieth century. The reasons for this massive rural-urban migration include factors that tend to push people out of the countryside and others that pull people to the cities (**Figure 8.24**). People leave rural areas because wages are low; because services such as safe drinking water, health care, and education are absent or limited; or because they do not have access to land to produce food for home consumption or for sale. Unemployment as a result of agricultural mechanization, price increases for agricultural inputs, and the loss of crop and food subsidies have also driven people from rural areas to the cities. Other push factors have been environmental degradation and natural disasters, such as Hurricane Mitch in Honduras, as well as long-running civil wars or military repression of rural people as in Guatemala.

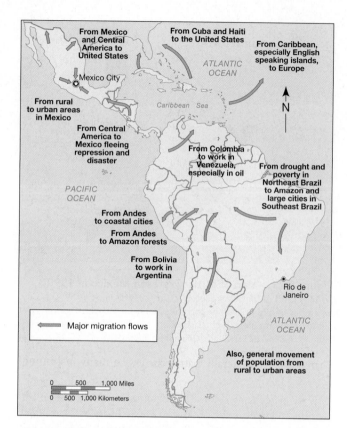

Figure 8.24 Major migration streams in Latin America and the Caribbean There are major migration streams within countries and between countries in Latin America and to the rest of the world from Latin America. The most significant overall trend is rural to urban migration, but poor people are also moving into frontier regions in the Amazon and southern Mexico, from the Andes to work in Argentina and Venezuela, and from political unrest and natural disasters. The Caribbean has flows to Europe, especially from the English-speaking islands, and to the United States, especially from Cuba, Haiti, the Dominican Republic, and Puerto Rico.

Cities pull migrants because they are perceived to offer high wages and more employment opportunities, as well as access to education, health, housing, and a wider range of consumer goods. Governments often have an urban bias in providing services and investment to cities that are seen as the engines of growth and the locus of social unrest. Social factors that encourage migration to the cities include the promotion of urban lifestyles and consumption habits through television and other media and long-standing social networks of friends and families that link rural communities with people in cities who can provide housing, contacts, and information to new migrants.

Although most people have migrated to cities within their own country, there are several other important migration flows within the Latin American and the Caribbean region. Several countries have encouraged the colonization of remote frontier regions by providing cheap land and other incentives to migrants. For example, the building of roads and availability of land in the Amazon created a stream of migrants from coastal regions of Brazil to the interior, and the development of irrigation in Mexico and Chile attracted migrants to desert

regions. People have moved between countries in Latin America and the Caribbean in search of work or fleeing from war and repression, with major population movements out of the Andes to work in mining, agriculture, and oil in Argentina and Venezuela, and out of Central America to Mexico either as refugees or workers seeking higher wages. Some of the smaller migrant streams have included better-off sectors of society—for example, many intellectuals left Chile, Argentina, and Brazil for Mexico, Venezuela, and Costa Rica during times of repression of leftists and students by military governments.

The Latin American and Caribbean Diaspora

Latin American and Caribbean people have also left the overall region in considerable numbers, creating a global Latin American and Caribbean diaspora. The United States hosts the largest number of people who define themselves as being of Latin American or Hispanic heritage. Many Mexican families became part of the United States when the land they lived on became U.S. territory following the U.S.-Mexican War in 1848. They use the phrase "the border crossed us, we didn't cross the border" to emphasize that they are not migrants but long-standing residents. Between 1900 and 1930, 1.5 million Mexicans (10 percent of the total population) migrated to the United States to escape the chaos of the Mexican Revolution and partly to fill labor shortages created by the First World War. Although 400,000 Mexicans (some of them U.S. citizens) were deported during the Great Depression in the early 1930s, the growth of the U.S. economy from about 1940 on and the Second World War created such a demand for low-cost labor, especially in agriculture, that the U.S. and Mexican governments introduced a formal guest farm worker program, called the *Bracero* program, which distributed 4.6 million temporary permits for Mexicans to work in the United States between 1942 and 1964. Many **braceros** never returned to Mexico, and migration continued after the program ended even as U.S. immigration restrictions were tightened. Migrants are still drawn to the United States by higher wages, by jobs for women in the service sector, and by strong social networks that link communities in Mexico to family and friends in the United States.

In the last 50 years, Latin American and Caribbean migration to the United States has been dominated by Mexicans (about 40 percent of the total), but the total includes large numbers of people from Cuba (15 percent) and Central America (10 percent). Significant Latin American populations can also be found in Canada and Europe (especially in Spain). The Caribbean diaspora includes migration to the United States (mainly from Cuba, Jamaica, and Puerto Rico), but because of colonial links to Britain, large numbers of Caribbeans have migrated to Europe and British Commonwealth countries, especially from Jamaica and Barbados to Britain and Canada (**Figure 8.25**).

The money that is sent back to Latin America and the Caribbean from people working temporarily or permanently in other countries is called *remittances* and can be an important contribution to national and local economies. Many com-

(a)   (b)

Figure 8.25   Baseball and cricket players in the Latin American and Caribbean diasporas (a) Baseball is one route for young Dominican baseball players to migrate to the United States, although some, such as Sammy Sosa, have maintained strong links to their home country, funding schools and helping with disaster relief when hurricanes hit. (b) In former British colonies, cricket has provided a similar diaspora with players such as Sir Gary Sobers, who is ranked among the best in history and was knighted by Queen Elizabeth II.

munities in the Caribbean and Mexico rely on these funds to build houses, purchase agricultural inputs, or educate their children. They are one of the new but informal flows of international financial capital in the global economy.

Language and Cultural Traditions

The mixed racial and ethnic composition of Latin America is echoed in many aspects of cultural heritage and social practices in the region. Indigenous culture, including traditional dress, crafts, ceremonies, and religious beliefs, persists in regions such as highland Guatemala and Peru, partly as a result of colonial policies that kept Indian communities separate while demanding tribute and labor and partly due to resistance to the adoption of European culture by conservative indigenous religious and political leaders. Cultural traditions are now promoted to tourists and revalued through indigenous social movements seeking political rights and recognition. For example, indigenous Mayan centers such as Quetzaltenango in Guatemala are promoted as tourist destinations where traditional crafts may be purchased and photos taken (often for a price) of women and children in traditional colorful woven garments.

Indigenous languages endure in several regions of Latin America (**Figure 8.26**). The most widely spoken languages are *Quechua* in the Andean region (spoken by 13 million people), English Creole and French Creole in the Caribbean (10 million), *Guarani* in Paraguay (4.6 million), *Aymara* in the Andes (2.2 million), *Mayan* in Guatemala and southern Mexico (1.7 million), and *Nahuatl* in Mexico (1.3 million). Spanish is the dominant language across most of Latin America, except for Brazil (Portuguese), Belize and Guyana (English), Suriname (Dutch), and French Guiana (French).

Certain cultural views of the family and gender roles are characteristic of Latin America. Multiple generations often live and work together, individual interests are subordinated to those of the family, and the traditions of machismo and marianismo define gender roles within the family and the society. *Machismo* constructs the ideal Latin American man as fathering many children, dominant within the family, proud,

and fearless. *Marianismo* constructs the ideal woman in the image of the Virgin Mary as chaste, submissive, maternal, dependent on men, and closeted within the family. Latin American society is generally patriarchal, with many institutions that have prohibited or limited women's right to own land, to vote, to get a divorce, and to secure a decent education. These stereotypes are, of course, contradicted by individual cases and are breaking down in the face of new geographies and global cultures. Family links are weakened through migration and the isolation of many living spaces from each other and from those of other family members in urban environments. Men's and women's roles are changing as fertility rates decline and women enter the workforce and politics. Latin American and Caribbean feminists have organized to obtain the right to vote; to effect changes in divorce, rape, and property laws; to gain access to education and jobs; and to elect women to political office.

The foods of Latin America and the Caribbean blend indigenous crops such as corn or potatoes with European influences, especially from Spain. Although Mexico is associated with spicy dishes that include *chile*, the food is quite mild in the rest of Latin America. In livestock-producing areas, such as Argentina, grilled meat is extremely popular, but in much of Latin America, the poor eat simple meals of rice, corn, potatoes, and beans for protein. Modified versions of Mexican cuisine have diffused throughout North America and include many chain restaurants. Foods in the Caribbean reflect the medley of cultures in the region with African, Asian, and European influences combining to create dishes of fish, chicken, pork, and a range of vegetables, fruits, and starch crops (rice, potatoes, and yucca).

Latin American and Caribbean art and literature have incredible variety and regional specialization. Traditional textiles, pottery, and folk art are sold to tourists and by import stores in North America and Europe. Literary traditions include magical realism (where authors such as Gabriel García Márquez blend imaginary and mystical themes into their fiction), and Latin American and Caribbean authors have won six Nobel prizes for literature. Works of noted Mexican

Figure 8.26 Languages of Latin America and the Caribbean This map shows the major regions of living indigenous languages in Latin America and the Caribbean. Most of mainland Latin America uses Spanish as an official language, except for Brazil, which uses Portuguese; French Guiana using French; Suriname with Dutch; and Belize and Guyana with English. The Caribbean has more variation, and the dominant European languages depend on the colonial histories of the islands.

muralists, including Diego Rivera, and the complex paintings of his companion Frieda Kahlo, are numbered among the masterpieces of twentieth-century art.

In the 1990s several strands of Latin American and Caribbean music became popular. Traditional music, such as the Andean pipes of the groups Inti Illimani and Los Incas, has formed the soundtracks of documentary films and is available in music stores worldwide. The music of *Nueva Canción*, or New Song movement, with its social conscience, and singers such as Mercedes Sosa who have fled repression in their home countries, has become a part of the global folk music scene. Caribbean global influences include the reggae of Jamaica (for example, Bob Marley) and steel-drum bands of Trinidad, which resonate with the rhythms of Africa. But the biggest boom was in Latin pop and rock music, where stars such as Ricky Martin and Gloria Estefan produced worldwide hits from companies based mainly in Miami, Florida, Latin America's business capital in the United States, and in the revival in the late 1990s of the Cuban dance hall music such as that performed by the Buena Vista Social Club. Elements of Latin popular music derive from the traditional and contemporary dance rhythms of salsa, merengue, and tango, the latter closely associated with the nation of Argentina (**Figure 8.27**).

Religion

One of the main objectives of Spanish and Portuguese colonialism was the conversion of indigenous peoples to Catholicism. While some indigenous people fiercely resisted missionary efforts, others found ways to blend their own traditions with those of the Catholic Church. The process of conversion was facilitated by the reported appearance of the Virgin Mary of Guadalupe to an Indian convert in Mexico on December 9, 1531, leaving behind her brown-skinned image on his mantle, and by the efforts of some priests to protect local communities from the Spanish efforts to obtain land, tribute, and labor by force.

The slave trade brought African religious traditions to Latin America and the Caribbean, and these eventually merged with indigenous and Catholic beliefs to construct contemporary rituals of Candomble and Umbanda in Brazil, Voodoo in Haiti, and Santería in Cuba and other islands. Candomble and Umbanda are both sects of the Macumba religion, with rituals that involve dances, offerings of candles and flowers, sacrifice of animals such as chickens, and mediums and priests who use trances to communicate with spirits that include several Catholic saints. Voodoo (also spelled Voudou) rituals include drumming, prayer, and animal sacrifice to important

(a)

(b)

(c)

Figure 8.27   The culture of Latin America and the Caribbean   The popularity of Latin American and Caribbean music, literature, and food is growing around the world. (a) Argentine tango. (b) Gloria Estefan (originally from Cuba). (c) Bob Marley (Jamaica).

spirits based on traditional African gods and Catholic saints and are led by priests who act as healers and protectors against witchcraft. Santería, which is closely connected to the Yoruba religion of West Africa, blends saints with African spirits associated with nature, using rituals similar to other Latin American and Caribbean religions.

The emergence of a new form of Catholic practice, **liberation theology,** focused on the poor and disadvantaged, informed by the perceived preference of Jesus for the poor and helpless and by the writings of Karl Marx and other revolutionaries on inequality and oppression. This new orientation to the poor was espoused by the Second Vatican Council, called by Pope John XXIII in 1962. Priests preached grass-roots self-help to organized *Christian base communities* and often spoke out against repression and authoritarianism. In some cases, they were murdered by powerful interests who saw liberation theology as revolutionary and communistic, as happened to Archbishop Oscar Romero in El Salvador, who was shot to death while saying Mass on March 24, 1980.

In recent decades evangelical Protestant groups with fundamentalist Christian beliefs have grown and spread rapidly in Latin America and the Caribbean (**Figure 8.28**). Their message of literacy, education, sobriety, frugality, and personal salvation has become very popular in many rural areas and estimates suggest that up to 40 million Latin American and Caribbean people are now members of such churches.

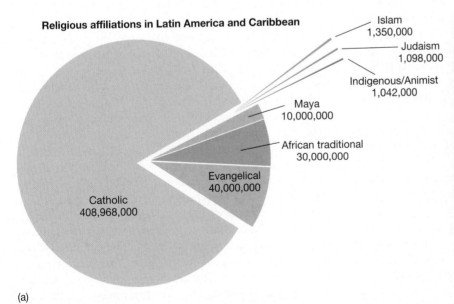

Religious affiliations in Latin America and Caribbean

Islam
1,350,000

Judaism
1,098,000

Indigenous/Animist
1,042,000

Maya
10,000,000

African traditional
30,000,000

Evangelical
40,000,000

Catholic
408,968,000

(a)

(b)

Figure 8.28   Latin American and Caribbean religions   (a) Although the majority of Latin Americans are Catholics, evangelical and African traditional religions, such as Santeria, are also important to millions of people. (b) This altar in Brazil illustrates how Santeria combines animistic elements with Catholic religious symbols.

Regional Change and Interdependence

Latin America and the Caribbean is a dynamic world region where economic, political, and social changes have been rapid in the twentieth century and have varied in their nature and impact among and within countries. Latin American and Caribbean countries took divergent political paths that have included socialist and military governments; authoritarian, single-party and multiparty systems; and highly centralized and localized administration. The challenges of creating functioning national governments and promoting economic growth dominated the post-independence period in the nineteenth century. The twentieth century saw regional factions, the working class and the poor demanding reform through revolution and populism, and threats to the distribution of wealth and elite power met by military and authoritarian rule. One of the most dramatic shifts has been from a continent dominated by military and authoritarian governments in 1970 to almost region-wide democratic systems in 2000.

The dual threats of economic instability and communist ideas contributed to a rise in authoritarianism and military governments in the 1960s and 1970s. Seeking financial order and control of socialist movements, the military took control of government in Brazil in 1964, Chile and Uruguay in 1973, and Argentina in 1976. These governments have been termed *bureaucratic authoritarian* states because they were based on an alliance of the military, professionals, and international business who sought to promote economic growth, often through opening markets and scientific management by the civil service and by severe control of opposition and unrest. While central authoritarian control certainly provided some degree of economic stability and growth, the military governments aggressively kept social order by repressing dissent, especially among students and workers perceived as having socialist ideals. In Argentina, the military government's so-called *Dirty War* is alleged to have killed 15,000 people and forced many others to leave the country. In Chile, the military government of General Augusto Pinochet has been accused of similar disappearances and human rights abuses.

Public and foreign outrage at authoritarian repression and human rights violations, the inability of military governments to solve economic problems, the end of the Cold War, and international and internal pressures that linked economic globalization to democratic governance resulted in gradual transitions to democratic governments in Argentina in 1983, Brazil in 1985, and Chile in 1989. In Argentina the departure of the military government was hastened by the loss of a war with Britain when Argentina invaded the Falkland Islands in 1982 (called the *Islas Malvinas* in Latin America).

Political opposition and activism has often taken the form of organized *social movements* that have also pressured for specific resources and issues, such as housing, water, human rights, or environmental protection. Geographer Anthony Bebbington argues that social movements have also filled a gap left in service provision and local administration created by the economic crisis and neoliberal policies that have shrunk government in many Latin American and Caribbean countries.

Green Revolution and Land Reform

For the first part of the twentieth century the yields of most agricultural crops in Latin America and the Caribbean were very low (less than 1 ton per hectare) and farmers with small plots of land could not produce enough to feed themselves, let alone sell in the market. As population and urban consumption demands increased, countries such as Mexico and Brazil had to import basic food crops such as wheat and corn. The legacy of large landholdings from the colonial period was compounded by the accumulation of land by the wealthy and by foreign companies in the late nineteenth century. This led to widespread rural poverty, landlessness, and frustration that aided uprisings such as the Mexican and Cuban revolutions and the election of socialist governments in Chile and Guatemala. In addition, many large landholdings were being used for extensive ranching or for export or low-productivity crops and were not contributing to the food needs of the growing urban populations.

Land reform—the redistribution of land with a goal of increasing productivity and reducing social unrest—was seen as a solution and was implemented by revolutionary governments and others seeking to reduce the risk of rural uprising. Mexico's post-revolutionary land reform redistributed expropriated and government lands to 52 percent of rural households between 1917 and 1980. In many cases, the land was distributed in the form of **ejidos,** communal lands given to groups of landless peasants who could farm collectively or as individuals but could not rent or sell the land outside the ejido.

Bolivia redistributed land to 79 percent of rural households between 1953 and 1975. The socialist governments of Guatemala (1952), Chile (1972), and Nicaragua (1979) distributed land to at least 20 percent of rural households, but some of these lands were subsequently returned to large landholders under military or more conservative governments. Pressure for land reform continues throughout the region. For example, in Brazil, the landless movement *Movimiento sim terra* has forced land redistribution by occupying more than 20 million hectares (nearly 50 million acres) and then demanding legal rights and political change with considerable public support. The question of whether land reform in Latin America has been successful or not is hotly debated, with some believing that the reform sector is inefficient and that communal lands should be privatized and others arguing that land reform has increased rural stability and agricultural production. Most have recognized that land reform on its own can be ineffective unless it is part of an overall agrarian reform package that also provides technical advice, inputs, credit, and market access to the new landowners.

A second solution to low productivity and poverty in rural areas was the **Green Revolution**—the process of agricultural modernization that used a technological package of irrigation, high-yielding seeds, fertilizers, pesticides, and mechanization to increase crop yields in several world regions. Mexico was a global center for Green Revolution technology, hosting the International Center for Improvement of Maize and Wheat (CIMMYT) near Mexico City. Scientists at the center, funded by the Rockefeller Foundation as well as Mexican and U.S. governments, used advanced plant-breeding techniques to produce new varieties of grains that resisted disease and responded to fertilizer and irrigation with very high yields. Norman Borlaug, who led the plant-breeding effort, was awarded the Nobel Peace Prize in 1970 for his efforts to end world hunger. Farmers, especially in irrigation districts in northern Mexico, were quick to adopt the new crop varieties, and national production of corn and wheat soared, turning Mexico into a major grain exporter by the 1970s (**Figure 8.29**). Other Latin American countries such as Argentina and Brazil also promoted Green Revolution agricultural modernization, including key crops such as rice and soybeans.

Although the Green Revolution increased crop production in many parts of Latin America, it was not an unqualified success because of its role in increasing inequality and in environmental degradation. The Green Revolution has been criticized because it increased the dependence on the imports of chemicals and machines from foreign companies and thus contributed to the debt problem. The benefits tended to accrue to wealthy farmers who could afford the new inputs and to irrigated regions, while poorer farmers on land watered only by rainfall fell behind or sold their land. In some cases, such as in southeastern Brazil, machines replaced workers, thus leading to unemployment, and Green Revolution technology and training also tended to exclude women, who play important roles in food production. The new agricultural chemicals, especially pesticides, contributed to ecosystem pollution and

worker poisonings, and the more intensive use of irrigation created problems of salt buildup in soils (*salinization*) and water scarcity. The most serious criticism of the Green Revolution was that it contributed to the worldwide *loss of genetic diversity* by replacing a wide range of local crops and varieties with a narrow range of high-yielding varieties of a few crops. Planting single varieties over large areas (*monocultures*) also made agriculture vulnerable to disease and pests.

A second Green Revolution is now under way, involving crops engineered using biotechnology to resist pests and diseases and to produce even higher yields. This research is opposed by some who fear unanticipated consequences from such efforts, exemplified by concern that bioengineered corn pollen was negatively affecting monarch butterflies in Mexico.

Economic crisis, the reduction of government programs, and opening of trade have slowed the progress of the Green Revolution in many countries. Fertilizer use in countries such as Brazil and Mexico has declined with high prices, fewer subsidies, and increased competition from imported corn and wheat, especially from the United States. Many governments have shifted from giving top priority to self-sufficiency in basic grains to encouraging crops that are apparently more competitive in international trade, such as fruit, vegetables, and flowers. These **nontraditional agricultural exports (NTAEs)** have become increasingly important in areas of Mexico, Central America, Colombia, and Chile, replacing grain production and traditional exports such as coffee and cotton (**Figure 8.30**). These new crops do obtain high prices but also require heavy applications of pesticides and water to meet export quality standards and fast refrigerated transport to market. They are vulnerable to climatic variation and to the vagaries of the

Figure 8.30 Crop production trends Production of fruit and vegetables has increased in Latin America and the Caribbean partly as a result of export of nontraditional crops. Tomatoes, cucumbers, and lettuce have grown rapidly in response to demand from North America and Europe. However, the high value and consumer demand for unblemished fruit and vegetables has increased the use of pesticides in countries such as Chile, where pesticide use doubled between 1984 and 1996, from just under 6000 metric tons to more than 12,000 metric tons. (*Source:* Data from the U.N. Food and Agriculture Organization online statistical database, available at http://www.fao.org)

Figure 8.29 Green Revolution in Mexico The International Center for Improvement of Maize and Wheat (CIMMYT) in Mexico. The center is responsible for plant breeding and research and developed the high-yield seeds that contributed to the Green Revolution.

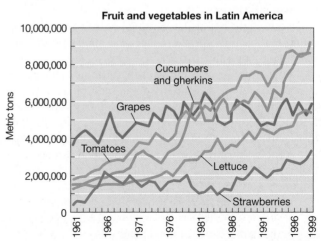

Fruit and vegetables in Latin America

international market, including changing tastes for foods and health scares about pesticide or biological contamination.

Fisheries are another important component of Latin American and Caribbean food and export systems, and activities range from subsistence fisheries in small coastal villages to large-scale commercial exploitation of offshore fisheries. The overall catch was more than 10 million metric tons in 1994, contributing on average about 10 percent of overall food supply and making a significant contribution to exports in Chile, Ecuador, and Costa Rica. Aquaculture (the cultivation of fish and shellfish under controlled conditions, usually in coastal lagoons) has been growing rapidly and has resulted in the clearing of coastal mangroves and an increase in exports, especially of shrimp from countries such as Honduras.

Continuing Inequality Between Social Classes

Political change and economic restructuring have not necessarily resulted in reductions in poverty and inequality in Latin America and the Caribbean. In fact, some indicators of inequality show some Latin American and Caribbean countries as having the greatest concentrations of wealth and land in the hands of an elite few, with the vast majority of the population remaining poor and landless. The highest average national incomes in Trinidad and Tobago and Venezuela are about 50 percent of those in the developed world, whereas in the poorest countries, including Haiti and Nicaragua, average incomes are less than 10 percent of the average in the developed world.

Recent reports from the Inter American Development Bank suggest that 25 percent of all income in Latin America and the Caribbean is received by only 5 percent of the population, compared to 16 percent in Southeast Asia and 13 percent in developed countries. At the other end of the scale, the poorest 30 percent of Latin Americans and Caribbeans receive only 7.5 percent of total income, compared to more than 10 percent in the rest of the world. Although income distribution became more equal from 1960 to 1982, conditions became more unequal during the late 1980s and have not recovered significantly since. These high levels of inequality are associated with high levels of poverty, with more than 150 million Latin Americans and Caribbeans earning less than a subsistence income of $2 U.S. per day and more than 20 percent of the population earning less than $1 per day in countries such as El Salvador, Guatemala, and Honduras (see World Data Appendix, p. 639).

The Gini index also indicates that income is unequally distributed and highly concentrated in most of Latin America. The Gini index of income inequality compares how equally income is distributed in a population. Lower values indicate that income is more equally distributed. Higher values suggest greater inequality between richer and poorer groups. The highest index levels occur in Brazil, Colombia, and Guatemala (**Figure 8.31**).

The vast gap between the richest 10 percent and the poorest 30 percent is also reflected in other social measures. For example, the richest heads of households average 11.3 years of education and the poorest 4.3 years, with even larger gaps of nine years in Mexico and Brazil. Those in the richest 10 percent tend to work in professional and technical occupations or to own their own businesses, whereas the majority of the poorest 10 percent work in the informal sector. The informal sector, or **informal economy**, in Latin America and the Caribbean comprises a variety of income-generating activities of the self-employed that do not appear in standard economic accounts including street selling, shoe shining, garbage picking, street entertainment, prostitution, crime, begging, and guarding or cleaning cars.

Social and health conditions are often considered a better measure of overall inequality within and between countries than are economic measures. National improvements in life expectancy, infant mortality, and literacy, for example, tend to reflect improvements at the lower end of the scale rather than for the better-off segments of the population. Latin America and the Caribbean tend to compare more favorably to the rest of the world on social and health indicators than on measures of income and income inequality. For example, life expectancy averages 70 years, higher than any other region in the developing world, and compared to a world average of 66.9 years. Literacy rates are also relatively high, averaging 88 percent, compared to 59 percent in Africa, 54 percent in South Asia, and 79 percent worldwide. According to the United Nations International Children's Fund (UNICEF), only 10 percent of children under age 5 are defined as underweight in Latin America and the Caribbean, compared to 30 percent worldwide.

However, there are wide gaps in social and health conditions within Latin America and the Caribbean. Haiti, Central America, and the Andes tend to have much worse conditions

Figure 8.31   Mayan family from Guatemala The Calabay Sicay family lives in the community of San Antonio de Palopó, Guatemala, on the shores of Lake Atitlán. Both parents work in the fields and weave colorful traditional fabrics on the looms displayed here with the family's possessions in front of their one-room house. Guatemala is one of the poorest countries of Latin America, with many families earning less than $1000 a year and continuing tensions between indigenous groups and the government in the aftermath of civil war and repression.

than do Argentina, Chile, Uruguay, northern South America, Costa Rica, Mexico, and the English-speaking Caribbean. For example, the average Haitian lives only 54 years, and the literacy rate is only 48 percent, compared to a life expectancy of 73 years in Argentina with literacy reaching 97 percent.

The United Nations Human Development Index combines measures of life expectancy, infant mortality, education, and income into an overall index of socioeconomic conditions (with a scale from 0.0 to 1.0 corresponding lowest and highest measures of human development). In Latin America and the Caribbean it ranges from a high of 0.84 in Argentina to a low of 0.44 in Haiti. Infant mortality is a key indicator of overall social conditions and ranges from a high of more than 100 infants dying per 1000 that are born in Haiti to less than 20 deaths per 1000 in Costa Rica, Cuba, and Chile.

These national indicators also hide large variations in economic and social conditions within Latin American and Caribbean countries. In Mexico, the southern regions of the country have lower incomes and life expectancy than do the northern and central areas of the country. In Brazil, the northeastern and Amazon zones have higher infant mortality and lower life expectancy and average monthly incomes than do the southern parts of the country. Each Latin American and Caribbean country has its own geography of inequality, with the more rural regions generally having lower social and economic conditions.

Gender inequality is also widespread in Latin America and the Caribbean. Women tend to earn much less on average than men. Female literacy, on average, is 2 to 15 percent less than that of male populations. In Ecuador, for example, female GDP per capita was $1173 in 1998 compared to $4818 for men. This inequality has been associated with systematic institutional biases that denied women in many countries the right to vote or right to marital property until the 1950s, with cultural traditions that discourage more than a few years of education for women, and with employment structures that pay women less than men or pay less for traditionally female work such as domestic service work and food processing.

Drugs in Latin America

Throughout the world certain plants have traditionally been used to reduce pain, as relaxants and stimulants, and as intoxicants and hallucinogenics. Used in both traditional and modern medicine, drugs such as opium, marijuana, and coca have been used to treat pain and anxiety for centuries, but because they can also induce relaxation, feelings of euphoria, and hallucinations, they have been used for recreation. These drugs also have chemical properties that make them highly addictive for people, especially to the opium and coca derivatives heroin and cocaine. Because of their medicinal and recreational uses, opium, marijuana, and cocaine have been transplanted around the world and are traded as controlled pharmaceutical products as well as illegal substances. The global drug trade links different regions through flows of the drugs and of the money that is gained from their sale.

Latin America produces drugs that are illegal in many countries, including cocaine, heroin, and marijuana. Cocaine derives from the coca plant, the leaves of which have been chewed by Andean residents for centuries to provide energy and alleviate the effects of high altitude. Peru, Colombia, and Bolivia produce about 98 percent of world cocaine supplies and Jamaica, Colombia, and Mexico are major producers of marijuana (cannabis). The coca-growing regions in Colombia are mostly located south of Bogota and near the border with Ecuador and Peru, and in Peru and Bolivia coca has been grown on the eastern slopes of the Andes. The main distribution centers are Cali and Medellín in Colombia.

Many Latin American farmers are swayed to grow drugs because of their high price compared to other agricultural products, and because of the power of the drug cartels in remote rural regions. In regions where crop yields are low, where people have only small plots of land, and where market prices for legal agricultural crops do not cover production costs, drug production is an attractive or even necessary survival option. The farmers receive only a fraction of the street value of the drugs when they are sold. In Bolivia a farmer might get $610 per kilogram (2.2 pounds) of coca leaves, whereas cocaine is sold for more than $100,000 per kilogram in the United States. Most of the drug exports are controlled by powerful families in Colombia and Mexico who manage the transport systems from rural Latin America by land, air, and boat into the main distribution and consumption centers in the United States, such as Los Angeles and Miami.

In some areas the drug trade has exacerbated political conflicts. The situation is most severe in Colombia, where several guerrilla movements control large zones despite opposition by government military units. The groups are alleged to have links with powerful drug lords and to offer protection to farmers involved in drug production. Because the conflicts are fueled by the drug economy, the United States is at risk of becoming involved in the strife through its support for the Colombian government's anti-drug activities. The drug economy threatens to destabilize other Latin American countries such as Bolivia, Peru, and Mexico because the drug traffickers increasingly control production areas and influence the police, the army, and political leaders through intimidation or bribery. Because of aggressive eradication efforts, including herbicide spraying funded by the United States, the area in coca cultivation has fallen from almost 300,000 hectares (741,000 acres) in 1990 to 180,000 hectares (445,000 acres) in 1998 and has shifted from Peru to Colombia. But production techniques have improved and the overall production has remained around 364,000 tons; the distribution is increasingly controlled by organized crime throughout the global system.

Analysts contend that farmers will continue to produce drugs until they can obtain a better living from other crops or other means of employment or until the demand is controlled in the United States. They argue that the United States should be focusing on the control of demand within its own borders or even on limited legalization of consumption, rather than on fighting a "war on drugs" in Latin America.

Core Regions and Key Cities of Latin America and the Caribbean

The core regions and key cities of contemporary Latin America and the Caribbean have been centers of production and political power since the colonial period and have emerged to become important players in the new global economy. The most important regions are those with high concentrations of population, industry, and services and the agricultural heartlands that produce for domestic and global markets. They include central Mexico with Mexico City, southeastern Brazil, including the cities of São Paulo and Rio de Janeiro, the U.S.–Mexico border region, and central Chile.

Central Mexico

The landscape of central Mexico is an elevated plateau dotted with volcanoes and well-watered basins where agriculture benefits from seasonal rainfall and cooler temperatures. The architecture and land use reflect the succession and merging of indigenous, Spanish colonial, and global cultures with Aztec pyramids adjacent to colonial cathedrals and high-rise corporate headquarters (see Figure 8.2). Central Mexico is the contemporary cultural and economic core of Mexico. More than a quarter of Mexico's population lives in this region, which is also a destination for tourists and international business.

Central Mexico had important functions at the time of the Aztec Empire as a center of trade and political control. Spanish colonialism maintained some of these core functions, using Mexico City to control the viceroyalty of New Spain but also to reorient production toward export to Europe. By 1800, the central region was producing sugar and silver for export to Spain and wheat and livestock on large haciendas to feed the mining and administrative centers. After the Mexican Revolution, about half of the large landholdings were converted to ejidos (communal lands) and these ejidos now grow both traditional crops of maize, beans, and wheat as well as vegetables for export and sale to urban markets. The most productive cropland is found in the Bajio region to the north and in the valley of Puebla to the southeast of Mexico City.

Mexico City   Mexico City is the heart of the dynamic agricultural and industrial zone of central Mexico located in the highland basins of the Mesa Central (**Figure 8.32a**). Mexico City is the economic, cultural, and political center of Mexico and is one of the largest urban complexes in the world. The city contains almost 20 percent of the country's population, hosts most government functions, and produces 40 percent of the gross national product. On a clear day, the view from the top of the modern high-rise office building that dominates the urban skyline—the Torre Latina—includes elegant colonial plazas and administrative buildings, modern skyscrapers owned by international corporations, and the gleaming snowy peaks of the volcanoes that ring the basin (**Figure 8.32b**).

As in most colonial cities, Mexico City's colonial center was designed around a main square or plaza formed by government buildings and the main Catholic cathedral and surrounded by the villas of the colonial elite and small specialized commercial zones for artisans. The nineteenth century saw the construction of wide avenues and elegant parks that were influenced by French urban design. The city grew very rapidly throughout the twentieth century, from a population of 500,000 in 1900 to almost 20 million in 2000, expanding from 27 square kilometers (10.4 square miles) to 1000 square kilometers (386 square miles) as the city embraced many satellite communities. Migrants who were pushed out of rural areas and attracted by the opportunities of the city drove most of the growth. Although the business core emerged in the central city, and the main industrial plants were located to the northwest, there has been a spatial deconcentration of manufacturing in recent years, with factories locating in neighboring cities such as Puebla and Cuernavaca. Business centers have grown up along the main north–south road and subway route of *Insurgentes* and around the *periférico* highway that rings the city.

As with other large Latin American cities, many of the new migrants to Mexico City could not afford to rent or purchase homes and settled in irregular settlements, or *barrios*, that surround the city (**Figure 8.32c**). As much as 50 percent of the housing stock is defined as self-help construction, ranging from cardboard and plastic shanties to sturdier wood and brick structures with aluminum or tile roofs. Many of these settlements occupy steep hillslopes, valley bottoms, and dry lake beds that are vulnerable to flooding, landslides, and dust storms. The *barrio* of Netzahualcóyotl houses more than 3 million people on the shores of Lake Texcoco and has been acknowledged and regularized by the government through land titling, provision of electricity and water, and even the recent construction of a subway line.

Mexico City has always faced water-management challenges because of highly seasonal rainfall that regularly floods the basin. The Aztecs constructed a sophisticated system of drainage and dams, as well as the agricultural system of the *chinampas* to feed the city. In the rainy season, enormous pumps now drain the city, but the major problem is now water scarcity for the large population, and drinking water is now pumped from 100 kilometers (62 miles) away over considerable physical barriers. The pumping of the groundwater reservoir (or aquifer) under the city has caused serious subsidence of parts of the city, including a drop of several meters under the magnificent opera house of "Bellas Artes."

The location of Mexico City on a former lake bed, with unconsolidated sediment, adds to the risks from earthquakes in this seismically active zone. The earthquake that woke residents in the early hours of the morning on October 19, 1985, killed as many as 10,000 people and destroyed more than 100,000 homes and other buildings (see Figure 8.8). Public outrage at shoddy construction of public housing and at the government's slow response to those left homeless contributed to growing political opposition and mobilization of city residents to press for better services and political attention. In addition to earthquakes, the city is at risk from the 5450-meter

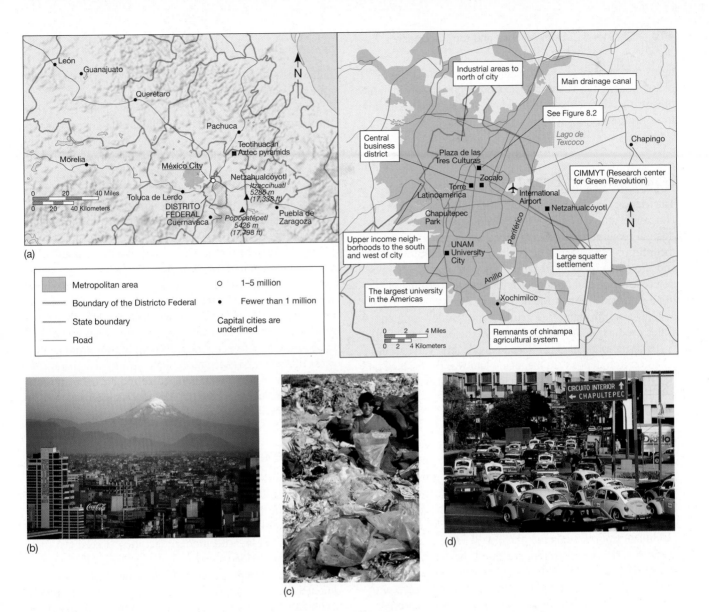

Figure 8.32   Mexico City   (a) Map of central Mexico and Mexico City. (b) Mexico City is located in a high basin surrounded by mountains, including the volcanic pair of Popocatépetl on the left (5450 meters, 17,887 feet) and Iztaccihuatl on the right (5288 meters, 17,343 feet). (c) The growing city has inadequate garbage collection and the poor pick through dumps in search for food and items to sell in the informal sector. (d) The green taxis use nonleaded fuel as one of the environmental policies designed to reduce the air pollution that causes environmental health problems and obscures the view of the volcanoes for most of the year.

(17,887-foot) volcano Popocatépetl, which overlooks the southern part of the basin.

Mexico City's most infamous environmental problem is its air pollution, which currently reaches levels dangerous to human health on more than 100 days a year. Thousands of automobiles, trucks, and buses, many with inadequate emission controls, are responsible for about 75 percent of the air pollution, with dust, fires, industrial plants, and miscellaneous energy use responsible for the remainder. The location of the city adds to the pollution problem because polluted air is often trapped in the basin by the surrounding mountains and by inversions where warm air traps cold air near the ground. The high altitude of the city at more than 2000 meters (6000 feet) means that fuel burns less efficiently and that humans must breathe more air because of the lower oxygen levels. Public protest and media attention to the air pollution crisis prompt-

ed the Mexican federal government to introduce a number of policies in the last 20 years, including lead-free gasoline, emissions testing, closure of a major oil refinery, and a program called *"hoy no circula,"* which requires city residents to leave their cars at home at least one day a week (**Figure 8.32d**). Unfortunately the continuing growth of the city and of car ownership has prevented any significant decline in pollution levels.

Migration rates to Mexico City have declined slightly in the 1990s, and the most rapid urbanization and industrialization is now occurring at nearby locations in central Mexico. For example, the popular tourist and language-school destination of Cuernavaca now has a population of almost a million people and is a major industrial zone producing chemicals, electronics, and other manufactured products. Other important industrial cities in central Mexico include Puebla, with a major automobile manufacturing plant and steel and textile industries, and Toluca, with an automobile plant. This growth of secondary cities in central Mexico has placed considerable pressure on the rich agricultural land that has fed the region for centuries.

Southeastern Brazil

The triangle that encompasses the cities of Rio de Janeiro, Belo Horizonte, and São Paulo is the powerhouse of a Brazilian industrial economy that ranks as the eighth largest in the world, the location of a dynamic global business and cultural center, and the home of more than 30 million people. This important core region had its origins in the founding of a major Portuguese colonial port at Rio de Janiero, the discovery of gold and silver in the eighteenth century, and the development of coffee production around São Paulo in the nineteenth century. The establishment of the Volta Redonda steelworks in 1946 relied on iron deposits in the region and was the basis of import-substitution projects to develop the Brazilian automobile and aircraft industries.

São Paulo   Brazilian geographer Milton Santos reports that São Paulo now employs more than 2 million manufacturing workers and produces 30 percent of Brazil's gross national product, having moved from a commercial center to a manufacturing hub to a service and information core for the global economy. Located on a high plateau about 50 kilometers (30 miles) from the Atlantic, the city of São Paulo has wide avenues and many skyscrapers around the central business district (**Figure 8.33**). It has also become a major financial center for Brazilian and international banks and has recently developed a large telecommunications and information sector with more than a million technical and scientific workers. Geographer Ane Schjolden has shown how the Brazilian telecommunications sector, initially protected by import tariffs and supported by government investment in research and development, has been taken over by global firms in the aftermath of trade liberalization and privatization.

Rio de Janiero   While Rio de Janiero has been overshadowed by the economic growth of its rival São Paulo, 250 kilometers (155 miles) to the southwest, it continues to be the cultural and media center of Brazil. Rio was the capital of Brazil from 1822 to 1960, and the urban structure includes an older city center with a wealthier residential zone and beaches such as Copacabana toward the south and a poorer more industrial zone to the north. The magnificent landscape of Rio's harbor and beaches draws worldwide attention during the festival of Carnival, a major tourist destination where the influence of African traditions emerges in music and dance. The commercial harbor is now a center for shipbuilding and for agricultural exports from the southeast of Brazil, including soybeans and orange juice in specially constructed tankers. In both Rio and São Paulo, massive football (soccer) stadiums holding up to 200,000 people are a focus of city pride and entertainment.

Rio and São Paulo followed the same pattern as other Latin American cities, attracting millions of migrants who settled in informal settlements around the urban core. The **favelas,** as they are called in Brazil, lack good housing and services. In São Paulo, 28 percent of residents have no drinking water and 50 percent have no sanitation. The crowding, high land costs, violent crime, poverty, and pollution of the city are starting to cause economic development to shift to smaller neighboring cities.

In 1956 concern about the concentration of population and development in this core region led the Brazilian government to move the nation's capital from Rio to Brasilia, a new city located 600 kilometers (370 miles) inland. This relocation is an example of so-called *growth pole development policies* by which government invests in a region to stimulate economic growth or to encourage decentralization from central cities. This policy has been moderately successful in that Brasilia has grown to almost 2 million people and houses many government bureaucracies, but the Rio–São Paulo axis continues to be the hub of the economy.

U.S.–Mexico Border Region

American tourists used to cross the border south into Mexico expecting a landscape of underdevelopment, subsistence agriculture, exotic food, and culture. Now, northern Mexico looks more and more similar to parts of U.S. border cities like San Diego and El Paso, with modern factories, hotels, fast-food restaurants, and video arcades.

Northern Mexico, especially the cities that border the United States, has become an increasingly important economic zone that is often used to exemplify the impacts of free trade and foreign investment in Latin America. The border region includes major cities, such as Tijuana and Ciudad Juárez with populations of more than 1 million people, and irrigated agricultural regions that produce a large percentage of Mexico's domestic and export crops. The industrial and commercial center focused on Monterrey can also be seen as part of this core region, although it is located much farther from the border. Monterrey emerged as part of an important industrial region including a major steel industry that used local coal and iron resources and has now grown to become a center for business, services, and high technology with a population of 1.5 million. The agricultural regions, such as the irrigation districts of Sonora and the lower Rio Grande valley (called the Río Bravo in Mexico), are consummate cases of Green Revolution technology and of new agricultural exports, with intensive use of pesticides and new seed varieties.

The border region is closely associated with **maquiladora** manufacturing, a system whereby companies can produce goods free of customs tariffs for export to the United States and elsewhere. In many cases the basic components are imported, and then the products themselves are just assembled in Mexico using low-cost labor. More than 500,000 people are now employed in 2500 maquiladoras in northern Mexico,

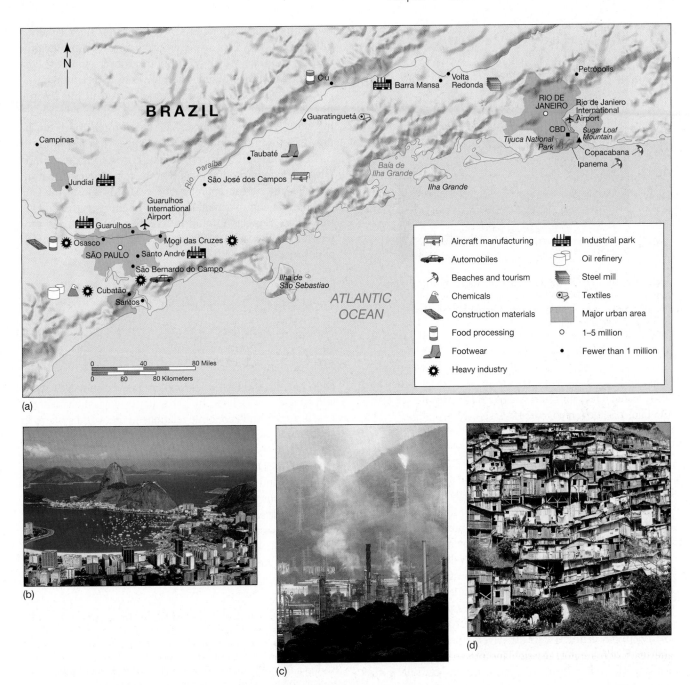

Figure 8.33 Rio de Janeiro and São Paulo (a) Map of the core region that links Rio and São Paulo. (b) Rio de Janeiro has a stunning location with a harbor overlooked by Sugar Loaf Mountain and the beaches of Copacabana. (c) Both Rio and São Paulo have serious urban problems of pollution such as in this region south of São Paulo called Cubatão. (d) Thousands of residents are living in poor-quality housing called *favelas* such as these in Rio. (*Source:* (a) Adapted from G. Knapp and C. Caviedes, *South America*. Englewood Cliffs, NJ: Prentice Hall, 1995, pp. 208–209.)

working in industrial plants that produce everything from clothes to computers (**Figure 8.34**).

Geographers have analyzed the social and environmental impacts of the maquiladoras. For example, Altha Cravey has described the new employment opportunities for young women, but under poor working conditions of low pay, ex-

posure to toxics, and restrictions on unionization. The problems of pollution and waste associated with urban and industrial development along the border between the United States and Mexico have led a variety of nongovernmental organizations and community groups to demand improved environmental protection. In Nogales, Mexico, women in informal

Figure 8.34   U.S.–Mexico border Rapid industrialization along the U.S.–Mexico border, especially the growth of the maquiladoras. In this photo, a woman assembles electronic components at a factory in Tijuana.

colonia settlements have organized to demand safe drinking water and the cleanup of wastes from factories, and they have also created a recycling and tree planting program (see Day in the Life: Yesenia, on page 414).

Central Chile

Central Chile is one of the most important agricultural export zones in Latin America and is increasingly compared to the U.S. state of California, with which it shares a moderate Mediterranean climate of warm, wet winters, and moderate summer temperatures (**Figure 8.35**). Spanish land grants distributed the fertile land of the Central Valley into large haciendas that produced wheat and raised cattle in the colonial period, especially to supply the growing cities of Santiago, Valparaíso, and Concepción. Wheat became an important export after independence but the major boost to exports came in the twentieth century with the development of refrigerated transport and shipment by air. This allowed Chile to take advantage of the hemispheric contrast in seasons, selling fruit and vegetables grown in the Chilean summer (November to March) to North American winter markets.

Production and export of fruit and vegetables grew dramatically, especially fresh grapes, apples, peaches, and berries, together with fruit packing and processing industries that employed thousands of people. Agricultural expansion has been aided by the Green Revolution package of technologies and by some land reforms that provided plots of land to those who will farm it intensively. Chile has also developed a wine industry that exported more than $500 million worth of wine in 1999.

Agricultural exports have only partially compensated for Chile's economic dependence on exports of copper and other minerals. Chile is the world's largest producer of copper and is extremely vulnerable to variations in the prices. The main copper area is north of the central zone, with gigantic mines owned by multinational corporations producing more than $5 billion of copper a year. This area was also a global center for production of nitrate, which was used in fertilizers, until the First World War, when synthetic alternatives were discovered.

The coastal zones of central and southern Chile are also important producers of timber for world, especially Japanese, markets. The forest industry has started to encroach on the groves of towering old-growth alerce trees that are similar to the redwoods of the western United States. Conservationists have been able to obtain some protection through environmental legislation and purchase of remaining forest land. One of the more controversial conservation efforts is that of U.S. millionaire Douglas Tompkins, who has purchased and protected a swath of 700,000 acres of forest land that cuts across almost the entire width of Chile.

Distinctive Regions and Landscapes of Latin America and the Caribbean

Latin America and the Caribbean have many distinctive regions and landscapes, defined by particular ecological characteristics, histories, and economic relations to the world system. In this section, we will focus on four regions: the Amazon basin, the Andes, the Caribbean islands, and Central America.

The Amazon basin and the Andes are distinctive for their vast forests and lofty mountains, with indigenous cultures who adapted to the challenges of the humid Tropics and the thin, cool air of the altiplano. Both of these regions were transformed by colonialism and export-oriented development in the search for minerals, other raw materials, and crops. In the twentieth century, the Amazon basin and the Andes developed reputations as regions of conflict associated with deforestation or

Figure 8.35   Central Valley of Chile The Central Valley of Chile has a lush agricultural landscape that benefits from a Mediterranean climate with warm temperatures and winter rainfall. With a growing season opposite to that of California, its twin in the Northern Hemisphere, the Central Valley has developed export agriculture focused on the production of grapes, wine, and fresh vegetables.

drugs and insurgency and thus drew international political and environmental attention.

The Caribbean islands and Central America share the legacy of exploitation and transformation by European colonial powers, especially plantation economies and slavery. Yet within these two regions, politics and economics have diverged to the point where they contain some of the richest and poorest countries and places in Latin America and the Caribbean and a wide range of social conditions. Latin America and the Caribbean is so vast that these four regions provide only a sample of the richness and diversity of landscapes.

Amazon Basin

The vast forests of the Amazon basin are perhaps Latin America's most distinctive regional landscape. Covering more than half a billion hectares (about 1.2 billion acres), the Amazon basin contains water, forest, mineral, and other resources of great value, yet has had relatively low population density until recent years. Although most people tend to associate the Amazon basin only with Brazil, the river basin and forests also include large parts of Bolivia, Peru, Ecuador, Colombia, Venezuela, and the Guianas (Guyana, Suriname, and French Guiana).

The ancestral economy of the Amazon is based on extractive processes, such as the collection of plants, animals, and products from the forest, including fish, nuts, and traditional medicines. These resources, which have been used for centuries by Amazonian indigenous groups such as the Yanomami, are renewable, and their collection does not destroy the forest.

The colonial image of the Amazon basin varied from a vision of a tropical Eden with untapped resources to an impenetrable disease-ridden jungle hell of savage tribes. The region was of botanical interest, but it held little economic interest until the late nineteenth century when the development of the automobile industry in the United States and Europe exploded the demand for rubber, a product obtained by tapping the latex sap of scattered rubber trees in the forest. Local rubber tappers, or *seringuieros,* sold the rubber to middlemen. They in turn traded with the "rubber barons," who constructed enormous mansions and a magnificent opera house in the Amazonian port of Manaus. The end of the rubber boom is said to have occurred when Henry Wickham shipped thousands of rubber tree seeds from Brazil to Kew Gardens, in England, where they were cultured and the seedlings exported to Southeast Asia. The success of the more efficient Asian plantations, especially in Malaysia, drove the Amazon into decline because Brazilian trees were too susceptible to disease when grown on plantations.

Amazonian development became a focus of Brazilian government policy in the 1970s because it was seen as a pressure valve for landless and impoverished peasants in other regions and as a way of securing national territory through settlement. Several highways were built across the Amazon, including the Trans-Amazon from Recife to the Peruvian border and the Polonoreste from Brasilia to Belém at the mouth of the Amazon.

Government policy was specifically designed to colonize the Amazon. They saw it as a *frontier region* similar to that of the western United States in the nineteenth century. Landless peasants were given title to plots of land if they promised to develop them productively, and peasants migrated in thousands along the new roads. As geographer Susannah Hecht has shown, much of the land was actually acquired by large landholders who took advantage of favorable incentives and tax breaks to develop ranches for speculation and tax havens. Hecht's fieldwork also found that when both small holders and large ranches cleared the land of forest, often by burning, soil fertility declined rapidly and this led to further deforestation as farms and pastures were abandoned.

Satellite images of the region show the process of Amazonian deforestation; the networks of new roads and associated forest clearance can be clearly seen (**Figure 8.36**). Satellites also show the thousands of fires that are set each year to clear land—these fires produce a dense layer of smoke that closes airports and chokes local residents. But the photos also show that the pattern of development and deforestation varies spatially, with some remote areas still relatively untouched and others along roads and around cities almost totally transformed to agriculture.

Estimates of the rate of deforestation of the Amazon basin do not always agree because of differences in the way in which forests are defined and satellite images are analyzed, and because clouds and smoke prevent accurate assessments in some regions. But the general consensus is that perhaps 15 percent of the Amazon forest has been cleared and that the current rate is about 130,000 square kilometers (50,000 square miles)

Figure 8.36   Satellite image of deforestation This satellite image from the southern region of the Brazilian Amazon in the state of Rondonia shows how the forest is cleared as roads and people move into the Amazon area. The cleared areas are shown in lighter brown and the forest in green. A distinctive grid pattern is clear as people farm rectangular plots along roads.

A Day in the Life

Yesenia

Yesenia is an 11-year-old girl who lives in Nogales, Sonora, Mexico, a city on the U.S.–Mexico border (**Figure 1**). When she was five years old, her family moved there from the agricultural state of Sinaloa, so her parents could find steady work.

Since the mid-1960s, when the Mexican government initiated an industrialization program to rehabilitate the northern economy, hundreds of maquiladoras—foreign-owned assembly plants—have located on the Mexican side of the border. Today about 80 factories operate in Nogales. In them, laborers assemble, among other things, suitcases, television remote controls, microchips, trombones, hospital supplies, auto parts, and computers for the world market.

While the early phase of maquiladora industry brought mostly single migrants to the border—mostly young women who ventured away from home and into the factories—today many neighborhoods in Nogales are filled with young families. Like Yesenia's parents, they came primarily to provide a better life for their children.

Yesenia's family, like other newcomers, built a house with whatever they could afford on land that was up for grabs. Yesenia lives with her parents, both of whom work in the maquiladoras, and her four younger brothers and sisters in a three-room house made of scrap wood and discarded shipping pallets. One cold Christmas Eve several years ago, Yesenia built a bonfire outside to try warm up. The flames grew too big too quickly and Yesenia was severely burned on the back of her right leg. Yesenia spent the holidays in the health clinic, where care is free for families of maquiladora workers. Before, when Yesenia's parents worked in the fields in Sinaloa, benefits like this were virtually nonexistent.

With such rapid growth (the population of Nogales increased fivefold between 1950 and 1980), housing shortages have forced new residents to build their own homes in marginalized neighborhoods, which lack basic services such as running water, drainage, and regular trash collection.

Children such as Yesenia are aware of the dangerous environmental conditions around them. In front of her house, a large accidental lake forms every summer when rainwater collects in the sloping intersection of two unpaved streets. "That's contaminated water," Yesenia says. "Nothing but infection there. When the big puddle forms, lots of us get sick from the flies."

Yesenia is also concerned about the quantity of garbage in her neighborhood. Without regular trash collection, refuse piles up in front of houses, in the streets, and in the schoolyard.

Figure 1   Yesenia's life Yesenia lives in Nogales, Mexico, adjacent to the border and to maquiladora manufacturing plants. She lives with her parents and her four younger brothers and sisters in a three-room house similar to those shown here. Many of these are constructed from scrap wood and discarded shipping pallets.

Yesenia's optimism has led her to persuade the director of the school to put up signs in the schoolyard that say, "Take care of the plants." Another time she gathered her neighbors together one morning to collect trash. Though difficult, these are tasks Yesenia feels she can accomplish.

For the future Yesenia says, "I wish for many good things. That people are good, that my family continues to be happy. I want to excel in the future. I like skating, I want to be a skater, but here in Sonora there is no skating and my parents don't have money, but this is my greatest dream, skating."

Contributed by Kimi Eisele, an independent writer from Tucson, Arizona.

a year. The fate of the Amazon has attracted global attention, led by scientists and environmental organizations that are concerned about the impacts of such large-scale forest loss on biodiversity and climate. The Brazilian government has responded by removing some of the tax breaks for development, by intensifying monitoring and control of deforestation, and by establishing parks and reserves. One of the best-known reserves is named for Chico Mendes, a rubber tapper, who organized resistance to deforestation by large ranchers and was murdered in 1988. He pushed for the establishment of areas that were protected for appropriate extractive uses. Other parks and reserves, such as Manu in Peru and Cuyabeno in Ecuador, are becoming important ecotourism destinations where international tourists stay in jungle lodges and are able to observe the rich bird and animal life of the forest.

The Brazilian government has argued that it is not appropriate for other countries like the United States and many in Europe, who cleared and developed their own territory in previous centuries, to criticize Brazil now for doing the same thing in trying to grow its own economy. National and international campaigns have also sought to protect the indigenous peoples of the Amazon who have lost their traditional hunting and gathering lands to development and who are vulnerable to the diseases and cultures of new immigrants to the region. The international musician Sting has held annual "rainforest concerts" to draw attention to this problem and to raise money for the plight of the Amazon environment and peoples.

The military government also sponsored mineral exploration in the Amazon basin, and rich resources of iron were discovered around Carajás and gold at nearby Serra Pelada. By 1982 more than 100,000 people had moved to the area to work in the new mines. Other controversial developments include large dam projects, such as that at Tucuri, and projects that replace diverse ecosystems with monocrop plantations (**Figure 8.37**). The discovery of oil in the western Amazon basin has increased the significance of the region to national economies but has also degraded the forests of Ecuador, Colombia, and Peru and led to boundary conflicts between countries and with indigenous groups.

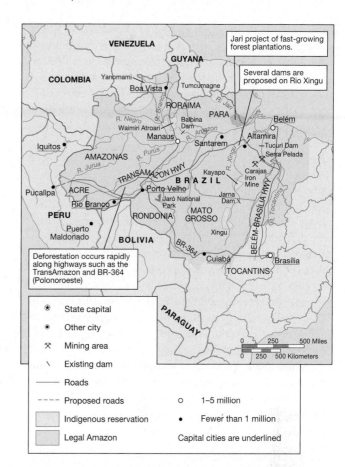

Figure 8.37 Map of development in Amazon This map shows some of the development projects that are causing deforestation and other environmental problems in the Amazon basin, including major roads such as the Trans-Amazon highway, dams such as Tucuri, and agricultural development around cities such as Acre. The basin also has many indigenous groups, some of whom have gained reserve status for their lands. (*Source:* Adapted from G. Knapp and C. Caviedes, *South America.* Englewood Cliffs, NJ: Prentice Hall, 1995, p. 233, and from "Controversial Infrastructure Projects Proposed or Underway in the Amazon Basin," available at http://www.amazonwatch.org)

The Andes

Its high elevations, indigenous cultures, distinctive agriculture, mining economy, and relative poverty characterize the Andean region. Mountains and volcanoes, reaching more than 6000 meters (20,000 feet), overlook the high plateau, or altiplano, where Latin America's largest lake, Lake Titicaca, lies at the border of Peru and Bolivia at more than 3800 meters (12,500 feet). The Andean residents include large populations of indigenous Aymara and Quechua speakers who have maintained traditional rituals, clothing, and crafts.

The traditional crops domesticated in the Andean region include the potato and a high-protein grain *quinoa*. The most important animals are the llama and alpaca, used for transport, wool, and milk. Cultural ecologists have described the vertical zonation of Andean agriculture (see Figure 8.10). Communities have fields scattered at different elevations to take advantage of different climatic and soil conditions. Grazing animals and potatoes are found at higher altitudes, and then lower down there are grains such as wheat and corn, and finally vegetables and fruits at lower levels with more tropical climates. Natural ecosystems also vary with elevation and include the *paramo* of the northern Andes, with unusual cold-adapted plants called *frailejón*.

Most agricultural production is for subsistence or local markets except for coca, a traditional crop with leaves that are chewed by Andean residents to alleviate the effects of high altitude. Coca is now exported to Colombia, where it is turned into cocaine. Crop yields are low because of poor soils and cold, dry climates, but some groups still use the intensive raised field and terrace systems of the Incas to reduce agricultural risks. The latifundia (large landholdings), established by the Spanish, created a system where most indigenous people were given small plots of land around the large haciendas where they were forced, often through debt to the owner, to work much of the time. Despite some efforts at land reform, this legacy of inequality lingers in poverty and discrimination against many Andean Indians and resentment has fueled rural revolt, illegal coca production, and some support for guerrilla movements such as the Shining Path (*Sendero Luminoso*) in Peru. Many young people now choose to migrate to the coastal cities to seek work and opportunities still not available in more remote Andean communities.

The mining industry is important in many parts of the Andes and includes copper in Chile and Peru, tin in Bolivia, emeralds in Colombia, and silver and gold in several countries. Despite some unionization and attempts to improve technology and working conditions, the life of many miners is still very difficult, and they endure high levels of respiratory diseases and accidents. The benefits of mining have also tended to concentrate in multinational corporations or in a few families that control the major companies, and prices have been very volatile, recently falling to low levels on world markets.

Development efforts in the Andean region have included large-scale government projects of road building and tourist development. But the collapse of many national economies during the 1980s and the subsequent restructuring to reduce government spending mean that in many areas the only external assistance now comes from nongovernmental organizations (NGOs).

Caribbean Islands

The Caribbean islands are a diverse mix of cultural traditions, political systems, and environments and include some of the poorest and wealthiest countries in the region. The distinctive physical geographies include extensive coral reefs and mangrove forests, small islands dominated by active volcanic peaks, and vulnerability to the hurricanes that arrive each fall. The Caribbean is divided into several subregions: the *Greater Antilles,* which include the islands of Cuba, Hispaniola (divided into Haiti and the Dominican Republic), Jamaica, and Puerto Rico; and the *Lesser Antilles,* which include the British and U.S. Virgin Islands, the Windward Islands, the Leeward Islands, and the islands of the southern Caribbean Sea north of Venezuela. Political groupings and affiliations include the many islands of the Bahamas, the former British colonies that remain within the Commonwealth, the French protectorates of Guadelupe and Martinque, and the U.S. territory of Puerto Rico.

Caribbean culture is heavily influenced by the African traditions of the millions of slaves who were brought to the region as labor for the colonial plantations. Many countries, such as Haiti and Jamaica, have a predominantly black population. Although most countries use the colonial languages of English, Dutch, French, and Spanish as official languages, millions of Caribbean residents speak versions of Creole, languages that blend English or French with African or even indigenous words to create distinct languages, such as Haitian patois. Reverberations of Africa are also found in Caribbean foods and music and in spiritual traditions such as Voodoo, Santería, and Rastafarianism.

Some islands, such as Barbados, have many residents of European origin, and several have significant populations of Asian descent. After the end of slavery, workers were brought from South Asia as indentured labor and were contracted to work for a plantation for a number of years. For example, 145,000 workers were brought from India to Trinidad between 1838 and 1917, and the population is now 40 percent South Asian heritage, with Hinduism as an important religion. Hindi is often heard spoken, and elements of Indian cuisine are reflected in local foods. Most of the Commonwealth countries, such as Jamaica, Antigua, and Trinidad share a passion for the British game of cricket; in Cuba and the Dominican Republic baseball is seen as the route to fame, with many players emigrating to work in the United States. The Caribbean has produced world-famous writers, such as V. S. Naipaul, Jamaica Kincaid, and Derek Walcott.

The economy of the Caribbean still has strong echoes of the colonial past, with many islands specializing in plantation export crops, such as sugar, tobacco, and coffee. Sugarcane long dominated the land and economy of islands such as Cuba, Jamaica, and the Dominican Republic, but sugarcane has declined with competition from sugar beets, and in Cuba's case the disintegration of Soviet markets, although the production of rum is still extremely important. Coffee is grown in the cooler highlands of countries such as Jamaica, Puerto Rico, and Haiti, with Jamaica's Blue Mountain coffee receiving premium prices on world markets. Tobacco is important in Cuba for its renowned cigar industry. Bananas are grown in the British Caribbean, such as St. Vincent, St. Lucia, and Jamaica, because they have received preferential access to the European Union under the Lomé Agreements to assist former British colonies with trade. These preferences are likely to disappear with moves toward tariff removal and open markets within the European Union and under the World Trade Organization. In all cases, these primary exports are vulnerable to world market fluctuations and to the tastes of global consumers.

Jamaica and Trinidad and Tobago have economies for which mineral and energy resources are important, including bauxite in Jamaica (exported for aluminum production) and oil and gas in Trinidad and Tobago, where important exports include gas derivatives such as ammonia fertilizer and methanol.

The Caribbean has both benefited and suffered from its location in "America's Backyard" as the impact of the Monroe Doctrine brought the region within the U.S. sphere of influence. U.S. political goals have included protecting the route to the Panama Canal, preventing the spread of communism, maintaining stability for U.S. corporations with assets in the region, and aligning trade in U.S. interests. (See Geography Matters: The Panama Canal, p. 417.)

Using these and other goals as justification the United States has intervened repeatedly in the Caribbean since the Spanish-American War of 1898 that freed Cuba from Spain and made Puerto Rico a U.S. colony. The United States sent troops to the Dominican Republic, Cuba, and Haiti several times in the earlier part of the twentieth century; invaded Grenada in 1983 to oust a socialist government and rescue

unnamed Latin American country where guerrillas and the army are fighting.

Mosquito Coast. Directed by Peter Weir, 1986. An inventor takes his family to Central America to build an ice factory in the middle of the jungle.

Portrait of the Caribbean. Directed by Roger Mills, 1992. This six-hour series looks at the political, cultural, and social formation of the region.

Quilombo. Directed by Carlos Diegues, 1984. This historical saga of seventeenth-century Brazil documents the stories of groups of runaway black slaves who escaped to mountainous jungle strongholds, where they formed self-governing communities known as *Quilombos*.

Voices of Latin America. Directed by Andrian Malone, 1987. This video examines the cultural identity of Latin America through its writers and literature using dramatizations and interviews.

Music

Inti Illimani. *The Best of Inti Illimani*. Xenophile, 2000.

The Mexican Revolution. Folk Lyric, 1996.

Various Artists. *2000 Latin Grammy Nominees*. Sony/Columbia, 2000.

Various Artists. *Beleza Tropical*. Emd/Luaka Bop, 1988.

Various Artists. *Forever Tango*. Bmg/RCA Victor, 1998.

Various Artists. *Tocando Tierra: A Tribute to Latin American Music*. Latin World Entertainment, 1999.

Various Artists. *Tougher Than Tough: The Story of Jamaican Music*. Uni/Mango, 1993.

Books

Allende, Isabel. *House of the Spirits*. New York: Bantam Books, 1986. The story of three generations of a family who experience political change in a country similar to Chile. Written by the niece of former socialist President Allende.

Alvarez, Julia. *In the Time of the Butterflies*. New York: Penguin USA, 1995. A tale of three sisters in the Dominican Republic during the rise of the Trujillo dictatorship.

da Cunha, Euclides. *Rebellion in the Backlands*. Chicago: University of Chicago Press, 1944. Classic account of a rebellion at the end of the nineteenth century in rural Brazil.

Dorfman. Ariel. *Heading South, Looking North: A Bilingual Journey*. New York: Penguin USA, 1999. Personal biography of the Argentinian playwright, who fled from his homeland to Chile and the United States.

Fuentes, Carlos. *The Years with Laura Díaz*. New York: Farrar, Straus and Giroux, 2000. Follows one woman through the upheavals of twentieth-century Mexico.

García Márquez, Gabriel. *One Hundred Years of Solitude*. New York: Harper Perennial Library, 1967. Magical realism novel from the Colombian Nobel Prize winner about the history of a family in the mythical town of Macondo.

Kinkaid, Jamaica. *A Small Place*. New York: Farrar, Straus and Giroux, 2000. An invitation to understand the Caribbean island of Antigua and the impacts of colonialism.

Llosa, Mario Vargas. *The War of the End of the World*. New York: Penguin USA, 1997. Revolutionary community in the nineteenth-century Brazilian backlands.

Menchu, Rigoberta. *I Rigoberta Menchu*. New York: Verso, 1987. Testimony from a Guatemalan Mayan woman about Mayan life and military repression.

Naipaul, V. S. *A House for Mr. Biswas*. New York: Vintage Books, 2001. A comic novel about the efforts of a Trinidad resident of Asian-Indian descent to own his own home.

Paz, Octavio. *Labyrinth of Solitude*. New York: Grove Press, 1985. Exploration and explanation of Mexico and Mexicans by a Nobel Prize winner.

Rulfo, Juan. *Pedro Paramo*. New York: Grove Press, 1994. A haunting search for roots in rural Mexico.

Walcott, Derek. *Tiepolo's Hound*. New York: Farrar, Straus and Giroux, 2000. Nobel Laureate explores European domination of the West Indies.

SAYAN MOUNTAINS

YABLONOVYY RANGE

ALTAY MOUNTAINS

Ulan Bator

MONGOLIA

TIEN SHAN

INNER MONGOLIA

Huang He (Yellow R.)

PAMIRS

XINJIANG

NINGXIA

QINGHAI

GANSU

Xi'an

SHAANXI

C H I N A

Chengdu

XIZANG (TIBET)

SICHUAN

H I M A L A Y A S

Chang Jiang (Yangtze R.)

Chongqing

Mekong R.

GUIZHOU

YUNNAN

GUANGXI

9 East Asia

0   150   300 Miles
0   150   300 Kilometers

90°E

Figure 9.1

KHINGAN RANGE

HEILONGJIANG

Harbin

Changchun

JILIN

Shenyang

LIAONING

NORTH
KOREA

Pyongyang

Beijing
Tangshan

Dalian

Tianjin

HEBEI

SHANXI

SHANDONG

Qingdao

Seoul

SOUTH
KOREA

Taejon
Taegu

Kwangju

Pusan

Yellow
Sea

Huang He (Yellow R.)

HENAN

ANHUI

JIANGSU

Nanjing

Shanghai

HUBEI

Hangzhou
Ningbo

ZHEJIANG

East China
Sea

JIANGXI

HUNAN

FUJIAN

Taipei

TAIWAN

Xiamen

Tropic of Cancer

GUANGDONG

Xi
Guangzhou

Shenzhen
Hong Kong

Zhu Jiang
(Pearl River)

South China
Sea

HAINAN

Sea of
Japan

Hitachi

JAPAN

Kanazawa
Tokyo

Nagoya

Kyoto

Kobe
Osaka

Nagasaki

PACIFIC
OCEAN

N

150°E

40°N

140°E

30°N

20°N

120°E

◉	More than 5 million
○	1–5 million
•	Fewer than 1 million
	Capital cities are underlined

East Asia consists of China, Japan, Mongolia, North and South Korea, and Taiwan. This is the most populous of all world regions, containing 1.45 billion people. The region is dominated in both demographic and territorial terms by China, which occupies 9.6 million square kilometers (3.7 million square miles; about 6.5 percent of Earth's land surface) and has a population of 1.26 billion, about 22 percent of the world's population. A glance at **Figure 9.1** shows that East Asia is naturally bounded by mountain ranges and oceans. To the northwest are the Elburz Mountains of northern Iran and the Tien Shan ("Mountains of Heaven") along the southern borders of the Central Asian countries. To the north are the Altay and Sayan Mountains that separate Siberia from Mongolia, and the Yablonovyy and Khingan ranges that separate southeastern Siberia from northern China. To the southwest are the Pamir Mountains and the Himalayas, while to the south and east are the coastal seas—the South China Sea, the East China Sea, and the Yellow Sea—of the Pacific Ocean.

East Asia is a region of tremendous contrasts in natural environments and history, while the human geography of the region has been subject to radical change during the modern era. China's vast and diverse domain has been restructured by a communist revolution that has itself been redirected several times in its objectives. Japan, on the other hand, has been transformed by no fewer than two dramatic "economic miracles."

Traditional rural communities with a great variety of distinctive ways of life still characterize much of the region, while at the same time East Asia is one of the most rapidly urbanizing of all world regions. Japan is already highly urbanized (almost 80 percent of the population lives in urban areas), but the pace and scale of urbanization in China is truly phenomenal. Overall, only 34 percent of China's population lives in urban areas, but 25 years ago the figure was only 17 percent, and 50 years ago it was 12 percent. China now has 53 cities with a population of 1 million or more, including 9 with a population in excess of 5 million.

Environment and Society in East Asia

A great deal of East Asia consists of juxtaposed plateaus, basins, and plains separated by narrow, sharply demarcated mountain chains. These broad physiographic regions contain a great diversity of ecosystems, which in turn have evoked a variety of human responses and adaptations.

Landforms and Landscapes

The satellite photograph of East Asia in **Figure 9.2** suggests a broad, threefold physical division:

Figure 9.2   East Asia from space   This image shows that much of East Asia consists of vast areas of upland plateaus and mountain ranges. [*Source:* (inset) Map projection, Buckminster Fuller Institute and Dymaxion Map Design, Santa Barbara, CA. The word *Dymaxion* and the Fuller Projection Dymaxion™ Map design are trademarks of the Buckminster Fuller Institute, Santa Barbara, California, © 1938, 1967 & 1992. All rights reserved.]

1. *The Tibetan Plateau, in the southwest, an uplifted **massif** of about 2.5 million square kilometers (965,000 square miles).* With an elevation of at least 2500 meters (8202 feet), rising to between 4500 and 5000 meters (14,763 to 16,403 feet) in the northwestern part of the plateau, the Tibetan Plateau is often referred to as the "roof of the world." Within the plateau are several mountain ranges—including the Himalaya Mountains—with peaks of 7000 meters (22,964 feet) or more. The Himalayas contain the world's highest peak, Mount Qomolangma (Mount Everest), at 8848 meters (29,027 feet).

2. *The central mountains and plateaus of China and Mongolia, a checkerboard of mountain ranges, plateaus, basins, and plains.* The most important of these plateaus and basins include the Mongolian Plateau, the Ordos Plateau, the Loess Plateau, the Yunnan-Guizhou Plateau, and the Tarim, Sichuan, and Zunghaer basins. Most of these have elevations of 1000 to 2000 meters (3281 to 6562 feet) above sea level and are bordered by uplifted mountain ranges of about 3000 meters (9843 feet). This pattern is largely the consequence of the intersection of two sets of geological structures that underlie much of China: a series of roughly parallel upfolds that run south–southwest to north–northeast and a less-pronounced series of mountain ranges that run east–west (**Figure 9.3**).

3. *The continental margin of plains, hills, continental shelves, and islands.* The bulk of the population of East Asia lives in this continental margin, which includes the great plains of China: the Northeast China (Manchurian) Plain, the North China Plain, and the plains of the Middle and Lower Chang Jiang (Yangtze River) valley. Most of these plains lie below 200 meters (656 feet) in elevation. South of the Chang Jiang is hill country with elevations of around 500 meters (1640 feet) and along the coast, including the Korean peninsula, are uplifted hills and mountains of 750–1250 meters (2460 to 4100 feet). The Japanese archipelago of Hokkaido, Honshu, Shikoku, and Kyushu forms the outer arc of East Asia's continental margin. Its backbone of unstable mountains and volcanic ranges that project from the shallow sea floor extends to the island of Taiwan and is part of the **Ring of Fire** that girdles the Pacific Ocean.

This broad division is the product of a long and complex geological history. The key to understanding the basic physiography of the region is the plate tectonics that have had a dramatic influence in recent geological times. The entire Tibetan Plateau was uplifted between 65 million years ago and 2 million years ago as the Indian-Australian plate moved northward

Figure 9.3 East Asia's physiographic regions East Asia can be divided into three broad physiographic units, the Tibetan Plateau and central mountains and plateaus corresponding to Outer China and the continental margins approximating to Inner China, the Japanese archipelago, the Korean Peninsula, and Taiwan.

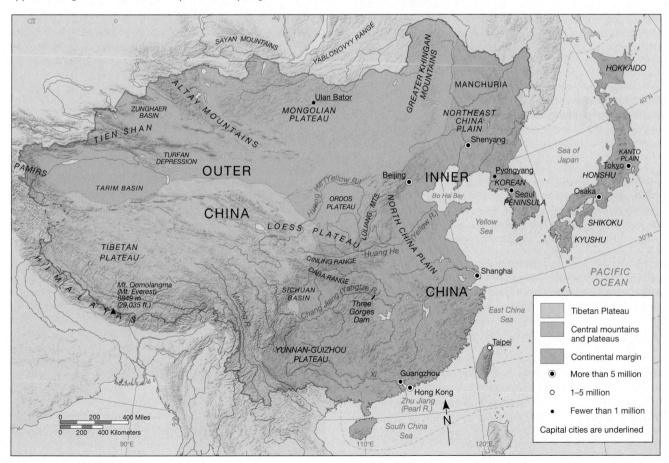

Geography Matters

Living with the Threat of Earthquake

Japan is situated within one of Earth's most geologically active zones. The Japanese archipelago is situated at the junction of three tectonic plates: the Eurasian, Philippine, and Pacific plates. In addition to volcanic activity—at least 60 volcanoes have been active within historic times in Japan—there is almost perpetual earthquake activity. Most of the 1000 or so earthquake events that occur in Japan each year are minor, but a few are serious enough to cause damage to property, and occasionally they can be devastating. In 1730 an earthquake in Hokkaido caused 137,000 deaths; in 1923 the Great Kanto earthquake resulted in 104,619 dead or missing. The most recent disaster was in Kobe in 1995, when a severe earthquake killed 6452 (**Figure 1**).

The strongest earthquakes seem to have occurred every 70 years or so, and experts agree that a major quake will almost certainly strike in Japan's highly urbanized Pacific Corridor—between Tokyo and northern Kyushu—in the relatively near future. Tokyo is particularly vulnerable to earthquake damage because much of the city, including most of the downtown and all of the waterfront districts, is built on loosely consolidated landfill. When severely shaken by an earthquake, this material behaves like liquid, causing the buildings on it to collapse. Tough building codes and sophisticated building technologies (putting buildings on rollers and installing long counterweights to offset shaking, for example) make it possible to erect structures that can withstand a severe earthquake, but 2 million of Tokyo's 2.6 million buildings were constructed before 1981, when building codes were significantly strengthened. In addition, the city's low elevation makes it vulnerable to tidal waves (*tsunami*) that would be triggered by an earthquake with an offshore epicenter.

The threat of a major earthquake is very much part of the background of living in Tokyo and a major shaper of both the built environment and the city's personality (**Figure 2**). To protect against *tsunami,* there is an extensive pattern of breakwaters in Tokyo Bay, sea walls and river walls at the waterfront, and massive gates that can close off river mouths to prevent a flood tide from surging in. To protect against the spread of fires that are likely to occur after a major earthquake, city planners have built continuous ribbons of fire-resistant apartment blocks. Nevertheless, emergency services anticipate a scenario in which 7000 will die and 4 million will be made homeless as 500,000 buildings collapse or burn. A control center with sophisticated geographic information system (GIS) displays is ready to keep track of casualties and damage, with huge video screens set up to display live images from cameras mounted on helicopters. Schoolchildren, shopkeepers, and office workers are regularly drilled in earthquake safety measures. Still, the ever-present threat of a serious earthquake has

Figure 1 Kobe, Japan The aftermath of the 1995 earthquake in Kobe, Japan.

engendered an air of fatalism. As geographer Roman Cybriwsky notes, "there is an attitude in Tokyo that time between disasters is limited, and that the city should therefore hurry forward at full steam with its business now, while it still can. Perhaps no other major city in the world defines itself as being so transitory as Tokyo; no other city views its own history so much as a series of urgent rebuildings between disasters."[*]

[*]R. Cybriwsky, *Tokyo: The Shogun's City at the Twenty-first Century,* New York: John Wiley & Sons, 1998, p. 31.

Figure 2 Earthquake planning Sinuous blocks of apartment buildings in the Tokyo metropolitan area are designed to limit the spread of fire after a serious earthquake.

and pushed up against the Eurasian plate, not only uplifting the Tibetan Plateau but also causing the mountain-building episode that resulted in the Himalayas. Subsequently, the weathering of the newly created mountains provided huge quantities of clay, silt, and other fluvial deposits that now blanket the plains. To the east, the movement of the Pacific plate toward the Eurasian plate caused the folding and faulting that has resulted in the mountains of the peninsulas and islands of the continental margin, including the Korean peninsula and the Japanese archipelago.

The geological uplift of the Tibetan Plateau also had important consequences for geomorphic processes. It increased the gradient of the rivers that flowed off the borders of the plateau, enabling them to incise into the plateau and produce a great number of gorges and canyons of considerable length and depth. As geographer Yi-Fu Tuan has noted, the canyons at the southeastern edge of Tibet, clothed in dense vegetation, provided an effective barrier between the Chinese and South Asian civilizations. The rivers are still incising rapidly, and the Tibetan Plateau itself is still not stable: Seismic disturbances continually occur along the Himalayan foothills, often bringing disaster to settlements there.

The physical environments of East Asia are, in fact, relatively hazardous. In addition to regular earthquakes along the Himalayan foothills, much of the North China Plain is subject to seismic activity. Here, earthquakes along fault lines beneath the silty soils have the effect of loosening the soil, causing major landslides. In eastern Shaanxi, a destructive earthquake occurs, on average, every 15 to 20 years. The most ruinous of these took place in December 1921, when more than a quarter of a million people perished. Fifty-five years later, almost as many (242,000) were killed when a strong earthquake devastated the city of Tangshan in Hebei Province. (In Imperial China, natural disasters were thought to presage the death of an emperor who had lost the Mandate of Heaven; Mao Zedong, the communist leader of the People's Republic of China, died in September 1976, just two months after the Tangshan disaster.) Much of Japan, too, is subject to earthquake hazards, with several major fault lines running along the Pacific coast of Honshu, between Tokyo and Hiroshima (see Geography Matters: Living with the Threat of Earthquake, p. 428).

Climate

The geological uplift of the Tibetan Plateau between 65 million and 2 million years ago is also key to understanding certain aspects of the climate of East Asia. The elevation of the Tibetan Plateau cut off the moisture that was formerly brought in to the interior of East Asia from the Indian Ocean by monsoonal winds. This contributed to the gradual dessication of the Tibetan Plateau, whose numerous lakes are now much reduced in size. Protected by mountain barriers and sheer distance from the coast, much of western and northwestern East Asia today averages less than 125 millimeters (5 inches) of rain a year. On the Tibetan Plateau, high elevations make for cool summers and extremely cold winters. Further north, in Xinjiang, Qinghai, Gansu, and Mongolia, summer temperatures

may be extremely hot: the Turfan Depression, some 154 meters (505 feet) below sea level, is one of the hottest places in East Asia, with recorded temperatures in excess of 45°C (113°F).

To the east of this vast arid region are two distinctive climatic regimes. The northern regime is subhumid. It is bounded to the west by the Lüliang Mountains and the Greater Khingan Mountains and extends southward as far as a latitude of about 35° N, encompassing the Northeast China Plain, the North China Plain, and the northern parts of the Korean peninsula and the Japanese archipelago. Winters here are cold and very dry. Summers are warm, with moderate amounts of rain from the southeasterly summer monsoon winds. Rainfall is, however, very variable, so that both drought and flooding are frequent occurrences. The southern regime is humid and subtropical. It extends west from the plains of the Middle and Lower Chang Jiang valley as far as the Sichuan basin and south as far as the southernmost coastlands of South China. Winters here are mild and rainy, and summers are hot with heavy monsoonal rains. Overall, annual rainfall is 1200 millimeters (47 inches) or more higher than it is in the north.

In the arid and subhumid regions of East Asia, drought is a critical natural hazard, causing widespread famine as a result of crop failures in drought years. In addition, the subhumid parts of East Asia tend to be prone to flooding. This is in part because the irregular summer monsoon rains can produce a sudden deluge. In addition, the rivers that flow through the plains carry a tremendous amount of silt, which is then deposited in their more sluggish lower reaches, building up the height of the river bed, and making their course unstable. The Huang He (Yellow River) is the largest and most notorious of these rivers, having changed its course several times in the last two centuries. In between these events, the Huang He regularly bursts its banks, flooding the farms and villages of the densely populated North China Plain.

Environmental History

Many of the landscapes of East Asia have been heavily modified by humankind. The principal human impact has been through the clearing of land for farming. Over the millennia, as the population grew and pre-modern civilizations flourished, much of humid and subhumid East Asia was cleared of its forest cover. At the same time, these landscapes were also modified by the Oriental passion for water control. Over the centuries, marshes were drained, irrigation systems constructed, lakes converted to reservoirs, and levees raised to guard against river floods (**Figure 9.4**). Even the topography was altered to suit the needs of growing and highly organized populations, with hills and mountainsides being sculpted into elaborate terraces in order to provide more cultivable land. Meanwhile, several of the world's most important food crops and livestock species were domesticated by the peoples of East Asia, beginning around 6500 B.C. Millet, soybeans, peaches, and apricots were domesticated in the more northerly subhumid regions, while rice, mandarin oranges, kumquat, water chestnuts, and tea were domesticated in the more humid regions to the south. Mulberry bushes were domesticated as

Figure 9.4 Levee construction China has a long history of labor-intensive projects that have modified and adapted the landscape.

silkworm fodder, and hemp was domesticated as a source of fiber and oil. Chicken and pigs were the most important live-stock species to be domesticated.

Only in the arid western parts of East Asia and the inhospitable Tibetan Plateau—a broad region often referred to as "Outer China" (**Figure 9.5**)—do contemporary regional landscapes reflect large-scale natural environments that are relatively unmodified by human intervention. Outer China is an outback region that contains barely 4 percent of the population of East Asia. Large areas are effectively uninhabited, and in the greater part of it, population density does not exceed one person per square kilometer. The scenery is often spectacular, encompassing snow-clad mountains, vast swamps, endless steppes, and fierce deserts. In broad terms, Outer China can be divided into three types of landscapes (Figure

Figure 9.5 Regional landscapes The landscapes of Outer China are relatively unmodified by human intervention. In contrast, the natural landscapes of Inner China and the continental margin have been greatly altered by human occupance. The Qinling and Daba ranges that extend eastward from the Kunlun mountains at the northern edge of the Tibetan Plateau represent a particularly important boundary within Inner China. These ranges mark one of the sharpest boundaries—floristic, faunal, and human—in the geography of East Asia. To the north are the landscapes of subhumid regions, while to the south are the landscapes of the humid and subtropical regions. (*Source:* Adapted from S. Zhao, *Geography of China.* New York: John Wiley & Sons, 1994, p. 31; and Y-F. Tuan, *China.* London: Longman, 1970, p. 25.)

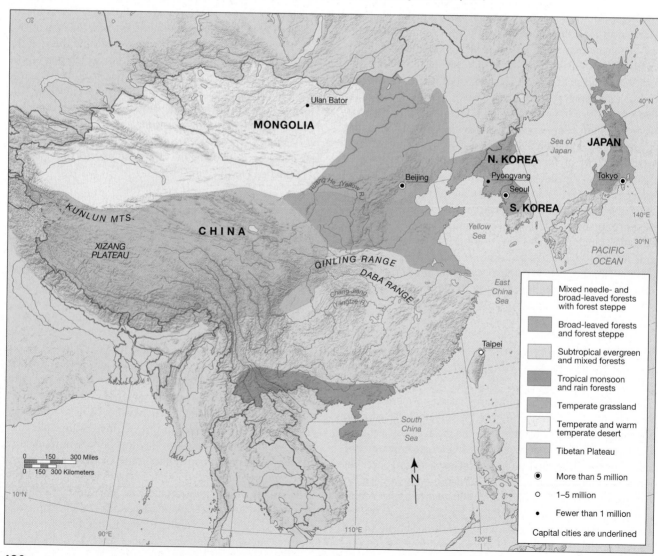

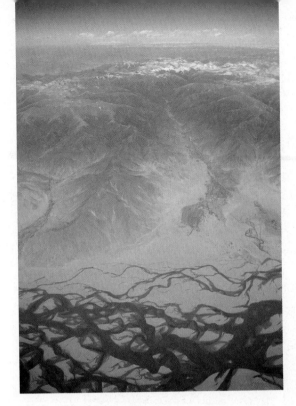

Figure 9.6   The Tibetan Plateau The highest and largest plateau in the world, the Tibetan Plateau occupies about 25 percent of China's land area but contains less than 1 percent of its population.

Figure 9.7   Mongolia The mountainous topography and arid climate of Mongolia and northwestern China support a very low density of population.

9.5). The Tibetan Plateau (**Figure 9.6**) is dominated by rugged mountains with intermontane basins of needle-leaf forests, alpine meadows, and marshes. Northwestern China and much of Mongolia (**Figure 9.7**) is a territory of temperate desert of dunes, scrub-covered hill slopes, and basins with plant communities that are tolerant of high levels of mineral salts in the soil, though dense forests of tall spruce clothe the cool wet flanks of the Tien Shan and Altay mountains. Inner Mongolia is mostly a mixture of steppe and semidesert.

The remainder of East Asia—"Inner China" plus the Korean peninsula, the Japanese archipelago, and Taiwan—can be divided broadly into the landscapes of the subhumid regions to the north of the Qinling and Daba ranges and those of the humid and subtropical regions to the south (Figure 9.5). The natural landscapes of the northern part of Inner China, along with the northern part of the Korean peninsula and the northern half of the Japanese archipelago, are dominated by mixed temperate broad-leaf and needle-leaf forests on the hills and mountains, with a forest-steppe mixture on the Loess Plateau. Higher elevations, and more northerly latitudes, are dominated by spuce, fir, and birch trees; lower elevations and more southerly latitudes are dominated by maple, basswood, and oak. As noted earlier, however, many of these natural landscapes have been heavily modified by humankind, and much of the forest has disappeared. Contemporary landscapes are dominated by dry-field farming, mostly of wheat and millet (**Figure 9.8**). To the south of the Qinling and Daba ranges, including Taiwan, the southern part of the Korean peninsula and the southern half of the Japanese archipelago, contemporary landscapes are dominated by a more intensive agriculture, with wet-field rice farming the dominant element (**Figure 9.9**). Historically, this area was dominated by evergreen broad-leaf forests of horsetail pine and yunnan pine, with tropical rain forests along the coastal margins.

East Asia in the World

East Asia has been the setting for some of the world's most sophisticated civilizations and most extensive empires. Imperial China and imperial Japan were both very inward-looking, but both eventually came into conflict with European and

Figure 9.9   Inner China's southern farming landscape The middle and lower Chang Jiang basin is the most productive grain region in China. Most of the flat land in the region is intensively cultivated under a double-cropping system, with rice, winter wheat, corn, and cotton as the chief crops.

Figure 9.8   Inner China's northern farming landscape The North China Plain, the initial core region of Chinese civilization, has the largest concentration of cultivated land in China. Winter wheat (shown here), corn, sorghum, millet, and cotton are the chief crops.

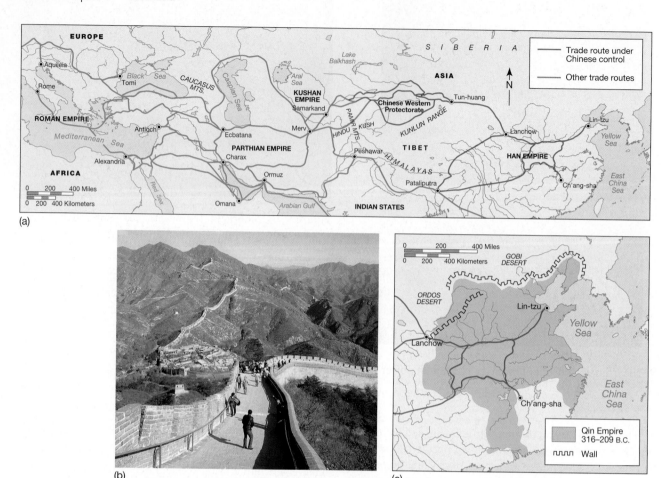

(a)

(b)

(c)

Figure 9.10   The Qin empire and the Silk Road   The Qin dynasty (316-209 B.C.) began the construction of the Great Wall. The wall, built with great hardship and much loss of life, was built by gangs of prisoners and forced laborers. The Wall that visitors see today dates from the Ming period and was built at various times between the late fourteenth and mid-sixteenth centuries. The political and economic stability created by the Qin dynasty was an important precondition for the subsequent success of the Silk Road, the overland system of trade routes that connected China with Mediterranean Europe. The map shows the trade routes of the Silk Road as they existed between 112 B.C. and A.D. 100. (*Source:* Adapted from I. Barnes and R. Hudson, *History Atlas of Asia.* New York: Macmillan, 1998, pp. 45 and 46–47.)

American imperialism. Both empires subsequently experienced revolutionary changes—though of a very different nature—before the modern states of China and Japan emerged to play key roles in the contemporary world economy.

Ancient Empires

China has had a continuous agricultural civilization for more than 8000 years. The first organized territorial state was that of the Xia dynasty, a Bronze Age state that occupied the eastern side of the Loess Plateau (in present-day Shanxi Province) and the western parts of the North China Plain (northwestern Henan Province) between 2206 and 1766 B.C. It was succeeded by the Shang dynasty (1766 to 1126 B.C.), during which walled cities appeared. The first unified Chinese empire, though, was that of the Qin dynasty (**Figure 9.10**). Emperor Shih Huang-ti (221–209 B.C.) established an imperial system that would last for 2000 years. He abolished feudalism, replacing the feudal hierarchy with a centralized bureaucratic administration, and had the Great Wall built to protect China from "barbarian" nomads.

The stability brought about by the Qin dynasty was an important precondition for the success of the Silk Road (Figure 9.10) as an economic and cultural link among the civilizations of China, Central Asia, India, Rome, and, later, Byzantium.

The history of the Chinese empire is complex, with constantly shifting territorial boundaries and successive dynasties that tended to move from vigorous beginnings, with power concentrated around a strong center, followed by a slow loss of control as regional centers gained more power, and a final collapse as a forceful new dynasty was established. Sometimes, the empire was fragmented or subdivided for extended periods, but over two millennia the overall trend was for a larger and more consolidated continental empire.

The first major dynasty to run through the cycle was the Han dynasty, from 206 B.C. to A.D. 220. Han emperors extended the Great Wall westward, allowing the Chinese to control more of the trade routes along the Silk Road. Another important economic development took place early in the seventh century, under the Sui dynasty, when the northern and southern regions of Inner China were linked by the first of a series of Grand Canals. Its purpose was to bring the plentiful rice of the south to the Sui capital in the northwest (present-day Xi'an) and to the armies stationed in the northeast. Over the next several centuries, the canal system was enlarged and modified. By the fifteenth century, China's canal system was more than 1000 kilometers (621 miles) in length. The imperial grain transportation system along the canals employed up to 150,000 soldiers to man its fleet and required the compulsory labor of many more to dredge and maintain the channels. (Canals of comparable scale only began to be cut in Europe—most notably in France—in the eighteenth century.) The infrastructure of China's canal system created a complementary relationship between the north and the south. The economic center of gravity was in the agriculturally prolific south, but the political center was almost always in the north. In 1279 the Mongols, under Kublai Khan, conquered China, establishing imperial rule for the first time over most of both Inner and Outer China. The Mongol dynasty, known as the Yuan dynasty, was expelled after less than a hundred years and was succeeded by the Ming (1368–1681) and the Qing (1681–1911) dynasties.

The Chinese concept of the centralization of power in an imperial clan spread to Japan in the sixth century A.D. and, as in China, a succession of dynasties maintained a rigid system in which a subjugated peasantry sustained a relatively sophisticated but inward-looking civilization. Japan's distinctive civilization was largely a result of the introduction of Buddhism, which arrived from India via China and Korea. Buddhist influence became so pervasive during the seventh and eighth centuries that the ruling elite established a new capital in A.D. 794 in order to make a clean break from the powerful temples in Nara, the old capital. The new capital, Kyoto, was to be the residence of Japan's Imperial family for more than 1000 years, during which it became the principal center of Japanese culture (see Sense of Place: Imperial Kyoto, p. 434).

For the last 200 years of this period, just as an industrial system was developing in the Western Hemisphere, the Tokugawa dynasty (1603–1868) strove to sustain traditional Japanese society. To this end, the patriarchal government of the Tokugawa family excluded missionaries, banned Christianity, prohibited the construction of ships weighing more than 50 tons, closed Japanese ports to foreign vessels (Nagasaki was the single exception), and deliberately suppressed commercial enterprise. At the top of the imperial social hierarchy were the nobility (the *shogunate*), the barons (*daimyos*), and the warriors (*samurai*). Farmers and artisans represented the productive base exploited by these ruling classes, and only outcasts and prostitutes ranked lower than merchants.

In terms of spatial organization, the Japanese imperial economy was built around a closed hierarchy of castle towns, each representing the administrative base of a local *shogun*. The position of a town within this hierarchy was dependent on the status of the *shogun*, which, in turn, was related to the productivity of their agricultural hinterland. As a result, the largest cities emerged among the rich alluvial plains and the reclaimed lakes and bay-heads of southern Honshu. Largest of all was Edo (known today as Tokyo), which the Tokugawa regime had

selected as its capital in preference to the traditional imperial capital of Kyoto, and which, bloated by military personnel, administrators, and the entourages of the nobility in attendance at the Tokugawa court, reached a population of around 1 million by the early nineteenth century. Kyoto and Osaka were next largest, with populations of between 300,000 and 500,000; and they were followed by Nagoya and Kanazawa, both of which stood at around 100,000.

Imperial Decline The dynasties of Imperial China and Imperial Japan both eventually succumbed to a combination of internal and external problems. Internally, the administration and defense of growing populations began to drain the attention, energy, and wealth of the imperial regimes. The imperial system itself also meant that cultural and social elites tended to be focused on the arts, humanities, and self-promotion at the imperial court rather than on economic or social development. As the economy stalled under these constraints, peasants were required to pay increasingly heavy taxes, driving many into grinding poverty and thousands to banditry. By the early nineteenth century, both China and Japan had moved into a phase of successive crises—famines and peasant uprisings—presided over by introverted and self-serving leadership. As in feudal Europe (see Chapter 3, p. 104), the peasantry began to flee the countryside in increasing numbers in response to a combination of rural hardship and the lure of the relative freedom and prosperity of the cities. Finally, the imperial courts of both China and Japan suppressed the spread of knowledge of modern weapons because they feared internal bandits and domestic uprisings. As a result, both empires relied on antiquated military technologies even as Europeans and Americans were racing ahead with new weapons developed through the new technologies of the Industrial Revolution.

This last point relates directly to the external problems that undermined both imperial regimes. European traders had been a growing presence in East Asia in the eighteenth century but had been restricted to a few ports. Initially, Europeans had bought agricultural produce (mainly tea) in exchange for hard cash (in the form of silver currency). Over time, this proved to be an unacceptable drain on European treasuries, and the British eventually provoked China into a military response by insisting on being able to trade opium (grown in India for export by the East India Company) for tea and other Chinese luxury goods. In 1839 the Chinese, having prohibited the sale of opium several decades before, destroyed thousands of chests of opium aboard a British ship. This was just the excuse the British needed in order to exercise their superior weaponry. The so-called Opium War (1839–1842) ended with defeat for the Chinese and the signing of the Treaty of Nanking, which ceded the island of Hong Kong to the British and allowed European and American traders access to Chinese markets through a series of **treaty ports** that included Amoy (now Xiamen), Canton (now Guangzhou), Ningpo (now Ningbo), Shanghai, and Tsingtao (now Qingdao). Shortly afterward, in 1853, U.S. Admiral Matthew Perry anchored his flagship in Edo Bay (now Tokyo Bay) to "persuade" the Japanese to open their ports to trade with the United States and

Sense of Place

Imperial Kyoto

Kyoto has a history that goes back to early in the seventh century. For more than a thousand years—from A.D. 794 until 1868—Kyoto was the home of the Japanese imperial family, the capital of a powerful empire and a sophisticated culture. Nevertheless, at first glance Kyoto is a modern Japanese city (**Figure 1**) with no striking features to reflect its long and glorious imperial past. The city was saved from the American firebombing that destroyed other Japanese cities toward the end of the Second World War, but postwar rebuilding and redevelopment has swept away a great deal of the historic fabric of the city. Between 1950 and 1995, more than 40,000 of the city's distinctive wooden townhouses (**Figure 2a**) were demolished to make way for modern apartment buildings and of-

fices. Today, the legacy of Kyoto's imperial past tends to be tucked away behind tall buildings and modern storefronts. There are, however, many treasures still intact. Kyoto has 1600 Buddhist temples, 400 Shinto shrines, several important palaces and castles, and dozens of gardens and museums. Thirteen of the Buddhist temples, together with three Shinto shrines and Nijo-jo castle (built in 1603 for the Tokugawa emperor), have been designated as World Heritage Sites by the United Nations Educational, Scientific, and Cultural Organization (UNESCO). With all of these assets, Kyoto still embodies the refined spirit of Japanese culture and is one of the most culturally rich cities in the world. More than 40 million visitors arrive each year to absorb the architectural, religious, and artistic heritage of the imperial past.

Two examples of the many single structures that make Kyoto such a powerful symbolic setting are the Katsura Imperial Villa and Kinkaku-ji. The Katsura Imperial Villa (**Figure 2b**) in the city's southwestern suburbs is considered to be one of the finest single examples of Japanese architecture, and its refined simplicity has been the source of inspiration for many leading modern architects in the West, including Frank Lloyd Wright. It was built in 1624 for the emperor's brother. Kinkaku-ji (**Figure 2c**) is one of Japan's most famous sights. Built in 1397 as a retirement villa for a shogun, it was subsequently converted into a Buddhist temple. Although it is these single structures and others like them that draw the tourists, the city's imperial past can also be seen in broader terms. The city's large contingent of *daimyos* (barons) and *samurai* (warriors), for example, supported a large and sophisticated geisha profession (**Figure 2d**), and the geisha quarter, Gion, survives as an entertainment district of upscale restaurants and bars.

Figure 1   Central Kyoto

---

other foreign powers (**Figure 9.11**). The Japanese quickly complied, and the lesson learned, on both sides, was East Asia's abject weakness in the face of superior Western military technology. In both China and Japan, the neocolonialist threat galvanized feelings of nationalism and xenophobia and precipitated a period of civil war among the feudal warlords that culminated in revolutionary change.

Japan's Revolutions

Japan was first to react to the humiliation of Western assertiveness. Japan's transition from feudalism to industrial capitalism can be pinpointed to a specific year—1868—when the Tokugawa dynasty was toppled by the restoration of the Meiji imperial clan by a clique of *samurai* and *daimyo* who were con-

vinced that Japan needed to modernize in order to maintain her national independence. Under the slogan "National Wealth and Military Strength," the new élite of ex-warriors set out to industrialize Japan as quickly as possible. A distinctive feature of the entire process was the very high degree of state involvement. Successive governments intervened to promote industrial development by supporting capitalist monopolies (called **zaibatsu**). In many instances, whole industries were created from public funds and, once established, were sold off to private enterprise at less than cost. Because early manufacturing was motivated strongly by considerations of national security, it was iron and steel, shipbuilding, and armaments that were prominent in the early phase of Japanese modernization. The latest industrial technology and equipment were purchased and brought in from overseas, and advisers (chiefly British) were

(a)

(b)

(c)

(d)

Figure 2   Kyoto   (a) Wooden townhouses.
(b) Imperial Gardens and pavilion, Katsura Imperial
Villa. (c) Kinkaku-ji (Golden Temple). (d) Geishas.

brought in to supervise the initial stages of development. Meanwhile, the state indulged in high levels of expenditure on highways, port facilities, the banking system, and public education in its attempt to "buy" modernization. Similarly, the Japanese railway system was financed by the state under British direction before being sold to private enterprise.

The Japanese financed this modernization by harsh taxes on the agricultural sector. As a result, there began a sharp polarization between the urban and the rural economies, characterized by the impoverishment of large numbers of peasant farmers. Yet the more productive components of the agrarian sector were able to contribute significantly to Japanese economic growth. Improved technology, better seed stock, and the use of fertilizers provided a 2 percent increase in rice production each year during the last quarter of the nineteenth century and the first

part of the twentieth century, thereby helping to feed the growing industrial workforce without great dependence on food imports. It is important to note that these increases in agricultural productivity were not absorbed by population growth, as has been the case for many late-developing countries. The Japanese demographic transition arrived later—after increases in agricultural productivity had helped to finance an emergent industrial sector but in time to provide an expanding labor force and market for industrial products.

Several other factors helped to foster rapid industrialization in Japan in the late nineteenth and early twentieth centuries. First was the cultural order that allowed the Japanese to follow government leadership and accept new ways of life: a recurring theme in modern Japanese economic history. Second was the success of educational reforms: by 1905, 95 percent of

Figure 9.11 Admiral Perry's visit to Edo Bay Japanese drawing of the U.S. naval squadron in Edo Bay, 1853, under the command of Commodore Matthew C. Perry.

Just as Japanese industry was becoming established, with a base in textiles and shipbuilding, the First World War provided a timely opportunity to expand productive capacity. With much European and American industry diverted to supply war materials, Japanese textile manufacturers were able to expand into Asian and Latin American markets. Meanwhile, with a large portion of the world's merchant ships destroyed by the hostilities, Japanese shipping industries took the opportunity to expand their merchant fleet. The profits from this commercial activity paid for the rebirth of the Japanese navy, which by 1918 had 12 battleships and battle cruisers, with 16 more under construction. The United States at the time had only 14 battleships, with three under construction. Within 50 years of the Meiji revolution, Japan had joined the core of the world-system.

This pattern of progress was halted, however, by the stagnation of international trade that followed the stock market collapse of 1929 and the subsequent Great Depression. Once again, state intervention provided a critical boost. A massive devaluation of the yen in 1931 allowed Japanese producers to undersell on the world market, while a Bureau of Industrial Rationalization was set up to increase efficiency, lower costs, and weed out smaller, less-profitable businesses.

all children of school age were receiving an elementary education. Third, Japanese sericulture (silk production) provided the basis for a lucrative export trade with which to help finance expenditure on overseas technology, materials, and expertise (see Geographies of Indulgence, Desire, and Addiction: Silk, p. 437). It has been estimated that between 1870 and 1930 the raw silk trade alone was able to finance as much as 40 percent of Japan's entire imports of raw materials and machinery. Finally, and most important, were the spoils of military aggression (**Figure 9.12**). Naval victories over China (1894–1895) and Russia (1904–1905) and the annexation of Taiwan (1895), Korea (1910), and Manchuria (1931) not only provided expanded markets for Japanese goods in Asia but also provided indemnities from the losers (which paid for the costs of conquest) and stimulated the armaments industry, shipbuilding, industrial technology, and financial organization in general.

By the early 1900s, a broad base of industries had successfully been established from what had been a feudal economy in 1868—Japan's first "economic miracle." Most were geared toward the domestic market in a kind of preemptive import-substitution strategy. The textile industry, however, had already begun to establish an export base. Unable to compete with Western nations in the production of high-quality textiles, the Japanese concentrated on the production of inexpensive goods, competing initially with Western producers for markets in Asia. Their success was based on labor-intensive processes in which high productivity and low wages were maintained through a combination of exhortations to personal sacrifice in the cause of national independence and strict government suppression of labor unrest.

Figure 9.12 Japanese expansionism Japan deliberately pursued a policy of military aggression in the late 1920s and early 1930s in order to secure a larger resource base for its growing military-industrial complex. (*Source:* Adapted from I. Barnes and R. Hudson, *History Atlas of Asia.* New York: Macmillan, 1998, p. 129.)

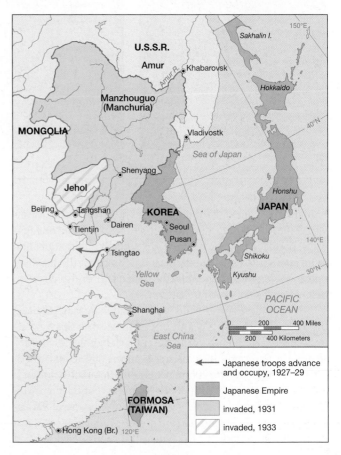

Silk

Silk has long been symbolic of opulence, luxury, and social status. From its origins in China, silk became one of the principal items of trade along the ancient trade routes between Asia and Europe. In Renaissance Europe, sericulture—the production of silk—and silk weaving were regarded as lucrative, leading-edge industries. Today, synthetic fibers have displaced lower-grade silks in most world markets, but good-quality silk fabrics remain highly desirable. Not only does their cost set them off as status symbols, but their light weight, sensuous feel, strength, and resistance to creasing make them a preferred indulgence of many of the world's more affluent consumers.

Silk is manufactured from the fibers of the cocoons produced by silkworms, which live on the leaves of mulberry bushes. Sericulture was first practiced in North China around 2700 B.C. and was well established in that region by the first century B.C. Silk was used as currency when Qin dynasty and Han dynasty envoys traveled in Central Asia and was the principal item carried by their merchants along the Silk Road (see Figure 9.10) through Central Asia and on toward India, Persia, and Rome. For centuries, China was the only country in the world capable of producing silk. The secrets of sericulture and silk manufacture were jealously guarded by early emperors: The penalty for attempting to export silkworm eggs or divulging the technique of silk weaving was death. Within China, though, silk fabrics came to be the embodiment of the refinement and sophistication of the elite.

Sericulture reached Japan through Korea by the third century, thus facilitating the first development of silk production outside China. Shortly afterward, sericulture was established in India. According to legend, two monks smuggled silkworms from China to Europe in the sixth century A.D. Promoted by the Byzantine Emperor Justinian I (A.D. 527–565), sericulture spread to Syria, Turkey, and Greece. The more prosperous cities of the Mediterranean, pursuing an import-substitution strategy, deliberately invested in silk manufacture, buying the raw silk from Greek and Arab merchants. By the end of the fourteenth century, silk production had been established in Bologna, Florence, Genoa, Lucca, and Venice. By the end of the fifteenth century, silk production played a vital economic role throughout the Italian peninsula, from the Alps to Sicily.

Chinese silk fabrics continued to be imported into Europe, however, because of their superior quality. Chinese governments under the Ming dynasty (1368–1681) vigorously promoted sericulture, and great quantities of Chinese silk were exported to Southeast Asia, Japan, and Spanish America as well as to Europe. By the eighteenth century, Britain led European silk manufacturing because of British innovations in silk-weaving looms, power looms, and roller printing. At the beginning of the nineteenth century, a Frenchman named Joseph Jacquard

Figure 1   Modern silk manufacture Silk factory in Suzhou, China.

developed a machine for figured-silk weaving that gave French manufacturers an edge. Then, with the opening of the treaty ports in the nineteenth century, exports of silk from China more than doubled. This was partly because of a silkworm disease that devastated sericulture in France and Italy in 1847, and partly because of the expansion of the world economy.

Silk was also important in Japan's entry to the modern world economy. With the Meiji revolution in 1868, the Japanese government organized a domestic silk industry, deliberately developing it with modern technologies as a major export sector. This enabled Japan to earn a significant portion of the foreign exchange it needed in order to purchase the machinery and raw materials necessary for its industrialization. Between 1870 and 1930, Japan's raw silk trade financed about 40 percent of its imports of foreign machinery and raw materials. Japan's success came at the expense of China. Silk continued to be China's major export until the 1930s, but the Chinese silk industry was undercapitalized and unresponsive to changing market conditions. Political instability, followed by Japanese invasions, further weakened the industry's effectiveness, and after 1949 it was relegated to a very low priority in Mao Zedong's Communist regime.

China's silk industry bounced back with the overall liberalization of the economy under Deng Xiaoping's leadership in the late 1970s. Since that time, China has once again become the world's main producer of silk (**Figure 1**), while world silk production has almost doubled, despite the fact that synthetic fibers have displaced silk for many uses. Together, China and Japan produce more than half the world's total annual production of silk fabrics.

Although these interventions helped to sustain Japanese industrialization and improve Japan's overall economic independence, they led directly toward crisis. Western governments—particularly the United States—began seriously to resist the purchase of Japanese goods. At home, the austerity resulting from devaluation and "industrial rationalization" precipitated social and political unrest. The government response was to increase military expenditures and to adopt a more aggressive territorial policy. In 1931 the Japanese army advanced into Manchuria to create a puppet state. In 1936 a military faction gained full political power and, declaring a Greater East Asian Co-Prosperity Sphere, began a full-scale war with China the following year. Japan attacked British colonies in the Far East in 1939, and by 1940 the Japanese had become heavily committed to an industrial empire based on war. By this time, as the rest of the world quickly realized, Japan had attained the status of an advanced industrial nation. The military leaders overplayed their hand, however, by attacking the United States at Pearl Harbor in December 1941. With the defeat of Japan in 1945, Japanese industry lay in ruins. In 1946, output was only 30 percent of the prewar level, and the United States, having begun to dismantle the *zaibatsu* and to impose widespread social and political reforms, was shaping up to impose punitive reparations.

Japan's Postwar Economic Miracle   Within five years, the Japanese economy had recovered to its prewar levels of output. Throughout the 1950s and 1960s the annual rate of growth of the economy held at around 10 percent, compared with growth rates of around 2 percent per year in North America and Western Europe. Having begun the postwar period at the bottom of the ladder of international manufacturing, Japan found itself at the top by 1963. By 1980, Japan had outstripped all of the major industrial core countries in the production of ships, automobiles, and television sets, and only the Soviet Union was producing more steel (**Figure 9.13**). The Japanese, in short, have not only achieved a unique transition directly from feudalism to industrial capitalism, they have also presided over a postwar "economic miracle" of impressive dimensions.

Explanations of this second economic miracle have identified a variety of contributory factors. Once under way, the reconstruction of the Japanese economy was able to draw on some of its previously established advantages: a well-educated, flexible, loyal, and relatively cheap labor force, a large na-

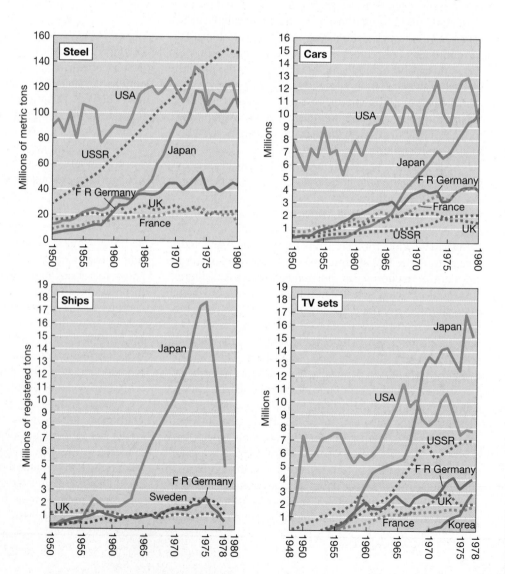

Figure 9.13   Japanese manufacturing growth   Between 1950 and 1980, Japanese manufacturing output grew at a phenomenal rate, outstripping other industrial countries by the early 1960s. The marked drop in Japanese shipbuilding in the mid-1970s was the result of the global economic recession that was caused by a fourfold increase in crude oil prices. (*Source:* P. Knox and J. Agnew, *Geography of the World Economy*, 3rd ed., London: Arnold, 1998, p. 175.)

tional market with good internal communications, a good geographical situation for trade within Asia, a high degree of cooperation between industry and government, and a mode of industrial organization—derived from the *zaibatsu*—big enough to compete with the multinational corporations of Western Europe and North America.

In addition, several new factors helped to transform reconstruction into spectacular growth. These included:

- Exceptionally high levels of personal savings (15 to 20 percent of personal disposable income through the 1960s and 1970s, compared with less than 5 percent in the United States), which helped to fund high levels of capital investment.

- The acquisition of new technology: Between 1950 and 1969 Japan was able to acquire, for around $1.5 billion in royalties and licenses, a body of thoroughly tested U.S. technology that had cost the United States $20 billion *per year* in research and development (R&D).

- New means of government support: On the one hand, a rigid and sophisticated system of import restrictions protected domestic markets from overseas competition. On the other hand, the growth of domestic industry was fostered by a multitude of tax concessions and by investment financing provided through the Japan Development Bank. Most important of all, however, was the orchestration of industrial growth by the Ministry of International Trade and Industry (MITI). In particular, MITI identified key recovery sectors (for example, steel and shipbuilding) and potential growth sectors (for example, automobiles, electronics, and computers) and facilitated their development by providing finance, ensuring protection from foreign competition, subsidizing technological development, and arranging corporate mergers.

The economic miracle achieved by Japan was remarkable not only for its overall success in terms of economic performance but also because it represented a unique pathway to development, one that was able to combine economic growth with income distribution. Real wages rose substantially, while income inequality was reduced to one of the lowest levels in the world. Equally remarkable was the interdependence of government and industry, characterized by some as "Japan, Inc." Orchestrated by the Ministry of International Trade and Industry (MITI), the state bureaucracy guided and coordinated Japanese corporations, organized in business networks (known as **keiretsu**), setting up favorable trade policies, technology policies, and fiscal policies to help Japanese industry compete successfully in the world economy. Also important was the exceptional degree of social stability and management-labor cooperation during this phase of tremendous change. This stability and cooperation was, like the interdependence of government and industry, a reflection of Japanese nationalism and people's commitment to rebuilding the nation. The same sense of national identity and purpose helped to foster people's adherence to traditional values and lifestyles, their willingness to work many more hours than their European

Figure 9.14 A Japanese family with their material possessions The Ukita family from Tokyo, photographed with their possessions outside their home in the mid-1990s, represents a statistically average Japanese family in terms of family size, residence, and income. Although Japan's overall economy has prospered remarkably since the Second World War, Japanese material consumption has been relatively modest, partly because of the cost of land and partly because of the Japanese habit of saving a relatively large portion of income.

and American counterparts, and their willingness to defer consumption (**Figure 9.14**), thus providing a pool of savings that could be invested in Japanese industry.

Revolutionary China

Revolution came to China in 1911, when the Qing dynasty was overthrown and replaced by a republic under the leadership of Sun Yat-Sen's Nationalist Party (or Kuomintang). The overthrow of the imperial government in 1911 was the outcome of a long economic decline and a series of military defeats. A rising generation of intellectuals believed that the "dead hand" of imperial bureaucracy had retarded Chinese responses to the West, and they searched for new philosophies from abroad—including capitalism, communism, liberal democracy, and socialism—as possible solutions to China's many problems. But before any progress could be made, the country fell into disarray, with contending warlords struggling for power. In the 1920s an alliance of China's fledgling Communist Party and the Nationalist Party, now led by Chiang Kai-shek, flourished for a while, but it ended abruptly in 1927 when Chiang Kai-shek, fearful of the growing power and ambitions of the Communist Party, attempted to quash their organization. The Communists organized a strategic retreat in what became known as the Long March (1934–1935). They traveled more than 9600 kilometers (5965 miles) from the southeast, through the rugged interior, to the plains of Yanan in northern China. During the march, in which more than 100,000 people perished, Mao Zedong emerged as the leader of the party. Mao devised a new strategy, aimed at gaining the support of China's rural peasantry—who made up 85 percent of China's population—as the major source of support for revolution. This approach was in direct contradiction to the preferred strategy of Soviet advisers, who had insisted that the key to revolution, as in Russia, must be the urban proletariat.

Japanese military advances into China in 1937 caused the Communists and the Nationalists temporarily to set aside their differences and resist the Japanese, but with the defeat of the Japanese in the Second World War, the Communists and the Nationalists resumed their internal conflict. It was the Communist forces who had fought hardest and suffered most against the Japanese, and their experience proved crucial in fighting the Nationalists. Having won the support of the vast peasantry, Mao Zedong's troops surrounded China's cities until, one by one, they fell to Communist control. By October 1949 the Communist Party had control over almost all of China except for the island of Taiwan, where the Nationalist leadership had retreated under U.S. protection.

Meanwhile, the end of the Second World War brought about an important geopolitical change to Korea. Under the sponsorship of Soviet troops that had moved south from Manchuria into the northern part of Korea, Kim Il Sung, an anti-Japanese Marxist-Leninist nationalist, was given power in 1948. Almost immediately, U.S. troops occupied the southern part of the country. Like East and West Germany, Korea was suddenly divided into two. The 38th parallel was agreed upon as the border, and both Soviet and U.S. military forces withdrew. But, in June 1950, just a few months after Mao Zedong's victory in China, Kim Il Sung's troops carried out a major attack on South Korea, seizing the capital Seoul and quickly extending control over almost the whole country. This led the United Nations to intervene, and U.S. forces rapidly rolled back the North Korean forces all the way to the Chinese border. This, in turn, prompted China to become involved on behalf of the North Koreans, and there followed a devastating war. Casualties were horrendous, with hundreds of thousands killed and much of the country destroyed. The 38th parallel was finally restored as the border in 1953, but the two Koreas have remained bitter rivals, and China's relations with most Western countries, especially the United States, were so seriously damaged that they did not recover until the 1970s. North Korea, meanwhile, has remained locked in a time warp of self-imposed isolation, with the first indications of a possible détente with South Korea appearing only in 2000 (see Sense of Place: North Korea, p. 441).

Back in China, Mao Zedong faced the task of reshaping society after two millennia of imperial control. The first years were tentative, with land reform and the formation of agricultural collectives as the principal objectives. In 1958, in what became known as the Great Leap Forward, Mao launched a bold scheme to accelerate the pace of economic growth. Land was merged into huge communes and an ambitious new Five-Year Plan was implemented. The fervor to industrialize led to the expectation that farmers would help to industrialize the countryside by building their own "backyard furnaces" capable of producing steel. The impact of the new Five-Year Plan on the landscape was dramatic. Whereas pre-Communist China had an average farm size of 1.4 hectares (3.5 acres), the new agricultural communes averaged 190 square kilometers (73 square miles) in size, with between 30,000 and 70,000 workers. Instead of a patchwork of fields, each with a different crop and presenting a rich palette of browns, yellows, and shades of green, there now appeared vast unbroken vistas, planted with crops dictated by the central planners.

China's planners, however, were concerned only with increasing overall production. They paid little or no attention to whether a need for the products existed, whether the products actually helped to advance modernization, or whether local production targets were suited to the geography of the country. The attempt to industrialize the countryside failed completely. Lacking iron ore, let alone any knowledge of how to make steel, peasants tore out metal radiators, pipes, and fences and sacrificed pots and pans in their zeal to produce steel in their backyard furnaces. Almost none of the final smelted product was usable. Meanwhile, several years of bad weather, combined with the rigid and misguided objectives of centralized agricultural planning, resulted in famine conditions throughout much of China. It is estimated that between 20 and 30 million people died from starvation and malnutrition-related diseases between 1959 and 1962. When the central economic leadership ordered all peasants to eat in large, communal mess halls, what was optimistically called the Great Leap Forward fizzled out. For people who valued family above all else, being deprived of time alone with their families for meals was the final disillusionment.

In an attempt to restore revolutionary spirit and to reeducate the privileged and increasingly corrupt Communist Party officials, Mao Zedong launched what he called a "Great Proletarian Cultural Revolution" in 1966. The Cultural Revolution brought a sustained attack on Chinese traditions and cultural practices and a relentless harassment of "revisionist" elites, the latter being defined broadly as anyone not belonging to the rural peasantry. Millions of people were displaced, tens of thousands lost their lives, and much of urban China was plunged into a terrifying climate of suspicion and recrimination. U.S. President Richard Nixon made a path-breaking visit to China in 1972 in an attempt to re-open China's relations with the Western world, but only with the death of Mao Zedong and the subsequent arrest of the politically extreme "Gang of Four" (which included Mao's wife) in 1976 did the Cultural Revolution came to an end. China's new leader, Deng Xiaoping, charted a more pragmatic course, gradually achieving stability and economic growth and opening China to Western science, technology, and trade.

In the aftermath of the Cultural Revolution, though, there remained little faith in Communist Party doctrine, while traditional values had been severely eroded. The result was that Western values of materialism, democracy, and individualism began to spread into the ideological vacuum along with a substantial revival of traditional Chinese culture. To combat the threat to established order, China's leadership therefore launched a series of mass campaigns—first against "spiritual pollution," then a repressive campaign against opponents of the ruling Communist Party (following the violent crackdown against protesters who challenged the legitimacy of the party in Beijing's Tiananmen Square in 1989), followed by a campaign against corruption, and then another to promote civil and respectful behavior. Despite all of this, China's leadership was able to keep the country on the path of economic liberalization, with the result that China's economy has been growing at double-digit rates for much of the past 20 years.

Under the leadership of Deng Xiaoping (1978–1997), China embarked on a thorough reorientation of its economy,

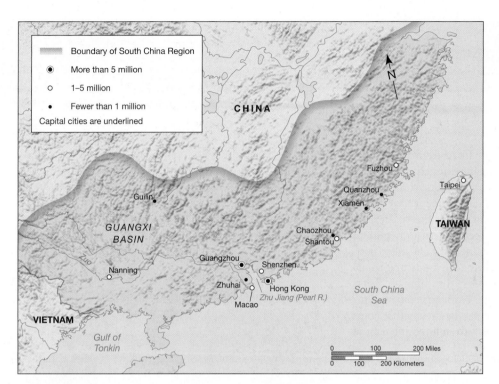

Figure 9.31 South China
Reference map showing principal physical features, political boundaries, and major cities.

The coastline of South China provides many protected bays suitable for harbors, and a series of important ports have developed, including Quanzhou, Shantou, Xiamen, and, on either side of the mouth of the Zhu Jiang, Macao and Hong Kong. These ports were a precondition for South China's emergence as a core region, providing an interface with the world economy. The established trade and manufacturing of Macao—a Portuguese colony that was returned to China in December 1999—and Hong Kong provided another important precondition for success. When Deng Xiaoping established his "open door" policy and set up the first Special Economic Zones, another important factor kicked in: capital investment from Hong Kong, Taiwan, and the Chinese diaspora. By 1993, more than 15,000 manufacturers from Hong Kong alone had set up businesses in neighboring Guangdong Province, and a similar number had established subcontracting relationships, contracting out processing work to Chinese companies. Today, the cities and special economic zones of South China's "Gold Coast" provide a thriving export-processing platform that has driven double-digit annual economic growth for much of the past two decades. Shenzhen (**Figure 9.32**) has grown from a population of just 19,000 in 1975 to 1.04 million in 2000, with an additional 2 million in the surrounding municipalities. The southern border of the Shenzhen Special Economic Zone adjoins the even more prosperous territory of Hong Kong, but the northern border is walled off from the rest of China by an electrified fence to prevent smuggling and to keep back the mass of people trying to migrate illegally into Shenzhen and Hong Kong.

Hong Kong   In 1950 the prospects for Hong Kong looked bleak. Japanese occupation during the Second World War had prompted 1 million of its prewar population of 1.6 million to disperse, along with many of the trading companies that had

been based there. The Communist revolution in China, the conflict in Korea, and Japan's derelict condition seemed to offer little prospect for the revival of a trading port. Yet today Hong Kong stands as a major world city, not only a thriving port city but a major manufacturing center and financial hub (**Figure 9.33**). Furthermore, it was able to grow into a metropolis of 6.1 million with few of the problems of inadequate infrastructure and illegal squatter housing that characterize the metropolises of the periphery.

One of the reasons for the transformation of Hong Kong was the arrival of refugee entrepreneurs from mainland China—and especially from Shanghai—as they fled the Communist takeover in 1949. With their expertise and their capital, together with Hong Kong's cheap labor, they established a thriving cotton textile industry, producing inexpensive jeans,

Figure 9.32 Shenzhen The population of the city of Shenzhen, just across the border from the Special Administrative Region of Hong Kong, grew from just 5000 in 1970 to more than 1 million in 2000.

Sense of Place

The Guangxi Basin

The Guangxi Basin merits attention for its distinctiveness as part of Inner China that has retained much of its traditional character. It is also a place where a unique combination of geomorphology and agricultural development have led to one of the world's most distinctive and spectacular landscapes. The Guangxi Basin is located in the southeastern part of Guangxi province, and is centered on the town of Guilin (population in 2000: 600,000). From classic old landscape paintings to contemporary tourist posters, the dramatic *karst* landscape (**Figure 1**) of the Guangxi Basin has come to embody a stereotypical traditional Chinese landscape that combines an intensive cultural landscape with an imposing natural landscape.

Karst landscape results from the work of underground water on the massive soluble limestone rocks of the region. Karst terrain is characterized by rocky outcrops, caves, potholes, sinkholes, and underground rivers. The term derives from the Karst region along the Dalmatian coast of Yugoslavia, but by extension has been applied to all areas of similar geomorphology. Around Guilin and in the Zuo River area to the south, near Nanning (population in 2000: 2.8 million), erosion has dissolved away areas of softer and more highly jointed limestone, leaving spectacular pinnacles several hundred feet tall. These improbably shaped hills contrast with the intermediate areas of intensely cultivated flatlands and winding rivers, making for some strikingly beautiful landscapes. The topography of the region, together with its relatively isolated lo-

cation toward the southern margin of China and the absence of important coal, oil, or mineral deposits, has left traditional ways of life relatively undisturbed. This, of course, adds to the scenic qualities of the area and has resulted in a significant volume of foreign visitors. The region is heavily populated and produces large crops of rice, together with cash crops of sugarcane and hemp.

Figure 1 Guilin landscape

T-shirts, and leisure wear for the expanding markets in the United States and Western Europe. The British governors of Hong Kong assisted this economic revival by keeping taxes and regulations to a minimum and by investing in the infrastructure of roads and port facilities. Meanwhile, in a rather unusual mixture of socialism with free enterprise, the government undertook a massive program of housing development. Squatter housing and slums were cleared, and hundreds of high-rise apartment blocks were built for the city's growing population, many of them with highly subsidized rents. Following the initial success of the housing program in the city itself, the government followed another British town-planning initiative, setting up several New Towns in the New Territories to the north of the city.

The revitalized Hong Kong proved attractive to investors, and the city's entrepreneurs moved from producing cheap apparel to subcontracting for designer wear and producing cheap electronics. As the Asian region began to prosper and markets expanded, Hong Kong became a major center for finance,

banking, and tourism. In the mid-1990s, there was some nervousness about the fate of the city on its return to China with the expiry of the 99-year lease that the British had signed with China in 1898. Nevertheless, it did not stop the development of a major new airport or significantly stem the pace of development. In the event, China saw the wisdom of maintaining Hong Kong's position as a capitalist dynamo and agreed to create a Special Administrative Region for Hong Kong. This arrangement preserves Hong Kong's legal system, guarantees the rights of property ownership, gives residents the right to travel, permits Hong Kong to continue independent membership in international organizations, and guarantees the democratic rights of assembly, free speech, academic research, the right to strike, and so on. The result is that Hong Kong continues to prosper, even after the 1998 financial crisis that affected most of East and Southeast Asia (and which sharply reduced the number of tourists from Japan and South Korea). Thousands of companies are located in Hong Kong for the purpose of doing business with China. Although most of its

(a)

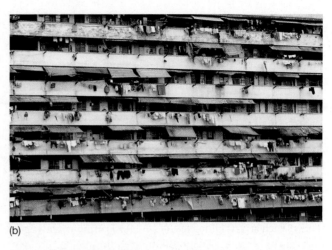

(b)

(c)

Figure 9.33   Hong Kong   Returned to China by the British at the expiration of their lease in 1997, Hong Kong remains a vibrant world city of critical importance to the economic development of South China. (a) The financial district and Hong Kong harbor. (b) Apartment building, Kowloon, Hong Kong. (c) Houseboats in Hong Kong harbor.

manufacturing has been transferred to neighboring Guangdong Province, where wages are much lower, Hong Kong remains the world's largest container port, the third-largest center for foreign exchange trade, the seventh-largest stock market, and the tenth-largest trading economy. Considering its size and its history, these are extraordinary achievements.

Japan's Pacific Corridor

Japan is notably lacking in natural resources for large-scale manufacturing industries and the country's "economic miracle" of the 1960s was founded on the import of raw materials and the manufacturing and export of finished products. Japan's resurgent industries thus flourished best in coastal locations, close to deep-water ports. The Pacific Corridor, between Tokyo and Kobe (**Figure 9.34**), developed into the core region of modern Japanese industrialization because it not only had several important deep-water ports but also large pools of skilled labor and relatively large amounts of flat land. The entire region has developed into a megalopolitan area, known as the Tokaido Megalopolis, that is comparable in size and scope to the megalopolitan area in the United States that stretches from Boston through New York and Philadelphia to Washington. The Tokaido Megalopolis contains more than 50 million people and accounts for more than 80 percent of Japan's total GDP. In 1999, Osaka alone accounted for a GDP that was greater than those of all but eight countries in the world.

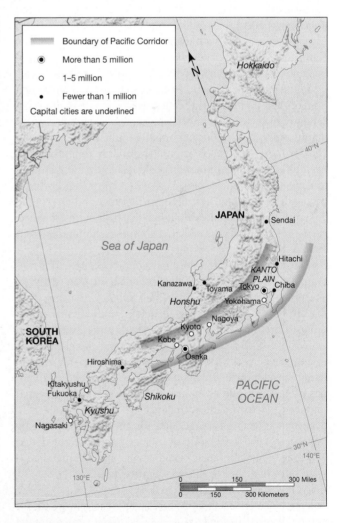

Figure 9.34   Japan's Pacific Corridor   Reference map showing principal physical features, political boundaries, and major cities.

Transportation has been a key factor in the successful development of the Pacific Corridor. Though port facilities were a precondition for the region's success, internal connections within the region were poor, making it difficult for manufacturers and suppliers to exploit agglomeration economies and restricting the movement of workers and consumers. In response, the Japanese government undertook a massive program of infrastructure investment. The showpiece of this program is the Shinkansen railway system (**Figure 9.35**). First opened in 1964 to coincide with the Tokyo Olympic Games, the "bullet trains" of the Shinkansen were for a long time the fastest in the world. The Shinkansen has turned the entire Pacific Corridor into a daily commuter belt, with trains carrying more than 1400 passengers each at 220 kilometers (125 miles) per hour, leaving at intervals of between 10 and 30 minutes throughout the day from each of the major cities of the Corridor.

The Pacific Corridor's prewar industrial base was dominated by cotton, silk, other textiles, toys, glass, and porcelain. These industries are still present in the region but their importance has been dwarfed by the growth of iron and steel, heavy metal products and machinery, shipping and shipbuilding, petrochemicals, paper products, ceramics, automobile and truck manufacture, cameras, scientific instruments, and electrical and electronics goods of all kinds. Tokyo has grown into a world city of the first rank, with a banking and financial sector that compares to those of London and New York. The population of the Tokyo metropolitan area in 2000 was 28.03 million. Nearby Yokohama (population 3.31 million) is first and foremost a port city and manufacturing center. Nagoya (3.38 million) is a center of heavy engineering, chemical, textile, and machinery manufacture and has an important commercial port; Kyoto (1.70 million) is a major cultural and tourist center but also has important electronics industries, including the manufacture of Nintendo games; Osaka (10.61 million) is a deep-water port with an industrial base of comparable size and diversity to Tokyo's but without Tokyo's international banking and finance; Kobe (1.42 million) is a center of shipping, shipbuilding, engineering, and manufacturing.

Most of these cities were almost entirely flattened by firebombing toward the end of the Second World War, with the exception of the old imperial capital, Kyoto. Unfortunately, it has not been possible to take advantage of building afresh in order to produce either more efficient or more livable cities. Growth has been so rapid, and Japanese land-use planning so weakly developed, that the entire region has the air of haphazard development. Tiny houses are wedged between tall hotels or modern offices or in the shadows of busy elevated highways; factories and warehouses sit amidst private residences or next to gleaming office towers; and the sprawling mix is festooned with ugly power lines, punctuated by netted golf ranges, tall smokestacks and cranes, and decorated with giant billboards and neon signs (**Figure 9.36**).

Tokyo The central city, or "ward area," of Tokyo extends for approximately 15 kilometers (9.3 miles) from the center of the city, covering an area of just under 600 square kilometers (230 square miles) and containing a population of more than

Figure 9.35 Shinkansen There are 115 round-trip journeys per day on the 515-kilometer (320-mile) Pacific Corridor route between Tokyo and Osaka.

8 million. Situated at the head of Tokyo Bay, the heart of the city is marked by the extensive grounds of the Imperial Palace, around which are clustered the main railway station and the office towers of the central business district (**Figure 9.37**). The scarcity of land, together with the expense of building safe high-rise buildings in an earthquake zone, has driven growth outward, with concentrations of offices, retailing, and industry along the major transport arteries and at nodal points where railways and subway systems intersect. The Tokyo metropoli-

Figure 9.36 Urban sprawl Japanese land-use planning has been notoriously weak, resulting in a high-density sprawl. Shown here is part of the Tokaido Megalopolis, with Mount Fuji in the background.

Figure 9.37   Tokyo With a population of more than 28 million in 2000, the Tokyo metropolitan area is highly congested. (a) Tokyo's central business district. (b) Commuter bicycles parked at a suburban railway station. (c) The Shinjuku district. (d) Apartment building in Yokohama.

tan area extends for up to 50 kilometers (31 miles) from the center of the city, containing a population of 28.03 million.

Geographer Roman Cybriwsky captures the magnitude of the city in his description of the view from Tokyo Tower:

> Perhaps the most striking impression . . . is of the immense size of the city. . . . The built-up area extends for well over 50 kilometers in some directions, much further than one can pick out in detail, and seems almost limitless. What is more, almost all of this enormous territory is extremely densely built up. A great many of the buildings seen are high- or mid-rise structures, at least five or six stories tall, and there is almost no empty space between them. . . . The buildings push right up against the edges of open spaces and form high, thick walls that define their boundaries pecisely and enclose them almost completely. . . .

> One is also taken aback by the incredible profusion of geometric shapes and building sizes that constitutes the rising mass of the new Tokyo. Some of the new developments are quite large and stand out as megastructures

that tower over their respective neighborhoods and impose on them a new authority. The Manhattan-like skyline of Shinjuku, seen in the middle distance to the northwest, is one example. . . .

> On the other hand, many of the other new buildings are so slim that they remind me of credit cards standing on end, or even pencils on end. Many are just barely wide enough to have elevators or stairways, and in quite a few structures the elevators are tiny and the stairways are affixed to the outside.[3]

Within this vast metropolis are all kinds of industries and occupations. Tokyo is important in almost every sphere of urban economic activity, and overwhelmingly dominant, within Japan, in most. The metropolitan area accounts for more than 25 percent of Japan's population but handles more than 80 percent of the country's imports and exports (by value), and accounts for about 60 percent of the country's employment in business services. In addition to the many government agencies, half of all

[3] R. Cybriwsky, *Tokyo,* New York: John Wiley & Sons, 1998, pp. 35–37.

Japan's businesses, cultural organizations, and media firms are located in Tokyo. More than 1 million college students study in the several dozen colleges and universities in the metropolitan area; and almost 90 percent of the foreign banks and transnational corporations have their offices in Tokyo. All this translates into an extremely high-energy environment. Crowds seem to be everywhere: even at 11 P.M. on a weekday evening, there is often standing room only on downtown subways. Throughout the central city, road traffic moves slowly, if at all.

If the modern city can be said to have an identifiable structure, then it is structured around the train and subway network that has evolved to cope with the millions of commuters who must pour into the central city each weekday. Where rail lines from the outer suburbs intersect with the circular Yamanote Line and the subway stations that serve the central city, nodes of office and retail development have sprung up, so that the central business district (CBD) is encircled by a series of secondary (but still very large) business districts, each at a distance of between 3 and 5 kilometers (1.9 to 3.1 miles) from the CBD. Among the more important are Shinjuku, Shibuya, Ikebukuro, Shinagawa, and Ueno. Many commuters cycle from their neighborhoods to a suburban train station (Figure 9.37) and change at one of these secondary centers to catch a subway to work. At the end of the day, they will likely break their return journey at one of these secondary centers to shop in one of the lavish department stores, to eat, or to relax in a bar. All of these centers are vibrant with people and blazing with neon long into the night.

In contrast, the suburbs, though the housing is tightly packed, are relatively quiet. The suburbs are, however, very diverse in character. Some are effectively industrial towns in their own right; some are dominated by universities; some are dominated by company housing or public housing; and some are planned new towns with high-tech industries and research institutes. Some are exurban settlements, their residents taking advantage of the high-speed bullet trains that allow them to purchase larger homes amid paddy fields, orchards, or tea plantations, with mountainous backdrops. Two things are particularly striking about all of these suburbs in comparison to the suburbs of North American metropolises: first, the small size and high density of people's homes; and second, the relative lack of spatial segregation along lines of income or class. Even at the scale of individual streets, housing is often very mixed, with very modest homes adjacent to those of more affluent families. In part, this is the result of the absence of the kind of land-use planning and zoning regulations that are common in North American cities. More important though, is a broader cultural difference: The refined understatement and strong sense of collective identity that have traditionally been valued within Japanese culture have led people to avoid ostentatious residential segregation.

Taiwan

Taiwan's growth as a core region within East Asia is a result of a special combination of factors. When the Communist revolution created the People's Republic of China in mainland China, the Nationalist government that established itself in Taiwan (then called Formosa) as the Republic of China took responsibility for orchestrating economic growth from the very beginning. Land reform was an important first step, along with the imposition of strict currency controls, the creation of government corporations in key industries, and strong trade barriers to protect domestic industries from foreign competition. An authoritarian regime was imposed on domestic society in order to suppress opposition to government policies. Economic aid from the United States, provided because of Cold War geopolitics, was also important in priming Taiwan's economy. By the early 1960s, Taiwan's political stability and cheap labor force provided a very attractive environment for export processing industries. Taiwan lost its full international diplomatic status in 1971, when U.S. President Richard Nixon's rapprochement with the People's Republic led to its entrance into the United Nations. In 1987 Taiwan's government lifted martial law, began a phase of political liberalization, and relaxed its rules about contact with mainland China. Meanwhile, mainland China still claims Taiwan as a province of the People's Republic and has offered to set up a Special Administrative Region for Taiwan, with the sort of economic and democratic privileges that it has given Hong Kong.

Taiwan's geopolitical problems, however, have not prevented it from achieving great economic success. With a land area of 36,000 square kilometers (13,900 square miles), only 25 percent of which is cultivable, and a population of 22 million, Taiwan would be the second smallest of China's provinces and seventh from the bottom in population. Yet Taiwan's per capita GNP in 1997, $13,230, was 15 times that of mainland China. In 1999 Taiwan was the world's thirteenth largest exporter, with a total export trade of $120 billion. Taiwan's economic growth rate over the past three decades has been phenomenal, averaging more than 8 percent per year. Sometimes referred to as "Silicon Island," Taiwan has some 1.2 million small- and medium-sized enterprises. Most make components, or entire products, according to specifications set by other, often well-known international firms, whose brand names go on the final product. Because Taiwan's firms tend to be small, they have been able to be flexible in responding to changes in technology. They have also been able to move quickly to take advantage of China's "open door" policy. Between 1990 and 2000, Taiwanese business people invested more than $100 billion abroad, the greater share of it in mainland China. In South China and in Fujian Province, across the Taiwan Strait from Taiwan, Taiwanese enterprises operate what amounts to a parallel economy.

Taiwan's economic success as one of the newly industrialized "Asian Tigers" is not without growing pains. In Taipei (population 2.89 million in 2000) and other cities, Taiwan's breathtakingly fast modernization has brought heavy pollution, acute housing shortages, and rampant corruption. The rapid acquisition of automobiles, scooters, and air conditioners has made the environment unbearable and transportation a nightmare. Growth has been so rapid that the government has been unable to solve the problems of water supply and waste disposal, and "garbage wars" over the issue of sanitary landfill placement have occasionally led to huge quantities of uncollected garbage. Within Taipei (see Sense of Place: Taipei, p. 468), the capital and the industrial and commercial heart of Taiwan, these problems have been intensified by an influx of young people from other parts of the island, drawn by the educational and economic opportunities and by the exciting sense of rapid change.

South Korea

Like Taiwan, South Korea has become one of the newly industrialized "Asian Tigers" through a combination of an authoritarian regime that orchestrated economic development through land reform, the protection of domestic industry, and the creation of state enterprises; massive inputs of foreign aid (again because of the country's geopolitical importance); and the presence of a disciplined and well-educated but low-wage workforce.

As in Taiwan, South Korea has moved through successive stages toward an increasingly balanced and liberalized economy. The initial emphasis was on import substitution, developing domestic industries, with government protection, to produce goods for the domestic market. The second stage, between the mid-1960s and mid-1970s, saw the growth of export-oriented, labor-intensive manufacturing. The South Korean government facilitated the development of these export industries by providing incentives, loans, and tax breaks to firms, and by encouraging the growth of giant, interlocking industrial conglomerates called **chaebol**. The success of this strategy was reflected in annual growth rates in the overall economy of more than 10 percent. In the late 1970s, South Korean economic planners decided to diversify the economy with greater emphasis on heavy industry, chemicals, automobile assembly, and shipbuilding. To do this, the government worked with the *chaebol*, which were best placed not only to maximize internal and external economies of scale but also to acquire new technology from transnational partners. By the mid-1980s, about a dozen *chaebol* had come to dominate the economy, employing the majority of the workforce, controlling the banking system, and dominating government economic policy. Manufactured goods accounted for 91 percent of total exports and more than half were in the form of ships, steel, and automobiles. During the late 1980s and 1990s, the economy diversified still further, with the manufacture of semiconductors and electronics and the emergence of telecommunications and information processing. Giant South Korean conglomerates like Samsung and Hyundai became household names around the world, and the economy once again grew at more than 10 percent each year. South Korea was hit particularly hard, however, by the Asian financial crisis of 1998: The preceding boom had led South Korean banks to make loans to their parent companies for investments throughout Asia and, when the bubble burst, many of the investments had to be written off. South Korea's economy had to be propped up with loans and guarantees from the International Monetary Fund, which in turn has required the liberalization of the economy, including curbs on the power and influence of the *chaebol*.

Figure 9.38   Seoul South Korea's capital city has more than doubled in size since the early 1970s, and now has a population in excess of 12 million.

In spite of the shock of the 1998 crisis, South Korea remains a core region within East Asia. With a total export trade of $135 billion in 1999, South Korea is the world's twelfth largest exporter. In terms of its overall economy, South Korea had a GNP of just under $500 billion, also ranking twelfth in the world. Seoul (**Figure 9.38**) has become one of the world's largest metropolitan areas (it was eleventh largest in 2000, with a population of 12.2 million) and is a fast-paced city with a broad economic base. Other important centers are Pusan (population 4.2 million in 2000), an international port and major industrial center specializing in automobile production, electronics, chemicals, iron and steel, and shipbuilding; Inch'on (2.8 million), an entrepôt for Seoul and a center of electronics and iron and steel manufacture; Taegu (2.6 million), a textile manufacturing center; and Kwangju (1.7 million) and Taejon (1.4 million). As in Taiwan, the pace of economic growth and urbanization has brought heavy pollution, acute housing shortages, and rampant corruption. Socioeconomic inequality is also a characteristic feature of South Korea, both within the cities and between the country's subregions. The government has attempted to improve equality among the subregions through spatial-planning policies that have directed industrial growth away from Seoul, toward provincial towns and cities. Currently, the major priorities of domestic policy are aimed at keeping up with the need for basic infrastructure improvements in order to maintain economic efficiency, and keeping up with educational spending in order to be able to further develop a high-tech sector. Government spending on social and environmental issues remains minimal, however, so that there is little immediate prospect of a reduction in social inequalities or of improvements in the quality of life for the many millions who live in cramped housing and degraded environments.

Distinctive Regions and Landscapes of East Asia

The most fundamental geographical division within East Asia is between those regions where natural landscapes are dominant and those where the imprint of human activity has significantly altered the vegetation and landscape. Natural

Sense of Place

Taipei

It is hard to believe that Taipei was once a small town surrounded by a wall, with gates that were closed at night. Two of the old city gates survive, but they are lost in a sprawl of hasty development. Swollen by refugees from the Communist revolution of 1949, the city already had a population of more than half a million in 1950. By 1975 the metropolitan area had grown to 1.84 million, and by 2000 it had grown to 2.89 million. It is a busy city, but in spite of the crowding, congested roads, and heavy pollution, it retains a friendly atmosphere (**Figure 1**).

The rapid growth of the city, coupled with Taiwan's equally rapid transition from an underdeveloped agrarian economy to a newly industrialized "Asian Tiger," has made for some sharp juxtapositions of land use, with luxury condominiums alongside squatter housing and modern offices next to a noisy, smoky factory. Often there is no separation of work and living space: People frequently use their homes as workshops, and many still work on the sidewalks right in front of their homes. Also characteristic is the intermingling of smells, cooking aromas from homes and street vendors compete with smelly pollution from traffic and frequent wafts of decaying sewage.

The built environment also reflects the consequences of Taipei's hasty growth. Amid the sprawl of the city there are no spacious parks and few imposing buildings. There are pockets of surviving wooden buildings from the nineteenth and early twentieth century, and there are some colonial-style buildings left over from the Japanese occupation of Taiwan

(1895–1945), most of them inspired by central European baroque and renaissance styles. After 1950, the rapid growth of the city was initially accommodated in 3- and 4-story steel-framed, brick-faced apartment houses. From the late 1970s, the pressure of growth resulting from Taiwan's economic success resulted in a building frenzy that produced medium-rise buildings with modern but unattractive facades. The result was to produce noisy canyons that tend to trap the pollution from the increasingly heavy traffic. In few cities around the world can so many people be seen wearing paper face masks in an attempt to avoid air pollution. Traffic congestion, along with an acute shortage of decent and affordable housing, are the city's two biggest problems, though the city also faces significant problems of water supply, drainage and flooding, sewage treatment and garbage collection, and ground subsidence. Nevertheless, the city carries a fast-paced air of energy and excitement. The city's younger and more affluent residents are sophisticated consumers of global products, and a significant proportion of their disposable income is spent on personal appearance: Exclusive European product lines are well represented in the city's department stores. Fast food franchises—including McDonald's, Hardee's, Kentucky Fried Chicken, Pizza Hut, and Baskin Robbins—provide hangouts for teenagers, but the city also has a broad range of restaurants with a more sophisticated international cuisine, and there remain a large number of traditional tea houses, many of which cater to customers who are connoisseurs.

Figure 1 **Taipei** (a) Pedestrian street, downtown Taipei. (b) Taipei inner city.

(a)

(b)

(a)

(b)

(c)

Figure 9.39 Mongolia A vast country, Mongolia has remained relatively untouched by modern economic development. (a) Pastoral landscape with traditional white *yurts*. (b) Ulan Bator. (c) Gobi Desert.

landscapes prevail in much of Outer China, where agricultural landscapes and the cultural imprint of villages, monasteries, shrines, and temples are small, widely scattered elements engulfed by overwhelming nature. Among the distinctive landscapes here are the high plateaus and deep gorges of the Tibetan Plateau, the expansive steppes of Mongolia, and the barren sand dunes of China's desert northwest. The imprint of humankind prevails in Inner China, the Korean peninsula, the Japanese archipelago, and Taiwan. The landscapes of these more populated parts of East Asia can, as we have seen, be divided broadly into the landscapes of the subhumid regions to the north of the Qinling and Daba ranges and those of the humid and subtropical regions to the south. In the north, the traditional landscapes of China were described by Yi-Fu Tuan as follows: ". . . large areas of level land, with hills; bare fields and brown, dust-laden scenes in winter; dry crops of wheat, millet, and sorghum; dirt roads for two-wheeled carts and draught animals; mud-walled houses; cities with wide streets, open spaces for agriculture and rectangular walls." In the south, Tuan described traditional landscapes with "a far more rugged topography; landscapes that are green at all seasons and watery scenes; evidences of intensive land use, the most prominent of which are the sculptural forms of wet-rice terraces; flagstone trails . . .; houses of brick and woven bamboo walls; crowded cities with narrow streets."[4] Most of these tra-

ditional landscapes have been further modified in recent decades by the imprint of industrialization and urbanization and have become absorbed into the orbit of the core regions and key cities described earlier.

Mongolia

The People's Republic of Mongolia is very sparsely populated, with a population of just 2.4 million spread over 1.57 million square kilometers (606,177 square miles). Almost 40 percent of the country's population lives in the capital, Ulan Bator (**Figure 9.39**), and another 19 percent lives in smaller towns and cities, leaving about 1.4 million scattered across the steppe landscapes that stretch for more than 1500 kilometers (932 miles) from east to west. As in the steppe lands of Central Asia (to the west of the Altay Mountains), there is a gradation of steppe landscapes that ranges from the wooded steppe through the steppe proper and desert steppe. Wooded steppe, where rolling grasslands are interspersed with stands of mixed woodland, accounts for 23 percent of Mongolia, mostly in valley bottoms and in the scenic basin complex of northern Mongolia known as the Great Lakes region. The wooded steppe quickly shades into the steppe proper, which accounts for a further 27 percent of the territory in Mongolia. Here, the landscape is flat, treeless, and overwhelmingly dominated by tall, hardy grasses. Here and there, small, stubby massifs mark the clearly discernible cones of extinct volcanoes. South of the steppe are equally extensive zones of desert steppe characterized by immense sweeps of boulder-strewn wastes and salt-pans, with patches of rough vegetation used by nomadic pastoralists. Most of the remaining 22 percent of Mongolia is taken up by the Gobi Desert (Figure 9.39), where bare rock and extensive sand dunes predominate. Several spectacular natural features are found in the Gobi, including huge basalt columns arranged in clusters resembling clusters of pencils.

For centuries, the steppes of Mongolia have been the realm of nomadic peoples. The current population consists

[1]Y-F. Tuan, *China*, London: Longman, 1970, p. 149.

mostly of Khalkh Mongols, significant numbers of whom still choose to pursue a nomadic lifestyle, living in felt tents called *yurts,* or *gers* (Figure 9.39). From 1924 until 1989, Mongolia had a Soviet-style economic system, under which the backward-looking, rigid, and hierarchical system (dominated by Mongolian Tibetan Buddhist religious leaders and various local khans and princes) was replaced by large state farms and co-operative trading enterprises. As agricultural production became more settled, clusters of *gers* became larger, concentrated around the more permanent dwellings and structures of the state farms. The landscape was also changed significantly by an extensive virgin lands program similar to that in the Soviet Union (see Chapter 4, p. 193), whereby hundreds of thousands of acres of virgin land were plowed up and large-scale irrigation systems were installed in order to allow for arable farming—especially wheat, root vegetables, and fodder crops. Meanwhile, Mongolia's extensive mineral resources (including oil, coal, copper, lead, and uranium) began to be exploited, and an industrial sector emerged, stimulating urban growth. In 1990 Mongolia abandoned communism, and subsequent governments have begun to establish links with Japan and other capitalist countries and to rehabilitate traditional Mongolian culture.

China's Desert Northwest

Northwestern China is a remote and desolate region that comprises some 22 percent of China's total land area but contains only about 2 percent of its population. Surrounded by mountain ranges, China's desert northwest includes all of the Xinjiang Uygur Autonomous Region, together with adjacent parts of Gansu Province and the Inner Mongolian Autonomous Region (**Figure 9.40**). Vast deserts stretch for thousands of kilometers before ending abruptly at the foot of towering mountain ranges, with most of the population concentrated in the cities and oases along the **spring lines** at the front margin of the higher plains.

A dominant feature of the region is the Tarim Basin, a huge depression whose center is occupied by the Taklimakan Desert. About half of the Basin's 331,000 square kilometers

(127,799 square miles) is yellowish sandy desert, and the other half is covered by a grayish depositional gravel desert. Summer temperatures in the Tarim Basin are exceptionally hot (35° to 45°C, or 95° to 113°F), while winter temperatures have been known to reach −51.5°C (−60.7°F). Strong winds are characteristic of the region in both summer and winter, making it extremely inhospitable. Most rivers and streams flowing from the surrounding mountains into the Tarim Basin die out in the desert floor of the Basin. The Tarim River (fed mainly by the Aksu River, which originates in the Tien Shan Mountains), used to terminate in a series of "wandering lakes"—Lop Nur, Karakoshun, and Taitema—all of which have dried up since 1972. The salt lake of Lop Nur is where the Chinese military tests its nuclear devices.

Beyond the Tarim Basin are extensive areas—more than 50 million hectares (193,050 square miles)—of desert pasture, montane grasslands and alpine meadows that have for centuries supported the region's nomadic populations of Uighurs, Kyrgyz, Uzbeks, Tajiks, and others. A marked vertical zonation means that transhumance dominates the lives of the nomadic pasturalists, with livestock (mainly sheep, goats, horses, and cattle) moving up the mountain slopes to make good use of alpine meadows and mountain grassland in the summer and fall, and moving back down to desert basins for winter and spring grazing (**Figure 9.41**). Around the oases that follow the spring lines, irrigation allows for a certain amount of commercial grain growing, together with market gardening and some fodder crops for livestock. The towns that sprang up at these oases include Gaochang, Kashgar, Korla, Kuqa, and Yining, all of which were important stops along the ancient Silk Road. Most of their past glory was stripped by nineteenth-century expeditions from Britain, France, Germany, Russia, and Japan and now resides in museums and private collections around the world. Gaochang itself is now a ruin, its massive walls the last remaining testament to its former importance.

Since 1949, the entire region has been flooded with Han settlers as China has sought to exploit the region's valuable oil, coal, nickel, and copper resources and to bring the region more closely under the control of the Beijing government. This has led to significant ethnic tensions (see page 447). About

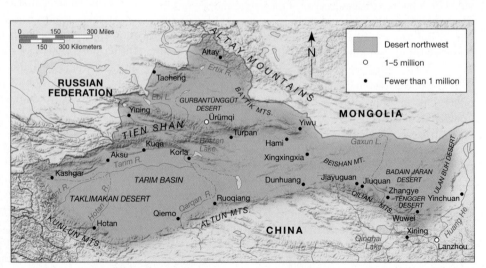

Figure 9.40 China's desert northwest Reference map showing principal physical features, political boundaries, and major cities. (*Source:* Adapted from S. Zhao, *Geography of China.* New York: John Wiley & Sons, 1994, p. 275.)

Figure 9.41 Nomadic pastoralists A goat herd on the Pamir Plateau, Tashikurgan, Xinjiang.

80 percent of the population of Urümqi, the capital of Xinjiang (population in 2000: 1.5 million), are Han Chinese. The city itself bears the imprint of Han colonization, its drab concrete-block architecture a mirror image of the industrial cities of Inner China. In contrast, towns such as Turpan (population 170,000 in 2000) that are dominated by Uighurs and other non-Chinese present very different townscapes. While town centers tend to be dominated by concrete-block architecture, most of the narrow streets that comprise the rest of the town are shaded by grapevine trellises and lined with mud-brick walls that enclose thatch-plaster houses.

The Tibetan Plateau

The Tibetan Plateau occupies about one-fourth of China's total land area. It is a unique physical environment, a vast area that has been violently uplifted in relatively recent geological time to produce the youngest, highest plateau in the world. The elevation of the entire plateau is at least 2500 meters (8202 feet), rising to between 4500 and 5000 meters (14,763 to 16,403 feet) in the northwestern part of the plateau. It is surrounded by a series of lofty mountains that tower to heights of between 6000 and 8000 meters (19,684 to 26,245 feet): the Himalayas on the southwestern border, the Kunlun and Karakoram mountains on the northwestern border, the Hengduan Mountains on the southeastern border, and the Altun and Qilian mountains to the northeast (**Figure 9.42**). In addition, there are many imposing mountain ranges inside the plateau. The climate is harsh, with extreme variations in seasonal and daily temperatures. Glaciers cover a total of more than 47,000 square kilometers (18,147 square miles), much of the landscape is shaped by the raw features of glacial erosion and deposition, and permafrost is widespread. This "roof of the world" is also the source area for many of the great rivers of East Asia and South Asia, including the Chang Jiang, the Huang He, the Mekong, the Indus, and the Yarlung Zangbo, which forms the upper reaches of the Brahmaputra.

Not surprisingly, the rugged terrain of the Tibetan Plateau is extremely sparsely inhabited. Parts of the region were occupied by nomadic peoples such as the Chang and the Ti as early as the second century B.C., while Han Chinese began to move into the margins of the region with the expansion and unification of the Chinese empire between the seventh and tenth centuries A.D. By that time, Tibet had developed a distinctive culture based on Tibetan Buddhism—developed from a fusion of the region's ancient animistic religion (known as Bon) with Tantric Buddhism imported from India. During the Ming (1368–1644) and Qing (1644–1911) dynasties, the entire Tibetan Plateau became part of the Chinese empire. Far away from the center of power, Tibet was divided into numerous small tribes and districts, with feudal chieftains ruling autocratically, generation after generation. As a result, Tibet became a highly repressive theocracy based on serfdom. In 1950 China seized on this fact as justification for its invasion of the strategically important region. Between 1950 and 1970 the Chinese "liberated" the Tibetans, drove their spiritual leader, the Dalai Lama, into exile, destroyed much of Tibetan cultural heritage, and caused the death of an estimated 1 million Tibetans. Today, about 60 percent of the population of the region is of Han Chinese descent. As we have seen (p. 447), these developments have given rise to significant ethnic tensions within the region, while the exiled Dalai Lama has acquired global celebrity as a result of his international tours promoting Buddhism and publicizing the "cultural genocide" committed by China in Tibet. The Dalai Lama does not want national independence for Tibet, but greater autonomy. For their part, the Chinese cannot understand the ingratitude of the Tibetans. As the Chinese see it, they have saved the Tibetans from feudalism and built roads, schools, hospitals, and factories.

Nomadic pastoralism remains the chief occupation within the Tibetan Plateau, the yak being the principal source of livelihood, along with sheep and goats. Yaks supply milk and

Figure 9.42 The Tibetan Plateau region Reference map showing principal physical features, political boundaries, and major cities. (*Source:* Adapted from S. Zhao, *Geography of China.* New York: John Wiley & Sons, 1994, p. 298.)

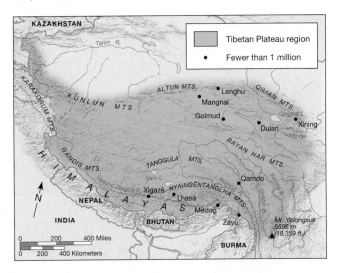

(a)

(b)

(c)

Figure 9.43 The Tibetan Plateau (a) A settled valley in the Tibetan Plateau. (b) Traditional Tibetan tent. (c) Potala Palace, the former seat of the Tibetan government and winter residence of the Dalai Lama.

wool as well as meat, and yak wool is the fabric used for the darkish, rectangular Tibetan tent that is a distinctive component of the Tibetan landscape (**Figure 9.43**) and can be distinguished significantly from the whitish, dome-shaped sheep-wool tents of the Mongolian Plateau (Figure 9.39). Chinese reforms have brought an expansion of farming to the more sheltered valleys of the region, with wheat, highland barley, peanuts, and rapeseed as the principal crops. The mixed forests of Gaoshan pine, Manchurian oak, and Himalayan hemlock are increasingly being brought into timber production, accelerating a long history of deforestation in the region and extending the imprint of humankind on this wild region. At the eastern edge of the plateau, overgrazing, combined with a decade of hotter, drier weather, has destroyed the thin topsoil, with the result that legendary horse country is now gradually turning into desert. Perhaps the most distinctive aspects of the human landscape, though, are the shrines, temples, and lamaseries (monasteries for lamas) of Tibetan Buddhism (Figure 9.43). Together with the spectacular scenery of the region, the picturesque, romantic, and mystical qualities of Tibet underpin a tourist industry that the Chinese authorities have allowed to develop, in controlled fashion, since the mid-1980s.

Summary and Conclusions

East Asia is the most populous of all world regions, containing 1.45 billion people. Though dominated in territorial, demographic, and geopolitical terms by China, it is Japan—the world's second most powerful single economy and one of the established nodes of the tripolar world-system—that is the key economic component of the region. In addition, Hong Kong, South Korea, and Taiwan in East Asia have developed as three of the Asian Tigers that have progressed from the periphery of the world-system to the semiperiphery. The geopolitics of the Cold War, coupled with significant flows of investment by transnational corporations, have helped both Taiwan and South Korea to become prosperous core regions within East Asia. Together with Japan, they have come to occupy specialized roles within the world economy and are highly interdependent with places and regions throughout the Pacific Rim and in Europe.

All of the core regions and key cities of East Asia are situated along the continental margin, where the bulk of the population lives. Within this continental margin are the most productive agricultural regions and the most prosperous cities. In Japan, the Pacific Corridor was developed as the geographic basis of the country's postwar "economic miracle." In China, which has made a significant impact on the world economy since the death of Mao Zedong and the adoption of an "open door" policy, the transition toward market economies has intensified the comparative advantages of its coastal regions, where economic and cultural globalization are swiftly becoming both cause and effect of striking changes in long-established patterns of production, trade, and culture.

Nevertheless, East Asia remains a region of tremendous contrasts. Traditional rural communities with a great variety of distinctive ways of life still characterize much of the region, while Outer China is a dramatic "outback" region with extensive uninhabited areas and spectacular scenery encompassing snow-clad mountains, vast swamps, endless steppes, and fierce deserts. In contrast, most of the rest of East Asia has been heavily modified by humankind. East Asia has been cleared of its forest cover, while marshes have been drained, irrigation systems constructed, lakes converted to reservoirs, levees raised to guard against river floods, and hills and mountainsides have been sculpted into elaborate terraces in order to provide more cultivable land. These changes reflect the imprint of some of the world's most sophisticated civilizations and most extensive empires. Today, East Asia is one of the most dynamic and most rapidly urbanizing of all world regions, but it still encompasses a rich mosaic of traditional landscapes and ways of life. As China—and perhaps even North Korea and Mongolia—become increasingly incorporated into the dynamics of global economic and cultural change, the pace of environmental, social, and political change is likely to increase, obliterating some elements of these long-standing local geographies while adding new elements and creating new regional patterns.

Key Terms

agglomeration (p. 451)
Asian Tigers (p. 442)
backwash effects (p. 451)
chaebol (p. 467)

counterurbanization (p. 446)
feng shui (p. 448)
geomancy (p. 448)

keiretsu (p. 439)
massif (p. 427)
Pacific Rim (p. 442)
pinyin (p. 448)

Ring of Fire (p. 427)
spring line (p. 470)
treaty ports (p. 433)
yurt (p. 470)
zaibatsu (p. 434)

Review Questions

Testing Your Understanding

1. When and why was the Great Wall of China built? Why were canal systems so important to Imperial China?
2. How did the Tokugawa Dynasty keep foreign influences out of Imperial Japan? How did Japan change following the Meiji Restoration of 1868?
3. What factors helped Japan industrialize rapidly before the Second World War? What factors helped Japan reconstruct and grow after the war?
4. After the fall of the last dynasty in 1911, which two groups fought for power in China? Who embarked on the Long March of 1934–35 and why was it a success by 1949?
5. What triggered the Korean War? What is the significance of the 38th Parallel? Who are Kim Jong-Il and Kim Dae-Jung?
6. In China, what were the results of the 1958 Great Leap Forward, the 1966 Cultural Revolution, and the 1979 Open Door Policy?
7. In 1970, how many children were born to the average Chinese family? By 1999, what was the average number of children per family? Who helps enforce China's one-child policy? Aside from lower birthrates, what are two main results of the one-child policy?
8. During the 1800s, why did so many people emigrate from East Asia? Where did they migrate to?
9. How do Confucianism and Shinto differ from each other, and how have they influenced the cultural geography of East Asia? What is animism?
10. How has China increased farm production as well as peasant incomes since 1980? Where and what are Special Economic Zones?
11. Why is clothing made in China produced more profitably than in Hong Kong or Canada?
12. What are some benefits and costs of the Three Gorges Dam project?

13. How do transportation routes help structure the city of Tokyo? How are Japanese suburbs different from American suburbs?

14. Compare Taiwan to South Korea with regard to their post-WWII economic development. What are *chaebol*?

Thinking Geographically

1. What are the three main physiographic provinces of East Asia? How do the Tien Shan and Himalaya mountains affect East Asia physically and culturally? How has human activity modified the landscape in East Asia?

2. Briefly discuss the territorial expansion of imperial Japan from 1890 to 1945. What motivated Japan's acquisition of new territory? Having lost all of its colonial possessions in 1945, how did Japan achieve its postwar economic miracle?

3. How does Star TV unify East Asia with the rest of Asia? Does Star TV help foster a unique Asian identity? How might Star TV influence younger generations of East Asian leaders?

4. Discuss which factors have led to urbanization and counterurbanization in East Asia during the twentieth century.

5. What are some dominant East Asian cultural traditions? How is geomancy (*feng shui*) used in everyday life? How have East Asian beliefs and traditions affected other world regions?

6. State-assisted capitalism occurs frequently in East Asia. Please explain how the central government involves itself in corporate activities in China, Japan, and South Korea.

7. How are private investors in Hong Kong, Taiwan, and elsewhere changing the geography of China, especially in the Special Economic Zones? How has the People's Republic of China's attempt to create a classless society been affected by capitalist investment in China?

8. Beijing, like other cities modified or created during periods of state socialism, has been described as "a city primarily designed to express the government's power and control." Compare Beijing to Moscow and to Tokyo. How is Beijing changing in response to foreign investment?

9. Shanghai exemplifies the dichotomy of modern China. Freed from strict economic control by becoming a Special Economic Zone, Shanghai is now more disorderly and less equitable but is also more productive and generates greater revenue. Compare Shanghai to Hong Kong—what factors make it possible for a city to have both strong social control and strong economic growth?

10. What physical and cultural factors make western China distinctive?

Further Reading

Abbas, A., *Hong Kong: Culture and the Politics of Disappearance.* Minneapolis: University of Minnesota Press, 1997.

Barnes, I., and Hudson, R., *The History Atlas of China.* New York: Macmillan, 1998.

Benewick, R., and Donald, S., *The State of China Atlas.* London: Penguin Reference, 1999.

Blunden, C., and Elvin, M., *Cultural Atlas of China.* New York: Checkmark Books, 1998.

Brugger, B., and Reglar, S., *Politics, Economy and Society in Contemporary China.* London: Macmillan, 1994.

Cannon, T. (ed.), *China's Economic Growth: The Impact on Regions, Migration, and the Environment.* New York: St. Martin's, 2000.

Cartier, C., *Globalizing South China.* Cambridge, MA: Blackwell, 2000.

Chai, J., Jackson, S., and White, H., *Economic History of Modern China.* New York: Routledge, 2000.

Chapman, G., and Baker, K. (eds.), *The Changing Geography of Asia.* New York: Routledge, 1992.

Craddock, S., *City of Plagues: Disease, Poverty and Deviance in San Francisco.* Minneapolis: University of Minnesota Press, 2000.

Cybriwsky, R., *Tokyo: The Shogun's City at the Twenty-First Century.* New York: John Wiley & Sons, 1998.

Dwyer, D. J. (ed.), *China: The Next Decades.* London: Addison Wesley Longman, 1994.

Eccleston, B., Dawson, M., and McNamara, D. (eds.), *The Asia Pacific Profile.* New York: Routledge, 1998.

Francks, P., *Japanese Economic Development,* 2nd ed. New York: Routledge, 1999.

Howard, K., *Korea: People, Country, and Culture.* London: School of Oriental and African Studies, 1996.

Kim, J., and Choe, S-C., *Seoul: The Making of a Metropolis.* New York: John Wiley & Sons, 1997.

Leeming, F., *The Changing Geography of China.* Cambridge, MA: Blackwell, 1993.

Lo, C. P., *Hong Kong.* London: Belhaven Press, 1992.

Mather, C., Karan, P. P., and Iijima, S., *Japanese Landscapes.* Lexington, KY: University Press of Kentucky, 1998.

McGrew, A., and Brook, C. (eds.), *Asia-Pacific in the New World Order.* New York: Routledge, 1998.

Olds, K., Dicken, P., Kelly, P., Kong, L., and Yeung, H., *Globalisation and the Asia Pacific.* New York: Routledge, 1999.

Pan, L. *Sons of the Yellow Emperor: A History of the Chinese Diaspora.* Boston: Little, Brown, 1990.

Preston, P. W., *Pacific Asia in the Global System.* Cambridge, MA: Blackwell, 1998.

Pyle, D. J., *China's Economy: From Revolution to Reform.* New York: St. Martin's Press.

Selya, R. M., *Taipei.* New York: John Wiley & Sons, 1995.

Sit, V. F., *Beijing: The Nature and Planning of a Chinese Capital City.* New York: John Wiley & Sons, 1995.

Smil, V., *China's Environmental Crisis: An Inquiry into the Limits of National Development.* New York: M. E. Sharpe, 1997.

Tuan, Y-F., *China.* London: Longman, 1970.

Watters, R., McGee, T., and Sullivan, G., *Asia Pacific.* Vancouver: UBC Press, 1997.

Wu, W., "Shanghai," *Cities,* 16, (1999), 207–16.

Zhao, S., *Geography of China.* New York: Wiley, 1994.

Film, Music, and Popular Literature

Film

China: Unleashing the Dragon. PBS Home Video Series, 1995. Documentary examines the social impact of China's experiment with a market economy.

Chungking Express. Directed by Kar-wei Wong, 1994. An acclaimed example of the Hong Kong movie industry.

Farewell My Concubine. Directed by Chen Kaige, 1993. The story of two men who met as apprentices in the Peking Opera and stayed friends for more than 50 years.

Great Wall Across the Yangtze. PBS Home Video Series, 1999. Documentary about the controversial Three Gorges Dam on China's Yangtze River.

Kokoro: The Heart Within. PBS Home Video series, 1998. A 10-part presentation of the land and people of Japan.

Not One Less. Directed by Yimou Zhang, 1999. Set in the remote, dry high plains of Hebei, China, the film depicts rural life in telling detail as it deals with the story of a young teacher in a village school.

Robert Thurman on Tibet. PBS Home Video series, 1999. Documentary reviews the history and culture of Tibet from ancient times to the present.

The Blue Kite. Directed by Tian Zhuangzhuang, 1993. The story of China's political upheavals is told from the point of view of a simple family trying desperately to survive.

The Silk Road. PBS Home Video series, 1992. Eighteen-part documentary series on the Silk Road across China and Central Asia.

The Story of Qiu Ju. Directed by Yimou Zhang, 1992. Gong Li, China's top actress, plays a naive, young married woman from a remote farming village who has to make repeated trips to the city to seek justice over a domestic dispute.

Music

Dadawa. *Sister Drum.* Wea/Elektra Entertainment, 1996.

Kodo. *Ibuki.* Tristar Music, 1997.

Shanghai Chinese Traditional Orchestra. *Chi Gong.* Wind Records, 1997.

Tomoko Sunazaki. *Tegoto.* Fortuna Records, 1989.

Various Artists. *Tea.* Wind Records, 1998.

Xiao-Lin, Yang. *I Take You There.* Kiigo, 2000.

Yuan, Lily. *Ancient Art Music Of China.* Lyrichord, 1991.

Popular Literature

Buck, P. *The Good Earth.* New York: Washington Square Press, 1999. A fictional account of life in nineteenth-century China.

Dutton, M. (ed.). *Streetlife China.* Cambridge: Cambridge University Press, 2000. A collection of pieces about life in contemporary China, focusing on the lives of ordinary people and the rules and rituals that govern their daily existence, including the impacts of the emergence of a consumer culture driven by market values.

Gao, M. C. F. *Gao Village: A Portrait of Rural Life in Modern China.* Honolulu: University of Hawaii Press, 1999. A good account of everyday life in a small Chinese village.

Guest, H. (ed.). *Travellers' Literary Companion—Japan.* Summary of literary works from and about Japan.

Hopkirk, P. *Foreign Devils on the Silk Road: The Search for the Lost Cities and Treasures of Chinese Central Asia.* Amherst: University of Massachusetts Press, 1984. Fascinating insights into the history of the ancient Silk Road as well as its later intersection with the "Great Game"—European imperial powers' jockeying for influence in Central Asia.

Kristof, N., and Wudunn, S. *China Wakes.* New York: Vintage Books, 1995. An account of contemporary politics and society in China.

Kuhn, L. *Made in China: Voices from the New Revolution.* Boston: PBS Books, 2000. Explores how the Chinese are adapting to their new hybrid system of personal and economic freedoms combined with rigid political controls.

Lever-Tracy, C., and Ip, D. *The Chinese Diaspora and Mainland China.* London: Macmillan, 1996. Describes the emerging economic ties between China and the Chinese diaspora.

Seagrave, S. *Lords of the Rim.* New York: Putnam, 1995. A fascinating insight into the world of the so-called "Overseas Chinese."

Seth, V. *From Heaven Lake: Travels through Sinkiang and Tibet.* New York: Vintage Books, 1987. An award-winning account of a hitchhiking trip from China into Tibet.

Weston, T. B., and Jensen, L. M. (eds.). *China Beyond the Headlines.* New York: Rowman and Littlefield, 2000. A series of essays that seek to dispel stereotypes about China by describing China's society, economy, and culture under the stresses of modernization.

Wong, J. *Red China Blues: My Long March from Mao to Now.* New York: Anchor, 1997. The detailed recollections of the journeys of an observant and engaged traveler over several extended visits to China.

Figure 10.1

10 Southeast Asia

Legend:
- ⊙ More than 1 million
- ○ 500,000–1 million
- • Fewer than 500,000

Capital cities are underlined

0 — 200 — 400 Miles
0 — 200 — 400 Kilometers

PACIFIC OCEAN

Ternate

Halmahera

Equator

Sorong

Moluccas

Jayapura

Ceram

Irian Jaya

Ambon

PAPUA NEW GUINEA

Arafura Sea

Merauke

AUSTRALIA

The continent of Asia is often divided into three large regions: East Asia, South Asia, and Southeast Asia. East Asia (Chapter 9) is dominated demographically and geographically by China, and South Asia (Chapter 11) is dominated demographically and geographically by India. Southeast Asia, the subject of this chapter, consists of a group of islands and peninsulas that lie between India and China (**Figure 10.1**). Although the region is strongly influenced by these two countries, it also has its own distinctive cultures and landscapes. The long coasts of Southeast Asia's peninsulas and islands, and the narrow marine channels between them, have made Southeast Asia an important crossroads for international maritime trade first with India, the Middle East, and China and later with Europe, Asia, and the Americas. The entire region has a tropical **monsoon** environment with warm temperatures and seasonal high rainfall, and the predominant land uses include irrigated rice cultivation in the lowlands and shifting agriculture in the highlands, as well as village gardens and plantation estate agriculture.

Southeast Asia is frequently divided into two main physical regions of approximately equal size: the mainland or peninsula region, which includes Burma, Cambodia, Laos, Thailand, and Vietnam, and the island or insular region, which includes Brunei, Indonesia, the Philippines, and Singapore. Malaysia includes a mainland peninsula and the northern region of the island of Borneo and is usually considered within insular Southeast Asia (**Figure 10.2**).

Mainland or peninsular Southeast Asia stretches from Burma, bordering Bangladesh along the Bay of Bengal, to the long peninsula of Malaysia through Thailand and Cambodia to Vietnam and a border with China on the South China Sea. Landlocked Laos lies inland surrounded by Thailand, Burma, China, Vietnam, and Cambodia. Prior to the Second World War, Vietnam, Laos, and Cambodia were called French Indochina because of their location between India and China, which has led to the mingling of cultures, and their shared history of French colonization. Island or insular Southeast Asia includes the island of Borneo, shared by the small nation of Brunei, Malaysia, and the Indonesian province of Kalimantan. It encompasses the arc of Indonesian islands from Sumatra in the west to Irian Jaya in the east and the new nation of East Timor, the island of Singapore, and the Philippine islands, which face Vietnam on the east side of the South China Sea.

The region covers an area of 4 million square kilometers (1.5 million square miles), about the size of Europe but consisting mainly of ocean, with a 2000 population of more than 500 million people. Indonesia dominates the region in both size and population, ranking as the fourth most populous nation in the world. The Association of Southeast Asian Nations (**ASEAN**) now links all of the countries in a regional security and economic alliance.

There is tremendous economic, ethnic, linguistic, and environmental diversity between and within the countries of Southeast Asia. Only Thailand avoided the direct European colonial control that grew to dominate the economies of the region from about 1500 to 1946. Colonialism in Southeast Asia built upon existing networks of trade and upon the labor and business expertise of residents as well as migrants from India and China. Because of the region's considerable geopolitical importance, Japanese forces invaded it during the Second World War, and U.S., British, and other Allied forces retook the region after a protracted war on land, air, and sea. At the end of the Second World War, the newly independent countries of the region faced Cold War tensions in the form of communist rebellions and various alliances with China, the Soviet

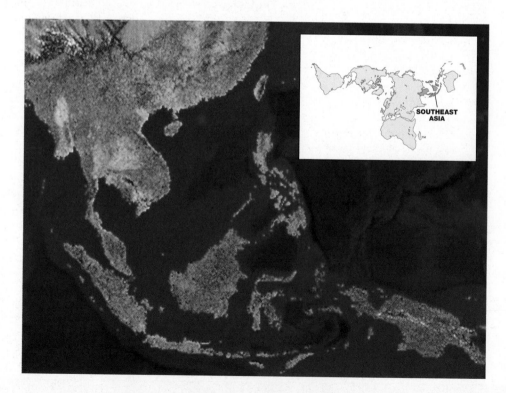

Figure 10.2 Southeast Asia from space This image clearly shows the many islands and major peninsulas that make up Southeast Asia. [*Source:* (inset) Map projection, Buckminster Fuller Institute and Dymaxion Map Design, Santa Barbara, CA. The word *Dymaxion* and the Fuller Projection Dymaxion™ Map design are trademarks of the Buckminster Fuller Institute, Santa Barbara, California, © 1938, 1967 & 1992. All rights reserved.]

Union, and the United States. The Vietnam War brought first France and then the United States into a regional conflict with the Chinese- and Soviet-supported forces of North Vietnam, a conflict that was based in South Vietnam and spilled over into Laos and Cambodia. Until relatively recently, communist and socialist governments of Vietnam, Laos, Cambodia, and Burma remained isolated within the region and from much of the Western world, with their economies focused on agriculture and rural lifestyles.

The socialist and communist experiences of Vietnam, Laos, Cambodia, and Burma contrast with the experience of countries such as Singapore, Thailand, Malaysia, Indonesia, and the Philippines, whose governments chose to open their economies to world capitalism in the latter half of the twentieth century. In these countries, existing trade in the traditional export commodities—such as rubber, rice, timber, and tin—expanded with the help of foreign investment to include textiles, clothing, and manufactured goods as well as new high-technology products for the telecommunications and computer industries. Agricultural intensification through new seeds and chemicals, as well as a growth in tourism, also contributed to economic prosperity, although the benefits of growth were not distributed equally.

Policies in both the socialist and capitalist economies of Southeast Asia were promoted through heavy state intervention in the economy and everyday life and included government subsidies, state-owned industry, and regulation of trade. Japan has become the most important foreign investor in the region, and there are also important flows of investment within Southeast Asia from countries such as Singapore and Malaysia.

The economic and industrial boom was seen as a model for development and used to illustrate the benefits of rapid globalization. Critics of globalization highlighted the exploitation of cheap labor in factories and sex workers in tourism, the grave environmental problems that accompanied development, and the serious social and political impacts of the financial crisis in 1997, which brought sudden recession to the region. Standard measures of economic development, such as GNP per capita, rank oil-rich Brunei and the international port and business center of Singapore among the most developed nations. Vietnam, Laos, Cambodia, and Burma are among the world's poorest nations on recent economic rankings, with Thailand, Malaysia, Indonesia, and the Philippines in the middle range of economies.

As in other regions, independence and subsequent changes in government have brought several changes in names of countries, territories, and key locations in Southeast Asia. Burma is called Myanmar by the current government, but we have decided to use the traditional name of Burma proposed by the United Nations and the country's pro-democracy movement and used by many of the country's residents. From 1976 to 1989 Cambodia was called Kampuchea, and North and South Vietnam were unified as Vietnam in 1976. The full name of Brunei is Brunei Darussalam, and many islands and places in Indonesia are also known by their Indonesian names and spellings. The large island of New Guinea was split between the independent nation of Papua New Guinea to the east and the Indonesian province of

Irian Jaya to the west. Some texts treat the whole island, with its tropical climate, lofty mountains, and diverse and scattered ethnic groups, as part of the Pacific island or Oceania region, but we include Irian Jaya in Southeast Asia. The newest nation in Southeast Asia will be the island of East Timor, currently transitioning to independence from Indonesia under U.N. administration.

Environment and Society in Southeast Asia

Landscapes and Landforms

The physical geography of Southeast Asia is shaped by processes of plate tectonics that have influenced the configuration of land and oceans, highlands and lowlands, geological hazards and resources, and the development of ecosystems. The mainland of Southeast Asia and the island of Borneo occupy the Sunda shelf of the Eurasian tectonic plate. The Sunda shelf is now flooded by the shallow South China Sea between Borneo and the mainland, but it was exposed during the ice ages, forming a **land bridge** to Asia. The Indonesian island arc from Sumatra to East Timor lies along the edge of the Eurasian plate. To the south and east, the Australian, Pacific, and Philippine plates are moving toward, and colliding with, the Eurasian plate, and are being forced downward in deep-sea trenches along major subduction zones. The Mariana deep-sea trench between the Pacific and Philippine plates in the Pacific Ocean southeast of the Mariana Islands is the deepest in the world at 11,034 meters (36,201 feet). The collision zones are also associated with mountain building, and volcanic activity and have created the thousands of islands that form the Indonesian and Philippine archipelagoes (**Figure 10.3**).

These relatively young islands have high mountains, steep slopes, and generally narrow coastal plains. Irian Jaya on the island of New Guinea has mountains reaching 5300 meters (16,000 feet); temperatures are so cold at these high elevations that these mountains are capped by permanent glaciers even at this tropical latitude. More than a dozen other mountains and volcanoes in insular Southeast Asia reach 3300 meters (10,000 feet). Many of the countries of Southeast Asia have very fragmented, elongated, or rugged geographies that have created immense challenges to national integration, transportation, and economic development. Indonesia has more than 13,600 islands (only half of them inhabited), and the Philippines is made up of more than 7000 islands. Burma, Thailand, and Vietnam all include long narrow segments less than 160 km (100 miles) wide.

The economies and lifestyles of insular Southeast Asia have been oriented to ocean fishing and maritime trade for centuries, and it has often been easier to use ocean or river transport rather than difficult overland routes. But the complexity of coastlines and the many islands of some nations, together with the economic significance of ocean fishery and oil resources, have created conflicts over maritime territorial jurisdictions and boundaries (see Geography Matters: The Spratly Islands, p. 482).

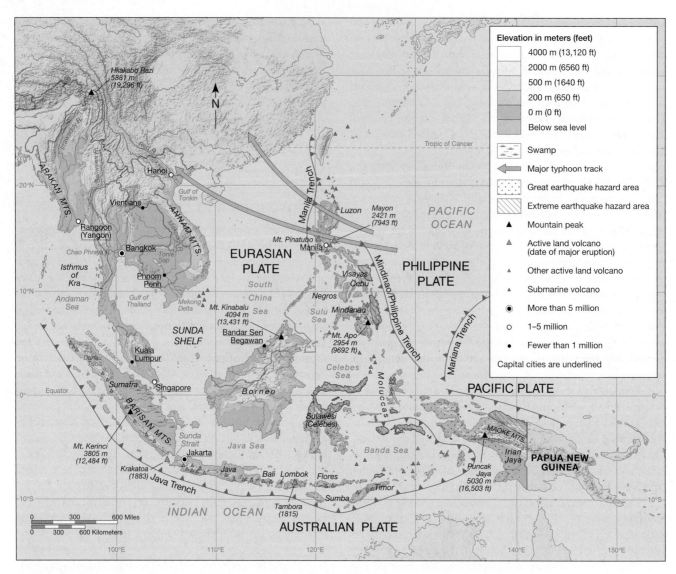

Figure 10.3 **Southeast Asia's physiographic regions** Southeast Asia, one of the most geologically active regions of the world, experiences many earthquakes and volcanoes because it lies at the intersection of active tectonic plates. The region is very mountainous except for the major river valleys of the Irrawaddy, Chao Phraya, Red, and Mekong rivers and some coastal plains. (*Source:* Map is based on information from T. R. Leinbach and R. Ulack, *Southeast Asia: Diversity and Development.* Upper Saddle River, NJ: Prentice Hall, 2000, Map 2.1; and H. C. Brookfield and Y. Byron, *South-East Asia's Environmental Future: The Search for Sustainability.* New York: United Nations University Press, 1993, Figure 13.1.)

The mainland is geologically older, but it is still very mountainous. Peaks in the highlands of Burma and Thailand reach 4000 meters (12,000 feet), and ridges are linked to the world's highest mountain range, the Himalayas. Between these north-south ridges lie areas of flatter land along the major river valleys and deltas of the Irrawaddy, Chao Phraya, Mekong, and Red rivers (the Red River is called the Yuan Jiang in China and Song Hong in Vietnamese). The Mekong River is Southeast Asia's longest waterway, flowing 4000 kilometers (2500 miles) from the highlands of Tibet to its delta in Vietnam and Cambodia. These river valleys, together with coastal plains, are the focus of human activity and settlement, although swamps and disease have posed problems for hu-

mans and animals in the lowland tropical environments. Two distinctive physical regions in mainland Southeast Asia are the Khorat plateau of Thailand, a sandstone upland region of poor soils that are suitable only for dryland crops and livestock, and the dry zone of Burma, where rainfall is significantly lower than it is in other areas.

The active volcanoes and earthquake zones of Southeast Asia pose great risks to the human populations of the region. The Indonesian and Philippine volcanoes are part of the **Ring of Fire** that surrounds the Pacific Ocean, linking this region with the seismically active regions of Japan, the western United States, and the Andes Mountains in South America. The island of Borneo sits on the Eurasian

Plate and is of older and more stable geological composition. In nineteenth-century Indonesia, the eruptions of Tambora in 1815 and Krakatoa in 1883 had severe local and global impacts. The Krakatoa volcano collapsed into the ocean during the eruption, leaving only the remnants of an island above sea level and causing a massive tidal wave (tsunami) that killed more than 35,000 people on the adjacent Indonesian islands of Java and Sumatra. Both eruptions ejected ash and soot high into the atmosphere, causing temperatures to fall worldwide and creating a "year without a summer" in 1884 that caused crop failures from cool temperatures in Europe and North America.

More recently, the eruption of Mount Pinatubo on the Philippine island of Luzon in June 1991 forced more than 100,000 people from their homes and destroyed 30,000 hectares (about 75,000 acres) of cropland (**Figure 10.4**). Scientific monitoring provided predictions that allowed warning and evacuations, limiting the loss of life to about 700 people. Pinatubo also caused a detectable cooling of global temperatures.

The risks of volcanoes are balanced by the benefits they provide in terms of soil nutrients. As in other parts of the world, volcanic eruptions have deposited ash that contributes to fertile, less acidic soils that can sustain high crop yields and associated population densities. In Java the rich volcanic soils have contributed to a productive agriculture and particularly dense populations. Soil fertility is also high in river valleys and deltas that receive regular replenishment of river sediment and nutrients in annual floods. Regions farther from volcanoes and rivers experience soil limitations typical of tropical climates around the world. Warm, wet climates mean that nutrients cycle rapidly back into forest vegetation as leaves and other organic debris decay, with most nutrients residing in the vegetation itself rather than in the soil. When forests are cleared, fertility declines rapidly as soil nutrients are washed out (leached) by heavy rains and the soil itself bakes in the hot tropical sun.

The geology and tectonics of Southeast Asia have promoted the formation and development of important mineral and energy resources, especially oil and tin, but generally the region is not as resource-rich as other regions, such as Africa or Latin America. Oil is the most important resource, contributing about 5 percent to global production. Indonesia, Malaysia, and Brunei are the most important producers, exporting oil mainly to Japan from major oil fields off the north coast of Borneo, Sumatra, and Java. With foreign assistance, Vietnam is starting to expand oil production, and Burma is reinvigorating its energy sector through exploitation of gas reserves for domestic use and for sale to Thailand. From colonial times, tin has been a major export from Malaysia, but exports have dropped in the last 20 years as easily worked deposits are exhausted and competition has lowered prices. Thailand and Indonesia also have significant tin reserves and production. The Philippines has developed a wide range of mineral resources, including copper, nickel, silver, and gold, and Indonesia is developing new gold, silver, and nickel mines in Irian Jaya despite opposition from environmentalists and indigenous groups.

Mountainous topography, rivers, and high rainfall favor hydroelectric development in Southeast Asia, and there are a number of proposals to expand hydroelectric capacity beyond the current modest facilities (see the discussion of the Mekong Basin, page 516). The World Resources Institute reports that only about 1 percent of the 1-million-megawatt estimated hydroelectric potential has been developed so far, with the greatest potential in Burma and Indonesia.

Figure 10.4 Eruption of Mount Pinatubo Mount Pinatubo erupted on the island of Luzon in the Philippines in 1991, forcing the evacuation of thousands of people and the closing of a major U.S. air base and cooling global temperatures.

Climate

Straddling the equator, most of Southeast Asia has a tropical **monsoon** climate, with warm temperatures all year-round and with high, but seasonal, rainfall. Temperatures average above 27°C (80°F) because the sun is high in the sky for most of the year, but it is cooler at higher elevations in the mountains. Annual rainfall totals more than 200 centimeters (80 inches) across much of the region and is strongly associated with seasonal monsoon winds.

As the sun heats inland Asia from May to October, low pressure builds over the continental interior, and winds start to flow in from cooler high-pressure regions over the Indian

Geography Matters

The Spratly Islands

The most disputed territories in Southeast Asia are the Spratly Islands, which lie between Vietnam and the Philippines in the South China Sea. The islands consist of several hundred treeless, sun-baked coral reef outcrops, many of them less than 1 square kilometer (less than half a square mile) in size. Brunei, China, Malaysia, the Philippines, Taiwan, and Vietnam each claim one or more of the Spratly Islands, and all except Brunei have established military posts on one or more of the islands (**Figure 1**). China's strategy includes the building of "fishing shelters" that other countries claim are actually military facilities (**Figure 2**). China has invaded the seven islands claimed by Vietnam and also threatened those claimed by the Philippines. The competition focuses on access to oil and gas resources and the potential for tourism.

Clashes occurred in 1988 when the Chinese evicted the Vietnamese from Johnson Reef and in 1995 when the Philip-

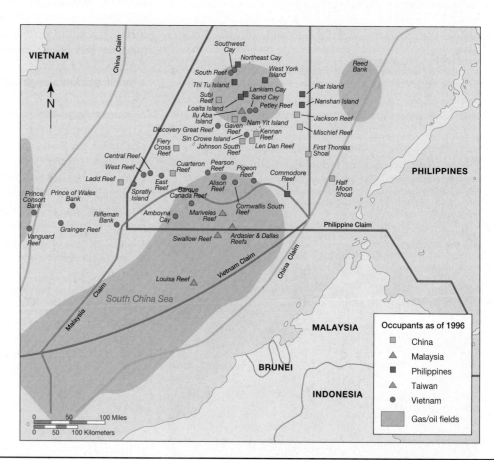

Figure 1 The Spratlys This map shows the multitude of contradictory claims in the South China Sea. (*Source:* Redrawn from M. J. Valencia, J. M. Van Dyke, and N. A. Ludwig, *Sharing the Resources of the South China Sea.* Honolulu: University of Hawaii Press, 1999, Plate 1.)

and Pacific oceans (**Figure 10.5**). These winds are laden with moisture and produce heavy rain over land, particularly where air currents rise and cool to produce orographic precipitation on highlands and south-facing island slopes. The onset of the monsoon is a momentous event in many places because it brings some alleviation from the very high and oppressive temperatures as rains cool the air, and because it brings water for crops and drinking water. But the monsoons can also cause severe floods, and the constant heavy rain strains nerves and promotes the growth of molds and fungus.

On mainland Southeast Asia, the November-to-March period brings cooler temperatures and higher pressures. The ocean temperatures become warmer relative to land, and lower pressure prevails over the oceans. During this period, the monsoon winds reverse, now blowing out of the continental interiors and across the ocean. On the mainland, this period is drier, resulting in lower overall annual rainfall totals, but the islands now receive a second sequence of monsoon rainfall, this time on the north-facing slopes. The Indonesian island of Sumatra, for example, receives the bulk of its rain on northern slopes

Figure 2   Chinese "fishing camp" on Mischief reef in the Spratly Islands Other countries claim that these buildings are military installations.

pines objected to Chinese construction on Mischief Reef. Indonesia has tried to mediate between the competing parties and to ensure stability in an area of the South China Sea that is crossed by internationally significant shipping lanes. One suggestion to improve relations included soccer or baseball matches between the troops stationed on different islands.

The Spratly dispute is an example of international conflicts over marine resources that usually center on rights for oil or fisheries. A 1958 U.N. conference began the process of allocating the oceans and their resources among countries, trying for years to reach agreement. The 1982 Convention on the Law of the Sea was signed by 117 nations, but not by several world powers, including the United States and Japan. But most countries have agreed to a territorial limit of sovereignty that extends 19.3 kilometers (12 miles) from the coast, and a 322-kilometer (200-mile) exclusive economic zone of rights to natural resources, including fish and oil.

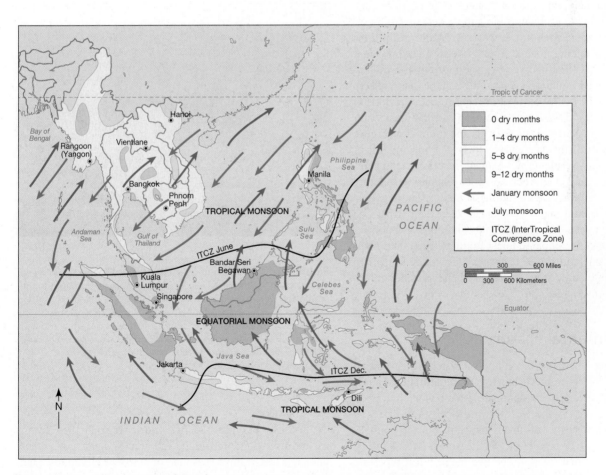

Figure 10.5   Climate map of Southeast Asia Southeast Asia has a hot, wet climate with seasonal monsoon winds and equatorial convergence that bring heavy precipitation. The mainland is drier with more seasonal rainfall, and regional climates can vary considerably from year to year as a result of El Niño conditions and the direction and intensity of tropical cyclones known as *typhoons*. (*Source*: Redrawn from R. Ulack and G. Pauer, *Atlas of Southeast Asia*. New York: Macmillan, 1988, p. 6.)

during the June-August period but on southern slopes between December and February. The monsoon was historically important, with summer onshore winds bringing traders and migrants to Southeast Asia, and with them their religious beliefs—Islam from Arabia and Hinduism from India. The shift in winds in winter carried the traders back home along with spices and other products.

The equatorial location of the Southeast Asian islands brings them within the influence of the **intertropical convergence zone (ITCZ),** where intense sun means high evaporation and a vigorous hydrological cycle associated with daily tropical thunderstorms throughout the year. Warm ocean temperatures from August to October combine with eddies in the trade winds to produce the large rotating storms known as *typhoons* in Asia and *hurricanes* in the Americas. Typhoons most commonly develop east of the Philippines and move eastward into the South China Sea. The combined effect of monsoons, the ITCZ, and typhoons means that the islands of Southeast Asia are among the wettest regions in the world with lush forest vegetation.

Located across the Pacific from Latin America, Southeast Asia, like Latin America, is affected by year-to-year variations in precipitation associated with **El Niño** (see Chapter 8, p. 382). Normally winds blow from east to west across the Pacific, bringing abundant rainfall to Southeast Asia. In El Niño years, the pattern reverses and winds blow from west to east, cooling ocean temperatures in the western Pacific, stabilizing clouds, and reducing rainfall. Fires associated with drought during the 1982–83 El Niño destroyed more than 6 million hectares (about 15 million acres) of forest on the island of Borneo. The 1997 El Niño brought drought to many regions of Southeast Asia and was associated with crop failures, water shortages, and severe fires in Malaysia and Indonesia (**Figure 10.6**). The world's media conveyed images of burning forests and choking city residents and blamed the crashes of a commercial airliner and of ferryboats on poor visibility. Respiratory illness increased not only in Indonesia but also in neighboring Singapore, and many tourists canceled their trips. The damage was intensified by economic conditions and policies that had relocated people to forest frontiers, promoted commercial logging, and placed pressures on swidden farmers to open up more land during dry times when it was difficult to control clearance fires.

Environmental History

Ecosystems The vegetation and ecosystems of Southeast Asia reflect the wet tropical climate, with a natural land cover of dense forests originally dominating the region. Indonesia is ranked second in the world in terms of its diversity of species, or *biodiversity*, with many native plants and animals, including at least 10 percent each of the world's forests, plants, birds, and mammal species. The two major forest types are evergreen (with leaves year-round) tropical forests in the wetter areas and tropical deciduous, or monsoon, forests where rainfall is more seasonal or lower and trees lose their leaves in

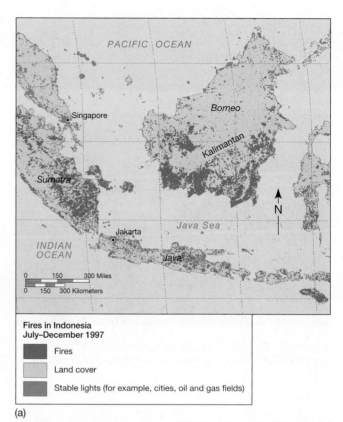

Fires in Indonesia
July–December 1997

■ Fires

■ Land cover

■ Stable lights (for example, cities, oil and gas fields)

(a)

(b)

Figure 10.6 El Niño in Southeast Asia, 1997 (a) El Niño conditions in Southeast Asia in 1997 brought severe drought to the region and were associated with devastating fires in Malaysia and Indonesia. (b) These fires caused serious air pollution and health problems. A soccer game in Indonesia continues despite the poor air quality. [*Source:* (a) National Geophysical Data Center, National Oceanic and Atmospheric Administration, 2001, available at http://www.ngdc. noaa.gov/dmsp/fires/indo.html]

the dry season (**Figure 10.7a**). In drier regions, these forests are less diverse and transition to savanna and grasslands where rainfall is less or more seasonal, or at higher elevations. Mangroves and marshes are found along the long mainland and island coastlines, together with a rich offshore marine ecosystem that includes coral reefs and productive fisheries. In-

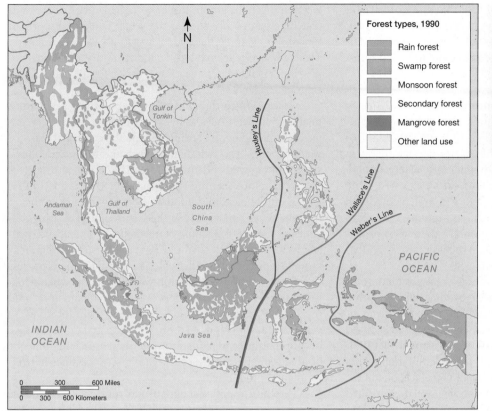

(b)

(a)

Figure 10.7 Southeast Asian vegetation and ecology (a) The natural land cover of most of Southeast Asia is forest—evergreen in the wetter areas and tropical deciduous in drier regions. This map shows the actual land cover in 1990 and the extensive conversion of forests to agriculture and other land uses. Wallace's Line, named after nineteenth-century naturalist Alfred Wallace, indicates the location of a deep-ocean trench that divided two sections of a great land bridge that rose above Southeast Asian oceans and seas 16,000 years ago. This line divides the two major biogeographical zones of Asia and Australia. (b) An orangutan in Tanjung Puting Park, Borneo. Species such as these are endangered as a result of habitat destruction. [*Source:* (a) Based on maps in R. Ulack and G. Pauer, *Atlas of Southeast Asia.* New York: Macmillan, 1988, p. 11; and T. R. Leinbach and R. Ulack, *Southeast Asia: Diversity and Development.* Upper Saddle River, NJ: Prentice Hall, 2000, Map 2.6b.]

donesia leads the world in mangrove area, with more than 4 million hectares (10 million acres).

Plate tectonics and climate change together produced a fascinating division in the ecology of the Southeast Asian region. The last ice age, when sea levels dropped as ice sheets locked up moisture, exposed the Sunda Shelf between the mainland and Indonesia until about 16,000 years ago, and many species migrated across this land bridge to the islands of Indonesia, including tigers, elephants, and orangutans. To the south, other animals such as kangaroos and opossums moved north across the Australian Plate to New Guinea and other islands. These mammals are called **marsupials,** because after birth the incompletely developed offspring continue to develop and nurse in an abdominal pouch on the mother (see Chapter 12). In contrast, most other mammals are *placental,* meaning offspring develop completely within the mother's uterus. Between Bali and Lombok there is a deep ocean trench that remained ocean, even during ice ages, and prevented these two very different types of species communities from mixing, thereby creating an

ecological division called **Wallace's Line,** after naturalist Alfred Wallace, who traveled extensively in the region in the nineteenth century and first noted this striking contrast. This boundary between Australian and Oriental or Asian ecosystems was later modified by naturalists Huxley and Weber.

As discussed later in this chapter, deforestation has destroyed or is threatening the habitat of many species in Southeast Asia, including charismatic animals, such as tigers and orangutans (**Figure 10.7b**). But the deep forests of mainland Southeast Asia are still relatively unexplored by biologists, and new species are still being discovered, including relatively large animals such as the Vu Quang ox and giant Muntjac deer, identified as recently as the 1990s. The tropical forests of Southeast Asia are of great cultural and economic significance in the history of the region, providing shelter, food, medicines, and marketable goods over the centuries.

The forests that covered much of the region have provided many useful functions to local communities, including materials for house and boat construction, traditional medicines, and a wide range of food, including fruit and game. As trade with other regions such as China developed, demand for specialized products, such as aromatic sandalwood and teak, grew and resulted in some increased forest exploitation.

Human Use of the Environment   About 5000 years ago the selective breeding of a grass with edible seeds produced rice, the basis of human diets in Southeast Asia. Domesticated in several regions of Asia, rice became the most important crop in the region, growing best where rainfall was evenly distributed through the year and totaled more than 120 centimeters (80 inches). Rice complemented fish and vegetables in a traditional diet, and the plant was also used for fodder and thatch in building. Other crops probably domesticated in Southeast Asia include taro, sago, bananas, mango, and sugar.

The modification of landscapes for rice production included the construction of terraces, paddies, and irrigation systems (**Figure 10.8a**). Terraces cut into steep hillsides provided level surfaces that facilitated water control and reduced erosion. The construction of dikes (ridges) around fields allowed them to be flooded, plowed, planted, and drained before harvest in a system called *paddy farming*. Rice, one of the few major crops that can grow in standing water, was suited to the flooding that accompanied heavy monsoon rains. Wet rice, or **sawah**, is the most important crop in Southeast Asia. Traditional rice production is highly labor intensive, with work throughout the season in preparing and maintaining fields, transplanting seedlings, weeding, and harvesting each stalk by hand. Women perform much of the labor.

The main adaptation to the seasonal and yearly variability of rainfall in Southeast Asia is irrigation, ranging from small canal and pump systems to large water-diversion systems and dams. Many of the more recently constructed dams are also used for electricity generation and have been controversial. Often they have been located in areas of high biodiversity, where indigenous groups use or own the land or where multilateral organizations such as the World Bank are a source of funds. For example, large dam proposals in Indonesia and Thailand have generated local, national, and international opposition, resulting in project cancellation for the Nam Choan Dam in Thailand.

Adaptation to the constraints of the tropical environment in Southeast Asia, as in other tropical regions, has centered on managing soil fertility and on the control of seasonal water supplies. The **swidden,** or **slash and burn,** agricultural system often used in tropical forests involves cutting trees and brush and burning them in order that crops can be grown on the cleared ground and benefit from nutrients in the ash. When soil fertility in these agricultural clearings declines as a result of leaching of nutrients by rainfall and damage from the relentless tropical sun, subsistence migratory farmers move on to another patch of forestland. This process of extensive shifting cultivation was sustainable when widely scattered villages controlled large areas of land and could use fields for a few years and then leave them for 10 to 30 years to recover (**Figure 10.8b**). If access to land decreases, or population and consumption increase, land may be cleared more frequently, causing long-term declines in environmental productivity.

The third traditional agricultural land use is the house garden where people plant vegetables and other crops, such as fruit trees, close to their homes.

Southeast Asia in the World

Humans probably reached Southeast Asia through migrations from Europe and mainland Asia and over land bridges and oceans about 60,000 years ago. These people formed populations that were the ancestors of contemporary indigenous groups in New Guinea and Australia. A later migration around 5000 years ago brought farmers and fisher

Figure 10.8   Traditional agriculture in Southeast Asia   (a) Irrigated rice fields such as these in Bali have supported the food needs of Southeast Asia for hundreds of years. (b) Another major land use in Southeast Asia is the clearing and burning of forestland for swidden agriculture, such as shown here near Dumai Riau, Sumatra, Indonesia, in July 2000.

(a)

(b)

people from southern China to the coast of Vietnam and to the islands of present-day Indonesia and the Philippines. Archaeological sites in Vietnam provide evidence of early hunting-and-gathering activities, including fishing and collection of forest products. Domesticated agriculture, including rice, cattle, pigs, and chickens, probably spread southward from China beginning in about 3000 B.C. By 700 B.C. the Dong Song culture had developed wet-rice cultivation and a metal industry that produced bronze artifacts.

Maritime trade brought merchants from the Middle East, China, and India about 2000 years ago, together with new religions, crops, and technologies. The most important interaction was with India, whose Hindu religion and related forms of government were attractive to local chiefs who saw advantage in the divine privilege they could be granted as god-kings (*deva raja*). To claim this power they needed the blessing of Hindu priests, who in turn demanded the construction of temples in the Indian architectural tradition. Two types of urban settlement emerged: port cities focused on trade, and sacred religious and ceremonial capitals focused on elaborate temples. The earliest kingdoms inspired by Indian models included Langkasuka on the Malay peninsula and Fu-nan, which was centered on present-day Cambodia and Vietnam (**Figure 10.9**). By the tenth century there were large kingdoms, influenced by Indian culture and religion, along the Irrawaddy River in Burma and in the Chen-la and Champa kingdoms in what is now known as Vietnam. These kingdoms relied on state taxation of rice production and power structures based on kinship, religious and royal bureaucracies, and they built spectacular capitals, such as Pagan in Burma (**Figure 10.10a**). In northern Vietnam, there were important Chinese influences on kingdoms at Annam on the Red River and Hanoi.

Two of the most powerful states to emerge in Southeast Asia were the mainland Khmer empire and the island Srivijaya culture of Sumatra. Srivijaya ruled Sumatra and the southern Malay peninsula from the seventh to twelfth centuries by controlling the region's long-distance maritime trade through the Strait of Malacca. This powerful state had a capital at Palembang and Buddhist religious traditions. The Khmer kingdom emerged in the ninth century near Tonle Sap Lake and built the magnificent twelfth-century temple compound of Angkor Wat, one of the largest religious structures ever built and a focus of tourism in contemporary Cambodia (**Figure 10.10b**).

An increasingly powerful Thai kingdom, centered on the city of Ayutthaya on the Chao Phraya River, conquered the Khmer empire in the fourteenth century and maintained control of the Malaysian peninsula for four centuries. A series of empires also flourished on Java and produced magnificent temples, such as Borobudur, the Buddhist temple (**Figure 10.10c**).

By the fifteenth century, the political geography of Southeast Asia had restructured around a series of sultanates such as Malacca (in present Malaysia) and Brunei (on the island of Borneo) and a new set of mainland empires. Malacca's strategic location controlling the important trading route between

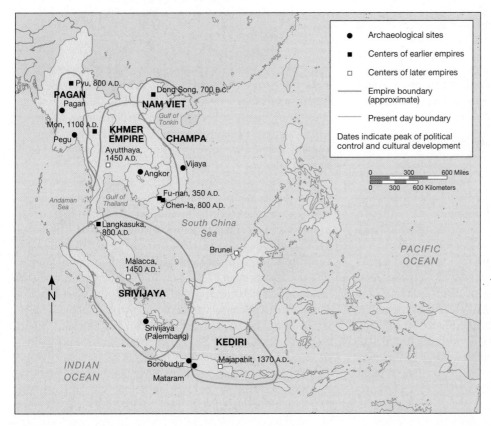

Figure 10.9   Early kingdoms and important cultural sites
Southeast Asia hosted a series of powerful kingdoms over the last 3000 years, including many important religious centers and ports. This map shows the major empires in the late twelfth century A.D., as well as the centers of some earlier and later empires and important archaeological sites. (*Source:* Based on R. Ulack and G. Pauer, *Atlas of Southeast Asia.* New York: Macmillan, 1988, pp. 16–18; and N. Tarling (ed.), *The Cambridge History of Southeast Asia*, vol. 1. Cambridge: Cambridge University Press, 1999.)

(a)

(b)

(c)

Figure 10.10   Pho Pagan, Angkor Wat, and Borobudur   (a) The ruins of the sacred capital at Pagan in present-day Burma. (b) The complex that includes the twelfth-century temple of Angkor Wat and sacred city of Angkor Thom was constructed by the Khmer empire just north of the Tonle Sap Lake in present-day Cambodia. The design represents symbols from the Hindu cosmos, including mountains, artificial lakes, and sculptures. After years of conflict, the site is being restored and promoted as a major tourist destination. (c) Statue of the Buddha at Borobudur, Indonesia, sacred city of the Mataram empire.

India and China gave it great commercial power as a sea-based trading state, and when the rulers enthusiastically adopted Islam, Malacca became the center of dissemination of Islamic beliefs and institutions in Southeast Asia.

European Colonialism

Scholars usually divide the colonial period in Southeast Asia into two periods. The mercantile period, based on trade, spanned from about 1500 to 1800. The industrial period concentrated on political and economic control of exports and lasted from about 1800 to 1945. A number of European powers sought to control Southeast Asia and its valuable trade, with Portugal, Spain, the Netherlands, Britain, France, and the United States controlling different parts of the region at different times and often in competition (**Figure 10.11**). Portugal dominated early mercantile trade with the region. The Portuguese sailed around Africa and established a headquarters at Goa in India and then moved on to gain control of the vibrant and strategic port of Malacca in 1511. To obtain commodities such as cloves, nutmeg, and pepper from local producers in the scattered islands of Indonesia, they relied on

indigenous and other Asian merchants. Other Europeans, such as the British, traded directly with local merchants on the outer islands who often wished to avoid trading through Malacca because by then it was dominated by Portuguese Catholics who criticized Islam, imposed heavy taxes, and monopolized markets, thereby keeping prices down for producers and middlemen.

Meanwhile, the Spanish sailed across the Pacific and sought to conquer the Philippines in order to spread Catholicism, expand the Spanish Empire, and gain access to the trade in spices and other commodities in Southeast Asia. The city of Manila, founded in 1571, became the center of trade with Latin America, with galleons sailing regularly to Acapulco, Mexico, Spain's main Pacific port in the Americas. As in Latin America, Spanish colonial rule resulted in the dominance of Catholicism, the consolidation of land ownership under Spanish landowners, and the reorientation of land use to export crops—especially sugar—through Spanish-owned plantations or systems of tribute.

The third colonial power to move rapidly into Southeast Asia was the Netherlands at the end of the sixteenth century. The Dutch dominated trade from about 1600 to 1750, with an initial focus on the Molucca Islands of Indonesia—the fa-

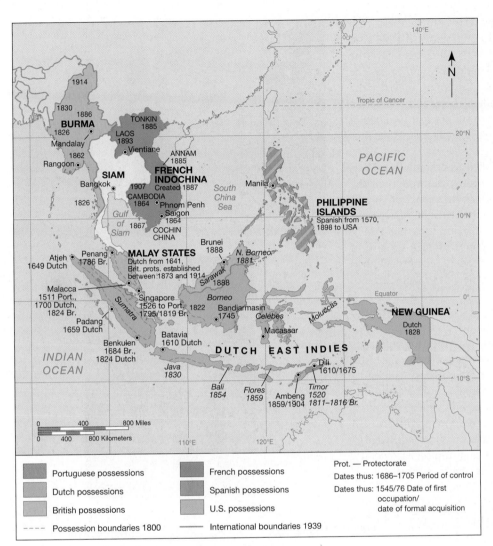

Figure 10.11   European expansion into Southeast Asia
This map show the areas controlled by the major colonial powers in Southeast Asia and shows the dominance of the Dutch in the island subregion, the British and French in the peninsula/mainland region, and the Spanish in the Philippines. (*Source:* Redrawn from B. Crown and A. Thomas (eds.), *Third World Atlas.* Milton Keynes: Open University Press, 1984, p. 39.)

mous **Spice Islands** that produced the valuable commodities of nutmeg, cinnamon, and cloves. The formation of the Dutch East India Company in 1602 consolidated private commercial interests from the Netherlands into a coordinated effort supported by the Dutch government and military to control trade in Indonesia (then called the East Indies). The privately chartered trading company was based in Batavia on the island of Java, a city that later became the capital of independent Indonesia and was renamed Djakarta (now Jakarta). The Dutch attempted to control markets and increase commodity prices by restricting the production of valuable spices such as pepper and destroying communities that ignored restrictions or participated in smuggling. By 1700 the Dutch had evicted the Portuguese from Malacca, subdued sultanates in Java, and eroded local political power by co-opting leaders to serve the Dutch East India Company and by encouraging Chinese migration to fill administrative and commercial posts.

The British colonial effort in Southeast Asia was an extension of their activities in India and focused on Burma, Malaya, and Borneo and on the control of strategic ports. As with the Dutch, the British colonial enterprise was led by a trading company, the British East India Company, formed in 1600 and focused for three centuries on British domination of South Asia (see Chapter 11, p. 534). For the most part the British waited until the nineteenth century to make a move from their empire in India into Southeast Asia, beginning with control of the strategic ports of Penang (1786), Singapore (1819), and Malacca (1824). They then fought two wars with the Burmese and used a protectorate system to gain control of the Malay peninsula. The British focused on reorienting the economies to exports of tin, rubber, and tropical hardwoods and on controlling trade routes between India and China. The most unorthodox British endeavor was the emergence of a personal kingdom, or *raj,* in Sarawak on the island of Borneo, operated by British administrator James Brooke from 1846.

The French influence originated in the eighteenth century with their missionaries in Indochina and their support of local emperors. It was consolidated in the nineteenth century in response to rivalry with Britain over commercial links to China. French Indochina brought together Cambodia, Laos, and the districts of Tonkin, Annam, and Cochin China in Vietnam as the Union of French Indochina.

The United States was a colonial power in Southeast Asia for less than 50 years, acquiring the Philippines in 1898 after victory in the Spanish-American War, which also gave the United States control over Guam, Puerto Rico, and Cuba. The legacy of this brief episode in the Philippines includes the widespread use of English, U.S. military bases and orientation to American popular culture, the growth of sugar production in response to preferential access to U.S. markets, and the promotion of education and elections.

As noted earlier, Thailand (called Siam until 1939) was able to maintain its political independence throughout the colonial period. It provided a buffer between British and French interests, losing territory to the British in Malaya and Burma and to the French in Cambodia and Laos, but able to maintain a core independent kingdom with tacit support of Britain. Although it remained independent, Siam was linked into colonial trading systems and vulnerable to the policies of the European powers.

The nineteenth century provides several insights into the processes and geography of colonial activities in Southeast Asia and the legacies of these activities for countries that became independent in the twentieth century. The first general process was the integration of Southeast Asia into a global trading system, building on long-standing traditions of trade with the Middle East, India, and China. Although spices (pepper, cloves, nutmeg, and cinnamon) were still important exports from the region in the nineteenth century, trade also included tortoise shell and tropical hardwoods, such as teak, and aromatic woods, such as sandalwood. Other important exports were tin, hemp, sugar, palm oil, tobacco, tea, and rubber. Economic developments in Europe and North America in the late nineteenth and early twentieth centuries increased demand for many of these products.

For example, in the late nineteenth century, the economy of Malaya, under British control, focused on the production of natural rubber and the mining of tin. Rubber production grew rapidly after 1876 when plants grown from seeds in Britain smuggled out of Brazil were introduced to Malaya. The area covered by rubber plantations grew from 2000 acres in 1898 to 2.1 million acres in 1920 as the explosion of automobile ownership in North America and Europe increased demand for rubber tires (**Figure 10.12a**). Rubber was initially grown by expatriate planters on plantations, encouraged by enthusiasts such as "Rubber Ridley," who walked around Malaya at the turn of the century evangelizing rubber and distributing seeds from his pockets. Even though there was a ban on production by local people in Malaya in order to protect the profits and monopoly of European planters, rubber quickly became popular with local farmers because of the high price and low labor its production demanded. Tin mines used imported Chinese labor because local people were reluctant to abandon subsistence rice production (**Figure 10.12b**). The Philippines produced sugar, tobacco, coconut, and pineapples.

During the nineteenth century, Dutch control over the Indonesian economy tightened with the introduction of the no-

(a)

(b)

Figure 10.12 Rubber and tin in Southeast Asia (a) Rubber plantation in Sumatra. Rubber was introduced into Malaya by the British in the nineteenth century and the area under production grew rapidly in response to demand from North America and Europe. Although demand dropped with the development of synthetic substitutes, there are still a large number of plantations, such as this one in Sumatra. (b) Overseas Chinese tin miner in Malaya. The mineral resources of tin in Malaya sparked British interest during the nineteenth century. They brought in thousands of Chinese to work in the mines, forming a large overseas Chinese population who continue to live in Malaysia. This miner carries his shovel in 1953.

torious **Culture System** into Java in 1830. The Culture System required Javanese farmers to devote one-fifth of their land and their labor to export crop production, especially coffee and sugar, with the profits going to the Dutch government. In many cases officials demanded more land and labor than specified, causing distress and food shortages in local populations. Abandoned in 1870 in the hope that farmers would produce more efficiently for export under a more lenient system, the Culture System has aroused considerable scholarly debate concerning its impact on the economy and peoples of Java. For example, anthropologist Clifford Geertz argued that the culture system promoted *agricultural involution,* a process of agricultural intensification to produce and share higher yields of rice in response to the loss of land to export crops and population growth. He highlighted this process of involution as a contrast to other regions of the world, such as Latin America and Africa where local people gave up food production, expanded agricultural lands, or were displaced by export cropping and population pressures.

The British and French promoted rice production to feed laborers and growing populations, and expansion was especially dramatic in the Irrawaddy delta of Burma and in the Mekong Delta of French Indochina. Local and imported labor was employed to clear forests and build irrigation and drainage systems in these vast deltas that became the rice bowls of Southeast Asia. According to geographer Raymond Bryant, rice production in the Irrawaddy delta grew from 700,000 acres in 1852 to more than 6 million acres in 1906.

The persistent stream of migrants from India and China accelerated in the nineteenth century. Thousands of workers were brought in to convert the deltas to rice production, to mine tin and other minerals, to manage the services in major ports, and to develop small businesses to serve colonial and local demands for consumer goods. In some cases, a shortage of European administrators was filled when rights to collect taxes and harbor duties, to market opium, or to operate gambling were auctioned by colonial governments and purchased by Chinese because such activities were valuable revenue generators. These migrants, who were seen as more entrepreneurial and fit to govern by some colonial administrators, became a core of the colonial economies and are the ancestors of the large populations of South and East Asians, and especially Chinese, that live in Southeast Asia today.

The raw materials of Southeast Asia were very important to the industrialization of the core European economies and brought Southeast Asia into a peripheral relationship within the world-system. The colonial systems created spatial inequalities. Key port and trading cities, such as Singapore, Batavia, and Manila, developed as regional cores and gateways to the world, with export agricultural regions oriented to these cores and with remote rural peripheries left in subsistence livelihoods with little investment in education or other services. In some cases minority and indigenous peoples were left isolated by export-oriented transport systems or by paternalistic colonial officials who decided to protect these "primitive" and exotic societies. In regions dominated by Buddhism, such as Burma, and Islam, such as Indonesia, Christian missionaries focused conversion efforts on remote hill tribes who followed animistic religions and were less resistant to Christianity than were the Buddhists and Muslims living in lowland areas.

Most of Southeast Asia remained under colonial control during the first part of the twentieth century. As in other regions with strong links to the world economic system, the Great Depression of the 1930s had serious economic impacts in Southeast Asia and was associated with several anticolonial uprisings. The price of rubber in 1932 was only 12 percent of what it was in 1928, tin was 20 percent, and sugar 26 percent, and many farmers shifted back into subsistence production.

The Second World War in Southeast Asia

The Japanese occupation of Southeast Asia during the Second World War had severe effects on economies and local peoples (**Figure 10.13**). Japan sought access to the natural resources of the region, claiming legitimacy from a shared set of cultural values with the regions that they captured under the slogan "Asia for the Asiatics." But the Japanese occupation cut off many regions from trade revenues, used forced labor, and diverted food and other resources to Japan at the expense of local economies and food security. The destruction of bridges and roads by the Allies and by native populations as a means of preventing Japanese advances and then by the Japanese themselves as they retreated left Southeast Asia's infrastructure in ruins, events documented in such classic films such as *The Bridge over the River Kwai*. In Vietnam the Japanese requisitioned rice for their own use and also forced farmers to grow jute fiber rather than rice, resulting in a famine that killed more than 2 million people in 1944–45. The Japanese were particularly harsh on the Chinese population of Southeast Asia.

Independence, the Cold War, and Vietnam

European colonial power in Southeast Asia was diminished by the Japanese invasion that had reduced the image of Western racial superiority and by the costs of the war. Hastened by post-war global calls for decolonization, independence came relatively peacefully to most of the region except for Indochina and Indonesia, where extensive French and Dutch settlement and investments made the Europeans reluctant to hand over power.

The Philippines was granted independence from the United States immediately after the war, in 1946. The British granted independence to Burma in 1948 after Indian independence and partition in 1947 and created the Federation of Malaysia in 1963. Singapore left the federation to become an independent country in 1965, and Brunei converted from a British protectorate to an independent nation in 1983. Indonesia was granted independence by the Dutch in 1949 only after a violent struggle following a declaration of independence in 1945, and western New Guinea, formerly Dutch New Guinea, became part of Indonesia as Irian Jaya in 1963.

Indonesia faced great challenges in forging a sense of national unity among scattered islands and diverse cultures. The concept of *Pancasila*—unity in diversity through belief in one God, nationalism, humanitarianism, democracy, and social justice—as the national ideology, and the promotion of a national Indonesian language were used to try to unify the country. However, fractures soon developed around religious differences and in response to government repression of criticism and political opposition.

Communist groups were part of the independence movements in several countries, and these groups continued to seek revolutionary change after independence. In 1965 a communist coup against Indonesian leader Sukarno was unsuccessful,

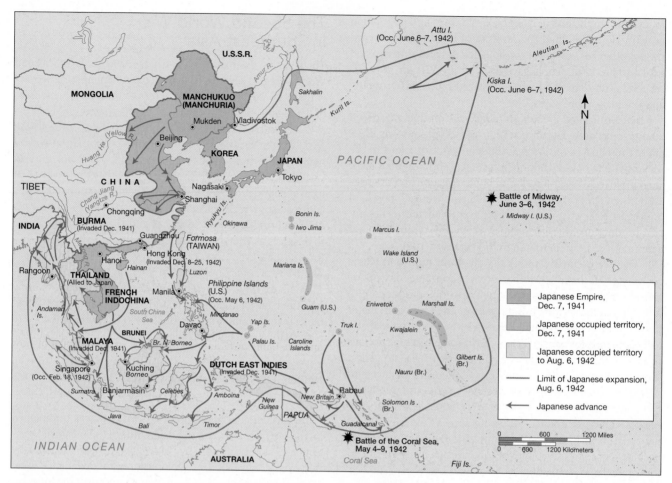

Figure 10.13 **Japan's occupation of Southeast Asia in the Second World War** Japanese expansion into Southeast Asia was rapid during the Second World War, and it occupied vast territory controlled by European powers. As Japan fought the Allied powers, including the British, French, and Americans, across the region, the conflict had devastating effects on local food supplies and infrastructure. (*Source:* Redrawn from M. Dockrill, *Atlas of 20th Century World History.* New York: Harper, 1991, pp. 74–75.)

and at least 500,000 people associated with the communist movement were massacred in the aftermath (one perspective on this period is portrayed in the film *The Year of Living Dangerously*). In Burma, a military coup in 1962 resulted in a socialist form of government that isolated Burma from the world and nationalized land and industry. Termed the Burmese Way of Socialism, the political philosophy combined elements of socialism and Buddhism.

But it was in Indochina that the intersection of independence, communism, and the Cold War had the most global and long-lasting impacts. Communist groups had led resistance against Japanese occupation during the Second World War and against French recolonization at the end of the war. Led by Ho Chi Minh, the communists fought for a base in northern Vietnam and established a separate government in Hanoi that supported guerilla war against the French in southern Vietnam with assistance from the Chinese and Soviet Union. When the French withdrew from Indochina after a devastating loss to Ho Chi Minh's forces at Dien Bien Phu in 1954, Laos and Cambodia became independent and the Gene-

va accords formally divided Vietnam into North Vietnam and South Vietnam. North Vietnam became an independent communist country, and thousands of refugees, especially Catholics, fled to South Vietnam, which had become an anticommunist partner of the United States with a capital in Saigon. Communist rebels in Cambodia and Laos joined with North Vietnamese forces (called the *Vietcong*) to try to bring the whole region under communist control, with considerable success in several regions of South Vietnam. Subscribing to a **domino theory** that held that the communist takeover of South Vietnam would lead to the spread of communism throughout Southeast Asia, the United States sent military advisers to South Vietnam in 1962. This was followed by the U.S. bombing of North Vietnam in 1964 and escalation to a full-scale land war with more than half a million U.S. troops in 1965.

The Vietnam War, together with its aftermath, was probably the most serious global manifestation of Cold War competition and wrought terrible social and environmental effects in Southeast Asia as well as on U.S. domestic and international

Figure 10.14 The Vietnam War U.S. troops waiting to be evacuated at Khe Sanh, Vietnam, in 1968—a year during which domestic opposition to U.S. involvement in Southeast Asia grew dramatically.

politics. More than a million Vietnamese people died, together with 58,000 Americans (**Figure 10.14**). U.S. forces sprayed 2 million hectares (5 million acres) of Vietnam with defoliants, such as Agent Orange, that poisoned ecosystems and caused irreparable damage to human health. Cambodia and Laos were also bombed with napalm and defoliated to disrupt communist supply lines and camps.

The media images of destruction, the loss of American lives, and the cost of the war created considerable opposition to the war in the United States, including protests on college campuses and marches on Washington. The United States gradually withdrew its forces in the early 1970s and left South Vietnam as the Vietcong approached Saigon in 1973; in 1975 Vietnam was unified under communist rule.

Two million people left South Vietnam fearing repression after unification, many (the so-called "Vietnamese boat people") sailing away in small, fragile boats. The communist government confiscated farms and factories from owners to create state and worker-owned enterprises, resettled hill tribes into intensive agricultural zones, and moved 1 million people into new economic development regions. But U.S.-led economic sanctions from 1973 to 1993 limited the potential for exports and restricted some critical imports such as medicines.

In Cambodia, the Khmer Rouge communist revolutionaries overcame the U.S.-backed military government in 1975 and instituted a cruel regime under the leadership of Pol Pot. The Khmer Rouge suspended formal education, emptied the cities, and set out to eliminate the rich and educated and to isolate themselves from the world, renaming the country Kampuchea. Mass murders in the so-called "killing fields" (depicted in the movie of the same name) and the brutal "death march" out of the capital Phnom Penh in 1975–76 killed at least 2 million people—a quarter of Cambodia's population, including most intellectuals and professionals—between 1975 and 1979, when Vietnam invaded and installed a new government. Conflict did not end even with the 1991 U.N. peace settlement and the return of the monarch Prince Sihanouk, because the Khmer Rouge retook power in a coup in 1997.

Economic Development and New Export Economies

The wars and instability in the countries of Indochina limited their economic development and trade with the world for more than 30 years. Meanwhile many other countries in Southeast Asia reoriented their economies, first of all to import substitution and later to export manufacturing. Malaysia, Singapore, and Thailand provide important regional examples of how government-led economic development policies brought Southeast Asia into a new relationship with the global system in the second half of the twentieth century. Southeast Asia is a classic case of how changes in the **international division of labor** fostered manufacturing in the global periphery.

In the early years after independence in the 1960s, governments pursued **import substitution** industrialization (ISI) policies that sought to develop a domestic capability in manufacturing rather than rely on imported goods. For example, in Malaysia tariffs on imports such as clothing and plastics were increased dramatically in order to protect locally produced goods from global competition. Dominated by non-Malay (especially Chinese) investment and producing low-value goods that quickly saturated domestic markets or competed, in the case of steel, with global overcapacity, the economic gains of ISI did not benefit much of the population or the national trade balance. Malaysia, Singapore, and Thailand all made decisions to shift toward export-oriented industrialization and to try to profit from their competitive advantages in the global economy, especially a low-cost but relatively well-educated workforce. Strong state involvement and incentives and high levels of foreign direct investment characterized economic development (**Figure 10.15**).

In 1971 Malaysia implemented a program for a New Economic Policy that set out to shift the economy away from a dependence on tin and rubber exports to higher value exports and to distribute the benefits of economic development to the ethnic Malay population called *Bumiputra* ("those of the soil"). This program explicitly discriminated against the ethnic Chinese populations. Incentives for foreign investment included tax breaks and freedom from customs tariffs when locating within a **Free Trade Zone (FTZ)** or Export Processing Zone (EPZ). From 1980 onward, foreign investment flowed into a range of Malaysian economic sectors and the share of manufacturing in exports increased from 20 to 70 percent.

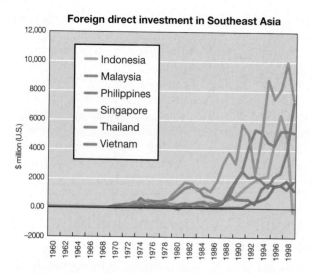

Figure 10.15 Foreign investment in Southeast Asia
Foreign direct investment increased during the 1980s and early 1990s in many Southeast Asian countries but dropped suddenly in 1997, precipitating a serious financial crisis. (*Source:* Data from World Bank, *World Development Indicators.* Washington, DC: World Bank, 2000.)

Manufacturing sectors that developed or relocated to Southeast Asia included automobile assembly, chemicals, and electronics. Japan was a major source of foreign investment in Southeast Asia, including the relocation of Japanese-owned firms seeking cheaper labor and land. This *offshore manufacturing* included 250 Japanese firms in Malaysia by 1990.

Thailand also attracted considerable Japanese investment and relocation of Japanese firms that were interested in the availability of cheap, well-educated labor, as well as comparative political stability and less discrimination against East Asians. Indonesia and the Philippines also pursued an export strategy, including electronics and apparel manufacturing in the Philippines and automobile manufacturing in Indonesia.

The rapid growth of these Southeast Asian economies, averaging 8 percent per year in the early 1980s, was seen as part of the larger East Asian economic miracle. The more successful countries—Thailand, Malaysia, and Singapore—joined the so-called **Asian tigers** of Hong Kong, South Korea, and Taiwan as the "little tigers" and were termed *Newly Industrializing Economies (NIEs)* rather than "developing" or "underdeveloped" countries. High rates of savings, balanced budgets, and low inflation were all indicators of a successful transition to modern industrial economies. Foreign investment flowed into Southeast Asia not only to benefit from cheap labor but also to take advantage of domestic markets for consumer goods, such as automobiles and soft drinks, and of valuable natural resources, such as oil and minerals.

The Crisis

The risks of such close financial and trade linkages to the global economy became dramatically evident in the 1997 collapse of Southeast Asian economies. The collapse was preceded by a slight decrease in competitiveness as wages increased in South-

east Asia compared to low-cost labor in China and by extensive international borrowing by governments and banks at low interest rates. Thailand was the first economy to fall when a decline in the value of the Japanese yen relative to the Thai baht (which was tied to the U.S. dollar) placed exports at a competitive disadvantage, and currency speculators on global markets began to bet on a currency devaluation. The Thai government raised interest rates and used up foreign exchange reserves in attempts to prop up the baht but was eventually forced to unlink from the dollar, causing a sharp fall in currency value. The high interest rates stopped economic development and the country fell into recession, with massive job layoffs.

In the 1997 panic, $12 billion was withdrawn from the region. Malaysian markets lost 80 percent of their value over a 2-year period. The crisis in the region was exacerbated by overinflation of stocks, speculative real estate development, and overlending by banks freed from government oversight. The impacts of the 1997 crisis were most severe in Indonesia, where the rupiah currency lost 80 percent of its value as the country coped with a severe drought caused by El Niño. Food prices increased, and government spending cuts resulted in the removal of subsidies on gas and kerosene. Public resentment focused on the ethnic Chinese, who fled the country, abandoning their shops and businesses and thereby creating further economic scarcities, and on President Suharto, who was forced to resign. Unemployment reached 40 percent, and those living below the poverty line increased from 10 percent to more than 50 percent of the population.

The International Monetary Fund agreed to help restructure debt and stabilize the economies so long as strict structural-adjustment policies were followed, including the reduction of corruption and removal of tariffs. Indonesia received the third largest bailout to date, requiring strict cutbacks in government spending and a reduction in corruption. Thailand and Malaysia responded to the crisis with austerity measures, including reductions in government spending and appeals to the public to accept reduced services and increased prices of basic goods.

The return of foreign investment has again brought the risks and benefits of globalization to the region. One overall indicator of linkage to the global economy is the relationship between the value of exports and the value of **gross domestic product (GDP)**. Singapore, with exports valued at 152 percent of GDP in 1998, and Malaysia, at 114 percent, are the most integrated, followed by Indonesia, the Philippines, and Thailand, with export value at about half the value of GDP. In terms of key global commodities, Southeast Asia produces more than half of the world's rubber, coconut, tin, palm oil, and hardwoods.

Peoples of Southeast Asia

The population of Southeast Asia was estimated at more than 500 million in 2000 (see World Data Appendix, p. 639). The country of Indonesia, with a population of 200 million, is the world's fourth largest country. Vietnam and the Philippines each have about 75 million people, followed by Thailand at 65 million, Burma at 45 million, and Malaysia at 22 million. Although there is archaeological evidence of early humans in Southeast Asia, the majority of contemporary Southeast Asians are descendants of migrants from East Asia.

Population growth and life expectancy in Southeast Asia surged with the eradication of malaria and improved medical care after about 1950. After several decades of growth at more than 2 percent a year, overall population growth has slowed, mainly as a result of significant declines in Indonesia, Thailand, Singapore, and Vietnam. In these countries the **total fertility rate** (average number of children born to a woman of childbearing age) has fallen from more than 6 to less than 3. In Singapore, where the fertility rate has fallen to 1.5—below the replacement level of 2.1—and population growth is negative, the government is now promoting marriage and childbearing, especially among the highly educated.

Fertility and population growth rates have remained higher in the Philippines, partly as a result of opposition to birth control by the Catholic Church, and in Malaysia, where the government encourages the ethnic and Muslim Malay population to have at least five children per married couple.

Fertility rates are also higher in the poorer countries of the region, including Cambodia and in Laos, where total fertility is 5.4 children per woman. Southeast Asian data support theories that fertility decline is associated with higher income, lower infant mortality, and higher status of women. Cambodia and Laos have lower GNP per capita, higher infant mortality, and lower levels of female literacy and schooling than do other countries in the region. In Cambodia, Laos, and Vietnam, overall population growth was until recently reduced by high death rates from war and famine.

Indonesia's population policy is often promoted as a model for noncoercive family planning. Indonesia promotes two-child families through advertising, grassroots leadership training, and free distribution of birth-control pills and condoms. Population growth rates in Indonesia have fallen from 3 percent to 1.5 percent.

The map of population distribution shows that people are concentrated in the river valleys and deltas and on the island of Java (**Figure 10.16**). The highest population densities (people per hectare) are in Singapore, the Philippines, and Vietnam. Levels of urbanization range from 100 percent in the city-state of Singapore, to more than 50 percent in Malaysia and the Philippines, to less than 25 percent in Cambodia, Laos, Thailand, and Vietnam.

Migration and the Southeast Asian Diaspora

There are three major types of migration within the countries of Southeast Asia. The first is the worldwide phenomenon of people flowing into the cities from rural areas driven by landlessness and agricultural stagnation or pulled by the attractions of urban areas, including job opportunities, education and health services, and access to consumer goods and pop-

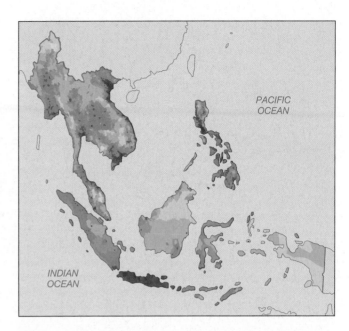

Figure 10.16   Population density in Southeast Asia 1995
The majority of the people of Southeast Asia live in or near river valleys and deltas, and on the islands of Java, the Philippines, and Singapore. For a view of worldwide population density and a guide to reading population-density maps, see Figure 1.40. (*Source:* Center for International Earth Science Information Network (CIESIN), Columbia University; International Food Policy Research Institute (IFPRI); and World Resources Institute (WRI). 2000. *Gridded Population of the World (GPW)*, Version 2. Palisades, NY: CIESIN, Columbia University. Available at http://sedac.ciesin.org/plue/gpw)

ular culture. The second set of flows arises from war and civil unrest within countries, such as the mass evacuations from cities in Cambodia under the Khmer Rouge and from war zones in Vietnam. The third, and distinctively Southeast Asian pattern, is the resettlement of populations from urban to rural areas, especially in Indonesia. The Dutch moved thousands of people from Java and Bali to Sumatra, Kalimantan, and Sulawesi to work on plantations in 1904, but this was dwarfed by massive relocation programs initiated by the Indonesian government beginning in 1950.

This **transmigration** program was designed to redistribute population from densely settled Java and the city of Jakarta to reduce civil unrest, increase food production in peripheral regions, and further goals of regional development, national integration, and the spread of the official Indonesian language. It is estimated that more than 4 million people moved to the Moluccas, Sulawesi, Sumatra, and Kalimantan, with 1.7 million of the migrants receiving official government sponsorship in the form of transport, land grants, and social services (**Figure 10.17**). Geographer Thomas Leinbach identifies a number of problems with the transmigration program. He includes lack of infrastructure in the new settlements, conflict between the settlers and indigenous groups, and the destruction of forests as peasant farmers accustomed to the fertile soils of Java cleared more rain forests as their crops failed because soil degraded after deforestation. Malaria, pests, and weed invasion also hindered the success of the program. The Philippines

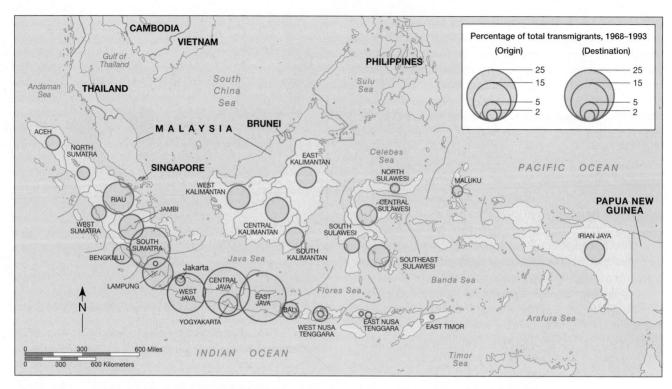

Figure 10.17   Transmigration flows in Indonesia The Indonesian government has relocated thousands of people from urban areas on Java to rural areas of the islands of the Moluccas, Sumatra, Sulawesi, and Borneo. (*Source:* Redrawn from T. R. Leinbach and R. Ulack, *Southeast Asia: Diversity and Development.* Upper Saddle River, NJ: Prentice Hall, 2000, Map 12.7.)

followed similar resettlement programs in subsidizing migration from Luzon to frontier regions such as Mindanao.

War, ethnic and religious conflict, and poverty within Southeast Asia forced thousands of people to move within the region between 1970 and 1995. Included in the total were more than 300,000 Laotians and 370,000 Cambodians to Thailand, about 300,000 Vietnamese to China and Hong Kong, 300,000 Muslims from Burma to Bangladesh, 200,000 Muslims from the Philippines to Malaysia, and 110,000 Burmese to Thailand. The U.N. High Commission for Refugees reports that in 1999 there were still 15,000 Cambodians and 100,000 Burmese in refugee camps in Thailand; 45,000 Filipinos in Malaysia; 22,000 Burmese still refugees in Bangladesh; 300,000 Vietnamese in China; 162,000 refugees from East Timor in Indonesia; and 8,000 Indonesians in Papua New Guinea. There is also a flow of labor migrants between countries, with more than 500,000 Indonesians working in the Malaysian construction industry and 400,000 Thais working in Singapore and Malaysia.

International migration to and from the region has a long history. The most important flow into the region has been the centuries of movement of Chinese into Southeast Asia beginning as early as the fourteenth century. Driven by civil wars, famine, and revolution in China, more than 20 million Chinese moved to Southeast Asia during the colonial period to work as contract plantation laborers to harvest rubber and to work in mines and railways. Many of these migrants then shifted into jobs in retail and trading and as clerical employees of the

colonial trading companies. These so-called **overseas Chinese** became essential to the success of the colonial economy, and upon independence they became the entrepreneurs who ran banks, insurance companies, and shipping and agricultural businesses. Explanations for the business success of the overseas Chinese include the Confucian tradition of family and ethnic business links and their providing their children with a professional education, sending them overseas and to private schools when local facilities were not accessible. As a result of this massive in-migration, ethnic Chinese make up a large percentage of the overall population in Singapore (78 percent), Malaysia (30 percent), and Thailand (11 percent). In other countries, such as Indonesia, ethnic Chinese make up a large percentage of the economic elite, even when they are a small percentage of total population (**Figure 10.18**). The overseas Chinese generally live in urban areas, in separate neighborhoods often known as Chinatowns, and with their own social clubs and schools.

The economic dominance and closed societies of the overseas Chinese are some of the reasons why they have been a target of discrimination in Southeast Asia. Singled out for persecution by the Japanese in the Second World War, overseas Chinese have also been vulnerable to communist groups who see them as a symbol of capitalism. One of the main reasons for Singapore independence from the Malay Federation was the desire of many Malays to detach from this city dominated by ethnic Chinese. In contemporary Malaysia, where ethnic Malays are favored, the Chinese must have a Malay business

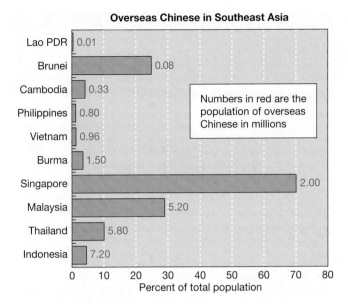

Figure 10.18 Overseas Chinese in Southeast Asia
Overseas Chinese make up a significant percentage of the population in Singapore, Brunei, Malaysia, and Thailand. (Source: Data from J. Rigg, *Southeast Asia: The Human Landscape of Modernization and Development.* New York: Routledge, 1997, p. 123.)

partner and are restricted by quotas in government jobs and higher education. In Indonesia, anger during the economic crisis of 1997–98 was focused on looting and murder of Chinese business owners, many of whom fled the country.

Out-migration from Southeast Asia includes large refugee flows and many thousands of labor migrants (**Figure 10.19a**). Since 1974, 1.5 million people have left Vietnam as refugees from war and communism. About half of the Vietnamese refugees went to the United States, and France, Canada, and Australia accepted others. U.S. cities such as San Francisco have large Vietnamese populations living in distinctive neighborhoods with a strong Vietnamese heritage. Some remained in refugee camps in Hong Kong. After Vietnam invaded Laos in 1975, more than 300,000 Hmong people fled to Thailand, fearing persecution because they had supported the Americans; they were subsequently resettled in the United States. Recently, with peace and reform, many Vietnamese are returning to Vietnam, and some bring capital for investment.

The largest numbers of labor migrants from Southeast Asia work in the Middle East, especially in Saudi Arabia, Kuwait, and Oman, and in Hong Kong and Japan. Many of the workers in the Middle East are Muslim women working in the service sector as nurses and maids. Philippine men have a tradition of joining the merchant marines of many countries and working as cooks, seamen, and mechanics. Thousands of Thais and Filipinos work in Hong Kong, where their wages are typically much lower than those paid to local laborers. Many Philippine women work as maids or nannies in North America, Europe, and Singapore (see Geography Matters: Chains of Love, p. 498).

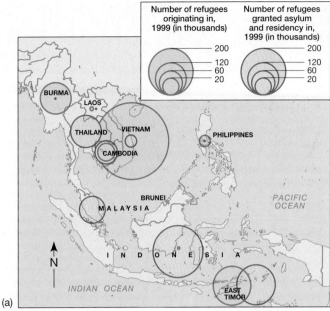

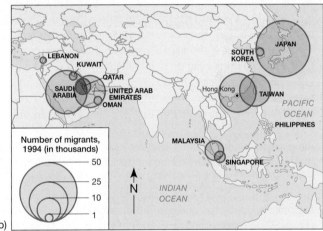

Figure 10.19 Migrations in Southeast Asia (a) Southeast Asia has experienced enormous refugee flows within and from the region, including out-migrations from Vietnam, Burma, Laos, and Cambodia to Thailand and beyond the region, from the Philippines to Malaysia, and from East Timor to Indonesia. In 1999 the total refugees registered with the U.N. in the region was more than 700,000 people but was less than in many previous years. (b) This map shows where Philippine labor migrants were moving in 1994 and illustrates the significance of remittances from workers in Japan, Hong Kong, and the Middle East to the Philippine economy. [*Source:* (a) U.N. High Commission on Refugees, *Refugees and Others of Concern to UNHCR 1999 Statistical Overview.* Geneva, Switzerland: Statistical Unit United Nations High Commissioner for Refugees, July 1998, Table I.2 and Table II.2; (b) Redrawn from T. R. Leinbach and R. Ulack, *Southeast Asia: Diversity and Development.* Upper Saddle River, NJ: Prentice Hall, 2000, Map 4.4.]

The money sent back by these labor migrants (*remittances*) is very important to national and local economies. For example, it is estimated that $10 billion was sent back to the Philippines in 1994, an amount equal to between 10 and 20 percent of the GNP and supporting more than 8 million jobs (**Figure 10.19b**). Thailand's remittances are worth more than $1.6 billion each year.

A Day in the Life

Chains of Love

Vicky Diaz is a 34-year-old mother of five. A former teacher in the Philippines, she migrated to the United States to work as a housekeeper and nanny to the two-year-old son of a wealthy family in Beverly Hills, Los Angeles. "Even now, my children are trying to convince me to go home," she says. "They weren't angry when I left because they were still very young. My husband couldn't get angry either because he knew this was the only way I could help him raise our children. I send them money every month." The Beverly Hills family pays Vicky $400 a week; Vicky pays her own family's live-in domestic worker back in the Philippines $40 a week. But living in this "global care chain" is not easy on Vicky and her family. "I'm paid well, but I'm sinking in work. Even while I am ironing the clothes, they can still call you to the kitchen to wash the plates. It is also very depressing. Away from my children, the only thing I can do is give all my love to Tommy [the two-year-old American child]."

Global care chains are usually made up of women. Most start in a poor country and end in a rich one. The chain Vicky Diaz finds herself in is an increasingly common one: (1) an older daughter from a poor family cares for her siblings while (2) her mother works as a nanny caring for the children of a migrating nanny who (3) cares for the child of a family in a rich country. Are first-world countries importing maternal love as they have imported copper, zinc, and gold from third-world countries in the past? Each person along the chain feels he or she is doing the right thing for good reasons. But transfer of care takes its toll both on the child left behind and on the mother. Sometimes the toll it takes is overwhelming. As one Filipino nanny said: "The first two years I felt I was going crazy. I would catch myself thinking about my baby. Every moment, every second of the day. When I receive a letter from my children, I cannot sleep. I cry." Given the depth of this unhappiness, one might imagine care chains are a minimal part of the global show. But this is not the case. Since the early '50s, for example, 55% of migrants from the Philippines have been women and, next to electronics manufacturing, their remittances make up the country's major source of foreign currency. Recent improvements in the economy have not reduced female emigration; rather, it continues to increase. In addition, migrants are not drawn from the poorest classes. In one study of migrant Filipino domestic workers, over half the nannies were graduates and most were married mothers in their 30s. Money provides a powerful incentive to work and the yawning global wage gap provides a powerful incentive to move. Migration is, simply, a ticket to a better life. Just as global capitalism helps create a third-world supply of mothering, so it creates a first-world demand for it. At the first-world end, there has been a huge rise in the number of women in paid work—from 15% of mothers of children aged six and under in 1950 to 65% today. Indeed, American women now make up 45% of the U.S. labor force; 75% of mothers of children aged 18 and under now work, as do 65% of mothers of children aged six and under. Thus, at the first-world end of care chains, we find working parents who are grateful to find a good nanny and able to pay more than the nanny could earn in her native country. Given the existence of this global care chain, and given the chain's growing scope, it is worth asking how we are to respond. One approach is to try to reduce incentives to migrate by addressing the causes of the migrant's economic desperation. Another would be to raise the value of caring work, such that whoever did it got more credit as well as money for it, and care wasn't such a "pass on" job. The value of the labor of raising a child—always low relative to the value of other kinds of labor—has, under the impact of globalization, sunk lower still. The declining value of childcare can be compared with the declining value of basic food crops relative to manufactured goods on the international market. If we want developed societies, with women doctors, political leaders, teachers, and computer programmers, we will need qualified people to help care for their children. And there is no reason why every society should not enjoy such loving paid childcare. For these days, the personal is global.

Source: Extracted from A. R. Hochschild, "Global Care Chains and Emotional Surplus Value," in A. Giddens and W. Hutton (eds.), *On the Edge: Globalization and the New Millenium* (pp. 130–146). London: Sage, 2000.

Urbanization

About half of Southeast Asia's people live in cities, most of which have grown rapidly since 1950. Most of this growth is driven by migration from rural areas. Bangkok and Kuala Lumpur jumped in size as a result of annexation of neighboring regions into the metropolitan areas. Urban growth in Cambodia and Vietnam was slower, or even reversed, under communist governments whose policy was to return people to rural areas to grow food or develop small rural industries. The region is dominated by several enormous cities with metropolitan-area populations estimated at about 10 million, among them Bangkok, Thailand, Manila, and Jakarta. These cities have grown so rapidly that they have serious problems of **overurbanization**, with insufficient employment opportunities, an inadequate water supply, sewerage problems, and inadequate housing. Several countries have two major urban centers. In the case of Burma, this resulted from the relocation of its capital, from Mandalay to Rangoon (now Yangon). The unification of North and South Vietnam left the new country with two major urban centers—Hanoi and Ho Chi Minh City (formerly Saigon). New administrative capitals have been established to promote decentralization in Quezon City (Philippines) and Putrajaya (Malaysia), but their proximity to Manila and Kuala Lumpur has created one large urban region in each case. Key secondary cities with more than half a million residents include Palambang, Medan, Bandung, Ujung Pandang, and Surabaya in Indonesia, and Cebu and Davao in the Philippines.

Cities in Southeast Asia mix traditional design with that of colonial and modern planning, but they are surrounded by unplanned growth, including desperately poor squatter settlements of recent migrants. The cities retain characteristics of earlier functions and include sacred cities such as Mandalay, trading/port cities such as Malacca, cities established as colonial centers such as Manila and Jakarta (formerly Batavia), the newer colonial port cities such as Ho Chi Minh City (Saigon) and Yangon (formerly Rangoon), and the postsocialist cities of Hanoi and Yangon. The traditional core of the city, symbolized by temples, was replaced by the fort or garrison under colonialism and by a high-rise central business district in modern times (**Figure 10.20**). The specific character and problems of Bangkok, Jakarta, Manila, and Singapore are discussed later in the chapter, as they are centers of core regions. Singapore is the only true **world city** in Southeast Asia, acting as a world financial and trade center and as a major gateway to Asia.

Geographer Terence McGee, who has written at length on cities in Southeast Asia, argues that the region has a distinctive urban landscape in the form of extended metropolitan regions (*desakota*), as cities have extended out into intensively farmed agriculture, especially rice paddies, and town, industry, and agricultural villages have become intermixed. McGee's work reminds us that boundaries between the city and the countryside in Southeast Asia are often blurred. There are many small industries, such as textiles, in rural areas, and considerable agricultural production in the cities.

Ethnicity, Language, and Cultural Traditions

Southeast Asia accommodates an incredibly diverse set of cultures, which is reflected in more than 500 distinct ethnic and language groups. Indonesia, for example, has more than 300 distinct ethnic groups and languages, and the population of Laos speaks more than 90 different languages. In general, the cultures have remained distinct from each other and from the cultures of migrants who arrived from India, China, Europe, and other world regions, and no language has unified the region in the way that Arabic has in the Middle East or Spanish has in Latin America. The most common dialects are versions of Malay spoken in Malaysia, Indonesia, and Brunei, and used as a *lingua franca* (trade language). The major language fami-

Figure 10.20   Petronas Towers Kuala Lumpur, in Malaysia, provides a good example of a Southeast Asian city with a mixture of ancient temples and British colonial buildings dominated by the modern Petronas Towers. Many poorer residents live on raft houses in rivers and bays and cultivate crops on the riverbanks.

lies are Austronesian (or Malayo-Polynesian), spoken in insular Southeast Asia and the Malay peninsula; Tibeto-Burmese in Burma; Tai-Kadai focused on Thailand; Papuan in New Guinea; and the Austro-Asiatic languages of Vietnam and Cambodia (**Figure 10.21**).

Some national boundaries of Southeast Asia do enclose relatively coherent cultural and language groups. Thailand (Thai), Burma (Burmese), Cambodia (Khmer), Vietnam (Vietnamese), Laos (Lao), and Malaysia (Malay) all have large majority populations that speak the same language and share cultural and religious traditions. The adoption of a common language in Indonesia and Pilipino in the Philippines as official national languages are attempts at integration of many diverse cultural groups. However, there are significant minority populations in many of these countries, including the Karen and Shan in Burma and the Hmong in Laos, as well as large populations of Chinese and Indians in many countries. Even within these groups there are distinct cultural and regional differences. For example, the Indian populations include Bengalis and Sikhs in Burma and Tamils in Malaya. The imprint of Indian and Chinese cultural traditions is strong in Southeast Asia. The monuments of the early Hindu kingdoms reflect the art and architecture of India, as do the crafts and rituals of contemporary Bali (see Sense of Place: Bali, p. 501). Chinese

Figure 10.21 Map of languages Southeast Asia has hundreds of distinct languages, which fall into five major language families and are dominated by versions of Malay. (*Source:* Redrawn from R. Ulack and G. Pauer, *Atlas of Southeast Asia.* New York: Macmillan, 1988, p. 27.)

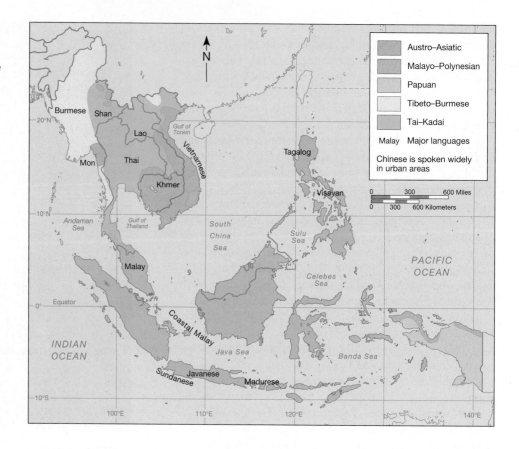

business traditions were spread by the millions of overseas Chinese who settled in the region.

The U.S. control of the Philippines and subsequent presence of major military bases has had a strong impact on local culture, and Thailand is also relatively open to U.S. and global culture. Official government criticism and popular resistance to U.S. culture and international media is more common in countries with communist regimes and where Islam is important.

The global influence of Southeast Asia is considerable, although the music, literature, and art of the region is not well known in the United States, and culinary influences are often confused with Western versions of Chinese food. However, Thai and Vietnamese restaurants are becoming more popular with the distinctive use of spices, lemon grass, and coconut milk, and tourism is increasing awareness about the region.

Religion The contemporary religious geography of Southeast Asia reflects centuries of evolution under Indian, Chinese, Arab, and European influence. Most generally, Buddhism tends to dominate the mainland region and Islam the islands (**Figure 10.22**). Buddhism includes both Theravada and Mahayana, the former more conservative and found mostly in Burma, Thailand, and Cambodia, and the latter associated with Vietnam. Islamic believers in Indonesia now outnumber those in the Middle East, although the practices are sometimes seen as more liberal than in the Middle East. Islam is also important in Malaya and is growing in the Philippines. The in-

tensification of fundamentalist Islamic belief has resulted in the seclusion and veiling of women and political conflict over the enforcement of Islamic law, especially in diverse populations. Hinduism is common in Burma and in Java and Bali. Christianity is the religion of 85 percent of the Philippine population as a legacy of Spanish colonialism and is also found among the hill tribes of Burma and in Vietnam, converted by French and British missionaries who could not make inroads into the Buddhist beliefs of lowland residents. Remote indigenous groups have maintained traditional animistic beliefs that imbue nature with spiritual meaning, especially in the mountains of Burma, Laos, and Vietnam, and in Borneo and Irian Jaya.

Family Life and Women's Roles In general, women have more power within Southeast Asian families and societies than in many other world regions. For example, couples often live with the wife's parents, with authority passing from father to son-in-law. This often gives women more power because they live with their own family rather than with in-laws. In Indonesia, women often manage the family money. The value of daughters is reflected in a more equal preference for female compared to male children than in either East or South Asia, where preferences are highly biased toward males. Women also have new employment opportunities in services and manufacturing and are preferred by many high-technology companies because they receive lower wages and are perceived as more careful and docile.

Sense of Place

Bali

With images of waving palm trees, soft breezes, towering volcanoes, and pristine beaches, Bali has been promoted to the world as a tropical paradise ideal for romantic vacations (**Figure 1**). This small island of 3 million residents in the Indonesian archipelago has become a major international tourist destination, with more than 4 million visitors a year. Bali has maintained Hindu traditions within an increasingly Muslim Indonesia and has a centuries-old landscape of rice terraces and temples. The government and the tourist industry promote a unique Balinese culture, tropical climate, and extensive beaches. The cultural attractions include Hindu temples dating from the fourth century A.D., crafts such as batik cloth and wood carving, dance, traditional *gamelan* orchestras, and the retelling of Hindu and other legends with intricately carved shadow puppets. More than 1 million tourists fly directly into Bali each year, with perhaps another million traveling from Java and neighboring countries, with tourism making up nearly 40 percent of the gross domestic product of the island economy. The standard of living in Bali is much higher on average than in other parts of Indonesia.

Tourism to Bali began to grow in the 1920s when Dutch steamships began to stop offshore and the luxurious Bali Hotel opened in the capital of Denpasar. An art colony grew up around the community of Ubud in the lush interior hills, and members of the international artistic and literary community who arrived to experience exotic culture introduced local artists to new techniques and opportunities. In 1937, Mexican author Miguel Covarrubias wrote a book on Bali in which he lamented that the island was "doomed to disappear under the merciless onslaught of modern commercialism and standardization." In 1972, the Master Plan for the Development of Tourism in Bali was drawn by the government of Indonesia to make Bali the "showcase" of Indonesia and to serve as the model of future tourism development for the rest of the country. The plan was financed by the U.N. Development Programme and carried out by the World Bank. The plan called for the focus of development in the southern peninsula of the island, Nusa Dua, with only day trips to the interior in order to protect cultural integrity. Bali has become a focus for research on the costs and benefits of tourism. Critics charge that the culture has lost its authenticity in the construction of images for international tourists who come especially from Australia and Europe and that economic development has placed unsustainable pressures on water supplies and coastal ecosystems (**Figure 2**). Coral reefs have been destroyed to build new tourist accommodations, and waste collection is inadequate. Some local people complain that the mass marketing of their culture is destroying their way of life and that communities are being forced to compete for tourists through lavish festivals and are ignoring the traditional laws (*adat*) that govern everything from land tenure to social relations. But geographer Geoffrey Wall, an expert on tourism in Bali, found that many residents view tourism positively and see it as fostering their culture, although those who live closest to their resorts had more concerns. Wall has become involved in the Bali Sustainable Development project that seeks to reduce the negative environmental and cultural impacts of tourism in Bali.

Figure 1   The Ulu Danu temple   The Ulu Danu temple on Lake Bratan in Bali is one of the cultural attractions often visited on day trips to Bali's verdant interior.

Figure 2   Tourists in Bali   Local people crowd a tourist bus in Bali trying to sell handiwork carvings based on traditional images.

Figure 10.22 Religion The geography of religious belief in Southeast Asia is dominated by versions of Buddhism on the mainland and Islam on the islands. Catholicism is important in the Philippines, and indigenous groups maintain animist beliefs in remoter mountain and island areas. (*Source:* Redrawn from R. Ulack and G. Pauer, *Atlas of Southeast Asia.* New York: Macmillan, 1988, p. 29.)

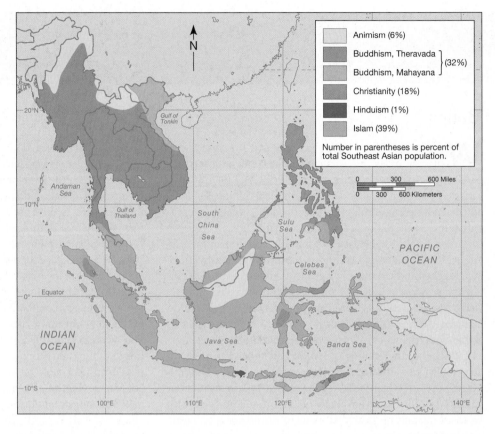

Nevertheless, cultures are patriarchal, and socioeconomic conditions are generally better for men than for women. Men tend to receive higher wages and more education. One of the more exploitative professions for women is as a sex worker in the tourist industry, most notably in Thailand, which is one of the world's centers of prostitution. Millions of sex tourists, especially Germans and Japanese, visit Thailand each year. Some of the sex workers are very young women sold into bondage by rural families or smuggled in slavelike conditions from Burma. Geographer Jonathan Rigg quotes studies suggesting that for women from poor rural families, sex work may appear as a ra-

tional choice for making a living, providing opportunities for them to send money back to their villages and families.

In recent years families in Southeast Asia have become more nuclear, in contrast to traditional extended links, and this has changed patterns of obligation from a focus on parents and older people to children (**Figure 10.23**). There are fewer arranged marriages or men with more than one wife, and there is less child labor. These changes are likely due to urbanization, the media, and higher education levels.

Religion and cultural tradition intersect in the reverence for the monarchy in countries such as Thailand and

Figure 10.23 A Thai family The Kuankaew family gather with their possessions, including a motor scooter, television, and video game, outside their house in Ban Muang Wa near Chiang Mai, Thailand. Their water buffalo, rice field, and several fruit trees can be seen in the background.

Sense of Place

Brunei

With its oil wealth, Islamic sultanate, and modern settlements, Brunei provides a dramatic contrast to the rest of Borneo. Brunei Darussalam (the full name of the country) is situated on the northwest of the island of Borneo and has a coastline of about 160 kilometers (100 miles) along the South China Sea. It is one of the smallest and wealthiest nations in the world, similar to Kuwait and Qatar in the Middle East in its reliance on oil profits. The country is split in two by the Malaysian state of Sarawak and consists of a low coastal plain with interior foothills. The discovery of oil in Brunei, with production beginning in 1932, transformed the former British protectorate at a time when it was vulnerable to the territorial claims of newly emerging independent regional powers of Malaysia and Indonesia.

The 350,000 people who live in Brunei benefit from a per capita GNP of more than $25,000 per year and heavy government investment in health and education. Brunei has what is called a *rentier* economy; national income is mainly derived from external sources, in this case oil exports, and not from domestic taxation.

The sultan of Brunei has a personal fortune of almost $40 billion and is reputed to own more than 150 cars, the Beverly Hills Hotel in Los Angeles, and a fleet of 200 Argentinean polo ponies. But residents of the nation have also benefited. They pay no taxes and are provided with free education and medical services, subsidies for consumer goods, and sponsorship for pilgrimages to Mecca, the center of Islam. The government has attempted to diversify away from oil by investing in agriculture

locally and internationally, including the purchase of a cattle ranch in Australia that is much larger than the country of Brunei itself. Recent reports claim that the royal family has squandered much of their fortune and had to auction items to pay their debts.

The landscape of the capital, Bandar (**Figure 1**), is dominated by the Sultan's palace and the magnificent Omar Ali Saifuddin mosque, but the neighborhood of Kampung Ayer preserves the traditional water villages of wooden houses built on stilts above the Brunei River and linked by bridges.

Figure 1   The Omar Ali Saifuddin mosque This mosque dominates the skyline of Bandar, the capital of Brunei Darussalam. Traditional water villages line the riverbank.

Cambodia. The Thai royal family is held in high, almost godlike esteem, with pictures of the king in the majority of homes. The royal family has selectively intervened during times of political crisis and plays an important balancing and leadership role.

Regional Change and Interdependence

Because Southeast Asia is so closely linked to the global economy, many changes, including economic development and crisis, are driven from outside the region. However, there are a number of shared regional dynamics and challenges that have originated or are sustained within the region, including patterns of inequality, political conflict and cooperation, agricultural development, and environmental degradation.

Inequality

There are large variations in economic and social conditions between Southeast Asia and other world regions and within Southeast Asia itself. The region includes countries that have some of the world's highest per capita gross national products (for example, in 1997 Singapore was at $32,960, and Brunei was at $27,270) but also some of the lowest (for example, Cambodia, with 1998 per capita GNP of only $260). The distribution of income tends to be most unequal in the strongest economies. For example, the wealthiest 10 percent control more than one-third of the wealth in Malaysia, the Philippines, Singapore, and Thailand, but generally levels of inequality are somewhat less than in other regions, such as Africa or Latin America. There are wide variations in how poverty is defined, but there do seem to be general improvements in overall conditions. For example, Malaysia reduced the proportion of people living in poverty from 60 percent in 1970 to 15 percent by 1990. Geographer Jonathan Rigg suggests that the poorest and

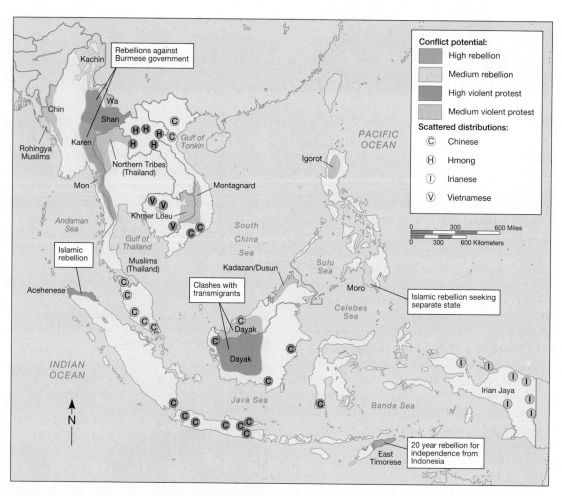

Figure 10.24   Map of conflict zones This map shows some of the major zones of conflict in Southeast Asia, including separatist movements in the outer islands of Indonesia and the Philippines. (*Source:* Adapted from T. R. Leinbach and R. Ulack, *Southeast Asia: Diversity and Development.* Upper Saddle River, NJ: Prentice Hall, 2000, Map 10.2)

most disadvantaged populations include those living in areas of extractive industry, such as timber and mining, indigenous groups, religious and ethnic minorities, guest workers, refugees, and the young and elderly. He points out that some are disadvantaged by who they are (ethnicity and religion) and others as a result of their occupation (urban informal sector such as sex worker or garbage picker) but that there is considerable variation within the region and within each group. Urban residents tend to be much better off than rural dwellers. In Malaysia and Thailand about a quarter of rural people live in poverty, compared to less than 10 percent of urban residents. Most Southeast Asian countries have considerable regional inequality.

Wealth concentration in Southeast Asia has been associated with *crony capitalism,* in which leaders allow friends and family to control the economy, and *kleptocracy,* in which leaders divert national resources for their personal gain. Evidence and public perception of such behavior by President Marcos in the Philippines led to uprisings and repression during his dictatorship.

Critics suggest that income and GNP per capita is an inadequate measure of living conditions and that indicators of basic human development, such as life expectancy, infant mortality, and literacy, provide more appropriate insights into levels of development. For all three of these indicators, Southeast Asia is above the world averages and has shown great improvements over recent decades. Brunei, Malaysia, Singapore, and Thailand show much better conditions than Burma, Cambodia, and Laos (see Sense of Place: Brunei, page 503). Indonesia, the Philippines, and Vietnam rank well above conditions in regions such as Sub-Saharan Africa and South Asia.

As noted earlier, women are more equal to men in Southeast Asia than in many other regions. Female literacy averages 85 percent, which is 92 percent of the male rate, compared to an average for women of 80 percent of the male rate in all developing countries; this rate is also better than that in both South and East Asia. In addition, women participate in the economy at higher rates than in other regions, at about three-quarters the rate of men.

Despite these relatively positive measures, many workers in Southeast Asia are exploited in terms of wages and labor conditions compared to workers producing similar goods in North America and Europe. Many work 12-hour days and seven days a week without benefits or unions, making products such as clothing and electronics for global markets. International companies such as Nike have come under criticism for sweatshop labor practices in countries such as Vietnam and Cambodia, where wages are less than $2 a day and workers earn less than 5 percent of what workers earn producing similar goods in the United States (see Day in the Life: Cambodia and Deth Chrib, p. 506).

Political Instability and Cooperation

Legacies of colonialism, territorial expansionism by some countries in the region after independence, and Cold War conflicts such as the Vietnam War contribute to continuing political instability in some regions of Southeast Asia. Some of the most long-lasting conflicts are those in areas ethnic minorities are seeking recognition and independence or where fractured physical and cultural geographies have made national unification more difficult (**Figure 10.24**).

In Indonesia, the outer islands have long resisted the dominance of Java, with the most serious insurrections in East Timor, Irian Jaya, and Aceh. East Timor is mostly Christian, a former Portuguese colony that occupies the eastern portion of the island of Timor, at the eastern end of the Indonesian archipelago. Initial hopes for independence in East Timor were dashed when Indonesia occupied the region as Portugal gave up colonial control in 1976. Twenty years of resistance and more than 200,000 deaths finally resulted in a local vote for independence in 1999. After the vote, anti-independence militias went on a violent rampage and thousands of refugees fled to West Timor. In 2000, East Timor was granted status as a newly independent country under initial U.N. administration. East Timor will become the eleventh nation in Southeast Asia.

In Irian Jaya, retained by the Dutch for more than a decade after Indonesian independence, a 1962 vote for union with Indonesia provoked a resistance movement because of suspicions of vote rigging by the Indonesian government, which then resettled many transmigrants on the island despite local opposition. A rise of Islamic fundamentalism in Aceh, the far west region of Sumatra, is associated with an armed rebellion with roots in decades of resistance to Dutch colonialism. In Kalimantan, indigenous peoples, especially the Dayak, have clashed with transmigrants from other regions of Indonesia.

Separatist movements are also found in the islands of the Philippines. The Islamic Moro liberation front on Mindanao has sought a separate state on religious differences and opposes resettlement of migrants on the island. Communist rebels, the New Peoples Army, were active during the 1980s. Tribal peoples in northern Luzon have rebelled against dams and deforestation. Repression under the Marcos government has been replaced with attempts at mediation and concession with these groups.

Ever since independence in 1948, the Burmese government has faced serious rebellion from hill tribes and religious minorities, who resist the political, linguistic, and religious domination of the Buddhist and Burmese-speaking authority. Muslims on the northern Arakan coast fled into neighboring Bangladesh. The Karen, who live along the border with Thai-

land and were favored by the British because of their Christian beliefs, were repressed by the independent Burmese government. With a population of 3.4 million, the Karen have been able to establish an insurgent state supported by smuggling, diamonds, and opium (see Geographies of Indulgence, Desire, and Addiction: Opium, p. 508). The Shan, with a population of 4.5 million in northern Burma, have similarly used drug profits to establish control of their territory and oppose the Burmese government.

The repressive authority of the Burmese military government is also opposed by Burmese who want a more democratic government. Popular protests in 1988 resulted in martial law and the establishment of the State Law and Order Restoration Council (SLORC—now known as the State Peace and Development Council). Elections were held in 1990 in which an opposition party, the National League for Democracy (NLD), won 60 percent of the vote, compared to 21 percent for the existing government party. SLORC cancelled the election results and remained in power. One of the leaders of the NLD is Aung San Suu Kyi, daughter of Burma independence hero General Aung San (**Figure 10.25**). Resolute in her quest for democracy and in her opposition to the authoritarian approach of SLORC, she was kept under house arrest by the government and constitutionally barred from leading the country because she had lived abroad. She was awarded the Nobel Peace Prize in 1991.

Defenders of more authoritarian styles of both military and elected governments in Southeast Asia claim that such systems are based on traditional *Asian values*. Among them are the Confucian idea of individual submission to authority, which emphasizes the importance of putting the nation and community before self, of consensus rather than conflict, of

Figure 10.25 Aung San Suu Kyi Nobel Peace Prize-winner Aung San Suu Kyi speaks to supporters while under house arrest in Burma.

A Day in the Life

Cambodia and Deth Chrib

Deth Chrib sits in front of a sewing machine 16 hours a day, seven days a week, in a Phnom Penh garment factory (**Figure 1**). It is the best job she has ever had. "It's pretty easy, compared to working on a farm," Deth says. "The only problem is that sometimes I think too much about my children and I can't focus on sewing." Now Deth is in danger of losing her job because her employer, June Textiles, a garment factory owned by a Singapore-based corporation, has been accused by Western media of using child labor. U.S. casual and sportswear giants Gap and Nike are threatening to cut off their contracts. Deth, 30, has little knowledge of the issues and forces that will determine her fate, but she is aware of what it could be. "I know Nike is a company that orders clothes from June, but I don't know who wears them," Deth says. "They only give me the cloth and I sew it. But now I've heard that maybe we won't have any more clothes to sew. I'm very worried because I don't want to be a prostitute again."

Deth is a pawn in the worldwide struggle over globalization. From Seattle to Phnom Penh, protesters are fighting the incursion of supposedly rapacious multinational corporations. American college campuses are abuzz with the dangers of free trade, which activists say allows rich companies like Nike to exploit workers like Deth. But Deth is just worried about losing her job. There is no doubt that working in a Cambodian garment factory is tough. The hours are long and the pace is relentless.

Deth's worries began in September with the production of a British TV documentary featuring an interview with a June Textiles worker who said she was just 14 years old. That is one year younger than Cambodian law allows. The company, the government, and even some union officials said that the report was not true, that the girl had lied, but Nike and Gap swung into damage control mode—and stopped their contracts with June. Together, the two labels account for three-quarters of the output of June's 3800 workers. The factory and the U.S. clients are still talking, but the outcome is uncertain. "If the situation continues, it will affect the workers," says C. K. Chang, June Textiles' Malaysian deputy general manager. "We won't be able to keep all of them."

The factory has been little short of salvation for Deth. Its existence is a sign of Cambodia's tentative opening to foreign investment that could help boost the economy. Deth lived through the dark days of the Khmer Rouge regime from 1975 to 1979—when two million Cambodians were killed through overwork, starvation, torture, and executions. She had to do heavy manual labor, despite being just five years old. Two brothers died. After the regime's ouster, Deth's family started a farm in Takeo province, southeast of Phnom Penh. The work took its toll, and Deth's parents died of illnesses when she was 19. At the time,

Figure 1   Deth Chrib and her children in Cambodia

Deth had two children, but no husband. She left the farm to seek a better life in the city. Arriving in Phnom Penh in 1992, she could only find a job as a café waitress, earning $12 a month. After several months, someone told Deth she could make more as a prostitute. She sent her children back to the family farm to be cared for by her sister and moved into a brothel.

"My living condition was very bad," Deth says, explaining her decision. "I had nothing and things kept getting worse and worse, so I became a prostitute. No one in my family ever did anything like that." It was a decision she regretted. Day and night, she was pressured to bring in more customers. "In the brothel, I worked 24 hours a day," she says. "We only slept when there were not guests." On average, she serviced four men a day. If she did not get at least three, she would be scolded and sometimes beaten. For each customer, she earned about 50 cents. After a year of selling her body and enduring physical abuse from both clients and the brothel owner, a cousin told Deth about the great job she had at a garment factory and urged Deth to apply.

Now, Deth earns about $60 a month, including over-time—more than triple what she used to—sewing T-shirts and shorts. Best of all, she no longer has to lie to her family about her job. "Garment factory jobs are very good for women because we would be prostitutes if we didn't have that opportunity," Deth says. "My condition is much better now. I have a good reputation and society doesn't consider me bad."

It is still not an easy life. Deth now has five children, three daughters and two sons. She had married, but her husband left her two years ago. However, she is now able to support her family on her own and lives with her children in a $10-a-month room in a wooden house near June Textiles. There is a mat on the floor, a hammock, and a few boxes and plastic bags filled with clothes. None of her children go to school. Her two oldest children, girls aged 10 and 8, help out by chopping wood, earning 18 cents for each pack of 100 pieces, and looking after the younger ones. Deth wants her sons to attend school, but cannot afford the 15 cents per term fee for each. She hopes her daughters can follow in her footsteps and become garment workers when they are old enough.

The approximately 200 garment factories in Cambodia employ some 150,000 workers or just over 1% of the population, and account for 90% of the country's export earnings. Earlier this year, thousands of workers from Phnom Penh garment factories went on strike, demanding a minimum wage of $70 a month. The then-prevailing level of $40 had been set in 1997, while the Cambodian Labor Organization, a local NGO, estimated that a family of five living in the capital spent $191 a month. After six days and several violent clashes, the strikers got a $5 raise. Other issues such as forced overtime and intimidation of union organizers remain unresolved.

But child labor is not a major problem in Cambodian garment factories. Quite simply, adult workers willing to take low-paid jobs are plentiful in this impoverished country. Factories want to keep clients like Gap and Nike happy, partly because those corporations pay more. Nike will pay $8 for a shirt, as opposed to the $6 paid by lesser-known buyers. "Working to improve labor conditions has already become a permanent part of the industry because of human rights groups, corporate buyers, and the U.S. quota system," Chang says. "We are aware of all of them when we are dealing with our workers." But improvements can carry a cost to workers. Officials from Gap recently came to town to go through employee documents at the 25 factories in Cambodia that make clothing for the company, check workers' ages and train factory managers on its code of conduct. It all sounds good. But at Best Honour International Garment, 37 workers have been fired so far for not being able to provide adequate proof that they are at least 18.

There are no easy answers. Unless powerful Western buyers, pushed by conscience-stricken customers and the media, play hardball with factory managers, workers may be subject to terrible abuses—use of children, forced overtime, limits on toilet breaks, locked fire escapes. But cracking down on bad practices can hurt innocent workers as well. For Deth, there is only one certainty. "I don't know what the fate of my children will be if I lose my job," she says. "But even if it's hard, even if I have to do something bad, I will always feed my children."

Source: Extracted from G. Chon, "Dropped Stitches." *Asia Week,* vol. 26, no. 50, Dec. 22, 2000.

---

hard work and discipline, and of racial and religious harmony. This philosophy has been used to justify control of the press and constraints on freedom of expression by dominant parties who control government, the media, and sometimes even election outcomes. Calls for more open democracy and freedom of speech have increased with economic development, education, and a more diverse and international media.

Despite these internal, regional, and international political conflicts, Southeast Asia provides a model for economic and political cooperation in the form of **ASEAN**—the Association of Southeast Asian Nations formed in 1967 (**Figure 10.26**). ASEAN was formed with the goal of encouraging intra-regional trade and reducing the political interference of both the United States and the Soviet Union. The initial alliance of Indonesia, Malaysia, Singapore, and Thailand was distinctly anticommunist and formed a defensive regional security group against China. The Asian Free Trade Association (AFTA) was

Figure 10.26   ASEAN ministers Foreign Ministers of the Association of Southeast Asian Nations (ASEAN) link arms on the occasion of Cambodia's entry into the organization in 1999.

Geographies of Indulgence, Desire, and Addiction

Opium

The mist-shrouded highlands of Burma, Laos, and Thailand contain much of mainland Southeast Asia's remaining forests, but they are also known for the fields of opium poppies that provide as much as half of the global supply of the addictive drug known as heroin. The drug economy has fueled rebellions and has even evolved a tourist industry in the exotically named Golden Triangle that centers on the region where Burma, Laos, and Thailand intersect (**Figure 1**).

The allure of the opium poppy as a narcotic is centuries old and was an important component of trade between China and Europe, with the exchange of opium for Chinese tea, silk, and spices, creating thousands of addicts in China (see Chapter 9, p. 433). The medical uses of opium expanded at the end of the nineteenth century with the isolation of morphine and heroin as powerful painkilling drugs. During the nineteenth century, literary figures such as John Keats and Elizabeth Barrett Browning popularized the recreational use of opium, and Britain and China went to war twice over control of the lucrative opium trade. Britain introduced opium production into its colonies in South and Southeast Asia to create an alternative supply. The U.S. Congress banned opium in 1905 as heroin addiction rose and eventually prohibited both opium and heroin sales. Black markets for these narcotics developed worldwide and expanded with the popularity of drug consumption in the 1960s in North America and Europe.

Large markets for heroin now exist in the United States, Europe, and Australia, as well as in Thailand. The U.N. Drug Report estimates opium production at 1303 tons in Burma, 124 tons in Laos, and only 8 tons in Thailand. Only Afghanistan is a larger global source, and Pakistan and Iran are also significant in opium and heroin production, with smaller amounts produced in Colombia and Mexico. The sap inside the opium poppy seed pods provides the raw opium, which is then sold to a refinery and concentrated into morphine. Morphine can then be combined with chemicals and processed to create highly potent and addictive heroin worth more than $100,000 a kilogram. The United Nations estimates that there are 8 million heroin and

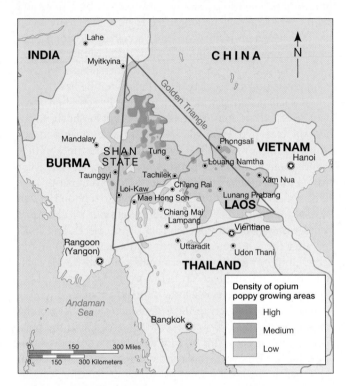

Figure 1   The Golden Triangle (*Source:* Redrawn from "The Opium Kings," *PBS Frontline,* 1998. Available at http://www.pbs.org/wgbh/pages/frontline/shows/heroin/maps/shan.html)

opium addicts worldwide, many of them unable to work productively and turning to crime in order to fund the regular purchase of heroin. Addiction can result in overdoses and death from respiratory failure, and the use of shared needles to inject heroin has contributed to the transmission of HIV/AIDS among addicts. Opium and heroin production and distribution has been monopolized by international crime syndicates and is an important source of funds for some rebel groups.

The opium poppies that are used to produce heroin are grown mainly by poorer farmers on small plots of land in the

created as part of ASEAN in 1993, reducing many tariffs within the region to less than 5 percent. A policy of constructive engagement with both military and socialist governments led to invitations to the rest of the region to join ASEAN (Brunei 1994, Vietnam 1995, Burma and Laos 1997, Cambodia 1999). Despite some resistance from Malaysia, ASEAN has generally been open to discussions with other nations and groups, including China and Australia.

Agricultural Development

Although agriculture has decreased in economic significance across most of the region, it is still an important employer and is essential to food security, employing more than 40 percent of the economically active population in Indonesia, Thailand, and the Philippines. Rice production continues to dominate the land area of Southeast Asia, but the traditional exports of

Figure 2   Opium harvest A woman of the Wa ethnic group harvests opium poppies in Burma.

Figure 3   The Golden Triangle Buddhist monks pose in front of tourist sign for the Golden Triangle.

warmer and drier highlands that stretch from Turkey through Afghanistan and Pakistan to the Golden Triangle (**Figure 2**). The Golden Triangle of Southeast Asia provides a powerful example of a landscape and society transformed by opium production, producing about two-thirds of the world's total. U.S. troops in Vietnam provided a major market for heroin produced in the Golden Triangle, and the Nationalist Chinese forces of the Kuomintang (K.M.T.), who had fled from Maoist China to Burma, fostered its production. The K.M.T. worked with indigenous groups rebelling against the Burmese government, including the Wa in northern Burma and the Shan on the northeastern plateau. The Shan taxed production and transportation of opium in order to purchase arms for their rebellion against the Burmese state. The Wa also produced millions of methamphetamine pills in remote laboratories.

An aggressive campaign to eradicate poppy production in northern Thailand was successful in persuading farmers to switch to production of alternative crops such as tobacco and soybeans by offering generous cash rewards. The opium crop in Thailand has decreased by 90 percent over a 10-year period, and what is now grown is mainly for consumption by hill tribes. However, Thailand is still a major smuggling route, with the newest group of smugglers being Nigerians. Thailand has also developed a tourist industry centered on the Golden Triangle that includes treks operated from the city of Chiang Mai to indigenous villages to view local cultures, the opium museum, and specimen poppy fields (**Figure 3**). Bringing $2 million to the region each year, tourism is proposed as a more sustainable activity for the Golden Triangle than opium or logging, although critics suggest that tourism exploits local culture and puts stress on local food supplies and is still focused on the mystique of opium.

spices and plantation crops, such as rubber, tea, coffee, and sugar, are still important. The area in plantations is only 15 percent of that in rice, with about half of the plantation area planted to rubber. Regional and international demand has increased the area planted to oil palm and pineapples, especially in Malaysia, the Philippines, and Thailand. Multinational corporations such as Del Monte and Dole are heavily involved in the production of pineapples and other fruits in Thailand and the Philippines. Southeast Asian agriculture was transformed by two major factors—the **Green Revolution** and land reform—in the last 50 years.

The Green Revolution (see Chapter 8, p. 404) in Southeast Asia focused on increasing yields of rice from a base at the *International Rice Research Institute* (IRRI), established in 1960 at Los Baños in the Philippines. Plant breeders from around the world, with support from the Rockefeller and Ford

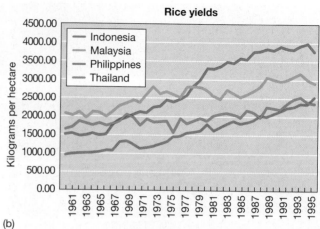

(a)

(b)

Figure 10.27 **Green Revolution** (a) The International Rice Research Institute at Los Baños in the Philippines, the agricultural research institution that supported the development of high-yielding varieties of rice in Southeast Asia. (b) Growth in rice yields as countries adopted the Green Revolution in Southeast Asia. [Source: (b) Data from U.N. Food and Agricultural Organization, available at http://www.fao.org]

foundations, the United States, and other national governments, focused on developing varieties of rice that would respond well to applications of fertilizer and would be resistant to disease. The most famous variety was IR-8, a short and stiff-stemmed "dwarf" rice that was responsive to fertilizer and had a short growing season. The Green Revolution package of technologies included high-yielding seeds, fertilizers, and agricultural machinery and contributed to dramatic increases in rice and other cereal production in Indonesia, the Philippines, and Thailand (**Figure 10.27**). As in other world regions, the benefits of the Green Revolution in Southeast Asia were unequally distributed. Communal land and rights to the rice harvest were lost in the new orientation to exports and wage labor, and the introduction of mechanical rice harvesting increased rural unemployment in countries such as Malaysia. The rate of successful adoption was higher among those with access to irrigation. When the new varieties were planted uniformly across large areas as monocultures, they became vulnerable to diseases and pests, such as the brown plant hopper. Rather than continue to use pesticides to combat such pests, the Indonesian government sponsored the use of Integrated Pest Management, which used less chemically intense techniques, by 2.5 million small farmers.

Land reform was an important policy of new socialist governments—for example, the one in Vietnam—and of others seeking to reduce rural unrest in countries such as the Philippines, where land ownership was highly concentrated. In North Vietnam more than 70 percent of the land was redistributed and collectivized in 1954, although many peasants continued to farm small individual plots within larger collectives. Land reform in the Philippines has been partial and slow, with 7 percent of the crop area distributed to 9 percent of the mainly tenant population between 1972 and 1983.

Landowners avoided reform by evicting tenants who might become eligible for land and by denying irrigation rights to new small holders.

Environment

Deforestation is the most significant region-wide environmental problem in Southeast Asia (**Figure 10.28**). The original land cover of the region was dominated by forests that provided habitat for a diverse ecology and food, medicines, fuel, fiber, and construction materials to local peoples. Swidden farming cleared patches of forest for crops, but widespread clearance began in the late 1800s with the expansion of rice production and export of tropical hardwoods under European colonial control. Most of the deforestation during this period was in lowland regions such as the Irrawaddy delta. After the Second World War, timber extraction expanded in the highlands, especially the cutting of teak in Thailand and Burma for export to the furniture industry. More recently, the growth of oil palm plantations, the pulp and paper industry, and the cutting of trees for plywood and veneers has placed even greater pressure on the forests.

The driving forces for contemporary deforestation include export agricultural plantations, logging for tropical hardwoods and pulp, and land clearing by frontier migrants and traditional swidden farmers. Swidden clearing has intensified in those areas where population has increased, land is limited, and chainsaws are available. Forest has also been cleared by those resettled under Indonesia's transmigration program and by Malaysia's Federal Land Development Authority (FELDA) programs to develop land for export crops. Japan is often blamed for deforestation in Southeast Asia because of the country's high demand for tropical hardwoods.

(a)

(b)

Figure 10.28   Deforestation in Southeast Asia   (a) An elephant hauls valuable teak wood near Pak Lay in Laos. (b) Trees have traditionally been robed to propitiate the spirits believed to dwell in them, an example of the Thai blend of animism and Buddhism, but in 1988, when loggers threatened to harvest a forest in northern Nan province, a monk named Phra Manasnati Pitak ordained trees in order to protect them. His action saved the forest, and soon other communities were following his example, such as in this region near Chiang Mai, Thailand.

Forest cutting increased as timber prices increased by 50 percent during the 1990s, at the same time that prices for many other commodities were falling. By 1996 more than half of the original forests had been cleared—Thailand and Vietnam had about 20 percent of their forests left, whereas Cambodia, Malaysia, and Indonesia retained about 65 percent. Only 40 percent of Burma's original forests remained. Timber exports are critical to the Burmese economy, comprising 40 percent of export earnings, including earnings from concession of more than 18,000 square kilometers (6950 square miles) for teak extraction by Thailand. The timber industry provides thousands of jobs and a significant amount of foreign exchange in Southeast Asia.

The impacts of deforestation include loss of species habitat and traditional cultures, flooding and soil erosion, and smoke and pollution from forest burning. Floods in Mindanao in 1981 killed 283 people and injured 14,000 below a region of major forest clearing in the Philippines, and deforestation has been associated with the near-extinction of the Manabo culture in northern Luzon.

Governments have responded by setting aside forest reserves and by banning logging in many regions, as well as by insisting on local processing rather than export of raw logs. But earlier patterns of crony capitalism that gave generous timber concessions to friends and relatives of government in Indonesia, Malaysia, and the Philippines have not been rescinded. Multinational companies such as Weyerhauser, Georgia Pacific, and Mitsubishi have also benefited from concessions.

Local people and international environmentalists have responded to deforestation in Southeast Asia by forming social movements and alliances to protect forests. In Thailand, Buddhist monks have helped to protect trees by wrapping them in saffron cloth and ordaining them, thus providing strong religious taboos against deforestation. The environ-

mental movement is one of a number of different organized grassroots efforts that also include action to support women, indigenous groups, factory workers, and religious minorities.

Core Regions and Key Cities of Southeast Asia

Perhaps the only place in Southeast Asia that merits status as an international core region and world city is Singapore, but there are a number of important cities and industrial regions that act as centers for the region and for some international economic activities. Each of the three megacities of Singapore, Bangkok, and Jakarta lie at the heart of metropolitan areas and hinterlands that are dynamic, productive, and closely tied to the international economic system. Singapore, while internationally significant in its own right as an industrial and service center, is located at the southern tip of the Malay peninsula and of a zone that includes the Malaysian cities of Johor Baharu, Kuala Lumpur, Malacca, and Penang along the Strait of Malacca. Bangkok dominates Thailand from its location in the rice bowl of the Chao Phraya basin. Jakarta is the major city on the island of Java, the economic and agricultural hub of Indonesia, and part of a metropolitan region called Jabotabek that combines the cities of Jakarta, Bogor, Tangerang, and Bekasi.

Singapore and the Strait of Malacca

Singapore is a small, flat, marshy island that has been drained and developed to become the most important port and business center in Southeast Asia and one of the ten wealthiest countries in the world in terms of gross national product per capita. Today, the international business executive flying into

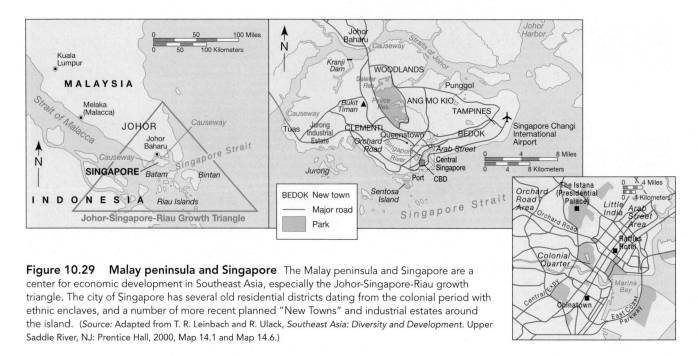

Figure 10.29   Malay peninsula and Singapore The Malay peninsula and Singapore are a center for economic development in Southeast Asia, especially the Johor-Singapore-Riau growth triangle. The city of Singapore has several old residential districts dating from the colonial period with ethnic enclaves, and a number of more recent planned "New Towns" and industrial estates around the island. (*Source:* Adapted from T. R. Leinbach and R. Ulack, *Southeast Asia: Diversity and Development.* Upper Saddle River, NJ: Prentice Hall, 2000, Map 14.1 and Map 14.6.)

Singapore's modern airport sees towering high-rise offices, luxury hotels, and large-scale industrial facilities. When the British colonized the so-called *Straits Settlements* along the shores of the channel between the Malay peninsula and Indonesia, Singapore became the jewel of their Southeast Asian colonies, with its strategic location on the major route from the Indian Ocean to China. Northward from Singapore, along the west coast of the peninsula, the cities of Malacca, Kuala Lumpur, and Penang became the core of the British colonial administration and economy in the region (**Figure 10.29**).

Although some equate Singapore and Hong Kong because they are both island Asian cities of great economic importance, Singapore appears much less chaotic and more modern than Hong Kong. It is more independent as a city-state, compared to Hong Kong's status as a British protectorate that recently became part of China.

The city of Singapore (effectively the same as the country) has a population of 3.5 million people and a high density of more than 500 people per square kilometer, the fourth highest population density of any country in the world (only Hong Kong, Macao, and Monaco are higher). The superior infrastructure—especially the excellent port and international airport—has made Singapore the import and transshipment center for the region. Containers are moved from smaller feeder ships and planes onto larger ocean and air transport. It is one of the world's largest oil refining centers, where crude oil is unloaded and refined before shipment to the rest of Asia. These functions are complemented by a large variety of maritime service activities, including banking, insurance, communications, and consulting.

When Singapore became independent from Malaysia in 1965, the government immediately made the attraction of foreign firms and investment a priority. It offered a low-cost, hard-work-ing, and compliant workforce for labor-intensive manufacturing and incentives to locate in new industrial parks. The country was successful in attracting companies that manufactured textiles and electronic equipment, such as stereos and TVs, and the economy began to grow at more than 10 percent a year. As wages began to increase, labor shortages developed, and trade barriers grew internationally in the late 1970s, these industries became less competitive. The government of Singapore made the astute decision to move away from labor-intensive manufacturing to higher-technology activities that required a skilled workforce, such as precision engineering, aerospace, medical instruments, specialized chemicals, and, most importantly, the emerging computer industry. To support this policy, the government invested in advanced education, especially engineering and computing, in high-technology industrial parks, and in some state-owned pilot companies such as Singapore Aerospace. The strategy was successful, and Singapore is now a center for the new information technology and aerospace, producing many of the world's computer hard drives and sound cards in companies like Seagate, as well as computer memory chips and software for Hewlett-Packard and Apple, among others.

Singapore has also diversified its service sector to include a wider range of financial, communications, and management activities and has attracted the regional headquarters of many multinational corporations. A number of Hong Kong businesspeople relocated to Singapore to avoid becoming part of China, and tourism has become a significant industry. This diversification, as well as high consumption levels in the domestic market, allowed Singapore to cope with the 1983 oil crisis and 1997 Asian financial crisis somewhat more easily than did other countries in the region.

Singapore has a population of diverse ethnicity and religion, but it is dominated (77 percent) by overseas Chinese, descendants of immigrants who moved to Singapore in the colonial period and followed the religions of Buddhism and Taoism. Other groups include Malays (about 15 percent, mainly Muslim) and Indians (about 7 percent, mainly Hindu). As noted earlier in the chapter, fertility rates are low in Singapore, and the government has tried to promote more births among the highly educated in order to reduce labor shortages and ensure a workforce to support the older population. Colonial Singapore was residentially segregated, with the British living next to the government buildings on the east shore of the Singapore River, Chinatown on the west riverbank, and Indian and Malay neighborhoods farther toward the east.

There are remnants of this structure in contemporary Singapore, but the city is now characterized by dozens of tall office buildings, housing complexes, and new towns, and industrial parks such as Jurong on the western end of the island. The container ports stretch along the south shore of the island (**Figure 10.30**).

Singapore has avoided the poor sprawling squatter settlements of other Asian cities through high wages and employment, careful urban planning, a public transit system, and the construction of thousands of publicly funded housing units. Compulsory but flexible savings plans have allowed most residents to purchase their apartments.

The government has made serious attempts to foster harmony between ethnic groups and to create a sense of national identity by designating four official languages (Mandarin Chinese, English, Malay, and Indian Tamil) and requiring a mix of groups in housing developments. They promote an Asian identity (as well as government authority and a cooperative labor force) through schools and national military service that emphasize hard work, community consensus, and respect for authority. They also enforce Singapore's image of a clean and crime-free environment through very strict rules against litter and graffiti, media censorship, and licensing of satellite dishes and street entertainers.

In recent years Singapore has looked beyond the city to development and investment in the wider region. Manufacturing

Figure 10.30 Singapore The Singapore skyline rises above hundreds of containers waiting transshipment from the international port that dominates trade in Asia.

has expanded to a new Growth Triangle (called SIJORI) that includes Johor Baharu in Malaysia and the Riau Islands (especially Batam) of Indonesia. Trade restrictions have been eased within the triangle so that labor- and land-intensive activities are carried out in the province of Johor and Batam, with value and business services added in Singapore in a flexible production system. Singapore also obtains half of its water supply from Johor and has developed agroprocessing industries in the Riau Islands that supply Singapore with pork, chicken, prawns, and flowers. Singapore investors are also active in Thailand and China.

Johor Baharu is the southernmost in a string of Malaysian cities that are strung along the western side of the Malay peninsula facing the Strait of Malacca. To the north, Penang shares many characteristics with Singapore in that it has become a center for high-technology manufacturing for the computer industry and has a population that is made up mostly of Chinese. Government investment in training skilled labor (many young women) and generous tax and free-trade incentives have attracted multinationals such as Intel, Sony, Philips, Motorola, and Hitachi to the island that is connected to mainland Malaysia by a bridge. Penang has become known as Silicon Island. Kuala Lumpur, located about halfway between Penang and Singapore, originally developed as a center for the colonial tin industry. Now, as the capital of Malaysia, it has a population of about 1.5 million and is the center of a manufacturing region called the Klang Valley Conurbation that includes a free-trade zone.

Bangkok and Central Thailand

The streets of Bangkok, Thailand (**Figure 10.31**), are crowded with people, street vendors, and a variety of vehicles ranging from hand-pulled rickshaws and tricycles to mini buses. The golden spires of religious buildings and soaring towers of secular buildings dot the skyline of the city, which is crossed by dozens of polluted canals, called *klong,* packed with houseboats and floating markets. Bangkok's problems include serious air and water pollution and an AIDS epidemic associated with a sex industry that has become a focus for international tourism (**Figure 10.32**).

Bangkok is located on the Chao Phraya River near its exit into the Gulf of Thailand and is surrounded by the rich agricultural land of the delta planted with rice. Rice yields have increased as a result of Green Revolution technology, and Thailand is a major rice exporter. However, the rice area around Bangkok has decreased as urban demand for fruit and vegetables has stimulated market gardens, small plots where produce is grown for sale to city residents. With a metropolitan population of more than 10 million people, Bangkok is the hub of mainland Southeast Asia and is a clear example of **primacy** (see p. 40). Bangkok dominates Thailand with 70 percent of the urban population (more than 30 times the size of the next largest city), 90 percent of the trade and industrial jobs, and 50 percent of the gross domestic product.

Until the 1970s, the Thai economy was focused on exports of rice, fruit, and seafood, including a local food-processing

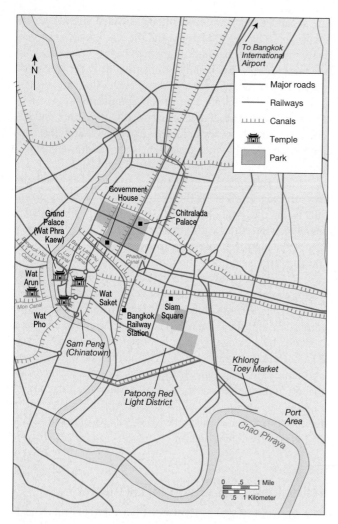

Figure 10.31   Bangkok, Thailand Bangkok developed along both sides of the Chao Phraya River in Thailand and is a city of many temples, palaces, and canals. (*Source:* Redrawn from T. R. Leinbach and R. Ulack, *Southeast Asia: Diversity and Development.* Upper Saddle River, NJ: Prentice Hall, 2000, Map 16.4.)

The origins of Bangkok's notorious sex industry lie to some extent in a traditional acceptance of prostitution but also in the role that the city played as a major R&R ("rest and recreation") center for U.S. troops during the Vietnam War. Sex tourism was constructed around an image of Asian women as passive and exotic, and by the 1990s tours were being heavily advertised in places such as Japan and Germany. Sex tourism is estimated to generate more than $10 billion, a significant source of national foreign exchange, and to involve about half of the almost 8 million tourists who arrived in Thailand each year.

According to U.N. estimates, at the end of 1999, 750,000 people in Thailand were infected with HIV/AIDS, about 2 percent of the adult population, and 66,000 people had died by 1998. This is by far the highest number of cases in Southeast Asia and is one result of the active sex industry. Local and international protests against the prevalence of child prostitution and a growing toll from HIV/AIDS prompted the Thai government to clamp down on the sex industry and to develop a strong anti-AIDS campaign that included distribution of condoms.

Java and Jabotabek

The island of Java is the most populous and economically significant of the islands of Indonesia, where Java, Bali, and Madura constitute the inner core of this island nation. Nearly 125 million people, 60 percent of Indonesia's population, and 25 percent of that of Southeast Asia, live on Java at high population densities and in four cities of more than 1 million people, including Jakarta with more than 10 million.

Java is famous for its fertile agricultural landscapes, carved with elaborate terraced rice paddies nurtured by centuries of human attention and by nutrient-rich volcanic soils. The island is about 1000 kilometers (650 miles) long from east to west, with a volcanic mountain chain creating the spine of the island. The landscapes around the active volcanoes, such as Mount Semeru (3676 meters, 12,060 feet) and Mount Bromo (2329 meters, 7639 feet), surrounded by a sea of sand, are particularly dramatic.

Early travelers from India brought both the Hindu and Buddhist religions to the island, producing the magnificent structures at Borobudur and a number of other temples. The Dutch established the headquarters of the Dutch East India Company in the seventeenth century at Batavia on the swampy northwest coast of Java, where there were several Muslim kingdoms and sultanates. Renamed Jakarta after independence, it became the capital of Indonesia and the focus of industrial development, especially an export textile industry and factories producing shoes, appliances, and lower-technology electronic goods.

Jakarta's downtown area includes the old city fort, the Portuguese church, many silver-roofed mosques reflecting Indonesia's Islamic tradition, a central business district, and spacious squares and parks, such as the Medan Merdeka (**Figure 10.33**). The Istiquial mosque (**Figure 10.34**) is one of the largest in the world, and the old Dutch port, Sunda Kelapa,

industry that produced canned goods. In the early 1980s the Thai government decided to promote export manufacturing through financial incentives, cheap labor, and the promise of political stability. They also advertised a business culture friendly to both Buddhist and Chinese investors from Hong Kong and Taiwan and to the United States, which had been closely allied to Thailand during the Vietnam War. A government-financed automobile manufacturing training institute helped attract a $750-million General Motors plant, as well as a Ford-Mazda truck manufacturing facility. There are more than 25,000 factories in the Bangkok metropolitan region, many of them labor-intensive textile producers.

Although manufacturing grew rapidly from 25 to more than 70 percent of exports between 1980 and 1998, the economic development of Bangkok has been constrained by inadequate infrastructure, including the lack of mass transit and the need to upgrade the international airport.

(a)

(b)

Figure 10.32   Life in Bangkok   (a) The city of Bangkok is crossed by many canals (or *klong*) crowded with houseboats and floating markets. Many of the canals are polluted. (b) Bangkok is notorious for the red light districts that are magnets for sex tourism and are centers for the spread of the HIV/AIDS epidemic.

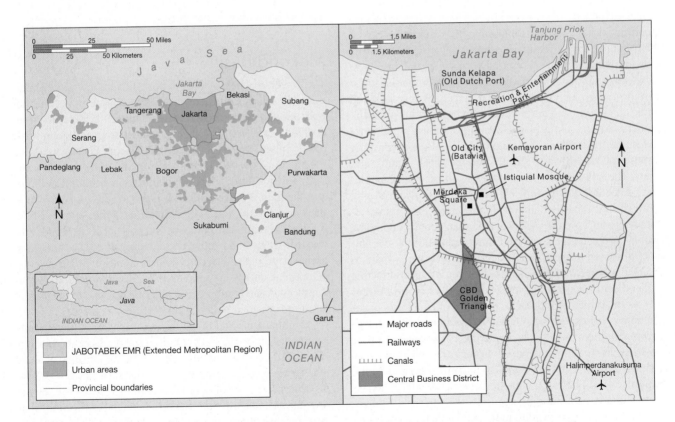

Figure 10.33   Jakarta   Jakarta is located along the Java Sea on the north coast of the Indonesian island of Java. The old Dutch port is now linked to the new harbor by a recreation and park development with avenues stretching inland to the colonial square of the Merdan Merdeka and the Central Business District known as the Golden Triangle. (*Source: Adapted from T. R. Leinbach and R. Ulack, Southeast Asia: Diversity and Development. Upper Saddle River, NJ: Prentice Hall, 2000, Map 12.8 and 12.9.*)

Figure 10.34   Jakarta's Istiquial mosque A view across the city of Jakarta with the enormous Istiquial mosque shown at right.

hosts dozens of sailing ships. Many houses are built of wood and bamboo mats. The expanded metropolitan region around Jakarta also includes the cities of Bogor, Tangerang, and Bekasi, with a total of 20 million people and thousands of factories representing 80 percent of Indonesia's industrial employment. The larger region combines the first two letters of each city into the name *Jabotabek* and has a population of more than 20 million.

The city is surrounded by unplanned squatter settlements that lack basic services, and it has a large informal sector. Informal sector activities include street selling, charging fees for guarding parked cars, assisting traffic flow, and filling cars to the three occupants required to travel on some streets. The growing population of Jakarta moves around the city in an increasing number of automobiles as well as on public transportation, such as buses and pedal taxis (called *betjol*), and traffic congestion and air pollution can be a serious problem.

The majority of the population of Jakarta use septic tanks and pit latrine facilities. A large proportion of the urban poor disposes of human waste directly into the canals, drains, and rivers. The rapid urbanization of the region has resulted in increased flooding downstream, especially in the low-lying areas of north Jakarta. The KIP (Kampung Improvement Programme) attempted to improve the welfare of the urban poor by upgrading their living environment and basic infrastructure. While the program improved the condition of large slum areas, a new pattern emerged as improvements increased property values, attracting higher-income groups and forcing the poor to create new slums.

As the capital of Indonesia, Jakarta is often a focus of social unrest. For example, in May 1998, after government budget cuts and price rises associated with the Asian economic crisis, rioters set fire to shops and other businesses owned by overseas Chinese and the family of President Suharto, who at the time had ruled for 32 years. Discontent was fueled by the cronyism of the Suharto family and by resentment of perceived Chinese wealth. Suharto was forced to resign, and in fiercely contested elections power passed through parliamentary selection to a moderate but unpredictable Islamic cleric, Abdurrahman Wahid. Criticism of his administration and protests in the streets of Jakarta eventually resulted in his dismissal and the

transfer of power to Megawati Sukarnoputri, daughter of Indonesia's first president at independence, and now the fifth president of the world's fourth largest country that now faces economic stagnation and regional conflict.

Distinctive Regions and Landscapes in Southeast Asia

The incredible physical and cultural diversity of Southeast Asia makes it hard to select just a few examples to illustrate the region's many distinctive landscapes, but we have chosen three: the Mekong Basin; the Indonesian islands of Borneo, Sulawesi, and Irian Jaya; and the Philippines. (The Golden Triangle, a fourth distinctive landscape within the region, is described in Geographies of Indulgence, Desire, and Addiction: Opium, p. 508.) The landscapes of the Mekong River and of the Golden Triangle link several countries of mainland Southeast Asia through the legacies of war and challenges of water development on the Mekong, the dilemmas of drug production, and tourism in the highlands of Thailand, Burma, and Laos that the Golden Triangle comprises. The islands of Borneo, Sulawesi, and Irian Jaya epitomize the frontier and indigenous landscapes of the region. Finally, we provide a brief overview of the diverse landscapes of the Philippines, with the distinctive legacy of Spanish colonialism and orientation to the Americas.

Mekong Basin

The lower reaches of the Mekong River flow slowly past thatched villages and rice paddies while the upper reaches rush through deep, forested gorges, linking the countries of Indochina in a legacy of past conflict and a promise of joint water development. With a source on the Tibetan Plateau in China, the Mekong flows 4200 kilometers (2600 miles) between and through Laos, Burma, and Thailand, to its delta in Cambodia and Vietnam (**Figure 10.35**). It is the twelfth longest river in the world and the tenth largest in terms of the volume of flow. It forms the main transportation and settlement corridor for the underdeveloped and conflict-torn countries of Laos and Cambodia. It provides water for irrigation and hydroelectric development and is a productive fishery. Until recently, the Southeast Asian section of the Mekong was one of the world's last large untamed rivers with no bridges or dams.

Vientiane, capital of landlocked Laos, and Phnom Penh, capital of Cambodia, both stand on the banks of the Mekong (**Figure 10.36**). The city of Vientiane has a population of about 400,000 people and still has the tree-lined boulevards and villas reminiscent of colonial French Indochina as well as numerous ancient temples. It sits on a bend in the Mekong, often vulnerable to floods, and is surrounded by fertile rice and vegetable fields on the floodplain. The city's central river port location in a country relying heavily on its rivers for transportation has made Vientiane the economic center of Laos. As in Vietnam, the Laotian communist government has

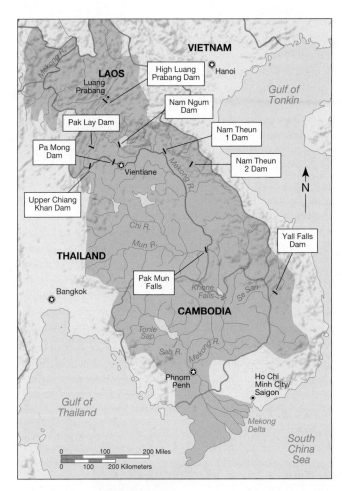

Figure 10.35   The Mekong River map   The Mekong River links the countries of mainland Southeast Asia as it flows from a source on the Tibetan Plateau to the delta in Vietnam and Cambodia. (*Source:* Redrawn from Probe International, Mekong Basin Project, available at http://www.probeinternational.org/pi/Mekong/images/mekongmap3.gif)

Figure 10.36   The Mekong River   The Mekong River is a major transportation route for Southeast Asia. As it flows into the delta near My Tho, Vietnam, boats move people across the river and transport rice for domestic use and export.

initiated reform, or "new thinking," and is expanding links with Thailand. Buddhism is very important in Laos, and many young men spend a period as a monk in the characteristic saffron robes.

One of the most unusual physical features in the Mekong Basin is the lake of Tonle Sap in Cambodia. The lake acts as a safety-valve overflow basin for flooding on the Mekong River. During the dry season, water flows out of the lake along the Sab River into the Mekong, and during flooding on the Mekong, water flows back up into the lake, sometimes more than doubling its area. Cambodia is still recovering from the aftermath of the Pol Pot/Khmer Rouge regime (see p. 493) and the city of Phnom Penh is starting to grow again after the forced depopulation by the Khmer communist government.

The Mekong provided a motive for cooperation among countries of the basin, even during times of tension, because of the potential for water resource development that could provide mutual benefits. The Mekong River Commission, established in 1957, has coordinated the planning of flood con-

trol and dam projects and has studied environmental issues within the basin with advice from geographer Gilbert White. The 1995 Chiang Rai accord provides mechanisms for resolving disputes within the basin. Several important dams have been constructed on the Mekong tributaries—among them the Nam Ngum in Laos, which generates several million dollars of hydroelectric sales to Thailand—and many others are proposed. Logging concessions granted in advance of dam construction are contributing to increased deforestation in the basin. Teak forests flooded by the Nam Ngum reservoir are now harvested in an unusual underwater logging operation. The construction of the Friendship Bridge between Thailand and Laos has improved transportation within the basin.

The lower Mekong landscape and delta still show the scars of war, with mosquito-infested bomb craters, ruined bridges and roads, residues of defoliants, and thousands of unexploded anti-personnel mines. Vietnam is attempting to restore the landscape with the support of foreign assistance for reforestation, rebuilding, reclamation, and nature conservation. The delta has recovered to become the national granary for rice production, with boats carrying the harvest to the coast for export and upriver to domestic markets.

Ho Chi Minh City (formerly Saigon), just to the north of the delta, the former capital of South Vietnam and base for the U.S. war effort, is undergoing economic revival and now has a population of 8 million. Seeking cheap labor and new markets, foreign investment is flowing into tourism, office construction, and textile manufacturing. A new Special Economic Zone to the south of the city has attracted shoe manufacturer Nike. Growth has been fostered by the Vietnamese government's policy of *doi moi* (renovation), which has promoted privatization, reduced central control, and foreign investment.

Borneo, Sulawesi, and Irian Jaya

The enormous islands of Borneo, Sulawesi, and Irian Jaya, the Indonesian portion of New Guinea, are frontier regions that have become a focus for resettlement, mineral development, forest exploitation, and resistance by indigenous groups (**Figure 10.37**). Borneo, a stable zone of old crystalline rocks, covers more than 750,000 square kilometers (290,000 square miles) and includes territory controlled by Indonesia (Kalimantan), Malaysia (Sarawak and Sabah), and the country of Brunei (see Sense of Place: Brunei, p. 503). Borneo, Sulawesi, and Irian Jaya have more than 25,000 species of plants, 10 percent of the world's biodiversity. Elephants, tigers, rhinos, and orangutans reside in the mountainous and forested interiors, with human populations concentrated in the coastal plains and river valleys around Hulu Sungai in southern Kalimantan and Pontianak in western Kalimantan. European explorers and writers such as Alfred Wallace and Joseph Conrad described the island as a luxuriant landscape inhabited by primitive tribes. This perception of primitive people and wild frontiers has persisted, with both Malaysian and Indonesian governments interested in developing the region that they portray as economically and politically isolated. Geograph-

er Harold Brookfield argues that many of the indigenous peoples of Borneo have actually traded with others in the region and the world for centuries, selling forest products such as rattan, pepper, and camphor to traders. Collectively called the Dayak by outsiders, there are actually many different groups with individual names, such as the Kenyah and Iban on Borneo.

Deforestation and the fate of indigenous populations are two of the critical geographic issues on the island of Borneo that have garnered the attention of international environmental and human rights groups.

Large-scale logging in Borneo increased after about 1960, when Japan shifted demand to Southeast Asia and altered preferences from teak on the mainland to hardwoods and trees used for plywood in the island regions. In addition to logging for export, Harold Brookfield identifies agricultural clearing for transmigration settlements and the introduction of chainsaws and heavy logging equipment as causes of deforestation that has moved from zones along rivers into more remote areas. By the late 1990s, Borneo was providing more than half of the world's tropical hardwoods, mainly through concessions given to multinational timber corporations. The damage from forest clearing included loss of biodiversity, erosion and flooding,

Figure 10.37   Borneo (a) Loggers fell trees for lumber in Kalimantan, Borneo. (b) Protest for independence from Indonesia by ethnic minorities from Irian Jaya in Jakarta in 2000. (c) Traditional tongkanans buildings in Tanatoraja, Sulawesi. (d) Transmigrants clash with Malays in Pontianak, Borneo, in 2000.

(a)

(b)

(c)

(d)

and loss of forest benefits to indigenous populations. Although bans on export of raw logs were put in place, the corresponding increase in wood processing on Borneo, including more than 50 plywood mills, kept the pressure on the forests.

The curiously shaped and mountainous island of Sulawesi (also called Celebes) is occupied by seven major indigenous groups with distinct cultures. One of the most distinctive cultural landscapes is associated with the Toradja. Their traditional houses (tonghanans) and rice barns have dramatic curved roofs that look like boats or buffalo horns, indicating the status of the family and attracting an increasing number of tourists.

Irian Jaya, the western part of the island of New Guinea, has become an important financial resource for the Indonesian government as a result of the development of several large mines by the multinational Freeport-McMoRan corporation since the late 1960s. Two mountains, Ertsberg and Grasberg, near Tembagapura have been almost removed in the mining of copper, gold, and silver with deposits valued at $50 billion. The Indonesian government has shares in the company and receives more than 5 percent of its tax revenue from the mines.

There is considerable local resistance to the mines and to the Indonesian government. Many residents of Irian Jaya see themselves as Papuans (Papua New Guinea occupies the eastern part of the island of New Guinea) rather than Indonesians. They claim that they opposed the Indonesian takeover of Irian Jaya in 1963 and that their desires for independence were ignored in the referendum to join Indonesia in 1969. Indigenous traditions of groups such as the Amingme, Dani, Komoro, and Ekari include hunting and collecting forest products, growing root crops, valuing pigs as a measure of wealth, and engaging in an economy based on gift exchange or ritual fighting. The Free Papua Movement (or OPM) has called repeatedly for independence and has focused their anger on the Freeport mines, including attacks in 1977 and 1996. They argue that land was expropriated without permission of local people, that too few locals are employed in the mines and are given low-skill jobs, and that the mines are causing water pollution in the form of acid drainage and silting of rivers. The Dani have asked for 1 percent of the total mine profits of $1.75 billion a year.

The Philippines

The Philippines are often seen as quite different from other parts of Southeast Asia because of specific geographic characteristics, including their distance from the Asian landmass, the long history of Spanish colonial rule, and subsequent control by the United States (**Figure 10.38**). The country comprises the two large islands of Luzon to the north and Mindanao to the south, as well as several other sizable islands, such as Cebu and Negros, and hundreds of smaller ones. The islands were originally covered with dense forests and were home to more than 75 distinct ethnic groups speaking hundreds of dialects.

Spain and the United States wanted to control the Philippines because of their location as a gateway to trade with Asia. The impacts of 400 years of domination from Spain and the

United States resulted in several distinctive characteristics, including a majority Catholic religion, mainly Spanish names, highly concentrated land ownership, an agriculture oriented to exports of sugar, tobacco, and pineapples to the Americas, the widespread use of English, and a general orientation to the west, especially the United States. As noted earlier, the United States acquired control of the Philippines as an outcome of the 1898 Spanish-American War. Even after the United States granted independence to the Philippines in 1946, it continued to treat the islands as an important strategic location, maintaining several enormous military installations such as Clark Air Force Base and Subic Bay Naval Base. The closure of the U.S. bases at Clark (because of the eruption of Mount Pinatubo) and Subic Bay in the 1990s was an initial blow to employment and the economy, but geographer Richard Ulack notes the rapid transition of Subic Bay, with its excellent infrastructure, into a new free-trade zone with 200 companies, including the main Asian hub for Federal Express.

Figure 10.38   Map of the Philippines The Philippines consists of dozens of islands, including the Luzon and Mindanao This map shows the major cities and industrial and agricultural regions. (*Source:* Adapted from R. Ulack and G. Pauer, *Atlas of Southeast Asia.* New York: Macmillan, 1988, p. 63.)

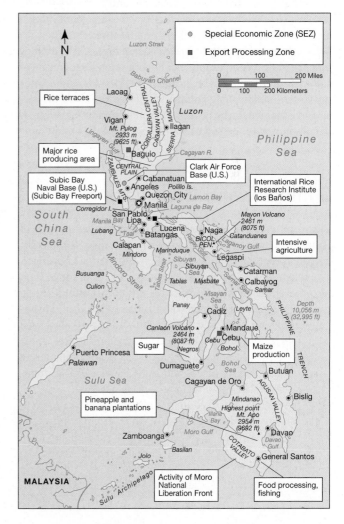

Figure 10.39 Manila This postcard of the Welcome Arch in Manila, the Philippines, draped in U.S. flags in 1915, symbolizes one aspect of the complex historical relationship between the Philippines and the United States.

The Spanish fostered the creation of an elite landholding group, many of mixed Spanish and local ethnicity, dominated by families who were designated as *principales*. This group came to control land and eventually the economy and politics after independence. Despite some attempts at land reform, Philippine land tenure still includes large export-oriented landholdings growing sugar, tobacco, and pineapples. The dominance of Catholicism has contributed to a higher birthrate and to Muslim efforts to seek autonomy through armed rebellion or by migrating to Malaysia as refugees. The Philippines provide a good example of how the global geography of colonialism influenced the local culture, politics, and economy so as to differentiate countries within a world region.

In other ways, the islands of the Philippines have much in common with the rest of Southeast Asia. Like other countries, the Philippines experienced a period of rapid integration into the global economy as a source of cheap labor for industries such as textiles, grew rice as the dominant crop, and faced a serious deforestation problem and the challenges of integrating different groups into a national identity. Manila, like Jakarta and Bangkok, has dozens of small export industries as well as several large squatter settlements (**Figure 10.39**). Industrial growth has also flourished in the secondary cities of Cebu, Davao, and Cagayan de Oro in the southern Philippines. Elsewhere in this chapter we highlight the Philippines as a case where labor migrants working in the Middle East, the United States, or on international shipping are contributing the equivalent of $10 billion a year to the Philippine economy.

Summary and Conclusions

This review of the geography and distinctive landscapes of Southeast Asia reveals a region of rapid change, complexity, and great economic, political, and environmental challenges. The region illustrates at least two major waves of integration into the global economic system—colonization by Europe (preceded by extensive trade with China and India) and twentieth-century economic restructuring toward exports associated with foreign investment and industrialization. Demographic linkages to the rest of the world include a large population of labor migrants and refugees and significant historical immigration from other Asian regions, especially China.

There can be few fixed images of Southeast Asia, as is evident in the sudden collapse of the thriving industrial economies of Thailand, Malaysia, Indonesia, and Singapore in 1997 and in the new reforms that have opened markets to foreign investment in the socialist and war-weary countries of Cambodia, Laos, and Vietnam. The geopolitical and economic relationships with Europe and the United States fashioned by colonialism and Cold War are being replaced by new linkages with Japan, China, and Australia and by the emergence of the independent regional network of ASEAN. The diversity of religious, ethnic, and political beliefs within Southeast Asian nations demands innovative approaches to national integration and recognition of local identity, but this diversity has also produced long-standing rebellions and demands for autonomy.

Future sustainable development in Southeast Asia must confront the impacts of urban growth and land-use change in much of the region, especially air pollution, spreading slums, and inadequate infrastructure in the cities, and deforestation of the highlands of the mainland and many of the islands such as Borneo. Managing the environment and land use more equitably and ecologically will contribute to maintaining and hopefully improving social and economic conditions in Southeast Asia.

Key Terms

ASEAN (p. 478)
Asian tigers (p. 494)
Culture System (p. 490)
domino theory (p. 492)
El Niño (p. 484)
Free Trade Zone (p. 493)
Green Revolution (p. 509)

gross domestic product (GDP) (p. 494)
import substitution (p. 493)
international division of labor (p. 493)
intertropical convergence zone (p. 484)

land bridge (p. 479)
marsupials (p. 485)
monsoon (p. 478)
overseas Chinese (p. 496)
overurbanization (p. 498)
primacy (p. 513)
Ring of Fire (p. 480)

sawah (p. 486)
Spice Islands (p. 489)
swidden (slash and burn) (p. 486)
total fertility rate (p. 495)
transmigration (p. 495)
Wallace's Line (p. 485)
world city (p. 499)

Review Questions

Testing Your Understanding

1. Southeast Asia is often divided into two main regions. What are these regions called and which countries are in each? What is the name of the newest independent nation that is emerging in Southeast Asia?
2. What is ASEAN and what role does it play in Southeast Asia?
3. Which countries in Southeast Asia have governments that follow socialist or communist ideas? Which countries have pursued more capitalist paths and open economies? How has the concept of "Asian values" been used to explain authoritarian tendencies in Southeast Asian politics?
4. What is the most important crop in Southeast Asia and the main systems by which it is produced? How and where did the Green Revolution affect this crop?
5. What were the five main European colonial powers in Southeast Asia and what regions were under their control between about 1500 and 1890? Which country remained independent during the colonial period? Identify four key mineral or agricultural products that the colonial powers exported from Southeast Asia and the regions they came from.
6. What factors led to the the Vietnam War and what were some of the impacts of the war on the peoples and environments of the region?
7. Which Southeast Asian countries participated in the so-called Asian economic miracle and what triggered the 1997 financial crisis? What were some of the responses to the crisis?
8. What is transmigration and how does it contribute to ethnic tensions and environmental degradation in Indonesia?
9. What is the current status of the Spratly Islands? Why are they so hotly contested?
10. What places are linked within (a) Jabotabek and (b) SIJORI, and what role are these groups intended to play within Southeast Asian economies?

Thinking Geographically

1. What is the monsoon and how does it influence the climate and vegetation of Southeast Asia?
2. What roles did India and China play in Southeast Asia prior to the colonial period and how did this influence culture and religion? What political and economic roles have Japan and the United States played in Southeast Asia during the twentieth century?
3. Who are the "overseas Chinese" and what are three Southeast Asian countries where they constitute more than 10 percent of the population? What are their traditional roles in the economy and how and where has this resulted in a backlash against ethnic Chinese?
4. In what ways are women in Southeast Asia connected to the economy and women in the United States through "global care chains," consumer behavior, or labor activism? What are some of the pros and cons of these relationships?
5. How do the crops and commercial transactions of the Golden Triangle affect Southeast Asia and other regions of the world?
6. When and why have Burma and Cambodia been a focus of international concern about human rights in the last three decades?

Further Reading

Atkinson, J. M., and Errington, S. (eds.), *Power and Difference: Gender in Island Southeast Asia.* Stanford, CA: Stanford University Press, 1990.
Brookfield, H. C., and Byron, Y., *South-East Asia's Environmental Future: The Search for Sustainability.* New York: United Nations University Press, 1993.
Brookfield, H. C., Lian, F. J., Kwai-Sim, L., and Potter, L., *Borneo and the Malay Peninsula.* In B. L. Turner II, W. C. Clark, R. W. Kates, J. T. Mathews, and W. B. Meyer (eds.), *The Earth as Transformed by Human Action* (pp. 495–511). Cambridge: Cambridge University Press, 1990.

Brown, I., 1997. *Economic Change in South-East Asia c. 1830–1980.* Kuala Lumpur: Oxford University Press, 1997.

Cox, C. R., *Chasing the Dragon: Into the Heart of the Golden Triangle.* New York: Holt/Marion Wood, 1996.

Dixon, C., and Drakakis-Smith, D. (eds.), *Uneven Development in South East Asia.* Brookfield, VT: Ashgate, 1998.

Dwyer, D. (ed.), *South East Asian Development: Geographical Perspectives.* Harlow, England: Longman Scientific & Technical, 1990.

Forbes, D. K., *Asian Metropolis: Urbanisation and the Southeast Asian City.* New York: Oxford University Press, 1996.

Hill, R. D. (ed.), *South-East Asia: A Systematic Geography.* New York: Oxford University Press, 1979.

Hitchcock, M., King, V. T., and Parnwell, M. J. G. (eds), *Tourism in South-East Asia.* London: Routledge, 1993.

Leinbach T. R., and Ulack, R., *Southeast Asia: Diversity and Development.* Upper Saddle River, NJ: Prentice Hall, 2000.

McGee, T. G., *The Southeast Asian City: A Social Geography of the Primate Cities of Southeast Asia.* New York: Praeger, 1967.

McGee, T. G., "Eurocentrism in Geography: The Case of Asian Urbanization." *Canadian Geographer 35* (1991), 332–344.

Murray, G., and Perera, A., *Singapore: The Global City State.* New York: St. Martin's Press, 1996.

Parnwell, M. J .G., and Bryant, R. L. (eds.), *Environmental Change in South-East Asia: People, Politics and Sustainable Development.* New York: Routledge, 1996.

Rigg, J., *Southeast Asia: A Region in Transition, a Thematic Human Geography of the ASEAN Region.* London: Unwin Hyman, 1991.

Rigg, J., *Southeast Asia: The Human Landscape of Modernization and Development.* New York: Routledge, 1997.

SarDesai, D. R., *Southeast Asia: Past and Present.* Boulder: Westview Press, 1997.

Schmidt, J. D., Hersh, J., and Fold, N. (eds.), *Social Change in Southeast Asia.* New York: Longman, 1997.

Scott, J. C., *The Moral Economy of the Peasant: Rebellion and Subsistence in Southeast Asia.* New Haven: Yale University Press, 1976

Spencer, J. E., and Thomas, W. L., *Asia: East by South: A Cultural Geography.* New York: John Wiley & Sons, 1971.

Tarling, N. (ed.), *The Cambridge History of Southeast Asia* (4 vols.). Cambridge: Cambridge University Press, 1999.

Ulack, R., and Pauer, G., *Atlas of Southeast Asia.* New York: Macmillan, 1988.

Wall, G., Mitchell, B., and Knight, D., "Bali: Sustainable Development, Tourism and Coastal Management," *Ambio 26* (1997), 90–96.

Watters, R. F., and McGee, T. G., *Asia-Pacific: New Geographies of the Pacific Rim.* London: Hurst & Co., 1997.

Films, Music, and Popular Literature

Films

Blue Collar and Buddha. Directed by Taggart Siegel, 1987. This documentary follows the mounting resentment toward the Laotian refugees who have settled in a small blue-collar American town.

Indochine. Directed by Regis Wargnier, 1992. A story of romance and separation during the transition from French colonial rule of Indochina to the U.S. presence in Vietnam.

The Killing Fields. Directed by Roland Joffé, 1984. The film traces the experiences of Cambodian Dith Pran, who was working for *The New York Times,* after he was forced by the Khmer Rouge to join the long march out of Phnom Penh into the countryside.

The Opium Kings. Directed by Adrian Cowell, 1998. This documentary, shot in the guerilla-held opium-producing regions of Burma, is a report on the black-market heroin that comes from the Shan States of Burma.

The Scent of Green Papaya. Directed by Tran Anh Hung, 1993. The interior life of a Vietnamese household in the 1950s, as seen through the eyes of a young servant girl.

The Year of Living Dangerously. Directed by Peter Weir, 1982. Set in Indonesia during the 1965 coup against President Sukarno, the film examines the chaos in Jakarta from the perspective of an Australian journalist.

Music

Bayanihan Dance Company. *Bayanihan Dance Company.* New York: Monitor, 1992.

Naam, Fong. *Ancient-Contemporary Music from Thailand.* Tucson, AZ: Celestial Harmonies Records, 1995.

Various Artists. *The Music of Vietnam* (3 vols.). Tucson, AZ: Celestial Harmonies Records, 1995.

Various Artists. *Rough Guide to the Music of Indonesia.* London: World Music Network, 2000.

Popular Literature

Ahmad, S. *No Harvest but a Thorn*. New York: Oxford University Press, 1972. Depicts the hardships of a farming family in Malaysia.

Ananta Toer, P. *The Buru Quartet*. New York: William Morrow, 1991. Pramoedya Ananta Toer's four-part epic encompasses the beginnings of the collapse of colonialism nearly 100 years ago.

Bnounyavong, O. *Mother's Beloved: Stories from Laos*. Hong Kong: Hong Kong University Press, 1999. This book of 14 deceptively simple stories tell of ordinary Lao people, their customs, traditions, and values.

Burgess, A. *The Long Day Wanes: A Malayan Trilogy*. New York: Norton, 1965. Three novels about the Chinese, Indian, Malays, Eurasians, and British colonialists who inhabited the pre-independence Malaya of the 1950s.

Godshalk, C. S. *Kalimantaan*. New York: Henry Holt, 1998. This novel describes the lives and ambitions of the English in colonial Borneo, including one man, Gideon Barr, who sets himself up as a rajah.

Lubis, M., *Twilight in Djakarta*. New York: Vanguard Press, 1964. A searing critique of corruption and poverty in 1950s Jakarta.

Orwell, G. *Burmese Days*. New York: Harcourt Brace, 1934. A novel about life and racism in a remote outpost of colonial Burma.

Rizal, J. *Noli Me Tangere (The Lost Eden)*. Bloomington, IN: Indiana University Press, 1961. A classic account of Philippine suffering under Spanish rule that inspired independence struggles.

Sudham, P. *Monsoon Country*. St. Johns, Newfoundland, Canada: Breakwater Books, 1990. This novel conveys the cultural tension between the East and West and the clashes between the new powers and the old values in Thailand.

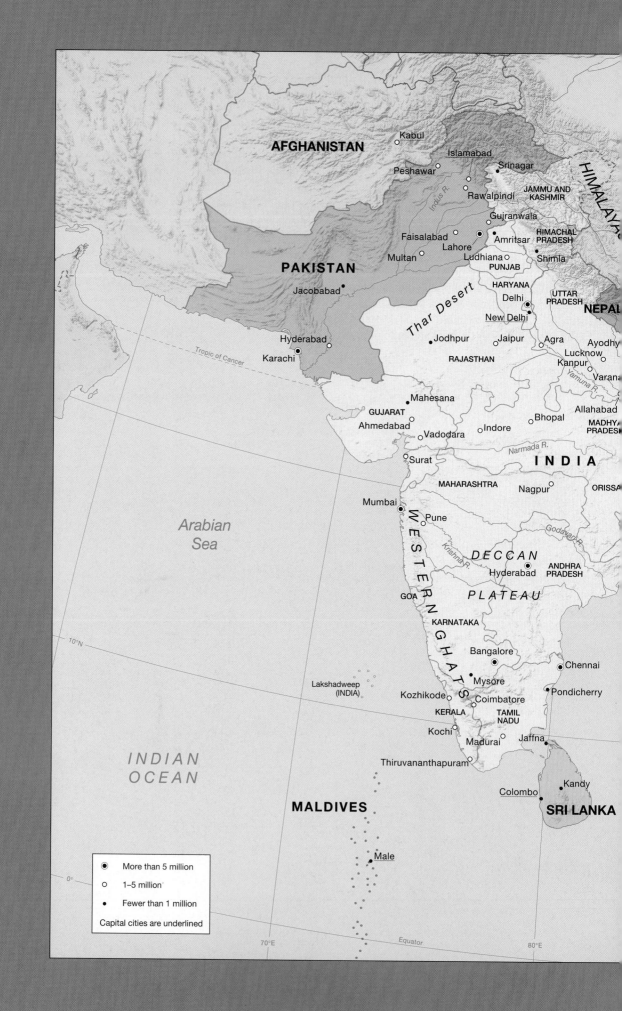

AFGHANISTAN

Kabul

Islamabad

Peshawar
Srinagar

Rawalpindi

JAMMU AND
KASHMIR

HIMALAYA

Gujranwala

Faisalabad
Amritsar

HIMACHAL
PRADESH

Lahore

Multan
Ludhiana

Shimla

PAKISTAN
PUNJAB

Indus R.

HARYANA

Jacobabad
Thar Desert
Delhi

UTTAR
PRADESH

NEPAL

New Delhi

Hyderabad
Jodhpur
Jaipur
Agra
Ayodhy

Karachi
Lucknow

Tropic of Cancer
RAJASTHAN
Kanpur

Varana

Yamuna R.

Mahesana

GUJARAT
Bhopal
Allahabad

Ahmedabad
Indore
MADHYA
PRADESH

Vadodara
Narmada R.

Surat
INDIA

MAHARASHTRA
Nagpur
ORISSA

Mumbai

Arabian
Sea
Pune

W
E
S
T
E
R
N

Krishna R.
Godavari R.

DECCAN

PLATEAU

Hyderabad
ANDHRA
PRADESH

GOA

G
H
A
T
S

KARNATAKA

10°N

Bangalore

Chennai

Lakshadweep
(INDIA)

Mysore

Pondicherry

Kozhikode
Coimbatore

KERALA
TAMIL
NADU

INDIAN
OCEAN

Kochi

Madurai
Jaffna

Thiruvananthapuram

Colombo
Kandy

MALDIVES
SRI LANKA

0°

Male

⊙ More than 5 million
○ 1–5 million
• Fewer than 1 million
Capital cities are underlined

70°E
Equator
80°E

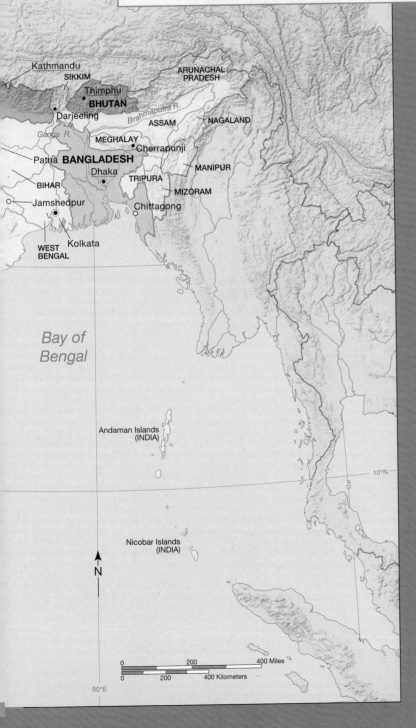

11 South Asia

Kathmandu
SIKKIM
Thimphu
BHUTAN
ARUNACHAL
PRADESH
Darjeeling
Brahmaputra R.
ASSAM
NAGALAND
Ganga R.
MEGHALAY
Cherrapunji
Patna
BANGLADESH
Dhaka
MANIPUR
BIHAR
TRIPURA
Jamshedpur
MIZORAM
Chittagong
Kolkata
WEST
BENGAL

Bay of
Bengal

Andaman Islands
(INDIA)

10°N

Nicobar Islands
(INDIA)

N

0   200   400 Miles
0   200   400 Kilometers

90°E

Figure 11.1

S outh Asia consists of Afghanistan, Bangladesh, Bhutan, India, the Maldives, Nepal, Pakistan, and Sri Lanka—a region that is naturally bounded by mountain ranges and seas (**Figure 11.1**). To the north is the almost impenetrable mountain rim of the Karakoram range and the Himalayas, while to the northwest are the forbidding ranges of the Sulaiman and the Hindu Kush. The Indian Peninsula and its large offshore island of Sri Lanka are girdled by the seas of the Indian Ocean—the Arabian Sea to the west, and the Bay of Bengal to the east. It is is a very heavily populated world region, with 1.35 billion people, more than a billion of whom live in India.

South Asia is semiperipheral or peripheral and politically volatile, though it is also a region of great potential. It is still a land of villages: Only about 10 percent of the population of Bhutan and Nepal live in urban settlements, while in the greater part of the region the urban population amounts to between 25 and 35 percent of the total. Not surprisingly, rural ways of life and traditional cultures remain extremely important throughout South Asia. Even broader as a defining characteristic of the region is poverty. The majority of South Asia's rural population is desperately poor, as are millions of the region's city dwellers. Hunger and malnutrition are widespread; barely half the adult population is literate; and only a minority of the population lives in sound housing with electricity and piped water. The abiding image, as described by Indian author Pankaj Mishra, is "the broken road, the wandering cows, the open gutter, the low ramshakle shops, the ground littered with garbage, the pressing crowd, the dust."[1]

However, another defining characteristic of the region derives from the stark contrasts that exist within and between places and subregions. Amid hierarchical traditions and social conservatism are deeply rooted ideals of equity and social justice. Amid predominantly rural settings are megacities like Delhi, Kolkata (formerly Calcutta), and Mumbai (formerly Bombay) in India, Dhaka in Bangladesh, and Karachi in Pakistan. Furthermore, amid extreme economic backwardness and widespread illiteracy there exists intellectual refinement and world-class technological innovation. To the stereotypical image of slow-moving lifestyles of dusty poverty we must add the legacies of sophisticated civilizations and the imprint of a growing middle class—computer programmers, managers, engineers, shop owners, media consultants, and so on—whose social practices and material consumption are in line with those of the middle classes in Europe and North America.

Environment and Society in South Asia

Two aspects of South Asia's physical geography have been fundamental to its evolution as a world region. First, as the satellite image reveals, South Asia is clearly set apart from the rest of Asia by a forbidding mountain rim (**Figure 11.2**). This arc

[1]P. Mishra, *Butter Chicken in Ludhiana: Travels in Small Town India,* London: Penguin, 1995, p. 93.

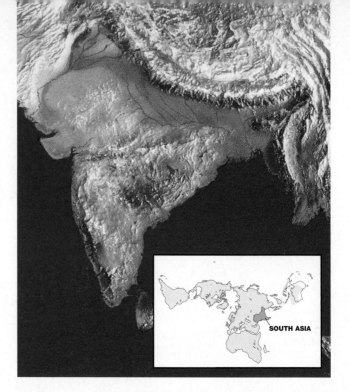

Figure 11.2 South Asia from space This image clearly shows how South Asia is naturally bounded by mountains and oceans. [*Source:* (inset) Map projection, Buckminster Fuller Institute and Dymaxion Map Design, Santa Barbara, CA. The word *Dymaxion* and the Fuller Projection Dymaxion™ Map design are trademarks of the Buckminster Fuller Institute, Santa Barbara, California, © 1938, 1967 & 1992. All rights reserved.]

of mountain ranges has isolated the peoples of South Asia, creating a large-scale natural setting in which distinctive human geographies have evolved. A second striking feature of the satellite image is the extent of the surrounding seas. Historically, the Arabian Sea provided a crucial routeway for trade between South Asia and the Middle East and the Mediterranean, while the Bay of Bengal gave access to (and from) Southeast Asia. These seas, together with the broader Indian Ocean, also produce the moisture for the summer monsoons, seasonal torrents of rain upon which the livelihood of the peoples of South Asia depends.

In geological terms, South Asia is a recent addition to the continental landmass of Asia. The greater part of what is now South Asia broke away from the coast of Africa about 100 million years ago and drifted slowly on a separate geological plate for more than 70 million years until it collided with the southern edge of Asia (see Figure 2.2, p. 53). The slow but relentless impact crumpled the sedimentary rocks on the south coast of Asia into a series of lofty mountain ranges and lifted the Tibetan Plateau more than 5 kilometers (3.1 miles) into the air. The Himalayas, which stand at the center of South Asia's mountain rim, are still rising (at a rate of about 25 centimeters—9.8 inches—per century) as a result of this geological event.

Landforms and Landscapes

Not surprisingly, the principal physiographic regions of South Asia—the Peninsular Highlands, the Mountain Rim, the Plains, and the Coastal Fringe—also reflect this major geological event. Between the mountain rim and the plateau highlands of peninsular India that stand on the ancient geological plate that drifted across from Africa are alluvial plains of young sedimentary rocks and material that has been washed down from the surrounding mountain rim and plateau. The coastal fringe of peninsular India, together with the coastal plains of Sri Lanka, the Maldives, and the Andaman and Nicobar islands, constitute a fourth physiographic region (**Figure 11.3**).

The Peninsular Highlands   The Peninsular Highlands of India form a broad plateau flanked by two chains of hills and uplands. The highlands rest on an ancient shield of granites and other igneous and metamorphosed sedimentary material, together with some very old sedimentary rocks. This shield

has remained a relatively stable landmass for much of the past 30 million years, but between 65 and 55 million years ago there occurred fissure eruptions of lava on an immense scale that buried the northwestern part of the peninsula beneath up to 3000 meters (9842 feet) of basalt, a dense volcanic rock. Today, this lava plateau—the Deccan Lava Plateau—covers about one-third of the Peninsular Highlands (Figure 11.3). To the north and southeast of the lava plateau, the ancient shield has been shaped into a broad area of plateaus, basins, and escarpments (steep slopes resulting from erosion or faulting). Surrounding this central body of the Peninsular Highlands is a series of hills and uplands. These include Aravalli Hills, the Western Ghats (the steep westward facing scarp of the eastward-tilted shield of the Peninsular Highlands), the Nilgiri

Figure 11.3   South Asia's physiographic regions   The physical geography of South Asia is framed by the highland plateaus of an ancient continental plate and a young mountain rim, separated by broad plains of alluvium washed down from both. (*Source:* Adapted from B. L. C. Johnson, *South Asia*, 2nd ed. London: Heinemann, 1982, p. 9.)

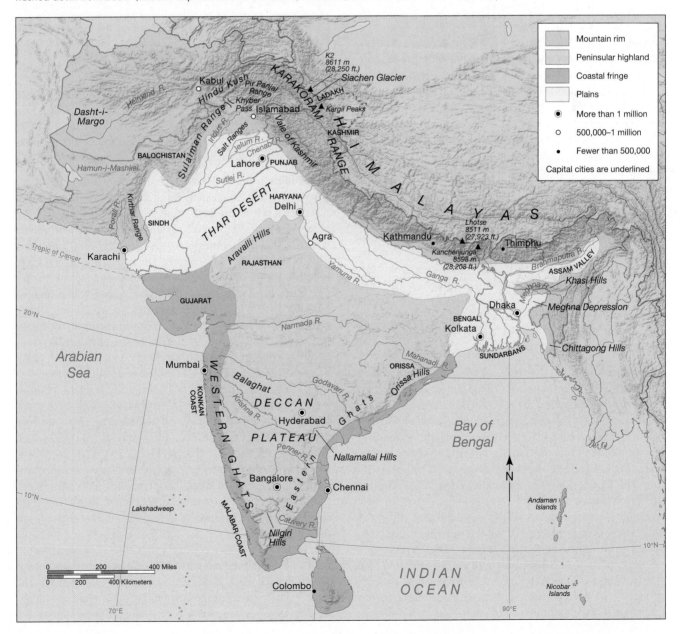

Figure 11.4 Deccan Lava Plateau Rice cultivation on the Deccan Lava Plateau near Hyderabad.

Figure 11.5 Kullu Vally, Himachal Pradesh, India The Kullu Valley is an extremely fertile valley tucked into the Himalayas and is known as "The Valley of the Gods" because of its fertility. Fields of rice paddies carpet the lower slopes. The homes built with alternating rows of wooden beams and stones, with slate roofs, are typical of the Himalayan foothills region.

Hills, the Eastern Ghats, the Orissa Hills, the Nallamallai Hills, and the Balaghat (Figure 11.3). These uplands are sparsely populated but have long provided refuge for many of India's tribal populations and for renegade princes, slopes on which to locate tea and coffee plantations, and shelter for wild game.

The rugged topography of the Peninsular Highlands as a whole has left the region somewhat isolated. Historically, it developed its own traditions and distinctive cultures, supporting a mosaic of local kingdoms. Because the only sources of water for farming are the seasonal and unpredictable annual rains of the monsoon, the peoples of the region have never enjoyed a rich agricultural economy, and subsistence farming and small villages still dominate the landscape throughout most of the region (**Figure 11.4**). There are, though, valuable deposits of iron ore, manganese, gold, copper, asbestos, and mica in the region, and thick seams of coal are present in the northeastern and east-central parts. Several of India's larger and most industrialized cities—including Ahmedabad, Bangalore, Bhopal, and Mysore—are located in the region.

The Mountain Rim At the heart of the Mountain Rim are the young but spectacular ranges of the Himalaya and Karakoram ranges. Arcing round to the west are the structurally complex ridges, ranges, and basins of the Hindu Kush, the Sulaiman range, and the Kirthar range, and behind these are several basins of inland drainage containing shallow **playa** lakes, such as Dasht-i-Margo and Hamun-i-Mashkel. Arcing round to the east is a series of parallel folded hills and valleys that reach the coast in the Chittagong Hills. Coal is found in workable quantities in parts of the mountain rim; oil-bearing ranges have been located in the Salt Ranges (also an important source of rock salt), the Assam Valley, and Gujarat, and natural gas is present in Bangladesh and the Sindh region. Still slowly rising, the Mountain Rim suffers from earthquakes, and severe shocks have occurred at various places along the entire length of the rim. Interspersed among the high peaks and the foothills are protected gorges and fertile valleys (**Figure 11.5**) that sustain isolated settlements of independent mountain peoples whose livelihood is based either on flocks of sheep, goats, and yak or on the tea plantations and orchards that cover the lower hills.

The Plains Broad plains of young sedimentary rocks and alluvium have been created by the deposition of material that has been eroded from both the Peninsular Highlands and the Mountain Rim. Three river systems—the Indus, the Ganga (Ganges), and the Brahmaputra—begin within 1600 kilometers (994 miles) of one another in the Himalayas but flow in three different directions through the mountains and into the plains. The Indus flows to the west, through Pakistan, to the Arabian Sea. The Brahmaputra flows eastward before doubling back through the Assam Valley and then flowing south to the Bay of Bengal. The Ganga flows southward before turning east, eventually merging with the Brahmaputra and forming a vast delta, more than 300 kilometers (186 miles) wide. All three river systems provide the Plains region with a steady, if uneven, flow of melting snow. As a result, the Plains region has long been widely irrigated and has supported a high density of population. The great dynasties of India—the Mauryan (320–125 B.C.), Gupta (A.D. 320–480), and Mughal (1526–1707)—all rose to prominence in this region, and the British moved their imperial capital here, to Delhi. Today, the Plains contain some of the most productive agricultural lands of South Asia (**Figure 11.6**). The cultivation of grains and rice is the predominant activity.

Not all of the Plains are so productive, however. Within the Plains are subregions with harsh environments for human settlement. The Thar Desert, between the Indus Valley and the Aravalli Hills, is one. At the other end of the Plains is the Meghna Depression, an immense backswamp of the Brahmaputra

Figure 11.6   Rice harvest on the Ganga Plains Although the Plains have been the hearth of successive empires because of their agricultural productivity, most of the Plains farmers rely on irrigation for their crops, and in some areas aridity is extreme.

system that provides a seasonal natural reservoir for the floods that regularly inundate lowland Bangladesh (see Figure 11.38). In the Assam Valley, the shifting, braided course of the turbulent Brahmaputra discourages settlement throughout much of the subregion.

The Coastal Fringe   The Coastal Fringe of South Asia consists of a rather mixed group of physical features. In places the Coastal Fringe is the product of marine erosion that has sawn into the edge of the ancient shield of the Peninsular Highlands; elsewhere it is the product of marine deposits; and in some places the plains consist of alluvial deposits in the form of deltas and mudflats. For the most part the Coastal Fringe is relatively narrow, and along India's western coast outlying spurs from the Western Ghats make it difficult to travel along the fringe by land. Far out into the Bay of Bengal are the Andaman Islands and the Nicobar Islands, both belonging to India but physiographically an extension of the Sumatran ranges of Southeast Asia (see Chapter 10, p. 479). About 650 kilometers (404 miles) to the southwest of India in the Indian Ocean are the Maldives, an independent state of 1190 tiny islands (only 200 of which are inhabited) grouped into 26 **atolls,** circular coral reefs that almost or entirely enclose a lagoon. Parts of the Coastal Fringe are occasionally subject to earthquakes, and in 2001 a major tremor centered on Bhuj, in western Gujarat, killed about 30,000 people and left hundreds of thousands homeless.

The Coastal Fringe is fertile in places and has been hospitable for human settlement. During the rainy monsoon seasons, the Coastal Fringe is filled with luxuriant growth, especially along the southwest Malabar Coast of India, where rich harvests of rice and fruit support rural populations of more than 1500 persons per square kilometer (3900 per square mile). Many of South Asia's largest and most prosperous cities developed from trading posts that were established along the Coastal Fringe in the seventeenth century, and the largest among them—Chennai (formerly Madras), Colombo, Karachi, Kolkata, and Mumbai—were great centers of commerce under British imperialism.

Climate

The Mountain Rim and the surrounding seas that we have noted as having been fundamental to South Asia's evolution as a world region are also strongly influential in shaping climate. The Mountain Rim effectively bars the movement of surface-level airstreams between South Asia and the rest of Asia, and vice versa. By contrast, no relief feature to the south of the Mountain Rim stands high enough to prevent the free flow inland of airstreams from the surrounding seas. Rather, all of the hills and uplands exert a strong **orographic effect,** causing moist air from the sea to lift and condense and producing heavy rainfall. Beyond these factors, the overall climates of South Asia are determined by global atmospheric circulation, as in other world regions. In South Asia, the dominant aspect of atmospheric circulation is the southwesterly summer monsoon. The word *monsoon* derives from the Arabic word *mausim,* denoting "season." Although now widely used to describe any seasonal reversal of wind flows in the lower-middle latitudes, *monsoon* was originally applied to the distinctive seasonal winds in the Indian Ocean that Arab traders relied upon to power their sailing ships on their annual voyages to and from the East Indies (the name formerly applied loosely to present-day Malaysia and Indonesia) in quest of spices, ivory, and fine fabrics.

In most of South Asia, the seasonal pattern of climate is as follows: a cool and mainly dry winter, a hot and mainly dry season from March or April into June, and a wet monsoon that "bursts" in June and lasts into September or later. In winter, a major branch of the jet stream tends to fend off the low-pressure systems of the circumpolar atmospheric whirl, helping to maintain stable high-pressure conditions over the Mountain Rim and the Tibetan Plateau. The prevailing winds are northeasterly, blowing from the interior toward the sea—these are the so-called "dry monsoon" winds (**Figure 11.7**). One exception to this pattern is in parts of Afghanistan, Pakistan, and Northwest India, where shallow low-pressure systems move through from the eastern Mediterranean, bringing light but useful rainfall in late winter. Another is northern and eastern Sri Lanka, where trade winds bring some winter rains. The southern part of Sri Lanka, together with the Maldives and the Nicobar Islands, are so far south as to be affected by intertropical convergence (see Chapter 2, p. 55), and so rarely have a dry month all year.

In terms of atmospheric circulation, similar conditions persist into the early summer. By May, daytime temperatures reach between 30°C and 40°C (between 86°F and 104°F) across most of South Asia. This is the season of heat and dust. By June, the earth is scorched, and farmers can grow nothing without irrigation. Everyone tends to become preoccupied with the discomfort of heat and humidity. One of the hottest places is Jacobabad, in Pakistan, where daytime temperatures in June average 45°C (113°F), with a maximum of 53°C (127°F) and a minimum of 29°C (84°F).

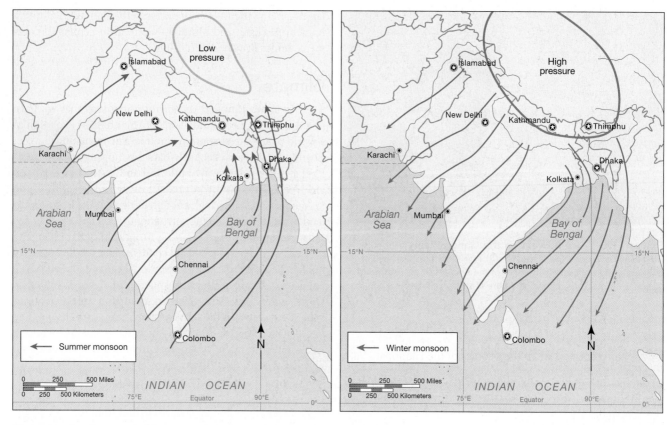

Figure 11.7   Summer and winter monsoons The wet summer monsoon "bursts" in June and lasts into September or later. In winter, a major branch of the jet stream tends to fend off the low-pressure systems of the circumpolar atmospheric whirl, helping to maintain stable high-pressure conditions over the Mountain Rim and the Tibetan Plateau. The prevailing "dry monsoon" winds are then northeasterly.

In mid-summer, the jet stream and the northern circumpolar atmospheric circulatory system move north of the Mountain Rim, allowing moist maritime air to invade the region. The southern circumpolar atmospheric circulation now produces strong trade winds that sweep north to become the southwest, or "wet" monsoon (Figure 11.7). The arrival of the wet monsoon season is announced by violent storms and torrential rain. This is the "breaking" or "bursting" of the monsoon (**Figure 11.8**). The monsoon season typically brings almost daily downpours, giving life to farms and fields. Where the low-pressure systems of the summer monsoon flow over hills and mountains, the monsoon rains are especially heavy. The Western Ghats and adjacent coastal plains typically receive between 2000 and 4000 mm (79 to 158 inches) of rainfall as the southwest monsoon winds meet the steep scarp slope at the edge of the Peninsular Highlands. Similar levels of annual rainfall are received in the central and eastern parts of the Mountain Rim. Cherrapunji, in the Khasi Hills south of the Assam Valley, boasts the world's average annual rainfall record of 11,437 mm (450.6 inches). It once recorded as much as 924 mm (36.4 inches) in one day at the onset of the monsoon season.

Figure 11.8   Breaking monsoon When the monsoon breaks, there is no escaping the torrential rains.

Occasionally, two streams of monsoon air, one moving up the Ganga Valley from the Bay of Bengal, the other across Rajasthan from the Arabian Sea, converge over the Himalayan foothills to produce abnormally heavy rainfall and floods. This happened in August 2000, leaving more than 500 dead and more than 4.5 million homeless in a 1500-kilometer (932-mile) long, 500-kilometer (311-mile) wide swath of the Himalayan foothills. It is in Bangladesh, however, that monsoon rains produce regular and widespread flooding. Swollen by monsoon rains, the **distributaries** of the Ganga and Brahmaputra river systems regularly spill over into the low-lying delta and plains areas. When monsoon rains are unusually heavy, flooding can be disastrous, inundating villages, drowning people and livestock, and ruining crops. When such conditions coincide with onshore typhoons, disaster can reach monumental proportions. In 1999 a particularly violent cyclone hit the low-lying coast of northeast India, pushing rivers backward and flooding much of the province of Orissa, killing an estimated 20,000 people and leaving 278,000 families homeless (see Figure 1.35, p. 28).

In contrast, for much of South Asia, there is a significant risk of drought and famine as a result of a late or unusually dry monsoon season. The Peninsular Highlands, the Indus Valley and adjacent plains, and the hills and basins of the northwestern portion of the Mountain Rim are especially prone to drought. In 2000 the late arrival of the summer monsoon left more than 50 million people in west and central India facing acute water shortages and widespread crop failure. In Afghanistan, several years of drought culminated in 2001 in disastrous crop failures that drove an average of 300 families into refugee camps each day.

Environmental History

The first extensive imprint of human occupation dates from at least 4500 years ago, when the peoples of the Harappan culture began to irrigate and cultivate large areas of the Indus Valley. Flourishing between 3000 and 2000 B.C., Harappan agriculturalists produced enough surplus, primarily in cotton and grains, to sustain an urban civilization. Harappan culture rivals its contemporary urban civilizations along the Euphrates and the Nile (see Chapter 5, p. 211) for the tag of "the cradle of civilization." Archaeological evidence shows that Harappans carried on trade with these civilizations in the Fertile Crescent, as well as with peoples in Southeast Asia and China. Soon after 2000 B.C., floods obliterated Harappan cities, leaving them covered in mud. It is possible that environmental degradation—especially the loss of the natural vegetative cover and the silting-up of irrigation and drainage channels—contributed to the flooding, though it may have been due entirely to climate change or tectonic movement.

The next significant impact on the natural environments of South Asia came with the incursion of tribal herdsmen from Central Asia into the Gangetic Plains. They were called Aryans (after the language that they spoke), and between 1500 and 500 B.C. they developed arable farming, assimilated or repulsed neighbors, adopted a settled life, organized into functional groups, opened trade links, built cities, and created a rich culture—the Vedic culture—whose mythic understanding of the world came to be a cornerstone of Hinduism. As their numbers grew, clans began to split away to annex new territory. In doing so, they began the transformation of the plains of the Ganga Valley from a moist green wilderness of forest and swamp to a dusty plain where tufts of trees survive only as shade for huddled villages and flocks of sheep (**Figure 11.9**). The deforestation of South Asia had begun. By 500 B.C the Vedic peoples had spread eastward into present-day Bihar and had learned to cultivate rice, clearing land along valley slopes in order to build terraces.

While the Plains region remained the most intensively developed, human settlement spread throughout South Asia, and a succession of kingdoms, sultanates, and empires slowly

Figure 11.9   Deforestation Although the Plains were originally covered with woodlands, 3000 years of agricultural clearances have left much of the region bereft of trees. This photograph was taken in the western plains, near Rajasthan.

brought more land under cultivation. For the most part, human occupance was sustainable. Subsistence farmers, herders, fisherfolk, and artisans drew on local resources for their food, traditional medicines, housing materials, and fuel, but apart from clearing part of the forest cover, their activities could be sustained from one generation to another without significant harm to the environment. It was the arrival of European traders, and especially British rule, that accelerated the deforestation of South Asia. In 1750, when the British were beginning their imperial conquests, more than 60 percent of South Asia was still forested, from the dry alpine forests of the Himalayan foothills and the mangrove forests of Bengal to the acacia forests of the Peninsular Highlands and the evergreen rain forests of the tropical coastal plains. British imperial rule brought the systematic clearance of land for plantations, and the methodical exploitation of valuable tropical hardwoods for export to Europe and North America. As railways opened up the interior, deforestation gathered pace. By 1900, only 40 percent of South Asia remained forested.

The most rapid period of change, however, has been the past 50 years, as the independent countries of South Asia have sought to modernize and expand their domestic economies. Between 1951 and 1976, for example, some 15 percent of India's land area was converted to cropland. Meanwhile, population growth in rural India has led to more and more wood being taken as fuel. Only 20 percent of India remains under forest today, and less than half of that is intact, natural forest—the rest consists of forest plantations, which have displaced natural ecosystems with monocrops. About one-third of the forest plantations in India consist of eucalyptus, a fast-growing, nonindigenous species that is very demanding of soil moisture. India's forests still represent a valuable trove of biodiversity (together, they house some 45,000 species of plants, 1250 species of birds, more than 350 species of mammals, and nearly 400 species of reptiles), but they are under severe pressure, along

South Asia's Disappearing Megafauna

In 2000 a World Conservation Union (IUCN) study identified 11,046 plant and animal species from around the globe as being at risk, including 180 species of mammals and 182 species of birds that are critically endangered. Emblematic of this acute problem are the so-called charismatic megafauna: large and exotic species like the rhinoceros, the elephant, whales, and tigers. In South Asia, some charismatic megafauna have already disappeared. Cheetahs disappeared from the wild in India more than 50 years ago, the last sighting being in 1948, when three young males were shot dead by a hunting party in the jungles of Bastar in Madhya Pradesh, central India. Today, the Bengal tiger (**Figure 1**) and the Asian elephant (**Figure 2**) are emblematic of critically endangered species in South Asia.

Experts put remaining numbers of Bengal tigers at somewhere between 3060 and 3985, although this could be an overestimate. Tiger populations are notoriously difficult to evaluate, despite recent technological advances that help keep track of their movements. What is clear is that the tiger is being lost. Animal by animal, its footprints are vanishing from South Asia's forests. The causes of the animal's decline are the same ones that are killing off many other species: loss of habitat and remorseless poaching.

In April 2000, the United Nations Convention on International Trade in Endangered Species (CITES) issued a critical report on how India is caring for its tigers, claiming that the Indian government has displayed a lack of concern and effort. Tiger losses to poachers, concluded CITES, are being covered up by officials, and figures for the remaining animals are deliberately inflated. Tigers are hunted by poachers in response to the huge demand for tiger body parts in traditional Chinese and Japanese medicine. A single tiger is worth more than $50,000 to poachers. The skin alone fetches an estimated $11,000, while a 10-gram tablet containing tiger bone sells for $25 and a bowl of tiger penis soup sells for $53.

The Asian elephant is also under great stress. In South Asia, elephants have long been important for cultural and religious reasons. In countries like Sri Lanka and Nepal, elephants take the lead in many festivals and ceremonies, while in India Ganesha is the elephant-headed Hindu god of wisdom. Stables of domesticated elephants are still kept in parts of South Asia for heavy work, but the region's herds of wild elephants are under stress. There are still between 17,000 and 25,000 wild elephants in South Asia, but as the region becomes more crowded, the elephants are rapidly running out of space. Farmers clear elephants' habitat to grow crops, while roads and urban development claim still more habitat. As elephants trespass on agricultural land, they are sometimes trapped, shot, or poisoned by farm managers, as in a case documented in 2000, when Pelwatte Sugar Industries in Sri Lanka killed elephants whose traditional migration routes pass through plantation land.

Asian elephants are also the target of poachers, who sell the animals' ivory tusks to make jewelry and their skin for shoes and bags. About 80 percent of the ivory goes to Japan, where it is used for making personal name seals (called *hankos*), which are considered an elegant substitute for a person's signature. Only male Asian elephants have tusks and, while the total population of wild Asian elephants is between 17,000 and 25,000, the number of tuskers of breeding age is just 1000 to 1200. It is this small fraction of the elephant population that is being targeted by poachers.

Figure 2 Asian elephant Like the Bengal tiger, the Asian elephant is endangered because of loss of habitat and indiscriminate killing by poachers.

Figure 1 Bengal tiger The most charismatic of all "charismatic megafauna," the Bengal tiger is one of the most seriously endangered species: Its habitat is fast being diminished, and it is hunted by poachers, with inadequate protection from government agencies.

with some of the key species that rely on them (see Geography Matters: South Asia's Disappearing Megafauna, p. 532).

The combination of population pressure and the desire on the part of newly independent countries to jump-start industry and agriculture has also put pressure on other natural resources, especially water. In parts of Punjab and Haryana, the "breadbasket" of India where almost a third of the country's wheat is grown, the water table has fallen more than 4 meters (13 feet) in the last decade. In the southern Indian state of Tamil Nadu, groundwater levels have fallen more than 25 meters (82 feet) in the last decade as a result of overpumping, leaving Chennai, like many other large cities, dependent on supplementary water supplies hauled in by tanker.

In parallel with the acceleration of resource depletion, there has been an acceleration in levels of environmental pollution. In India some 200 million people do not have access to safe and clean water, and an estimated 80 percent of the country's water sources are polluted with untreated industrial and domestic wastes. The Asian Development Bank has estimated that fewer than 1 in 10 of the industrial plants in South Asia comply with pollution-control guidelines. Only 10 percent of all sewage in South Asia is treated. India alone generates about 50 million tons of solid waste each year, most of which is disposed in unsafe ways: burned, dumped into lakes or seas, or deposited into leaky landfills. Air pollution has also become a serious issue. Little is known about the effects of acid rain on forests and cropland in South Asia, but the effects of air pollution in cities are clear. According to the Tata Research Institute in New Delhi, air pollution in India causes an estimated 2.5 million premature deaths each year. Motor vehicle emissions are a major contributor to urban air pollution. During the 1990s, the number of vehicles on Indian roads increased by 300 percent, and there has been a corresponding rise in rates of respiratory diseases.

Still, most people in South Asia—65 to 75 percent, in fact—are not city dwellers. Geographers Madhav Gadgil and Ramachandra Guha of the Indian Institute for Science estimate that some 400 to 500 million people remain what they call "ecosystem people," living at subsistence levels but in sustainable ways that have protected and preserved the environment. Increasingly, however, these ecosystem people are being pushed onto unproductive soils and arid hillsides as commercial forestry, mining, the construction of roads and dams, and the spread of industry limit their access to the land. As the environmental commons diminish and populations increase, a destructive cycle is set in motion. Ecosystem people are forced to use their limited resources in increasingly unsustainable ways, depleting sources of fuelwood, exhausting soils, and draining water resources.

South Asia in the World

South Asia has developed distinctive cultures and generated influential concepts and powerful ideals that have spread around the world. From the hearth areas of Harappan and Vedic civilization in the Plains, sophisticated cultures and powerful political empires spread across vast sections of the region. South Asia's resources and its geographic situation on sea-lanes between Europe and the East Indies made it especially attractive to European imperial powers, and in modern times it has become a pivotal geopolitical region with an emergent industrial and high-tech sector and an important market for core-region products.

The Mauryan Empire (320–125 B.C.) was the first to establish rule across the greater part of South Asia. By 250 B.C. the emperor Asoka had established control over all but present-day Sri Lanka and the southern tip of India. Securing control had wrought such havoc and destruction, however, that Asoka renounced armed conquest and adopted a policy of "conquest by *dharma*," that is, through the example of spiritual rectitude and chivalrous obligations. *Dharma* is a key concept of Buddhist teachings, and the spread of Buddhist principles of vegetarianism, kindness to animals, nonacquisitiveness, humility, and nonviolence are perhaps the most important legacy of Asoka's reign over South Asia.

After Asoka's death in 232 B.C., the Mauryan Empire fell into decline, and northern India soon succumbed to foreign invaders from Central Asia. After more than four centuries of division and political confusion, the Gupta Empire (A.D. 320–480) united northern India and came to control all but the northwestern hill country, the Peninsular Highlands, the southwestern coastlands (modern Kerala), and Sri Lanka. The Gupta period is generally regarded as the classical period of Hindu civilization. It produced the decimal system of notation, the golden age of Sanskrit and Hindu art, and important contributions to science, medicine, and trade. It was the Mughal Empire (1526–1707), however, that brought the most comprehensive and extensive economic, political, and administrative unification of South Asia, providing a framework that was absorbed into the British Empire in the eighteenth century.

Mughal India

Toward the end of the fifteenth century, a clan of militant Turks from Persia (now Iran) moved east in an attempt to evade the control of Tamerlane's Mongol empire. These Turks were the Mughals. Led by Babur, they conquered Kabul, in what is now Afghanistan, in 1504. By 1605, Babur's grandson Akbar had established control over most of the Plains, and in the next century Mughal rule had been extended to all but Sri Lanka and the southern tip of India (**Figure 11.10**). Akbar's rule was an extraordinary time, his achievements driven by his personal desire to synthesize the best of the many traditions that fell within his domain while maintaining strict control. Traditional kingdoms and princely states were kept intact, but they were integrated within a highly organized administrative structure with an equitable taxation system and a new class of bureaucrats. Persian became the official language, but Akbar abolished the tax on non-Muslims that had been instituted by his grandfather. Mughal rule did not seek to impose Islam on indigenous populations, but Mughal commitment to the religious precepts of Islam, together with the equitable system of Mughal governance, gave great stature to Islam. Over time,

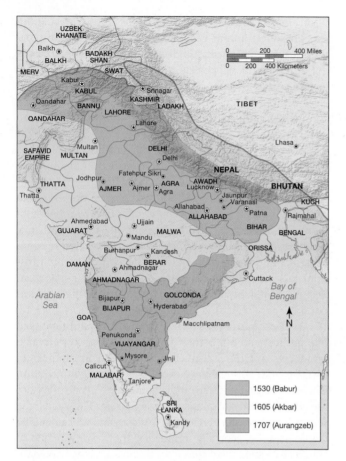

Figure 11.10 Mughal India Two centuries of Mughal rule began with the conquest of Kabul, in present-day Afghanistan, in 1504; by the time of Aurangzeb's death in 1707, Mughal rule had extended to almost all of the subcontinent. (*Source:* Redrawn from J. Keay, *India: A History.* New York: Atlantic Monthly Press, 2000, p. 314.)

Islam proved attractive to many, especially in the northwest (the Punjab) and the northeast (Bengal), both areas where Buddhism had previously been dominant. By 1700, mosques, daily calls to prayer, Muslim festivals, and Islamic law had become an indelible part of the social fabric of South Asian life.

Mughal rule was also characterized by a luxurious court and by extensive support for creativity in art, architecture, music, and literature. Akbar led the way by building Fatehpur Sikri (**Figure 11.11**) as the capital for his empire. The most fa-

mous legacy of Mughal architecture, however, is the Taj Mahal, built by Akbar's grandson, Shah Jahan, as a mausoleum for his wife Mumtaz Mahal, who died in 1631 while giving birth to their fourteenth child. Spectacular architecture became a signature of Mughal rule, the landscape of the northern regions becoming punctuated not only with lavish mosques and palaces but also with forts and citadels, towers, and gardens. Most ambitious of all was Shahjahanabad (now known as Old Delhi), which was designed to supersede Agra as the imperial capital. Here was a whole new city, complete with processional thoroughfares, spacious squares, bazaars, caravansaries, shaded waterways, and massive stone walls that were pierced by 11 gates and guarded by 27 towers. Its rigid geometry has long since been blurred but some of the walls and gates remain, as do the imperial complex known as the Red Fort and, nearby, the great Jama Masjid, which in its day was the largest mosque in India.

The last of the great Mughal emperors, Aurangzeb (1658–1707), provoked a series of rebellions and uprisings as a result of his anti-Hindu policies and his reinstatement of the tax on non-Muslims. At the same time he had to deal with raids from the Marathas, a warlike people from the Konkan coast, and was forced to conduct military campaigns on several fronts. With his death in 1707, the Mughal Empire collapsed, leaving South Asia open to the increasing interest and influence of European traders and colonists.

The *Raj*

European traders had been a regular presence along the coasts of South Asia long before the dissolution of the Mughal Empire. The Portuguese were the first, with the arrival of Vasco da Gama in India in 1498, but early in the seventeenth century the British East India Company (established in 1600) and the Dutch East India Company (established in 1602) set out in a deliberate attempt to contest the Portuguese monopoly of the Indonesian spice trade. South Asia, situated as it was on the route between Europe and the East Indies, provided an attractive array of intermediate stops at which to trade for the calicos, chintzes, taffetas, brocades, batiks, and ginghams of Gujarat, Bengal, Golconda, and the Tamil country. Soon, the armed ships of the two East India companies had pushed aside the Portuguese and had begun to tap into the ancient trade between India's east coast ports and Southeast Asia. In

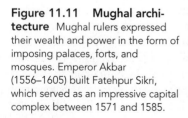

Figure 11.11 Mughal architecture Mughal rulers expressed their wealth and power in the form of imposing palaces, forts, and mosques. Emperor Akbar (1556–1605) built Fatehpur Sikri, which served as an impressive capital complex between 1571 and 1585.

the 1660s the French established the *Compagnie des Indies* and joined in the scramble for trade.

By the 1690s, European trading companies had established a permanent presence in several ports, though they had no interest in establishing colonies or exerting any kind of political authority, even when South Asia fell into disarray at the end of Mughal rule. But European wars in the eighteenth century (in particular the War of the Austrian Succession, 1740–1748, and the Seven Years' War, 1756–1763) spilled over into an imperial contest in South Asia. Initially, East India Company armies fought over territory simply to protect their trading hinterlands. With its own navy, backed by the powerful Royal Navy with its superior weapons technology, the British East India Company was most successful. In 1773 the British government transformed the Company into an administrative agency and, soon afterward, during the French Revolution and Napoleonic Wars, the British pushed ahead with aggressive imperialist policies in South Asia, using a mixture of force, bribery, and political intrigue to gain control over more and more of the region, which by this time was riddled with political and religious disunity.

The focus of British imperialism now shifted beyond trade and territorial control to social reform and cultural imperialism. In a famous memo written by East India Company Supreme Council member Thomas Macaulay in 1853, British administrators were urged to create a special class of South Asian people who would be "Indian in blood and colour, but English in taste, in opinions, in morals, and in intellect." One by one, the territories of native rulers were annexed or brought under British protection (**Figure 11.12**). Under the governor-generalship of Lord Dalhousie (1848–1856), there was a push to bring Western institutions and a modern industrial infrastructure to South Asia. Railroads, roads, bridges, and irrigation systems were built; and restrictions were placed on slave-trading, *suttee* (a widow's ritual suicide on her husband's funeral pyre), and other traditional practices. Western educational curricula flourished in private colleges, and British-style public universities were established. All of this provoked a conservative and anticolonial reaction, which came to a head in 1857, when an Indian Army unit rebelled because 85 of its soldiers were jailed for upholding their religious principles, refusing to use ammunition greased with animal fat. The incident quickly spread into a year-long civil uprising—the "Indian Mutiny"—throughout the north-central region. Massacres were carried out on both sides, but the mutiny was eventually quelled, and in 1858 the British Crown assumed direct control over India. In 1876 Queen Victoria was declared Empress of India.

Thus emerged the *Raj,* British rule over South Asia, which by 1890 extended to the entire region with the exception of present-day Afghanistan and Nepal (Figure 11.12). The British brought plantation agriculture to South Asia, producing food crops for the British domestic population and commodity crops for British industry and British merchant traders. Among the most important plantation crops were coconuts, coffee, cotton, jute, rubber, and tea (see Geographies of Indulgence, Desire, and Addiction: Tea, p. 537). The *Raj* also introduced Western industrial development and technology to South Asia,

displacing indigenous crafts and industries. Western political ideas of social reform, democracy, freedom of expression, and the materialism that accompanies free markets in land, labor, and commerce were also fostered by the *Raj,* along with the Western concept of national states, something that was to be an explosive legacy at the conclusion of the *Raj* in 1947, when Britain partitioned colonial India into separate independent national states.

Partition   Grassroots political resistance to British imperial rule had been institutionalized through the Indian National Congress Party, formed in 1887 to promote greater freedom and democracy, not only from imperial rule but from the traditional and autocratic rule of hereditary maharajas (leaders of the princely states that were under British protection). A leader and the inspirational figure of this movement was Mohandas Gandhi, whose vision of social justice and accountability and methods of nonviolent protest (including boycotts and fasting) were inspired by the ancient Buddhist concept of *dharma.* Under Gandhi's leadership, the case for national independence became irrefutable and, soon after the conclusion of the Second World War, the British set about withdrawing from South Asia altogether.

In creating new, independent countries, Britain sought to follow the European model of building national states on the foundations of ethnicity, with particular emphasis on language and religion. As a result, it was decided to establish a separate Islamic country, called Pakistan ("Land of the Pure"). Administrative districts under direct British control that had a majority Muslim population were assigned to Pakistan, together with those princely states whose ruling maharajas wished to join Pakistan rather than India. The result was that Pakistan was created in two parts, East Pakistan and West Pakistan, one on each shoulder of India, separated by 1600 kilometers (994 miles) of Indian territory.

In 1947, when the two national states were officially granted independence, millions of Hindus and Sikhs found themselves as minorities in Pakistan, while millions of Muslims felt threatened as a minority in India. Communal violence erupted across the region. In desperation, more than 12 million people fled across the new national boundaries—the largest refugee migration ever recorded in the world. As Hindus and Sikhs moved toward India and Muslims moved toward Pakistan in opposite directions between the two countries, many hundreds of thousands were senselessly killed. Hyderabad, in the Peninsular Highlands, which had a Hindu majority population but a Muslim leader who had volunteered Hyderabad to Pakistan, was quickly absorbed into India when riots broke out at the time of partition. In Kashmir the situation was reversed: a Hindu maharaja had elected to join India, but Pakistani forces intervened on partition to protect the majority Muslim population.

Having withdrawn from the greater part of South Asia, the British granted independence to the island of Ceylon as a Commonwealth dominion in 1948. In 1949 Britain handed to India its formal control over the external affairs of the kingdom of Bhutan, and in 1968 Britain granted independence to the

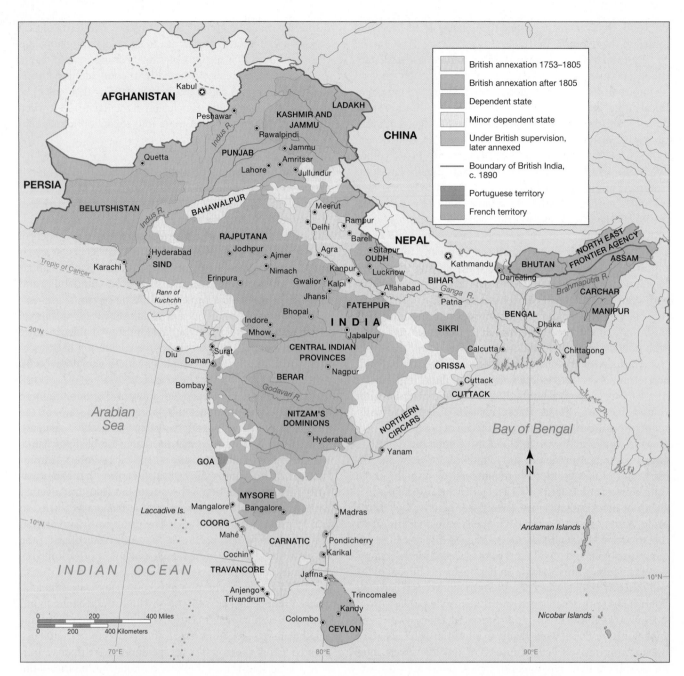

Figure 11.12 The British conquest of India British interest in India began as a consequence of merchant trade in the early eighteenth century, but by 1890 the British had come to control, directly or indirectly, most of South Asia. (*Source:* Adapted from I. Barnes and R. Hudson, *The History Atlas of Asia*. New York: Macmillan, 1998, p. 118–19.)

Maldives. The *Raj* was finally over, but the legacy of partition remains an important dimension of the geography of South Asia in today's world. In some ways, the states of South Asia are still adjusting to the partition of India and Pakistan in 1947. In Pakistan, divergent regional interests in East and West Pakistan quickly developed into regionalism, with East Pakistani leaders calling for secession. As a result, the country was split into two independent states in 1971: West Pakistan became Pakistan, and East Pakistan became Bangladesh. Meanwhile, neither India nor Pakistan could agree on the status of Kashmir, and the two countries briefly went to war over the region in 1948, in 1965, and again in 1971.

South Asia in Today's World

Kashmir is still a flashpoint, with heavily armed bands of Muslim militants ambushing and kidnapping Hindu victims in the mountain valleys, and Indian and Pakistani troops facing off against one another with sporadic skirmishes across the no-man's-land of the Siachen glacier. China is also involved in the dispute over Kashmir, with three separate fragments of the region claimed by India but controlled by China. The national boundaries drawn up by the British on partition also led to contention between India and Bangladesh because of a dispute over the distribution of the waters of the River Ganga.

Tea

Tea, a mild drug that makes a refreshing drink, was cultivated in China and Japan for centuries before becoming an important commodity in international trade. Thereafter, tea became a catalyst of economic, social, and political change. The Opium War (see Chapter 9, p. 433) was fought in large part over tea; the American Revolution was sparked by riots over a tax on tea; and the social and economic fabric of large parts of South Asia was destroyed by colonial mercantile forces that transformed them into tea plantations.

Tea, which was first brought to Europe by Portuguese traders, was initially marketed as a medicine. Tea drinking was adopted by the royal court of King Charles II in England, who reigned from 1660 until 1685, and tea quickly became an indulgence of the European bourgeoisie. By 1700 more than 17 million pounds of tea were being exported to Europe each year, mostly to Holland, Portugal, and England. As East Asian trade came to be dominated over the next century by the British East India Company, the cost of tea fell, and the taste for it spread. It soon became an addiction for the middle and working classes in Britain and a desirable indulgence for many in the American colonies. A significant market for tea had meanwhile developed in Russia and in many of the countries of the Middle East and North Africa.

South Asia became an important source of tea only in the nineteenth century, when the British sought to find commercial advantage in their newly acquired territories in Assam. In 1848 the Assam Company hired a botanist, Robert Fortune, to travel incognito to China to discover the secrets of cultivating and processing tea. After four expeditions, he brought to Assam not only the knowledge that the company needed but 12,000 seedlings, the specialized tools used in processing tea, and a skilled Chinese workforce. The British colonial government granted land on inexpensive leases to anyone who had the capital to establish a tea plantation. Inevitably, those with the capital were European settlers, and thousands of would-be planters flocked to Assam. Valuable timber was cut, forests were cleared, and the land not needed for tea plantations was rented to tribal peoples from Bihar and Orissa who were contracted to work—for miserably little pay and in dreadful conditions—as laborers in the plantations.

The opening of the Suez Canal in 1869 cut transport costs, made tea even cheaper, and allowed producers a bigger profit margin. Tea plantations in Assam and neighboring Bengal increased sixfold between 1870 and 1900, by which time there were half a million plantation workers in the region. In this same period, Ceylon—now Sri Lanka—emerged as a major tea-producing area. In the 1870s, a quarter of a million acres of coffee plantations in Ceylon were destroyed by blight, leaving coffee planters bankrupted and land extremely cheap. Thomas Lipton, a prosperous grocer from Glasgow, Scotland,

Figure 1　Tea pickers At the beginning of the economic chain that links South Asian producer regions with consumers in Europe and North America, tea pickers work long hours in harsh conditions for meager pay.

arrived in Ceylon in 1871 while on a world cruise, bought dozens of the plantations at bargain prices, and turned them over to the production of tea. Lipton was a pioneer of vertical economic integration, and by taking over all of the operations from growing tea to processing, management, transport, blending, packaging, and marketing, Lipton was able to cut the cost of tea to European consumers by 35 percent. Lipton and other tea planters brought in low-caste Tamil Hindus from famine-stricken areas of South India to work as laborers in the Ceylonese plantations. The conditions in which they were forced to work were atrocious, as bad as those of their counterparts in Assam and Bengal. Being low caste, low paid, isolated on estates in the highlands, and with no political voice, the "estate Tamils" were effectively enslaved on the plantations.

After independence in India, Bangladesh, and Sri Lanka, most of the European planters were forced to sell their tea plantations, either to the respective governments, under nationalization programs, or to indigenous business interests. The conditions for plantation workers have remained relatively unchanged, however. Plantation workers are still an impoverished group, a fact that is often belied by the image of smiling tea-pickers dressed in traditional costume, adorned with jewelry, working in sunshine amid beautiful scenery (**Figure 1**). Women, especially, are exploited cruelly. They are restricted to the lowest paid jobs: plucking, sifting, and weeding. Many start monotonous, repetitive, and back-breaking work in the fields at age 12 or younger. Fieldwork takes place in searing heat and in pouring rain, with no protective clothing except, perhaps, a fertilizer bag over her head and shoulders, and with hands bandaged because of sores caused by pesticides.

Afghanistan and Geopolitics

Afghanistan, known in ancient times as Gandhar, is situated pivotally between Central Asia and South Asia. Control of its mountainous terrain and the passes—such as the Khyber Pass—has often been of geopolitical significance. Time and again, Afghanistan has been fought over, more often than not leaving its villages and its countryside in ruins.

It is perhaps hard to believe that the strife-torn modern state of Afghanistan was once famous for its art and culture. Afghanistan's trading centers were important links on the emerging Silk Road between China and Mediterranean Europe (see p. 432), and their wealth soon attracted invaders. Alexander the Great swept into Afghanistan—then a part of the Persian Empire—in 329 B.C. This invasion paved the way for a cultural awakening and the emergence of the Gandhar school of art, known for its amalgam of Indian and Greek styles.

But invading the mountain passes of Afghanistan is easier than maintaining control of them. After Alexander the Great, the Scythians, White Huns, Turks, Genghis Khan, the British, and the Soviet Union all sent their armies into Afghanistan, only to mount a retreat. The problem, put simply, is physical geography. The country is dominated by the rugged Hindu Kush mountains that sweep from the west to the east, finally petering out near the northwestern city of Herat, where the mountains sink into the desert. Tens of thousands of square kilometers of the Hindu Kush form an intricate and seemingly endless maze of valleys and ravines. Jagged rock-strewn mountains and rugged valleys and caves provide ideal territory in which to fight a guerilla war against invaders or occupying forces. The problems of topography are compounded by the weather. By late October, swirling snow descends on the mountains, sealing off many of the passes, valleys, and high plateaus, and making the movement of troops almost impossible until late spring.

Nevertheless, Afghanistan's geopolitical significance has attracted one invader after another. In A.D. 642, Arabs invaded the region and introduced Islam. Arabs quickly gave way to Persians, who controlled the region until they were conquered by Turkic Ghaznavids in A.D. 998. The Turks turned Ghazni into a great cultural center and a base for frequent forays into India. In A.D. 1219 a Mongol invasion, led by Genghis Khan, resulted in the destruction of many cities, including Ghazni, Herat, and Balkh, and the despoliation of fertile agricultural areas. Following Genghis Khan's death in 1227, a number of petty chieftains and princes struggled for supremacy until late in the fourteenth century, when one of his descendants, Timur, incorporated Afghanistan into his vast Asian empire. Babur, a descendant of Timur and founder of the Mughal dynasty, made Kabul the capital of his principality at the beginning of the sixteenth century. Then, in 1747, Ahmad Shah Durrani forged the preconditions for modern Afghanistan by consolidating the many chieftainships, petty principalities, and fragmented provinces of the region into one country. His rule extended from Mashhad in the west to Delhi in the east, and from the Amu Darya River in the north to Arabian Sea in the south.

Late in the eighteenth century, Afghanistan's geopolitical significance increased dramatically. For the eastward-expanding Russian empire, Afghanistan represented the last barrier in a possible thrust toward the rich plains of India. For the British, who were establishing a hold on India, Afghanistan represented a bastion against Russian expansion. Both the Russians and the British desperately wanted to control Afghanistan, and so began the "Great Game"—the struggle between the two imperial powers for control of Afghanistan. The British were able to outmaneuver the Russians in the Great Game but were not able to establish territorial control. After three wars with the stubborn Afghans, the British finally granted Afghanistan independence in 1921. The first Anglo-Afghan war (1839–1842) resulted in the destruction of the British Army and is remembered as an example of the ferocity of Afghan resistance to foreign rule and the difficulty of moving troops and supplies through the difficult terrain of the Hindu Kush (**Figure 1**). The second Anglo-Afghan war (1878–1880) was sparked by the Afghans' refusal to accept a British mission in Kabul and resulted in the British and Russians together establishing the official boundaries of modern Afghanistan. The third Anglo-Afghan war began in 1919

Figure 1 The Anglo-Afghan War of 1839–1842 A nineteenth-century painting depicting British troops entering the Bolan Pass in 1839 on their way to Kabul. The Bolan and Khyber passes, it was feared, could also give Russian troops access to India.

with an attack by Afghans on India and ended later that year when the war-weary British relinquished their control over Afghan foreign affairs by signing the Treaty of Rawalpindi.

A brief period of Afghan independence followed, during which Mohammad Zahir Shah, who reigned from 1933 to 1973, established a relatively liberal constitution. Although Zahir's experiment in democracy produced few lasting reforms, it permitted the growth of extremist political movements, including the communist People's Democratic Party of Afghanistan (PDPA), which had close ideological ties to the Soviet Union, Afghanistan's new northern neighbor.

In 1973, following charges of corruption against the royal family and poor economic conditions caused by a severe drought the previous year, a military coup, led by former prime minister Sardar Mohammad Daoud, abolished the monarchy and declared Afghanistan a republic, with Daoud as its first president. His attempts to carry out badly needed economic and social reforms met with little success and were soon followed, in April 1978, by a bloody coup led by the PDPA. During its first 18 months of rule, the PDPA brutally imposed a Marxist-style reform program that ran counter to deeply rooted Islamic traditions. As a result, opposition to the PDPA's Marxist government emerged almost immediately.

Meanwhile, the Soviet Union moved quickly to take geopolitical advantage of the 1978 coup, signing a new bilateral treaty of friendship and cooperation with Afghanistan that included a military assistance program. Before long, as opposition insurgency intensified, the PDPA regime's survival was wholly dependent upon Soviet military equipment and advisers. In December 1979, faced with a rapidly deteriorating security situation, the Soviet Union sent a large airborne force and thousands of ground troops to Kabul under the pretext of a field exercise. A total of more than 120,000 Soviet troops were subsequently sent to Afghanistan, but they were unable to establish authority outside Kabul. An overwhelming majority of Afghans opposed the Communist regime, and Afghan freedom fighters (*mujahideen*) made it almost impossible for the regime to maintain a system of government outside major urban centers. Poorly armed at first, the *mujahideen* began receiving substantial assistance in the form of weapons and training from the United States, Pakistan, and Saudi Arabia in 1984. Trained and armed, the *mujahideen* exploited Afghanistan's terrain expertly. *Mujahideen* guerillas firing from surrounding ridges would disable the first and last vehicles in the Soviet columns and then slowly pick off the soldiers trapped in the center. Even with attack helicopters flying in support, Soviet columns were decimated. The guerilla war ended in 1989 with the withdrawal of the Soviet Union, but as the victorious *mujahideen* entered Kabul to assume control over the city and the central government, a new round of internecine fighting began between the various militia groups that had coexisted uneasily during the Soviet occupation.

With the demise of their common enemy, the militias' ethnic, clan, religious, and personality differences surfaced, and civil war ensued. Large-scale fighting in Kabul and in northern provinces caused thousands of civilian deaths and created new waves of displaced persons and refugees, hundreds of thousands of whom trekked across the mountains to Pakistan for sanctuary. Eventually, it was the hard-line Islamist faction of the *mujahideen,* the Taliban (**Figure 2**), that gained control of Kabul and most of Afghanistan. By 2001, the only remaining resistance to the Taliban regime was the Northern Alliance, a loose coalition of minority ethnic groups, including the Tajiks, who controlled between 5 and 10 percent of Afghan territory, including the beautiful Panjshir Valley 105 kilometers (65 miles) to the northeast of Kabul, and a small enclave around the far northern provinces of Badakhshan and Takhar. The Taliban regime not only imposed harsh religious laws and barbaric social practices on the Afghan population but also harbored the source of an entirely new geopolitical factor with worldwide implications: Osama bin Laden and his Al Qaeda terrorist network that the U.S. government asserts was responsible for the September 11, 2001, attacks on the Pentagon and the World Trade Center.

Figure 2 Taliban militiamen The militia that dubbed itself the Taliban (which translates as "Islamic students") emerged in 1994 from the rural southern hinterlands of Afghanistan under the guidance of the reclusive former village preacher Mullah Mohammed Omar. Fed by recruits from conservative religious schools across the border in Pakistan (most of whom were destitute refugees from the 1979–1989 war against the Soviet invasion), the Taliban won military and political support from Pakistan. It rose to power by promising peace and order for a country ravaged by corruption and civil war and the prospect of re-establishing the traditional dominance of the majority ethnic group, the Pashtun.

The boundary dispute was finally resolved in 1996, although tension persists over water control.

At a broader scale, South Asia quickly assumed a significant level of importance in global geopolitics. South Asia's strategic location meant that it was of great interest to the superpowers during the Cold War. India took full advantage of this, playing both sides against one another in seeking aid, while following a hybrid approach to domestic policy, with a democratic form of governance but a socialist-style approach to economic development. Post-colonial ties to Britain were maintained by Bangladesh, India, Pakistan, and Sri Lanka through the British Commonwealth, a loose association of countries tied together by patterns of trade, a shared language, and a similarity of institutions.

Afghanistan has long occupied a situation of geopolitical significance (see Geography Matters: Afghanistan and Geopolitics, p. 538). For much of the latter part of the twentieth century, Afghanistan had economic and cultural ties to the neighboring Islamic countries of Iran and Pakistan, and to the neighboring republics of Tajikistan and Turkmenistan in the Soviet Union. When a military coup established a dictatorship in Afghanistan in 1973, the country established close ties with the Soviet Union and began to pursue a Soviet-style program of modernization and industrialization. This provoked resistance from a zealous group of fundamentalist Islamic tribal leaders called the *mujahideen,* who were armed and trained by Pakistan. After several years of strife, the Soviet Union moved in to secure its geopolitical interests in the region. The Soviet military intervention was bloody but ineffective. It lasted for a decade, ending just before the collapse of the Soviet Union in 1990. Indeed, the military failure of the Soviet Union in Afghanistan is widely held to have contributed to its eventual collapse.

Ironically, South Asia has become even more of a geopolitical hot spot since the end of the Cold War. Both India and Pakistan have developed the capability of producing nuclear weapons. With the territorial dispute over Kashmir still simmering, India announced in May 1998 that five nuclear tests had been carried out in the Thar Desert close to the Pakistani border. This triggered a fervent bout of national pride within India, but it prompted Pakistan to respond within a few weeks with its own show of strength by carrying out a series of nuclear tests. India and Pakistan, together with Israel, are among the few states not to have signed the Nuclear Non-Proliferation Treaty that was formulated by the superpowers during the Cold War in 1968. In 2001, Afghanistan became the focus of operation "Enduring Freedom" in the wake of terrorist attacks on the Pentagon in Washington, DC, and the World Trade Center in New York.

On a more positive note, South Asia—and India, in particular—has come to play an increasingly important role within the world-system. India is the world's largest democracy and has maintained stable parliamentary and local government through elections and rule of law since the adoption of its Constitution in 1950. India has also developed a significant industrial base. Although two-thirds of the labor force is still engaged in agriculture, half of the country's GDP is accounted for by an industrial sector that is the tenth largest in the world, by value.

In 1992, after India lost its major trading partner with the collapse of the Soviet Union, India embarked on a series of reforms as a condition of a Structural Adjustment Program attached to a loan from the World Bank. Before these reforms, many key institutions—including banks, utilities, airlines, railways, radio, and television—were government owned and operated. High tariffs, restrictions on foreign ownership, high taxes, and widespread corruption all kept foreign investors away and suppressed the energy of Indian entrepreneurs. Although more than three-quarters of the economy remained in the private sector, government bureaucracy had developed a complex system of permits, licenses, quotas, and permissions that further restricted economic vitality. The reforms have created a more open and entrepreneurial economy. Key institutions have been privatized, and foreign investment has been flowing into the country, helping to generate exceptionally high economic growth rates. The fact that India's middle class conducts business in English gives India an important comparative advantage in today's world economy.

India now has an affluent middle class estimated at some 200 million—a huge, well-educated, and sophisticated consumer market that has become part of the "fast" world (see Chapter 1, p. 29) and an important agent of globalization. The rapid growth of India's affluent middle classes serves not only to accentuate the contrasts within South Asia between the traditional and the modern but also to highlight the desperate situation of an even larger group: the extremely poor. According to the United Nations Development Fund, some 53 percent of India's population (that is, more than half a billion people) live on less than a dollar a day—the World Bank's definition of dire poverty. In fact, most of India's poor (390 million of them) somehow exist with an income of a dollar a *week*.

The other countries of South Asia have not experienced the kind of economic boom enjoyed by India, though there have been attempts to foster regional economic integration. In 1985 the South Asian Association for Regional Cooperation (SAARC) was established. Although progress has been slow—mainly because of the friction between India and Pakistan—the member states did sign a South Asian Preferential Trade Agreement (SAPTA) in 1996, which established some modest mutual tariff concessions. Impatient with the pace of SAARC, India also set up two subregional cooperation groups: one with Nepal, Bhutan, and Bangladesh, and another with Sri Lanka and the Maldives. India has also signed a free-trade agreement with Sri Lanka and has pursued wider avenues of economic and diplomatic cooperation, becoming a "dialogue partner" in the Association of Southeast Asian Nations (ASEAN) and lobbying for a seat as a permanent member of an expanded United Nations Security Council.

Peoples of South Asia

South Asia has the second largest and the fastest-growing population of all world regions. The total population of South Asia in 2000 stood at 1.35 billion, with India accounting for just over 1 billion. With overall growth rates in the region of 1.9 percent per year (compared to 1.0 percent per year in China), South Asia is headed for a population of 1.63 billion by 2010. **Figure 11.13** shows the distribution of population within

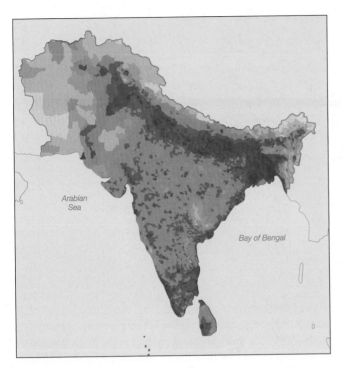

Figure 11.13   Population density in South Asia 1995
The density of population is very high throughout most of South Asia, but especially so in the Plains and in subregions with good soils and humid climates. For a view of worldwide population density and a guide to reading population-density maps, see Figure 1.40. (*Source:* Center for International Earth Science Information Network (CIESIN), Columbia University; International Food Policy Research Institute (IFPRI); and World Resources Institute (WRI). 2000. *Gridded Population of the World (GPW)*, Version 2. Palisades, NY: CIESIN, Columbia University. Available at http://sedac.ciesin.org/plue/gpw)

South Asia. The first thing to note about this map is the very high density of population throughout most of the region. The overall density of population in India is 316 persons per square kilometer (819 per square mile), compared to 131 persons per square kilometer (338 per square mile) in China and 29 persons per square kilometer (75 per square mile) in the United States. In detail, patterns of population density reflect patterns of agricultural productivity. The combination of good soils with a humid climate or with extensive irrigation supports densities of more than 500 persons per square kilometer (1300 per square mile) in a belt extending from the upper Indus plains and the Ganga plains through Bengal and the Assam Valley. Similar densities are found along much of the Coastal Fringe.

Urbanization

In comparison with other world regions, South Asia is still very much a land of villages. Approximately 65 percent of Pakistan's population and 73 percent of India's live in rural settings; in Afghanistan, Bangladesh, and Sri Lanka the rural population is around 80 percent; and the tiny state of Bhutan is 94 percent rural. Rural-to-urban migration is shifting the balance toward towns and cities, however. This is largely a result of population pressure in rural areas, where natural population increase has reduced the amount of cropland per person to half of what

it was in 1960. In that year, there were just 9 cities of 1 million or more in South Asia, and only one of these had more than 5 million inhabitants. In 2000 there were 50 cities of 1 million or more, including 9 of 5 million or more. Mumbai, the largest metropolis in South Asia, grew from 4.1 million to 18.0 million between 1960 and 2000; Dhaka, in Bangladesh, grew from less than 650,000 to almost 11 million; and in Pakistan, Karachi grew from 1.8 million to 11.8 million.

Population Policies

Both the overall rate of population growth and the rate of urbanization are cause for concern in South Asia. With hundreds of millions already living in extreme poverty, high rates of natural increase, intensified in urban areas by high rates of in-migration, bring the prospect of serious food and water shortages, mass starvation, and food riots. As a result, each of the countries of the region has developed policies to try to limit population growth, with varying degrees of success. India was the first country to establish such policies, announcing an official family-planning program as early as 1952. Little attention was paid to the program until the mid-1960s, when the government announced specific demographic targets and opened "camps" around the country for the mass insertion of intrauterine devices (IUDs). The program soon failed, mainly because of negative public reaction to the poor training of health workers and unsanitary conditions in the camps.

Next came vasectomy camps. More than 10 million men were coerced into being sterilized in the 1970s in an "Emergency Drive" for family planning that saw all kinds of government administrators—from police to teachers and railway inspectors—given monthly quotas to recruit "volunteers" for vasectomy camps. Bureaucrats in some Indian states sought to reinforce the sterilization drive with harsh penalties. In Bihar state, for example, families with more than three children were denied public food rations; and in Uttar Pradesh, teachers who refused to volunteer for sterilization were fined a month's salary. Not surprisingly, a popular backlash put an end to the vasectomy program. Today, the level of public mistrust remains high, the quality of family-planning services remains poor, and the demand for contraceptives is low. In 1998 the Indian government acknowledged the evidence of international experience—that female education is the single most influential determinant of lower birthrates—and finally abandoned targets for sterilization and contraception. Several Indian states are now following Kerala's successful example of emphasizing women's education and better infant and maternal care. While this policy shift seems likely to be successful in the long run, it means that India can expect population growth to continue for several decades before leveling off.

The South Asian Diaspora

The South Asian diaspora amounts to some 5 or 6 million people, most of them located in Europe, Africa, North America, and Southeast Asia (**Figure 11.14**). The origins of this diaspora can be traced to the abolition of slavery in the British Empire in 1833. The consequent demand for cheap labor in

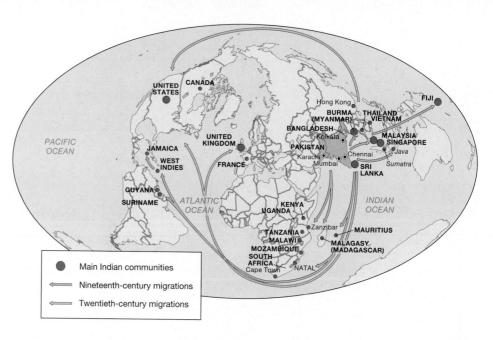

Figure 11.14 The South Asian diaspora Nineteenth-century migrations from South Asia were mainly to British colonies in East Africa, South Africa, and Southeast Asia, where there was a demand for cheap labor in plantations and on railways. In the twentieth century the principal flows were to factory and service jobs in Britain and the United States. (*Source:* Redrawn from G. Chaliand and J-P. Rageau, *The Penguin Atlas of Diasporas.* New York: Viking, 1995, p. 155.)

the plantations and on the railways of the British Empire was filled in part by emigrants from British India. In the mid-nineteenth century, thousands of Indians left for the plantations of Mauritius (in the Indian Ocean), East Africa, the West Indies, and South Africa. The stream of emigration intensified in the early twentieth century. By 1920 there were more than a million South Asian immigrants in Burma, about 600,000 in Malaya, 300,000 in the West Indies, 200,000 in South Africa, and 100,000 in East Africa, while about 20,000 South Asians had made their way to Britain to work in factories and another 5000 or so had found their way to North America to work in railway building, in sawmills, and as farm workers.

After the Second World War the pattern changed significantly. Independence in former British colonies led to the exclusion of South Asian immigrants, and both Burma and Uganda expelled most South Asian immigrants. But a new destination for South Asian emigrants opened as the postwar economic recovery in Europe resulted in a severe shortage of labor on assembly lines and in transportation. Britain received more than 1.5 million South Asian immigrants, whose permanent presence not only filled a gap in the labor force but has also served to enrich and diversify British urban culture. About 800,000 South Asians moved to North America, mainly to larger metropolitan areas where most found employment in service jobs. From the 1970s onward there has also been a steady stream of South Asian immigrants to the oil-rich Persian Gulf states, recruited on temporary visas to fill manual and skilled manual jobs.

South Asia has also experienced a "brain drain" of significant proportions over the past several decades. Beginning with the emigration of physicians and scientists to Britain in the 1960s, the brain drain accelerated as South Asian students, having completed their studies in British and American universities, stayed on to take better-paying jobs rather than return to South Asia. The idea of living abroad gained popularity among India's cosmopolitan and materialist middle classes as newspaper and television features publicized the global successes of Indian emigrants. Geographer Pamela Shurmer-Smith comments that ". . . it is sobering to know that virtually the whole of the youth of a social fraction in India is now craving

to live outside its own country and that this obsession has been largely constructed by the power of the international media."[2] In the 1990s the most distinctive aspect of the brain drain from South Asia was the emigration of computer scientists and software engineers from India to the United States and parts of Europe. By 2000, more than 2000 of the 15,000 employees on Microsoft's Redmond campus were South Asian immigrants.

Cultural Traditions

Diversity has to be the key word to describe the cultural geography of South Asia. The whole region has deep cultural roots, but these roots are often tangled. Even where traditions have not been mixed or hybridized, there are significant differences in the degree to which traditional cultures have accommodated or resisted globalization. In Afghanistan and Pakistan, powerful Islamist movements have resisted globalization, attempting to re-create certain aspects of traditional culture as the basis of contemporary social order. There is strong adherence in both countries, for example, to traditional forms of dress and public comportment (**Figure 11.15**). In Afghanistan the ultra-orthodox Taliban rulers who have controlled 90 percent of the country since 1996 have imposed a harsh version of Islamic law that follows a literal interpretation of the Muslim holy book, the Quran. Under Taliban laws, murderers are publicly executed by the relatives of their victims. Adulterers are stoned to death, and the limbs of thieves are amputated. In 1999 a woman accused of murdering her husband was shot to death in a stadium packed with thousands of men and women, many of whom had brought their children along to watch. Lesser crimes are punished by public beatings. The Taliban's long list of rules include not wearing shorts or short-sleeved shirts in public—even athletes are required to wear the traditional baggy pants and long tunic. In 2000, Taliban religious police interrupted a soccer game in the southern Afghan city of Kandahar to arrest 12 visiting Pakistani players for wearing shorts. The offenders were released after

---

[2]P. Shurmer-Smith, *India: Globalization and Change,* London: Arnold, 2000, p. 171.

Figure 11.15 Traditional dress There is strong adherence in many communities to traditional forms of dress and public comportment. These women in Peshawar, Pakistan, are wearing the typical everyday attire of Islamic women in the region.

their heads had been shaved in punishment. Under Taliban rule, taking photographs or painting pictures is considered impure. There is no television, no cinema, and newspapers do not print pictures. Music of any kind is forbidden. In February 2001, Mullah Mohammed Omar, fundamentalist leader of the Taliban, issued an edict ordering the destruction of all non-Islamic statues. Within a month, the Taliban militia had fully carried out the edict, including the destruction by artillery of two huge 1700-year-old sandstone Buddhas at Bamiyan in the Hindu Kush mountains of central Afghanistan.

These harsh cultural rules are symptomatic of deeper barbarities. Under the Taliban, Afghan women have been robbed of all of their rights, Afghan children have no schools or hospitals to go to, and thousands of opponents of the Taliban have been systematically massacred. Meanwhile, the entire country has suffered three successive years—1999, 2000, and 2001—of severe drought and crop failure. The resulting famine and poverty has seemingly helped turn many formerly peaceful farmers into desperate extremists, while hundreds of thousands have abandoned their homes and become refugees, joining the million or so Afghans who had fled to Pakistan during the Soviet military intervention of the 1980s. In the three years before Pakistan sealed its border with Afghanistan in September 2001, more than 2 million Afghan refugees crossed into Pakistan.

In contrast to Afghanistan, India has successfully fostered democracy after its introduction through British colonial governance in the nineteenth century. In India, contemporary culture is open to the economic and cultural flows of globalization. The result is that traditional cultures, still strong, are juxtaposed vividly against modern global culture. For example, the tradition of parents seeking marriage partners for their children through newspaper advertisements continues relatively undiminished; but those same advertisements often provide an e-mail address or even a Web site for replies.

Religion Tradition itself is very important in South Asian cultures. Many different indigenous cultural threads have evolved into a variety of regional patterns. Over the centuries much has been added, while little appears to have been lost. As a result, the regional cultural geography of South Asia is extremely complex. At face value, one of the most important bases for regional differences in cultural traditions is religion. The two most important religions in South Asia are Hinduism and Islam.

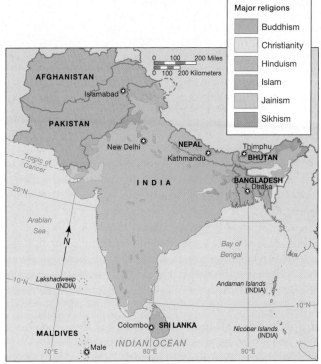

Figure 11.16 The geography of religion in South Asia Partition between India and Pakistan in 1947 resulted in mass migrations of Hindus from Pakistan to India and of Muslims from India to Pakistan. There remain, however, more than 80 million Muslims in India, concentrated in a number of subregions.

Hinduism is the dominant religion in Nepal (where about 90 percent of the population are Hindu) and India (about 80 percent). Islam is dominant in Afghanistan (99 percent), Bangladesh (more than 80 percent), the Maldives (100 percent), and Pakistan (about 80 percent). Buddhism, though it originated in South Asia, is followed by only about 2 percent of the region's population. It is the predominant religion in Bhutan and Sri Lanka, and there is an enclave of Buddhism in Ladakh, the section of Kashmir closest to China. Jains are another distinctive religious group whose origins are in South Asia. Jains, like Buddhists, trace their faith to Prince Siddhartha, a religious leader who lived in northern India in the sixth century B.C. and who came to be known as Buddha, that is, the Enlightened One. Sikhs, whose religion was founded by Guru Nanak in the sixteenth century A.D., are concentrated in the Punjab, which straddles the India-Pakistan border (**Figure 11.16**).

The broad regional patterns reflected in Figure 11.16 are much more complex when considered in any detail. Underlying much of this complexity is the fact that Hinduism is not a single organized religion with one sacred text or doctrine; it has no unifying organizational structure, worship is not congregational, and there is no agreement as to the nature of the divinity. Rather, Hinduism exists in different forms in different communities as a combination of "Great Traditions" and "Little Traditions." The Great Traditions derive from the *Rg Veda,* a collection of 1028 Vedic poems that date from the tenth century B.C. One of the key aspects of this Great Tradition is the belief that human lives represent an episode of cosmic existence, followed after death by the transmigration of the soul to some other form of life. The "Little Traditions" of Hinduism

Hinduism's Sacred Places

Most religions designate certain places as sacred, often because a special event occurred there. Sites are often designated as sacred in order to distinguish them from the rest of the landscape that is considered ordinary or profane. Sacred places are special because they are the sites of intense or important mystical or spiritual experiences. Sacred places include those areas of the globe recognized by individuals or groups as worthy of special attention because they are the sites of special religious experiences and events. Sacred space does not occur naturally; rather, it is assigned sanctity through the values and belief systems of particular groups or individuals. Geographer Yi-Fu Tuan insists that what defines the sacredness of a place goes beyond the obvious shrines and temples. Sacred places are simply those that rise above the commonplace and interrupt ordinary routine. In almost all cases, sacred places are segregated, dedicated, and hallowed sites that are maintained as such generation after generation. Believers—including mystics, spiritualists, religious followers, and pilgrims—recognize sacred places as being endowed with divine meaning.

In Hinduism, the number seven has special significance. There are seven especially sacred cities: Varanasi (**Figure 1**), associated with the god Shiva (the destroyer, but without whom creation could not occur); Haridwar (where the Ganga enters the Plains from the Himalayas); Ayodhya (birthplace of Rama, one of the incarnations of Vishnu, the preserver); Mathura (birthplace of Krishna, another incarnation of Vishnu, sent to Earth to fight for good and combat evil); Dwarka (legendary capital of Krishna thought to be located off the Gujarat coast); Kanchipuram (the great Shiva temple); and Ujjain (the site every 12 years of the Kumbh Mela, a huge religious fair). There are also seven sacred rivers: the Ganga, the Yamuna, the mythical Saraswati, the Narmada, the Indus, the Cauvery, and the Godavari. It is often the case that religious followers are expected to journey to especially important sacred places to renew their faith or to demonstrate devotion. A pilgrimage is a journey to a sacred place, and a pilgrim is a person who undertakes such

Figure 1   Varanasi Situated on the banks of the holy Ganga River, Varanasi is one of the most sacred places in India. Hindu pilgrims come to bathe in the waters of the Ganga, a ritual that is held to wash away all sins.

consist of the many local gods, beliefs, rituals, and festivals, and the sacred spaces that are associated with them (see Geography Matters: Hinduism's Sacred Places, above).

The complexity of Hindu traditions within India is compounded by the existence of a sizable minority population who adhere to other religions. The most important of these is Islam. Although several million Muslims migrated from India to Pakistan at the time of partition, more than 112 million still reside in India today. There are also almost 20 million Christians in India. According to legend, Christianity was first introduced to South Asia by the Apostle Thomas during the first century. Silk traders passing through northwest Pakistan to China during the second century encountered Christians, but the small Christian community did not increase significantly until the arrival of colonial powers. The Portuguese brought Roman Catholicism to the west coast of India in the late 1400s, and Protestant missions, under the protection of the British East India Company, began to work their way through the region in the 1800s. Christianity is most widespread in the state of Kerala, in southwest India, where nearly one-third of the population is Christian.

Language   A great diversity of languages is spoken in South Asia. In India alone there are some 1600 different languages, about 400 of which are spoken by 200,000 or more people.

a journey. Hindus visit sacred pilgrimage sites for a variety of reasons, including to seek a cure for sickness, wash away sins, or fulfill a promise to a deity. The Ganga is India's holiest river, and many sacred sites are located along its banks, including Haridwar and Varanasi. There are, however, many other important sacred sites (**Figure 2**) that attract Hindu pilgrims, and thousands of temples, shrines, and other sacred sites.

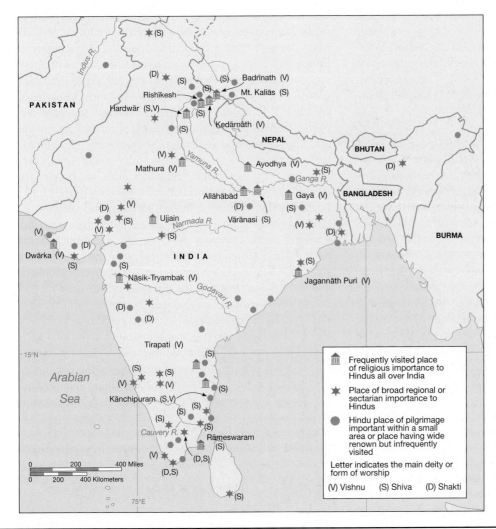

Figure 2   Sacred sites of Hindu India Many of the most sacred Hindu sites are located along the seven sacred rivers of Hinduism: the Ganga, the Yamuna, the mythical Saraswati, the Narmada, the Indus, the Cauvery, and the Godavari. (*Source:* Adapted from I. Ragi al Farugi and D. E. Sopher, *Historical Atlas of the Religions of the World.* New York: Macmillan, 1974.)

There is, however, a broad regional grouping of four major language families. The Indo-European family of languages, introduced by the Aryan herdsmen who migrated from Central Asia between 1500 and 500 B.C., is prevalent in the northern plains region, Sri Lanka, and the Maldives. This language family includes Hindi, Bengali, Punjabi, Bihari, and Urdu. Munda languages are spoken among the tribal hill peoples who still inhabit the remoter hill regions of peninsular India. Dravidian languages (which include Tamil, Telegu, Kanarese, and Malayalam) are spoken in southern India and the northern part of Sri Lanka. Finally, Tibeto-Burmese languages are scattered across the Himalayan region.

In India the boundaries of many of the country's constituent states were established after partition on the basis of language. Overall, no single language is spoken or understood by more than 40 percent of the people. Since the time when India became independent, there have been efforts to establish Hindi, the most prevalent language, as the national language, but this has been resisted by many of the states within India, whose political identity is now closely aligned with a different language. In terms of popular media and literature, there is a thriving Hindi and regional language press, while film and television are dominated by Hindi and Tamil, with some Telegu programming.

English, spoken by fewer than 6 percent of the people, serves as the link language between India's states and regions. As in other former British colonies in South Asia, English is the language of higher education, the professions, and national business and government. Without English, there is little opportunity for economic or social mobility. Most children who do attend school are taught only their local language, and so are inevitably restricted in their prospects. A guard, sweeper, cook, or driver who speaks only Hindi or Urdu will likely do the same work all his life. In contrast, those who can speak English—by definition, the upper-middle classes—are able to practice their profession or do business in any region of their country or in most parts of the world. English-language South Asian literature has produced many excellent novels. Among the most notable authors are Anita Desai, V. S. Naipaul, Arundhati Roy, and Salman Rushdie.

Caste   A very important—and often misunderstood—aspect of India's cultural traditions is that of **caste**. Caste is a system of kinship groupings, or *jati,* that are reinforced by language, region, and occupation. There are several thousand separate *jati* in India, most of them confined to a single linguistic region. Many *jati* are identified by a traditional occupation, from which each derives its name: *jat* (farmer), for example, or *mali* (gardener), or *kumbhar* (potter). Modern occupations such as assembly-line operators, clerks, and computer programmers, of course, do not have a traditional *jati,* but that does not mean that people doing these jobs cease to be members of the *jati* into which they were born. People within the same *jati* tend to sustain accepted norms of behavior, dress, and diet. They are also endogamous, which means that families are expected to find marriage partners for their children among other members of the *jati.*

In each village or region, *jati* exist within a locally understood social hierarchy—the caste system—that determines the accepted norms of interaction between members of different *jati.* In a normal village caste system, individuals will typically interact on a daily basis with others from about 20 different *jati.* Each individual person's *jati* is fixed by birth, but the position of the *jati* within the local caste system is not. Nevertheless, the broad structure of caste systems always places certain groups at the top and others at the bottom. Caste systems tend to hold in high esteem those who are religious and those who are especially learned. Those who pursue wealth or hold political power are typically less well regarded, but those who perform menial tasks are accorded least status of all. Priestly *jatis*—known as brahmins—are always at the very top of the caste hierarchy. Brahmins are expected to lead ascetic lives and revere learning.

At the opposite end of all caste systems are the so-called "untouchables"—*jatis* whose members deal with human waste and dead animals. Mohandas Gandhi, the inspirational leader of India before independence, crusaded to dissociate these *jatis* from the demeaning term *untouchable.* Gandhi called them Harijans, meaning "children of God," but today most people in these *jatis* prefer to be referred to as Dalits, meaning "the oppressed," and the Indian government refers to them as "Sched-uled Castes." Traditionally, the Dalits were forced to live outside the main community because they were deemed by the brahmins to be capable of contaminating food and water by their touch. They were denied access to water wells used by other *jatis,* refused education, banned from temples, and subject to violence and abuse. Although these practices were outlawed by India's constitution in 1950, discrimination and violence against Dalits is still routine in many rural areas.

Contemporary Culture   Contemporary culture provides many sharp contrasts with the deep-rooted traditions of South Asia, although there are places and regions (Afghanistan, in Bhutan, and many of the more remote rural areas of South Asia) where contemporary culture finds few expressions. The growth of a large and affluent middle class in India since the country's 1992 economic reforms has brought the sights and sounds of Western-style materialism to India's larger towns and cities: fast-food outlets, ATM machines, name-brand leisure wear, consumer appliances, video games, and luxury cars. Cricket, a legacy of British colonialism, has become the preeminent sport in both India and Pakistan (**Figure 11.17**). Long popular, the new affluence of the middle classes has taken cricket beyond a popular pastime with a passionate following to a sport that generates huge sums in betting and supports a star system to rival that of baseball in the United States.

Cable television arrived in India in the early 1990s, at about the same time that the government initiated its economic reforms. After years without access to popular Western culture, urban middle-class Indians could now watch, via Hong-Kong-based Star TV, programming that included MTV, "Baywatch," and "The Oprah Winfrey Show." The expectation among many was that such programming would quickly displace Indian culture, at least among the young and the middle classes. The sheer size and market power of India's middle classes, however, has meant that this scenario of an externally imposed global culture has not come about. Rather, Indian television and cable companies quickly began to produce films,

Figure 11.17   International cricket People in India and Pakistan are passionate about cricket. During international matches, interest is intense and emotions run high. Shown here is the cricket ground in Kolkata during the match between India and South Africa in 1991, which drew a record crowd of 95,000.

musical shows, sitcoms, and soap operas in Hindi, Tamil, and some other local languages. The only Hollywood-made programs that earn reasonable ratings are those that are dubbed, while the domestic Indian television and movie industry has quickly grown to major proportions (see Sense of Place: Bollywood, p. 548).

Just as the impact of globalization has been mediated and transformed by India's television and movie industry, other aspects of economic and cultural globalization have found mixed expression amid South Asia's traditional cultural patterns. Thus, for example, it is still common to see people dressed in traditional clothing—saris for women, dhotis (loin cloths) for Hindu men, turbans for Sikh men, and so on—often in combination with Nike or Adidas sneakers or some other nontraditional apparel. Similarly, although there has been a proliferation of fast-food outlets such as Domino's Pizza and vending machines selling soft drinks such as Pepsi and Coca-Cola, Western-style food retailing has little appeal to affluent households, most of whom still live in neighborhoods where street vendors sell high-quality fruits, vegetables, dairy products, and other basics door-to-door. Appliances such as washing machines, dishwashers, and power tools are also less prevalent than might be expected among South Asia's affluent middle classes, simply because of the millions of people available to undertake domestic labor at very low wages.

Meanwhile, as in other world regions, the cultural shifts involved in globalization flow out as well as in. South Asian mysticism, yoga, and meditation found their way into Western popular culture during the "flower power" era of the 1960s after the Beatles had visited India. South Asian cuisine, with its spicy curries and unleavened breads, found its way into Britain at about the same time and has since become an established item in restaurants and supermarkets in much of Europe and North America. Meanwhile, South Asian methods of nonviolent protest such as boycotts and fasting, inspired by the ancient Buddhist concept of *dharma* and developed in the twentieth century by Gandhi, have spread all around the world. Contemporary South Asian literature from writers such as Anita Desai, Vikram Seth, Arundhati Roy, and Salman Rushdie has found a global readership. South Asian art and music have been less influential, though Indian singers and musicians are well represented in the "international music" sections of Western record stores and some artists, such as Sheila Chandra, have crossed over into a broader international audience.

Ethnicity and Nationalism

The Western concept of nation-states did not transfer very well to South Asia, where tremendous cultural diversity means that national political boundaries tend to encompass diverse groups in terms of ethnicity, language, religion, and cultural identity, while at the same time dividing some groups, leaving some in one country and some in another. The partition of British India in 1947 demonstrated this in relation to Hindus and Muslims, as did the subsequent secession of Bangladesh from Pakistan in relation to Bengali ethnic and cultural iden-

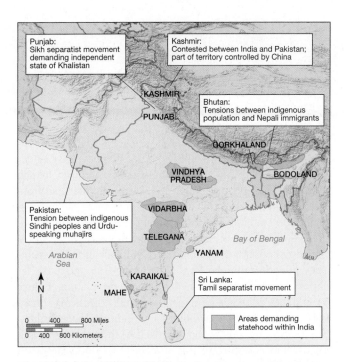

Figure 11.18   Regional and separatist movements in South Asia   The imposition of modern political and administrative boundaries on centuries-old patterns of cultural and ethnic differentiation has led to ongoing tensions and a number of cases of regionalism, separatism, and irredentism.

tity. But South Asia's cultural diversity, framed within national boundaries that have been relatively recently imposed, has also given rise to several other cases of regionalism, separatism, and irredentism (**Figure 11.18**) that are a continuing basis for political tension, social unrest, and, occasionally, outright rioting or armed conflict.

One of the most troubled areas is the Punjab, a region that was divided in two by partition. In the 1980s the Sikh population in the Indian portion of the Punjab developed a nationalist movement, demanding a separate state of Khalistan, under the leadership of a Sikh holy man called Bhindranwale. A series of terrorist attacks and kidnappings led to the occupation, in 1984, of the Golden Temple in Amritsar, the most sacred shrine of the Sikh community. In response, Indian troops stormed the Golden Temple, killing Bhindranwale and many of his supporters. This, in turn, led to the assassination of Indian Prime Minister Indira Gandhi by two of her bodyguards who were Sikh. Outraged Hindus immediately turned on Sikh communities throughout northeast India, killing more than 3000 in riots and vengeful attacks. Since the mid-1990s, levels of violence in the Punjab have diminished, though ethnic tensions still simmer.

Neighboring Kashmir, whose predominantly Muslim population found itself isolated as a minority within India at partition (see p. 535), has three times been the cause of war between India and Pakistan (in 1948, in 1965, and 1971). Kashmir remains a contentious and complex arena. Kashmir's northern border is not an accepted international border—it is a "line of control" established after the 1971 war. Pakistan controls the northwestern portion of what India claims as

Sense of Place

Bollywood

When Mumbai was still known as Bombay, the city developed a huge Hindi-language film industry, which acquired the nickname "Bollywood." Although revenues from the 700 or so movies that are produced in India each year—about $850 million—do not compare to those of Hollywood, they nevertheless represent a significant industry within India. Equally important, they represent a unique cultural element. They provide a popular form of escapism from the harsh realities of daily life for the majority of the population (**Figure 1**), and they do so in a form that is culturally distinctive, drawing on classical Hindu mythology and traditional social values. The roots of the Bollywood approach lie in traditions of folk theater and performance that stretch back 2000 years, with familiar themes: good triumphing over evil, the struggle of the poor, the sins of the big city, and the melodrama of family life. Hindi-language movies have been a potent force in shaping Indian ideas of nationhood. At the same time, many of the Bollywood movies and TV productions deal with themes such as caste and modernization in ways that relate directly to the lives of Indian viewers. The potency of Bollywood movies as cultural agents is underscored by the case of a figure created in the film *Jai* (directed by Chandrakant): the character of Santoshi Mata, supposedly a descendant of the Hindu gods Shiva and Parvati. Despite having no basis in the scriptures, Santoshi Mata has acquired a genealogy and been absorbed into the pantheon of Hindu deities as a genuine goddess; women appeal to her for success in the modern urban world.

Although India has produced avant-garde movies that have been recognized for their artistic and dramatic content, most Bollywood products are exuberant, spectacle-driven entertainment: melodramatic fantasies that mix action, violence, romance, music, dance, and moralizing into a distinctive, formulaic form that has been called "Busby Beserkeley"—a reference to the Hollywood musicals of Busby Berkeley in the 1930s. Bollywood stars such as Sridevi, Twinkle Khanna, Dimple Kapadia, Karisma Kapoor, and Chunky Pandey have their careers and private lives monitored by adoring fans with an intensity that Hollywood agents would envy. The majority of Bollywood films have some sort of musical content, and the songs (lip-synched by the actors but sung by "playback artists" who are also stars) dominate Indian pop charts. Soundtracks include sitars, synthesizers, pianos, and violins to provide a score that moves effortlessly from classical Indian ragas to Mozart to hip-hop and rap music. Every taste is catered to, making a bridge between the traditional and the modern, and between East and West.

There is a large export market for Bollywood films among the Indian diaspora. In 1998 the Bollywood hit *Kuch Kuch Kota Hai* (directed by Karan Johar) was one of the top 10 grossing movies of the year in the United Kingdom. In 1999, India earned $100 million in film exports, which have been growing by 50 percent a year. In addition to expatriate Indian markets, Bollywood movies are successful in the Persian Gulf states and in Russia. Nevertheless, Bollywood has its problems. Indian filmgoers, especially those in cities and those with access to satellite television, have become more and more difficult to satisfy with the standard Bollywood recipe. In 2000, Bollywood produced 40 percent fewer movies than it had in 1990. Meanwhile, India has become one of the largest markets for Hollywood films. In 2000, 1000 of India's 14,000 movie theaters, mostly in the larger cities, were showing U.S. films.

Figure 1 Bollywood escapism These billboards hint at the escapist themes that are common to Bollywood movies.

Kashmir, and China controls the northeastern corner. In 1986 there began a renewed campaign of insurgency by Muslim separatists in the Indian-controlled portion of Kashmir. Since 1989, more than 30,000 people—separatist guerillas, policemen, Indian army troops, and civilians—have died in a guerilla campaign aimed at the incorporation of Kashmir into Pakistan as part of a larger Islamic state. The Pakistani-backed campaign culminated in Pakistan sending its own forces across the border into the Kargil Peaks district in 1999. Pakistani troops were withdrawn after India launched a full-scale military offensive to evict them and U.S. President Clinton put pressure on the Pakistani government, which subsequently fell to a military coup d'état, the fourth such coup since Pakistan became independent in 1947.

Within Pakistan, meanwhile, ethnic tensions have developed around linguistic differences. Most indigenous Pakistanis speak Punjabi or Sindhi, but families who migrated from India at the time of partition—known in Pakistan as *muhajirs*—have tended to retain Urdu as their language. In order to protect and maintain their distinctive identity, the *muhajirs* formed a political party, the Mohajir Quami Movement. This attracted a great deal of resentment among indigenous Pakistanis, and in 1995 more than 1800 people were killed in riots in Karachi. In 1998 continuing tensions led the government to the imposition of martial law and to the exile, in London, of the leadership of the Mohajir Quami Movement.

In the small Himalayan state of Bhutan, there have been tensions between the indigenous population and Nepali immigrants, whose number has grown to more than one-fourth of the Bhutanese population. The government of Bhutan has formally adopted the traditional language, Dzongkha, as the official language, mandated the wearing of Bhutanese national dress for formal occasions, and restricted Bhutanese citizenship to Nepalis who could prove residency in the country since 1958. A census undertaken in 1988 in order to enforce this residency law led to civil disorder within Bhutan and to tens of thousands of Nepali refugees who made their way to refugee camps set up by the United Nations in eastern Nepal.

Sri Lanka's ethnic tensions involve both language and religion. The majority population is Buddhist and Sinhalese-speaking. In the northeastern part of the country, however, the majority population is an enclave of Tamil-speaking and Hindu population that represents about 17 percent of Sri Lanka's total population. Ever since independence from Britain in 1948, the Sri Lankan government has pursued a nationalistic posture that has resulted in the oppression of this Tamil population. The first casualties were 600,000 descendants of Tamil plantation workers who had been brought to Ceylon (as it was then called) from southern India to work in tea plantations. The deportation of these "plantation Tamils" led to the formation in the 1980s of a militant and bloody Tamil separatist movement that crystallized in 1983 into the "Tamil Tigers"—the Liberation Tigers for Tamil Eelam. In the early 1990s more than a million Tamil villagers were displaced by fighting between the Tamil Tigers and the Sri Lankan army, becoming refugees in their own land. Since the mid-1990s the level of conflict has diminished, but Sinhalese and Tamil nationalism continues to result in sporadic terrorist attacks and outbreaks of violence.

India's ethnic tensions, in addition to those in the Punjab and Kashmir, include those related to separatist movements in Jharkand (an alliance of tribal peoples in southern Bihar, western Orissa, and eastern Madhya Pradesh), Vidarbha and Telegana (involving tribal peoples of the Peninsular Highlands), and Assam, where the Assamese-speaking indigenous population has long been resentful of the Bengali administrators and business elite (Figure 11.18). These tensions have provoked a strong reaction within India's majority Hindu population. A nationalistic form of Hinduism, *Hindutva*, emerged in the late 1980s, fanned by an epic television series (more than 100 30-minute episodes) based on the classic Hindu story, the *Ramayana*. This coincided with the emergence of a new political party committed to Hindu nationalism, the Bharatiya Janata Party, or BJP. The BJP quickly attracted popular support, which became focused on the BJP leadership's campaign to build a temple to Rama (recognized by Hindus as an incarnation of the supreme god Vishnu) on the site of his legendary birthplace in the small town of Ayodhya (population 70,000) in Uttar Pradesh. The campaign was acutely sensitive because the site was already occupied by the Babri Masjid, a mosque. In 1992 a crowd estimated at almost three-quarters of a million gathered in Ayodhya and, despite the presence of 15,000 government troops, succeeded in scaling the mosque and demolishing it. The incident unleashed ethnic tensions and latent feelings of fear and hatred that erupted into a spasm of communal riots, killings, and looting throughout India. After this, the BJP moderated somewhat its stance on *Hindutva*, and in 1999 became the key partner in a 24-party coalition that came to power after national elections. The new prime minister was Atal Vajpayee of the BJP.

Regional Change and Interdependence

South Asia is at a critical juncture in its development. On the one hand, it has considerable potential in its human and natural resource base and emerging new economy. During the 1990s, the region was able to sustain a 3 percent annual growth rate when other regions such as Africa and Latin America were posting negative growth rates. Food production in South Asia has shown a significant increase, and the region as a whole is now a net exporter of food (which is not to say that there are no food shortages). Most South Asian countries are opening up their economies, introducing financial discipline, and attempting to build up technological capability that will permit them to compete in the global economic system. Meanwhile, there is a 1-billion-strong domestic market waiting to be fully developed. India's middle class, at 200 million or more, is the largest in the world and a major consumer market in its own right. More important, perhaps, is India's strong tradition of democracy. India, with a billion people, is the world's largest democracy and a critical element in the region's overall stability.

On the other hand, the region is facing a potentially deep and multifaceted crisis that could well undermine this potential. In marked contrast to India, other countries in the region have struggled to sustain democracy, sometimes failing altogether. Throughout South Asia, poverty threatens to swamp the gains of economic development, while extreme inequality threatens to undermine political stability. Meanwhile, both poverty and economic development pose a serious threat to South Asia's resource base and its fragile ecological system.

Democracy and Political Freedom

Four of the South Asian states that were formerly part of British India—Bangladesh, India, Nepal, and Pakistan—have constitutions that use the British parliamentary form of government, led by a prime minister who is elected by legislators who are themselves voted into office to represent local electorates. Sri Lanka replaced its parliamentary system in 1978 with a presidential system similar to that of the United States.

In India, every adult, male and female, has the right to vote, and there has developed a deep-seated sense of democracy that extends to the legislative assemblies of the country's 31 states and to local governments within each state. For the most part, India's democratic framework has worked extraordinarily well, given the sheer size and diversity of the country. In addition to the machinery of democratic government, India has a free and lively press and an independent judiciary. There was a brief period between 1975 and 1977 when the country's democratic machinery and civil rights were suspended by Prime Minister Indira Gandhi in order to protect herself from a legal challenge to her office. India's democracy has also been flawed (as elsewhere in South Asia) by endemic corruption. Companies have become used to buying favors in order to do business with politicians and bureaucrats, while citizens have become used to having to pay "facilitation fees" to the police and petty officials to get access to services. Much of this corruption was generated by the complex system of permits, licenses, and permissions that developed during India's socialist political economy, between 1947 and 1992. The hope is that India's market reforms of the 1990s will not only promote economic vitality but also will reduce the need and opportunities for bribery. Nevertheless, a national poll in 1997 found that people still regard corruption as the greatest national evil, far above unemployment or poverty.

Bangladesh and Pakistan have fared less well, each having fallen under military rule: Bangladesh between 1975 and 1989 and Pakistan between 1958 and 1971, 1977 and 1988, and from 1999 until the time of this writing. In the Maldives and Bhutan, traditional hierarchical systems are only slowly evolving toward democracy. Afghanistan introduced democratic elections in the 1920s, but democracy and political freedoms were resisted by the regional warlords who had traditionally enjoyed autocratic rule. In the 1970s the same traditional forces resisted the military dictatorship that established close ties with the Soviet Union, eventually prompting a Soviet invasion (in 1979) that led to

a decade of fierce guerilla fighting. With the collapse of the Soviet empire in 1989, there was a period of intense infighting among rival groups within Afghanistan, with a new, militant Islamist revolutionary force, the Taliban, gaining control over most of the country by 1996. As we have seen, the Taliban's regime has reduced Afghanistan to a state of oppressive religious autocracy.

India's Economic Transition

Since India's market reforms of the early 1990s, the country's economic development has triggered an important transformation that has important implications for regional change and interdependence within South Asia. Since 1992, India's government has built on its structural economic reforms, bringing in a series of second-phase market reforms that have made it easier for free-enterprise capitalism to flourish. India's manufacturing productivity has increased and the amount of foreign direct investment flowing into the country has increased dramatically: from $76 million in 1991 to $1.61 billion in 1999. The results of this investment are most visible among India's newly affluent middle class. Market reforms have meanwhile triggered an associated cultural change: flaunting success is no longer frowned upon, and so India's expanding middle class is increasingly unabashed about its cars, Palm Pilots, mobile phones, and vacations in Phuket and Singapore.

Breaking with socialist principles of centrally planned development and social and regional equality has unleashed the spatially uneven economic development processes of capitalism. The growth and the wealth has not been evenly distributed throughout India. There has been dramatic growth in certain industries and certain places and regions, while elsewhere there has been disinvestment and recession: the classic "creative destruction" of capitalism. One of the most dramatic examples of regional growth is that of the software industry in Bangalore and Hyderabad (see Sense of Place: Bangalore's High-Tech Fast World, p. 551). More generally, the growth has been centered in larger metropolitan areas and preexisting industrial centers, again following classic principles of capitalist economic development. Places and regions with an initial advantage in terms of factories, skilled labor, specialized business services, and affluent markets can attract more investment, faster, through "cumulative causation," the self-reinforcing spiral of regional growth. The corollary is that places and regions with a weak industrial base, with a weak or obsolescent infrastructure, and with an unskilled or poorly educated workforce tend to experience a downward spiral of recession. In India today, it is the remoter rural regions that are experiencing most acutely the negative consequences of the country's economic reforms.

There have also been less predictable consequences of the liberalization of India's economy. As geographer Pamela Shurmer-Smith has noted, the lifting of export controls has enabled farmers with access to large amounts of capital to reorganize their production toward lucrative overseas markets, with the result that domestic consumers have to pay more for traditional staples. Thus, for example, many farmers are

Bangalore's High-Tech Fast World

First impressions of Bangalore are similar to those of other major cities in India: a sprawl of decaying single-story houses and shops, cramped apartment buildings, crumbling colonial offices, mile after mile of squatter slums, and the pervasive sights and smells of poverty. Yet within Bangalore is a parallel universe of high-tech industry and the Western-style materialism and fast-world lifestyles of its workers. Koramangala, a south Bangalore neighborhood, is home to a wide variety of software companies, from local start-up operations to subsidiaries of Compaq, Motorola, Nexus, Oracle, Texas Instruments, and Verizon. Infosys Technologies, a Bangalore-based firm that writes and maintains software for hundreds of corporations worldwide, employs 6000 people, has a modern 17-hectare (42-acre) campus with a decidedly Silicon Valley feel, and was the first Indian company to be listed on the U.S. Nasdaq exchange. Another Bangalore firm that is now listed by Nasdaq is Wipro Limited, which through its information-technology division employs 6500 software workers worldwide and grossed $310 million in revenues in 1999. Daimler-Chrysler has a small center in Bangalore that produces communications technology for automobiles. Philips Research, a branch of the $34 billion Philips Electronics N.V., has established a research center for embedded and software technology. Altogether, Bangalore is now home to more than 300 high-tech companies that employ 40,000 people.

Sabeer Bhatia, a young Bangalore software engineer, brought attention to the city's software industry in 1997 when he invented the world's first Web-based e-mail service, Hotmail, which he subsequently sold to Microsoft for $400 million. But the origins of Bangalore's boom go back for decades. Bangalore is home to the Indian Institute of Science, a world-renowned technical school that has produced top scientists and engineers since 1911. In 1958, Texas Instruments set up a successful design center in the city. In the 1960s the Indian government chose Bangalore as a site for one of its weapons and aeronautics laboratories, fostering a skilled labor force. After the structural economic reforms of the Indian government in the early 1990s, free-enterprise capitalism was able to flourish, and in Bangalore there was a pool of Indian programmers who had become experts at writing concise, elegant code on their rather limited hardware. When American software companies began to encounter rising costs in Silicon Valley, they found a large pool of highly trained, English-speaking, ambitious, and inexpensive software engineers. Transnational corporate investments, combined with local entrepreneurialism, have enabled Bangalore to become a world-class engine of high-tech development.

Occasional citywide power outages and woefully inadequate Internet infrastructure remain limiting factors on the city's ability to take full advantage of its capabilities. But already the success of Bangalore's high-tech companies and their workers have brought a distinctive fast-world dimension to the city. In Koramangala, landscaped corporate campuses are equipped with upscale cafés and restaurants, exercise centers, tennis and basketball courts, libraries, and theaters. In nearby suburbs are spacious suburban homes whose residents drive Mercedes, BMWs, Jaguars, and Audis. Downtown, on Mahatma Gandhi Road, the city's successful young men and women can be seen in designer clothes, talking into cell phones, shopping in expensive boutiques, and eating in Western-style pubs and restaurants (**Figure 1**; see also Day in the Life: Meenakshi Nagarajan, p. 565).

Figure 1   Downtown Bangalore Early evening rush-hour brings crowds of shoppers to Bangalore's downtown area.

switching from growing grains for local consumption to cash crops like cotton and tobacco, while others are turning to the specialist cultivation of flowers and strawberries to be shipped to newly affluent urbanites or to be air-freighted abroad. Now that a global market has become aware of high-quality local specialties such as the fragrant Basmati rice of the Himalayan foothills and the short-season Alphonso mangoes of Maharashtra, their price within India has put them in the luxury class, out of reach of many of the consumers who have traditionally regarded them as occasional treats.

Table 11.1	Indicators of Poverty in South Asia				
	Percentage of Total Population Not Expected to Survive to Age 40	Adult Illiteracy Rate	Percentage of Total Population without Access to Safe Water	Percentage of Total Population without Access to Sanitation	Percentage of Population with an Income of Less Than $1 a Day
Afghanistan	No data	31.5	88	92	No data
Bangladesh	21.5	61.1	5	57	28.5
Bhutan	20.2	55.8	42	30	No data
India	16.1	46.5	19	71	52.5
Maldives	13.5	4.3	40	56	No data
Nepal	22.5	61.9	29	84	53.1
Pakistan	14.7	59.1	21	44	11.6
Sri Lanka	6.4	32.9	15	22	No data

Poverty and Inequality

Against the background of acute and chronic poverty that have been ever-present in the landscapes of South Asia, the material wealth and Western lifestyles of the growing middle classes serve to highlight the extreme inequality that is also characteristic of the region. Official statistics reveal that hundreds of millions in South Asia live not just in poverty, but in ignorance and destitution (**Table 11.1**). If anything, poverty and inequality are increasing.

In rural South Asia, scarcity is the norm (**Figure 11.19**). Illiteracy is commonplace, and even the most basic services and amenities are lacking. Life expectancy is low, and hunger and malnutrition are constant facts of life. Cow dung is used

for fuel (**Figure 11.20**), and most villagers brush their teeth with sticks from neem trees (which have a natural antibacterial sap). In urban areas, poverty is compounded by crowding and unsanitary conditions. In South Asia's largest cities, a third or more of the population live in slums and squatter settlements, and hundreds of thousands are homeless. In Kolkata alone, it is estimated that more than 700,000 people sleep on the streets each night (**Figure 11.21**). Clean drinking water is

Figure 11.20   Rural poverty   Grinding poverty is the norm in rural South Asia, where dried cow dung is collected from fields to provide fuel. While this provides a sustainable source of cooking fuel, it robs the fields of natural fertilizer.

Figure 11.19   An Indian family with their material possessions   The Yadav family from Ahraura village, in Uttar Pradesh, photographed with their possessions outside their home in the mid-1990s, represent a statistically average Indian family in terms of family size, residence, and income.

Figure 11.21   Urban poverty   For many of the urban poor, poverty means homelessness. In Kolkata entire families dwell on the sidewalk, and an estimated 700,000 people are forced to sleep on the streets each night.

limited, and most poor households do not have access to a latrine of any kind.

The worst concentrations of poverty are characterized by overcrowding, a lack of adequate sanitation, shockingly high levels of ill health and infant mortality, and rampant social pathologies. Consider, for example, the squatter settlement of Chheetpur in the city of Allahabad, India. The settlement's site is subject to flooding in the rainy season, and a lack of drainage means stagnant pools for much of the year. Two standpipes (outdoor taps) serve the entire population of 500, and there is no public provision for sanitation or the removal of household wastes. In this community, most people have food intakes of less than the recommended minimum of 1500 calories a day; 90 percent of all infants and children under age 4 have less than the minimum calories needed for a healthy

diet. More than half of the children and almost half the adults have intestinal worm infections. Infant and child mortality is high—though nobody knows just how high—with malaria, tetanus, diarrhea, dysentery, and cholera the principal causes of death among children younger than age 5.

A great deal of this poverty results from the lack of employment opportunities in cities that are swamped with people. In order to survive, people who cannot find regularly paid work must resort to various ways of gleaning a living. Some of these ways are imaginative, some desperate, some pathetic. Examples include street vending, shoe-shining, craftwork, street-corner repairs, and scavenging on garbage dumps (**Figure 11.22**). This informal economic sector consists of a broad range of activities that represent an important coping mechanism. For too many, however, coping means resorting to begging, crime, or prostitution. More than a half a billion people in South Asia must feed, clothe, and house themselves entirely from informal sector occupations.

Women and Children   Among South Asia's poor, women bear the greatest burden and the most suffering. South Asian societies are intensely patriarchal, though the form that patriarchy takes varies by region and class. The common denominator among the poor throughout South Asia is that women not only have the constant responsibilities of motherhood and domestic chores but also have to work long hours in informal-sector occupations (**Figure 11.23**). In many poor communities, 90 percent of all production is in the informal sector, more than half of which is the result of women's efforts. In addition, women's property rights are curtailed, their public behavior is restricted, and their opportunities for education and participation in the waged labor force are severely limited. Women's subservience to men is deeply ingrained within South Asian cultures, and it is manifest most clearly in the cultural practices attached to family life, such as the custom of providing a dowry to daughters at marriage (see Day in the Life: Bibi Gul, p. 555). The preference for male children is reflected in the

Figure 11.22   Informal economic activities   In cities where jobs are scarce, people have to cope through the informal sector of the economy, which includes a broad variety of activities, including agriculture (backyard hens, for example), manufacturing (craft work), and retailing (street vending). (a) Collecting cans for recycling, Mumbai. (b) Street dentist, Jaipur.

(a)

(b)

(a)   (b)

Figure 11.23   Women's work In most households, women must not only raise children, prepare food, and do most of the domestic chores but also work in informal-sector activities or as unskilled laborers. (a) Women construction workers, Deghradun, India. (b) Woman collecting water, Tangalla, Sri Lanka.

widespread (but illegal) practice of selective abortion and female infanticide. Within marriages, many (but by no means all) poor women are routinely neglected and maltreated. More extreme are the cases—usually reported only when they involve middle-class families—of "bride burning," whereby a husband or mother-in-law fakes the accidental death (kitchen fires are favored) or suicide of a bride whose parents had defaulted in their dowry payments. Several thousand such deaths are reported in India each year, and this is almost certainly only a fraction of the real incidence.

The picture is not entirely negative, however, and one of the most significant developments has been the emergence of women's self-help movements. Perhaps the best known of these is the Grameen Bank, a grassroots organization formed to provide small loans to the rural poor in Bangladesh (see Geography Matters: the Grameen Bank, p. 556). In India, the Self-Employed Women's Association (SEWA) has made a major contribution to building self-confidence and self-reliance among poor working women by mobilizing and organizing them. SEWA was formed in 1972 in Ahmedabad in the state of Gujarat. It evolved from a trade union of textile workers, but, unlike conventional trade unions, SEWA organizes women workers in the informal sector: vegetable vendors, rag and paper pickers, bamboo workers, cart-pullers, and garment workers. SEWA has given its members a degree of independence from middlemen and, consequently, an invaluable sense of independence. Following the example of the Grameen Bank in Bangladesh, SEWA has also established its own bank in order to finance income-generating projects for small groups of women, helping them to meet the emergencies that would otherwise drive them to money-lenders. A third example of women's self-help movements comes from Rajasthan, in India, where the Women's Development Programme, sponsored by the government, organizes rural women as volunteers to counter the deep-seated patriarchy of the region. Community-based groups, coordinated by these volunteers, disseminate information on women's legal rights, health and literacy programs, and income-generating schemes, and oc-

casionally organize campaigns to protest particularly extreme injustices to individual women.

Children in impoverished settings are even more vulnerable than women. Throughout South Asia, the informal labor force includes children (**Figure 11.24**). In environments of extreme poverty, every family member must contribute some-

Figure 11.24   Child labor The exigencies of poverty mean that children are required to contribute to the household economy, often from a very early age. Here a girl from Khulna, Bangladesh, is making matchboxes.

A Day in the Life

Bibi Gul

Bibi Gul is 35 years old. She was born and grew up in the rough tribal area of Baluchistan. When she was 9 years old, her mother died, her father remarried, and she was placed in the care of her grandparents, who lived in poverty. At age 13, for a bride price that was common in the area, Bibi's grandparents assured their own financial future by passing Bibi on to a man more than twice her age.

For the next 15 years Bibi's life was a cycle of repeated pregnancies. She had six children who survived and three miscarriages. No health or education centers were available in Bibi Gul's village, because of opposition from the men of the community, who, however, allowed an embroidery center to be established to revive the dying art of traditional stitchcraft.

Bibi Gul was encouraged by the embroidery trainer, who was quick to appreciate her intelligence and reading ability, and she rapidly became a competent teacher herself. She not only helped to provide other girls and women with an income-generating activity but also established a forum for discussing women's issues.

In recognition of her emerging organizational abilities, Bibi Gul was invited to a leadership workshop in 1989. It was the first time that she had left her hometown. She did not remove her burga (the local veil) throughout the eight-day workshop, even though only women were present, and she did not speak during the sessions. But when she returned to her community, a great change came over her.

She began to visit other women in her village and talked with them about health, immunization, education, family planning, and income-generating opportunities, and she encouraged them to participate in the women's program. She was gradually given responsibility for conducting forums on women's issues in three of the villages in her area. This entailed traveling by public transport to visit other house-bound women. She slowly shed not only her fears and inhibitions but also her burga when in female company. She developed into an active member of the community.

Bibi Gul had to face her husband's initial resentment of her newfound independence, and she also became the object of adverse comment from the conservative community in which she lives. Nevertheless, she has worked with quiet determination, teaching her husband to care for their children when she is away on training courses and gradually earning the respect of the female community as a whole.

Source: Adapted from "Three Women from Pakistan," *UNESCO Courier,* September 1995, p. 17.

---

thing, and so children are expected to do their share. Industries in the formal sector often take advantage of this situation. Many firms farm out their production under subcontracting schemes that are based not in factories but in home settings that use child workers. In these settings, labor standards are nearly impossible to enforce.

The International Labour Office has documented the extensive use of child labor in South Asia, showing that many of the children involved in a great variety of work—tending animals, carpet-weaving, stitching soccer balls, making bricks, handling chemical dyes, mixing the chemicals for matches and fireworks, sewing, and sorting refuse—are less than 10 years old, most of them working at least six and as much as 12 hours a day. A particularly cruel type of exploitation of child labor is **bonded labor.** This kind of bondage occurs when persons needing a loan but having no security to back up the loan pledge their labor, or that of their children, as security for the loan. In addition, there are many street children, some of whom do casual work and beg but return to their families at night, while others live on the street and effectively have no families. UNICEF has estimated that there are more than 11 million of these street children in India. Finally, perhaps the cruelest and most reprehensible exploitation of children is as sex workers. In parts of India—notably in the small towns of rural regions—there are prostitute *jati*, where the cycle of recruitment into sex work is an unavoidable legacy from mother to daughter. Meanwhile, in the red-light districts of every large city there are hundreds of young bonded or kidnapped rural girls who have been sold into brothels.

Environmental Issues

As we have seen, South Asia's environmental history has left a legacy of serious environmental issues that include deforestation, water shortages, and air and water pollution (see p. 533). Given the acute problems of population pressure and poverty in South Asia, it is not surprising that concepts of sustainable development and social responsibility for environmental protection are very weakly developed. Each country in the region has a set of environmental laws and regulations, but they are routinely flouted and only weakly enforced. Corruption is one factor that contributes to this, but another reason is that governments simply do not have the institutional apparatus or the funds to enforce environmental laws. More important still, perhaps, is the short-term perspective that derives from the high priority given to economic

Grameen Bank

The Grameen Bank runs completely against the established principles of banking by lending to poor borrowers who have no credit. The idea for the bank came from its founder, Muhammad Yunus, an economics professor who, as part of his research, surveyed villagers in Jobra, Bangladesh, in 1974 and discovered that only $27 would be enough to release all of the 42 debt-encumbered villagers from the clutches of the local money-lender. Yunus lent the villagers the $27 himself and was duly repaid, a few cents at a time. In 1976 he undertook a pilot project that gave small-business loans to the poorest of the rural poor. Because they had no collateral and were illiterate, the borrowers did not have access to conventional sources of credit. The loans were given on a group liability basis, and the accrued interest was repayable in small weekly installments over a period of one year, during which Yunus provided financial advice to his clients and helped them generate savings. But he was unsuccessful in persuading the banks that they should follow the example of his successful pilot project. So in 1983 Yunus founded the Grameen Bank of Bangladesh as a specialized microcredit institution (*Grameen* means "village," or "of the village"), with the express purpose of providing small sums of credit to people with no collateral. The group-liability concept means that peer pressure works as a driving force for the borrowers to repay the loan on time. In case of default, the borrowers as well as other members of the group become ineligible for further loans.

Since 1983, more than 2.3 million Bangladeshis, spread over 38,000 villages, have borrowed from Grameen Bank, which now claims to be a financially sustainable, profit-making venture with 12,000 employees. Cumulatively, the bank has loaned more than $2.5 billion, 98 percent of which has been repaid, according to Grameen, a rate that is far higher than that for any conventional financial institution operating in the country. The average size of a Grameen loan is about $120, typically enough to purchase a cow, a sewing machine, or a silkworm shed. The most distinctive feature of the Grameen Bank is that 95 percent of its borrowers are women (**Figure 1**). Studies have shown that the bank's operations have resulted in improvements in nutritional status, sanitation, access to food, health care, pure drinking water, and housing, and that more than one-third of all borrowers have risen above the poverty line, with another third close to doing so.

Grameen programs have been replicated around the world. In 2000 there were 241 international programs in 58 different countries, including China, Colombia, Indonesia, Kenya, Mexico, Nepal, Nigeria, Papua New Guinea, Sri Lanka, and Tanzania. Meanwhile, Yunus has begun to explore the possibilities of using the Grameen system to provide opportunities for poor villagers to take advantage of innovations in energy, communications, and information technology. In 1997, Grameen Telecom, a nonprofit company, was established to launch cellular telephone operations in rural areas. Telephones are greatly needed in Bangladesh, where the telephone density is one of the lowest in the world. Twenty-eight Grameen Bank borrowers were given loans of approximately $350 each, which covered the cost of the telephone, the hook-up, training, and repair services. As the "wireless women" of their villages, they purchase air time at wholesale prices from Grameen Telecom and sell the service to their neighbors at the market rate. Basically, they act as human pay phones in places where virtually no one has even seen a telephone or made a phone call. By 1999 the telephone operators were earning net profits of approximately $2 a day—more than $700 a year, significantly in excess of the $250 average annual per capita income in Bangladesh.

Figure 1   Rural enterprise Microcredit programs such as those pioneered by the Grameen Bank have enabled tens of thousands of rural women to begin small businesses.

development: Enforcing environmental laws would wipe out a significant part of South Asian countries' competitive advantage in world markets.

The long-term costs of this situation are certain to be measured in serious environmental degradation and loss of biodiversity. Meanwhile, the short-term costs are significant. A 1998 World Bank study estimated that India loses $13.8 billion every year—equivalent to 6.4 percent of the country's gross domestic product (GDP)—as a result of environmental degradation. The largest share of this cost—$8.3 billion—is associated with health impacts resulting from water pollution. The health impacts and consequent loss of productivity of urban air pollution account for an estimated loss of $2.1 billion. Soil degradation and the consequent loss of agricultural output is estimated to cost $2.4 billion a year; and rangeland degradation, resulting in a loss of livestock carrying capacity, costs $417 million each year. Deforestation is estimated to cost $244 million annually.

Such estimates do not always take account of the disastrous effects of environmental problems on peoples' lives or, indeed, of the raw cost in human lives of disasters such as flooding or the release of untreated toxic waste. One of the most horrific disasters of all time took place in Bhopal, India, in 1984, when lethal methyl isocyanate leaked overnight from a Union Carbide plant, killing more than 6000 people in nearby neighborhoods and permanently damaging the health of hundreds of thousands more. The exact causes and responsibility for the event have still not been settled conclusively, though the Bhopal disaster has been interpreted by many as being emblematic of the potentially disastrous effects of lax attitudes on the part of plant owners and managers toward environmental planning and regulation. Most of the time, such laxity does not involve loss of human life. Nevertheless, the results can be calamitous both to communities and to the environment.

Take, for example, the consequences of poor environmental planning in the case of the dams and irrigation schemes along the Porali River in Pakistan. The depth and spread of the river's delta, with its extensive mangrove swamps, made it a haven and breeding ground for fish. For centuries, local villages had earned their living from this natural bonanza. But the river and its delta began to silt up due to a combination of upstream dams and badly applied irrigation techniques, all installed as part of an economic development program. In particular, the huge Tarbela Dam (which is itself suffering from sedimentation because of deforestation in the mountains) has been a major cause of silting by preventing the otherwise natural scouring out of mud during the rainy season. The result is that the rich ecology of the mangrove forests of the coastal belt of Sindh and Balochistan is dwindling, and the future of the coastal villages is seriously threatened. As fish habitat has shrunk and stocks fallen, fishermen have switched from traditional techniques—catching large specimens with long lines—to using fine-mesh nets. This quickly depleted stocks still further, reducing the average catch to small immature fish. Affluent fishing communities that used to pay their taxes in gold now find it a challenge to feed themselves.

The most dramatic case of poor environmental planning came to light in 2000, when it was discovered that millions of tube wells in Bangladesh are drawing arsenic-contaminated water. Tube wells are water wells that are lined with a durable and stable material, usually cement, which makes it possible to sink wells to a greater depth than traditional water wells. They were installed throughout the country as a result of a campaign in the 1970s by UNICEF, the United Nations children's fund. The purpose of the wells was to provide drinking water free of the bacterial contamination of the surface water that was killing more than 250,000 children each year in Bangladesh. Unfortunately, the well water was never tested for arsenic contamination, which occurs naturally in the ground-water, and for many years the well water was believed to be completely safe. By the 1990s, high rates of certain types of cancer throughout much of Bangladesh led researchers to investigate, resulting in the identification of the cause as arsenic-contaminated water from tube wells. Medical statistics indicate that 1 in 10 people who drink such water over a prolonged period will ultimately die of lung, bladder, or skin cancer. The World Health Organization, in a 2000 report, described the crisis as the largest mass poisoning of a population in history. The scale of the environmental disaster far exceeds those of Bhopal or Chernobyl (see p. 180): As many as 85 million people still draw arsenic-contaminated water from their local wells, and although the technology is available to purify Bangladesh's plentiful supplies of surface water, it will take many years to replace the estimated 6 million tube wells that are affected.

Core Regions and Key Cities of South Asia

South Asia, like other peripheral world regions, does not contain industrial regions that drive the global economy. Nevertheless, there are within South Asia several historically important regions that have developed concentrations of industry, each with one or two metropolises that act as second- or third-level world cities: centers of industry and commerce that provide important nodes in the flows of goods, services, information, and capital not only within South Asia but also around the world. The irrigated plains of the Upper Ganga and the Indus were both cultural hearth regions with agricultural productivity that has long sustained a high density of population. The importance of the Damodar Valley and Hooghlyside, in contrast, is relatively recent and is based on heavy industry that has concentrated around coalfields. Cotton textiles are the basis of core industrial regions in Eastern Gujarat and the Mumbai-Pune corridor, while South India is characterized by a series of semiautonomous industrial subregions.

The Upper Ganga Plains

The Upper Ganga Plains have historically constituted the most prominent region of India, and today they are the most heavily populated region of the country. The great empires of India rose to power here. The Ganga is the sacred river of Hinduism, and four of Hinduism's seven holy towns are located in the region, including Varanasi, the holiest of them all. More than 2000 years ago the Upper Ganga Plains were part of Asoka's great Buddhist empire, based at Patna (**Figure 11.25**). Muslim raids from the northwest began in the eleventh century, and by the sixteenth century the plains were the seat of the great Mughal Empire, the capital of which was for some time located at Agra and nearby Fatehpur Sikri. Following the decline of the Mughal Empire, the region became a flourishing center for the arts under the Kingdom of Oudh (or Avadh), whose capital was in Lucknow. New Delhi, in the northwestern corner of the region, became the national capital of British India in 1911 and remains the capital of India.

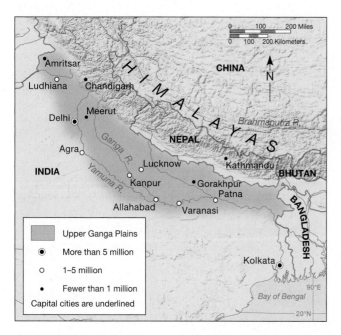

Figure 11.25 Upper Ganga Plains Reference map showing principal physical features, political boundaries, and major cities.

There are no significant mineral resources in the region, which is an immense plain built up from detritus eroded from the Himalayas. The monotony of the plains is broken up only by minor physical features: the *bhabar,* a tract of boulders and coarser gravels that skirts the hills to the north of the plains; the marshy *terai* areas that collect the drainage that falls freely through the *bhabar;* shallow salt-pans, known as *usar* plains, where chemical soil constituents have been deposited in dazzling sheets as water has evaporated from short-lived lakes and ponds; and occasional patches of low, sandy, undulating uplands known as *bhur.* Otherwise, most of the plains are very much as they were described almost 100 years ago in a government survey:

" . . . a level plain, the monotony of which is broken only by the numerous village sites and groves of dark-olive mango-trees which meet the eye in every direction. The great plain is, however, highly cultivated, and the fields are never bare except during the hot months, after the spring harvest has been gathered, and before the rainy season has sufficiently advanced for the autumn crops to have appeared above the ground. . . . With the breaking of the monsoon in the middle or end of June the scene changes as if by magic; the turf is renewed, and tall grasses begin to shoot in the small patches of jungle. Even the salt *usar* plains put on a green mantle, which lasts for a very short time after the close of the rains. A month later the autumn crops—rice, the millets, and maize [corn]—have begun to clothe the naked fields. These continue to clothe the ground until late in the year, and are succeeded by the spring crops—wheat, barley, and gram [a kind of chick-pea]. In March they

ripen and the great plain is then a rolling sea of golden corn [wheat], in which appear islands of trees and villages. . . ."[3]

The wealth of the Upper Ganga Plains came from this agricultural productivity, carefully nourished by irrigation canals and wells. It has proved sustainable, but it can carry only a certain density of population, and that density may well have been reached or surpassed in many parts of the plains. Industry has developed throughout the region, but for the most part on a relatively small scale, involving agricultural processing, textiles, glassmaking, crafts, and carpet weaving. Ludhiana (population 1.7 million in 2000), in the northwestern corner of the region, is a textile center and the location of the world's largest bicycle manufacturer, Hero Bicycles, which produces 3 million bicycles annually. Agra (population 1.2 million), Allahabad (1.1 million), Lucknow (2.6 million), and Patna (1.3 million) are all important textile and light-engineering centers. Kanpur (formerly Cawnpore, population 2.7 million), the region's most important industrial city, rose to prominence as a textile center at the time of the U.S. Civil War, which created a sudden demand for Indian cotton just as the city had been linked by rail to Kolkata. Today, it is one of the most heavily polluted cities in the world. By far the largest and most important city of the Upper Ganga Plains, however, is Delhi.

Delhi Delhi is situated at a great crossroads, an important strategic location at the narrowest point of the Indo-Gangetic plains, the most productive agricultural regions of South Asia. For centuries, Delhi provided an essential base for controlling access to and from South Asia's northwestern frontier, and thereby the key overland routes to Central Asia and the Middle East. As a result, Delhi has been the site of the capital of at least eight different empires. Equally, Delhi has seen many different invaders throughout the ages. Timur (Tamerlane) plundered it in the fourteenth century; the Afghan Babur occupied it in the sixteenth century; and in 1739, Nadir Shah, the Persian emperor, sacked the city and made off with the famous Peacock Throne and with the 186-carat Kohinoor diamond (which was cut down to 108.93 carats and is now in what is known as Queen Mary's crown, part of the British royal family's crown jewels).

Delhi's golden age was in the seventeenth century under the Mughal emperor Shah Jahan (1628–1658), when it was known as Shahjahanabad. Shah Jahan built the famous Red Fort (**Figure 11.26**) with its palace and city walls, as well as the imposing Jama Masjid (Friday Mosque). Today, Shahjahanabad is part of Old Delhi, whose central focus is Chandni Chowk—"Silver Street," the bazaar of goldsmiths and jewelers—which runs west from the Red Fort on the right bank of the River Yamuna. Old Delhi is characterized by narrow streets and alleys, low-rise buildings, outdoor markets, bazaars, mosques, temples, and crowds. To the north and west of Old Delhi, the modern metropolis (population 11.7 million in

[3]*United Provinces Gazeteer,* Vol. 1, Calcutta: United Provinces, 1908, p. 8. Quoted in O. H. K. Spate and A. Learmonth, *India and Pakistan,* London: Methuen, 1972, p. 549.

(a)

(b)

(c)

Figure 11.26   Delhi   The capital of India, its third-largest city, and north India's industrial hub, Delhi still has elements of its Mughal and colonial past, though both are swamped by the slums and squatter settlements resulting from over-urbanization. (a) The Red Fort and Palace, Old Delhi. (b) Outdoor market, New Delhi. (c) Squatter housing along a large water pipe, New Delhi.

2000) has spilled out into a sprawl of industrial suburbs and high-density slums and squatter settlements.

To the south is New Delhi, the planned capital of British India. Government administrative functions were moved to Delhi from Calcutta (now Kolkata) in 1912, but New Delhi was completed only in 1931. Designed by British architect-planner Edwin Lutyens, New Delhi was laid out with spacious roads and an impressive ensemble of imposing (but rather ungainly) government buildings in a modernistic geometric street pattern. The site originally stood separate from the unsanitary and congested environments of Old Delhi, but both Old and New Delhi are now engulfed within the vast sprawl of metropolitan Delhi. There is, however, a marked contrast between the suburbs that surround Old Delhi and those that surround New Delhi. The southern suburbs around New Delhi have very little industry and are dominated by the middle-class neighborhoods of civil servants, interspersed with the spacious suburbs of New Delhi's diplomatic sector and with shopping centers and office complexes.

The Indus Plains

Like the Ganga Plains, the Indus Plains have a long history of agricultural productivity that has supported a succession of empires. Harappan agriculturalists, flourishing between 3000

and 2000 B.C., produced enough surplus to sustain a civilization that was centered in the cities of Kot Diji (near present-day Sukkur), Moenjodaro (near Larkana), and Harappa (near Sahiwal; **Figure 11.27**). The center of gravity of later empires shifted north as it became more important to command access to the overland routes to Central Asia. For a thousand years, from the sixth century B.C. until around A.D. 450, the northern Indus plains took over as the core area of civilization, where a striking fusion of Greek, Central Asian, and Indian art and culture developed. Taxila (near present-day Islamabad) was of particular importance.

From these earliest times, the region's productivity was dependent on irrigation, for the climate is hot, the rainfall irregular, and the soils sandy (**Figure 11.28**). The plains consist of a great mass of alluvium brought down by the Indus and its five tributaries (from west to east, these are the Jhelum, Chenab, Ravi, Beas, and Sutlej) that flow across the Punjab (*panj ab* means "five rivers"). The river floodplains are naturally fertile, but the interfluves (areas of runoff between river valleys, known in this region as *doabs*), though they have good soils, are semi-arid, and require irrigation. There were two traditional methods of irrigation that established the plains as the granary of successive empires. One was a series of inundation canals—channels that were constructed to carry the floodwaters of the monsoon season beyond the regular floodplains of the rivers.

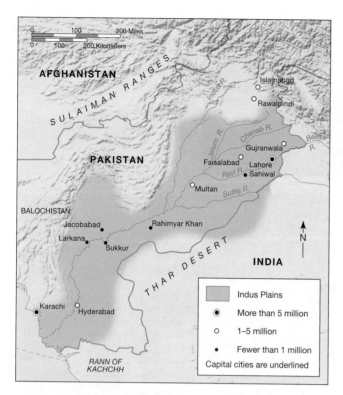

Figure 11.28 **Indus Plains agriculture** Centuries of labor-intensive investment in irrigation systems have made productive agriculture possible in the sandy soils and semiarid climate of the Indus Plains.

Figure 11.27 **Indus Plains** Reference map showing principal physical features, political boundaries, and major cities.

The other was the tube well, a simple shaft sunk to the level of the water table, from which the water is raised by a variety of means, the most common being the "Persian Wheel," driven by bullocks or other farm animals (**Figure 11.29**). British colonial engineers extended the inundation canals in the nineteenth century and in the early twentieth century added a series of dams and irrigation schemes that extended irrigation to a greater portion of the doabs in the Punjab and to the lower Indus plains known as the Sindh. The new farmlands created by these irrigation schemes came to be called the Canal Colonies, and today they are distinctive within the landscape of the plains for their severely rectilinear field patterns, in contrast to the small and irregular-shaped field systems of the older-established areas.

With irrigation, farmers on the Indus Plains can grow two sets of crops. The first set of crops, sown to take advantage of the monsoon rains and harvested by early winter, includes rice, millet, corn, and cotton. The second, sown at the start of the cool, dry season and harvested in March or April, includes wheat, barley, rapeseed, mustard, and tobacco. This productivity supports a high density of population, mostly in large, nucleated villages. As in the Ganga Plains, craft industries are present throughout the towns and villages of the region. In the Indus Plains these are dominated by the traditional manufacture of homespun and woven fabrics in cotton, silk, and wool, carpets, footwear, pottery, and metalworking.

Modern manufacturing industry is dominated by cotton textiles, and woolen knitwear and is concentrated in the larger cities. Lahore (population 6.0 million in 2000) is the cultural, educational, and artistic capital of Pakistan and has a

Figure 11.29 **Irrigation** Raising water from tube wells in the Punjab is usually accomplished by "Persian Wheels," driven by bullocks or other farm animals.

relatively large engineering and electrical goods sector. Multan (population 1.5 million) is situated at the center of the country's most important cotton-growing region, and its manufacturing sector is dominated by cotton textiles. Gujranwala (population 2.0 million) is an engineering and metalworking center. Hyderabad is a tobacco-processing and textile manufacturing center whose economy has been particularly hard hit by the ethnic tensions between the indigenous (Sindhi) population and the *muhajirs,* Muslim refugees from India who settled in the Sindh in large numbers at partition. Similar tensions exist in Karachi (population 11.8 million), though it is large and cosmopolitan enough that these tensions have not adversely affected its economy. Like all cities in this region, Karachi has a large informal economic sector and corre-

A Day in the Life

Meenakshi Nagarajan

Welcome to Purple Haze, one of the 150 or so pubs that make up the central nervous system of Bangalore's very-much-happening party scene. Places like this are a nightly pit stop for the high-tech elite: Hundreds of programmers, programmer wannabes, low-level managers, and occasionally even a CEO or board chair stop by to down a couple of Kingfisher beers as they listen to blaring rock from the '70s and '80s. In Bangalore there's a huge selection of places like this to match your tastes and lifestyle—from Purple Haze to the '50s-style Black Cadillac to NASA, a futuristic spot with a space-shuttle interior and waitresses dressed like airline flight attendants.

I'm at Purple Haze with three coders from Infosys, all in their early twenties: two women—Meenakshi Nagarajan and Roopa Gosain—and their male pal, Mushtaq Ahmed. They each make more than Rs200,000 ($5000), and as Infosys employees they're looking at a bright future. Everyone knows that Infosys people have some of the sweetest stock-option programs on the [Deccan] Plateau, so they'll have no problem finding desirable mates when the time comes. Right now, though, partying is job one. Ahmed, dressed in khakis and oxford cloth, is making frantic pub-crawl plans on his cell phone.

A few days after my night out at Purple Haze, I catch up again with Nagarajan—Meena, to her friends—at the bus stop outside the Infosys world headquarters. To pass the time while she waits for the bus, she's thumbing through a tattered copy of *The Fountainhead,* Ayn Rand's capitalist philoso-drama. "This Ayn Rand, she has good ideas that are in this book," she says. Nagarajan once told me that she's "180 degrees different" from her parents. "The biggest change," she said, "is that we don't think twice about buying something if we want it. With them, providing for your family took every rupee you had. Personally, I'm working on building a large CD collection and a library full of good books."

It's Friday, and Nagarajan says it's been an exhausting 60-hour week, but she's not slowing down. Flipping her black ponytail, she recites a long list of activities: friends over after work for MTV and some snacks, then out to dinner with her "batchmates" at an Italian restaurant downtown. Then, to-

Figure 1   Meenakshi Nagarajan Meenakshi Nagarajan (left) with co-workers at Bangalore's Purple Haze pub. As coders for Infosys, they earn more than $5000—more than 20 times the average Indian's annual income.

morrow, back to work to tie up a few loose ends. Then a shopping trip to pick out a motor scooter. Then off to see *The Matrix* with her roommate, Gosain, and her Purple Haze pal Ahmed, who always seems to be raiding her cabinets at home for potato chips. And then, Sunday. "Sunday, I'm not even going to get out of bed!" she says with enthusiasm.

What high tech means for Nagarajan is a level of independence that was unimaginable for her mother. She and Gosain live together in Koramangala in a one-bedroom apartment full of college-kid furniture. Most nights they're at a pub or a movie, or home watching MTV India till 1 A.M.

Source: Extracted from B. Wetzler, "Boomgalore," *Wired,* March 2000, pp. 152–71.

---

30 highest peaks. In addition to these high peaks are parallel ranges of lower but still impressive mountains and bands of deeply incised, rugged foothills. The higher ranges are bare rock with glaciated features, but some of the lower ranges and foothills are forested with "chir" (*Pinus longifolia*), while the low outer ridges carry a sparse dry scrub. The Mountain Rim can be traversed only in a few key passes. These include the famous Khyber Pass in Pakistan's North-

west Frontier and the Rohtang La and Kunzum La between northern India and Tibet.

Because of the high altitude and barren terrain, much of this region is uninhabitable. But interspersed among the high peaks and the foothills are protected gorges and fertile valleys that sustain isolated settlements of mountain peoples, most of whom tend flocks of sheep, yak, and goats, or work the tea plantations and orchards that cover the lower

Figure 11.34   Mountain Rim monastery The mountainous landscape of this region is distinctive for its many isolated monasteries, temples, and shrines. Shown here is the Lamayuru monastery in Ladakh.

Figure 11.35   Vale of Kashmir Human settlement within the Mountain Rim is largely confined to sheltered valleys, many of which are quite productive. Shown here is part of the Liddar Valley in the Vale of Kashmir.

hills. The physical geography of the region, with its mosaic of remote valleys and basins, provided the framework for the territories of tiny feudal states. While these societies have been incorporated politically into modern national states, their landscapes and ways of life remain largely unchanged. The great variety of peoples within this mosaic is reflected by distinctive and colorful folk traditions and a broad range of ethnic, linguistic, and religious attributes. The peoples of the region are also among the most intractable in South Asia. The Mountain Rim is a region of strained geopolitics, ethnic tensions, ferocious independence movements, and political cultures that are steeped in guns and violence. The Pashtuns (Pathans) of Pakistan's Northwest Frontier region are the world's largest autonomous tribal society and have a long history of fierce independence. Many of the hundreds of tribes in the Mountain Rim remain antagonistic both to neighboring groups and to their national governments. Parts of the Mountain Rim remain off-limits to outsiders as national governments continue to struggle to subdue tribes that persist in pursuing independent ways of life.

Though local populations have resisted integration with broader empires and economies, they have absorbed a rich variety of cultural influences, including Hinduism, Buddhism, and the Mahayana Buddhism of Tibet and Ladakh that has mingled with Hinduism to produce the singular culture of Nepal. The temples and shrines of these and other religions are found scattered throughout the region, along trails and at the entrance to passes (**Figure 11.34**), but human settlement is for the most part concentrated in sheltered valleys and basins. The largest of these is the Vale of Kashmir (**Figure 11.35**), a fertile and verdant basin some 130 kilometers (81 miles) long and between 30 and 40 kilometers (19 to 25 miles) wide that is enclosed by the Himalayas to the east and the snow-capped ridges of the Pir Panjal range to the

west. In this and other valleys it is possible to grow rice, with corn and wheat at higher elevations and orchards—especially of apricots and walnuts—on the valley slopes. Herds of sheep, yak, and goats are kept on the higher slopes of the valleys and in the more arid valleys and basins. The moderate microclimate of these valleys also proved attractive to the leadership of the *Raj,* who established **hill stations** as rest and recreation centers for their troops and bureaucrats, away from the heat and dust of the plains. Among the more important of these hill stations were Darjeeling (**Figure 11.36**), Shimla, and Srinagar. The British also found the climate of the foothills of the eastern Himalayas well suited to the cultivation of tea and established extensive tea plantations to supply the seemingly inexhaustible demand in Britain for tea.

Figure 11.36   Darjeeling Hill stations such as Darjeeling were established by British colonial administrators as summer refuges for bureaucrats and troops. After independence, Darjeeling became a popular mountain resort; the city now has a population of less than 100,000 but boasts more than 100 tourist-class hotels.

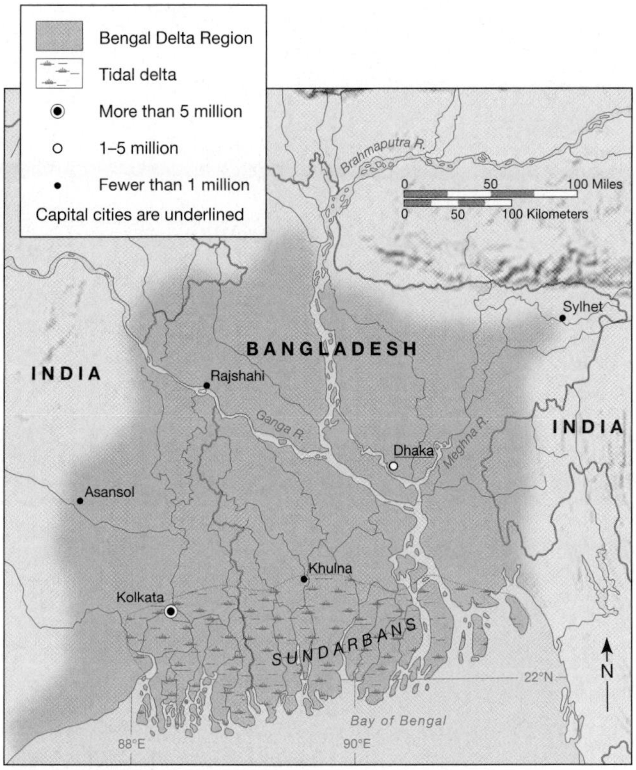

Figure 11.37 Bengal Delta Reference map showing principal physical features, political boundaries, and major cities. (*Source:* After B. L. C. Johnson, *South Asia*, 2nd ed. London: Heinemann, 1982, Fig. 15.2.)

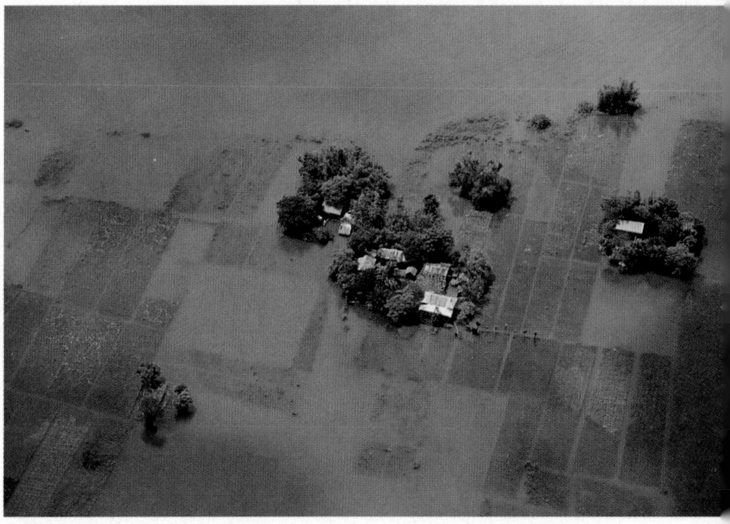

Figure 11.38 Monsoon season Flooding is a regular occurrence in the monsoon season in the Bengal Delta. Though the floods are troublesome, the silt deposited during the floods helps to replenish soil nutrients. Serious problems occur, though, when the rains are unusually heavy, and when high tides and/or exceptionally high rainfall are accompanied by strong onshore winds.

The British found the fiercely independent local tribal peoples unwilling to work in the plantations and so contracted hundreds of thousands of tribal peoples from Bihar and Orissa to work as laborers in the plantations. Their descendants are known today as the "tea tribes" of Assam.

The Bengal Delta

The Bengal Delta covers a large portion of Bangladesh and extends into India (**Figure 11.37**). The unique landscape of the region is dominated by water. The delta country is the product of three major rivers, the Ganga, the Brahmaputra, and the Meghna and their distributaries, along with a number of lesser rivers and their tributaries that sluice down to the Bay of Bengal, creating a vast web of waterways. Because these rivers lie in deep deposits of sand and clay and carry such enormous quantities of water, especially in flood stage, they are almost impossible to control with engineering works; as a result, flooding is a normal event. In the monsoon season, about 70 percent of the delta region is flooded up to a meter or two (3 to 6 feet) in depth (**Figure 11.38**). The entire region is flat and low-lying, the land never rising above 10 meters (33 feet). Entering this environment, the rivers meander and braid and often burst their banks to shift course. The history of the region is full of thriving towns that were permanently abandoned because the rivers on which they were situated silted up or changed course.

Nevertheless, human occupance has adapted to the environment, and the region is rather densely populated (about 900 persons per square kilometer; 2340 persons per square mile). Annual floods are a routine part of life, and farmers rely on floodwaters to water and fertilize the land. Occasionally, however, the flooding goes beyond routine inconvenience and hardship to reach disastrous levels. In 1988 all three of the major rivers reached flood stage at the same time, with the result that floods drowned more than 2000 people. In 1970, 1991, and 1999, cyclones hit the delta area during especially heavy flooding and exceptionally high tides, leading to devastating damage and widespread loss of life. The 1970 cyclone killed between 300,000 and 500,000 people. Some geographers have suggested that annual flooding is getting more pronounced, pointing to deforestation in India and Nepal as the cause of increased runoff.

The natural landscape of the delta country is dominated by marshlands and dense jungle that is a haven for wildlife. There are, however, certain differences that relate to the different parts of the delta itself. Farther inland, the older parts of the delta are more stable, less prone to flooding, and more widely adapted to agriculture, including jute, the principal commercial crop of the region. In contrast are areas of "moribund delta," formerly active floodplain areas where river flow is now at a minimum, river channels are choked with water-hyacinth, and soils, without the benefit of regular flooding by nutrient-rich rivers, have become leached and barren. Parts of the older, inland, delta have been affected by subsidence that has resulted in extensive marsh areas, some of them hundreds of square kilometers in extent. On the younger parts of the delta there are also marshy areas called *backswamps.* These are the result of the levees that build up along river banks, which subsequently tend to trap floodwater, preventing it from

Figure 11.39 Swamp ecology The Sundarbans occupy the youngest part of the Bengal delta, where saltwater penetrates the river channels, creating a distinctive ecology of mangrove and tropical swamp forest that is home to a great variety of species, including crocodiles, Bengal tigers, and Chital deer. The tall trees in this photograph are the damaged remnants of a dense, mature forest that was decimated by a severe cyclone in 1991.

Figure 11.40 Deccan Lava Plateau The vast flat landscape of the Deccan is interrupted only by wide, gently-stepped valleys cut by rivers.

draining back into the river channel. Finally, much of the youngest part of the delta is tidal, and in the southernmost reaches of the delta saltwater penetrates the deltaic distributary channels, creating a distinctive ecology of untouched mangrove and tropical swamp forest—the Sundarbans ("beautiful forest")—that is home to crocodiles, Bengal tigers, and Chital deer (**Figure 11.39**).

The Deccan Lava Plateau

The most physiographically distinctive region of Peninsular India is the Deccan Lava Plateau (see Figure 11.3), a great expanse (approximately 300,000 square kilometers; 780,000 square miles) of basalt rock into which rivers have cut a landscape of wide, gently stepped valleys (**Figure 11.40**). The basalt rock is solidified lava that poured from earth fissures between 55 and 65 million years ago, covering this part of the peninsula with up to 3000 meters (9842 feet) of lava. The result is a landscape so flat, as geographers O. H. K. Spate and Andrew Learmonth once observed, "as to make one believe in the flat-earth theory." In detail, the features of the landscape are also distinctive, as the same authors noted: "mesas and buttes, their tops remarkably accordant, often as if sliced off with a knife. . . . The flanks of the hills are often stepped by

the great horizontal lava flows . . ., and the whole country looks ridiculously like a relief model so badly constructed that the cardboard layers show through the modeling."[4]

The Deccan Lava Plateau is also culturally distinctive as the hearth area of the Maratha people. The Marathas resisted Muslim, Moghul, and British power with varying degrees of success, defending their fortified towns and villages with fast, light cavalry that was well suited to the terrain. The martial culture of the Marathas also incorporated *thuggees,* ritual murderers and bandits from whom the word *thug* is derived. The *thuggees* were eventually subdued by the British in the nineteenth century. Today, the military heritage of the Marathas is marked by the citadels that dominate the larger towns and the forts and gates that can be seen in some of the villages. The towns and villages of the lava plateau are compact and widely spaced, the low and unreliable rainfall supporting a relatively low density of population. In terms of agriculture, the region is distinctive for the absence of rice cultivation. Most farmers rely on a mixture of wheat and millet, with a few cattle, or specialize in cotton or sugar beet as cash crops.

---

[4]O. H. K. Spate and A. Learmonth, *India and Pakistan: A General and Regional Geography,* London: Methuen, 1972, p. 693.

Summary and Conclusions

Isolated and protected by an arc of mountain ranges, the peoples of South Asia have developed distinctive cultures and generated influential concepts and powerful ideals that have spread around the world. The Harappan culture that flourished between 3000 and 2000 B.C. was one of the world's hearth areas of urban civilization. South Asia's resources and its geographic situation on sea-lanes between Europe and the East Indies made it especially attractive to European imperial powers from the sixteenth century, and in the latter part of the twentieth century, South Asia's strategic location—between the Middle East and Southeast Asia, and adjacent to China—meant that it was of great interest to the superpowers during the Cold War.

Today, South Asia has the fastest-growing—and the second largest—population of all world regions, with an important diaspora that extends to Europe, Africa, North America, and Southeast Asia. In comparison with other world regions, South Asia is still very much a land of villages, though it contains several metropolises of global importance. The region has become even more of a geopolitical hot spot since the end of the Cold War as India and Pakistan, both of them now with access to nuclear weapons, continue to struggle to come to terms with partition. Meanwhile, South Asia is at a critical juncture in its development. On the one hand, it has considerable potential in its human and natural resource base and emerging new economy. India, in particular, is opening up its economy through reforms that have created a more open and entrepreneurial economic climate. Key institutions have been privatized, and foreign investment has flowed into the country, helping to generate exceptionally high economic growth rates. India now has a huge, well-educated, and sophisticated consumer market of more than 200 million that has become part of the "fast" world and an important agent of globalization. With the world's largest democracy and a significant industrial base, India has come to play an increasingly important role within the world-system, and the fact that India's middle class conducts business in English gives the country an important comparative advantage in today's world economy.

On the other hand, the whole of South Asia is facing a potentially deep and multifaceted crisis that could well undermine this potential. In marked contrast to India, other countries have struggled to sustain democracy, sometimes failing altogether. In Islamic Afghanistan and Pakistan, powerful fundamentalist movements have resisted globalization, attempting to re-create certain aspects of traditional culture as the basis of contemporary social order. Throughout South Asia, poverty threatens to swamp the gains of economic development, extreme inequality threatens to undermine political stability, and both poverty and economic development pose a serious threat to South Asia's fragile ecological system.

Key Terms

atoll (p. 529)
bonded labor (p. 555)
caste (p. 546)
hill station (p. 566)
distributary (p. 530)
orographic effect (p. 529)
playa (p. 528)
Raj (p. 535)

Review Questions

Testing Your Understanding

1. What are the three major river systems in the Plains region?
2. Who were the Harappans and the Aryans/Vedics?
3. Emperor Asoka of the Mauryan Empire helped spread which tenets of Buddhism? How did Mughal rule further transform the people of South Asia?
4. How did Imperial Britain change South Asia? How did tea cultivation affect the population of Sri Lanka (Ceylon)?
5. What were the principal consequences of the 1947 partition of India and Pakistan?
6. What is the Grameen Bank? What is Grameen Telecom? How do these services operate and are they successful?
7. What is the role of women in India's economy? How does the success of women's self-help movements transform Indian society?
8. Why is child labor so pervasive in South Asia?
9. Briefly discuss the environmental damage and risks to human health caused by the industrial chemical spill in Bhopal, India, the high sedimentation rates at Porbela Dam, Pakistan, and arsenic in groundwater from tube wells in Bangladesh.
10. Why did the Soviet Union intervene militarily in Afghanistan from 1979 to 1989? How did the Afghan War affect the Soviet Union and Afghanistan after 1989? Who controls Afghanistan today and how do they enforce their interpretation of Islamic law?
11. India is the world's largest democracy yet until recently its family-planning practices have been coercive. With public mistrust high and rates of sexually transmitted diseases climbing, how is India trying to move toward more successful contraception programs?
12. Define caste, jati, brahmin, and Dalit.

13. Who are the Sikhs, Mujahirs, Tamil Tigers, Sinhalese, and the BJP?

14. The 2001 earthquake in Gujarat killed more than 20,000 people and rendered more than a million homeless. Why is eastern Gujurat such a densely populated area?

15. How does demand for medicinal products in Southeast Asia and for ivory in Japan affect tiger and elephant populations in South Asia?

Thinking Geographically

1. How do the Himalyan Mountains and monsoon season help define the character of South Asia? What is the orographic effect?

2. What challenges face the people of South Asia with regard to water and air pollution?

3. How did Britain transform the physical and cultural geographies of South Asia with plantations, manufacturing, preferential education, and cultural exchange?

4. Compare poverty in rural and urban South Asia. With malnutrition and illiteracy rates high, how do the poor manage to survive?

5. India provides outsourcing for core regions, especially in the high-tech industry where computer programmers at companies like Infosys create a large amount of computer code for far less money than it would cost in North America or Europe, yet for relatively high salaries in South Asia. How does that change the lives of programmers in the United States and in India?

6. How did the U.S. Civil War (1861–1865) affect agricultural and industrial development in India? Which cities in India benefited most, and how?

7. Why is India so successful at retaining many of its local cultural traditions in the face of rising globalization since the 1992 market reforms?

8. As South Asians migrated worldwide, which occupations did they take? Which cultural traditions did they bring with them?

9. Why do China, India, and Pakistan all have a claim to the Kashmir region?

10. Bhutan strictly controls the number of visitors to the country in order to protect fragile ecosystems. Nepal, home to Mount Everest and other popular climbing destinations within the Himalayan Mountains, welcomes tourist revenue. Does this have any bearing on relations between the countries or on the way each country treats its minority populations?

Further Reading

Allen, D., (ed.), *Religion and Political Conflict in South Asia.* Westport, CT: Greenwood Press, 1992.

Barnes, I., and Hudson, R., *The History Atlas of Asia.* New York: Macmillan, 1998.

Chapman, G. P., *The Geopolitics of South Asia.* London: Ashgate Publishing, 2000.

Corbridge, S., and Harriss, J., *Reinventing India: Liberalization, Hindu Nationalism, and Popular Democracy.* Cambridge: Polity Press, 1999.

Dutt, A., and Geib, M., *An Atlas of South Asia.* Boulder: Westview Press, 1987.

Farmer, B. H., *An Introduction to South Asia,* 2nd ed. New York: Routledge, 1993.

Fielding, M., "The Geopolitics of Aid: Provision and Termination of Aid to Afghan Refugees in North West Frontier Province, Pakistan." *Political Geography,* 17, 1998, 459–487.

International Labour Office, *Child Labour: Targeting the Intolerable.* Geneva: ILO, 1996.

Jha, V., Hewison, G., and Underhill, M., *Trade, Environment, and Sustainable Development: A South Asian Perspective.* New York: St. Martin's Press, 1997.

Johnson, B. L. C., *Bangladesh,* 2nd ed. London: Heinemann, 1982.

Johnson, B. L. C., *South Asia,* 2nd ed. London: Heinemann, 1981.

Keay, J., *India: A History.* New York: Atlantic Monthly Press, 2000.

Khandker, R., Khalily, B., and Khan, Z., "Grameen Bank: Performance and Sustainability." Washington, DC: World Bank *Discussion Papers,* #306, 1995.

Mitra, S. K. (ed.), *Subnational Movements in South Asia.* Boulder: Westview Press, 1994.

Norton, J., *India and South Asia,* 4th ed. Guilford, CT: Dushkin/McGraw-Hill, 1999.

Rawling, M., *Commodities: How the World Was Taken to Market.* London: Free Association Books, 1987.

Rothermund, D., *An Economic History of India.* London: Routledge, 1993.

Shurmer-Smith, P., *India: Globalization and Change.* London: Arnold, 2000.

Spate, O. H. K., and Learmonth, A. T. A., *India and Pakistan: A General and Regional Geography.* London: Methuen, 1967.

Thomas, G. C., "Competing Nationalisms: Secessionist Movements and the State," *Harvard International Review,* Summer 1996.

Wetzler, B., "Boomgalore," *Wired,* March 2000, 152–71.

Wolpert, S., *India,* rev. ed. Berkeley: University of California Press, 1999.

Film, Music, and Popular Literature

Film

Bhutan: The Last Shangri-La. PBS Home Video, 1999. Documentary about life, landscapes, and ecology in the Buddhist Himalayan kingdom.

City of Joy. Directed by Roland Joffé, 1992. This movie depicts life in a Kolkata slum; gives a good flavor of India and the monsoon season.

Gandhi. Directed by Richard Attenborough, 1982. Hit movie biography of the lawyer who became the famed leader of the Indian revolt against the British through his philosophy of nonviolent protest.

The Great Indian Railway. PBS Home Video, 1998. Documentary that describes journeys on India's vast railway network.

Heat and Dust. Directed by James Ivory, 1982. Based on the book of the same title by Ruth Jhabwala, this movie depicts backpacker's India.

Kandahar. Directed by Mohsen Makhmalbaf, 2001. An independent film about a young journalist based in Canada who returns to her native Afghanistan to find the country in the grip of the Taliban.

Mission Kashmir. Directed by Vidhu Vinod Chopra, 2000. Good example of a Bollywood movie in the thriller genre.

My Beautiful Laundrette. Directed by Stephen Frears, 1985. Based on the novel by Hanif Kureishi, this movie tells the story of a romance between an Anglo-Pakistani and a London skinhead.

Phantom India (L'Inde Phantôme). Directed by Louis Malle, 1969. This two-part movie provides a fascinating, in-depth look at post-Independence India.

Salaam Bombay. Directed by Mira Nair, 1988. Portrays the life of street children in Mumbai.

Music

Chandra, Sheila. *Weaving My Ancestors' Voices.* Real World Records, 1993.

Jhaveri, Shweta. *Anahita.* Intuition Records, 2000.

Khan, Nusrat Fateh Ali. *Dust to Gold.* EMD/Real World, 2000.

Musafir. *Dhola Maru.* Sounds True, 1999.

Shankar, Ananda. *Walking On.* EMD/Real World, 2000.

Shankar, Ravi, and Glass, Philip. *Passages.* BMG/Private, 1990.

Subramaniam, L. *Global Fusion.* Wea/Atlantic/Atrium, 1999.

The Sabri Brothers. *Greatest Hits.* Shanachie, 1997.

Popular Literature

Denker, D. *Sisters on the Bridge of Fire.* Los Angeles: Allstory.com, 1993. An account of solo travel through Pakistan, sharing the lives of local women.

Grewal, R. *In Rajasthan.* London: Lonely Planet Publications, 1997. A personal journey that takes in the traditional villages, palaces, and personalities of one of India's most exotic regions.

Hasan, M. "Partition: The Human Cost," *History Today*, September 1997. An account of the trauma and tragedy of partition through literature and personal histories.

Kipling, R. *Kim.* New York: Doubleday, 1966. A classic epic novel of India under the British by one of the great storytellers.

Mehta, G. *Karma Cola: The Marketing of the Mystic East.* London: Jonathan Cape, 1991. Describes the collision between India's desire for Western technology and methods and the West's fascination with India's ancient wisdoms.

Mishra, P. *Butter Chicken in Ludhiana: Travels in Small Town India.* London: Penguin, 1995. A popular account of travel through the "real" India.

Nasrin, T. *Lajja.* New York: Prometheus, 1997. The title of this novel means "shame"; it describes the 1992 Muslim-Hindu clash over the destruction of the Babri Masjid in India.

Newby, E. *Slowly Down the Ganges.* London: Lonely Planet Publications, 1998. Captures the sights and sounds of travelling down the river.

Novak, J. *Bangladesh: Reflections on the Water.* Dhaka: University Press, 1994. A penetrating overview of the country and its people.

Rushdie, S. *Midnight's Children.* New York: Penguin USA, 1995. A novel that tells of the lives of the children who were born, like modern India itself, in 1947.

Rushdie, S., and West, E. (eds.). *Mirrorwork: 50 Years of Indian Writing, 1947–1997.* London: Henry Holt, 1997. A celebration of Indian writing in English.

Sarkar, T. "Women in South Asia: The *Raj* and After," *History Today*, September 1997. Examines the evolving position of women.

Seth, V. *A Suitable Boy.* New York: HarperPerennial, 1993. An epic novel about post-Independence India.

Tharoor, S. *India: From Midnight to the Millennium.* New York: HarperPerennial, 1998. A passionate but often agonized account of the history of India since independence that attempts to define what makes India one country and Indians of various ethnic, religious, and cultural backgrounds one nationality.

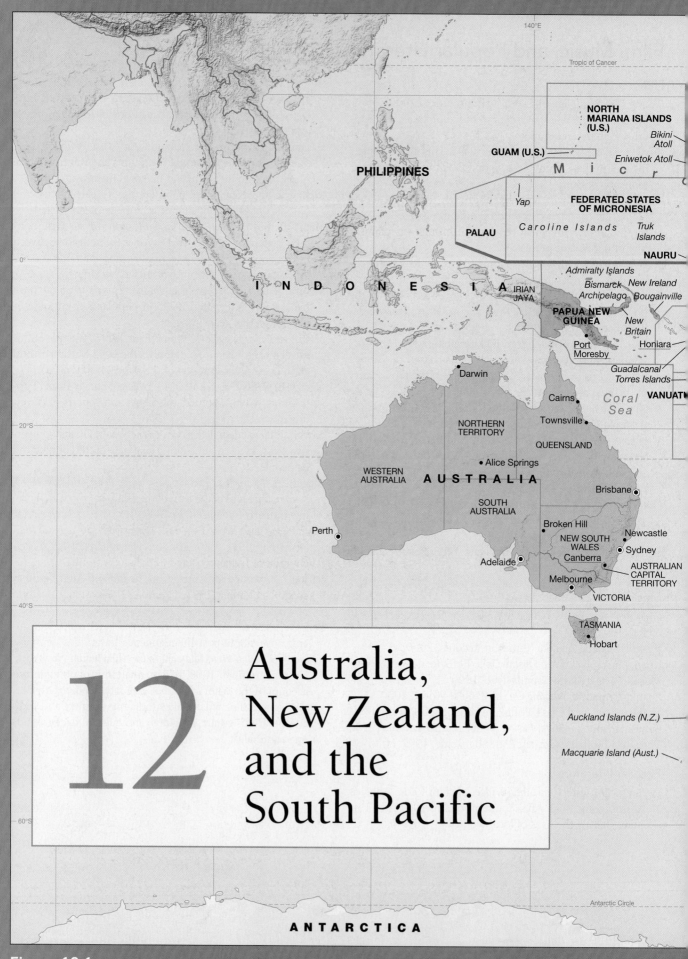

Figure 12.1

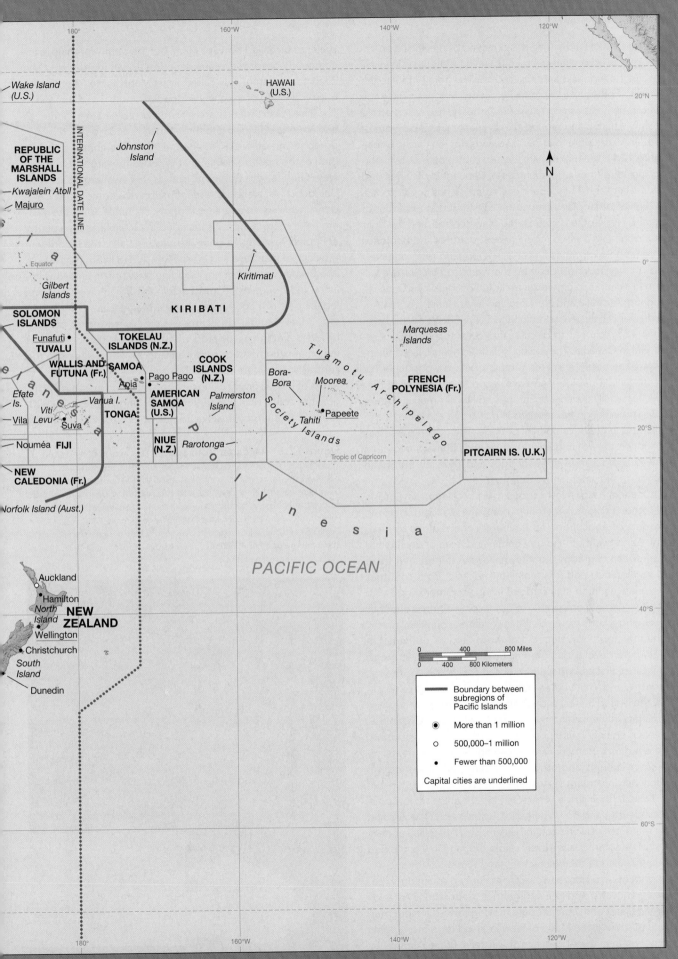

One-third of Earth's surface is occupied by the Pacific, an immense ocean that spans 16,000 kilometers (10,000 miles) between the Americas and Asia and reaches from Alaska to the Antarctic. Portuguese explorer Ferdinand Magellan gave the Pacific its name, meaning "peaceful," after he had sailed across a stormy Atlantic and around the tempestuous Cape Horn of South America. More than 20,000 islands dot the Pacific Ocean (**Figure 12.1**). The majority are located in the southeastern portion, including the large countries of Papua New Guinea and New Zealand and 11 independent nation islands or island clusters. Eleven other island groups are constituted as territories of Australia, France, the United States, New Zealand, and the United Kingdom. As a world region, these countries and territories are commonly known as *Oceania*, a shorthand term originally used by the French to describe the world region that we will use in this chapter. Some texts exclude Australia from Oceania, designate Australia and New Zealand as Australasia or the Antipodes, or call this entire region the South Pacific. In this book we have also chosen to discuss Antarctica as a distinctive region within Oceania. The region is defined by its shared orientation to the ocean realm, its comparatively low population of only 31 million people, and its isolation from other world regions, especially the core of Europe and North America (**Figure 12.2**). This isolation has resulted in distinctive ecosystems and challenges in the development of trade. The region possesses a lengthy history of migration, first by early humans migrating from Asia thousands of years ago, and then by European colonists in the nineteenth century.

Oceania has played various roles in relation to Europe and other regions. Australia and New Zealand became providers of agricultural and mineral commodities to nineteenth- and twentieth-century Britain and a major destination for European migration, but they have now shifted their trade and immigration patterns more toward Asia. Both countries have relatively high standards of living, commensurate with average incomes comparable to other core countries and thus challenging a north–south divide that defines countries "north" of the equator as developed and those "south" of the equator as underdeveloped. The Pacific islands rose to strategic significance first as stopovers on ocean trading routes, then as Second World War battlegrounds between the Allies (especially the United States) and Japan, and most recently as bases for military influence and marine territorial claims. Oceania as a whole is emerging as a popular destination for tourists drawn by the region's dramatic landscapes and tropical retreats, indigenous cultures and exotic ecosystems, and the modern, mellow urban centers of Australia and New Zealand.

Despite their shared characteristics of isolation, low population density, and British heritage, Australia and New Zealand have many physical and cultural differences. Australia's ancient physical geography is relatively stable, flat, and low compared to New Zealand's tectonic activity and mountain ranges. Although tension exists between the majority population of European ancestry and the indigenous populations in both countries, and both countries are moving toward a more multicultural society, the Maori of New Zealand have more rights and power than the Aborigines of Australia. Some Aborigines prefer to be called *Koori* or *Murri*, but the term *Aborigine* is still commonly used as a self-description and in legislation. New Zealand is alternatively called by its Maori name "Aotearoa," meaning "land of long white clouds."

The Pacific islands are also marked by many differences and are often divided into three broad groups, defined by geography and ethnicity (see Figure 12.1). **Melanesia** is the region of the western Pacific that includes the largest islands of New Guinea, the Solomon Islands, Fiji, Vanuatu, and New Caledonia. All are independent nations except for New Caledonia, which is an overseas territory of France. Only the eastern half of New Guinea is included in Oceania as the country of Papua New Guinea since the western half is the Indonesian province of Irian Jaya and is usually included in Southeast Asia. The Bismarck island chain, including New Britain and New Ireland, and Bougainville and Buka in the Solomon Islands, are legally part of Papua New Guinea. Melanesia—as its meaning, "dark islands," indicates—is mainly an ethnic or race-based definition associated with the darker skin color of its inhabitants compared to those living on other islands in the central and eastern Pacific. The region is better defined in terms of its geographical location in the southwestern Pacific or in the confluence, evident mainly in language, of a 40,000-year-old Papuan culture associated with New Guinea and a

Figure 12.2 Oceania from space This image conveys the importance of the Pacific Ocean to Oceania and the contrast between the large island continent of Australia, the larger islands of New Guinea and New Zealand, and the scattering of smaller islands across the southern Pacific. [*Source:* (inset) Map projection, Buckminster Fuller Institute and Dymaxion Map Design, Santa Barbara, CA. The word *Dymaxion* and the Fuller Projection Dymaxion™ Map design are trademarks of the Buckminster Fuller Institute, Santa Barbara, California, © 1938, 1967 & 1992. All rights reserved.]

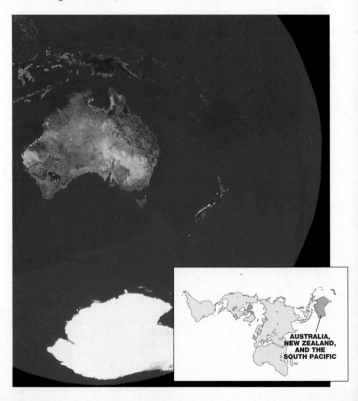

AUSTRALIA,
NEW ZEALAND,
AND THE
SOUTH PACIFIC

more recent migration from Southeast Asia. Even this classification overlooks strong links to other regions of the Pacific.

Micronesia (meaning "small islands") is the northern group of islands at, or north of, the equator, including the independent countries of Nauru, Kiribati, Palau, the Marshall Islands, the Federated States of Micronesia, the U.S. territory of Guam, and the Commonwealth of the Northern Mariana Islands in association with the United States. **Polynesia** (meaning "many islands") comprises the western Pacific islands and includes the independent countries of Samoa, Tonga, and Tuvalu; the U.S. territory of American Samoa; the French overseas territories of French Polynesia (including the island of Tahiti, the Society Islands, and the Marquesas Islands) and Wallis and Futuna; the New Zealand territory of Tokelau; the British territory of the Pitcairn Islands, and Niue and the Cook Islands, which maintain close ties with New Zealand. The U.S. state of Hawaii is also sometimes considered part of Polynesia because of ethnic and linguistic links between the indigenous Maori, Hawaiiana, and people of the eastern Pacific islands. This larger region can be defined by a triangle that joins New Zealand, Hawaii, and Easter Island. Easter Island is administered by Chile, in Latin America, just as the Galapagos, also in the far eastern Pacific, are administered by Ecuador.

This chapter examines the environment, history, and peoples of Oceania, highlighting the core region of southeastern Australia and the distinctive landscapes of the Australian outback, Pacific islands, and Antarctica. Oceania provides an opportunity to highlight some important global issues and their regional impacts, including climate change and ozone depletion, marine territorial disputes, indigenous rights and multiculturalism, and the development of nuclear energy.

Environment and Society in Australia, New Zealand, and the South Pacific

Environmental conditions in Oceania have posed many challenges and offered many opportunities to the region's human residents, who have adapted to the constraints of the physical environment and transformed it in dramatic ways. In this chapter we develop the environmental context for human activities in Oceania by examining the physical environments of several large subregions. Australia, the largest landmass in the region, has several contrasting environments, including the dry interior and western plateau, and an eastern coastal lowland, backed by uplands. The climate ranges from tropical conditions on the northern Queensland coast to cooler conditions in the southeast state of New South Wales and on the island of Tasmania. Australia is the only country in the world that also forms an entire continent, and it is the driest and flattest of all the continents. New Zealand has two major environmental regions, the more tropical North Island and the cooler South Island. The Pacific islands can be classified into two physical types, the higher volcanic islands and the low coral atolls.

Landforms and Landscapes

Geological history, especially tectonic activity, influences the nature of many of the major environmental regions of Oceania. Australia is a very old and stable landmass that was part of the Southern Hemisphere supercontinent called Gondwanaland (see Chapter 2, p. 52), which broke away from Pangaea 200 million years ago and moved toward the South Pole. The Indo-Australian Plate separated about 50 million years ago and began moving northeast until it collided with the Pacific Plate. New Zealand now sits at the plate boundary where the Pacific Plate is moving under (subducting) the Indo-Australian Plate and producing high levels of volcanic and earthquake activity.

Australia The Australian landmass lies at the core of the Indo-Australian Plate and forms a continental shield of ancient stable rock with very little volcanic, earthquake, or other mountain-building activity. Australia, which has an area of 7.7 million square kilometers (3 million square miles), is divided into three major physical regions by a series of interior, low-lying basins that divide the western Australian Plateau from the uplands of eastern Australia (**Figure 12.3**).

The *eastern highlands* of Australia are the remnants of an old folded mountain range with a steep escarpment on the eastern flanks that was created when New Zealand broke away from Australia about 80 million years ago. The highland crest is often called the Great Dividing Range because it separates the rivers that flow to the east coast from those flowing inland or to the south. The highest mountain in the eastern uplands is Mount Kosciusko (2228 meters, 7310 feet) in the Snowy Mountains of New South Wales. These highlands are very important to water resources and agriculture and serve as recreation areas for the residents of Australia's coastal cities. The east coast is distinguished by the offshore presence of the world's largest coral reef, the Great Barrier Reef (see Sense of Place: The Great Barrier Reef, p. 577).

The *interior lowlands* were once flooded by a shallow ocean and now contain the Lake Eyre basin, filled only occasionally by inland-draining rivers. Waters in the Lake Eyre basin are very salty because of the high levels of evaporation in the hot, dry climate. A large part of the lowlands is also called the **Great Artesian Basin** because it is underlain by the world's largest groundwater aquifer, a reservoir of underground water in porous rocks. The basin is artesian because overlying rocks have placed pressure on the underground water so that when a well is drilled, the water rises rapidly to the surface and discharges as if from a pressurized tap (**Figure 12.4**). The wells that tap the Great Artesian Basin are critical to the human settlements and livestock of arid east-central Australia, although the cost of drilling is high and the water is often very warm and salty. The southern part of the interior lowlands is drained by the Murray and Darling river systems, which flow west out of the eastern highlands toward the southern coast. These river basins have been transformed by irrigation projects.

Two-thirds of Australia is occupied by the *western plateau* of old shield rocks with a few low mountains, such as the Macdonnell and Hamersley ranges, and large areas of flatter desert

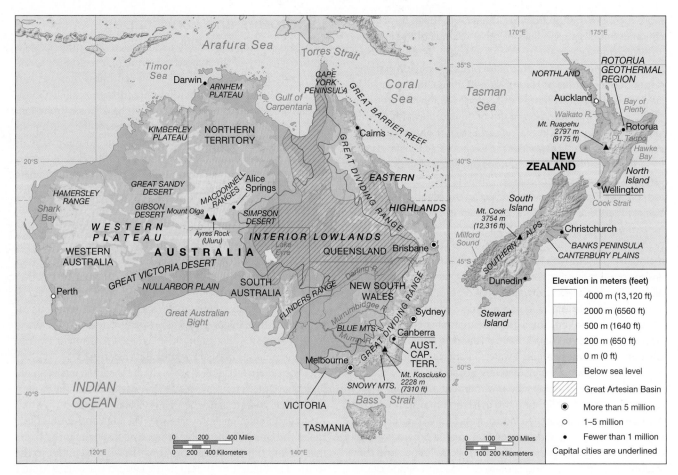

Figure 12.3   The physical landscape of Australia and New Zealand Major physiographic regions and distinctive landforms of Australia and New Zealand are shown on these maps. For Australia, key features include the eastern highlands, central deserts, Great Artesian Basin, and Great Barrier Reef and for New Zealand, the Southern Alps and west coast fjords such as Milford Sound.

plains and plateaus, such as the Simpson and Great Sandy Deserts and the Kimberly and Arnhem plateaus. This region has numerous mineral deposits—which lay the foundation for Australia's mining industry—and old, weathered soils that are too nutrient-poor or salty for agriculture. Western Australia

Figure 12.4   Well in the Great Artesian Basin Water rises under artesian pressure from the enormous underground reservoir that underlies east-central Australia and sustains livestock in the arid interior. Windmills provide supplemental energy for pumping.

and the interior lowlands contain impressive examples of desert landforms, including the wind-shaped undulating ridges of sand dunes, stony plains with varnished rock fragments called *desert pavement*, and dry interior drainage basins called *playas*. Centuries of erosion by wind and water have left erosion-resistant domes of rocks standing above the surrounding landscape. Extremes of heat and cold, combined with moisture and various chemical processes, create weathering processes that can varnish the rocks or peel them off in concentric layers. The most famous of these isolated rock domes are Ayres Rock (called *Uluru* by the Aborigines) and the Olgas (called *Kata Tjuta*); both are popular tourist destinations that bring thousands of visitors to isolated regions of central Australia (**Figure 12.5**).

New Zealand   In contrast to Australia, New Zealand is a much younger and more tectonically active landscape, located where the Pacific Plate is moving under the Indo-Australian Plate, thrusting land upward into high relief and promoting volcanic activity. The New Zealand physical landscape includes two major islands spanning 1600 kilometers (976 miles) from north to south, with a combined area of just over 268,000 square kilometers (103,000 square miles). The South Island is about 25 percent larger than the North Island, from which it is separated by the narrow Cook Strait.

Great Barrier Reef

The Great Barrier Reef is considered one of the natural wonders of the world (**Figure 1**). Fringing the northeast coast of Australia, the reef is more than 2000 kilometers (1250 miles) long and easily visible from space. The view from the air does not reveal the real beauty of the landscape because the most attractive parts of the reef are underwater. Diving below the surface, the visitor encounters intricate and colorful corals and waving sea grasses that are home to millions of brilliant fish and majestic marine animals, such as turtles, whales, and dugong (large marine seal-like mammals).

The Great Barrier Reef actually consists of 3400 individual and fringing coral reefs, incorporating more than 300 species of coral and hosting more than 1500 species of fish and 4000 different types of mollusks (**Figure 2**). The reefs were formed over millions of years from the skeletons of marine coral organisms in the warm tropical waters of the Coral Sea.

The reef is Australia's second most important foreign tourist destination after Sydney, and the most important for domestic tourists. Tourists visit the reef by boat or stay at resorts on the reef and participate in activities that include fishing, diving, snorkeling, and reef walking. Some resorts have constructed viewing areas below the surface and offer submarine tours. Tourism generates almost $1 billion a year.

Today the reef is under pressure from trawling in the lagoon, chemical and sediment pollution, climate change, and coastal development and has attracted worldwide attention from conservation organizations because of its global biological significance. Overgrazing and land clearing on the Queensland mainland has sent sediments and nutrients equivalent to more than 3.5 million dump trucks a year into the lagoon, causing increases in populations of algae and turning the clear waters into murky soup. Fertilizers and pesticides used on sugar cane, banana, and cotton plantations are also washing into the sea and poisoning marine life. One of the most serious threats to the reef in addition to pollution reducing the clarity of the water is the coral-eating crown-of-thorns starfish that is destroying the reefs.

Most of the reef is protected within the Great Barrier Reef Marine Park, and the reef is the largest unit on the list of United Nations World Heritage sites. Efforts to protect the reef include restrictions on fishing by trawlers, limits on tourism permits in some sensitive or heavily used areas, an environmental management charge on tourist operators, and detailed contingency plans in the case of oil spills.

Figure 1 The reef from the air The Great Barrier Reef fringes the northeastern coast of Australia and is the largest of the United Nations World Heritage sites.

Figure 2 The marine landscape The undersea landscape of the Great Barrier Reef has become a major destination for tourists who swim among the corals and grasses to view the colorful tropical fish.

The South Island has rugged mountains rising to more than 3500 meters (11,500 feet) in the Southern Alps, dominated by Mount Cook at 3754 meters (12,316 feet). The South Island is far enough south (41 to 47 degrees south latitude) to have extensive permanent snowfields and more than 300 glaciers, some flowing almost to sea level. The southern portion of the west coast is penetrated by magnificent fjords, such as Milford Sound, created when the sea flooded the deep valleys cut by glaciers (**Figure 12.6a**). The spectacular alpine scenery is a major draw for tourists. The east coast of the South Island has much gentler relief, with rolling foothills, long valleys with braided rivers and freshwater lakes, and alluvial plains (formed

(a)

(b)

Figure 12.5   The desert landforms of Australia (a) Ayres Rock in Uluru National Park is an erosion-resistant isolated dome of rock that stands 348 meters (1143 feet) above the desert and has a circumference of 9 kilometers (6 miles). Of spiritual significance to the Aborigines, the rock has also become a major tourist destination. (b) The Olgas, also in Uluru National Park, are even more dramatic examples of rock formations left after centuries of erosion by wind and water.

from stream deposits) such as the Canterbury Plains, which comprise an important agricultural area.

The North Island has much more volcanic activity than does the South Island, and many volcanoes, craters, and lava flows dot the central region. Recent eruptions, such as that of Mount Ruapehu in 1995–96, have ejected ash over a large area and caused damage to agriculture and property. Volcanically heated water that emerges from hot springs, geysers, and steam vents is captured in geothermal facilities and provides an important energy resource (**Figure 12.6b**). The North Island also has extensive areas of rolling hills and valleys where the warmer climate and rich volcanic and river-deposited soils nourish a productive agriculture.

New Guinea   The island of New Guinea is the second largest island in the world (after Greenland, because Australia is considered a continent rather than an island) and is much larger than the country of New Zealand at more than 800,000 square kilometers (309,000 square miles). The western portion is administered by Indonesia, but more than 57 percent of the island forms the independent nation of Papua New Guinea and is considered to be part of Oceania and Melanesia. The island is separated from Australia to the south by the Torres Strait. The mountain spine of the island of New Guinea rises to more than 4000 meters (13,000 feet), with many extinct volcanoes and high isolated basins. The southern part of the island has a large marshy lowland plain drained by the Fly River.

Figure 12.6   New Zealand landscapes (a) The fjord of Milford Sound is about 3 kilometers (2 miles) wide, extending inland for almost 20 kilometers (12 miles) and with steep slopes falling from Mitre Peak (1695 meters, or 5560 feet, high) to the bottom of the flooded fjord, 512 meters (1680 feet) deep. Milford Sound is located within Fiordland National Park on New Zealand's South Island, where mountains in the Southern Alps rise to more than 3500 meters (11,500 feet). (b) The region around Rotorua, on the North Island of New Zealand, has numerous hot springs, geysers, and steam vents associated with volcanic activity. Geothermal energy sources have been developed and contribute 10 percent of New Zealand's electricity production, and the region is also popular with tourists attracted by the landscape, spas, and thriving Maori culture.

(a)

(b)

(a)

(b)

Figure 12.7   Pacific islands (a) The higher islands of the Pacific are mostly volcanic in origin and rise steeply from the sea to forested mountain slopes. The island of Moorea in French Polynesia is a spectacular example of a volcanic high island. (b) The lower islands include atolls such as Tetiorua in the Society Islands of French Polynesia, where coral reefs protect shallow lagoons.

Pacific Islands   The Pacific islands can be classified into the high volcanic islands and the low coral islands called *atolls* (**Figure 12.7**). The high islands, which are mostly volcanic in origin, rise steeply from the sea and have very narrow coastal plains and deep narrow valleys. Many high islands, such as those of Hawaii and Samoa, are created in linear chains when tectonic plates move over hot spots where molten rock reaches the surface. Others, such as the Marianas and Vanuatu islands, form island arcs along the edge of tectonic plates. The heights of the islands promote heavy rainfall and are often capped by clouds, creating spectacular landscapes such as those of Tahiti, where the mountains rise to 2100 meters (6900 feet), Bougainville (3000 meters or 9840 feet), or Hawaii, where the Mauna Loa volcano reaches 3900 meters (12,800 feet).

The low islands are mostly coral islands, or **atolls,** created from the buildup of skeletons of coral organisms that grow in shallow tropical waters. Atolls are usually circular, with a series of coral reefs or small islands ringing and sheltering an interior lagoon that may contain the remnants of earlier islands or a volcanic island that has sunk below the surface. Although in some cases, such as the islands of Nauru and Guam, tectonic activity may uplift coral reefs to create higher elevation limestone plateaus, many of these islands are very low lying with most of the land within a meter of sea level, making them vulnerable to storms, tidal waves, and sea-level rise. The majority of the low islands in the Pacific are atolls, including Kiribati, the Marshall Islands, and Tuvalu. The Kwajalein Atoll in the Marshall Islands is the largest in the world, with 90 small coral islands circling a 650-square-kilometer (251-square-mile) lagoon.

Climate

Much of Oceania lies within the Tropics and is dominated by the warm seas and moisture-bearing winds of these latitudes. Australia and New Zealand reach farther south and thus have climates that range from tropical to the cooler temperate climates of the southern westerly wind belts, with annual average temperatures declining from north to south. Australian climate is dominated by the very dry and hot conditions that parch the interior and define it as one of the most arid regions on Earth, with two-thirds of the country receiving less than 50 centimeters (20 inches) of rainfall a year (**Figure 12.8**). The dryness is created by the great distances from moisture sources and the dominance of the subtropical high-pressure zone with descending dry air. The harsh climate limits human activity and has required complex adaptations from both people and ecosystems.

Most of the coastal perimeter has higher precipitation. For example, northeastern Australia has heavy rainfall from the southeasterly trade winds that rise over the eastern uplands with associated orographic rainfall (see Chapter 2, p. 57). The rainfall at Cairns in Queensland averages more than 460 centimeters (180 inches) a year. Northern Australia receives most of its rain from monsoon winds, drawn inland by the southward shift of the intertropical convergence zone and the heating of the landmass in the Southern Hemisphere summer from November to February. The rainy season is called the *Big Wet,* and it sometimes brings tropical cyclones, such as Cyclone Tracy, which devastated the city of Darwin in 1974. Southern Australia receives rainfall from storms associated with the westerly wind zone, especially during the winter (June to August) when the storm tracks shift northward. The southern coast, specifically the regions around the cities of Adelaide and Perth, has the mild temperatures and winter rainfall associated with the Mediterranean climate type that is also found at this latitude in California, Chile, southern Europe, and South Africa. The southernmost part of New South Wales, Victoria, and the island of Tasmania are wetter as a result of exposure to westerly rain-bearing winds for most of the year and have considerable snowfall in the mountains.

Griffith Taylor, the founder of academic geography in Australia, wrote a series of books and articles between 1920 and 1950 on the environmental opportunities and constraints posed by the Australian climate. He produced numerous maps showing relationships between climate and agricultural potential and argued against the settlement of migrants in the desert interior.

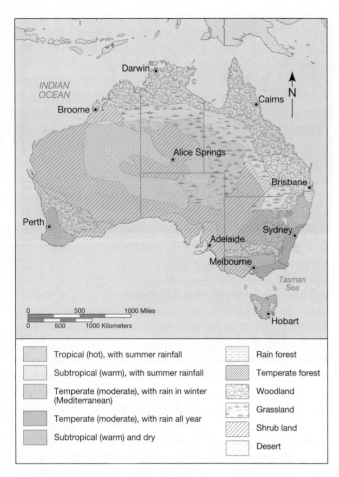

Figure 12.8 The climate and vegetation of Australia The climate of Australia is determined by proximity to the ocean and by latitude, with cooler temperatures in New South Wales, Victoria, and Tasmania, tropical warm and wet conditions in Queensland, and dry conditions in the Australian interior. Vegetation patterns are closely associated with climate and include rain forest in the northeast and desert vegetation in the interior. (*Source:* Based on T. McKnight, *Oceania: The Geography of Australia, New Zealand and the Pacific Islands.* Englewood Cliffs, NJ: Prentice Hall, 1995, Figure 2.2e; and G. M. Robinson, R. J. Loughran, and P. J. Tranter, *Australia and New Zealand: Economy, Society and Environment.* New York: Oxford University Press, 2000, Figure 3.1.)

New Zealand sits in the middle of the westerly wind belt, and frequent storms release heavy rain on the west coast as they rise over the high mountain ranges. The eastern coasts of New Zealand are much drier because they lie in the rain-shadow to the east of the mountains, and this area can be sunny in summer when subtropical highs move southward and create clear and stable conditions. The North Island is generally much warmer than the South, but mountain climates are cooler and wetter throughout the country, often with heavy snow that favors the promotion of ski tourism in the June-October period, drawing many skiers from North America.

The Pacific islands have warm temperatures associated with year-round high sun and the warmth of the tropical ocean. Those islands with higher elevations experience substantial rainfall as moist winds rise over the island, cooling and releasing their moisture. Clouds of condensing moisture cap many high islands. Islands throughout the region receive rainfall from the towering cumulus clouds that occur at the intertropical convergence zone as intense heating creates rising air (convection) and heavy afternoon rains. The lower islands are much drier because they do not benefit from orographic uplift over mountain ranges and are small enough to elude the convective downpours. As a result, many low-lying islands experience near desert conditions and shortages of fresh water.

Oceania extends into the region of the Pacific Ocean that is directly connected to **El Niño,** the periodic change in ocean temperature off the coast of Peru that affects weather worldwide (see Chapter 8). It is in the tropical Pacific that an El Niño first develops and where scientists monitor its intensity using systems of floating buoys and an index called the Southern Oscillation, which measures the air pressure difference between Tahiti and Darwin in northern Australia. Normally, pressure is lower at Darwin than at Tahiti, and winds blow westward across the Pacific, bringing rain to Australia and Indonesia. As an El Niño develops, the pressure rises at Darwin and drops in Tahiti as the winds slow or shift direction away from Australia, bringing drought conditions to northern Australia and many western Pacific islands. The 1997–98 El Niño caused severe drought in Papua New Guinea, Australia, and Micronesia, with crop failures and food shortages in New Guinea, expensive shipments of drinking water to smaller islands, and serious livestock loss and wildfires in Australia.

Environmental History

The long isolation of Oceania has contributed to the development of some of the world's most unique, diverse, and vulnerable ecosystems. Many of the species that evolved from the isolated populations have remarkable adaptations to the physical environment and are found only in that locality. Australia is beloved by biogeographers and conservationists for its great biodiversity and unusual species (**Figure 12.9**). There are more than 20,000 different plant species, 650 species of birds, and 380 different species of reptiles.

Australia has several remarkable species of marsupials and of monotremes, animals descended from primitive mammals that died out on most other continents. A **marsupial** gives birth to a premature offspring that then develops and feeds from nipples in a pouch on the mother's body. Australian marsupials range in size from the large gray kangaroo (with adults reaching 2 meters, 6 feet, tall) to koalas, wombats, and ferocious Tasmanian devils, to small mice and voles. **Monotremes** are very unusual mammals in that they lay eggs rather than gestate their young within the body but then nurture the young with milk from the mother. They include the platypus, with a ducklike bill and tail like a beaver, and the spiny anteater. Other Australian fauna are adapted to extreme drought—for example, the water-holding frog, which burrows into the soil and saves water in a cocoonlike sac around its body that is sometimes an emergency water source for humans. Australia is also notable for the large number of dangerous or poisonous species, especially in the northern region, including the man-eating saltwater crocodile, the deadly taipan and tiger snakes, and the venomous funnel-web spider. The presence of

(a)

(b)

(c)

Figure 12.9   Unusual Australian animals (a) This road sign on the Nullarbor Plain warns drivers to beware of two of Australia's marsupials—the kangaroo and the wombat—as well as feral camels, an introduced domestic animal that now runs wild. (b) A koala munches on the leaves of one of Australia's many species of eucalyptus (or gum) trees. (c) The fierce and rare marsupial called the Tasmanian devil, whose populations declined with the introduction of the dingo, the Australian wild dog.

the marine jellyfish known as the sea wasp, which has millions of lethal stingers, requires beach closings from November to March on the Queensland and Northern Territory coast.

Australian ecosystems are defined by their growth form and dominant species and include several types of forest and shrub land with species well adapted to dry climates and frequent fires. The most important type of tree is the **eucalyptus,** commonly known as the *gum tree.* Eucalyptus trees have waxy leaves that hang vertically to reduce evaporation and moisture loss and contain strong-smelling oils that are used to treat respiratory illness. The release of an oily mist from eucalyptus led to the naming of the Blue Mountains in eastern Australia. Although they grow rapidly and provide valuable timber, pulp, fuel, oils, and food to wildlife, eucalyptus is very flammable and increases the intensity of forest fires. Eucalyptus has been introduced into many other world regions for reforestation and pulp plantations. Acacia, commonly called *wattle,* is another dominant tree and shrub type, especially in the drier woodlands. The densest forests are found along the wettest portion of the east coast, with the largest remnants in Queensland composed mainly of trees of Southeast Asian origin interspersed with woody vines (*lianas*). The forests of southeastern and southwestern Australia are dominated by drought-adapted eucalyptus, with more open woodlands in the transition to the dry interior. The northwestern and southern regions have a shrub vegetation called *mallee* that includes low, multibranching eucalyptus and plains covered with bushes similar to the sagebrush found in North America. The northern interior has extensive but sparse grasslands, and the driest zones have scattered grasses and shrubs characteristic of desert ecosystems (but without cacti).

New Zealand was heavily forested prior to the arrival of humans about 1000 years ago, with towering *kauri* conifers in the north and beech in the cooler south. The remaining one-third of the land was covered with scrub, with grasses at drier, lower elevations, and with alpine grasslands (tundra) at high altitudes. Faunal evolution in New Zealand produced no predators or carnivores, and several birds remained flightless, including the now extinct moa, which was 3 meters (nearly 10 feet) tall, and the kiwi bird. The kiwi is the national emblem of New Zealand, and it has given its name to a popular fruit

(formerly known as the Chinese gooseberry) and has become a nickname for the New Zealand people.

The Pacific islands have also developed many different ecosystems and species, with the smaller islands generally less diverse than the larger ones in accordance with the theory of **island biogeography.** Those plant species that can be easily transported by ocean (for example, coconuts) or air (for example, fruit seeds eaten and excreted by birds) are more widely distributed, and the variety declines as one moves eastward, away from the larger landmasses. Luxuriant rain forests are found on the wetter and higher islands, and marshes and mangroves thrive along the coastal margins. The larger islands, such as New Guinea and Hawaii, also have extensive middle-elevation grasslands. The smaller and drier coral islands have much sparser vegetation, but coconut palms are ubiquitous, are a basis of human subsistence, and grow along many beaches. There are few native mammals on the Pacific islands (with the exception of New Guinea), and the richest fauna include the birds that have been able to fly from island to island and marine organisms, especially those of reefs and lagoons, including turtles, shellfish, tuna, sharks, and octopus.

The early human history of Oceania is usually divided into two main phases: the migration of humans from Southeast Asia into Australia, New Guinea, and nearby islands about 40,000 years ago, and a second dispersal to more remote Pacific islands such as Fiji, Tonga, and Samoa about 3500 years ago. The expansion eastward required difficult voyages against the trade winds, and Hawaii was settled only about 1500 years ago. Recent biological evidence from genetic DNA analysis confirms Maori legends that suggest that migrants from Polynesia settled New Zealand about 1000 years ago.

The early inhabitants of Australia are ancestors of today's Aborigines who still preserve some traditions that reflect the early adaptations and modifications of the Australian environment. These traditions include a complex spiritual relationship to the land, a nomadic lifestyle, and the use of fire in hunting. Gathering of roots, seeds, grubs, insects, and lizards contributed essential calories and proteins to traditional diets, and in some coastal areas traps for stonefish and eel were constructed. As Aborigine populations grew, they may have reduced local populations of major game species such as kangaroos, but the most significant environmental change was the transformation of vegetation through the use of fire to improve grazing for game and to drive animals to hunters. Ecologists believe that over thousands of years, some Australian vegetation became more resistant to these fires. The only domesticated animal was the dingo, a dog that was probably imported from Southeast Asia. The Aborigine worldview linked people to each other, to ancestral beings, and to the land through rituals, art, and taboos. Their worldview is associated with the **Dreamtime,** a concept that joins past and future, people and places, in a continuity that ensures respect for the natural world.

The Maori arrived in New Zealand much more recently but are believed to have caused much more widespread environmental transformations than have the Australian Aborigines. They hunted the enormous moa bird to extinction, cleared as much as 40 percent of the original forests, and practiced agriculture based on the shifting cultivation of sweet potatoes. The Maori migrated from Polynesia, where subsistence was based on fishing and the cultivation of root crops such as taro and yams and tree crops such as coconuts and breadfruit, all originally domesticated in Southeast Asia.

The next major stage in the environmental history of Oceania was associated with the incorporation of the region into the world economy during the later phases of European colonialism. This brought massive transformations of the environment through newly introduced species and new economic practices, including mining, extractive forestry, livestock raising, and large-scale crop production.

The Introduction of Exotics and Ecological Imperialism

Beginning with the introduction of the dingo from Southeast Asia by Australian Aborigines about 3500 years ago, foreign species have proceeded to devastate the native species of Oceania. The dingo, a canine similar to a coyote, probably out-competed and out-hunted the marsupial predators, such as the now-extinct Tasmanian wolf, as well as rodents. But it was the **ecological imperialism** of species introduced by the Europeans that led to the endangerment and extinction of numerous other native species through hunting, competition, and habitat destruction. These species are also called "exotics" because they come from elsewhere.

The flightless birds of Oceania were the most vulnerable to introduced predators such as rats, cats, dogs, and snakes. After centuries of such predation, several birds have become extinct and others are in danger. Conservation efforts to protect birds today include the establishment of reserves and the careful monitoring and elimination of predators. On the Pacific islands, there are great efforts to contain the spread of the introduced brown tree snake, the mongoose (a very aggressive small mammal), and a carnivorous snail, all of which prey on local species. Other changes included the introduction of European weeds, pests, and crops and the escape of domesticated livestock. These "feral" animals of Australia include horses, cattle, sheep, goats, and pigs, as well as camels that were introduced to provide transportation across vast deserts. The

overall escaped population may total 3 million animals, the largest population of feral domestic animals in the world.

Another ecological disaster was the introduction of the European rabbit to Australia in 1859. Over the next 50 years, the rabbit population exploded to plague proportions that devastated pasturelands. The rabbit population was partially eradicated in the 1950s by the introduction of a disease called *myxomatosis.* Later, the introduction of the prickly pear cactus in the 1920s to create hedges led to the infestation of more than 20 million hectares (50 million acres) of Australian pasture before the cactus was eradicated by introducing a beetle that fed on the plants. The cane toad, introduced to control pests in the sugarcane district of Queensland, has destroyed frogs, reptiles, and small marsupials, is highly toxic to predators, and has spread to northern Australia. The eucalyptus has become a widespread exotic on other continents, out-competing other species in places such as California and contributing to fire hazards.

Australia, New Zealand, and the South Pacific in the World

The story of Oceania's integration into the world system is dominated first by the British colonization and settlement of Australia and New Zealand in the nineteenth century and the orientation of their economies to the export of agricultural products and minerals. In the late twentieth century, a political and economic shift away from Britain and toward Asia and the United States occurred, along with a shift from protectionist and state-managed economies to free trade, privatization, and foreign investment. The Pacific islands grew in global significance after they were parceled out among European powers in the late nineteenth century, and some were targeted for export agriculture and mineral development. Their critical strategic significance in the Second World War increased U.S. influence in the Pacific islands, later reinforced through the growth of international tourism.

European Exploration and Early Settlement

Although Oceania had limited contact with other regions of the world, especially Southeast Asia, it was not until the mid-1700s that European explorers opened up the area for trade and eventually colonization. Spanish and Portuguese sailors controlled Guam as a stop for galleons traveling from Manila to Acapulco and may have encountered Australia. The Dutch explored the west coast and south coasts of Australia and claimed Van Diemen's Land in 1642, later named Tasmania after the Dutch explorer Abel Tasman (**Figure 12.10**). Tasman also made contact with the Maori in 1642, but the encounter was violent and the Dutch did not land, calling the region *Nieuw Zeeland* (after a region of the Netherlands). It was more than a hundred years later that the most enduring claim was advanced by explorer Captain James Cook, who landed at Botany Bay in 1770, claiming the land for Britain

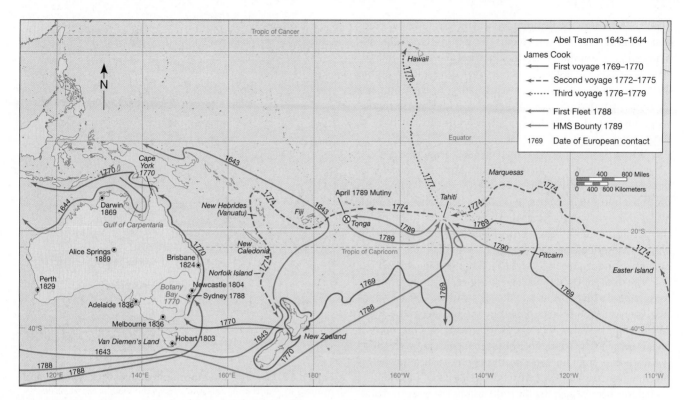

Figure 12.10   The European exploration of Oceania This map shows the sequence of European exploration and settlement of Oceania, including the voyages of Abel Tasman, Captain James Cook, and of the HMS *Bounty*. Cook made three voyages to Oceania, landing at Botany Bay and claiming Australia for Britain; he was ultimately killed in Hawaii. (*Source*: Based on C. McEvedy, *The Penguin Historical Atlas of the Pacific*. New York: Penguin Books, 1998, pp. 49, 63, 65, 90.)

and calling the new territory New South Wales. Cook also explored New Zealand at this time and established harmonious relationships with the Maori.

Based on Cook's reports, the British government decided to people New South Wales by using it as a penal colony, sending boatloads of convicts in order to relieve pressure on British prisons but also to reinforce territorial claims and provide cheap labor for economic development. The *First Fleet* of 11 ships with about 730 male and female convicts arrived in 1788, settling at the site of the city of Sydney. From 1800 to 1830, the British established several settlements around the coast, including Hobart and Launceston in Tasmania; in mainland Australia, Newcastle and Brisbane were established on the east coast, and Melbourne, Albany, and Perth along the southern coast. Eighty percent of Australians still live within 35 kilometers (22 miles) of the coast. More than 160,000 convicts were eventually transported to Australia by 1868, half to New South Wales and half to Van Diemen's Land (Tasmania). Most were not serious criminals and many have been identified as petty thieves and Irish political activists. It has been argued that a strong Irish Catholic element among these early immigrants explains the more anti-British and anti-establishment tradition in Australia compared to New Zealand. The convicts worked for the government or were assigned to private employers. Many eventually gained freedom through pardons, often prompted by the desperate need for more farmers to produce food. Many other free settlers arrived in Australia, and some were given cash rewards for emigrating and were assigned convicts as laborers.

The initial goal of the settlements was self-sufficiency, but many, including Sydney, were unable to produce an adequate range of foodstuffs because of poor soils, plant disease, and a variable climate, even after convicts were released to increase the number of farmers. As a result, many settlements depended on imports of food and other goods from Britain. The most successful agricultural areas were in coastal valleys, such as the Derwent valley in Tasmania, which produced wheat for Sydney. The main exports from the initial settlements were associated with the whaling trade, such as whale oil and seal pelts, with Hobart as the main base of the South Pacific whaling fleet.

Merinos and Mines in Australia

A momentous shift took place with the import of the first livestock to Australia, especially the Merino sheep, which thrived in central New South Wales and Tasmania. The first wool shipment to Britain took place in 1807, and high prices encouraged further expansion of the sheep industry. By 1831 more than 1 million kilograms (2.2 million pounds) of wool were being exported. By 1860 exports totaled 16 million kilograms (35 million pounds), and there were 21 million sheep in Australia. The demand was driven by the success of the British textile industry, including technological improvements associated with the Industrial Revolution; the demand tied the pastoral economy of Australia and New Zealand to the industrial core. The Australian Agriculture Company was established in 1824 and invested 1 million pounds to promote wool and other agricultural exports. Although the British government was initially reluctant to permit frontier settlements away from the closely supervised port communities, stockmen began to move inland, especially to the grasslands, such as the Bathurst Plain on the western slopes of the Great Dividing Range. As the European frontier expanded, it came into conflict with the Aborigines who defended their traditional lands and lost their lives in the process.

Wheat production was centered in southeastern Australia, especially on the red-brown soils around Adelaide. Wheat production expanded, with the use of fertilizers to compensate for phosphorus deficiencies, to occupy a belt from New South Wales through Victoria and Adelaide that came to be called the "fertile crescent." The potential area for wheat production was circumscribed by "Goyder's line" of rainfall that separated arable land from semidesert to the north (**Figure 12.11**).

The mid-1850s brought a great degree of self-government to Australia, with two-thirds of the legislatures elected by popular vote (and the rest appointed by the British) in the states of New South Wales, Victoria, South Australia, Queensland, and Tasmania. The boundaries between these states were mostly drawn as straight lines irrespective of physical features or indigenous land rights. The convicts' slavelike labor for colonists ended in eastern Australia by 1840 and in the west by 1868. The next major transformation of the Australian economy occurred with the discovery of gold in 1851, first in New South Wales, next in Victoria, then in Queensland and the Northern Territory, and finally and most dramatically in Western Australia in the 1890s. Gold diggers were drawn by the thousands from all over Australia, England, and China in a *gold rush* fueled by rumors of giant nuggets weighing 30 kilograms (100 pounds) or more. The town of Broken Hill at the border of New South Wales and South Australia became one of Australia's most important mining communities, producing lead, silver, and zinc that were exported to Europe to sustain industrial development. The Australian frontier based on livestock raising and mining development parallels that of western North America and has given rise to folklore similar to that of the U.S. Wild West. Stories of rural outlaws (called *bushrangers*) such as Ned Kelly, leader of a gang, captured the public imagination during the 1870s and continue to contribute to the image of the Australian frontier today.

Important to the development of the Australian economy was the construction of railroads, which radiated from ports to terminuses at livestock yards, grain elevators, and mines. This transportation pattern increased the importance of the port cities but hindered later national integration, especially when it became evident that three different rail gauges had been selected by different colonies.

The frost-free climate of the central coast of eastern Australia allowed for the development of tropical agriculture around Brisbane beginning in the 1860s, especially sugar plantations, using indentured laborers brought in from Vanuatu and the Solomon Islands. Cattle were also introduced into the warmer and drier regions of central and northern Australia after wells were drilled into the Great Artesian Basin in the 1880s, where they grazed more lightly than sheep on the sparse vegetation.

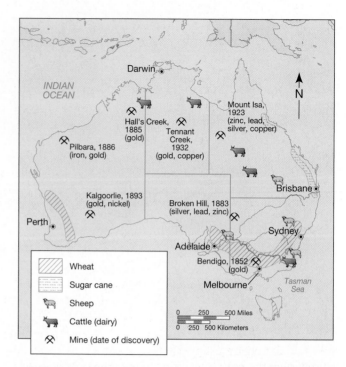

Figure 12.11 The development of livestock, wheat, and mining in Australia This map shows the main areas of settlement, agriculture, and mining development in nineteenth-century Australia, including the development of the "fertile crescent" of wheat production in southeastern Australia, the sugarcane region along the tropical east coast, and the important mining centers of Broken Hill and Kalgoorlie. (*Source:* Based partly on T. McKnight, *Oceania: The Geography of Australia, New Zealand and the Pacific Islands.* Englewood Cliffs, NJ: Prentice Hall, 1995, Figures 4.1 and 4.2; J. M. Powell, *An Historical Geography of Modern Australia: The Restive Fringe.* New York: Cambridge University Press, 1988, pp. 3–55; and G. M. Robinson, R. J. Loughran, and P. J. Tranter, *Australia and New Zealand: Economy, Society and Environment.* New York: Oxford University Press, 2000, Figure 6.5.)

New Zealand

The history of European settlement and economic development in New Zealand is closely tied to that of Australia. A sealing station was established on the South Island of New Zealand in 1792, but British sovereignty and the first official settler's colonies were not established until the 1840s, initially as part of New South Wales. Small whaling settlements had established trading relationships with the Maori, who through this and subsequent contacts, obtained access to firearms, were exposed to disease, and were brought into a capitalist economy. Missionaries had also established settlements in the early 1800s and contributed to the transformation of Maori culture. The British annexed New Zealand in 1840 through the **Treaty of Waitangi**, a pact with 40 Maori chiefs on the North Island. This treaty purported to protect Maori rights and land ownership if the Maori accepted the British monarch as their sovereign, granted a crown monopoly on land purchases, and became British subjects. At the last minute before the treaty was signed, land agents purchased large areas of land around the Cook Strait, often without identifying the true Maori owners. This land was held by the private New Zealand Association and included the sites of the cities of New Plymouth, Wellington, and Nelson.

Alarmed by European settlement, some Maori resisted the British and waged warfare for several years until suppressed in 1847. The introduction of sheep prompted further settler expansion in search of pastures and a renewal of hostilities with the Maori during the Maori wars of the 1860s. The discovery of gold in 1861 on the South Island in the Otago region made Dunedin the largest settlement in New Zealand (with a population of 60,000) within 10 years and fostered the development of wheat production on the Canterbury Plains.

The next stage of the integration of Australia and New Zealand into a global economy was driven by a technological innovation, the development of refrigerated shipping, after 1882. This allowed both economies to expand or shift from producing nonperishables such as wool, metals, and wheat to the more valuable export of meat and dairy products. New Zealand became Britain's "farm in the South Pacific," trading its high-quality agricultural products for imported manufactured goods from Britain. Trade was facilitated by the opening of the Suez Canal in 1867 and the Panama Canal in 1914, which reduced the time and cost of ocean transport to Europe. Australia and New Zealand, like Canada, became staple economies that were dependent on the export of natural resources.

Independence, Involvement in War, and Import Substitution

Independence was granted to Australia in 1901 when the Commonwealth of Australia was established with a federal structure governing the six states of Western Australia, South Australia, Queensland, New South Wales, Victoria, and Tasmania, and administering the Federal territories of the Northern Territory and (after 1908) the Australian Capital Territory around Canberra. New Zealand declined to join this new nation and chose dominion status as a self-governing colony of Britain in 1907. Both countries soon gained their own colonies in the Pacific. In 1901, Britain turned over Papua New Guinea to Australia and the Cook Islands to New Zealand. New Zealand was granted a mandate over Samoa in 1920 and jurisdiction over Tokelau in 1925.

Thousands of Australians and New Zealanders fought for the Allies in the First World War in the Australian and New Zealand Army Corps (Anzac), most notably at Gallipoli, Turkey, in 1915, when more than 33,000 of them died. The war contributed to the growth of industry, with the development of a steel and auto industry in Australia, and to increased agricultural exports at higher prices because of difficulties the war created in international trade.

The Second World War influenced Australia's geopolitical orientation. Although Australia fought in defense of Britain in Europe and North Africa, the rapid advance of the Japanese in Asia and the Pacific, including the capture of 15,000 Australians in Singapore and the bombing of the northern Australian city of Darwin, caused Australia to look to the United States. American and Australian troops fought together in New Guinea and the Pacific islands, forging bonds that endured in subsequent years and were formalized in the ANZUS security treaty between Australia, New Zealand, and the United States, which was signed in San Francisco in 1951.

As the Cold War deepened after 1950, Southeast Asia became the focus of U.S. concern about Communist expansion from China. As the United States became increasingly involved in Indochina (see Chapter 10, p. 491), both Australia and New Zealand sent troops to Vietnam.

From the 1920s, policies to substitute expensive imports with domestic production resulted in heavy subsidies of manufacturing and high tariffs on imported goods, especially in Australia. The goal was to create new jobs and diversify the economy and reduce the sensitivity to global demand for wool and minerals. As in other regions, these *import-substitution* policies had mixed success. Although Australia and New Zealand had middle-class populations with a demand for manufactured goods, the overall market was small, and labor costs were high as a result of a strong tradition of labor union activism. The new industries were often inefficient because they were protected from competition from the world market. New Zealand had a particularly high level of government involvement in the economy, with state-run marketing boards controlling the export of commodities, such as wool, meat, dairy products, and fruit, and government ownership of banking, telecommunications, energy, rail, steel, and forest enterprises.

Reorientation to Asia and the Impacts of Global Economic Restructuring

Although Australia protected manufacturing and subsidized agriculture, there were few barriers to foreign investment. Many sectors had high levels of foreign ownership, including minerals and land, especially by British firms. During the 1970s, investment patterns began to change, with Asian, especially Japanese, capital starting to flow into Australia. Another major change occurred when Britain decided to enter the European Community in 1973 and was forced to end preferential trade relationships with **British Commonwealth** nations, including Australia and New Zealand. New Zealand was particularly hard hit by the loss of guaranteed markets, but the shock provided an impetus to seek new markets in Asia.

Both Australia and New Zealand decided to reduce government intervention and regulation of the economy in the 1980s. The Australian government deregulated banking, reduced subsidies to industry and agriculture, and sold off public-sector energy industries beginning in about 1983. The move to the neoliberal policies of free trade and reduced government was even more dramatic in New Zealand. There, sweeping policy reversals eliminated agricultural subsidies, removed trade tariffs, reduced welfare spending, and privatized government-owned enterprises, including airlines, postal services, and forests. Although these policies did succeed in reducing the national debt, they also increased economic inequality and unemployment.

The most rapidly growing sector during the 1980s and 1990s was services, with employment in this sector increasing from 58 percent in 1980 to 71 percent in 2000 in Australia and from 58 percent to 66 percent in New Zealand (**Figure 12.12**). Finance, tourism, and business services expanded the most and were associated with an increase in female employment and considerable foreign investment. The region has become a major international tourist destination with more than 5 million visitors in 1999, more than half of them from Asia, especially Japan. Asia has also invested in hotels and other tourist facilities in the region.

The fate of agriculture in Australia and New Zealand is unclear, as competition from Latin America and Asia, the high cost of inputs, the loss of subsidies, and the changing structure of demand create a new geography of agricultural trade (see Geography Matters: The New Geography of Food and Agriculture, p. 588).

Colonization of the Pacific Islands

The Pacific islands were of less initial interest to the European explorers, and they were drawn more slowly into the world economy than were Australia and New Zealand. The Spanish and Portuguese explorers encountered a few islands on their ocean voyages. The Portuguese explorer Fernando de Magalhães (Magellan) landed on Guam in 1519, and the Spanish came upon the Caroline, Solomon, and the Cook islands. The Dutch voyaged from their base in Indonesia to the Tuamotu archipelago and Tonga. In most cases visits were brief, with a small amount of bartering for food. As was the case in other regions of the world, the explorers brought European diseases to local populations who had no resistance to them, so the most serious impact was mortality from diseases. Guam, Palau, parts of the Federated States of Micronesia, and the Mariana Islands became Spanish colonies.

As Britain and France rose to power in Europe in the eighteenth century, a series of explorations set out for the Pacific. British explorer Samuel Wallis and Luis Antoine de Bougainville of France were made welcome by the people of Tahiti in the 1760s, and their reports of friendly people and abundance cultivated the myth of a tropical paradise. Ambitious chiefs, especially the Pomare family, were happy to trade with the Europeans, welcoming Captain James Cook on his Pacific explorations (**Figure 12.13**) as well as subsequent visits from Protestant missionaries as a way to enhance their own power and control.

One of the most well-known South Pacific voyages was that of the H.M.S. *Bounty* in 1789. Captain William Bligh went to Tahiti to collect breadfruit trees for transport to the Caribbean, where they could be grown as a basic food source for slaves working on plantations. After visiting Tahiti, the *Bounty* began its return voyage, but near Tonga there was a mutiny, led by Fletcher Christian and sailors who wanted to return to the idyllic life of Tahiti. Bligh and some loyal crewmembers were set adrift in a small boat and somehow made it 5800 kilometers (3600 miles) southwest to Timor. Some of the mutineers from the *Bounty* eventually settled, with

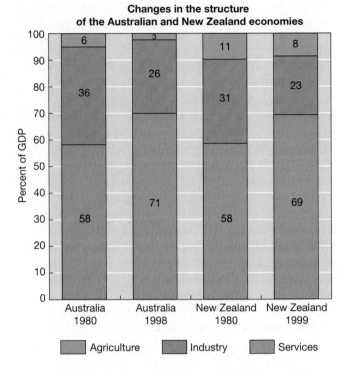

Changes in the structure of the Australian and New Zealand economies

Figure 12.12 **The structure of the Australian and New Zealand economies** Agriculture is a more important component of the New Zealand economy than in Australia, whereas industry is more important to Australia than to New Zealand. Between 1980 and the late 1990s the share of agriculture and industry in the Australian and New Zealand economies declined and the importance of services increased considerably. (*Source:* Data from the World Bank Group, April 2001, available at http://www.worldbank.org/data/countrydata/countrydata.html)

several Tahitian women, on the isolated Pitcairn Island, where their descendants live today.

At the beginning of the nineteenth century, the Pacific islands were brought into contact with missionaries, whalers, and traders. The London Missionary Society was very active in the Pacific, seeking conversions in Tahiti, Samoa, Tuamotu, Tuvalu, and the Cook Islands. The Methodists focused on Tonga and Fiji. The missionaries often worked through native chiefs who were advised to alter local laws and traditions to conform to European principles, sometimes provoking local rebellions. Missionary activity altered traditional social ties, beliefs, and political structures. Hundreds of whaling ships called regularly at islands such as Tahiti, Fiji, and Samoa for supplies and maintenance. The discovery of sandalwood, a valued aromatic wood, attracted traders to Fiji, Hawaii, and New Caledonia, which were ruthless in their treatment of local people.

Coconut, the staple product of the Pacific, became part of European trade from about 1840 in the form of copra, dried coconut meat used to make coconut oil for soaps and food. Its importance was such that some Europeans sought to establish plantations on islands such as Fiji and Samoa. A scarcity of cotton during the U.S. Civil War prompted the establishment of cotton plantations on Fiji in the 1860s and increased the need for laborers for the cotton plantations in Queens-

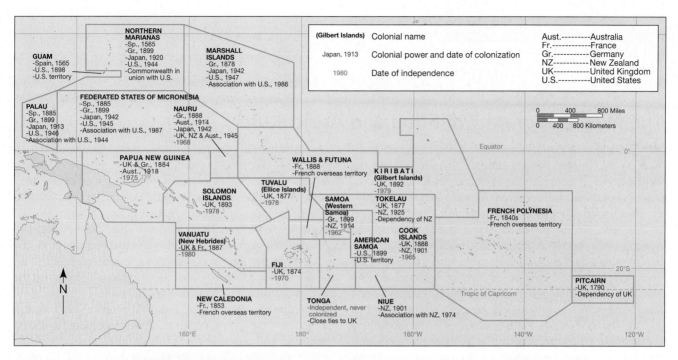

Figure 12.13   The colonization and independence of the Pacific Many Pacific islands are now independent from their former colonial rulers. This map shows the pattern and dates of colonial control, dates of independence of Pacific island countries, and current affiliations of territories. Note that only Tonga resisted colonization and that some countries, such as Nauru and Palau, were handed from one European power to another. (*Source:* Data from Australian Broadcasting Company, available at http://www.abc.net.au/ra/carvingout/maps/statistics.htm)

land, Australia. Beginning in the 1840s, Pacific islanders had been kidnapped and enslaved in a process called *blackbirding* that brought thousands of laborers, collectively called *kanakas* (because many of them were of Kanak origin from the islands now known as Vanuatu), to Australian cotton and sugar plantations. This practice continued until 1904.

Formal colonial rule was established in the Pacific by France, using the need to protect missionaries and other French nationals as a pretext for the takeover, beginning with the Marquesas in 1842 and Tahiti and the other Society Islands in 1843. France also seized New Caledonia in 1853 and shipped convicts, ranchers, and coffee farmers to the island. Britain, Germany, and the United States then began to compete to control islands where they had trade interests or significant expatriate populations. By 1900, Germany had acquired Western Samoa, Northern New Guinea, Nauru, and most of the islands in Micronesia. Britain took control of Fiji, southern New Guinea, and the Solomon Islands. Tonga was able to maintain its independence through an alliance with Britain. The United States held Hawaii, Eastern Samoa, and Guam, the latter taken from Spain together with the Philippines after the Spanish-American War in 1898. German colonies in the Pacific were transferred to Japan, New Zealand, Britain, and Australia after the First World War.

The islands were governed according to the different colonial styles of European nations modified to local conditions. Britain ruled through governors who incorporated native leadership into their administrations in a form of indirect rule (see Chapter 6, p. 276). The Germans administered their Pacific colonies through commercial companies, and the French practiced direct rule and assimilation into French culture and institutions.

The colonial powers restructured the economies and peoples of the Pacific islands in ways that left enduring legacies. The British brought large numbers of contract workers from India to work on plantations in Fiji, creating a divided society of Asians and Pacific islanders that produces significant political tensions even today. On the islands of Nauru and Banaba, imported labor was used by the British to mine phosphate ore for Australian agriculture.

The imposition of colonial rule did meet with resistance in the form of armed uprisings, alternative trading networks, political movements, and defiant behavior. For example, in Papua New Guinea, locals mocked their Australian rulers by shaking in their presence and rocking in chairs on verandas, sipping drinks, and discussing the laziness of the white men in a subtle parody of colonial attitudes.

The Second World War and Independence in the Pacific

The Second World War marked a critical turning point in the history of the Pacific, with thousands of foreign soldiers fighting across the islands and constructing military bases. In the process, many islanders lost their lives. The most significant impacts were in the western Pacific, where the Japanese advanced from their colonies in Micronesia (such as Palau) to Guam, New Guinea, and the Solomon Islands and then attacked the U.S. base at Pearl Harbor in Hawaii in 1941

Geography Matters

The New Geography of Food and Agriculture

Geographers who study agriculture and food are using several new approaches and concepts to understanding the way in which the restructuring of international trade, the activities of multinational corporations, and rapid shifts in government policies are affecting agriculture and the rural landscape in countries such as New Zealand and Australia. Agriculture has been transformed through horizontal integration in which smaller enterprises are merged to create larger units (for example, when adjacent farms are consolidated into one large landholding, resulting in the disappearance of small family-run farms) and through vertical integration, in which a single firm takes control of several stages in the production process (when a company owns the fertilizer and seed companies as well as the food-processing plant and supermarkets). For example, the international corporation ConAgra owns grain companies, feedlots, meat-processing, and wholesale distribution in Australia. As such, ConAgra is an example of an agribusiness that organizes food production from the manufacturing of chemical inputs and the genetic manipulation of animal breeds or crop varieties to the processing, retailing, and consumption of the agricultural product.

Geographers such as Richard Le Heron and Guy Robinson have written extensively about how New Zealand agriculture has changed in response to the restructuring of the global food system. They document how New Zealand's agricultural system evolved during the nineteenth century into a food regime with an orientation to exports of wool and lamb based on a pastoral landscape and a guaranteed market in the core economy of the United Kingdom. After the Second World War, a second regime developed that included dairy cows on small farms and processing of products such as butter for export using refrigerated shipping. By the mid-twentieth century, the New Zealand government was heavily involved in the agricultural system through "marketing boards" that mediated farmers' relationships with international markets through quality controls, price supports, and marketing. In the 1970s, the shock of the oil crisis (increasing the cost of agricultural inputs) and the loss of the imperial preference market when Britain joined the European Community resulted in further state support for producers with price supports, incentives, and subsidies for inputs, such as fertilizers providing more than a third of farm revenues. Even these institutional supports could not fully buffer farmers against the increasing cost of inputs and loss of markets for the staples of wool, meat, and dairy, and some farmers began to diversify into nontraditional exports such as venison, produced on deer farms, and fruit such as kiwi and Asian pears, responding to a new global food regime of specialty foods and the export of fruit and vegetables (**Figure 1**).

A dramatic change in agricultural policies in 1984 abruptly removed most price supports, trade protections, and farm subsidies and required farms to pay for extension services, water, and quality inspections. Farm incomes fell by up to half, debt increased, 10 percent of farms were sold, herd sizes were significantly reduced, and 10,000 farmers protested in front of parliament. New Zealand agriculture was thrown into a global free market and the full impact of what has been called the "international farm crisis," while most other developed countries, including the United States, Canada, and Europe maintained considerable state regulation and support for their agricultural systems. Although New Zealand farmers coped by adjusting their herd sizes and changing their crop mix, some farmers went out of business and their properties were horizontally integrated into larger farms. The landscapes of important agricultural regions such as the Hawkes Bay region of the North Island began to change as farms switched into fruit production, often associated with increases in pesticide use (**Figure 2**). But New Zealand was also one of the first countries to adopt certification for organic agricultural products, and there is a thriving domestic market for sustainably grown foods. Transnational agribusiness firms such as H. J. Heinz purchased New Zealand agricultural processing enterprises with the goal of supplying growing Asian markets. Michael Moore, the New Zealander who had spearheaded his country's plunge into the free market as minister of trade, became the head of the World Trade Organization, charged with reducing barriers to trade worldwide.

(**Figure 12.14**). Three years of intense and bitter warfare on land, air, and sea included many famous battles such as those of Guadalcanal, Tarawa, and Saipan in which the Allies, especially the United States, retook the islands from Japan. The damage from bombing was extensive, and exposure to Western values and goods also transformed the islands. For example, canned foods such as Spam™ and corned beef, as well as U.S.-style music, clothing, and sports, became popular.

The strategic and economic significance of the Pacific islands slowed their decolonization after the war. The United States was determined to maintain military bases and control of the Pacific, especially in response to Cold War competition with China and the Soviet Union. High prices for primary products such as copra and sugar brought profits to colonial powers. Self-government began with elected governments and small independence movements. Western Samoa became in-

ward the eastern Pacific. The missionaries were successful in converting most of the Pacific islanders to Christianity, with a range of Protestant denominations and Catholicism prevalent in French Polynesia. Hinduism is important in Fiji among the Asian Indian population.

Traditional beliefs are still important in more remote islands and regions, especially in Papua New Guinea, and some communities have fused traditional and Christian practices by combining harvest rituals with the celebration of Christmas, for example. Melanesia is associated with a set of religious movements that have been called **cargo cults,** where the dawn of a coming new age was associated with the arrival of goods brought by spiritual beings or foreigners. In some cases, symbolic ship piers or airstrips for the arrival of the cargo of goods were constructed. Anthropologists interpret these cults in several ways, including protests against European colonialism and oppression or as demands for a return to the tradition of reciprocal gift-giving. More recently, critical perspectives suggest that the idea of a cargo cult overgeneralizes many different kinds of movement and reflects a Western obsession with commodities and lack of understanding of local peoples.

Culture and Society

The traditional cultures of Australia, New Zealand, and the Pacific islands have been disrupted by contact with the rest of the world and have sometimes become transformed into societies oriented to cultures based on the demands of tourism and the need to strengthen political identity. In Australia, Aboriginal cultures, based on strong spiritual ties to land, were marginalized and homogenized when Aborigines were removed from their ancestral lands and resettled on reserves. Aboriginal art, often based on rock art designs in dotted forms, silhouettes, and so called X-ray styles, has become very popular in contemporary markets (**Figure 12.16a**), and native dances and songs, such as those that are part of the social gathering known as *corroboree*, are performed for tourists. Aboriginal symbols

have been appropriated for major events such as the Sydney 2000 Olympic Games. Maori tradition is celebrated as integral to New Zealand's official bicultural identity, including the welcome ceremonies that include the *hongi* (pressing noses together) and the *haka* war dance now performed at international sporting events. Maori architecture includes distinctive carved and decorated meeting houses called *whare*. Artistic expressions include intricate carvings such as those found on war canoes and decorative masks and tattoos (*moko*), which are also found on other Pacific islands (**Figure 12.16b**).

Overall, society in Australia and New Zealand is still influenced by British legacies, including the significance of sports such as cricket and rugby, where the national teams, such as the New Zealand All Blacks rugby team, have gained international renown and have enthusiastic, if not fanatic, local support. Australia and New Zealand have also produced a series of award-winning films and novels, and two of the world's most popular opera singers—Australian Joan Sutherland and New Zealander and part-Maori Kiri Te Kanawa. The echoes of the colonial relationship with Britain provoke considerable ambivalence, and there have been strong attempts to establish distinct identities by embracing indigenous traditions, new immigrant cultures such as those from Asia, or the particular livelihoods and landscapes of the outback frontier or the surfing beach (**Figure 12.17**).

Gender roles in Oceania are influenced by stereotypes, but roles are as rapidly changing and complex as in any other world region. In Australia, the image of the frontier rancher or miner is associated with heavy drinking, gambling, male camaraderie, and a tough, laconic attitude epitomized by movie characters such as those in *Crocodile Dundee* or the lone outlaw portrayed in *Mad Max*. But other films, such as *Priscilla, Queen of the Desert*, celebrate an alternative image of Australia as a center for gay and other transsexual identities. Sydney's Mardi Gras parade has become a celebration of gay culture and a major tourist attraction. Many women in Australia and New Zealand have shifted from a role as traditional housewives into a multitude of

Figure 12.16 Aborigine and Maori culture (a) Bark paintings by the Australian Aboriginal artists Bunangur, Dhartaugull, Malangi, and Milpurrurru use traditional symbols. (b) Maori and other Polynesian warriors, some with facial *moko* (tattoos), perform a traditional *haka* dance in Gisborne, New Zealand, at the dawn of the new millennium.

(a)

(b)

Figure 12.17 Sports in Oceania Australia is known for its beach culture, especially surfing, such as here on Bondi Beach in Sydney.

careers and to senior political positions supported by a strong feminist movement.

It is nearly impossible to generalize about the cultures of the Pacific islands, which are as varied as the many languages. Some of the most distinctive social and material forms include the strict separation of the male and female in much of New Guinea and Melanesia, the tradition of ritual warfare and reciprocal gift exchange, the importance of local leaders, or "big men," and close links with kin and extended families. Pigs are considered an important measure of wealth on many islands, including New Guinea, and a traditional plant, *kava,* is consumed as a recreational and ritual relaxant on many islands. Polynesian cultures have a strong orientation to the ocean, and natives have been stereotyped by some explorers, anthropologists, and tourists as sexually promiscuous and living an easy life of tropical abundance. Well-known anthropologist Margaret Mead explored the complexity of Pacific cultures. In her book *Coming of Age in Samoa* Mead described a relatively unproblematic and sexually liberal process of adolescence and analyzed the elaborate nature of sex roles in different cultures in New Guinea. Her work has been criticized by some as taking a romantic and deterministic view of Pacific island culture and for an uncritical acceptance of stories told to her by informants. Contemporary cultures on the Pacific islands reflect the tensions between the maintenance and revival of traditional cultures, their selective construction for the tourist industry, and the widespread penetration of global culture, especially formal education, television, and processed foods.

Regional Change and Interdependence

Oceania has seen some important regional changes, restructuring, and reorientation in recent years, including shifts in alignment from Europe to North America and Asia discussed earlier and the challenges of coping with smaller markets and geographic isolation within a global economy. Within the region the stability of some independent democracies and de-

pendencies in the Pacific have been threatened by internal tensions, while political and economic integration has been sought through regional cooperation agreements. Glaring inequalities within generally wealthy countries such as Australia and New Zealand have highlighted the fate of indigenous groups, while at the same time the countries have embraced multicultural identities. Oceania is particularly vulnerable to human activities, such as the use of fossil fuels, marine overharvesting, and pollution, which are altering global climate, sea levels, and fishery resources. The region also encapsulates critical examples of the impacts of international tourism and of the risks of nuclear energy (see Geographies of Indulgence, Desire, and Addiction: Uranium in Oceania, p. 596).

Political Stability

The most serious recent political conflicts in Oceania have involved encounters between ethnic groups in Fiji and demonstrations by independence or secessionist movements in New Caledonia and Papua New Guinea. But the region has also seen renewed attempts at regional integration and peacekeeping through a variety of regional and subregional cooperation agreements.

The conflict in Fiji is a legacy of British colonial policies that brought Asian Indians as indentured workers for sugar plantations from 1879 to 1920. By the 1960s Indo-Fijians almost outnumbered the ethnic Fijian population, dominating commerce and urban life, and maintaining a separate existence with little intermarriage and continued cultural and religious differences. Although the indigenous Fijians took over government at independence in 1970, the victory of Indo-Fijian backed parties in elections in 1987 resulted in military coups to ensure indigenous political dominance. After many attempts at a compromise constitution and economic decline associated with drought and reduced tourism, new elections in 1999 resulted in another Indo-Fijian victory. Only a year later, a group of ethnic Fijian nationalists led by George Speight took the prime minister and parliament hostage, prompting riots and looting of Indo-Fijian shops. They claimed that the Indo-Fijian government was threatening the land rights of ethnic Fijians and placing Indo-Fijians in all major government positions. After weeks of negotiations, with the involvement of the Australians and others seeking stability in the Pacific, the hostages were released and the rebels were arrested by the military. Many Indo-Fijians fled the country, despite promises of new elections during the summer of 2001.

In the nickel-rich islands of New Caledonia, the indigenous Melanesian population, known as *Kanaks*, has been pressing for independence for years but has been outvoted by those of French descent (called *demis*), who prefer to remain part of France. This has led to riots and disruption by dissatisfied militants seeking an independent and indigenous nation. In the Solomon Islands, residents of Bougainville are trying to secede from Papua New Guinea, claiming ethnic affiliation with the independent Solomon Islands and complaining that they do not receive a fair share of the profits from mining. Within the Solomon Islands there are conflicts between ethnic groups over land, such as those between long-time residents of Guadalcanal and immigrants from the neighboring island of Malaita.

Regional cooperation agreements include the South Pacific Commission, founded in 1947, which focuses on social and economic development and includes 21 island nations and territories, Australia, New Zealand, the United States, France, and the United Kingdom. The **South Pacific Forum,** established in 1971, excludes France, the United Kingdom, the United States, and their colonies and promotes discussion and cooperation on trade, fisheries, and tourism between all of the independent and self-governing states of Oceania. It has supported maritime territorial rights and a nuclear-free Pacific as well as the independence goals of French Polynesia and New Caledonia. But there are also dozens of nongovernmental organizations and intergovernmental agencies that operate region-wide, especially among the smaller Pacific islands. For example, the University of the South Pacific fosters higher education across 12 countries through distance education and three main campuses in Fiji, Samoa, and Vanuatu.

Australia and New Zealand are members of larger economic and political alliances such as the Asia-Pacific Economic Cooperation group (APEC), which also includes New Guinea, and which focuses on improving transportation links and liberalizing regional trade. Both Australia and New Zealand have been able to take advantage of APEC to increase exports to Asia, especially to Japan. In attempts to foster regional markets as global trade liberalizes and restructures around them, Australia and New Zealand created the **Closer Economic Relations** (CER) agreement in 1983, which built upon an earlier New Zealand-Australia Free Trade Agreement (NZAFTA) and set out to remove all tariffs and restrictions on trade between the two countries. The increased trade between the two countries has been especially beneficial to New Zealand, with its small domestic market, which has doubled its exports to Australia. Many manufacturing firms now operate in both countries.

Poverty and Inequality

Although Oceania has generally higher incomes and better living conditions than many other world regions, there are significant differences between and within countries of the region. Australia and New Zealand are distinctive for their very high average incomes as expressed by the per capita annual gross domestic product (GDP) at more than $22,000 for Australia and $17,000 in New Zealand in the late 1990s. Some Pacific islands have average GDP near or above $10,000 per person as a result of their associations with the United States (American Samoa, Guam, and Palau), France (French Polynesia, New Caledonia), or as a result of mineral wealth (Nauru). Others, especially Kiribati, Tuvalu, and Vanuatu, have incomes below $2000 per capita per year. Although there is little published information on the distribution of incomes within most of the smaller countries, and although inequality is apparently less than in many other world regions, there is persistent poverty throughout the region, including Australia and New Zealand, where the Aborigines and Maori populations are particularly disadvantaged. Although Australia and New Zealand were for many years reputed to have strong welfare systems and equitable societies, at least for the nonindigenous populations, income inequality has increased in the last two decades, and the government has reduced or privatized social services, especially in New Zealand. Larger populations of single parents, the elderly, refugees, and workers in low-paid service sector jobs are also diminishing the overall ranking of Australia and New Zealand as places where everyone can make a good living.

Monetary measures such as GDP are of limited use where many people are living in economies based on exchanges and barter or on subsistence. The concept of **subsistence affluence** has been used to describe Pacific island societies where although monetary incomes may be low, local resources such as coconut and fish provide a reasonable diet, and extended family and reciprocal support prevent serious deprivation. Adequate diets and relatively effective health and education systems contribute to comparatively high life expectancies and literacy and low infant mortality throughout the Pacific (**Figure 12.18**). Life expectancies range from 62 years in Kiribati and Vanuatu to more than 75 years in American Samoa, Australia, French Polynesia, Guam, the Northern Marianas, and New Zealand. Women live 2 to 4 years longer than men, and literacy is above 90 percent for both men and women in much of the region. Papua New Guinea, Kiribati, and Vanuatu have higher infant mortality and lower literacy than other parts of Oceania.

Some of the more serious social problems in the region include alcohol abuse and high levels of domestic violence. Papua New Guinea has some of the highest levels of violence against women in the world.

Figure 12.18   A Samoan family's possessions The Lagavale family sits in front of their open-structure house, or *fale*. Their possessions include mats made from the Pandanus and coconut palms, an outrigger canoe, and cows, pigs, and chickens, as well as a radio and sports equipment. Being an average Samoan family, the Lagavale family's per capita income is $930 per year, of which half is spent on food.

Uranium in Oceania

L ike oil in the Middle East, the extraction and use of uranium links Oceania to the global hunger for cheap energy and to geopolitical conflicts beyond the region. Uranium is a radioactive element that can be split in a process of nuclear fission to produce a chain reaction that releases large amounts of energy. Controlled reactions can be used to generate electricity in nuclear power plants, whereas uncontrolled reactions can be used in atomic bombs that release enormous amounts of thermal energy and radioactivity. Uranium became a desirable commodity after the Second World War demonstrated the power of atomic weapons at Hiroshima and Nagasaki and the potential of nuclear-powered electrical generation became apparent.

This interest in uranium had important impacts on several regions of Oceania. The United States, Britain, and France all joined the Cold War arms race and the effort to develop even more powerful weapons based on uranium and related elements such as plutonium. They chose to test many of the atomic weapons in the Pacific, with devastating implications for local residents and environments. The United States tested its bombs in the Marshall Islands, relocating the residents of Bikini and Enewetak atolls to other islands, and exploding several different types of nuclear weapons between 1946 and 1958 (**Figure 1**). Although the prevailing winds were supposed to carry the radioactive fallout from bomb testing away from inhabited islands, in 1954 radioactive ash dusted the island of Rongelap and its almost 100 residents, including several relocated from Bikini. Radioactive exposure can have serious short- and long-term effects, including acute poisoning, leukemia, and birth defects, so the U.S. government evacuated the residents of Rongelap at short notice with little information about the hazard they had been exposed to or warning that they would not be able to return to their homelands. Years later, in 1968, the residents of Rongelap and Bikini were told it was safe to return, but those on Bikini later had to be re-evacuated when scientists discovered that dangerous levels of radioactivity persisted in food gathered on the islands. Although the United States has monitored the health of the islanders and established a $90-million trust fund, many residents of the islands are angry and resentful about the experiments that disrupted their lives.

France conducted more than 150 bomb tests on the tiny atolls of Moruroa and Fangataufa in French Polynesia begin-

Figure 1   Bikini Atoll The atomic bomb test at Bikini Atoll in the Marshall Islands on July 25, 1946. Fallout from this and subsequent tests posed serious health risks to Pacific islanders and resulted in the evacuation of several atolls.

ning in 1966. The first bombs showered the surrounding regions with radioactivity, reaching as far as Samoa and Tonga hundreds of miles to the west. Opposition from other Pacific islands, including New Zealand and Australia, culminated in boycotts of French products, including wine and cheese, during the 1970s. France moved to underground testing and refused to release information about accidents and monitoring of radioactive pollution or health in French Polynesia. While locals use the bomb tests as a reason to seek independence from France, international activists have tried to stop the French bomb tests. In 1985 the environmental group Greenpeace planned to protest tests by sailing their ship *Rainbow Warrior* to Moruroa, but French intelligence agents scuttled the ship while it was moored in the harbor of Auckland, New Zealand (**Figure 2**).

The international scandal prompted New Zealand to take a strong stand against nuclear proliferation, banning all nuclear-powered and nuclear-armed vessels from its harbors, breaking off diplomatic relations with France, and taking a leadership role in the antinuclear movement in the Pacific. This created a long-term strain on relations between New Zealand and the United States because U.S. military vessels, which will not admit nuclear capability, were therefore banned from New Zealand. However, New Zealand's actions contributed to the announcement by France in 1996

Indigenous Issues and Multiculturalism

Some of the most passionately debated issues in contemporary Australia and New Zealand relate to the rights of their indigenous peoples and the creation of a multicultural society and national identity. The countries share a history of British

colonialism and dispossession of indigenous lands and cultures but have distinctly different contexts and contemporary approaches to intercultural relationships.

In New Zealand, Maori rights are framed by the 1840 Treaty of Waitangi. Although the Maori interpreted the treaty as guaranteeing their land and rights, the century that fol-

Figure 2 *Rainbow Warrior* The Greenpeace ship *Rainbow Warrior* scuttled by French agents in the harbor of Auckland, New Zealand, because of its role in protesting nuclear testing in French Polynesia. New Zealand has declared its ports as nuclear-free zones and has joined Pacific island nations in strongly protesting nuclear activities in the Oceania region.

Figure 3 **The Ranger uranium mine** The Ranger uranium mine is located in Kakadu National Park in Australia's Northern Territory. The area is sacred to the Aborigine population and has striking landscapes and ecosystems. Moves to expand mining activity have resulted in protests by Aborigine groups and environmental activists.

that it would end nuclear testing after riots in Tahiti and declines in tourism.

The British tested their bombs on Christmas Atoll, now within the nation of Kiribati, and also at several locations in Australia, including the Monte Bello islands off the coast of Western Australia and Maralinga in South Australia. Critics now claim that neither the Australian government nor its people were made fully aware of the risks of these tests and that the local Aborigines were heavily exposed to radiation and continue to wander into the contaminated test sites.

The consumption of uranium has also threatened Australian Aborigines through the mining of uranium on or near their lands in northern Australia. The Ranger mine commenced operations in the Northern Territory in 1980 within the boundary of Kakadu National Park, a region of great natural beauty listed as a World Heritage site for both natural and cultural values (**Figure 3**). The mine has produced more than 16 million tonnes (35 billion pounds) of radioactive mine waste and has created serious water-pollution problems in the area. Australia produces 27 percent of the world's uranium, exported to fuel nuclear power stations in the United States, Japan, Europe, Canada, and South Korea, even though Aus-

tralia itself does not produce electricity from uranium. Great controversy has arisen over proposals to open another mine at Jabiluka on land belonging to the Mirrar Aboriginal Group; activists have blockaded the mine road, and protests have occurred around Australia.

Thus, uranium links the countries of Oceania to the global geography of energy consumption and to the desire for geopolitical supremacy in a multitude of ways. Though nuclear testing has been halted in the Pacific, radioactivity persists for thousands of years and will continue to pose risks to people and ecosystems. Although uranium prices are currently low because few new nuclear power stations are under construction in the aftermath of the Chernobyl accident (see Chapter 4, p. 180), many countries are reconsidering the nuclear energy options and seeking to purchase uranium as other supplies become scarce or create environmental problems.

lowed saw large-scale dispossession of Maori land and disrespect for Maori culture. Maori landholdings were reduced from 27 million hectares (100,000 square miles), to only 1.3 million hectares (5000 square miles) or 3 percent of the total area of New Zealand. A series of protests, court cases, and reawakening to Maori tradition led in 1975 to the establishment of the

Waitangi tribunal, which eventually reinterpreted the Treaty of Waitangi as more favorable to the Maori and investigated a series of Maori land and fishery claims. The Maori were established as *tangata whenua* (the "people of the land") and Maori was recognized as an official language of New Zealand (**Figure 12.19**). Some land claims were settled or compensated

Figure 12.19 Maoris demand that New Zealand honor the Treaty of Waitangi Maori activists protest during a visit of Britain's Queen Elizabeth in 1989, asking the New Zealand government to honor the Treaty of Waitangi.

through money or grants of government land but others are too large or threatening to private interests to be easily recognized. A bicultural Maori and *Pakeha* (a Maori term for whites) society has been adopted rather than a multicultural policy that would encompass other immigrant groups, such as Pacific islanders and Asians, or recognize the differences within Maori and other cultures.

New Zealand's recognition of Maori rights and culture as part of a national identity has not solved some of the deeper problems of racism toward the Maori or of their poverty and alienation. Maori unemployment is twice that of white residents; average incomes, home ownership, and educational levels are less than half; and welfare dependence is much higher.

Australian Aborigines, in contrast, have had no recourse to a treaty to assert their rights. The European colonists saw the indigenous peoples as primitive and their land as *terra nullius*, owned by no one, and therefore freely available to settlers. Only in the 1930s were reserves set aside for Aboriginal populations but mostly in very marginal environments with little autonomy or access to services. In many ways the Aboriginal population had been made "invisible," not counted in the census or allowed to vote until the 1960s and stereotyped as a primitive and homogeneous nomadic culture when in fact there were many different cultures. One of the most misguided programs set out to assimilate the Aboriginal population by forcibly removing their children from their families and communities and placing them in white foster homes and institutions from 1928 to 1964. This **stolen generation** of as many as 100,000 Aboriginal children was given voice and officially acknowledged by the Australian government in a national inquiry in the 1990s:

> Our life pattern was created by the government policies and are forever with me, as though an invisible anchor around my neck. The moments that should be shared and rejoiced by a family unit, for my brother and mum

and I are forever lost. The stolen years that are worth more than any treasure are irrecoverable.[1]

Indigenous Australians are disadvantaged on almost all economic and social indicators, with unemployment at four times the national average, much lower average incomes, housing quality, and educational levels, as well as higher levels of suicide, substance abuse, disease, and violence.

Growing awareness and regret at the treatment of Aborigines has led to efforts at apology and reconciliation by many white Australians. In 2000 more than 300,000 people marched across the Sydney Harbor Bridge in a walk of reconciliation, and more than one million Australians signed "Sorry Books" that stated, "we stole your land, stole your children, stole your lives. Sorry," as a way of apologizing for the treatment of Aboriginal peoples (**Figure 12.20**). The Aboriginal Land Rights Act of 1976 gave Aborigines title to almost 20 percent of the Northern Territory and opened government land to claims through regional land councils. The states of South Australia and Western Australia have also handed over land to Aboriginal ownership or leases. The more contentious claims surround land with valuable mineral resources, especially uranium, or where development threatens spiritual sites.

Several landmark lawsuits have also addressed native land claims, including the case brought by Eddie Mabo, an indigenous activist from the Torres Strait Islands off the north coast of Australia, who claimed that his people (the Meriam) had never surrendered their land rights. In 1992 the High Court decided in his favor, effectively overruling the doctrine of *terra nullius,* which assumed that Australian land was unoccupied when the British arrived, and catalyzing Aboriginal claims for land and compensation. Claims have been filed for land in the city of Brisbane and for one-quarter of the state of New South Wales. A second decision in 1996—the Wik judgment concerning northern Queensland—ruled that Aboriginal rights to traditional lands could coexist with grazing leases. The 1993 Native Title Act, which passed with a narrow national majority, recognized native title and established processes for making claims based on customary use of land rather than legal title. Aboriginal control now extends over about 15 percent of Australia, with claims to at least another 20 percent.

Australian Aborigines still have much less power and recognition than the Maori of New Zealand, and this emerges in Australia's adoption of a multicultural rather than bicultural policy of national integration. Multiculturalism emerged in the 1970s and set out to embrace the distinctive cultures of many different ethnic and immigrant groups. The National Agenda for a Multicultural Australia (published in 1989) set out to promote tolerance and cultural rights and to reduce discrimination, while maintaining English as the official language and avoiding special treatment for any one group. In contrast to New Zealand, where Maori language and culture is an essential component of the new national bicultural identity, Abo-

---

[1]Human Rights and Equal Opportunity Commission, *Bringing Them Home: Report of the National Inquiry into the Separation of Aboriginal and Torres Strait Islander Children from Their Families.* Sydney: Australian Government Printing Office, April 1997, Confidential Submission #338, Victoria, p. 3.

(a)

Figure 12.20   Aboriginal issues in Australia (a) Millions of Australians have apologized to the Aborigines for discrimination and the damage to the "stolen generation" of Aboriginal children who were taken forcibly from their homes. (b) This map shows the current status of Aboriginal lands in Australia, with the majority in the Northern Territory. Key legal settlements such as the Mabo and Wik judgments have prompted Aborigine land claims to more than 35 percent of the country [*Source:* (b) Adapted from G. M. Robinson, R. J. Loughran, and P. J. Tranter, *Australia and New Zealand: Economy, Society and Environment.* New York: Oxford University Press, 2000, Figure 5.5.]

rigines are just one of many ethnic groups in a multicultural society, and some have resented this status. There has also been considerable opposition to immigration, Aboriginal rights, and multiculturalism in the last decade. The government of Prime Minister John Howard declined to apologize formally for past offenses against Aborigines and set limits on land claims. The One Nation political party, led by Pauline Hanson, opposes immigration, multiculturalism, and any special preferences for Aborigines, and rose in popularity among those who wanted a return to a more monocultural (essentially European) society.

Climate Change and Ozone Depletion

Oceania is especially vulnerable to global environmental changes, particularly ozone depletion, global warming, and sea-level rise. Ozone gas provides a protective layer in the upper atmosphere against the ultraviolet component of solar radiation that can cause damage to living organisms. When scientists first noticed a dramatic drop in the amount of ozone over the South Pole in the 1980s, the so-called "hole" in the ozone layer was linked to the global emission of manufactured chemicals, especially chlorofluorocarbons (CFCs) used in air conditioning and refrigeration that can persist for many decades in the atmosphere. The particular dynamics and chemistry of the Antarctic atmosphere meant that ozone losses were higher in this region. **Ozone depletion** can result in higher levels of ultraviolet radiation and associated increases in skin cancer, cataracts, and damage to marine organisms. The skin cancer risks were highest in regions near the southern Antarctic zone where ozone depletion was concentrated and where people spent a lot of time exposed to the sun. Australia, with its southern latitude location, sunny days, and traditions of beach-going and sunbathing, was especially

vulnerable with growing rates of skin cancer. International concern about ozone depletion eventually led to the signing of the Montreal Protocol in 1997, by which countries committed to reducing their use of CFCs and other ozone-depleting chemicals, but their use continues in many regions, and they persist in the atmosphere. The continued risk of skin cancer led to campaigns to use protective sun lotions, wear sunglasses, and reduce exposure to the sun in countries such as Australia.

Human activities have also caused another change in the global atmosphere—an increase in carbon dioxide associated with the burning of fossil fuels and forests. Although carbon dioxide is an essential component of the atmosphere in that it traps the sun's energy and helps keep Earth at a livable temperature, recent increases in carbon dioxide are likely to result in an increase in temperatures that has been called **global warming** (see Chapter 2, p. 60). Carbon dioxide levels have increased since the Industrial Revolution of the eighteenth century and are predicted to double by 2050. Other gases, such as methane, also act to warm the atmosphere in the so-called **greenhouse effect.** The resulting change in the balance of incoming and outgoing solar energy is projected to increase global temperatures by 2 to 5 degrees Celsius (5 to 8 degrees Fahrenheit), with the greatest relative increases toward the North and South poles. Global warming is also likely to change the pattern and strength of winds, and the intensity of evaporation, resulting in shifts in precipitation and availability of water resources. In Oceania the impacts of global warming may include drier conditions in the already drought-prone interior of Australia, increased risk of forest fires, and the melting of New Zealand's magnificent glaciers. Many scientists suggest that the impacts of global warming are already evident, but others suggest that the evidence is still uncertain and the predictions exaggerated.

A rise in global temperatures is also likely to produce a significant rise in sea levels, mainly because a warmer ocean takes up slightly more volume than a cooler one and secondarily because global warming may melt glaciers and ice sheets such as those in the Antarctic. Sea-level rise is of urgent concern to many Pacific islands, especially those on low coral atolls where any increase in sea level may result in the disappearance of the land below the sea or an increased vulnerability to storms. In response to the threat of global warming and sea-level rise, the Pacific islands were early members of the **Alliance of Small Island States (AOSIS),** which maintains a sustained voice in international negotiations to reduce the threat of global climate change. For example, the AOSIS countries have lobbied hard for major carbon dioxide emitters such as the United States and China to sign the Kyoto protocol, an international agreement to reduce carbon dioxide and other greenhouse gas emissions. There are already reports of sea-level increases in many areas of the world—for example, an increase of 13 centimeters in 7 years on Tonga—and several small but uninhabited Pacific islands have disappeared beneath the waves. One of the more vociferous members of AOSIS is the country of Tuvalu, where the islands are less than 4.5 meters (15 feet) above sea level and spring tides recently rose to 3.2 meters (11 feet) and flooded large areas.

Marine Disputes and Fisheries

Oceania is defined by its marine environment, and it is not surprising that the region shares many concerns about fisheries, territorial claims over ocean boundaries, and pollution of marine ecosystems. Pacific islanders eat more fish per person than any other population, and fishing and coastal tourism are critical to the majority of smaller island economies. Marine resources include not only fish and shellfish but also valuable exports such as pearls and shell (mostly for shirt buttons), as well as products made from whales, the species that initially attracted many Europeans to the Pacific. Because of the warm water, the southern Pacific is actually less biologically productive than the colder water that wells up near the continental shelves of Australia and New Zealand. The preponderance of reefs also limits the harvest of fish and other marine resources because coral organisms use up most of the nutrients and the coral reefs snag nets. Island societies consume a very wide range of fish species as well as other marine organisms, such as sea cucumbers (exported to China) and giant clams. They use many traditional methods to harvest marine resources, including spear diving, fish traps, and traditional poisons.

The main fishery is tuna that live near the surface of the western Pacific, with most of the remaining commercial fish caught in the coastal waters of New Zealand and Australia. The Pacific tuna harvest is more than 1 million tons (2 billion pounds) of fish a year, of which 90 percent is caught in large nets towed by massive boats from Japan, Korea, China, and the United States. New Zealand has an economically important cool-water fishery that includes squid and roughy as well as coastal shellfish resources under Maori control.

The declaration of the international 200-nautical-mile (370-kilometer) **Exclusive Economic Zone (EEZ)** in the 1970s was of tremendous significance to Oceania because it allowed countries with a small land area but many scattered islands such as Tonga and the Cook Islands to lay claim to immense areas of ocean. The pattern of Pacific islands is such that most of the region is now covered by their EEZs, with very little unclaimed ocean (**Figure 12.21**), and these island nations can demand licensing fees from the international fleet that seeks to catch tuna within their zones.

Oceania provides many interesting examples of how communities manage renewable resources such as fisheries. Fisheries are often thought of as **common property resources,** which are managed collectively by a community who has rights to the resource rather than owned by individuals. Strategies for the management of common property resources in the Pacific include traditional moratoriums, called *tabu* in the Pacific (periods or places where fishing is not permitted), and recognition of family or group access based on customary rights to harvest a resource. More recently some countries have brought fisheries under government control or have regulated harvesting through permits and quotas. Beyond the Exclusive Economic Zones, fisheries are open to all and are vulnerable to the so-called *tragedy of the commons* in which the common resource is overexploited by individuals who do not recognize how their own use of the resource can aggregate with that of many others to degrade the environment—for example, overfishing a given species to the point of extinction. One of the greatest challenges in the sustainable management of fisheries is the lack of information about fish numbers, movement, and reproduction, especially in the Pacific.

Core Regions and Key Cities of Australia, New Zealand, and the South Pacific

The isolation, smaller populations, and scattered geographies of the nations of Oceania produce fewer core regions, key cities, or distinctive landscapes than do some other world regions. Even Sydney, Australia, the region's largest and most important city, may not rank as a true world city compared to global trade and service centers such as Singapore or Hong Kong. But within the region there are certainly regions and settlements that act as important centers for their areas, linking them to the global economy and acting as the source of economic and cultural development. We have identified southeastern Australia, centered on the cities of Sydney, Canberra, and Melbourne, as the core region of Oceania, and selected a small number of distinctive landscapes to illustrate the geography of the region, including the Australian Outback and the Pacific islands, as well as Antarctica, not yet discussed in this chapter but included under Oceania in this book.

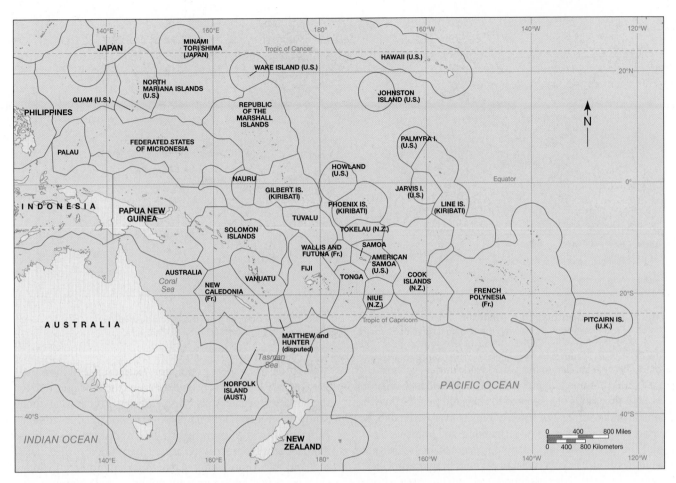

Figure 12.21   Marine territorial claims in the Pacific The Exclusive Economic Zones of Pacific nations and territories leave very little unclaimed ocean in the South Pacific. [*Source:* Redrawn from M. Rapaport (ed.), *The Pacific Islands: Environment and Society.* Honolulu: The Bess Press, 1999, Figure 30.3.]

Southeastern Australia

The core region of southeastern Australia stretches from the Gold Coast in southeastern Queensland south along the coast to the cities of Newcastle, Sydney, Wollongong, Melbourne, and Adelaide and inland to Australia's capital city of Canberra and the important agricultural regions of the Darling Downs, Murrumbidgee Irrigation Area, and the Barossa Valley (**Figure 12.22**). A moderate climate, a coastline of safe harbors and world-class beaches, and rich mineral and agricultural resources all make southeastern Australia an important center of historical and contemporary economic development that is the home to more than 60 percent of Australia's population (more than one-third the population of Oceania).

The rich agricultural lands of southeastern Australia host an important livestock industry of milk, beef, and lamb production, with animals that graze on pastures improved by fertilizer (especially phosphate) and introduced grasses. Southeastern Australia is also the heart of the Australian wheat industry in a "fertile crescent" that stretches with rolling farmland from the Darling Downs of southern Queensland to central South Australia. Wheat and other grains are often grown in rotation with pasture and sheep raising on larger farms. Wheat from this area combines with that from Western Aus-

tralia to constitute an export wheat industry that ranks with that of Canada and France, with only the United States ahead of this group. The irrigated farms of the Murrumbidgee Irrigation Area and the grazing lands of the Riverina are also very important to agriculture. Cotton from New South Wales contributes to Australia's dominance in world cotton exports. Southeastern Australia also includes several areas of intensive horticulture (fruit and vegetables) and vineyards. Australian wine has gained an excellent reputation in world markets, especially the wines of the Barossa Valley north of Adelaide and the Hunter Valley of New South Wales. Australian wine exports are the fourth largest in the world (behind France, Italy, and Spain). Wine is produced with the most advanced technology and innovative marketing and supports a thriving tourist industry similar to that of the Napa valley in the U.S. state of California.

Sydney   The city of Sydney, the capital of the state of New South Wales, has a spectacular location on the shore and low hills that surround an extensive natural harbor with many inlets (**Figure 12.23**). Two architectural symbols dominate the landscape—the Sydney Harbor Bridge completed in 1932 as one of the longest steel arch bridges in the world, and the Sydney Opera House with its brilliant white sail- or shell-shaped roof. These

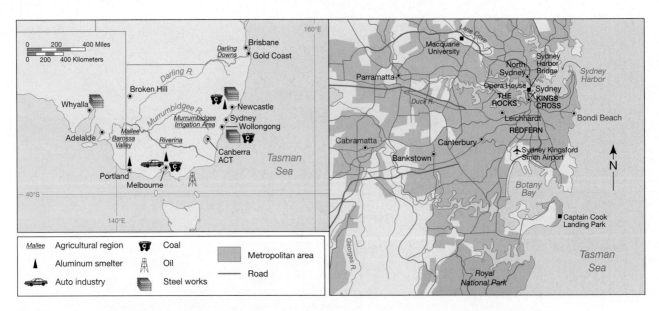

Figure 12.22 Southeastern Australia The most economically significant region of Oceania is the agroindustrial region of southeastern Australia with the dynamic cities of Sydney and Melbourne and the federal capital of Canberra.

images became familiar to many worldwide during the broadcasts of the 2000 Summer Olympics, which were held in Sydney and brought thousands of visitors to the city. Although Sydney has some manufacturing, the city economy is overwhelmingly service oriented, focusing on trade, banking, and tourism.

The city center has many commercial high-rise office buildings as well as government buildings from the nineteenth century, and the older neighborhoods often consist of small one-story bungalows. Sydney, with a population of almost 4 million people, has high levels of car ownership and sprawls into surrounding suburbs. The wealthier areas of the city are to the east, with poorer residents concentrated in the western suburbs. As immigrants from different regions settled in groups of similar origin, Sydney developed several ethnic neighborhoods, including Greek and Italian (Leichhardt), and Vietnamese (Cabramatta). Many Aborigines who migrated to the city settled in the suburb of Redfern, and their poverty and discrimination has made this neighborhood a focus of indigenous action and social programs. A number of older downtown neighborhoods, such as Paddington, have been renovated by higher income groups in a process of **gentrification.** Some of the older warehouse and manufacturing districts near the harbor have been redeveloped into shopping, museum, convention, and entertainment structures, such as the Darling Harbor district. Sydney is surrounded by parks and protected areas and by magnificent beaches that encourage an outdoor lifestyle that includes swimming, surfing, and sailing. The major industrial cities of Wollongong and Newcastle with coal mines and steelworks lie to the south and north of Sydney, and Newcastle also has two major aluminum smelters in the Hunter Valley that cuts inland from the coast.

Melbourne and Canberra The manufacturing center of southeastern Australia is Melbourne, the capital of the state of Victoria. Located on a sheltered harbor on Port Phillip Bay,

Melbourne has a local population exceeding 2 million people. The city first developed when it was the transport hub for the nineteenth-century gold rush and grew further when thousands of refugees and migrants were sponsored to come to Australia after the Second World War and were sent to work in Melbourne's industrial sector, which included textiles, clothing, and metal processing. Contemporary industry includes chemicals, food processing, automobiles, and computers.

Canberra, the federal capital of Australia, lies inland, about 150 miles (240 kilometers) southwest of Sydney. It was chosen as the site for the capital in 1909 and is a relatively successful example of a city planned and built according to principles of urban design and as a compromise location for a national capital between other larger cities. Federal government agencies, the Australian National University, and parliament are all located on wide avenues near the artificial lake that was part of the new city landscape design, and more than half of all jobs are in government. The National Capital Planning Authority is one of the strictest urban-planning authorities in the world, with residential and commercial land leased from the government.

Distinctive Regions and Landscapes of Australia, New Zealand, and the South Pacific

The Outback

The **Outback** is the term generally applied to the remote and drier inland areas of Australia and carries an image of a reddish, dusty landscape occupied by a few cows, Aborigines, and mining enterprises (**Figure 12.24a**). Although low rainfall and frequent drought are common in inland Australia, the exploitation

(a)

(c)

(b)

Figure 12.23   Sydney, Canberra, and Melbourne (a) This view of Sydney features the Sydney Opera House (middle left) and Sydney Harbor Bridge with the modern office skyline across the harbor. (b) Australia's federal capital, Canberra, is one of the world's most carefully planned cities. (c) The central business district of Melbourne.

of underground water from the Great Artesian Basin has allowed the development of scattered homesteads that raise livestock on sheep and **cattle stations.** Cattle are left to fend for themselves for the most part and are only brought into the stations once or twice a year. Sheep are raised where rainfall is higher to the east and west. Life on the cattle and sheep stations is often tough, exposed to the hazards of drought and wildfires and remote from

schools, shops, and hospitals. Distance education was pioneered in these remote settlements, with children taught through radio broadcasts; and the ill receive emergency medical care from the Flying Doctor service (see A Day in the Life: The McSporran Family and Anna Creek Station, p. 603).

Many Aborigines still live somewhat traditional lives in the Outback, although many work on stations and mines or

(a)

(b)

Figure 12.24   The Outback (a) Herding cattle by helicopter across the dry landscape of an Outback cattle station. The distances involved are so vast that helicopters and motorbikes are used to herd cattle, in addition to the traditional horses and dogs. (b) A massive transport truck called a "road train" carries freight at high speeds across the remote road network of the Australian Outback.

The McSporran Family and Anna Creek Station

It has a clothesline, a barbecue, and a shed. A gum tree and a rusting basketball hoop make do as goal posts for family football games, and from the kitchen door you see green lawn and petunias. But this is a backyard like no other: it is also the biggest cattle station in the world, running flat and dry across 26,000 square kilometers of far northern South Australia toward a fence somewhere beyond the horizon (**Figure 1**). Mobs of cattle wander across it under a sun that seeps rose and lilac as it sets. The fine dust of thousands of kilometers of Outback country swirls everywhere, and when clothes are washed, the water runs deep red.

Thirteen people live at Anna Creek station: Grant McSporran, who runs it for the Kidmans pastoral empire, his wife Tracey, and their two sons, the head stockman and his partner, three stockmen ringers, the cook, the teacher, the pilot, and a handyman. The nearest neighbors are 17 km away at the William Creek pub; beyond that it's 160 km the other way on dirt roads to the tough opal-mining town of Coober Pedy. The station's white-painted houses face in toward each other as if in defense against the vastness. Behind them rise row on row of low sand dunes where snakes and lizards bask as the wind whines past.

The country beyond has been farmed for more than a century, first with sheep and then, when dingoes decimated the flocks, with cattle. There are now 16,000 head of cattle feeding on Anna Creek, moving from dam to dam in their restless search for food. Every year between March and late November, before the heat and the flies drive men and cattle crazy, the herd must be mustered. With the cattle scattered across an area bigger than Wales, the job takes Grant and his ringers months.

The land they move through looks bare as it rolls on out of sight. "You look out and you can't see a thing," says Pat Fogg, who teaches Grant and Tracey's sons. But there are details of beauty in the emptiness: shards of black rock like small, sharp teeth, and the tiny white and purple flowers that tremble in the breeze. And if you know where to look, there are memorials to human toil—like the weathered posts of the Overland Telegraph line, laid across the 3,200 km from Port Augusta to Darwin by hardy men with camels in 1872.

The public phone rings outside the station's classroom, sending white cockatoos wheeling from the gums. Inside, 13-

Figure 1   Anna Creek The Anna Creek cattle station is the largest in Australia.

year-old Michael McSporran does his math homework while his brother Greg, 18 months younger, reads Aboriginal Dreamtime stories. They have 11 horses between them and want little else. "Very rarely do they say they're bored," says Miss Pat, as her students call her. Not many teachers have lasted more than a year; most are driven by the isolation back to crowds and the coast. From the two-way radio come the crackling voices of children on remote stations around the country singing the School of the Air's song: "There is a school where no child goes/They sit at home by their radios." Distance means something else in country like this. "It turns you different," says Nick Edwards, a 24-year-old ringer. "You can't handle cities anymore." Here, the length of a road is far less important than the intervals between rains and good cattle prices. The driver of the road train packed with jostling cattle cheerfully describes his 13-hour journey from Anna Creek to the slaughter yards as a "suburban trip." Someone suggests chartering the local light plane to fly to the annual Coober Pedy horse races. "But it's only 150 km away," is the incredulous reply.

Source: Excerpted from "A Splendid Isolation," *Time South Pacific,* November 23, 1998.

reside in small settlements known as outstations and often receive government support. As discussed earlier, some areas of inland Australia, especially in the Northern Territory and Western Australia, have been set aside for Aborigines.

Some of the earliest transportation routes into the Outback were roads and railways that connected important mining centers with the coast. Mining centers developed across the inte-

rior at Broken Hill in New South Wales and Kalgoorlie (gold) in Western Australia, and more recently at new finds near Mount Isa (lead, zinc, copper) in Queensland, and Pilbara (iron ore) in Western Australia. Uranium is also mined in the Outback. At the town of Coober Pedy, which is a center for the mining of opals valued for jewelry, the temperatures are so intense that much of the town has been built underground.

These mining communities are the most important settlements in the Outback, together with the town of Alice Springs, which developed on the telegraph line and later the Stuart Highway from Adelaide in the south of the country to Darwin in the north. Goods are transported across the vast interior using "road trains," enormous trucks that pull a chain of trailers at rapid speeds (**Figure 12.24b**).

Life in the Outback is vulnerable to several natural hazards, including frequent droughts that can decimate livestock herds and exacerbate problems of soil erosion and overgrazing. Droughts are also associated with severe wildfires that can race across the tinder-dry bush vegetation, especially where oily eucalyptus fuels the fire. Livestock operations are also at risk from doglike dingoes, which are called *Warrigal* by the Aborigines, who sometimes tame them. Dingoes hunt alone or in small groups, preying on kangaroos, rabbits, sheep, and cattle. The world's longest fence was built in order to keep the dingoes out of southern Australia. The so-called Dog Fence is a straight wire-and-post fence 5322 kilometers (3307 miles) long, twice as long as the Great Wall of China. Another irritant in the Outback are the hordes of flies that land on any sign of moisture, especially the faces of humans and animals.

The Islands of the Pacific

The islands of the Pacific form a distinctive image in the minds of most of the world—tropical paradises where local people fish, collect coconuts, and make crafts while tourists relax on beautiful beaches and swim in peaceful lagoons fringed with coral reefs. The reality of the Pacific islands is, as we have already seen, far more complex and in some cases conflict-ridden and difficult than the popular image suggests. This, together with the islands' far-flung geography and isolation from each other, makes it difficult to treat the islands as a coherent distinctive landscape. Even the geographical classification into Melanesia, Micronesia, and Polynesia, or into high and low islands, clusters islands that are very different into subregional groups. Another way to understand the distinctive landscapes of the Pacific islands is to focus on groups of islands that share distinct forms of integration into the global economy that have then shaped their social and physical environments in similar ways. Two such categories are islands transformed by mineral extraction and islands transformed by international tourism. One further group, the islands transformed by nuclear testing, is described in Geographies of Indulgence, Desire, and Addiction: Uranium in Oceania, on p. 596.

Mining Landscapes of the Pacific   One distinctive set of islands are those where the economy is almost wholly dependent on exports of mineral resources, where mining has destroyed the landscape and created social tensions over the wealth that flows from exports. Perhaps the most dramatic case is that of Nauru, an oval island with an area of only 21 square kilometers (8 square miles) consisting of an uplifted coral platform about 30 meters (100 feet) above sea level (**Figure 12.25**).

Centuries of roosting by sea birds covered most of Nauru with deep deposits of **guano** (bird droppings) that have created the highest-quality phosphate rock in the world. Exploita-

Figure 12.25   Nauru The landscape of Nauru has been devastated by the mining of phosphate rock that accumulated from the guano of roosting seabirds. This shows the dock where phosphate is loaded onto ships for export.

tion of the phosphate for use as a fertilizer began in 1906, and Nauru phosphate was especially valuable in making the phosphorus-poor soils of Australia productive for crops and pasture. When Nauru became independent from Australia in 1968, the government took over control of the mining industry. Phosphate dominates the economy and is now strip-mined, crushed, and sent by conveyor belts to ships that anchor outside the reef that surrounds the island. The profits are divided between the government, local landowners, and a long-term trust fund that will become more important because the phosphate is expected to run out within the next few years. The gross national product per capita of the 10,000 or so citizens (including many non-residents) is high, at more than $10,000 per year. Many locals choose not to work, and the mining is done by temporary migrant contract workers from other islands. Nauru has invested in real estate in Australia and in other enterprises as alternative sources of revenue after the phosphate income disappears and provides generous social security to all citizens.

The landscape of Nauru is now a desolate wasteland stripped of vegetation and soil, with cavernous pits dotting a rainless rocky plateau. Drinking water comes from an aging desalination plant or is shipped in from Australia. High levels of consumption and sedentary lifestyles among local people have brought the diseases of affluence to the island, including obesity, diabetes, and heart disease and a loss of traditional culture.

Banaba, an even smaller island north of Nauru that is part of Kiribati, had similar resources of guano-derived phosphate deposits that were exhausted by 1980, with some money set aside in a reserve fund that now supports the government of the sparsely inhabited and ecologically devastated island.

Mineral extraction also dominates the landscapes and livelihoods of larger islands in the Pacific. The main island of New Caledonia, located northwest of New Zealand, has one of the largest nickel reserves in the world, extensive mines, and an enormous and polluting nickel refinery located in the capital city of Nouméa. Nickel is used not only in coins but in many important industrial alloys, and its value is one of the reasons that France has held on to New Caledonia as an Overseas

Territory, supported by the descendants of French immigrants, but opposed by the indigenous Melanesian Kanaks, who seek independence. Mining also exacerbates problems on the island of Bougainville, which is controlled by Papua New Guinea but is geographically and culturally part of the Solomon Islands. The giant Panguna copper mine owned by the multinational company Rio Tinto was one of the world's largest open-pit mines and contributed as much as a quarter of Papua New Guinea's export earnings. The mine was developed in the forests without the participation of the resident indigenous Nasioi, and it has polluted several rivers that provided fish and drinking water to other groups. Local people seeking a share of the copper revenues, concerned about the mine's environmental impacts, and demanding independence from Papua New Guinea have joined a rebel movement. The rebels have closed the mine on several occasions and come into violent conflict with authorities. More than 10,000 people have died from the civil war, and an air and sea blockade imposed by Papua New Guinea beginning in 1989 prevented medicines and food supplies from reaching Bougainville. The blockade ended only in 1998 with a cease-fire and beginnings of peace negotiations. The conflict over mining on indigenous lands in Bougainville parallels that surrounding the Ok Tedi copper and gold mine on the mainland of Papua New Guinea, where the pollution of rivers by mine tailings has prompted international environmental concern (**Figure 12.26**). The Ok Tedi case provides another example of how mineral exports and mining by foreign-owned companies can transform local environments, cultures, and politics in Oceania.

Tourist Landscapes of the Pacific   The widespread image of the Pacific islands as a vacation paradise has origins in the accounts of the first European visitors, who described tropical abundance and peaceful locals, including the romantic depiction of island women as exotic and available partners. The island of Bora-Bora in French Polynesia uses these images to advertise its luxury five-star resorts, describing the location as the "islands of dreams," an "emerald in a setting of turquoise, encircled by a necklace of pearls." Tourists are invited to purchase crafts and to view the traditional dances of "beautiful Polynesian women." Bora-Bora, a small mountainous island surrounded by coral reefs and lagoons, certainly has a stunning location, but its beauty has attracted so many luxury hotels that it is rapidly becoming an expensive caricature of the typical Pacific island paradise (**Figure 12.27**).

The notion of the Pacific as a tourist destination grew as international air and cruise routes included stops at island groups such as Hawaii and Fiji, often en route to Australia or Asia. But the big boom in Pacific tourism occurred from about 1980 onward as air travel became more accessible, and increased numbers of North Americans and Asians (especially Japanese) sought luxury and exotic vacations. The most popular Pacific island destinations, after the U.S. state of Hawaii, are Guam, Fiji, and Tahiti, and the total number of tourists to the Pacific islands (Hawaii excepted) reached more than 3 million a year in the late 1990s. The significance of international tourism to individual countries is tremendous and is the

Figure 12.26   Mining in New Guinea The Ok Tedi mine in Papua New Guinea has polluted the Fly River with mine waste and affected the lives of local indigenous people.

major source of foreign exchange for the Cook Islands, Fiji, French Polynesia, Samoa, Tonga, Tuvalu, and Vanuatu. Some of the challenges faced by these tourism-dependent economies include vulnerability to international trends in tourism and political unrest that dissuades tourists and the need to ensure

Figure 12.27   Bora-Bora Luxury tourist resorts dot the coastline of Bora-Bora in French Polynesia. Visitors to these resorts bring valuable foreign exchange to the economy, but the resorts put pressure on the local resource base and encourage local peoples to modify their livelihoods and rituals to please foreign visitors.

that the benefits of tourism reach throughout the population and that the negative impacts on local cultures and environments are minimized.

Samoa is trying to build an image as the *ecotourism* capital of the South Pacific, promoting trips to waterfalls, lava flows, snorkeling, sea kayaking, rugby football matches, and accommodation in the traditional open thatched hut, or *fale*. The councils of village chiefs have decided that large-scale tourist development is inconsistent with Samoan values and have promoted eco-villages that commit to protect their wildlife and culture and to adopt sustainable development policies. More than 85,000 people visited Samoa in 1999, and employment is rapidly shifting from agriculture to the service sector.

Antarctica

Antarctica once lay at the heart of a single supercontinent called Gondwanaland (see Chapter 2, p. 52). Fossils of marine animals and forests found in Antarctica suggest that the Antarctic seas cooled about 65 million years ago, and most trees and reptiles disappeared by 5 million years ago. Today, Antarctica covers an area of 8.8 million square kilometers (5.5 million square miles) and contains 70 percent of the world's freshwater and 90 percent of the world's ice. It is the highest, coldest, and windiest of all the continents. Temperatures here average −51°C (−60°F) during the six-month Antarctic winter, when the sun does not rise above the horizon and the continent is in perpetual twilight. The world's lowest temperature was recorded on July 21, 1985, at the Russian Vostok base at −89°C (−129°F). By September each year, half the surrounding ocean is frozen over, creating a vast mantle of Antarctic pack ice with an area of 20 million square kilometers (32 million square miles) and a thickness of more than 2 meters (6.6 feet). The average elevation is above 2300 meters (7000 feet) and the highest mountains reach almost 5000 meters (16,000 feet).

It is always difficult to categorize Antarctica within the normal groupings of world regions, but in some ways it fits well within Oceania because of the importance of the marine environment, its isolation, the links to New Zealand and Australia as well as the United Kingdom and the United States, and the growing importance of tourism.

Antarctica's landscapes are beautiful but sterile. There are few growing plants, no surface lakes, and no running water. The vast polar ice cap, more than 3 kilometers (nearly 2 miles) thick in places, is pierced by mountain peaks called *nunataks*. Glaciers and snowfields cover most of the rest of Antarctica, though in the brief summer the snow in some coastal regions melts to reveal a lunar landscape of rock, boulders, and volcanic ash. The largest glaciers, such as the Lambert and the Beardmore glaciers, are so massive that they force great quantities of ice beyond the landmass onto the surrounding seas, where it floats as an ice shelf, terminating in towering ice cliffs. Antarctica's landscapes are also striking for the near-silence. Because absolutely nothing grows in the region, there is no sound of wind in the vegetation. There is very little animal life or human habitation, no buzzing insects, and the crying of birds is localized in a few of the relatively more hospitable

Figure 12.28 Antarctic landscape Antarctica is an ice-covered landscape populated by penguins and other species adapted to the cold temperatures. The penguins feed on the abundant fish in the oceans surrounding the icecaps.

coastal locations. Because it is frozen for much of the year, even the sea is silent. As the ice melts along the coasts, the most active and noisy residents of Antarctica's landscapes are millions of seals and penguins and the occasional giant blue whale surfacing offshore (**Figure 12.28**).

Visually, the most striking aspect of Antarctica's landscapes is the sheer scale. In the clear, bright light of unpolluted air, the unbounded snow- and icefields seem endless. The distances are indeed immense, and the atmosphere is so clear in Antarctica that one can easily be deceived: mountains that seem no more than 20 kilometers (12.4 miles) away may in fact be 80 kilometers (49.7 miles) distant. And in the absence of haze, the colors of distant objects are different: In Antarctica, mountains seen at a distance seem yellow rather than blue. In detail, the snowy, icy landscape is subject to constant change as ice features move and new snow blankets old features. Along parts of the coast, stark granite promontories provide fixed landmarks, but much of the coastline of Antarctica is ephemeral. The latest maps of Antarctica are always out of date because of the changing configuration of glaciers as they reach the sea. In summer, the Antarctic pack ice breaks up and the coastal glaciers calve huge icebergs that shift, drift, and change shape by the hour. Scientists are concerned that global warming may be melting both the sea ice and the ice sheets, and this could lead to the collapse of large areas of ice such as the Ross Ice Shelf and a resulting rise in sea levels.

Although scarcely inhabited by humans, there is a great deal of interest in Antarctica, both from the point of view of scientific research and from the point of view of the potential exploitation of reserves of natural resources such as deposits of iron ore, coal, gas, and oil that may lie beneath Antarctica and under the seas around it. The 1911–12 race to the South Pole between Britain's Captain Robert F. Scott and Norwegian explorer Roald Amundsen has gained mythic proportions and inaugurated further scientific exploration and territorial claims

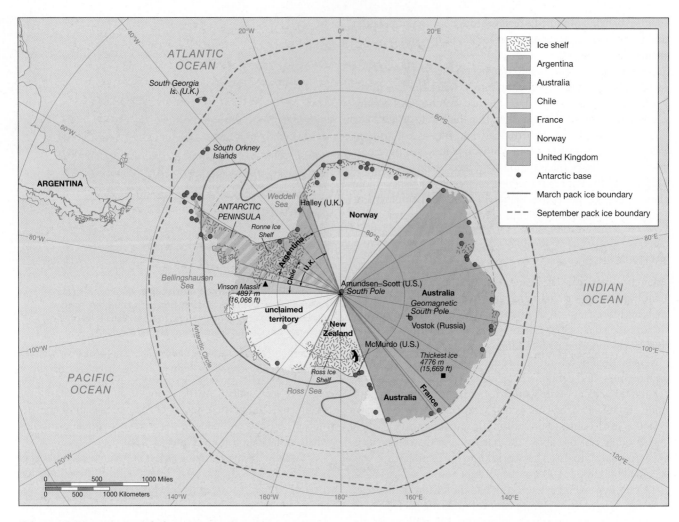

Figure 12.29 Territorial claims in Antarctica Seven countries have territorial claims in Antarctica but more than 40 have signed the Antarctic Treaty. The treaty bans nuclear tests and the disposal of radioactive waste, ensures that the continent can be used only for peaceful purposes and mainly for scientific research, and includes a 50-year ban on mineral and oil exploration. Sixteen countries have maintained scientific research bases on the Antarctic continent, including the U.S. Amundsen-Scott base at the geographical South Pole and the Russian Vostok base at the magnetic South Pole. [*Source:* Based on G. Lean and D. Hinrichsen (eds.), *Atlas of the Environment.* Santa Barbara, CA: ABC-CLIO, 1994, pp. 182–183; and Terraquest, Virtual Antarctica Expedition, available at http://www.terraquest.com/va/expedition/maps/cont.map.html]

on the continent. Amundsen reached the pole in December 1911, and Scott arrived on January 18, 1912. On their return trip, Scott and his companions perished when they were slowed by bad weather and they ran out of food.

International relations on the continent are governed by the **Antarctic Treaty,** which covers the area south of 60° S. The treaty, created in 1958 and now signed by 44 countries, bans nuclear tests and the disposal of radioactive waste, and ensures that the continent can be used only for peaceful purposes and mainly for scientific research. In 1991 the treaty added a 50-year ban on mineral and oil exploration. Nevertheless, several countries—Australia, Argentina, Chile, France, New Zealand, Norway, and the United Kingdom—still claim specific slices of the Antarctic pie, hoping to be able to assert rights to offshore fisheries and onshore resource exploitation (**Figure 12.29**).

Small groups of military personnel, scientists, and support personnel have been based in Antarctica in pursuit of

these scientific and national interests. There are no airports whatsoever in the entire continent and only one hotel (on King George Island), and so these specialized personnel are flown in on aircraft equipped with skis and accommodated in camps and research institutes called *stations.* Odd-shaped buildings, domes, and antennas, seemingly arranged haphazardly, dot the landscape of these stations, where snowmobiles and vehicles with monster truck tires provide the only sign of outdoor activity. The stations are supplied by specialized icebreaking ships that are also used for scientific research. When the Soviet Union collapsed in 1989, most of the former Soviet Union's specialized ships were suddenly idle. But not for long: Entrepreneurial Western travel companies hired the ships and their crews, turning high-tech research vessels into ecotourist cruise ships. About 17,000 travelers now visit Antarctica each summer. Cruises are by far the most popular tourist activity in Antarctica, but organized tours also offer mountain climbing, kayaking, sailing, camping, and even scuba diving.

Though these numbers are minuscule on the overall scale of world tourism, the environmental impact is already giving cause for concern. Bird species such as petrels, penguins, and albatrosses are declining in number. Petrels are long-lived birds—the oldest on record survived 50 years—and they feed on a variety of foods, making them an ideal Antarctic "indicator species." Many ecologists believe that the increased human presence in Antarctica may be disturbing these sensitive birds so much that they fail to breed. Another possible culprit is commercial fishing, which is booming, often illegally, throughout the southern oceans. Meanwhile, the efficiency, persistence, and greed of Russian, Japanese, and Norwegian whaling fleets have led to the near-extermination of the blue whale and the decimation of other species.

Summary and Conclusions

Oceania is the most geographically isolated of world regions, yet it is also closely connected by migration, trade, environment, politics, and tourism to the world-system. Our analysis of the region has highlighted some important themes, commonalties, and differences across the region. The historical geography of Oceania illustrates how ecological imperialism—especially the introduction of exotic species, together with European immigration and trade in staple agricultural and mineral commodities—transformed Australia, New Zealand, and the Pacific in the nineteenth century. Australia and New Zealand became two of the world's wealthiest countries in the twentieth century, with preferred trading links to Britain succeeded by new export and investment relations with Asia. Immigration patterns paralleled this shift, with increased numbers of migrants from Asia and the Pacific arriving, especially in Australia. These new immigrants, together with increased awareness of the rights of indigenous Aborigines and Maori, challenged Australia and New Zealand to rethink their national identities, with Australia promoting a multicultural society and New Zealand a bicultural approach of a Maori-Pakeha (white) society.

The Pacific islands were strongly affected by the Second World War, and they continue to play a role in world politics. Countries such as France and the United States maintain territories and military bases in the region and tested nuclear weapons there until relatively recently. The world's last frontier, Antarctica, is protected under international agreements from militarization. The Pacific has also become the testing ground for the impacts of global pollution. It was the first world region to experience the effects of ozone depletion, and some of its low-lying islands have already disappeared as a result of sea-level rise associated with global warming.

Within the region there are great contrasts—between the densely populated core region of southeastern Australia and the remote, rural Outback; between the high-income districts of modern cities such as Sydney and their poorer neighborhoods; between the semitropical North Island and alpine South Island of New Zealand; between the high volcanic islands and low coral atolls of the Pacific; and between the Pacific islands dependent on mining, military bases, remittances, and tourism.

While there are many problems in the Pacific, including ethnic strife and struggles for greater autonomy, the region is peaceful and prosperous compared to many other world regions. Regional cooperation includes agreements on trade, environment, and security, and the promotion of a Pacific way of self-determination, cultural esteem, and peace through consensus.

Key Terms

Alliance of Small Island States (AOSIS) (p. 600)	**Dreamtime** (p. 582)	**guano** (p. 605)	**stolen generation** (p. 598)
Antarctic Treaty (p. 608)	**ecological imperialism** (p. 582)	**island biogeography** (p. 581)	**subsistence affluence** (p. 595)
atoll (p. 579)	**El Niño** (p. 580)	**marsupials** (p. 580)	**Treaty of Waitangi** (p. 584)
British Commonwealth (p. 585)	**eucalyptus** (p. 581)	**Melanesia** (p. 574)	**White Australia policy** (p. 592)
cargo cult (p. 593)	**Exclusive Economic Zone** (p. 600)	**Micronesia** (p. 575)	
cattle station (p. 603)	**gentrification** (p. 602)	**monotreme** (p. 580)	
Closer Economic Relations (p. 595)	**global warming** (p. 599)	**Outback** (p. 602)	
common property resources (p. 600)	**Great Artesian Basin** (p. 575)	**ozone depletion** (p. 599)	
	greenhouse effect (p. 599)	**Polynesia** (p. 575)	
		South Pacific Forum (p. 595)	

Review Questions

Testing Your Understanding

1. Define: (a) Oceania; (b) Melanesia; (c) Micronesia; (d) Polynesia. What are three factors that make Oceania a distinctive region?

2. What are some of the physical differences between high volcanic islands and low coral atolls? (Discuss how they formed and what they look like, whether they receive a lot of rainfall, what the soils and vegetation are like, and so forth.)

3. How are marsupials and monotremes different from other mammals?
4. What is ecological imperialism and how did it transform the environment of Oceania?
5. Which European powers colonized which sections of Oceania and what are the legacies of the colonial period (for example, in terms of the official languages)?
6. Currently, why and where do the United States, Britain, France, and New Zealand maintain dependent territories in the Pacific?
7. How and where has nuclear weapons testing transformed the people and places of Oceania? Is nuclear weapons testing still going on today?
8. How does the term MIRAB summarize the basis of many modern Pacific island economies?
9. Why have the Maori generally been more successful than the Aborigines in reasserting their land rights and cultural heritage? What aspects of Maori and Aborigine culture are now part of mass-marketed popular culture?
10. What are some of the factors that created recent political instability in Fiji, Bougainville, and New Caledonia?
11. What is Antarctica's unique political status? Why do some states maintain claims to the region? Why do scientists and environmentalists value Antarctica?

Thinking Geographically

1. What are some of the ways in which the large expanse of ocean has defined culture and economy in Oceania?
2. How, why, and when did Australia and New Zealand start to shift their economic and political orientations away from Great Britain toward Asia and North America?

3. What are the differences between the ways in which Australia and New Zealand have dealt with their indigenous peoples, immigration, and multiculturalism in general? Include a discussion of the White Australia policy, the stolen generation, the Treaty of Waitangi, and *terra nullius*, and any changes over time.
4. Why was Tahiti a key location in the colonial history of Pacific island states? How did the perceptions of early colonial explorers influence the image of the Pacific in other parts of the world? Is Tahiti an independent country today?
5. How did the Second World War affect the Pacific? Which regions of the Pacific are still strongly influenced by international geopolitical activities (for example, through military bases, bomb testing)?
6. What are the ozone hole and the greenhouse effect and how do they differ in their impact? Why is Oceania particularly vulnerable to these environmental changes and what is the name of the organization that helps some of the smaller countries respond to these changes?
7. Using specific islands as examples, how does resource extraction and tourism affect Pacific island states?
8. In what ways does southeastern Australia operate as a core region for Oceania yet still not rank as a core region within the global economy?
9. In what ways do livestock, mining, and indigenous livelihoods make the Outback a distinctive but contested landscape?

Further Reading

Brookfield, H. C., and Hart, D., *Melanesia: A Geographical Interpretation of an Island World*. London: Methuen, 1971.

Bunge, F. M., and Cooke, M. W., *Oceania, A Regional Study*. Washington, DC: U.S. Dept. of the Army, 1985.

Cameron, I., *Lost Paradise: The Exploration of the Pacific*. Topsfield, MA: Salem House Publishers, 1987.

Campbell, I. C., *A History of the Pacific Islands*. Berkeley: University of California Press, 1989.

Colbert, E. *The Pacific Islands*. Boulder: Westview Press, 1997.

Crocombe, R., *The New South Pacific. An Introduction*. Suva: The University of the South Pacific, 1990.

Diamond, J., *Guns, Germs, and Steel: The Fate of Human Societies*. New York: W. W. Norton, 1997.

Firth, S., *Nuclear Playground*. Honolulu: University of Hawaii Press, 1987.

Freeman, O. W., *Geography of the Pacific*. New York: Wiley, 1951.

Horne, D. *The Lucky Country*. Ringwood, Victoria: Penguin, 1998.

Howe, K. R., Kiste, R. C., and Lal, B. (eds.), *Tides of History: The Pacific Islands in the Twentieth Century*. Sydney: Allen & Unwin, 1994.

Hughes, R. *The Fatal Shore*. London: Pan Books, 1988.

Le Heron, R. B., *Globalized Agriculture: Political Choice*. New York: Pergamon Press, 1993.

Lockwood, V., *Tahitian Transformation: Gender and Capitalist Development in a Rural Society*. Boulder and London: Lynne Reiner, 1993.

Manderson, L., and Jolly, M. (eds.), *Sites of Desire, Economies of Pleasure. Sexualities in Asia and the Pacific*. Chicago: The University of Chicago Press, 1997.

McEvedy, C., *The Penguin Historical Atlas of the Pacific*. New York: Penguin Books, 1998.

McKnight, T., *Oceania: The Geography of Australia, New Zealand and the Pacific Islands*. Englewood Cliffs, NJ: Prentice Hall, 1995.

Mead, M. *Coming of Age in Samoa*. New York: Quill, 1928.

New Zealand, Government of, *Atlas of the South Pacific*. Wellington, New Zealand: Government Printing Office, 1986.

Powell, J. M., *An Historical Geography of Modern Australia: The Restive Fringe*. New York: Cambridge University Press, 1988.

Rapaport, M. (ed.), *The Pacific Islands: Environment and Society*. Honolulu: The Bess Press, 1999.

Rickard, J., *Australia: A Cultural History.* London: Longman, 1996.

Robinson, G. M., Loughran, R. J., and Tranter, P. J., *Australia and New Zealand: Economy, Society and Environment.* New York: Oxford University Press, 2000.

South Pacific Regional Environment Programme, *Environment and Development: A Pacific Island Perspective.* Manila: Asian Development Bank, 1992.

Stanley, D., *South Pacific Handbook,* 6th ed. Chico, CA: Moon Travel Handbooks, 1996.

Stratford, E. (ed.), *Australian Cultural Geographies.* Melbourne: Oxford University Press, 1999.

Taylor, G. T., *Australia: A Study in Warm Environments and Their Effect on British Settlement.* London: Methuen, 1940.

Vayda, A., *Peoples and Cultures of the Pacific.* Garden City, NY: American Museum of Natural History, Natural History Press, 1968.

Waddell, E., Naidu, V., and Hau'ofa, E. (eds.), *A New Oceania: Redefining Our Sea of Islands.* Suva: University of the South Pacific, Beake House, 1993.

Watters, R. F., and McGee, T. G. (eds.), *Asia-Pacific: New Geographies of the Pacific Rim.* London: Hurst and Co., 1997.

Film, Music, and Popular Literature

Films

Black Harvest. Directed by Robin Anderson and Bob Connolly, 1992. Presents the troubled relationship between the Ganiga tribe and Joe Leahy, the mixed-race owner of a coffee plantation built on land sold cheaply by the tribe.

Cannibal Tours. Directed by Dennis O'Rourke, 1987. Depicts the interaction between tourists on a luxury cruise in the South Pacific and the aboriginal people of Papua New Guinea.

Half Life: A Parable for the Nuclear Age. Directed by Dennis O'Rourke, 1985. Documents the official government cynicism behind U.S. nuclear testing in the Pacific and led to open debate on the morality of exposing the Marshall Islanders to fallout.

Mad Max. Directed by George Miller, 1979. Conflict between a police officer and biker gangs in a bleak and violent vision of future Australia.

Mutiny on the Bounty. Directed by Lewis Milestone, 1962. Fictional account of the voyage of the *Bounty* to Tahiti and the famous mutiny.

Once Were Warriors. Directed by Lee Tamahori, 1995. A family descended from Maori warriors is devastated by a violent father and the societal problems of being treated as outcasts.

The Piano. Directed by Jane Campion, 1993. A young woman is sent from Scotland to a remote area of New Zealand's South Island in the mid-nineteenth century for an arranged marriage.

Picnic at Hanging Rock. Directed by Peter Weir, 1975. Mysterious disappearance of Australian schoolgirls during a picnic in rural Victoria.

Priscilla, Queen of the Desert. Directed by Stephan Elliott, 1994. A comedy that follows a road trip by three drag queens through the landscape of Australia.

Music

Kiri Te Kanawa. *Maori Songs.* Emd/EMI classics, 1999.

Various Artists. *Dance Music of Tonga.* Pan, 1994.

Various Artists. *Music of Oceania: Samoan Songs.* Musicaphon, 1999.

Various Artists. *Oceania.* Uni/Point Music, 2000.

Various Artists. *Rough Guide: Australian Aboriginal Music.* World Music Network, 1999.

Various Artists. *Spirit of Polynesia.* Saydisc, 1993.

Various Artists. *Tuvalu.* Pan, 1995.

Popular Literature

Chatwin, Bruce. *The Songlines.* New York: Penguin, 1988. An exploration of Australian Aboriginal culture and the invisible pathways or Dreamtime tracks that are mapped and communicated through song.

Eri, Vincent. *The Crocodile.* Auckland: Longman Paul, 1970. One of the first published novels from Papua New Guinea.

Grace, Patricia. *Potiki.* Honolulu: University of Hawaii, 1995. A novel by a New Zealand Maori writer regarding Maori-Pakeha relations that explores how Maori are adapting to change while attempting to preserve a sense of cultural identity.

Hau'ofa, Epeli. *Tales of the Tikongs.* Honolulu: University of Hawaii, 1994. Short stories of Polynesian life by a Tongan writer.

Hulme, Keri, *The Bone People.* New York: Viking Press, 1986. A dreamlike novel about relationships between three individuals and the Maori culture in New Zealand.

Wendt, Albert. *Leaves of the Banyan Tree.* Honolulu: University of Hawaii, 1994. A novel that explores colonialism and independence through three generations of Samoans.

White, Patrick. *The Tree of Man.* New York: Viking Press, 1955. An epic family saga, by a Nobel Prize Prize-winning author, depicting an ordinary couple at the beginning of the twentieth century who establish a farm in the Australian wilderness. They raise their children, have grandchildren, and eventually see their land engulfed by suburb.

Figure 13.1
Visualization of global Internet traffic flows

13 Future Regional Geographies

It is, of course, impossible to accurately predict the future of the very complex and diverse regions we have discussed in this text. Yet, for all sorts of reasons, it is important to be able to envisage what the future might hold for them (**Figure 13.1**). We have to live in the world for a while, and naturally we want to leave it in good shape for future generations. We therefore need to be able to identify the key changes that the future might bring so that we can work toward enabling the most desirable outcomes for future residents of planet Earth. To attempt to look into the future means we must ask some very difficult questions. Will globalization undermine regional cultures? Will technology and human determination be able to cope with the environmental stresses that industrialization in the periphery and rapid population increases will inevitably create? Will peripheral and semiperipheral regions be able to develop sustainably into core regions? Will new regions emerge based on new types of connectivity such as trade, the Internet, or any number of political movements such as mobilizations against globalization or the human rights movement? In this chapter, we attempt to address these and related questions in order to get a sense of the future of the ten world regions we have described.

Globalization and the Future of Regions

The most effective way to approach the questions listed above is to try to get a sense of how different aspects of globalization are changing the world and how they might continue to do so. As we discussed in Chapter 2, the globalization of the capitalist world-system is a process that has been occurring for at least 500 years. But since the Second World War, world integration and transformation has been remarkably accelerated and dramatic. Among the forces driving integration and transformation are the strengthening of regional alliances such as the European Union and the Organization of Petroleum Exporting Countries, the increasing connectivity of the most remote regions of the world due to telecommunications and transportation linkages, the emergence of the new economy in the core countries, and the rise of global institutions such as the World Trade Organization. How will the forces of broadening global connectivity—and the popular reactions to them—change the fates and fortunes of world regions whose current coherence owes more to eighteenth- and nineteenth-century European colonialism than it does to forces of integration or disintegration in the twenty-first century? To answer this and related questions, we need to understand what the experts think about the processes behind globalization and how they might affect its future po-

tential. But we also need to understand the very risky issue of predicting the future, how predictions are made, and how useful predictive exercises can be.

Predicting the Future

Predicting the future can be a tricky business. The uncertainties of geopolitical transformations, the unexpected impacts of technological breakthroughs, and the complexity of environmental change all conspire to make our future seem, at first glance, highly unpredictable. Nevertheless, there is no shortage of visionary projections (**Figure 13.2**). Broadly speaking, these can be divided into two sorts of scenarios: optimistic and pessimistic. Optimistic futurists stress the potential for technological innovations to discover and harness new resources, to provide faster and more effective means of transportation and communication, and to enable new ways of living. This sort of futurism is characterized by science-fiction cities of mile-high skyscrapers and spaceship-style living pods, by bioecological harmony, and by unprecedented social and cultural progress through the information highways of cyberspace. It projects a world that will be stabilized and homogenized by supranational or even "world" governments. The sort of geography implied by such scenarios is rarely spelled out. The relevance of the region, we are led to believe, will be transcended by technological fixes.

To pessimistic futurists, however, this is just "globaloney." They stress the finite limitations of Earth's resources, the fragility of its environment, and population growth rates that exceed the capacity of peripheral regions to sustain them. This sort of doomsday forecasting is characterized by scenarios that include irretrievable environmental degradation, increasing social and economic polarization, and the breakdown of law and order. The sort of geography associated with these scenarios is rarely explicit, but it usually involves the probability of a sharp polarization between the haves and have-nots at every geographical scale.

Fortunately, we do not have to choose between these two extreme scenarios. In order to arrive at a more grounded understanding of future geographies, however, we must first glance back at the past. Then, looking at present trends and using what we know about processes of geographic change and principles of spatial organization, we can begin to map out the kinds of geographies that the future most probably holds.

Looking back at the way that the geography of the world-system has unfolded, we can see now that a fairly coherent period of economic and geopolitical development occurred between the outbreak of the First World War (in 1914) and the collapse of the Soviet Union (in 1989). Some historians refer to this period as the "short twentieth century." It was a

Figure 13.2   Future scenarios In some ways, the future is already here, embedded in the world's institutional structures and in the dynamics of its populations. Still, there are some aspects of the future that we can only guess at. Six different scenarios are presented here, two each for three types of futures: Conventional Worlds, Barbarism, and Great Transitions (*Source:* Global Scenario Group, Stockholm Environment Institute, Boston, MA). (a) Homeless in Tokyo; (b) California windmills; (c) teenage guerillas in the Balkans; (d) global Mobil (Redrawn from *The Guardian*, Tuesday, April 17, 2001, p. 27.); (e) Slobodan Milosovic on trial for warm crimes in the Netherlands; (f) electric car recharging.

Conventional Worlds

The two scenarios presented here—the Reference Scenario and the Balanced-Growth Scenario—have a common vision of a world in which development is governed by gradual and steady industrial growth. While population grows, world economic output expands indefinitely as consumption and production practices in peripheral and semi-peripheral regions converge toward those of the increasingly richer core. The regions of the world become progressively more interdependent as the competitive private market remains the main engine for economic growth. Transnational corporations increase their role as the dominant economic node of the global, borderless economy. The liberal state persists as the dominant unit of governance. The two variations on the Conventional World futures differ in terms of policy intervention and impacts.

(a)

Reference Scenario

Economic growth is the central force motivating change as it assumes that most of the world's regional economies will open and that largely unregulated markets will expand internationally. New and expanded markets foster rapid technological development. Population growth increases in the peripheral and semi-peripheral regions of the world as most of the core regions grow slowly or not at all. As the core gets richer, the marginalized become increasingly poorer and inequality increases (Figure 13.2a). Environmental quality improves in some of the core regions and gets worse in the periphery and overall global environmental conditions deteriorate. Social justice issues intensify.

Balanced Growth Scenario

This scenario differs from the Reference Scenario by assuming the implementation of policy reforms that guide economic growth and sustain the environment. Balanced growth results from the introduction of better technology, reduction of subsidies for natural resource use, pollution taxes, land reform, development incentives, increased foreign aid for education and health, and broader economic opportunities in the periphery and semi-periphery (Figure 13.2b). The gap between the elite and the marginalized is less than in the Reference Scenario, and, as a result, the poorer regions of the world are stabilized and widespread social conflict is avoided.

(b)

Barbarism

The two scenarios that underpin the Barbarism future assume that the contemporary negative stresses present in Conventional Worlds scenarios overwhelm the coping capacity of markets and institutions. The world veers toward Barbarism as regions with declining physical amenities experience breakdown. Growing populations, persistent poverty, and increasingly disastrous environmental problems lead to the barbarization of the marginalized regions. In a future of Barbarism, social welfare policies are increasingly abandoned in favor of productivity and competitiveness policies as states lose relevance and power compared to large multinational corporations. The marginalized experience growing environmental pollution and natural resource constraints. Rather than full-scale wars, the result is small-scale armed conflicts and violence.

(c)

Social Breakdown

In this scenario, chaos ensues as random violence diverts resources from economic growth to security concerns. Civil order breaks down as states become too weak or too fragile to set the global economy back on track (Figure 13.2c). As refugees fleeing from disaster help to destabilize neighboring regions, states pour their resources into police powers, border fences, guards, and monitoring of the movements and activities of citizens, and ultimately limiting or obstructing trade and travel. The result is the collapse of globalization, overburdened by rising unemployment, depressions, political instability, and outbreaks of civil disorder in elite, marginalized, and embattled regions as everyone gets poorer.

Fortress World

In this scenario, the core regions recognize the crisis that is mounting and create alliances among themselves to protect their own interests. With the multinational corporations as their agents, the elite rich and powerful become entrenched, surrounded by oceans of misery (Figure 13.2d). The result is a society of elites and marginalized with entry into the elite by birth only. Strategic reserves of fossil fuels, minerals, freshwater, and genetic diversity are put under military control by the rich groups. While the impacts of pollution are limited for the elite, pollution and its noxious health effects increase for the marginalized.

ExxonMobil has business interests in 200 countries

(d)

Great Transitions

The most optimistic future is that of the Great Transition, in which the world's regions evolve to a higher stage. Although these scenarios may seem naive and improbable, they are not impossible and may even be necessary to achieve the goals of sustainability and equity. While there is a range of possible Great Transitions scenarios, we consider two: Global Governance and the New Sustainability paradigm, which differ in their means but not in their ends.

(e)

Global Governance

This scenario is built upon a growing collective realization that individuals, institutions, and states must restrict certain activities and undertake others for the common global good. Cooperation is essential. The leadership for this effort comes from multinational and transnational corporations, intergovernmental global organizations, and nongovernmen-

New Sustainability

In this scenario, increases in technological and economic growth are concentrated in core regions and dominated by transnational corporations. The gap between the elite and the marginalized is extreme. Environmental problems in the periphery grow. Migration flows to the core increase. The result is a rise in global social move-

(f)

tal organizations. Acting in concert, these three entities are a counterforce to states and are given limited regulatory power to enforce voluntary guidelines and to tax (but not to restrict) international flows of currencies, goods, and telecommunications. In addition, international courts are strengthened, mediation bodies flourish, and the dispute resolution capability of international organizations is greatly enhanced (Figure 13.2e).

ments opposing high-consumption lifestyles. Corporations follow the market and alter what they produce, how they produce it, even how they market it (Figure 13.2f). Sustainability becomes an economic and environmental goal. A global civil society is born based on social justice and open mechanisms for decision making and consensus seeking.

period when the modern world-system developed its triadic core of the United States, Western Europe, and Japan, when geopolitics was based on an East–West divide, and when geoeconomics was based on a North–South divide. This was a time when the geographies of specific places and regions within these larger frameworks were shaped by the needs and opportunities of technology systems that were based on the internal combustion engine, oil and plastics, electrical engineering, aerospace industries, and electronics and by the political struggles of the Cold War. In this short century, the modern world was established, along with its now-familiar landscapes and spatial structures: from the industrial landscapes of the core to the unintended metropolises of the periphery and from the voting blocs of the west to the newly independent nation-states of the south.

Looking around now, much of the established familiarity of the modern world and its geographies seems to be disappearing, about to be overwhelmed by a series of unexpected developments—or obscured by a sequence of unsettling juxtapositions. The United States is giving economic aid to Russia and Vietnam; Eastern European countries are on a schedule to become part of the European Union over the next 5 to 8 years; Germany has unified, but Czechoslovakia and Yugoslavia have disintegrated; apartheid South Africa has been transformed, through an unexpectedly peaceful revolution, to black majority rule. Meanwhile, in Zimbabwe, blacks are violently seizing white farms in a kind of reverse apartheid; former communist Russian ultranationalists have become comradely with German neo-Nazis; Hindus, Sikhs, and Muslims are in open warfare in South Asia; and Sudanese military factions steal food from aid organizations in order to sell it to the refugees for whom it was originally intended.

In short, we entered a dramatic period of transition in 1989 that is continuing to unfold in the new century. The economic and cultural flux of the world has provided some very colorful examples:

McDonald's, Pizza Hut, and American dollars are everywhere. Overnight jet flights and international direct dialing to North America afford the basic infrastructure for South American narcocapitalism. . . . Parts of Africa are returning to a hunting and gathering economy. Russia's markets are often empty and its factories are idle, but billions in oil, metals, lumber, and weapons are smuggled, like dope from Bolivia or Burma, to foreign markets through Kaliningrad. Moscow's GUM department store has a Benetton, while the city's nouveau riche mafia "entrepreneurs" ostentatiously zoom around in German BMWs and Mercedes Benzes, courtesy, in many cases, of lucrative car theft rings operating in Western Europe. Bloomingdale's sells Red Army watches at the costume jewelry counter. . . . Communist China's "military" industries are making millions selling knock-off running shoes to Singapore traders and reverse-engineered strategic rockets to Saudi princes.

. . . Karaoke machines offer ancient Motown hits to American corporate managers, Hong Kong entrepreneurs, and German sex tourists in Thailand. . . . Some of Iran's, Pakistan's, Libya's, and Mexico's largest urban populations are located in Paris, London, Milan, and Los Angeles. Disneyland now claims territory in Europe, Asia, and North America.[1]

These examples show that we cannot simply project our future geographies from the regions and landscapes of the past. Rather, we must map them out from a combination of existing structures and budding trends. We have to anticipate, in other words, how the shreds of tradition and the strands of contemporary change will be rewoven into new landscapes and new spatial structures.

As we look ahead to the future, we can appreciate that some dimensions of future geographies are more certain than others. In some ways, the future is already here, embedded in the world's institutional structures and in the dynamics of its populations. We know, for example, a good deal about the demographic trends of the next quarter-century, given present populations, birth- and death rates, and so on (**Figure 13.3**). We also know a good deal about the distribution of environmental resources and constraints, about the characteristics of local and regional economies, and about the legal and political frameworks within which geographic change will probably take place.

However, we can only guess at some aspects of the future. Two of the most speculative realms are those of politics and technology. While we can foresee some of the possibilities (maybe a spread and intensification of ethnonationalism; perhaps a new railway era based on high-speed trains), politics and technology are both likely to spring surprises at any time. The September 11, 2001, terror attack in the United States is a painful example of the sort of political surprises that are likely to occur. In a matter of about 100 minutes, terrorists who crashed two commercial jetliners into each of the towers of the World Trade Center and one into the Pentagon were able to deliver a crippling blow to the U.S. economy, draw the country into full military alert, and profoundly transform the conduct of daily life in the United States (**Figure 13.4**). Other events—such as a political revolution in China, war between India and Pakistan, or unanticipated breakthroughs in biotechnology—can cause geographies to be rewritten suddenly and dramatically. As we review the prospects for regional transformations, therefore, we must always be mindful that our prognoses are all open to the unexpected. But, as painful as such surprises can be, we must also be aware that the unexpected can also be our biggest source of optimism.

In the following section we examine the debates about contemporary globalization in order to arrive at a clearer sense

---

[1]G. ÓTuathail and T. Luke, "Present at the (Dis)integration: Deterritorialization and Reterritorialization in the New Wor(l)d Order," *Annals, Association of American Geographers*, 84 (1994), 381–398.

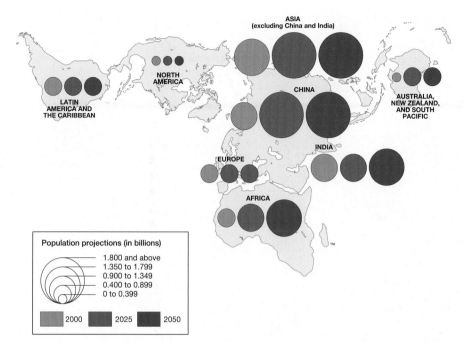

Figure 13.3   Population geography of the future Population projections for 2025 and 2050 show a very marked disparity between world regions, with core countries and core regions growing very little in the next half-century compared with the periphery and semiperiphery.

Population projections (in billions)

1.800 and above
1.350 to 1.799
0.900 to 1.349
0.400 to 0.899
0 to 0.399

2000   2025   2050

of how globalization might affect the future of the world's regions. An important aspect of globalization is the widespread perception that the world, through economic and technological forces, is increasingly becoming one shared political and economic space where the events in one part of the globe will have repercussions for all other parts, whether near or far. The regionalization scheme we described in the previous chapters implies that flows, interactions, and networks are strongest *within* the geographical groupings of states we have identified as Europe; Russia, Central Asia, and the Transcaucasus; the Middle East and North Africa; Sub-Saharan Africa; North America; Latin America and the Caribbean; East Asia; Southeast Asia; South Asia; and Australia, New Zealand, and the South Pacific. Given that economic and technological forces are breaking down the barriers within and between both near and distant places, will the most recent phase of globalization strengthen some regional connections, weaken others, or make regions altogether irrelevant? Alternatively, will globalization enable some regions—core regions, for instance—to create even greater differences of wealth and power than already exist in the world-system? Understanding what the experts believe about globalization will give us a better glimpse of the future it is likely to make possible.

The Globalization Debates

The number of books published on globalization and its impact on the world's regions has grown at a tremendous rate over the last 15 years. Check out the shelves in any bookstore and you will find hundreds of them, with whole sections devoted to globalization within political science, sociology, geography, economics, media studies, and business management. Yet, while the literature on globalization is extensive, it is pos-

sible to group the main participants in the contemporary debates about globalization into three general camps: the hyperglobalists, the skeptics, and the transformationalists. While these three viewpoints do not exhaust the range of the debates, they do provide a clear sense of the issues on which the experts on globalization agree and disagree.

The Hyperglobalists One group of experts—the hyperglobalists—believes the current phase of globalization signals the beginning of the end for the nation-state and the

Figure 13.4   The destruction of the World Trade Center Completed in 1973, the 110-story twin towers of the World Trade Center dominated the New York City skyline. The terror attack that caused their complete destruction resulted in an instantaneous transformation of the physical and psychological landscape of New York City.

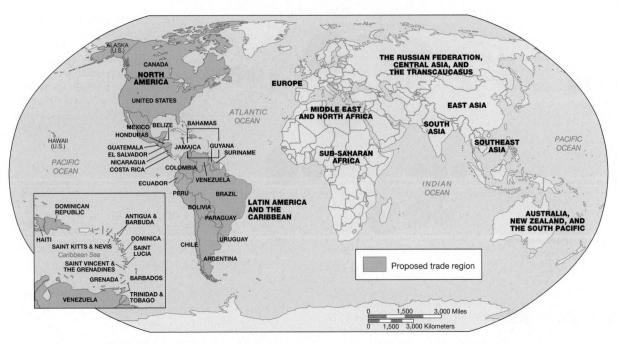

Figure 13.5 The Free Trade Area of the Americas This map shows the new free-trade zone proposed by U.S. President George W. Bush at the Quebec summit in April 2001. If actually enacted, FTAA would become the world's largest free-trade zone, linking 34 countries (from North America, Latin America, and the Caribbean, excluding Cuba) in a market of 800 million people. The project was first proposed in 1990 by former President George Bush but it will be up to his son to finish the treaty's negotiation by January 1, 2005, at the latest. Countries are expected to ratify the treaty by the end of that year. The treaty will give North American businesses duty-free access to Latin American and Caribbean markets, and those countries can export tariff-free to North America. The deal covers nearly all goods traded. Opponents of FTAA, who staged protests in Québec, believe that the treaty will allow corporations to bypass democratically adopted environmental and worker protection laws.

"denationalization" of economies, meaning that national boundaries will become irrelevant with respect to economic processes and that national governments will not control their once geographically bounded economies but will instead facilitate connections among and between different parts of the world. The Free Trade Area of the Americas is one such possible large-scale trade area that could have the effect of weakening boundaries and strengthening economic ties among North America and Latin America and the Caribbean (**Figure 13.5**). FTAA, still in the proposal stage, was presented by President George W. Bush at the recent world trade summit on expanding NAFTA held in Québec. If FTAA is ratified by all 34 of the possible member nations by 2005, it will be at least another five years after that before its effects can be assessed. The wider implications of the hyperglobalist position is that the world will become borderless as national governments become increasingly meaningless or as they function as merely the facilitators of global capital flows and investments. Hyperglobalizers believe that the nation-state, as the primary political and economic unit of contemporary world society, will eventually be replaced by institutions of global governance in which individuals claim transnational allegiances that are founded upon a commitment to neoliberal principles of free trade and economic integration. Politically, the global spread of liberal democracy will reinforce the emergence of a global civilization with its own mech-

anisms of global governance, replacing the outmoded nation-state with global institutions like the International Monetary Fund (IMF) or the World Trade Organization (WTO).

The Skeptics A second broad argument within the globalization literature belongs to the skeptics, those who believe that contemporary levels of global economic integration represent nothing particularly new and that globalization is an exaggerated myth. The skeptics look to the nineteenth century and draw on statistical evidence of world flows of trade, labor, and investment to fortify their position. Pointing to these data, they argue that contemporary economic integration is actually much less significant than it was in the late nineteenth century, when nearly all countries shared a common monetary system known as the gold standard. The skeptics are also dismissive of the idea that the nation-state is in decline. They argue that national governments are essential to the regulation of international economic activity and that the continuing liberalization of the world economy can only be facilitated by the regulatory power of national governments.

The skeptics assert that their analysis of nineteenth-century economic patterns demonstrates that we are today witnessing not globalization but rather "regionalization," as the world economy is increasingly dominated by three major regional financial and trading blocks: Western Europe, North America, and Japan. The skeptics understand regionalization

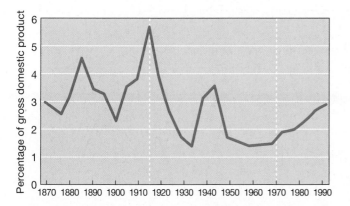

Figure 13.6   International capital flows among the core economies, 1870–1995 The position taken by the skeptics on globalization is highly controversial. Their argument, that the world was more integrated at the end of the nineteenth century than at the end of the twentieth, is based on statistical evidence of flows of capital, trade, and people over a 125-year period, as shown in this graph. The data represented on this graph certainly do show higher capital flows in earlier periods, but what the skeptics' argument fails to take into account are other important indices of globalization, such as increasing integration through faster transportation and higher speed and more efficient communication linkages. (*Source:* Data from P. Hirst and G. Thompson, *Globalization in Question: The International Economy and the Possibilities of Governance.* Cambridge: Polity Press, 1999, p. 28.)

and globalization to be contradictory tendencies. They believe that because of the dominance of these three major regional blocks, the world is actually less integrated than it once was because Western Europe, North America, and Japan control the world economy and limit the participation of other regions in that economy (**Figure 13.6**).

The Transformationalists   According to the transformationalist view, contemporary processes of globalization are historically unprecedented as governments and peoples across the globe confront the absence of any clear distinction between what is global and what is local, between what are domestic affairs and what are international ones. Like the hyperglobalists, this group understands globalization as a profound *transformative* force that is changing societies, economies, and institutions of government—in short, the world order. In contrast to the hyperglobalists and the skeptics, however, the transformationalists make no claims about the future trajectory of globalization nor do they see present globalization as a pale version of a more "globalized" nineteenth-century past. Instead, they see globalization as a long-term historical process underlain by crises and contradictions that are likely to shape it in all sorts of different and unpredictable ways. Moreover, unlike the skeptics, the transformationalists believe that the historically unprecedented contemporary patterns of economic, military, technological, ecological, migratory, political, and cultural flows have functionally linked all parts of the world into a larger global system (**Figure 13.7**). In fact, they see regionalism as a stepping stone to more global connectedness in which free trade agree-

ments such as FTAA help to draw regions in to a global neoliberal economic framework.

In this book, we come closest to the transformationalist position and believe that this approach to globalization can tell us a great deal about what we might expect the future—especially the future of world regions—to be. The transformationalists, mindful that no one can accurately predict the future, no matter how reasonable and rational their theories, suggest that we are heading toward a period where world regions will experience a wide range of internal changes at the same time that the strength of their connections with other parts of the world will increase. Whether those internal adjustments and external linkages will signal the wholesale transformations of the ten regions we have discussed in this book is very difficult to determine. What we can identify, however, is tendencies and how those tendencies might reshape the current world of regions.

What is most disturbing about the transformationalist view of globalization is their identification of global patterns of increasing disparities in wealth. Transformationalists believe that globalization is leading to increasing social stratification in which some states and societies are more tightly connected to the global order while others are becoming increasingly marginalized. They contend that there is no evidence to sustain the hyperglobalist claim that the new global social structure is tending toward a global civilization in which equality among individuals will eventually prevail. They argue the opposite: that the world will increasingly consist of a three-tiered system—comprising the elites, the embattled, and the marginalized—that cuts across national, regional, and local boundaries. Thus within nations and places, disparities of wealth will increase just as they will increase across the world-system itself. The economic projections discussed in the following section can provide us with some sense of how and why this will be so.

Global Stratification, Regional Change

For many years now, organizations such as the United Nations and the World Bank have prepared forecasts of the world economy. These forecasts are based on models that take data on economic variables (for example, trends in countries' gross domestic product, or GDP; their imports and exports; their economic structure; their investment and saving performance; and their demographic dynamism) and use known relationships between and among these variables to predict future outcomes. The problem is that economic projection is an inexact science. Economic models are not able to take into account the changes brought about by major technological innovations, significant geopolitical shifts, governments' willingness and ability to develop strong economic policies, or the rather mysterious longer-term ups and downs that characterize the world economy. Current forecasts of economic growth through 2010 are decidedly modest, with GDP per

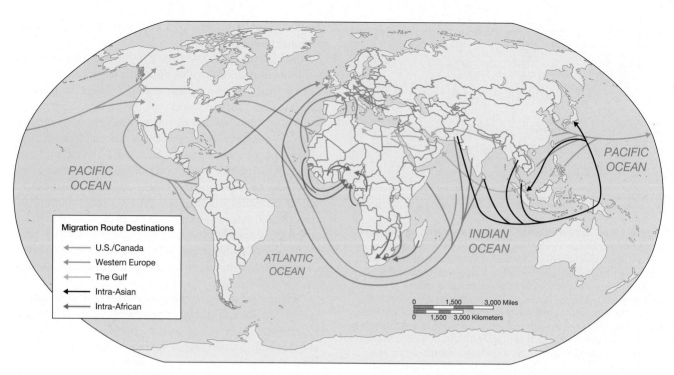

Figure 13.7 Major global migrations, 1945–1995 While migration has been a fact of life for most of the twentieth century, more recently the world has opened up even more. From the late nineteenth to the mid-twentieth century, most migratory flows originated in Europe and Asia and were destined for core countries. But since the end of the Second World War and the decline of the European colonial empires, new regions have become the target of migratory flows. Current figures show that although many migrants are still moving to core regions, not all are. There are also major patterns of migration within Southeast Asia, the Middle East and North Africa, Sub-Saharan Africa, and Latin America. Thus, at the same time that there are global flows of migrants, there are also significant regional flows. (*Source:* Redrawn from S. Held, A. McGrew, D. Goldblatt, and J. Perraton, *Global Transformation*. Cambridge: Polity, 1999, p. 298.)

capita in core economies growing only 1.6 to 2.3 percent each year, while parts of the periphery may experience a decline (**Figure 13.8**).

Such forecasts, of course, are made only for the short term. In overall terms, the global economy is vastly richer, more productive, and more dynamic than it was just 15 or 20 years ago. Every prospect exists that, in the longer term, the world economy will continue to expand. Geographer Brian J. L. Berry has analyzed the timing of past long-wave economic cycles and on that basis has calculated that the next sustained boom might begin around 2010 (**Figure 13.9**). Should this in fact take place, suggests Berry, it will probably be based on the currently emerging infrastructure of information technologies, which will form the platform for a new technology system that will exploit further developments in electronics (semiconductors, superconductors, artificial intelligence), bioengineering, and materials technology.

It is when we focus on the prospects of particular regions and places within those regions that things begin to look less encouraging. Future geographies seem likely to be structured by an even greater gap between the haves and have-nots of the world. The Mexican Zapatista guerilla leader, Subcommandante Marcos, has described the opening up of the global economy as a death sentence for the poor. The gap between the world's core areas and the periphery has already begun to

widen significantly. Even in the United States there is evidence of growing disparity. The Census Bureau, for example, reported in mid-2000 that the gap between the rich and poor in the United States was the widest it had been since the Second World War. The United Nations has calculated that the ratio of GDP per capita (measured at constant prices and exchange rates) between the developed and developing areas of the world increased from 10:1 in 1970 to 12:1 in 1985, and 13:1 in 2000.

Little hope exists that any future boom in the overall world economy will reverse this trend. In spite of the globalization of the world-system (and in many ways *because* of it), much of the world has been all but written off by the bankers and corporate executives of the core. In many peripheral regions, 20 percent or more of all export earnings are swallowed up by debt service—the annual interest on international debts (**Figure 13.10**). In 1998, for example, 29 peripheral countries, including Angola, Congo, Ethiopia, Guyana, Jordan, Nicaragua, Sierra Leone, Sudan, Syria, and Zambia—had total debts so large that they owed more than they produced (**Figure 13.11**). Nicaragua, with a debt of $5.9 billion, a population of only 4.8 million, and a total GNP of $1.8 billion, had the greatest debt burden per person of any country. Although the per capita GNP is only $375, the per capita debt is more than $1200. Practically none of this money is ever likely to be repaid, but so long as the U.S., Eu-

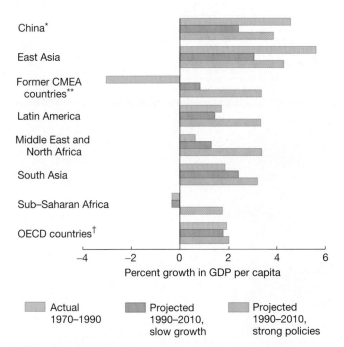

Actual 1970–1990	
Projected 1990–2010, slow growth	
Projected 1990–2010, strong policies	

* China includes Hong Kong.

** Former CMEA countries (CMEA: Council for Mutual Economic Assistance. Member countries were Bulgaria, Cuba, Czechoslovakia, East Germany, Hungary, Mongolia, Poland, Romania, USSR, Vietnam, and Yugoslavia.)

† OECD data include Australia, Canada, European Union, Japan, New Zealand, and United States only.

Figure 13.8   Forecasts of regional economic growth through 2010   This chart shows the actual GDP per capita between 1970 and 1990, together with the World Bank estimates of growth from 1990 through 2010. One forecast is based on the continuation of past trends of slow growth and regional divergence, the other on an assumption of the successful development of strong economic policies in all parts of the world, together with deepening international integration of trade. (*Source:* Data from "Workers in an Integrating World," The World Bank, *World Development Report 1995.* Washington, DC: The World Bank, 1995.)

simply threadbare. They face unprecedented levels of demographic, environmental, economic, and societal stress. In the worst-off regions, such as some parts of Sub-Saharan Africa, for example, the events of the next 50 years are likely to be played out from a starting point of limited basic resources, serious environmental degradation, disease, unprovoked crime, and refugee migrations (**Figure 13.12**). Some African countries will be further disadvantaged because the prices of commodities produced there and in other peripheral regions have been dropping, while imported goods from the core have become more expensive. What this means is that these countries now have, and will probably continue to have, reduced purchasing power in the global marketplace because of the decline in the value of their exports. Today, the combined effects of an external debt crisis, dwindling amounts of foreign aid, insufficient resources to purchase technology or develop indigenous technological innovations, and the high costs of marketing and transporting commodities continue to disable the peripheral regions' full participation in the global economy—especially

Figure 13.9   The long-wave "clock"   In the past, long-wave economic cycles have been fairly regular, with phases of depression occurring at 50- to 55-year intervals, interspersed with episodes of "stagflation"—a combination of economic stagnation and rapid price inflation. At that rate, the next economic boom should be around 2010, after which we can expect a phase of inflationary speculation to lead to the next stagflation crisis, sometime in the 2030s. (*Source:* B. J. L. Berry, *Long-Wave Rhythms in Economic Development and Political Behavior.* Baltimore: Johns Hopkins University Press, 1991, Fig. 70. Copyright © 1991 The Johns Hopkins University Press.)

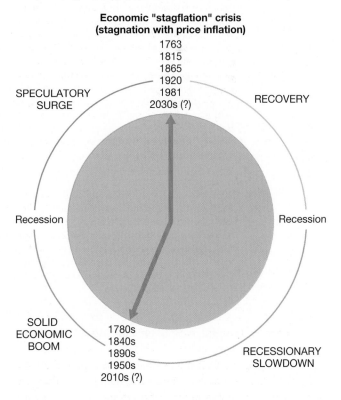

ropean, and Japanese banks continue to receive interest on their loans, they will be satisfied. Indeed, the core regions are doing extremely well from this aspect of international finance. In 1998 the core of the world economy took in about $275 billion in debt servicing, while paying out a total of less than $85 billion in new loans. There is a danger to the future well-being of the better-off regions, however, in that debtor countries might act together in deliberately defaulting on their debts, which would cause a major disturbance to the global financial system. In fact, critics of globalization have charged that the mid-1990s "bailout" of the Mexican economy was really a rescue for Wall Street and the International Monetary Fund, both of which have enormous financial investments in Mexico.

The Marginalized

As for the marginal regions, their future could be very bad indeed. It is not just that they have already been dismissed by investors in the core, nor that their domestic economies are

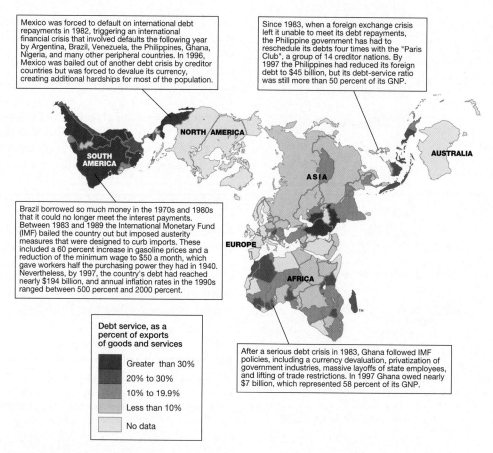

Mexico was forced to default on international debt repayments in 1982, triggering an international financial crisis that involved defaults the following year by Argentina, Brazil, Venezuela, the Philippines, Ghana, Nigeria, and many other peripheral countries. In 1996, Mexico was bailed out of another debt crisis by creditor countries but was forced to devalue its currency, creating additional hardships for most of the population.

Since 1983, when a foreign exchange crisis left it unable to meet its debt repayments, the Philippine government has had to reschedule its debts four times with the "Paris Club", a group of 14 creditor nations. By 1997 the Philippines had reduced its foreign debt to $45 billion, but its debt-service ratio was still more than 50 percent of its GNP.

Brazil borrowed so much money in the 1970s and 1980s that it could no longer meet the interest payments. Between 1983 and 1989 the International Monetary Fund (IMF) bailed the country out but imposed austerity measures that were designed to curb imports. These included a 60 percent increase in gasoline prices and a reduction of the minimum wage to $50 a month, which gave workers half the purchasing power they had in 1940. Nevertheless, by 1997, the country's debt had reached nearly $194 billion, and annual inflation rates in the 1990s ranged between 500 percent and 2000 percent.

After a serious debt crisis in 1983, Ghana followed IMF policies, including a currency devaluation, privatization of government industries, massive layoffs of state employees, and lifting of trade restrictions. In 1997 Ghana owed nearly $7 billion, which represented 58 percent of its GNP.

Debt service, as a percent of exports of goods and services

- Greater than 30%
- 20% to 30%
- 10% to 19.9%
- Less than 10%
- No data

Figure 13.10   The debt crisis At the beginning of the twenty-first century, peripheral and semiperipheral countries owed $2.17 trillion to banks and governments in the core regions of the world-system. In some countries, the annual interest on international debts (their "debt service") accounts for more than 20 percent of the annual value of their export goods and services. Many countries first got into debt trouble in the mid-1970s, when Western banks, faced with recession at home, offered low-interest loans to the governments of peripheral countries rather than being stuck with idle capital. When the world economy heated up again, interest rates rose, and many countries found themselves facing a debt crisis. The World Bank and the International Monetary Fund (IMF), in tandem with Western governments, worked to prevent a global financial crisis by organizing and guaranteeing programs that eased poor countries' debt burdens. Western banks were encouraged to swap debt for equity stakes in nationalized industries, while debtor governments were persuaded to impose austere economic policies. These policies have helped to ease the debt crisis, but often, as discussed in Chapter 2, these policies have resulted in severe hardship for ordinary people. In dark humor, the IMF became known among radical development theorists as "imposing misery and famine." In 1999 an international organization, sponsored by many churches and nongovernmental organizations (NGOs), was formed in Europe, under the name Jubilee 2000 (see also Chapter 6), with the objective of campaigning for the cancellation of international debts, the idea being to give debtor countries a new start for the new millennium. Spreading to North America as well, Jubilee 2000 was successful in encouraging the government of Denmark to cancel international debts worth $635 million, and some other core countries forgave debt arising from earlier aid support. Jubilee 2000, intended as a 1-year campaign, was dissolved in 2001. Neither the United States nor the IMF forgave any debt during the Jubilee 2000 campaign. (*Sources:* Organization for Economic Cooperation and Development, the World Bank, and the United Nations Population Division. Map projection, Buckminster Fuller Institute and Dymaxion Map Design, Santa Barbara, CA. The word *Dymaxion* and the Fuller Projection Dymaxion™ Map design are trademarks of the Buckminster Fuller Institute, Santa Barbara, California, © 1938, 1967 & 1992. All rights reserved.)

in Latin America and the Caribbean, Sub-Saharan Africa, and parts of North Africa.

At the moment, post-independence ideals of modernization and democracy now seem remoter than ever in these regions. Corrupt dictators, epitomized in the Democratic Republic of the Congo by former presidents Mobutu and Ka-bila, and in Nigeria by former President General Abacha, have created "kleptocracies" (from *kleptomania*: an irresistible desire to steal) in place of democracies. Of the estimated $12.42 billion in oil revenues that came to Nigeria as a result of the Persian Gulf crisis of 1990–91, for example, $12.2 billion seems to have passed into the hands of corrupt government

Figure 13.11   Civil war in Sierra Leone In addition to crushing levels of debt, countries like Democratic Republic of Congo, Sudan, and Sierra Leone are also involved with civil wars that suck millions of dollars each day from the domestic economy in order to supply the military. Pictured here are "Kamajor" militia fighters, part of a pro-government civil defense movement in Sierra Leone.

leaders and their friends. Among the regimes that were widely regarded as kleptocracies during the 1990s were those of Cambodia, Côte d'Ivoire, Haiti, Indonesia, Pakistan, Russia, Venezuela, and Yugoslavia.

Amid this political chaos, disease has prospered. Parts of Sub-Saharan Africa may be more dangerously unhealthy today than they were 100 years ago. Malaria and tuberculosis are out of control over much of Sub-Saharan Africa, while AIDS is truly epidemic. In Uganda, where annual spending on health is only $3 per person (compared with debt repayments of $17 per person), one in five children die before his or her fifth birthday, and there is an AIDS epidemic among young adults. More than 22 million people in Sub-Saharan Africa in 2000 were living with the HIV virus (the human immunodeficiency virus that causes AIDS)—one in every 12 people. Almost 30 percent of Zimbabwe's population aged 15 to 49 is infected with the HIV virus, followed by Botswana and Zambia, with rates of 25 percent and 19 percent, respectively.

The worst of the epidemic, however, may have passed in Sub-Saharan Africa. For the first time, there are signs that HIV incidence—the annual number of new infections—may have stabilized there (**Figure 13.13**). In 2000, new infections totaled an estimated 3.8 million, compared to a total of 4 million in 1999. It is not clear why the incidence rates have begun to drop. It may be that the disease has gone on for so long that there are now few new populations to affect. It is also likely that prevention programs have started to take effect in a few African countries, especially in places like Uganda.

While the number of new cases of HIV infection is down in Africa, however, it is up in Russia. In 2000, more cases of HIV infection were reported there than in all the previous years of the HIV epidemic combined. If the number of HIV cases for the Ukraine is added to those reported in the Russian Federation, conservative estimates put the number of people at the end of 2000 living with HIV or AIDS in Eastern Europe and Central Europe at 700,000, an increase of 280,000 over the previous year. The main driving force behind the dramatic increase is unsafe drug-injecting practices.

For the marginalized, the challenges of the contemporary global economy are many and complex. In most instances the marginalized appear to lack many of the key resources, and in some instances the political capacity, necessary to meet those challenges. The prospects for the embattled are severely limited, and solutions to the very difficult obstacles they face are likely to require radical social, political, and economic change.

The Elite

At the other end of the spectrum, the prospects for the core regions of the world-system are bright, especially for large, core-based transnational corporations. With the end of the Cold War, new markets in Eastern and Central Europe have opened up to capitalist industry, along with more resources and a wider range of skilled and disciplined labor. New transport and communications technologies have already facilitated the beginnings of the globalization of production and the emergence of a global consumer culture. Top companies have reorganized themselves to take full advantage of this globalization. Reforms to the ground rules of international trade have removed many of the impediments to free-market growth, and a new, global financial system is now in place, ready to service the new global economy. An example of this is the World Trade Organization (WTO), an institution created in conjunction with GATT (the General Agreement on Trade and Tariffs), which is supranational in its scope. The WTO has begun to provide a system of regulations that supersedes national-level regulations and laws. What this seems to mean in practice is that national restrictions over foreign corporations are increasingly subordinated to the new rules of the WTO. Proponents of the WTO argue that without such an organization the terms of international trade are more likely to be set by powerful countries and transnational corporations, at the expense of the weak. Subordinating powerful national interests to WTO-enforced free trade, they argue, will benefit less-affluent countries by giving them free access to core-country markets and by requiring core countries to stop dumping the products of their subsidized agricultural sectors in peripheral markets. Critics of the WTO argue that the aims and area of responsibility of the WTO have been shaped by international business, and that the way that WTO negotiations take place—with dispute resolution panels made up of unelected bureaucrats rendering decisions in closed sessions—advances the interests of business in general and transnational corporations in particular (**Figure 13.14**).

For the world-system core regions, therefore, the long-term question is not so much one of economic prosperity but of relative power and dominance. The same factors that will consolidate the advantages of the core as a whole—the end of the Cold War, the availability of advanced telecommunications, the transnational reorganization of industry and finance, the liberalization of trade, and the emergence of a global culture—

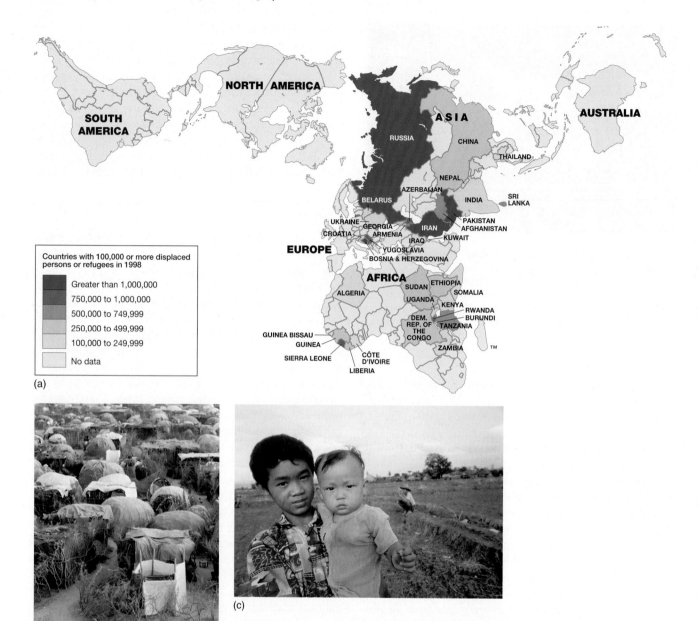

(a)

Countries with 100,000 or more displaced persons or refugees in 1998

- Greater than 1,000,000
- 750,000 to 1,000,000
- 500,000 to 749,999
- 250,000 to 499,999
- 100,000 to 249,999
- No data

(b)

(c)

Figure 13.12   Global civil strife and refugee populations (a) A CIA report estimated that about 40 million people were at risk of malnutrition or death in the late 1990s because of war, famine, disease, and criminal anarchy. Recently, according to estimates supplied by the U.S. Commissioner for Refugees, there were more than 21 million refugees, asylum seekers, and displaced persons in the world—more than 6 million of them in Africa. Eighteen of the world's poorest countries are in Africa. The map shows countries with 100,000 or more displaced persons or refugees in 1998. (b) Many refugees, such as the Somalis shown here in Kenya, spend years, even generations, in makeshift camps far away from their places of origin. About 45,000 refugees live in this camp on the Somali/Kenya border. The camps may be funded by international organizations, such as the United Nations, or by national governments. Often there is little or no support for refugees who must find their own way in their new surroundings, often lacking everything except the clothes on their back. (c) Although refugee children sometimes grow to adulthood in refugee camps, the repatriation of Burmese refugees has been occurring. Fleeing Burma in 1991 and 1992, some 250,000 Muslim Burmese who crossed the border into Bangladesh largely for political reasons were housed in camps supported by the government of Bangladesh and the United Nations High Commission on Refugees (UNHCR). By 1998 approximately 230,000 Burmese refugees had been repatriated and provided with various forms of economic assistance, including technical support and education, by UNHCR. The final 20,000 Burmese refugees have yet to be repatriated. [*Source:* (a) Map projection, Buckminster Fuller Institute and Dymaxion Map Design, Santa Barbara, CA. The word *Dymaxion* and the Fuller Projection Dymaxion™ Map design are trademarks of the Buckminster Fuller Institute, Santa Barbara, California, © 1938, 1967 & 1992. All rights reserved. Data from *Refugees and Others of Concern to the UNHCR,* 1998. Geneva: United Nations High Commission for Refugees, December 12, 1999, available at http://www.unhcr/ch/statist/98oview/tab1_1.htm]

Figure 13.13   Protesters during the XIII International AIDS Conference In summer 2000, these demonstrators in Durban, South Africa, demanded that pharmaceutical companies make AIDS drugs more affordable and thus more available to some of the world's poorest victims of the disease. Several major pharmaceutical companies responded by slashing the prices of AIDS drugs for African countries.

Figure 13.14   Anti-globalization protests, Genoa, 2001 Anti-globalization protesters have been active in Europe as well as the United States. Genoa, Italy, was the site of violent protests in July 2001, when more than 50,000 protesters staged demonstrations over several days. One protester was killed by the Italian police. The protest against the globalization of the economy held in Genoa was directed against the G8-summit attended by President George W. Bush and core country leaders.

will also open the way for a new geopolitical and geoeconomic order. This is likely to involve some new relationships between places, regions, and countries.

As we suggested in Chapter 2, the old order of the "short" twentieth century (1914–89), dominated both economically and politically by the United States, is rapidly disappearing. In our present transitional phase, the new world order is still evolving. This does not necessarily mean, however, that the United States will be unable to renew its position as the world's dominant power. Britain had two consecutive stints as the dominant world power—the hegemon that was able to impose its political view on the world and set the terms for a wide variety of economic and cultural practices (see Chapter 2).

Alternatively, we may not see the same kind of hegemonic power in the new world order of the twenty-first century—there may not be a new hegemon at all. Instead, the globalization of economics and culture may result in a polycentric network of nations, regions, and world cities bound together by flows of goods and capital. Order may come not from military strength rooted in national economic muscle but from a mutual dependence on *trans*national production and marketing, with stability and regulation provided by powerful international institutions (such as the World Bank, the IMF, the WTO, the European Union, NATO, and the United Nations).

The Embattled

In between the elite and the marginalized are the embattled middle regions that possess some of the structural strengths of the elite regions and yet also are burdened with many of the weaknesses of the marginalized ones. For the embattled regions in the middle, there are reasons to be hopeful, yet there are also very clear signs of significant and extremely difficult challenges to be overcome. In order to understand some of

these hopes and challenges, it is important to appreciate the dramatic social and economic transformations that have come about recently in the global economy.

Globalization experts not only divide economic activities into primary, secondary, tertiary, and quaternary activities as we have done in Chapter 1, but more recently they have come to recognize two new categories that supersede these older ones. The first category of economic activity characteristic of the new global economy is "real-time" activities. These are economic activities in which distance and location no longer determine how economic operations will occur. An example of real-time economic activities are those undertaken by an emerging and rapidly growing class of businesspeople who are not tied to any particular place but conduct their work through travel or via strong and widely available telecommunications connections (**Figure 13.15**). Key members of the new economy—marketing experts, computer consultants, legal specialists, financial accountants, and top managers—are representative of real-time economic actors. Regions whose economic activities are predominantly of the real-time sort are those that are more economically flexible and able to respond to new impulses and trends. Individuals in the real-time economic category are a highly mobile group, able to go wherever their services are needed and wherever the highest price for their services can be obtained. Elite regions and individuals are heavily involved in real-time economic activities.

The other new category of economic activity identified by globalization experts is known as "material" activities because material things like tools and equipment, upon which this sort of economic activity is based, require a location to which workers must travel in order to do their work (**Figure 13.16**). Whereas workers in the real-time economy can go where the wage rates are highest, workers in the material economy are not mobile and must submit to whatever the wage rate is in their location. In contrast to regions that

Figure 13.15 Businessman in airport with laptop computer and mobile phone Members of the real-time economic group are highly mobile and are able to conduct business as they move from one geographical location to another, thanks to advances in telecommunications and electronic technologies, with tools that are easily transportable.

Figure 13.16 Factory workers in the periphery Much of the core's industrial processes have been exported to the periphery where labor is cheaper and environmental restrictions are looser than in many core countries. Pictured here are Chinese workers in a factory that contracts to produce Reebok athletic shoes. These workers are clearly involved in the "material" economy rather than the "real time" economy and are far less mobile and subject to local economic conditions. While economic opportunity is improving for these sorts of workers in the embattled category, it must be recalled that industry can always go elsewhere should local labor or political conditions prove unacceptable to corporations like Reebok that will simply offer their contracts to other more appealing bidders.

emphasize real-time economic activities, regions that are dominated by material economic activities are less flexible and therefore less competitive in the global economy. Whereas regions possessing material economic activities control jobs and the potential for wealth generation from production, they also are based on tools and equipment that wear out, become outdated, break down, and must be replaced, all at a significant cost. The more geographically footloose services provided by real-time economies and workers generally involve technological designs, instructions, and advice, all things that can be replaced or updated relatively more easily than can equipment. The point is that these two new categories of economic activity have begun to transform the global organization of work and workers, with the effect that some regions and groups of workers are heavy participants in material activities while others participate more strongly in the real-time economic activities. Those regions that are heavily committed to material activities tend to be part of the embattled middle.

The embattled middle regions and social groups do not enjoy the same high levels of wealth and living standards that are available to elite groups and regions, though it is also true that those in the embattled middle are certainly better off than the marginalized. In short, they are caught in the middle between the elites and the marginalized. Consider the substantial differences that exist between elite workers' salaries, benefits, and overall standard of living as compared to the embattled middle workers. Social groups tied to the material activities are unlikely to have stock options as part of their employment package. They are not likely to be covered by company- or government-sponsored health and unemployment insurance nor are they likely to have the opportunity to use sick leave. In short, their place in the labor market is structurally insecure, and they are often working part-time,

are self-employed, are employment agency workers, or are contract workers.

But compared to the truly marginalized regions (and workers) of the world, the embattled middle regions must consider themselves to be a lot better off. The embattled regions are able to trade with both the elite and the marginalized, exchanging different kinds of products with each, at the same time that they achieve modest profit and intermediate wage levels. The embattled social groups are more than likely to have at least some form of work—part-time, casual, or full-time. And even though these jobs may not be highly remunerated or carry a wide range of benefits, it is certainly better than being unemployed, which is often the situation for the marginalized. The embattled middle can easily see that they are better off than the marginalized at the same time that they are hopeful that one day, they too can gain elite status.

The most significant point to keep in mind about the stratification of the world economy through recent globalization is that the elite, the marginalized, and the embattled are not concentrated in any one particular region or regions but are rather spread out across the globe. It is no longer accurate to see Brazil, for instance, as a wholly embattled region or its population as exclusively embattled. Instead, it must be seen that Brazil, like Great Britain or Nigeria, contains the range of stratified regions and groups within its national boundaries such that elite, embattled, and marginalized regions and social groups are all part of the larger whole. Moreover, the elite regions and social groups should be seen as having more in com-

(a)   (b)

Figure 13.17   Global social hierarchy These photographs demonstrate that material possessions and levels of wealth and poverty connect different social groups around the world. The new social hierarchy that has emerged around globalization has resulted in a situation where the elites, the embattled, and the marginalized tend to have more in common with their own social stratum wherever they live than with members of the other stata, even in their own cities and regions. Pictured here are members of the marginalized strata in (a) Bhutan and (b) Ethiopia. Although the particular circumstances in which they live may be different (perhaps a large level of international debt in one or little connection to the global economy in another), the chances for an overall improvement in standard of living are slim.

mon with elite regions and groups in other parts of the world than they do with the embattled and marginalized groups within their own national boundaries. The same is true of the embattled and the marginalized such that marginalized social groups or regions within the United States are likely to have more in common in terms of their relative well-being and chances for betterment with the same regions and social groups in Nigeria than they do with the elites and the embattled in their own country (**Figure 13.17**).

It is clear that at the global level there is sufficient wealth to provide for basic human needs for all the world's regional populations. Unfortunately however, that wealth is unevenly distributed geographically. Moreover, as national states cede more and more responsibility for managing the global economy to supranational organizations and as the welfare state disappears, there are fewer standard mechanisms to redistribute wealth within or across countries. While the hyperglobalists believe that globalization will ultimately raise all social groups to the same level of well-being and wealth, both the transformationalists and the skeptics are less optimistic as hard data show the current gap between rich and poor to be widening. Indeed, there is also a great deal of skepticism among globalization experts that the levels of economic development enjoyed by the core of the world economy can be sustainable throughout the globe. We turn to the complex questions of sustainability in the following section.

Sustainability and Regional Change

The increasing development of the world economy that is heralded by the hyperglobalists is not necessarily welcomed by other globalization experts. Although there is a general

desire to improve the livelihoods of all people throughout the globe who are currently living in poverty or distress, there are serious reservations about the aim of raising all living standards to the level that core regions and the elite enjoy. In fact, one set of globalization experts firmly believes that some of the highest standards of living around the globe will have to be lowered in order to raise others because the widespread economic development such a goal would require is not environmentally sustainable. In short, because of the limitations set by Earth's resources—both in terms of amount of key resources such as energy or soil and quality of the environment with respect to air and water—improvements in social well-being at the level now enjoyed by the elite could not be sustained over the long term and may not even be possible to begin with. As a result, the future of economic globalization is increasingly being accompanied by a widespread discussion about the sustainability of such a process. We turn to a discussion of sustainability after we provide a general review of the critical link between resources and development.

Resources and Development

The expansion of the world economy and the globalization of industry will undoubtedly boost the overall demand for raw materials of various kinds, and this will spur the development of some previously underexploited but resource-rich regions in Africa, Europe, and Asia. Raw materials, however, will be only a fraction of future resource needs. The main issue, by far, will be energy resources. World energy consumption has been increasing steadily over the recent past (**Figure 13.18**). As the periphery is industrialized and its population increases further, the global demand for energy will expand rapidly. Basic industrial development tends to be highly energy-intensive,

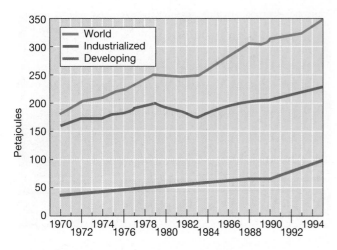

Figure 13.18 Trends in energy consumption Global commercial energy consumption rose from less than 200 petajoules—1 quadrillion joules—in 1970 to nearly 350 petajoules in 1997. (A joule is a unit of energy about equal to the force with which a grapefruit hits a table after being dropped from a height of about 10 centimeters, or 4 inches.) (*Source:* Data from the United Nations Statistical Division, *1995 Energy Statistics Yearbook.* New York: United Nations, 1997.)

however. The International Energy Agency, assuming (fairly optimistically) that energy in peripheral countries will be generated in the future as efficiently as it is today in core countries, estimates that developing-country energy consumption will more than double by 2010, lifting total world energy demand by almost 50 percent.

It has been predicted that in 2010, peripheral and semi-peripheral regions will account for more than half of world energy consumption. Much of this will be driven by industrialization geared to meet the growing worldwide market for consumer goods, such as private automobiles, air conditioners, refrigerators, televisions, and household appliances. Current trends give some indication of this. For example, since 1978 the number of electric fans sold in China has increased almost twentyfold, and the number of washing machines has increased from virtually zero to 97 million. Trends such as this caused China's oil consumption to leap by 11 percent in 1993, turning the world's sixth-largest oil producer into an oil importer.

While world reserves of oil, coal, and gas are, in fact, relatively plentiful, without higher rates of investment in exploration and extraction than at present, production will be slow to meet the escalating demand. The result might well be a temporary but significant increase in energy prices. This would have important geographical ramifications: Companies would be forced to seriously reconsider their operations and force core households into a serious reevaluation of their residential preferences and commuting behavior; and peripheral households would be forced further into poverty. If the oil-price crisis of 1973 is anything to go by (after crude oil prices had been quadrupled by the OPEC cartel), the outcome could be a significant revision of patterns of industrial location and a substantial reorganization of metropolitan form.

Significantly higher energy costs may change the optimal location for many manufacturers, leading to deindustrialization in some regions and to new spirals of cumulative causation in others. Higher fuel costs will encourage some people to live nearer to their place of work, while others will be able to take advantage of telecommuting to reduce personal transportation costs. It is also relevant to note that almost all of the increase in oil production over the next 15 or 20 years is likely to come from outside the core economies. This means that the world economy will become increasingly dependent on OPEC governments, which control more than 70 percent of all proven oil reserves, most of them in the Middle East.

On a more optimistic note, some potential efficiencies exist that might mitigate the demand for raw materials and energy in many parts of the world. Chinese industry, for example, currently uses 35 percent more energy per ton of steel than U.S. industry, mainly because it has small, inefficient plants with outdated, energy-guzzling capital equipment. New plants and equipment would allow significant energy savings. Considerable efficiencies could also be made through changes in patterns of consumption. In former socialist states, for example, years of price subsidies have made consumers (both industries and households) profligate with energy. East European countries currently consume four or five times as much energy per capita as do countries of the same income levels elsewhere.

In countries that can afford the costs of research and development, new materials will reduce the growth of demand both for energy and for traditional raw materials such as aluminum, copper, and tin. Japan, for instance, may be able to reduce motor vehicle fuel consumption by 15 percent (and thereby reduce its total fuel oil consumption by 3 percent) by using ceramics for major parts of engines. It may also be possible to substitute ceramics for expensive rare metals in creating heat-resistant materials. Improved engineering and product design will also make it possible to reduce the need for the input of some resources. U.S. cars, for example, were almost 20 percent lighter in 2000 than they were in 1974. In addition, of course, the future may well bring technological breakthroughs that dramatically improve energy efficiency or make renewable energy sources (such as wind, tidal, and solar power) commercially viable. As with earlier breakthroughs that produced steam energy, electricity, gasoline engines, and nuclear power, such breakthroughs would provide the catalyst for a major reorganization of the world's economic geographies.

The past 25 years have seen the growing awareness of the impacts that continued globalization will have on the regions in which we live. Increasingly citizens, nongovernmental organizations, and environmental policy makers have expressed growing concern over the negative outcomes of rapid and enduring global economic growth. However, because growth is so critically tied to improving the lives of poor people around the world, governments are reluctant to limit it. The response from the global community, hammered out during the course of international meetings, through academic publications, and in response to social protest, is a different approach to development than previously existed. This approach is known as sustainable development (see Chapter 1).

Sustainability

The world currently faces a daunting list of environmental threats: the destruction of tropical rain forests and the consequent loss of biodiversity; widespread, health-threatening pollution; the degradation of soil, water, and marine resources essential to food production; stratospheric ozone depletion; acid rain, and so on. And, as we have shown throughout this text, most of these threats will have their greatest impact in the world's periphery, where daily environmental pollution and degradation cannot be mitigated through expensive technological solutions as they are in the core.

In these peripheral regions there is, simply, less money to cope with environmental threats. In addition, the very poverty endemic to peripheral regions also adds to environmental stress. In order to survive, the rural poor are constantly impelled to degrade and destroy their immediate environment, cutting down forests for fuelwood and exhausting soils with overuse. In order to meet their debt repayments, governments feel compelled to generate export earnings by encouraging the harvesting of natural resources. In the cities of the periphery, poverty encompasses so many people in such concentrations as to generate its own vicious cycle of pollution, environmental degradation, and disease. Even climate change, an inherently global problem, seems to pose its greatest threats to poorer, peripheral regions.

Future trends will only intensify the contrasts between rich and poor regions. We know enough about the growth of population and the changing geography of economic development to be able to calculate with some confidence that the air and water pollution generated by low-income countries will more than double in the next 15 years. By 2010, China will probably account for one-fifth of global carbon dioxide emissions. We know, in short, that environmental problems will be inseparable from processes of demographic change, economic development, and human welfare. In addition, it is becoming clear that environmental problems are going to be increasingly enmeshed in matters of national security and regional conflict. The prospect of civil unrest and mass migrations resulting from the pressures of rapidly growing populations, deforestation, soil erosion, water depletion, air pollution, disease epidemics, and intractable poverty is real (**Figure 13.19**). These specters are alarming not only for the peoples of the affected regions but also for their neighbors.

Such images are also alarming for the elite and embattled regions and social groups whose continued prosperity will depend on processes of globalization that are not disrupted by large-scale environmental disasters, unmanageable mass migrations, or the breakdown of stability in the world-system as a whole. Over a decade ago the U.S. security agenda was cast in terms of environmental awareness. In an address to the United Nations in 1993, for example, former President Bill Clinton noted that global environmental issues "threaten our

children's health and their very security" and promised that the United States will "work far more ambitiously to fulfill our obligations as custodians of this planet not only to improve the quality of life for our citizens . . . but also because the roots of conflict are so often entangled with the roots of environmental neglect and the calamities of famine and disease." Unfortunately, after 9 years of international negotiations, in mid-2001, President George W. Bush announced that the United States would no longer honor its commitment to the 1997 Kyoto agreement. The implications of this decision are that the United States will continue to be the world's largest single generator (more than 25 percent) of greenhouse gases—emissions that are leading to worldwide rises in temperature. Such temperature increases—known as global warming—have more potential to damage Earth's web of life than any other factor outside of nuclear war or a collision with an asteroid. In addition to causing a rise in sea level throughout the world (which could result in widespread loss of property and livelihoods), global warming is also likely to contribute to increases in heat-related deaths (especially respiratory illnesses) and a widening of the range of disease-carrying rodents and insects (which would cause an increase in malaria, dengue fever, and Lyme disease, among other afflictions).

Despite the fact that the United States appears to be excluding itself from efforts to solve one of Earth's most pressing environmental problems, the notion of sustainable development continues to gain momentum around the globe.

Figure 13.19   Air pollution in Guangzhou, China China has embraced industrialization in a new way, experiencing explosive economic growth and the negative impacts on the environment that unregulated growth brings. The Guangzhou Province, visible from Hong Kong, and at the mouth of the Pearl River Delta, is home to more than 60 million people. The region has grown rapidly, largely through international manufacturing investment, so much so that it triples its economic output every 8 years. Pictured here is downtown Guangzhou (formerly Canton), which is currently experiencing a huge development boom. Notice the air pollution hanging over the area. The pollution is generated by the burgeoning industrial economy operating in the absence of any national or regional environmental restrictions.

In practice, sustainable development means economic growth and change should occur only when the impacts on the environment are benign or manageable and the impacts (both costs and benefits) on society are fairly distributed across classes and regions. Sustainable development is geared to meeting the needs of the present without compromising the ability of future generations to meet their needs. It envisages a future when improvements to the quality of human life are achieved within the context of local and regional ecosystems.

Sustainable development means using renewable natural resources in a manner that does not eliminate or degrade them—by making greater use, for example, of solar and geothermal energy, and by greater use of recycled materials. It means managing economic systems so that all resources are used optimally. It means regulating economic systems so that the benefits of development are distributed more equitably (if only to prevent poverty from causing environmental degradation). It also means organizing societies so that improved education, health care, and social welfare can contribute to environmental awareness and sensitivity and an improved quality of life.

A final and more radical aspect of sustainable development is a move away from wholesale globalization toward increased "localization": a desire to return to a more locally based economy where production, consumption, and decision making can be oriented to local needs and conditions. Localization also means basing livelihoods on the resources available in the local area. Thus peripheral regions, as well as workers and citizens throughout many parts of the core, are demanding a reinstatement of control over the economic events and institutions that directly shape their lives.

Put this way, sustainable development sounds eminently sensible yet impossibly utopian. The first widespread discussion of sustainability took place in the early 1990s and focused on the "Earth Summit" (the United Nations Conference on Environment and Development) meeting in Rio de Janeiro in 1992. Attended by 128 heads of state, it attracted intense media attention. At the conference, many examples were described of successful sustainable development programs at the local level. Some of these examples centered on the use of renewable sources of energy, as in the creation of small hydroelectric power stations to modernize Nepalese villages. Some examples centered on ecotourism in environmentally sensitive areas, as in the trips organized in Thailand's Phang Nga Bay (**Figure 13.20**). There tourists visit spectacular hidden lagoons by sea canoe under strict environmental regulation (no drinking, littering, eating, or smoking, and limits on the number of visitors allowed each day) and with a high level of social commitment (emphasizing respect for staff, local culture and family, proper training, and good pay). Most examples, however, centered on sustainable agricultural practices for peripheral countries, including the use of intensive agricultural features such as raised fields and terraces in Peru's Titicaca Basin—techniques that had been successfully used in this difficult agricultural environment for centuries, before European colonization. After the U.N. conference, however, many observers commented bitterly on the deep conflict of interest between core countries and peripheral countries that was exposed by the summit. Without radical and widespread changes in value systems and unprecedented changes in political will, "sustainable development" will remain an embarrassing contradiction in terms.

We cannot just wait to see what the future will hold. If we are to have a better future (and if we are to *deserve* a better future), we must use our understanding of the world—and of geographical patterns and processes—to work toward more desirable outcomes. No discipline is more relevant to the ideal of sustainable development than is geography. Where else, as British geographer W. M. Adams has observed, can the science of the environment (physical geography) be married with an understanding of economic, technological, social, politi-

Figure 13.20 Ecotourism in Phang Nga Bay, Thailand Spectacular and scenic Phang Nga Bay in Thailand has become the site of ecotourism development. Located in Phang Nga National Park, the bay is home to floating fishing villages arrayed around beautiful limestone "sea mountains" that rise from the ocean as vertical walls, some up to 300 meters (nearly 1000 feet) high. These sea mountains are actually islands that are largely uninhabited. Hidden among them are numerous caves, tunnels, lagoons, and collapsed caverns eroded by natural forces over millions of years. These *hongs* (the Thai word for "room") can be entered via the tunnels only when the tide is right. The most popular ecotourism activity in the bay is sea kayaking and canoeing, as both enable easy entrance into these unusual formations. A number of small-scale ecotourism companies operate in the bay with the aim of exposing visitors to the area's natural beauty without contaminating or disturbing the natural environment.

cal, and cultural change? What other discipline offers insights into environmental change, and who but geographers can cope with the diversity of environments and the sheer range of scales at which it is necessary to manage global change?

Adjusting to the Future

The immediate future will be characterized by a phase of geopolitical and geoeconomic transition; by the continued overall expansion of the world economy; and by the continued globalization of industry, finance, and culture. The processes of change involved in shaping this kind of future will inevitably bring critical issues, changes, conflicts, and threats. We can already identify what several of these might involve: fault lines of cultural dissonance, the changing configuration of state power, and regional change.

Globalizing Culture and Cultural Dissonance

At one level, globalization has brought a homogenization of culture through the language of consumer goods. This is the level of "Planet Reebok," where material cultures are enmeshed by 747s, CNN, music video channels, cell phones, and the Internet; and where they are swamped by Coca-Cola, Budweiser, McDonald's, Gap clothing, Nikes, Walkmans, Nintendos, Hondas, Disney franchising, and formula-driven Hollywood movies. Furthermore, sociologists have recognized that a distinctive culture of "global metropolitanism" is emerging among the transnational elite. This is simply homogenized culture at a higher plane of consumption (French wines instead of Budweiser, Hugo Boss clothes instead of Levi's, BMWs instead of Hondas, and so on). The members of this new culture are people who hold international conference calls; who send and receive faxes and e-mail; who make decisions and transact investments that are transnational in scope; who edit the news, design and market international products, and travel the world for business and pleasure.

These trends are transcending some of the traditional cultural differences around the world. We can, perhaps, more easily identify with people who use the same products, listen to the same music, and appreciate the same sports stars that we do. At the same time, however, sociocultural cleavages are opening up between the elites and the marginalized. By focusing people's attention on material consumption, they are also obscuring the emergence of new fault lines—between previously compatible cultural groups, and between ideologically divergent civilizations.

Several reasons account for the appearance of these new fault lines. One is the release of pressure brought about by the end of the Cold War. The evaporation of external threats has allowed people to focus on other perceived threats and intrusions. Another is the globalization of culture itself. The more people's lives are homogenized through their jobs and their material culture, the more many of them want to revive subjectivity, reconstruct we/us feelings, and reestablish a distinc-

tive cultural identity. For the marginalized, a different set of processes is at work, however. The juxtaposition of poverty, environmental stress, and crowded living conditions alongside the materialism of "Planet Reebok" creates a fertile climate for crime. The same juxtaposition also provides the ideal circumstance for the spread and intensification of religious fundamentalism. This, perhaps more than anything else, represents a source of serious potential cultural dissonance.

The overall result is that cultural fault lines are opening up at every geographical scale. This poses the prospect of some very problematic dimensions of future regional geographies from the scale of the city to the scale of the globe. The prospect at the metropolitan scale is one of fragmented and polarized communities, with outright cultural conflict suppressed only through electronic surveillance and the "militarization" of urban space via security posts and the "hardened" urban design using fences and gated streets (**Figure 13.21**). This, of course, presupposes a certain level of affluence in order to meet the costs of keeping the peace across economic and cultural fault lines. In the unintended metropolises of the periphery, where unprecedented numbers of migrants and refugees will be thrown together, there is the genuine prospect of anarchy, and intercommunal violence exists—unless, that is, intergroup differences can be submerged in a common cause such as religious fundamentalism.

At the subnational scale the prospect is one of increasing ethnic/racial rivalry, parochialism, and insularity. Examples of these phenomena can be found throughout the world. In North America, we see increasing ethnic rivalry represented by the secessionism of the French-speaking Quebeçois, and the

Figure 13.21   Gated community in Shanghai, China
While gated communities with closed-circuit television cameras have become routine phenomena in much of the core, it is important to recall that the core exists in the periphery just as the periphery is part of the core. Shown here are armed guards at a gated community in the city of Shanghai in China. Behind the gates of this community live members of the international elite who, though clearly members of the Chinese nation, are also connected to elites throughout the world in terms of the their strong attachments to the global economy. The appearance on the landscape of the gated and guarded community in China is particularly disturbing given that throughout much of the twentieth century China appeared to be committed to an egalitarian social order.

insistence of some Hispanic groups in the United States on the installation of Spanish as an alternative first language. In Europe, examples are the secessionism (from Spain) of the Basques, the separatist movement of the Catalans (also in Spain), the regional elitism of northern Italy, and, recently, outright war between Serbs and Croats. In South Asia, examples are provided by the recurring hostility between Hindus and Muslims throughout the Indian subcontinent, and between the Hindu Tamils and Buddhist Sinhalese of Sri Lanka. In Africa, examples include ethnic rivalry, parochialism, and insularity such as the continuing conflict between the Muslim majority in northern Sudan and the Christian minority in the south of the country as well as widespread unrest in northeast Democratic Republic of Congo. Where the future brings prosperity, tensions and hostilities such as these will probably be muted; where it brings economic hardship or decline, they will undoubtedly intensify.

The prospect at the global scale is of a rising consciousness of people's identities in terms of their broader historical, geographical, and racial "civilizations": Western, Latin American, Confucian, Japanese, Islamic, Hindu, and Slavic-Orthodox. According to some observers, deepening cleavages of this sort could replace the ideological differences of the Cold War era as the major source of tension and potential conflict in the world. The potential for such a scenario was sharply underscored by terrorist attacks on the Pentagon and the World Trade Center in September 2001, which were interepreted by governments around the world as a declaration of war by Islamist extremists against Western values and geopolitics.

Globalization, Transnational Governance, and the State

As we have already noted, globalization has been as much about restructuring geoeconomics as it has been about reshaping geopolitics. In fact, the hyperglobalists believe that the impact of globalization on politics has been so profound that it is leading to the diminution of the powers of the modern state, if not its ultimate disappearance. The hyperglobalists believe that because the modern state is organized around a bounded territory and because globalization is creating a new economic space that is transnational, the state is increasingly unable to respond to the needs of the new transnational economy. Although we in this text do not subscribe to such a position, we do recognize that the state is undergoing dramatic changes that are restructuring its role with respect to both local, domestic concerns as well as global, transnational ones. Moreover, these changes are very much part of recent history.

In the twentieth century, from the end of the Second World War until 1989 when the Berlin Wall was dismantled, world politics had been organized around two superpowers. The capitalist West rallied around the United States and the communist East gathered around the Soviet Union. But with the fall of the Berlin Wall signaling the end of communism, a bipolar world order came to an end and a new world order, organized around global capitalism, emerged and has become increasingly solidified around a new set of political powers and institutions that recast the role of the state in a new context. In short, a new world order predicated on the fall of communism and enabled by the technological (particularly communication and transportation) changes of the last 25 years is restructuring the architecture as well as the conduct of contemporary politics at both the international and the domestic level.

Throughout the text, we have referred to the importance for the contemporary global economy of such regional and supranational organizations as the European Union, the North American Free Trade Agreement (NAFTA), the Association of Southeast Asian Nations (ASEAN), the Organization of Petroleum Exporting Countries (OPEC), and the WTO (see Chapter 2 for a more extended discussion). We have also noted that these organizations are unique in modern history as they aim to treat the world or their respective regions as seamless trading areas not hindered by the rules that ordinarily regulate national economies. The increasing importance of these trade-facilitating organizations is the most telling indicator that the world, besides being transformed into one global economic space, is also experiencing global geopolitical transformations. Another example of growing transnational governance is provided by the international response to the emergence of global environmental problems such as fisheries depletion, global warming, and the loss of biodiversity, in which the United Nations connects with other international organizations to develop treaties and agreements that seek to manage resources and pollution at a global level. But rather than disappearing altogether, the powers and roles of the modern state are changing as it is forced to interact with these sorts of organizations as well as with a whole range of other political institutions, associations, and networks (**Figure 13.22**).

The point is not that the state is disappearing but rather that it must now contend with a whole new set of processes and other important political actors, not only on the international stage but also within its own territory. For instance, geographer Andrew Leyshon has shown how the financial revolution of the 1980s established a transnational financial network that is far beyond the control of any one state, even a very powerful state like the United States, to regulate effectively. In fact, what the increasing importance of transnational flows and connections—from flows of capital to flows of migrants—indicates is that the state is less a container of political or economic power and more a site of flows and connections.

As we discussed in Chapter 2 (see Figure 2.38 in particular), the increasing importance of flows and connections means that contemporary globalization has enabled an increasingly shrinking world. Geographers call this phenomenon **time-space convergence,** which is a decrease in the friction of distance between places. In addition to allowing people and goods to travel farther faster and to receive and send information more quickly, time-space convergence, or the smaller world that globalization has enabled, means that politics and political action have also become global. In short, politics is able to move beyond the confines of the state out into the global political arena, where rapid communications enable complex supporting networks to be developed and deployed, thereby facilitating interaction and decision making. A good

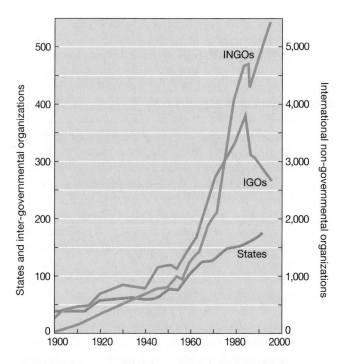

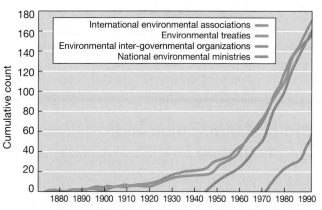

Figure 13.23   Growth of international environmental organizations   Since the Second World War, there has been a dramatic increase in global environmental problems as well as the development of national, international, and voluntary organizations to assess and address those problems. (*Source:* Data from S. Held, A. McGrew, D. Goldblatt, and J. Perraton, *Global Transformation.* Cambridge: Polity, 1999, p. 388.)

Figure 13.22   Growth of states and international governmental and nongovernmental organizations in the twentieth century   While the number of states has grown steadily over the twentieth century, intergovernmental organizations (IGOs) and international nongovernmental organizations (INGOs) have also experienced dramatic growth, particularly since the 1960s. In 1909 there were 37 IGOs and 176 INGOs. In 1996 there were 260 IGOs and 5472 INGOs. Another important feature of the internationalization of governance is the number of international treaties in force between states. The number of international treaties increased from 6351 to 14,061 between 1946 and 1975. While states continue to persist as the main forms of national government, they have shed many of their previous governing responsibilities as they have been taken over by international governing organizations as well as by nongovernmental organizations. (*Source:* Data from Union of International Associations, *Yearbook of International Organizations, 1996-1997.* Munich: K. G. Saur, 1996, p. 54.)

example of this is the protests that occurred in Seattle, Washington, in 1999 over the scheduled meeting of the WTO (see Chapter 1). Telecommunications, and especially the Internet, enabled the protest leaders to organize and deploy their actions with the participation of interested groups all over the world. These Seattle protests were an expression of truly global politics that matched the global politics of the WTO itself, which had come to Seattle, after years of planning, to hold its 1999 meeting. Thus both institutionalized politics as well as popular political movements are able to be truly global in their reach. One indication of the increasingly global nature of politics, outside of formal political institutions, is the increase in environmental organizations whose purview and membership are global (**Figure 13.23**).

What has been most interesting about the increasing prominence of the institutionalization of global politics is that it has been less involved with the traditional preoccupations of relations between states and military security issues and more involved with issues of economic, ecological, and social

security. The massive increase in flows of trade, foreign direct investment (FDI), financial commodities, tourism, migration, crime, drugs, cultural products, and ideas has been accompanied by the emergence and growth of global and regional institutions, the role of which is to manage and regulate these flows. The impact of these twin forces on the modern state has been to draw it increasingly into this complex of global, regional, and multilateral systems of governance. And as the state has been drawn in to these new activities, it has shed, or de-emphasized, some of its previous responsibilities, such as social welfare provision and military buildup. The involvement of the state in these new global activities, the growth of supranational and regional institutions and organizations, the critical significance of transnational corporations to global capital, and the proliferation of transnational social movements and professional organizations is captured by the term **international regime.** The term is meant to convey the idea that the arena of contemporary politics is now international, so much so that even city governments and local interest groups—from sister city organizations to car clubs—are making connections and conducting their activities both beyond and within the boundaries of their own states. An example of this is the human rights movement that has gained ascendancy over the last four to five decades or so. Until the Second World War, safeguarding human rights was the provenance of states whose rules and regulations legislated the proper treatment of its citizens, from prisoners to schoolchildren. Since the late 1940s and 1950s, nearly all states have come to accept the importance of a comprehensive political and legal framework that focuses on human rights and that allows an international organization to intervene in the operations of a sovereign state that is in violation of the International Bill of Human Rights.

The point of this discussion of the changing role of the state and the emergence of new regimes of international governance is to better appreciate them as primary forces shaping the future of world regions. The contemporary world political order, therefore, is one in which the state system persists at the same time that a growing number of international, supranational, and regional authority structures are emerging and gaining strength. Such a new political order can be seen to have both fragmenting and integrating influences on the present organization of world regions.

Regional Integration and Fragmentation

Will globalization bring about increased regional integration so that our current regions will remain the building blocks of the world order? Or will overall fragmentation lead to the emergence of new regions? The answers to these two questions are not easy because for different parts of the world, we have evidence to support both successful integration and continued fragmentation. If we look to Europe, for instance, regional integration has been remarkably successful. Since 1957, the various states of Europe have become increasingly integrated around political and economic concerns, so much so that the European Union is a shining global example of a coherent and prosperous world region. Elsewhere, however, especially in Africa and Latin America, attempts at

regional integration have been far less successful, and subnational pressures have, in fact, lead to more fragmentation than integration. A helpful way to think about the possibilities for future regional integration is to consider why regional integration might be attractive for some states and troubling for others.

Probably the most important reasons to pursue regional integration revolve around strategic and security issues. As we have already discussed, regional trading blocks are created because they are understood to enable economic and trading advantages. And the one variable that demonstrates the foundation of contemporary globalization is the increase in global trade linkages (**Figure 13.24**). Regional trading blocks are also believed to provide security for weaker states against economic and political threats. Following the devastation wreaked on Europe by the Second World War and compounded by the loss of their colonies, the member states of the European Union (organized in 1957 as the European Economic Community, or EEC) viewed it as a way of increasing political and economic leverage in world affairs. It was widely understood that no single state could have exercised the power that a cooperative union could. The perceived military threat of the Soviet Union was also an

Figure 13.24   World trade interconnectedness Since the Second World War, a number of agreements have been put into place that have helped to foster the multilateral trading patterns shown in this map. Bretton Woods and GATT were particularly important in providing the basis for a multilateral trading order. In 1995 the World Trade Organization transformed and superseded GATT as the primary international trade agency. (*Source:* Data from S. Held, A. McGrew, D. Goldblatt, and J. Perraton, *Global Transformation.* Cambridge: Polity, 1999, p. 166.)

Number of states that each state trades with

Greater than 120
91 to 120
61 to 90
31 to 60
0 to 30
No data

important factor in the decision to establish a regional bloc among the founding, Western European, members of the EEC. Finally, since its founding, the still-developing European Union has been seen by smaller and later joining states—such as Denmark and Ireland—as a way of improving their political and market status, something they would lack if they remained isolated.

But just as there are strong incentives for forming regional blocks, there are also substantial obstacles that make it difficult for states to cooperate and intensify any blocks that already exist. These obstacles tend to revolve around difference and the desire by states to maintain control over their own affairs. The issue of difference centers on the fear that integration can undermine the cultural traditions of any particular nation because cooperation would necessarily require standardization of practices. Because nation-states are rarely homogeneous, the cultural and economic diversity that already exists is often used as a basis for demanding more local control rather than ceding a portion of control to an even more remote political unit. It is commonly perceived that integration is likely to exacerbate the tensions surrounding difference and increase social problems. These are just the sorts of challenges that the Latin American and African trading blocks have faced. But it is also important to recognize that even the very successful EU is not immune to these sorts of counterpressures to regional integration either. Britain is currently reluctant to enter into the single EU currency scheme, for instance, and attempts to expand the Asian trading blocks such as ASEAN as well as NAFTA have also raised fears about protecting local interests and maintaining local control.

Because there are both pros and cons for regional integration and mindful that predicting the future is a tricky business, our view is that the contemporary world order of regions is likely to continue over the next decade with some possibility for surprise. We are convinced that the contemporary triadic core is likely to continue to dominate the global economy and, moreover, is more than likely to increase its strength over the first decade of the twenty-first century, with some important additions and transformations.

As you will recall from Chapter 2, the triadic core consists of Europe, North America, and East Asia—even more precisely, the EU, the United States, and Japan. So while one world region is a dominant player on the world stage, the other two are actually states, which, although involved in regional alliances (NAFTA in the case of the United States and ASEAN in the case of Japan), are at the moment the strongest, most aggressively controlling players in their respective regions. Still, we expect the triadic core to maintain its hegemony for the next 10 years. The important additions and transformation we anticipate include the possibility that Mexico may become increasingly integrated into the North American region through continued growth and expansion of NAFTA. While Mexico will not achieve the level of wealth of either Canada or the United States over this decade, it will likely increase its position in the world economy and alter and add to the strength of the current North American region. Because the Free Trade Areas of the Americas (FTAA) proposal is still in its infancy, it is too soon to tell what impact that might have on creating a mega-trading zone among the regions of North America and Latin America and the Caribbean. Europe, through the EU, will also continue to develop and gain strength as the current applicant states to the EU (Estonia, Latvia, Lithuania, Poland, Czech Republic, Slovakia, Hungary, Slovenia, Romania, Bulgaria, and Cyprus) become more economically strong and politically stable and are allowed to enter the EU. And while there will still be an uneven distribution of wealth and development in the transforming EU, the new states will be better off than they were 10 years ago, and Europe overall will be economically and politically stronger. East Asia is also likely to gain strength both politically and economically, as Japan climbs out of a recession and China continues to develop its economic base and open its markets to the West. Although China, too, has an uphill battle in raising its level of wealth to that of Japan, its enormous market and the industriousness of its population suggest that China has a bright economic future in store.

Yet, if we are fairly certain about the continued integration and growth of Europe, North America, and East Asia, we must also acknowledge that regions outside the core are not likely to experience any dramatic economic transformations that will catapult them into core status. Latin America, Southeast Asia, South Asia, and Russia, Central Asia, and the Transcaucasus, Sub-Saharan Africa, and the Middle East and North Africa will continue to press for special trading relationships with the core, and though there will certainly be changing fields of influence and allegiance, the relative economic disadvantage of these regions with respect to the core will continue. The Australia, New Zealand, and South Pacific region is likely to maintain its position as economically part of the core, though politically not particularly central to it.

Summary and Conclusions

In this chapter we discussed the future of contemporary world regions in light of the likely transformations that globalization may bring about. We reviewed the three positions in the globalization debate as a way of understanding the different approaches to globalization's possible effects. As the global integration of the world economy continues, we should expect to see that social stratification will continue to increase the gap between the haves and the have-nots of the world, and a three-tiered hierarchy that has already become established will deepen. This hierarchy consists of the elite, the marginalized, and the embattled who have more in common with their counterparts in other parts of the world than with the other social strata of their own region.

One of globalization's impacts that is most widely debated is the issue of sustainability. If current practices persist, the continued development of the world's regions will require massive inputs of new resources. Because of this, there is widespread concern about the impact this will have on the environment, society, and culture. Social movements and international organizations have begun to address the question of sustainable global economic development, though there is little widespread agreement on how to distribute the economic and social costs of sustainability and how to protect local cultures from its negative impacts.

The contemporary world-system has been dramatically transformed by the pressures and impacts of the globalization of the capitalist economic system. One effect has been the changing role of the nation-state, which, though once the primary political actor on the world stage, must now contend with new transnational and international organizations and institutions in negotiating the fortunes of its national space. In attempting to predict some of the other important changes that globalization will mean for world regions, we advanced the position that we are likely to see more of the same and perhaps a bit more of it. In other words, the world economy will continue to be dominated by the current triadic core, though some small changes will occur within the triad that will strengthen the hegemonic position of Europe, North America, and East Asia while the other world regions will continue to be largely peripheral.

Key Terms

international regime
(p. 633)

time-space convergence
(p. 632)

Review Questions

Testing Your Understanding

1. What was the "short twentieth century"?
2. What are the main arguments of the hyperglobalists; skeptics; and transformationalists?
3. How does national debt affect the economic development potential of poor countries?
4. Who are the: marginalized; elite; and embattled?
5. What is the difference between "real time" and "material" economic activities?
6. What is the link between poverty and pollution?
7. What is sustainable development?
8. What is global metropolitanism?
9. What trends characterize the modern state at the start of the twenty-first century?
10. Why is state fragmentation often a more likely outcome than regional integration?
11. If current trends continue, what does the near future hold for the geography of world regions?

Thinking Geographically

1. What are some of the key factors behind the regional geographies that developed during the "short twentieth century"?
2. What may be some of the worldwide demographic trends of the next 25 years?
3. Why does Subcommandante Marcos of Mexico describe current trends in the global economy as "a death sentence for the poor"?
4. Why is sustainable development a global as well as a local issue?
5. Why is global warming considered not just an environmental issue but a security threat to countries like the United States?
6. How is the human rights movement an example of an "international regime"?

Further Reading

Agnew, J., and Corbridge, S., *Mastering Space: Hegemony, Territoriality, and International Political Economy*. New York: Routledge, 1995.

Allen, J., and Hamnett, C., *A Shrinking World: Global Unevenness and Inequality*. Milton Keynes: The Open University Press, 1995.

Anderson, J., Brook, C., and Cochrane, A., *A Global World? Re-ordering Political Space*. Milton Keynes: The Open University Press, 1995.

Berry, B. J. L., *Long-Wave Rhythms in Economic Development and Political Behavior*. Baltimore: Johns Hopkins University Press, 1991.

Chatterjee, P., and Finger, M., *The Earth Brokers: Power, Politics, and World Development*. London: Routledge, 1994.

Cohen, R., and Kennedy, P., *Global Sociology*. London: Macmillan, 2000.

De Alcantara, C. H. (ed.), *Social Futures, Global Visions*. Oxford: Blackwell, 1996.

Hammond, A., *Which World? Scenarios for the 21st Century*. Washington, DC: Island Press, 1998.

Held, D., McGrew, A., Goldblatt, D., and Perraton, J., *Global Transformations: Politics, Economics and Culture*. Cambridge: Polity, 1999.

Hirst, P., and Thompson, G., *Globalization in Question: The International Economy and the Possibility of Governance* (2nd ed). Cambridge: Polity, 1999.

Holdgate, M., *From Care to Action: Making a Sustainable World*. London: Earthscan, 1996.

Hoogvelt, A., *Globalization and the Postcolonial World: The New Political Economy of Development*. Baltimore: Johns Hopkins University Press, 1997.

Johnston, R. J., Taylor, P. J., and Watts, M. (eds.), *Geographies of Global Change*. Cambridge, MA: Blackwell, 1995.

O'Riordan, T., *Environmental Science for Environmental Management* (2nd ed.). Harlow: Longman, 1998.

ÓTuathail, G., and Luke, T., "Present at the (Dis)integration: Deterritorialization and Reterritorialization in the New Wor(l)d Order," *Annals of the Association of American Geographers* 84 (1994), 381–398.

"The World in 2001." London: The Economist Newspaper Ltd., 2000.

World Data Appendix

The appendix on the following pages contains geographic and demographic data for major world states and some smaller dependencies. The data are grouped into the world regions that are the subjects of chapters 3 to 12, and the regions are arranged in the order in which they are discussed in this book. A more complete table of world data can be found at the Web site for this book (www.prenhall.com/Marston). There are some data set references in this book that are available only at the Web site table of world data.

World Data Appendix[a]

Country Name	Population in millions, 1999[b]	Projected population, 2015 (millions)	Surface area (1000 sq km)	Population density, 1999 (people per sq km)	Average annual population growth (%), 1965–1990	Urban population as percent of total population, 1998	Total fertility rate, 1998[b]	Infant mortality, 1998[c]	Life expectancy at birth, 1998[d]	Life expectancy at birth, female-male difference[e]
Europe										
Albania	3.34	3.93	28.75	121.9	1.8	40	2.5	25.1	71.6	5.94
Andorra	0.07		0.45	144.4	1.2	93	1.3	195.8	83.5	6.00
Austria	8.08	7.99	83.86	97.6	0.3	65	1.3	4.9	77.7	6.20
Belgium	10.20	10.24	32.82	310.9	0.2	97	1.6	5.6	77.8	6.27
Bosnia and Herzegovina	3.77	4.29	51.13	73.9	0.2	42	1.6	12.6	71.5	5.54
Bulgaria	8.26	7.34	110.91	74.7	0.0	69	1.1	14.4	70.9	7.20
Croatia	4.50	4.30	56.54	80.5	0.1	57	1.5	8.2	73.7	8.56
Cyprus	0.75	0.90	9.25	81.5	0.6	66	1.4	8.1	77.7	4.67
Czech Republic	10.29	9.88	78.86	133.2	0.2	75	1.2	5.2	74.5	7.00
Denmark	5.30	5.34	43.09	124.9	0.3	85	1.8	4.7	76.5	5.04
Estonia	1.45	1.33	45.1	34.3	0.4	69	1.2	9.3	69.5	11.05
Finland	5.153	5.26	338.15	16.9	0.4	66	1.8	4.2	77.4	7.30
France	58.85	61.13	551.5	106.9	0.6	75	1.8	4.8	78.8	7.60
Germany	82.05	78.67	356.98	234.9	0.2	87	1.4	4.9	77.4	6.35
Greece	10.52	10.29	131.96	81.6	0.2	60	1.3	6.1	78.4	5.40
Hungary	10.11	9.44	93.03	109.5	0.0	64	1.3	9.7	71.4	9.04
Iceland	0.27	0.30	103	2.7	0.6	92	2.0	3.6	79.1	4.58
Ireland	3.71	4.11	70.28	53.8	0.8	59	1.9	6.2	76.8	5.24
Italy	57.59	54.37	301.27	195.8	0.3	67	1.2	5.4	79.0	6.78
Latvia	2.45	2.15	64.6	39.5	0.2	69	1.1	14.9	68.4	11.40
Liechtenstein	0.032		0.16	200.0	1.0	23	1.5	5.1	78.8	7.31
Lithuania	3.70	3.61	65.2	57.1	0.7	68	1.4	9.2	69.1	10.40
Luxembourg	0.43	0.50	2.6	165.0	1.3	88	1.7	4.8	76.7	6.79
Macedonia, FYR	2.01	2.17	25.71	79.0	0.0	61	1.8	16.3	73.8	4.32
Malta	0.38	0.40	0.32	1178.1	0.7	91	1.9	5.9	77.2	5.13
Moldova	4.30	4.18	33.7	130.4	0.8	46	3.0	17.5	64.5	7.40
Monaco	0.032			16410.0	0.5	100	1.8	5.9	78.8	8.12
Netherlands	15.70	16.30	40.84	462.8	0.7	89	1.6	5.0	78.3	5.76
Norway	4.43	4.66	323.88	14.4	0.5	75	1.8	4.1	78.7	5.56
Poland	38.67	38.87	323.25	127.0	0.6	65	1.4	9.5	73.2	8.40
Portugal	9.97	9.83	91.98	108.9	0.3	61	1.5	8.4	75.8	7.16
Romania	22.50	21.33	238.39	97.7	0.5	56	1.3	20.5	69.9	7.80
San Marino	0.0027		60.5		1.5	81	1.3	6.3	81.1	7.45
Slovak Republic	5.39	5.47	49.01	112.1	0.6	57	1.4	8.8	73.7	8.09
Slovenia	1.98	1.93	20.25	98.5	0.6	50	1.2	5.0	74.9	7.60
Spain	39.37	38.13	505.99	78.8	0.6	77	1.2	5.4	78.8	7.00
Sweden	8.85	8.64	449.96	21.5	0.4	83	1.5	3.6	79.6	5.28
Switzerland	7.11	7.03	41.29	179.7	0.6	68	1.5	4.4	79.6	6.16
Ukraine	50.30	44.00	603.7	86.8	0.3	68	1.3	13.9	66.0	11.05
United Kingdom	59.06	59.24	244.88	244.4	0.3	89	1.7	5.8	77.7	5.26
Yugoslavia, FR (Serb./Mont.)	10.62	10.71	102.17	104.1	0.7	52	1.7	12.6	72.39(Serb.) 75.46(Mont.)	5.06
Russian Federation, Central Asia, and Transcaucasus										
Armenia	3.80	4.06	29.8	134.6	1.6	69	1.3	14.7	66.4	7.30
Azerbaijan	7.91	9.27	86.6	91.3	1.7	57	2.0	16.6	62.9	7.10
Belarus	10.24	9.43	207.6	49.3	0.5	71	1.3	11.3	68.0	11.70
Georgia	5.44	5.35	69.7	78.1	0.6	60	1.3	15.2	64.5	8.30
Kazakhstan	15.59	16.31	2717.3	5.8	0.8	56	2.0	21.6	63.2	11.08
Kyrgyzstan	4.70	5.63	198.5	24.5	1.8	34	2.8	26.2	63.4	8.10
Russian Federation	146.91	137.64	17075.4	8.7	0.4	77	1.2	16.5	67.2	11.60
Tajikistan	6.12	7.89	143.1	43.5	2.7	28	3.4	23.4	64.1	5.72
Turkmenistan	4.72	6.04	488.1	10.0	2.8	45	2.9	33.2	60.9	7.02
Uzbekistan	24.05	30.32	447.4	58.1	2.6	38	2.8	22.5	63.7	6.32
Middle East and North Africa										
Algeria	29.92	39.80	2381.74	12.6	2.8	59	3.5	34.8	69.7	3.20
Bahrain	0.64	0.80	0.69	931.9	1.8	88	2.8	20.5	72.9	4.87
Egypt, Arab Rep.	61.40	78.66	1001.45	61.7	2.2	45	3.2	49.1	63.3	3.12

[a]With the exception of some of the data for infant mortality, life expectancy, and adult literacy, which are taken from the CIA World Factbook Online 2001, and the two United Nations Development Programme (UNDP) development indexes, all data are from the World Bank 2000 World Development Data Disk.

[b]Total fertility rate is the number of children who would be born to a woman if she were to live to the end of her childbearing years (approximately ages 14–45) and bear children in accordance with current age-specific fertility rates.

[c]Infant mortality rate is the number of deaths of infants younger than 1 year of age during the indicated year per 1000 live births in the same year.

[d]Life expectancy at birth is the number of years a newborn would live if prevailing patterns of mortality at the time of its birth were to stay the same throughout its life.

[e]A positive value for life expectancy represents female advantage in years.

[f]Adult illiteracy rate is the percentage of adults aged 15 and older who cannot, with understanding, read and write a short, simple statement about their everyday life.

Crude death rate per 1000 people, 1998	Crude birthrate per 1000 people, 1998	Adult illiteracy, 1998 (%)[f]	Child malnutrition (weight for age) as percent of children under age 5, 1992–1998[g]	Prevalence of HIV as percent adult population, 1997[h]	Energy use per capita, 1997 (commercial kg of oil equivalent)	Total external debt, 1998 ($ millions)[i]	Gross national product per capita, 1998 ($)	Gross domestic product, 1998 ($ million)	UNDP Human Development Index, 1998[j]	UNDP Gender Development Index, 1998[k]
7	18	7.0	8.10	0.01	317.19	820.60	810	3047.00	0.713	0.708
5	11	0.0						1200.00		
10	10	2.0		0.18	3439.10	16000.00	26830	211857.59	0.908	0.901
10	11	2.0		0.14	5610.94	28300.00	25380	248184.41	0.925	0.921
7	13			0.04	479.30	3400.00		6200.00		
14	8	2.0		0.01	2480.25	9906.80	1220	12257.58	0.772	0.769
12	11	3.0	0.70	0.01	1686.51	8296.70	4620	21751.87	0.795	0.790
8	13	3.4					11920	9000.00	0.886	0.877
11	9	0.1	1.00	0.04	3937.85	25300.50	5150	56378.89	0.843	0.841
11	13	0.0		0.12	3994.35	21700.00	33040	174869.74	0.911	0.909
13	9	0.0		0.01	3810.73	781.70	3360	5201.91	0.801	0.798
10	11	0.0		0.02	6435.03	30000.00	24280	123502.35	0.917	0.913
9	13	1.0		0.37	4223.62	106000.00	24210	1426966.97	0.917	0.914
10	10	1.0		0.08	4231.36		26570	2134205.40	0.911	0.905
10	9	5.0		0.14	2434.60	57000.00	11740	120723.81	0.875	0.869
14	10	1.0		0.04	2492.49	28580.00	4510	47807.39	0.817	0.813
7	15	0.1				2600.00	27830	6420.00	0.927	0.925
9	15	2.0		0.09	3411.91	11000.00	18710	81949.25	0.907	0.896
10	9	2.0		0.31	2839.13		20090	1171864.81	0.903	0.895
14	8	0.0		0.01	1806.40	755.90	2420	6395.80	0.771	0.770
7	12	0.0						730.00		
11	10	2.0		0.01	2376.41	1949.70	2540	10736.12	0.789	0.785
9	12	0.0					45100	14700.00	0.908	0.895
8	15			0.01		2392.30	1290	2492.42	0.763	
8	13	8.5				130.00	10100	5300.00	0.865	0.848
9	10	4.0		0.11	1028.76	1034.70	380	35546.20	0.589	0.570
13	10	1.0						870.00		
9	12	1.0		0.17	4799.77		24780	381818.60	0.925	0.919
10	13	0.0		0.06	5500.78		34310	145892.08	0.934	0.932
10	10	1.0		0.06	2720.70	47708.10	3910	158574.49	0.814	0.811
11	12	12.6		0.69	2051.28	13100.00	10670	106697.06	0.864	0.858
12	11	3.0	5.70	0.01	1956.86	9513.30	1360	38157.50	0.770	0.767
8	11	4.0						500.00		
10	11			0.01	3198.04	9892.60	3700	20361.59	0.825	0.822
9	9	1.0		0.01	3212.55	6200.00	9780	19523.81	0.861	0.857
9	9	3.0		0.57	2729.40	90000.00	14100	553230.27	0.899	0.891
11	10	1.0		0.07	5868.62	66500.00	25580	226491.81	0.926	0.923
9	11	1.0		0.32	3698.93		39980	263630.21	0.915	0.910
15	9	2.0		0.43	2959.86	12718.30	980	43614.77	0.744	0.740
11	12	1.0		0.09	3863.43		21410	1357197.48	0.918	0.914
10	11		1.60	0.10		13742.10		20600.00		
6	11	1.0	3.30	0.01	476.49	799.70	460	1900.48	0.721	0.718
6	16	3.0	10.10	0.01	1529.34	693.40	480	3925.58	0.722	
13	9	2.0		0.17	2448.58	1119.70	2180	22554.65	0.781	
7	9	1.0		0.01	422.57	1674.40	970	5128.89	0.762	
10	14	2.0	8.30	0.03	2439.08	5713.70	1340	21978.93	0.754	
7	22	3.0	11.00	0.01	602.59	1147.70	380	1704.13	0.706	
14	9	2.0	3.00	0.05	4018.78	183601.30	2260	276611.17	0.771	0.769
5	21	2.0		0.01	562.41	1069.50	370	2163.66	0.663	0.659
6	20	2.0		0.01	2615.07	2266.40		2367.00	0.704	
6	23	1.0	18.80	0.01	1797.99	3162.00	950	20383.99	0.686	0.683
6	26	38.4	12.80	0.07	904.32	30665.40	1550	47346.77	0.683	0.661
4	21	13.5				2700.00	7640	8600.00	0.820	0.803
7	24	48.6	11.70	0.03	655.88	31964.10	1290	82709.56	0.623	0.604

[g]Prevalence of child malnutrition is the percentage of children under five whose weight for age and height for age are less than minus two standard deviations from the median for the international reference population aged 0–59 months.

[h]Prevalence of HIV refers to the percentage of people aged 15–49 who are infected with HIV.

[i]Total external debt is debt owed to nonresidents repayable in foreign currency, goods, or services. It is the sum of public, publicly guaranteed, and private nonguaranteed long-term debt, use of IMF credit, and short-term debt.

[j]An index, calculated by the United Nations Development Programme, that combines weighted life expectancy, adult literacy, school enrollment, and gross domestic product per capita into a ranking that ranges up to 1.0 (highest level of development).

[k]An index, calculated by the United Nations Development Programme, that combines the differences between male and female weighted life expectancy, adult literacy, school enrollment, and gross domestic product per capita into a ranking that is adjusted for gender inequalities and ranges up to 1.0 (highest level of development).

World Data Appendix[a]

Country Name	Population in millions, 1999[b]	Projected population, 2015 (millions)	Surface area (1000 sq km)	Population density, 1999 (people per sq km)	Average annual population growth (%), 1965–1990	Urban population as percent of total population, 1998	Total fertility rate, 1998[b]	Infant mortality, 1998[c]	Life expectancy at birth, 1998[d]	Life expectancy at birth, female-male difference[e]
Iran, Islamic Rep.	61.95	82.11	1633.19	38.2	2.8	61	2.7	26.0	69.7	1.74
Iraq	22.33	31.29	438.32	51.0	3.1	71	4.6	103.0	66.5	2.08
Israel	5.96	7.58	21.06	289.2	2.6	91	2.7	5.7	78.6	3.95
Jordan	4.56	6.73	89.21	51.3	4.2	73	4.1	27.1	77.4	3.34
Kuwait	1.87	2.88	17.82	104.7	4.2	97	2.8	11.9	76.1	5.92
Lebanon	4.21	5.18	10.4	411.6	1.9	89	2.4	27.2	71.3	3.60
Libya	5.30	7.43	1759.54	3.0	3.6	87	3.7	23.2	75.5	4.14
Morocco	27.78	35.32	446.55	62.2	2.2	55	3.1	49.4	69.1	3.76
Oman	2.30	3.35	212.46	10.8	3.9	81	4.6	18.0	71.8	3.00
Qatar	0.74	0.70	11	67.5	3.4	91	3.3	22.1	74.5	5.02
Saudi Arabia	20.74	33.67	2149.69	9.6	4.4	85	5.7	20.2	67.8	3.50
Sudan	28.35	40.55	2505.81	11.9	2.5	34	4.6	69.1	56.6	2.78
Syrian Arab Republic	15.28	21.80	185.18	83.1	3.2	54	3.9	28.0	68.5	4.54
Tunisia	9.34	11.50	163.61	60.1	2.1	64	2.2	28.1	73.7	3.69
Turkey	63.45	77.85	774.82	82.4	2.2	73	2.4	37.9	71.0	5.16
United Arab Emirates	2.72	3.74	83.6	32.6	9.5	85	3.4	7.8	74.1	2.64
West Bank and Gaza	2.73	4.97			3.4		5.9	23.7	72.08 (70.82-Gaza)	3.21
Western Sahara	0.251		2660.00		2.3	95	6.6	133.6	49.8	2.68
Yemen, Rep.	16.60	26.59	527.97	31.4	3.2	24	6.3	82.0	59.8	0.98
Sub-Saharan Africa										
Angola	12.00	19.45	1246.7	9.627	2.5	33	6.7	123.6	38.3	3.20
Benin	5.95	9.07	112.62	53.8	2.8	41	5.7	86.7	50.2	3.50
Botswana	1.56	1.83	581.73	2.8	3.2	49	4.2	61.8	39.3	1.90
Burkina Faso	10.73	15.89	274	39.2	2.3	17	6.7	104.0	46.7	1.54
Burundi	6.55	9.17	27.83	255.0	2.2	8	6.2	118.2	46.2	2.74
Cameroon	14.30	20.34	475.44	30.7	2.7	47	5.0	76.6	54.8	3.00
Cape Verde	0.42	0.60	4.03	103.2	1.0	53	4.2	54.6	68.5	6.66
Central African Republic	3.48	4.57	622.98	5.6	2.2	40	4.8	98.4	44.0	3.84
Chad	7.28	11.61	1284	5.8	2.4	23	6.4	98.9	50.5	3.40
Djibouti	0.64	0.70	23.2	27.4	1.5	83	5.8	103.3	49.7	3.67
Comoros	0.53	1.10	2.23	238.0	3.1	29	5.4	86.3	60.2	4.43
Congo, Dem. Rep.	48.22	79.14	2344.86	21.3	3.1	30	6.3	90.3	48.8	3.08
Congo, Rep.	2.78	4.33	342	8.1	2.8	61	6.0	89.5	47.4	4.42
Côte d'Ivoire	14.49	19.07	322.46	45.6	3.5	45	5.0	87.6	45.2	1.10
Equatorial Guinea	0.43	0.70	28.05	15.4	2.5	37	4.9	94.8	50.3	4.12
Eritrea	3.88	5.75	117.6	38.4	2.7	18	5.7	60.7	55.8	3.08
Ethiopia	61.27	87.60	1104.3	61.3	2.7	17	6.4	106.8	45.2	1.80
Gabon	1.18	1.71	267.67	4.6	2.6	79	5.1	85.5	50.1	2.70
Gambia, The	1.22	1.77	11.3	121.6	3.3	31	5.6	76.4	53.2	3.52
Ghana	18.46	26.81	238.54	81.1	2.6	37	4.8	64.6	57.4	3.48
Guinea	7.08	10.00	245.86	28.8	2.1	31	5.5	118.3	45.6	1.00
Guinea-Bissau	1.16	1.59	36.12	41.3	2.4	23	5.4	128.4	49.0	2.80
Kenya	29.29	39.07	580.37	51.5	3.3	31	4.6	76.2	48.0	1.68
Lesotho	2.06	2.71	30.35	67.8	2.3	26	4.6	93.0	50.8	2.48
Liberia	2.96		111.37	30.7	1.9	45	6.4	30.1	47.2	2.89
Madagascar	14.59	22.92	587.04	25.1	2.6	28	5.7	92.0	55.0	3.04
Malawi	10.53	15.29	118.48	112.0	3.0	22	6.4	133.8	37.6	0.32
Mali	10.60	16.72	1240.19	8.7	2.4	29	6.5	116.5	46.7	3.82
Mauritania	2.53	3.71	1025.52	2.5	2.5	55	6.3	90.0	50.8	3.24
Mauritius	1.16	1.35	2.04	571.3	1.3	41	2.0	19.3	71.0	8.04
Mozambique	16.95	23.78	801.59	21.6	2.1	38	5.2	134.5	37.5	2.72
Namibia	1.66	2.23	824.29	2.0	2.6	30	4.8	67.5	42.4	1.64
Niger	10.14	16.96	1267	8.0	3.1	20	7.3	118.0	41.3	3.50
Nigeria	120.82	184.70	923.77	132.7	2.9	42	5.3	76.4	51.6	3.46
Rwanda	8.11	11.78	26.34	328.5	2.8	6	6.1	123.1	39.3	2.42
São Tomé and Principe	0.14		0.96	147.6	3.2	44	6.1	50.4	64.3	2.86
Seychelles	0.08		0.45	174.8	0.5	63	1.9	17.7	71.7	11.25
Senegal	9.04	13.34	196.72	46.9	2.8	46	5.5	68.6	62.2	3.64

[a]With the exception of some of the data for infant mortality, life expectancy, and adult literacy, which are taken from the CIA World Factbook Online 2001, and the two United Nations Development Programme (UNDP) development indexes, all data are from the World Bank 2000 World Development Data Disk.

[b]Total fertility rate is the number of children who would be born to a woman if she were to live to the end of her childbearing years (approximately ages 14–45) and bear children in accordance with current age-specific fertility rates.

[c]Infant mortality rate is the number of deaths of infants younger than 1 year of age during the indicated year per 1000 live births in the same year.

[d]Life expectancy at birth is the number of years a newborn would live if prevailing patterns of mortality at the time of its birth were to stay the same throughout its life.

[e]A positive value for life expectancy represents female advantage in years.

[f]Adult illiteracy rate is the percentage of adults aged 15 and older who cannot, with understanding, read and write a short, simple statement about their everyday life.

Crude death rate per 1000 people, 1998	Crude birthrate per 1000 people, 1998	Adult illiteracy, 1998 (%)[f]	Child malnutrition (weight for age) as percent of children under age 5, 1992–1998[g]	Prevalence of HIV as percent of adult population, 1997[h]	Energy use per capita, 1997 (commercial kg of oil equivalent)	Total external debt, 1998 ($ millions)[i]	Gross national product per capita, 1998 ($)	Gross domestic product, 1998 ($ million)	UNDP Human Development Index, 1998[j]	UNDP Gender Development Index, 1998[k]
5	22	27.9	15.70	0.01	1777.31	14390.60	1650	113140.36	0.709	0.691
10	32	42.0	11.90	0.01	1240.03	139000.00		59900.00	0.583	0.548
6	22	5.0		0.07	3014.22	38000.00	16180	100524.84	0.883	0.877
4	31	13.4	5.10	0.02	1080.72	8484.50	1150	7392.99	0.721	
2	23	21.4	1.70	0.12	8935.88	6900.00		25171.45	0.836	0.827
6	21	13.6	3.00	0.09	1264.98	6725.10	3560	17228.71	0.735	0.718
4	29	23.8	4.70	0.05	2908.97	4100.00		39300.00	0.760	0.738
7	25	56.3	9.50	0.03	339.62	20687.10	1240	108000.00		
3	29	20.0	23.30	0.11	3003.10	3628.90		14962.03	0.730	0.697
4	16	19.6	6.00			13100.00		12300.00	0.819	0.807
4	34	37.2		0.01	4906.34	16300.00	6910	128891.85	0.747	0.712
11	33	53.9	33.90	0.99	413.89	16843.00	290	10366.13	0.477	0.453
5	29	29.2	12.90	0.01	983.03	22435.30	1020	17411.70	0.660	0.636
6	18	33.3	9.00	0.04	738.47	11077.90	2060	19955.74	0.703	0.688
6	21	17.7	10.30	0.01	1140.19	102073.90	3160	198843.73	0.732	0.726
3	17	20.8	7.00	0.18	11966.67	12600.00	17870	47234.06	0.810	0.927
5	42					108.00	1560	3588.98		
16	45									
12	40	62.0	46.10	0.01	207.88	4138.00	280	4318.16	0.448	0.389
19	48	58.0	42.00	2.12	587.40	12172.80	380	7472.37	0.405	
13	41	63.0	29.20	2.06	377.08	1646.80	380	2306.36	0.411	0.391
16	33	30.2	17.00	25.10		548.20	3070	4876.48	0.593	0.584
19	44	80.8	32.70	7.17		1399.30	240	2580.56	0.303	0.290
20	42	64.7	37.00	8.30		1118.70	140	885.21	0.321	
12	38	36.6	22.20	4.89	413.43	9828.80	610	8701.21	0.528	0.518
7	30	27.1				260.00	1200	618.00	0.688	0.675
19	37	40.0	23.20	10.77		921.30	300	1056.53	0.371	0.359
16	45	51.9	38.80	2.72		1091.40	230	1693.87	0.367	
15	41	53.8				356.00		550.00	0.447	
10	40	41.5	17.00			197.00	370	410.00	0.510	0.503
15	46	22.7	34.40	4.35	311.26	12929.20	110	6964.36	0.507	0.499
16	43	25.1	17.00	7.78	458.71	5119.00	680	1960.67	0.430	0.418
17	37	51.5	23.80	10.06	393.85	14851.80	700	11005.26	0.420	0.401
13	38	18.9				290.00	1110	960.00	0.555	0.551
12	40	75.0	43.70	3.17		149.30	200	649.91	0.408	0.394
20	45	64.5	47.70	9.31	286.71	10351.80	100	6543.84	0.309	0.297
16	36	36.8		4.25	1418.59	4424.60	4170	5518.36	0.592	
13	42	61.4	26.20	2.24		477.00	340	415.70	0.396	0.388
9	35	35.5	27.30	2.38	383.44	6883.50	390	7500.78	0.556	0.552
17	41	64.1		2.09		3545.90	530	9200.00	0.394	
21	41	46.1	23.00	2.25		964.40	160	3598.27	0.331	0.298
12	35	21.9	22.50	11.64	494.14	7009.80	350	11578.58	0.508	0.503
13	35	28.7	16.00	8.35		692.10	570	792.42	0.569	0.556
17	47	49.4				3000.00		2850.00		
11	41	80.0	40.00	0.12		4394.10	260	3748.63	0.483	0.478
23	47	42.0	29.90	14.92		2444.00	210	1687.62	0.385	0.375
16	47	69.0	26.90	1.67		3201.50	250	2694.68	0.380	0.371
13	40	62.3	23.00	0.52		2588.60	410	4900.00		
7	17	17.1	14.90	0.08		2481.60	3730	4198.96	0.761	0.750
20	41	59.9	26.10	14.17	460.85	8208.30	210	3893.07	0.341	0.326
13	35	62.0	26.20	19.94		217.00	1940	3092.18	0.632	0.624
18	52	86.4	49.60	1.45		1659.40	200	2047.80	0.293	0.280
12	40	42.9	39.10	4.12	753.33	30314.90	300	41353.43	0.439	0.425
21	46	39.5	29.40	12.75		1225.90	230	2023.92	0.382	0.377
8	43	27.0	16.00			268.00	270	169.00	0.547	
7	18	42.0	6.00			240.00	6420	590.00	0.786	
13	39	66.9	22.30	1.77	314.96	3861.40	520	4681.92	0.416	0.405

[g]Prevalence of child malnutrition is the percentage of children under five whose weight for age and height for age are less than minus two standard deviations from the median for the international reference population aged 0–59 months.

[h]Prevalence of HIV refers to the percentage of people aged 15–49 who are infected with HIV.

[i]Total external debt is debt owed to nonresidents repayable in foreign currency, goods, or services. It is the sum of public, publicly guaranteed, and private nonguaranteed long-term debt, use of IMF credit, and short-term debt.

[j]An index, calculated by the United Nations Development Programme, that combines weighted life expectancy, adult literacy, school enrollment, and gross domestic product per capita into a ranking that ranges up to 1.0 (highest level of development).

[k]An index, calculated by the United Nations Development Programme, that combines the differences between male and female weighted life expectancy, adult literacy, school enrollment, and gross domestic product per capita into a ranking that is adjusted for gender inequalities and ranges up to 1.0 (highest level of development).

World Data Appendix[a]

Country Name	Population in millions, 1999[b]	Projected population, 2015 (millions)	Surface area (1000 sq km)	Population density, 1999 (people per sq km)	Average annual population growth (%), 1965–1990	Urban population as percent of total population, 1998	Total fertility rate, 1998[b]	Infant mortality, 1998[c]	Life expectancy at birth, 1998[d]	Life expectancy at birth, female-male difference[e]
Sierra Leone	4.85	6.75	71.74	67.8	2.1	35	6.0	169.0	45.3	2.90
Somalia	9.08		637.66	14.5	2.9	28	7.2	125.8	47.6	3.19
South Africa	41.40	49.39	1221.04	33.9	2.2	53	2.8	51.5	51.1	5.32
Swaziland	0.99	1.00	17.36	57.5	2.0	25	5.9	109.0	56.3	1.83
Tanzania	32.13	44.78	945.09	36.4	3.0	31	5.4	85.0	52.3	2.12
Togo	4.46	6.29	56.79	82.0	3.1	32	5.1	78.2	54.7	2.42
Uganda	20.90	30.75	241.04	104.7	2.9	14	6.5	100.7	42.9	−0.77
Zambia	9.67	12.97	752.61	13.0	3.0	39	5.5	113.7	37.2	−0.12
Zimbabwe	11.69	14.08	390.76	30.2	2.9	34	3.7	72.7	37.8	2.84
North America										
Canada	30.30	33.67	9970.61	3.3	1.3	77	1.6	5.2	79.4	6.03
Greenland	0.06		341.7	0.2	0.1		2.5	18.3	68.1	7.17
United States	270.30	304.86	9363.52	29.5	1.0	77	2.0	7.0	77.1	6.04
Latin America and the Caribbean										
Antigua and Barbuda	0.07		0.44	152.0	0.7		1.9	23.1	74.8	4.65
Argentina	36.13	42.77	2780.4	13.2	1.5	89	2.6	18.6	75.1	7.18
Aruba	0.09		0.19	494.7	0.7		1.8	6.2	78.4	6.90
Bahamas, The	0.29		13.88	29.4	1.0	84	2.3	17.0	73.5	5.69
Barbados	0.27	0.30	0.43	617.7	0.6	38	1.7	12.4	75.9	5.17
Belize	0.24	0.30	22.96	10.5	2.8	47	4.1	26.0	74.6	4.62
Bolivia	7.95	10.92	1098.58	7.3	2.3	61	4.1	60.4	63.7	3.48
Brazil	165.87	200.01	8547.4	19.6	2.0	80	2.3	33.1	62.9	7.86
Chile	14.82	17.74	756.63	19.8	1.7	85	2.2	10.2	75.7	6.00
Colombia	40.80	51.43	1138.91	39.3	2.2	73	2.7	23.4	70.3	6.18
Costa Rica	3.53	4.40	51.1	69.1	2.7	47	2.6	12.6	75.8	4.62
Cuba	11.10	11.61	110.86	101.1	1.1	75	1.5	7.0	76.2	3.84
Dominica	0.07		0.75	97.3	−1.1	71	2.1	17.1	76.1	5.86
Dominican Republic	8.25	10.36	48.73	170.6	2.3	64	2.9	39.5	73.2	4.14
Ecuador	12.18	15.60	283.56	44.0	2.6	63	2.9	32.2	71.1	5.11
El Salvador	6.06	7.99	21.04	292.4	2.1	46	3.3	31.0	69.7	5.86
French Guiana			91000		2.9	79	3.2	14.0	76.1	6.83
Guadeloupe			1780		1.1	48	1.9	9.8	66.2	5.43
Guatemala	10.80	15.50	108.89	99.6	2.6	39	4.4	42.1	66.2	5.78
Guyana	0.85		214.97	4.3	−0.1	36	5.6	39.1	64.1	6.07
Grenada	0.10		0.34	282.9	−0.4		1.9	14.6	72.0	3.57
Haiti	7.65	10.04	27.75	277.5	1.9	34	4.3	70.5	49.2	4.74
Honduras	6.16	8.80	112.09	55.0	3.1	51	4.2	35.9	69.9	4.88
Jamaica	2.58	2.98	10.99	237.9	1.2	55	2.6	21.2	75.2	3.96
Martinque			1100		1.0	93	1.8	8.0	78.3	−1.57
Mexico	95.85	120.83	1958.2	50.2	2.4	74	1.7	30.2	71.5	6.06
Montserrat			100		2.1		1.9	9.1	78.0	4.45
Netherlands Antilles (Curaçao, Bonaire, Saba, Sint Eustatius, and Sint Maarten)	0.21		0.8	265.6	1.0	70	2.1	11.7	75.7	
Nicaragua	4.79	6.93	130	39.5	3.0	55	3.7	35.7	68.7	4.76
Panama	2.76	3.41	75.52	37.1	2.3	56	2.6	20.6	75.5	4.64
Paraguay	5.22	7.26	406.75	13.1	2.8	55	3.9	24.4	73.7	4.52
Peru	24.80	31.80	1285.22	19.4	2.3	72	3.1	39.6	70.0	4.76
Puerto Rico	3.86	4.38	8.95	435.2	1.2	74	1.9	10.0	75.6	8.90
St. Kitts and Nevis	0.04		0.36	113.4	−0.2	43	2.4	16.7	70.3	5.73
St. Lucia	0.15		0.62	249.2	1.2	30	2.4	15.6	71.9	5.73
St. Vincent and the Grenadines	0.11		0.39	290.3	0.4	44	2.1	17.1	73.1	3.46
Suriname	0.41		163.27	2.6	0.7	69	2.5	25.1	70.1	5.43
Trinidad and Tobago	1.29	1.46	5.13	250.5	1.1	73	1.8	15.9	68.0	4.60
Turks and Caicos Islands			430				3.3	18.7	73.3	4.36
Uruguay	3.29	3.63	177.41	18.8	0.6	91	2.4	16.4	75.2	7.94

[a]With the exception of some of the data for infant mortality, life expectancy, and adult literacy, which are taken from the CIA World Factbook Online 2001, and the two United Nations Development Programme (UNDP) development indexes, all data are from the World Bank 2000 World Development Data Disk.

[b]Total fertility rate is the number of children who would be born to a woman if she were to live to the end of her childbearing years (approximately ages 14–45) and bear children in accordance with current age-specific fertility rates.

[c]Infant mortality rate is the number of deaths of infants younger than 1 year of age during the indicated year per 1000 live births in the same year.

[d]Life expectancy at birth is the number of years a newborn would live if prevailing patterns of mortality at the time of its birth were to stay the same throughout its life.

[e]A positive value for life expectancy represents female advantage in years.

[f]Adult illiteracy rate is the percentage of adults aged 15 and older who cannot, with understanding, read and write a short, simple statement about their everyday life.

Crude death rate per 1000 people, 1998	Crude birthrate per 1000 people, 1998	Adult illiteracy, 1998 (%)[f]	Child malnutrition (weight for age) as percent of children under age 5, 1992–1998[g]	Prevalence of HIV as percent adult population, 1997[h]	Energy use per capita, 1997 (commercial kg of oil equivalent)	Total external debt, 1998 ($ millions)[i]	Gross national product per capita, 1998 ($)	Gross domestic product, 1998 ($ million)	UNDP Human Development Index, 1998[j]	UNDP Gender Development Index, 1998[k]
25	45	68.6	29.00	3.17		1243.10	140	646.63	0.252	
19	48	76.0				2600.00		4300.00		
9	25	18.2	9.20	12.91	2636.32	24711.50	3310	133461.48	0.697	0.689
20	41	21.7	10.00			281.00	1400	4200.00	0.655	0.646
16	41	32.2	30.60	9.42	455.29	7602.60	220	8016.20	0.415	0.410
16	40	48.3	25.10	8.52		1448.40	330	1509.65	0.471	0.448
20	47	38.2	25.50	9.51		3935.20	310	6775.15	0.409	0.401
19	42	21.8	23.50	19.07	634.00	6865.30	330	3351.56	0.420	0.413
13	31	15.0	15.50	25.84	865.51	4716.10	620	6338.22	0.555	0.551
7	12	3.0		0.33	7929.86	1900.00	19170	580623.40	0.935	0.932
8	17							945.00		
9	14	3.0	1.40	0.76	8075.59	862000.00	29240	8230396.76	0.929	0.793
6	20	11.0	10.00				8450	524.00	0.833	
8	19	3.8	1.90	0.69	1729.93	144050.19	8030	298131.00	0.837	0.824
6	13	3.0						1600.00		
7	20	4.5				385.00		5580.00	0.844	0.842
9	14	2.6	5.00			425.00		2900.00	0.858	0.788
5	32	29.7	6.00				2660	740.00	0.777	0.754
9	32	16.9	7.60	0.07	547.70	6077.50	1010	8586.39	0.643	0.631
7	20	16.7	5.70	0.63	1050.96	232004.21	4630	778208.87	0.747	0.736
5	18	94.8	0.80	0.20	1573.79	36302.30	4990	78737.65	0.826	0.812
6	24	8.7	8.40	0.36	761.23	33263.30	2470	102895.70	0.764	0.760
4	22	5.2	5.10	0.55	768.76	3970.70	2770	10479.12	0.797	0.813
7	13	4.3	9.00	0.02	1290.62	11100.00		18600.00	0.783	
7	18	6.0	5.00			109.00	3150	225.00	27.000	
5	25	17.9	5.90	1.89	672.60	4451.20	1770	15852.55	0.729	0.720
6	24	9.9	17.00	0.28	713.16	15140.00	1520	18360.17	0.722	0.701
6	27	28.5	11.20	0.58	690.82	3632.70	1850	11870.19	0.696	0.693
5	22	17.0						1000.00		
6	17	10.0						3700.00		
7	33	44.4	26.60	0.52	535.61	4565.00	1640	18941.82	0.619	0.610
8	18	1.7	12.00			1100.00	780	205.66	0.709	0.698
8	21	2.0				183.00	3250	360.00	0.785	
13	31	55.0	27.50	5.17	237.45	1047.50	410	3870.95	0.440	0.436
5	33	27.3	25.40	1.46	531.56	5002.30	740	5371.39	0.653	0.644
6	23	15.0	10.20	0.99	1551.68	3994.60	1740	6418.31	0.735	0.732
6	16	7.0						4240.00		
5	28	10.4	14.00	0.35	1501.06	159958.60	3840	1615.36	0.700	0.697
7	17	3.0						31.00		
6	17	3.6						2400.00		
5	31	34.3	12.20	0.19	550.87	5968.20	370	2007.42	0.631	0.624
5	22	9.2	6.10	0.61	856.20	6688.60	2990	9143.80	0.776	0.770
5	30	7.9	4.00	0.13	824.19	2304.50	1760	8608.18	0.736	0.723
6	25	11.3	7.80	0.56	620.70	32397.00	2440	62744.71	0.737	0.723
7	15	11.0								
9	19	3.0				115.00	6190	244.00	0.798	
5	22	33.0				131.00	3660	656.00	0.728	
6	18	4.0				99.00	2560	309.00	2.000	
6	21	7.0				512.00	1660	1480.00	0.766	
6	15	2.1	7.00	0.94	6414.45	2193.30	4520	6382.09	0.793	0.784
		2.0						117.00		
9	17	2.7	4.40	0.33	882.82	7600.20	6070	20578.02	0.825	0.821

[g]Prevalence of child malnutrition is the percentage of children under five whose weight for age and height for age are less than minus two standard deviations from the median for the international reference population aged 0–59 months.

[h]Prevalence of HIV refers to the percentage of people aged 15–49 who are infected with HIV.

[i]Total external debt is debt owed to nonresidents repayable in foreign currency, goods, or services. It is the sum of public, publicly guaranteed, and private nonguaranteed long-term debt, use of IMF credit, and short-term debt.

[j]An index, calculated by the United Nations Development Programme, that combines weighted life expectancy, adult literacy, school enrollment, and gross domestic product per capita into a ranking that ranges up to 1.0 (highest level of development).

[k]An index, calculated by the United Nations Development Programme, that combines the differences between male and female weighted life expectancy, adult literacy, school enrollment, and gross domestic product per capita into a ranking that is adjusted for gender inequalities and ranges up to 1.0 (highest level of development).

World Data Appendix[a]

Country Name	Population in millions, 1999[b]	Projected population, 2015 (millions)	Surface area (1000 sq km)	Population density, 1999 (people per sq km)	Average annual population growth (%), 1965–1990	Urban population as percent of total population, 1998	Total fertility rate, 1998[b]	Infant mortality, 1998[c]	Life expectancy at birth, 1998[d]	Life expectancy at birth, female-male difference[e]
Venezuela, R.B.	23.24	30.21	912.05	26.3	2.8	86	2.9	20.6	73.1	5.72
Virgin Islands, British			352		2.3		1.7	21.1	75.4	1.78
Virgin Islands, U.S.	0.12		0.34	347.9	1.1		2.3	9.6	77.0	8.05
East Asia										
China	1238.60	1388.47	9596.96	132.8	1.7	31	1.9	31.1	71.4	3.25
Japan	126.41	124.42	377.8	335.7	0.7	79	1.4	3.7	80.7	6.62
Korea, Dem. Rep.	23.17	26.18	120.54	192.4	2.0	60	2.0	54.4	70.7	4.28
Korea, Rep.	46.43	51.07	99.26	470.3	1.5	80	1.6	8.7	74.4	7.16
Mongolia	2.58	3.31	1566.5	1.6	2.6	62	2.4	50.3	67.3	3.04
Taiwan	22.19		35980		0.8		1.8	7.1	76.4	5.70
Southeast Asia										
Brunei	0.31	0.40	5.77	59.7	2.2	67	2.5	14.8	75.8	4.83
Burma	44.46	53.83	676.58	67.6	1.8	27	3.1	78.2		3.34
Cambodia	11.50	14.76	181.04	65.1	1.9	15	4.5	101.6	56.5	2.82
East Timor						8				
Indonesia	203.68	250.46	1904.57	112.4	2.0	39	2.7	43.0	68.0	3.72
Lao PDR	4.97	7.21	236.8	21.5	2.2	22	5.5	95.7	53.1	3.10
Malaysia	22.18	29.18	329.75	67.5	2.6	56	3.1	8.3	70.8	5.00
Philippines	75.17	100.03	300	252.1	2.6	57	3.6	32.2	67.5	3.72
Singapore	3.16	3.75	0.62	5186.1	1.9	100	1.5	3.8	80.1	4.10
Thailand	61.20	71.00	513.12	119.8	2.1	21	1.9	28.6	68.6	5.00
Vietnam	76.52	94.37	331.69	235.1	2.1	20	2.3	33.6	69.3	4.77
South Asia										
Afghanistan	25.05		652.09	38.4	3.5	20	5.9	149.3	45.8	-1.52
Bangladesh	125.63	161.79	144	965.1	2.3	23	3.1	72.8	60.2	0.14
Bhutan	2.0	3.10	47	16.2	2.2	6	5.1	111.0	61.1	-0.80
India	979.67	1224.40	3287.59	329.5	2.1	28	3.2	69.8	62.5	1.52
Maldives	0.26	0.50	0.3	875.3	3.1	25	5.6	65.5	67.4	2.35
Nepal	22.85	32.55	147.18	159.8	2.4	11	4.4	77.2	57.8	-0.32
Pakistan	131.58	194.65	796.1	170.7	2.8	36	4.9	91.5	61.1	1.88
Sri Lanka	18.78	22.62	65.61	290.5	1.6	23	2.1	16.4	71.8	4.44
Australia, New Zealand, and the South Pacific										
Australia	18.75	21.51	7741.22	2.4	1.5	85	1.8	5.0	79.8	5.60
Fiji	0.79	0.90	18.27	43.3	1.4	41	2.9	14.5	72.7	4.91
French Polynesia	0.23		4	62.1	1.8	54	2.3	9.3	72.2	4.75
Kiribati	0.09		0.73	117.8	2.3	37	4.4	55.4	60.9	5.93
Marshall Islands	0.06		0.2	342.0	3.9	65	5.4	41.0	65.5	3.64
Micronesia, Fed. Sts.	0.11		0.7	162.0	3.3	27	2.5	33.5	67.3	3.95
Nauru	0.001		21		2.1	100	3.7	10.9	60.8	7.15
New Caledonia	0.21		18.58	11.3	1.5	71	2.5	8.6	72.8	
New Zealand	3.79	4.10	270.53	14.2	1.1	86	1.9	5.3	77.8	5.42
Palau	0.02		0.5	40.2	1.8	71	2.5	17.1	71.0	6.41
Papua New Guinea	4.60	6.22	462.84	10.2	2.3	17	4.2	59.4	63.1	1.54
Samoa	0.17	0.20	2.84	59.8	-0.2		3.5	32.8	68.7	5.58
Solomon Islands	0.42		28.9	14.9	3.0	17	4.8	25.3	70.8	4.95
Tonga	0.10		0.75	137.1	1.9	32	3.2	14.5	70.6	4.91
Tuvalu	0.0001		0.0026	416.8	1.4	18	3.1	23.3	66.3	4.32
Vanuatu	0.18		12.19	15.0	1.7	21	3.3	62.5	65.0	2.75

[a]With the exception of some of the data for infant mortality, life expectancy, and adult literacy, which are taken from the CIA World Factbook Online 2001, and the two United Nations Development Programme (UNDP) development indexes, all data are from the World Bank 2000 World Development Data Disk.

[b]Total fertility rate is the number of children who would be born to a woman if she were to live to the end of her childbearing years (approximately ages 14–45) and bear children in accordance with current age-specific fertility rates.

[c]Infant mortality rate is the number of deaths of infants younger than 1 year of age during the indicated year per 1000 live births in the same year.

[d]Life expectancy at birth is the number of years a newborn would live if prevailing patterns of mortality at the time of its birth were to stay the same throughout its life.

[e]A positive value for life expectancy represents female advantage in years.

[f]Adult illiteracy rate is the percentage of adults aged 15 and older who cannot, with understanding, read and write a short, simple statement about their everyday life.

Crude death rate per 1000 people, 1998	Crude birthrate per 1000 people, 1998	Adult illiteracy, 1998 (%)[f]	Child malnutrition (weight for age) as percent of children under age 5, 1992–1998[g]	Prevalence of HIV as percent adult population, 1997[h]	Energy use per capita, 1997 (commercial kg of oil equivalent)	Total external debt, 1998 ($ millions)[i]	Gross national product per capita, 1998 ($)	Gross domestic product, 1998 ($ million)	UNDP Human Development Index, 1998[j]	UNDP Gender Development Index, 1998[k]
4	25	8.9	5.10	0.69	2525.79	37003.10	3530	95022.79	0.770	0.763
5	16	2.8						287.00		
5	16							1800.00		
8	16	18.5	15.80	0.06	907.00	154599.19	750	959030.03	0.706	0.700
7	10	1.0		0.01	4083.54		32350	3782963.63	0.924	0.916
9	20	1.0	32.20	0.01		12000.00		22600.00	0.854	0.847
6	14	2.0		0.01	3834.47	139097.41	8600	320748.49		
7	21	17.1	8.60	0.01		738.80	380	6100.00	0.569	0.566
6	14	6.0						357000.00		
3	21	9.3	52.00					5600.00	0.848	0.843
10	26		42.90	1.79	296.21	5680.40			0.585	0.582
12	33	65.0		2.40		2209.70	260	2870.94	0.512	0.534
8	23	16.2	34.00	0.05	692.54	150874.60	640	94156.16	0.670	0.664
13	38	43.0	40.00	0.04		2436.70	320	1260.94	0.484	0.463
5	25	16.5	20.10	0.62	2237.18	44773.10	3670	72488.99	0.772	0.762
6	28	5.4	29.60	0.06	520.23	47817.10	1050	65106.61	0.744	0.739
5	13	8.9		0.15	8660.54	9700.00	30170	84378.59	0.881	0.876
7	17	6.2	19.00	2.23	1319.48	86171.90	2160	111327.32	0.745	0.741
6	21	6.3	39.80	0.22	520.89	22359.10	350	27184.18	0.671	0.668
18	42	68.5				5500.00		21000.00		
10	28	61.9	56.30	0.03	196.77	16375.60	350	42701.83	0.461	0.441
14	36	57.8	38.00			120.00	470	2100.00	0.483	
9	27	48.0	53.20	0.82	479.06	98232.30	440	430024.20	0.563	0.545
8	39	4.0	43.00			237.00	1130	540.00	0.725	0.720
11	34	72.5	57.10	0.24	320.77	2645.70	210	4783.02	0.474	0.449
8	35	62.2	38.20	0.09	442.31	32228.60	470	63369.27	0.522	0.489
6	18	9.8	37.70	0.07	385.89	8526.00	810	15706.83	0.733	77.000
7	13	0.0	0.00	0.14	5483.75	220000.00	20640	361722.48	0.929	0.927
6	23	7.8	8.00			193.00	2210	5900.00	0.769	0.755
4	19	2.0						2600.00		
9	32					10.00	1170	74.00		
6	45	7.0					1540	989.44	0.451	
6	27	11.0					1800	1041.66	0.628	
7	28					33.00		100.00		
6	21	9.0						3000.00		
7	15	1.0		0.07	4434.61	30800.00	14600	52845.07	0.903	0.900
7	20	8.0						160.00		
10	32	27.8	30.00	0.19		2692.20	890	3745.69	0.542	0.536
6	16	3.0				180.00	1070	485.00	0.711	
4	35					152.00	760	1210.00	0.614	
6	25	1.5				62.00	1750	238.00		
8	22							7.80		
9	26	47.0	20.00			48.00	1260	245.00	0.623	

[g]Prevalence of child malnutrition is the percentage of children under five whose weight for age and height for age are less than minus two standard deviations from the median for the international reference population aged 0–59 months.

[h]Prevalence of HIV refers to the percentage of people aged 15–49 who are infected with HIV.

[i]Total external debt is debt owed to nonresidents repayable in foreign currency, goods, or services. It is the sum of public, publicly guaranteed, and private nonguaranteed long-term debt, use of IMF credit, and short-term debt.

[j]An index, calculated by the United Nations Development Programme, that combines weighted life expectancy, adult literacy, school enrollment, and gross domestic product per capita into a ranking that ranges up to 1.0 (highest level of development).

[k]An index, calculated by the United Nations Development Programme, that combines the differences between male and female weighted life expectancy, adult literacy, school enrollment, and gross domestic product per capita into a ranking that is adjusted for gender inequalities and ranges up to 1.0 (highest level of development).

Glossary

Aborigines: indigenous peoples of Australia.

acid rain: precipitation that has mixed with air pollution to produce rain that contains levels of acidity—often in the form of sulfuric acid—that are harmful to vegetation and aquatic life.

afforestation: converting previously unforested land to forest by planting trees or seeds.

agglomeration: the clustering together of economic activities at the scale of metropolitan areas or industrial sub-regions.

agglomeration economies: cost advantages that accrue to individual firms because of their location among functionally related activities.

Alliance of Small Island States (AOSIS): an association of more than 40 low-lying, mostly island, countries that have formed an alliance to combat global warming, which threatens their existence through sea-level rise.

altiplano: the high-elevation plateaus and basins that lie within even higher mountains, especially in Bolivia and Peru, at more than 3000 meters (9500 feet) in the Andes of Latin America.

altitudinal zonation: a vertical classification of environment and land use according to elevation based mainly on changes in climate and vegetation from lower (warmer) to higher (cooler) elevations.

americanization: process by which a generation of individuals born in the British American colonies felt less loyalty and fewer cultural ties with Great Britain and developed a new indigenous ethos.

Antarctic Treaty: international agreement to demilitarize the Antarctic continent, delay mineral exploration, and preserve it for scientific research.

apartheid: South Africa's policy of racial separation that prior to 1994 structured space and society to keep black, white, and colored populations apart.

archipelago: a group of islands or expanse of water with many islands.

aridity: a climate with insufficient moisture to support trees or woody plants.

ASEAN: the Association of Southeast Asian Nations, an international organization of the nations of Southeast Asia established to promote economic growth and regional security.

Asian Tigers: the newly industrialized territories of Hong Kong, Taiwan, South Korea, and Singapore that have experienced rapid economic growth and become semiperipheral within the world-system.

assimilation: the process by which peoples of different cultural backgrounds who occupy a common territory achieve sufficient cultural solidarity to sustain a national existence.

atoll: low-lying island landform consisting of a circle of coral reefs around a lagoon, often associated with the rim of a submerged volcano or mountain.

azimuthal projection: a map projection on which compass directions are correct only from one central point.

backwash effects: the negative impacts on a region (or regions)—including outmigration and the loss of capital—due to the economic growth of some other region.

Balfour Declaration: a 1917 British mandate that required the establishment of a Jewish national homeland.

balkanization: the division of a territory into smaller and often mutually hostile units.

banana republics: term used to describe small tropical countries, often run by a dictator, dependent on the export of a few crops such as bananas.

barchan: crescent-shaped sand dune, concave on the side sheltered from the prevailing wind.

Berlin Conference: a meeting convened by German chancellor Bismark in 1884–1885 to divide Africa among European colonial powers.

biogeography: the study of the spatial distribution of vegetation, animals, and other organisms.

biome: the largest geographic biotic unit, a major community of plants and animals or similar ecosystems.

bioprospecting: search for plants, animals, and other organisms that may be of medicinal value or may have other commercial use.

bonded labor: labor that is pledged against an outstanding debt.

bracero: guest worker from Mexico given temporary permit to work as a farm laborer in the United States between 1942 and 1964.

British Commonwealth: group of former British colonies and other countries allied with Britain that cooperate on political, economic, and cultural activities.

buffer zone: a group of smaller or less powerful countries situated between larger or more powerful countries that are geopolitical rivals.

bush fallow: modification of shifting cultivation where crops are rotated around a village and fallow periods are shorter.

canton: a small territorial administrative division of a country.

cargo cult: Pacific island religious movements in which the dawn of a coming new age was associated with the arrival of goods brought by spiritual beings or foreigners.

cartogram: a map projection that is transformed in order to promote legibility or to reveal patterns not readily apparent on a traditional base map.

cartography: the body of practical and theoretical knowledge about making distinctive visual representations of Earth's surface in the form of maps.

caste: a system of kinship groupings that is reinforced by language, religion, and occupation.

cattle station: livestock enterprises where cattle (or sheep) are raised on large grazing leases in the remote regions of Australia.

central business district (CBD): central nucleus of commercial land uses in a city.

chador: a loose, usually black robe worn by Muslim women that covers the body, including the face, from head to toe.

chaebol: South Korean term for the very large corporations in that country that, with government help, control numerous businesses and dominate the national economy.

chernozem: thick, dark grassland soils (also called Black Earths) that are neutral in terms of acidic content and rich in humus.

circle of poison: use of imported pesticides on export crops in developing countries that then export back the contaminated crops to the regions where the pesticides were manufactured.

circular migration: traditional and long-standing population movements that respond to seasonal availability of pasture, droughts, and wage employment.

civil society: a network of social groups and cultural traditions that operate independently of the state and its political institutions.

climate: typical conditions of the weather expected at a place often measured by long-term averages of temperature and precipitation (e.g., a rainy place).

Closer Economic Relations: agreement in 1983 that built upon an earlier New Zealand-Australia Free Trade Agreement (NZAFTA) and set out to remove all tariffs and restrictions on trade between the two countries.

cognitive image (mental map): psychological representation of locations that are made up from people's individual ideas and impressions of these locations.

colonialism: the establishment and maintenance of political and legal domination by a state over a separate and alien society.

Columbian Exchange: the interchange of crops, animals, people, and diseases between the Old World of Europe and Africa and the New World of the Americas beginning with the voyages of Christopher Columbus in 1492.

command economy: a national economy in which all aspects of production and distribution are centrally controlled by government agencies.

common market: a market in which internal restrictions on the movement of capital, labor, and enterprise are also removed from the basic framework of a customs union.

common property resources: resources such as fish or forests that are managed collectively by a community that has rights to the resource rather than it being owned by individuals.

comparative advantage: principle whereby places and regions specialize in activities for which they have the greatest advantage in productivity relative to other regions—or for which they have the least disadvantage.

conformal projection: a map projection on which compass bearings are rendered accurately.

continental drift: slow movement of the continents over long periods of time across Earth's surface (*see* plate tectonics).

continental shield: the stable cores of continents comprising old (500 million years or older) crystalline rocks often containing minerals.

contract farming: method by which farmers sign contracts with companies to produce crops to certain production and quality standards in return for a guaranteed price.

convergent plate boundary: area of Earth's crust where plates are moving toward each other.

core regions: regions that dominate trade, control the most advanced technologies, and have high levels of productivity within diversified economies.

counterurbanization: net loss of population from cities to smaller towns and rural areas.

creative destruction: the withdrawal of investments from activities (and regions) that yield low rates of profit, in order to reinvest in new activities (and new places).

cultural ecology: study of the relationship between a cultural group and its natural environment.

culture: a shared set of meanings that are lived through the material and symbolic practices of everyday life.

Culture System: Dutch colonial policy from 1830 to 1870 that required farmers in Java to devote one-fifth of their land and their labor to production of an export crop.

customs union: an international association organized to eliminate customs restrictions on goods exchanged between member nations and to establish a uniform tariff policy toward nonmembers.

deindustrialization: a decline in industrial employment in core regions as firms scale back their activities in response to lower levels of profitability.

demographic collapse: after about 1500, the rapid die-off of the indigenous populations of the Americas as a result of diseases introduced by the Europeans to which residents of the Americas had no immunity.

demographic transition: replacement of high birth and death rates by low birth and death rates.

desertification: the process by which arid and semiarid lands become degraded and less productive, leading to more desertlike conditions.

development theory: an analysis of social change that assesses the economic progress of individual countries in an evolutionary way.

diaspora: the spatial dispersion of a previously homogeneous group.

disinvestment: the sale of assets such as factories and equipment.

distributary: a river branch that flows away from the main stream.

divergent plate boundary: area of Earth's crust where plates are moving apart.

division of labor: the separation of productive processes into individual operations, each performed by different workers or groups of workers.

domestication: adaptation of wild plants and animals through selective breeding by humans for preferred characteristics into cultivated or tamed forms.

domino theory: The domino theory (also known as the domino effect) suggests that political unrest in one country can destabilize neighbors and start a chain of events like the fall of a stack of dominos. The U.S. was concerned that communist rebellions would spread, especially from Cuba into Latin America and from China and the Soviet Union via North Vietnam into Southeast Asia.

Dreamtime: Aboriginal worldview that links past and future, people and places, in a continuity that ensures respect for the natural world.

dry farming: arable farming techniques that allow the cultivation of crops without irrigation in regions of limited moisture (50 centimeters, or 20 inches per year).

Earth system science: an integrated approach to the study of Earth that stresses investigation of the interactions among Earth's components in order to explain Earth and atmosphere dynamics, Earth evolution and ecosystems, and global change.

ecological imperialism: a concept developed by historian Alfred Crosby to describe the way in which European organisms, including diseases, pests, and domestic animals, were able to take over the ecosystems of other regions of the world, often with devastating impacts on local peoples, flora, and fauna.

economic union: a union that provides for integrated economic policies among member states in addition to the characteristics of a common market.

economies of scale: cost advantages to manufacturers that accrue from high-volume production, since the average cost of production falls with increasing output.

ecosystem: the complex of living organisms, their physical environment, and all their interrelationships in a particular place.

ecotourism: environmentally oriented tourism designed to protect the environment and often to provide economic opportunities for local people.

ejidos: Mexican system of distributing communally held land to peasants after the Mexican Revolution.

El Niño: the periodic warming of sea surface temperatures in the tropical Pacific off the coast of Peru that results in worldwide changes in climate, including droughts and floods.

enclave: a culturally distinct territory that is encompassed by a different cultural group or groups.

encomienda: system by which groups of indigenous people were "entrusted" to Spanish colonists who could demand tribute in the form of labor, crops, or goods and in turn were responsible for the indigenous groups' conversion to Catholic faith and for teaching them Spanish.

Enlightenment: eighteenth-century movement marked by a belief in the sovereignty of reason and empirical research in the sciences.

entrepôt: a seaport that is an intermediary center of trade and transshipment.

environmental determinism: a doctrine holding that human activities are controlled by the environment.

equal-area (equivalent) projection: a map projection that portrays areas on Earth's surface in their true proportions.

equidistant projection: a map projection that allows distance to be represented as accurately as possible.

ethnic cleansing: the systematic and forced removal of members of an ethnic group from their communities in order to change the ethnic composition of a region.

ethnocentrism: the attitude that one's own race and culture is superior to that of others.

eucalyptus: fast-growing trees native and widespread in Australia, the leaves of which produce a strong-smelling oil and which have been introduced into many other world regions.

europeanization: highly selective process involving the mixings of native and imported practices that eventually created distinct colonial cultures and societies.

exclave: a portion of a country or a cultural group's territory that lies outside its contiguous land area.

Exclusive Economic Zone: agreement within International Law of the Sea Convention to set aside a 370-kilometer (200-nautical-mile) limit around islands and coasts where a country has the sole right to exploit or rent its resources such as fish or oil.

farm crisis: the financial failure and foreclosure of thousands of family farms across the U.S. Midwest.

fast world: people, places, and regions directly involved, as producers and consumers, in transnational industry, modern telecommunications, materialistic consumption, and international news and entertainment.

faulting: fracturing in the rocks of Earth's crust where they are displaced relative to each other in a vertical or horizontal direction.

favela: Brazilian term for informal settlements lacking good housing and services that grow up around the urban core.

federal state: a form of government in which power is allocated to units of local government within the country.

feminization of poverty: women are more likely to be poor, malnourished, and otherwise disadvantaged because of inequalities within the household, the community and the country.

feng shui: the application of a collection of ancient principles of geomancy that are believed by adherents to ensure health, wealth, happiness, long life, and healthy offspring through the spatial organization of cities, buildings, and furniture.

feudal systems: regional hierarchies of territorial lords composed, at the bottom, of local nobles and at the top of lords or monarchs owning immense stretches of land. Landowners delegated smaller parcels of land to others in return for political allegiance and economic obligations in the form of money dues or labor.

fjord: a steep-sided, narrow inlet of the sea, formed when deeply glaciated valleys are flooded by the sea.

flexible production region: a region within which there is a concentration of small- and medium-sized firms whose production and distribution practices take advantage of computerized control systems and local subcontractors in order to quickly exploit new market niches for new product lines.

formal region: a region with a high degree of homogeneity in terms of particular distinguishing features.

free trade association: an association whose member countries eliminate tariff and quota barriers against trade from other member states but continue to charge regular duties on materials and products coming from outside the association.

Free Trade Zone: area within which goods may be manufactured or traded without customs duties.

functional region: an area characterized by a coherent functional organization of human occupancy.

gender and development (GAD): an approach to development that links women's productive and reproductive roles and an approach to understanding the gender-related differences and barriers to better lives of both men and women.

gender division of labor: the separation of productive processes based on gender.

gentrification: the invasion of older, centrally located working-class neighborhoods by higher-income households.

geographic information systems (GIS): integrated computer tools for the handling, processing, and analyzing of geographical data.

geomancy: belief that the physical attributes of places can be analyzed and manipulated in order to improve the flow of cosmic energy, or *ch'i.*

ghetto: an urban residential district almost exlusively the preserve of one ethnic or cultural group.

global positioning system: a system of satellites that orbit Earth on precisely predictable paths, broadcasting highly accurate time and locational information.

global warming: an increase in world temperatures and change in climate associated with increasing levels of carbon dioxide and other gases resulting from human activities such as deforestation and fossil-fuel burning.

globalization: the increasing interconnectedness of different parts of the world through common processes of economic, environmental, political, and cultural change.

Great Artesian Basin: the world's largest reserve of underground water located in central Australia and under pressure so that water rises to the surface when wells are bored.

Green Revolution: the technological package of higher yielding seeds, especially wheat, rice, and corn, that in combination with irrigation, fertilizers, pesticides, and farm machinery was able to increase crop production in the developing world after about 1950.

greenhouse effect: trapping of heat within the atmosphere by water vapor and gases, such as carbon dioxide, resulting in the warming of the atmosphere and surface.

gross domestic product (GDP): an estimate of the total value of all materials, foodstuffs, goods, and services that are produced in a country in a particular year.

gross national product (GNP): similar to GDP, but also includes the value of income from abroad.

guano: phosphate-rich bird droppings that often accumulate on islands where seabirds roost.

guest worker: a foreigner who is permitted to work in another country on a temporary basis.

hacienda: large agricultural estate in Latin America and Spain that grows crops mainly for domestic consumption (for example, for mines, missions, and cities) rather than for export.

hajj: the pilgrimage to Mecca required of all Muslims.

harmattan: a hot, dry wind that blows out of inland Africa.

hate crime: act of violence committed because of prejudice against women; ethnic, racial, and religious minorities; and homosexuals.

hegemony: domination over the world economy, exercised through a combination of economic, military, financial, and cultural means, by one national state in a particular historical epoch.

hill station: administrative center situated on higher land in order to take advantage of the cooler weather at higher altitudes.

hinterland: sphere of economic influence of a town or city—the tributary area from which it collects products to be exported and through which it distributes imports.

homelands: areas set aside in South Africa for black residents as tribal territories where they were given limited self-government but no vote and limited rights in the general politics of South Africa.

imperialism: the extension of the power of a nation through direct or indirect control of the economic and political life of other territories.

import substitution: the process by which domestic producers provide goods or services that were formerly bought from foreign producers.

indentured servant: an individual bound by contract to the service of another for a specified term.

informal economy: those economic activities that take place beyond official record and not subject to formalized systems of regulation or remuneration (e.g., street selling, petty crime).

inselberg: isolated, domed hill left by erosion in a desert or semiarid region.

intermontane: lying between or among mountains.

internal migration: the movement of populations within a national territory.

international division of labor: the specialization of different people, regions, and countries in certain kinds of economic activities.

International Monetary Fund: an organization that provides loans to governments throughout the world.

international regime: the orientation of contemporary politics around the international arena rather than the national one.

intertropical convergence zone: region where air flows together and rises vertically as a result of intense solar heating at the equator, often with heavy rainfall, and shifting north and south with the seasons.

intifada: the violent uprising of Palestinians against the rule of Israel in the occupied territories.

irredentism: the assertion by the government of a country that a minority living outside its borders belongs to it historically and culturally.

Islam: a religion that is based on submission to God's will.

Islamism: an anticolonial, anti-imperialist, and overall anticore political movement.

island biogeography: theory that smaller islands will generally be less biologically diverse than larger ones.

jihad: a sacred struggle or striving to carry out God's will.

keiretsu: Japanese business networks facilitated after the Second World War by the Japanese government in order to promote national recovery.

kinship: a shared notion of relationship among members of a group often but not necessarily based on blood, marriage, or adoption.

La Niña: the periodic abnormal cooling of sea surface temperatures in the tropical Pacific off the coast of Peru that results in worldwide changes in climate, including droughts and floods that contrast with those produced by El Niño.

land bridge: a dry land connection between two continents or islands, exposed, for example, when sea level falls during an ice age.

land reform: a change in the way land is held or distributed, such as the division of large private estates into small private farms or communally held properties.

language family: a collection of individual languages believed to be related in their prehistoric origin.

language group: a collection of several individual languages that are part of a language branch, share a common origin, and have similar grammar and vocabulary.

latifundia: large rural landholdings or agricultural estates.

latitude: the angular distance of a point on Earth's surface, measured north or south from the equator, which is 0°.

law of diminishing returns: the tendency for productivity to decline, after a certain point, with the continued addition of capital and/or labor to a given resource base.

leadership cycles: periods of international power established by individual states through economic, political, and military competition.

liberation theology: a Catholic movement, originating in Latin America, focused on social justice and on helping the poor and oppressed.

loess: a surface cover of fine-grained silt and clay deposited by wind action and usually resulting in deep layers of yellowish, loamy soils.

longitude: the angular distance of a point on Earth's surface, measured east or west from the prime meridian (the line that passes through both poles and through Greenwich, England, and is given the value of 0°).

lost decade: the 1980s, when debt and economic crisis led to setbacks in economic development and social conditions in many developing countries.

Main Street: the dominant urban corridor of Canada, extending from Québec City to Windsor, Ontario.

maquiladora: industrial plant in Mexico, originally within the border zone with the United States and often owned or built with foreign capital, that assembles components for export as finished products free from customs duties.

maroon communities: settlements of runaway slaves in Brazil and elsewhere in the Caribbean.

Marshall Plan: strategy of quickly rebuilding the West German economy after the Second World War in order to prevent the spread of socialism or a recurrence of fascism. Named after U.S. Secretary of State George Marshall.

marsupial: Australian mammal such as the kangaroo, koala, and wombat that gives birth to premature infants that then develop and feed from nipples in a pouch on the mother's body.

massif: a mountainous block of Earth's crust bounded by faults or folds and displaced as a unit.

Megalopolis: the dominant urban corridor of the United States that extends along the eastern seaboard from Boston to Washington, DC.

Melanesia: the region of the western Pacific that includes the westerly and largest islands of Papua New Guinea, the Solomon Islands, Fiji, Vanuatu, and New Caledonia.

mercantilism: an economic policy where government controls industry and trade.

mestizo: term used in Latin America to identify a person of mixed white (European) and American Indian ancestry.

microfinance programs: programs that provide credit and savings to the self-employed poor, including those in the informal sector, who cannot borrow money from commercial banks.

Micronesia: the northern group of islands at, or north of, the equator in the western Pacific, including the independent countries of Nauru, Kiribati, Palau, the Commonwealth of the Northern Marianas, the Marshall Islands, and the Federated States of Micronesia (including Yap, Pohnpei, Chuuk, and Kosrae of the Caroline Islands), and the U.S. territory of Guam.

mikrorayon: neighborhood-scale planning unit of the Soviet era.

minifundia: very small parcels of land farmed by tenant farmers or peasant farmers.

minisystem: a society with a single cultural base and a reciprocal social economy.

Modernity: a forward-looking view of the world that emphasizes reason, scientific rationality, creativity, novelty, and progress.

monotreme: an animal that lays eggs but then nourishes the newborn with mother's milk.

Monroe Doctrine: proclamation of U.S. President James Monroe in 1823 stating that European military interference in the Western Hemisphere, including the Caribbean and Latin America, would no longer be acceptable.

monsoon: seasonal reversal of wind flows in parts of the lower to middle latitudes. During the cool season, a dry monsoon occurs as dry offshore winds prevail; in hot summer months a wet monsoon occurs as onshore winds bring large amounts of rainfall (e.g., moist winds that blow from the south into Asia between April and October bringing the monsoon rains, and out of Asia between November and March).

moraine: an accumulation of rock and soil carried forward by a glacier and eventually deposited at its frontal edge or along its sides.

mulatto: term used in Latin America to identify a person of mixed white (European) and black (African) ancestry.

multiculturalism: a process of immigrant incorporation in which each ethnic group has the right to enjoy and protect their officially recognized "native" culture.

Muslim: a member of the Islamic religion.

nation: a group of people often sharing common elements of culture, such as religion, language, a history, or political identity.

nationalism: the feeling of belonging to a nation as well as the belief that a nation has a natural right to determine its own affairs.

nationalist movement: organized groups of people, sharing common elements of culture, such as language, religion, or history, who wish to determine their own political affairs.

nationalization: the process of converting key industries from private to governmental organization and control.

nation-state: an ideal form consisting of a homogeneous group of people governed by their own state.

Near Abroad: independent states that were formerly republics of the Soviet Union.

neocolonialism: economic and political strategies by which powerful states in core economies indirectly maintain or extend their influence over other areas or people.

neoliberalism: a reduction in the role and budget of government, including reduced subsidies and the privatization of formerly publicly owned and operated concerns, such as utilities.

neotropics: ecological region of the Tropics of the Americas.

new international division of labor: the decentralization of manufacturing production from core regions to some semiperipheral and peripheral countries.

nontraditional agricultural exports (NTAEs): new export crops, such as vegetables and flowers, that contrast with the traditional exports such as sugar and coffee, and often require fast refrigerated transport to market.

North American Free Trade Agreement (NAFTA): 1994 agreement among the United States, Canada, and Mexico to reduce barriers to trade between the three countries, through, for example, reducing customs tariffs and quotas.

oasis: a spot in the desert made fertile by the availability of surface water.

offshore financial services: provision of banking, investment, and other services to foreign nationals and companies who wish to avoid taxes, oversight, or other regulations in their own countries.

ordinary landscapes: the everyday landscapes that people create in the course of their lives.

orographic effect: the influence of hills and mountains in lifting airstreams, cooling the air, and thereby inducing precipitation.

Outback: the dry and thinly populated interior of Australia.

overseas Chinese: migrants from China who settled in Southeast Asia as early as the fourteenth century, but mainly during the period of European colonialism, when they arrived as contract plantation, mine, and rail workers and then moved into clerical and business roles.

overurbanization: a condition in which cities grow more rapidly than the jobs and housing can sustain.

ozone depletion: the loss of the protective layer of ozone gas that prevents harmful ultraviolet radiation from reaching Earth's surface and causing increases in skin cancer and other ecological damage.

Pacific Rim: a loosely defined region of countries that border the Pacific Ocean.

pastoralism: system of farming and way of life based on keeping herds of grazing animals—cattle, sheep, goats, horses, camels, yaks, and so on, depending on the environment.

peripheral regions: regions that are characterized by dependent and disadvantageous trading relationships, by inadequate or obsolescent technologies, and by undeveloped or narrowly specialized economies with low levels of productivity.

permafrost: permanently frozen subsoil, which may extend for several meters below the surface layer and may defrost up to a depth of a meter or so during summer months.

petrodollars: revenues generated by the sale of oil.

physiographic region: a broad region within which there is a coherence of geology, relief, landforms, soils, and vegetation.

pinyin: system of writing Chinese language using the Roman alphabet.

place: a specific geographic setting with distinctive physical, social, and cultural attributes.

plantation: large agricultural estate that is usually tropical or semitropical, monocultural (one crop), and commercial- or export-orientated, most of which were established in the colonial period.

plate tectonics: theory that Earth's crust is divided into large solid plates that move relative to each other and cause mountain building, volcanic, and earthquake activity when they separate or meet.

playa: flat bottom of an undrained desert basin that becomes at times a shallow lake.

polder: an area of low land reclaimed from a body of water by building dikes and draining the water.

Polynesia: the central and southern Pacific islands that include the independent countries of Samoa, Tonga, the Cook Islands, Niue, and Tuvalu; the U.S. territory of American Samoa; the French overseas territories of Wallis and Fortuna and French Polynesia (including the island of Tahiti, the Society Islands, the Tuamotu archipelago and the Marquesas Islands); the New Zealand territory of Tokelau; and the British territory of the Pitcairn Islands. Polynesia sometimes includes New Zealand and the Hawaiian islands.

primacy: a condition in which the population of the largest city in an urban system is disproportionately large in relation to the second- and third-largest cities in that system.

primary activity: economic activity that is concerned directly with natural resources of any kind.

pristine myth: erroneous belief that the Americas were mostly wild and untouched by humans prior to European arrival. Large areas had been cultivated and deforested by indigenous populations.

quaternary activity: economic activity that deals with the handling and processing of knowledge and information.

Raj: the rule of the British in India.

region: larger-sized territory that encompasses many places, all or most of which share similar attributes in comparison with the attributes of places elsewhere.

regional geography: the study of the ways in which unique combinations of environmental and human factors produce territories with distinctive landscapes and cultural attributes.

regionalism: strong feelings of collective identity shared by religious or ethnic groups that are concentrated within a particular region.

remote sensing: the collection of information about parts of Earth's surface by means of aerial photography or satellite imagery designed to record data on visible, infrared, and microwave sensor systems.

Richter scale: a logarithmic scale ranging from 1 to 10 used to measure the amount of energy released by an earthquake.

rift valley: block of land that drops between two others, forming a steep-sided trough, often at faults on a divergent plate boundary.

Ring of Fire: chain of seismic instability and volcanic activity that stretches from Southeast Asia through the Philippines, the Japanese archipelago, the Kamchatka Peninsula, and down the Pacific coast of the Americas to the southern Andes in Chile. It is caused by the tension built up by moving tectonic plates.

salinization: salt deposits caused when water evaporates from the surface of the land and leaves behind salts that it has drawn up from the subsoil.

satellite state: a national state that is economically dependent and politically and militarily subservient to another—in its orbit, figuratively speaking.

savanna: grassland vegetation found in tropical climates with pronounced dry season and periodic fires.

sawah: irrigated or wet rice cultivation; term from Indonesia used in Southeast Asia.

secondary activity: economic activity involving the processing, transformation, fabrication, or assembly of raw materials, or the reassembly, refinishing, or packaging of manufactured goods.

sectionalism: extreme devotion to local interests and customs.

semiperipheral regions: regions that are able to exploit peripheral regions but are themselves exploited and dominated by the core regions.

sense of place: feelings evoked among people as a result of the experiences and memories that they associate with a place and to the symbolism that they attach to it.

shifting cultivation: agricultural system that preserves soil fertility by moving crops from one plot to another.

Silk Road: ancient east-west trade route between Europe and China.

site: the physical attributes of a location: its terrain, its soil, vegetation, and water sources, for example.

situation: the location of a place relative to other places and human activities.

slash and burn: agricultural system often used in tropical forests that involves cutting trees and brush and burning them so that crops can benefit from cleared ground and nutrients in the ash.

slow world: people, places, and regions whose participation in transnational industry, modern telecommunications, materialistic consumption, and international news and entertainment is limited.

social capital: networks and relationships that encourage trust, reciprocity, and cooperation.

South Pacific Forum: an institution that promotes discussion and cooperation on trade, fisheries, and tourism between all of the independent and self-governing states of Oceania.

sovereignty: the exercise of state power over people and territory, recognized by other states and codified by international law.

spatial diffusion: the way that things spread through geographic space over time.

spatial justice: the fairness of the distribution of society's burdens and benefits, taking into account spatial variations in people's needs and in their contribution to the production of wealth and social well-being.

Spice Islands: islands in eastern Indonesia, especially the Moluccas, that became the heart of the valuable trade in spices such as cloves, cinnamon, and nutmeg before and during the early period of Southeast Asian colonialism.

spring line: line of springs at the intersection of the water table with a sloping ground surface, usually resulting from the presence of less-permeable rock strata.

staples economy: an economy based on natural resources that are unprocessed or minimally processed before they are exported to other areas where they are manufactured into end products.

state: an independent political unit with territorial boundaries that are internationally recognized by other states.

state socialism: a form of economy based on principles of collective ownership and administration of the means of production and distribution of goods, dominated and directed by state bureaucracies.

steppe: semiarid, treeless, grassland plains.

stolen generation: Aboriginal children that were forcibly removed from their homes in Australia and placed in white foster homes or institutions.

structural adjustment policies: economic policies, mostly associated with the International Monetary Fund, that required governments to cut budgets and liberalize trade in return for debt relief.

subduction: the sinking of one plate under another as plates are converging or moving toward each other.

subsistence affluence: achievement of a good standard of living through reliance on self-sufficiency in local foods and with little cash income.

suburbanization: growth of population along the fringes of large metropolitan areas.

superfund site: locations in the United States officially deemed by the federal government as extremely polluted and requiring extensive, supervised, and subsidized cleanup.

supranational organization: a collection of individual states with a common economic and/or political goal that diminishes, to some extent, individual state sovereignty in favor of the collective interests of the membership.

sustainable development: a vision of development that seeks a balance among economic growth, environmental impacts, and social equity.

swidden: *See* slash and burn.

symbolic landscapes: representations of particular values and/or aspirations that the builders and financiers of those landscapes want to impart to a larger public.

taiga: ecological zone of boreal coniferous forest.

technology system: a cluster of interrelated energy, transportation, and production technologies that dominates economic activity for several decades.

territorial production complex: regional groupings of production facilities based on local resources that were suited to clusters of interdependent industries.

territory: the delimited area over which a state, an individual, or a group exercises control and that is usually recognized by other states, individuals, or groups.

tertiary activity: economic activity involving the sale and exchange of goods and services.

time-space convergence: the rate at which places move closer together in travel or communication time or costs.

total fertility rate: average number of children a woman will bear throughout her childbearing years, approximately ages 15 through 49.

trade creation effects: the positive economic effects of transnational integration.

trade diversion effects: the negative economic effects of transnational integration.

transform boundary: the process of plates sliding past each other horizontally.

transhumance: the movement of herds according to seasonal rhythms: warmer, lowland areas in the winter, and cooler, highland areas in the summer.

transmigration: policy of resettling people from densely populated areas to less populated, often frontier regions.

transnational corporation: a corporation that has investments and activities that span international boundaries, with subsidiary companies, factories, offices, or facilities in several countries.

Treaty of Tordesillas: agreement made by the Pope Alexander VI in 1494 to divide the world between Spain and Portugal along a north-south line 370 leagues (about 1800 kilometers, or 1100 miles) west of the Cape Verde Islands. Portugal received the area east of the line, including much of Brazil and parts of Africa, and Spain received the area to the west.

Treaty of Waitangi: an 1840 agreement in which 40 Maori chiefs gave the Queen of England governance over their land and the right to purchase it in exchange for protection and citizenship. Reinterpreted by the Waitangi tribunal in the 1990s, it provides the basis for Maori land rights and New Zealand's bicultural society.

treaty ports: ports in Asia, especially China and Japan, that were opened to foreign trade and residence in the mid-nineteenth century because of pressure from powers such as Britain, France, Germany, and the United States.

tribe: a form of social identity created by groups who share a common set of ideas about collective loyalty and political action.

tundra: an Arctic wilderness where the climate precludes any agriculture or forestry. Permafrost and very short summers mean that the natural vegetation consists of mosses, lichens, and certain hardy grasses.

unitary state: a form of government in which power is concentrated in the central government.

viceroyalty: largest scale of Spanish colonial administration leading to the emergence of Mexico City and Lima as the headquarters of the viceroyalties of New Spain and Peru, respectively.

visualization: computer-assisted representation of spatial data, often involving three-dimensional images and innovative perspectives, in order to reveal spatial patterns and relationships more effectively.

Wallace's Line: division between species associated with the deep-ocean trench between the islands of Bali and Lombok in Indonesia that could not be crossed even during low sea levels of the ice ages.

weather: instantaneous or immediate state of the atmosphere (e.g., it is raining).

White Australia policy: Australian policy, until 1975, that restricted immigration to people from northern Europe through a ranking that placed British and Scandinavians as the highest priority, followed by southern Europeans, with the goal of attaining a homogenized-looking and culturally similar population.

World Bank: a development bank and the largest source of development assistance in the world.

world city: city in which a disproportionate share of the world's most important business—economic, political, and cultural—is transacted.

world-empire: minisystems that have been absorbed into a common political system while retaining their fundamental cultural differences.

world region: a large-scale geographic division based on continental and physiographic settings that contain major clusters of humankind with broadly similar cultural attributes.

world religion: a belief system that has adherents worldwide.

world-system: an interdependent system of countries linked by political and economic competition.

xenophobia: a hate and/or fear of foreigners.

yurt: circular tent used by nomadic groups in Mongolia, northwest China, and Central Asia, constructed from collapsible wooden frames and felt made from sheep's wool.

zaibatsu: a large Japanese conglomerate corporation.

zambo: term used in Latin America to identify a person of mixed African and American Indian ancestry.

zionism: a movement whose chief objective has been the establishment for the Jewish people of a legally recognized home in Palestine.

Photo Credits

Contents

TOC-1: © Sandro Vannini/CORBIS   TOC-2: © David Samuel Robbins/CORBIS   TOC-3: Corbis Digital Stock   TOC-4: Dave Hamman; Gallo Images/CORBIS   TOC-05: © CORBIS   TOC-6: Nik Wheeler/CORBIS   TOC-7: Owen Franken/CORBIS   TOC-8: Owen Franken/CORBIS   TOC-9: © Liba Taylorr/CORBIS   TOC-10: Kevin Fleming/CORBIS   TOC-11: Juda Ngwenya/Getty Images, Inc.

Chapter 1

1-2: NASA/Jet Propulsion Laboratory   1-3: Paul L. Knox   1-5: Saul Steinberg, "View of the World from 9th Avenue", 1975. Wax crayon and graphite on paper, 20 x 15". Photograph by Ellen Page Wilson, courtesy of Pace Wildenstein. ©2002 The Saul Steinberg Foundation/Artists Rights Society (ARS), New York   1-18: Minosa-Scorpio/Corbis Sygma   1-20: Chromosohm/Sohm/Photo Researchers, Inc.   1-21: Paul L. Knox   1-23: Sarah Leen/Matrix International, Inc.   1-24: Raymond Gehman/CORBIS   1-25: Paul L. Knox   1-26: AP/Wide World Photos   1-27: Paul L. Knox   1-30: Sallie A. Marston   1-32a: David Arnold/NGS Image Collection   1-32b: Paul L. Knox   1-33: Paul L. Knox   1-34: Mike Yamashita/Woodfin Camp & Associates   1-35: AP/Wide World Photos   1-36: Patrick Doherty/Getty Images, Inc.   1-B03-01: Agence France Presse/CORBIS   1-B03-02: Loren Callahan/REUTERS/CORBIS

Chapter 2

2-4: Raid Planete Poussiere/Getty Images, Inc.   2-6c: © Alison Wright/CORBIS   2-14: ©National Maritime Museum Picture Library, London, England   2-17: Georg Gerster/Photo Researchers, Inc.   2-18: Gigli/F.A.O. Food and Agriculture Organization of the United Nations   2-20: Victor Englebert/Englebert Photography, Inc.   2-23: Palazzo Vecchio, Florence, Italy/Canali PhotoBank Milan/Superstock   2-24: Robert Holmes/CORBIS   2-25a: Library of Congress   2-25b: Peter Vadnai/Corbis/Stock Market   2-31: Martyn Austin/CORBIS   2-39: Philippe Brylak/Getty Images, Inc.   2-B01-01: © National Maritime Museum Picture Library, London, England   2-B01-02: © National Maritime Museum Picture Library, London, England   2-B02-03: Library of Congress   2-B02-04: Library of Congress   2-B02-05: The London News/Library of Congress   2-B02-06: Library of Congress   2-B02-09: Vince Streano/CORBIS   2-B02-10: Library of Congress   2-B02-11: Library of Congress

Chapter 3

3-2a: Tom Van Sant, Geosphere Project/Planetary Visions/Science Photo Library/Photo Researchers, Inc.   3-4: © Paul A. Souders/CORBIS   3-5: © Sandro Vannini/CORBIS   3-6: Paul L. Knox   3-7: © Vince Streano/CORBIS   3-12: Paul L. Knox   3-20: Peter Turnley/CORBIS   3-21: Peter Ginter/Material World   3-24a,b: ©Adam Woolfitt/CORBIS   3-25: Hans Wolf/Getty Images, Inc.   3-26: Joe Cornish/Getty Images, Inc.   3-27: Sandy Stockwell/CORBIS   3-28: © Yann Arthus-Bertrand/CORBIS   3-29: © Craig Aurness/CORBIS   3-31a: © Adam Woolfitt/CORBIS   3-31b: © Vince Streano/CORBIS   3-32a: © Vittoriano Rastelli/CORBIS   3-32b: © Robert Holmes/CORBIS   3-35a: © Bryan Pickering/Eye Ubiquitous/CORBIS   3-35b: © CORBIS   3-36a: © Yann Arthus-Bertrand/CORBIS   3-36b: © Chris Rainier/CORBIS   3-37: © Kevin Schafer/CORBIS   3-40a: Dr. Robert Muntefering/Getty Images, Inc.   3-40b: Vittoriano Rastelli/CORBIS   3-42a: © John Heseltine/CORBIS   3-42b: © Michael Busselle/CORBIS   3-43: © David Cumming/Eye Ubiquitous/CORBIS   3-44: © Yann Arthus-Bertrand/CORBIS   3-45a: © Michael S. Yamashita/CORBIS   3-45b: © Bill Ross/CORBIS   3-45c: Paul L. Knox   3-B01-01: The Granger Collection   3-B01-02: The Granger Collection   3-B02-01: AP/Wide World Photos   3-B03-02: Agence France Presse/CORBIS   3-B04-02: © Pressens Bild AB/The Liaison Agency Network   3-B05-02: © Bill Ross/CORBIS   3-B06-02: Sandy Stockwell/London Aerial Photo Library/CORBIS   3-B07-01: © 2001, SZTE EK/Karoly Kokas

Chapter 4

4-2a: WORLDSAT International and J. Knighton/Science Photo Laboratory/Photo Researchers, Inc.   4-4: © Roger Tidman/CORBIS   4-5: © Dean Conger/CORBIS   4-6: © Kevin Schafer/CORBIS   4-7: © Marc Garanger/CORBIS   4-8: © Wolfgang Kaehler/CORBIS   4-9: © Nevada Wier/CORBIS   4-10: © Staffan Widstrand/CORBIS   4-11: © Dean Conger/CORBIS   4-12: Russia and Eastern Images   4-15: © Wolfgang Kaehler/CORBIS   4-23: Louis Psihoyos & John Knoebber/Material World   4-24: © David Samuel Robbins/CORBIS   4-25: © Yann Arthus-Bertrand/CORBIS   4-27b: Mark Wadlow/Russia and Eastern Images   4-27c: © Wolfgang Kaehler/ CORBIS   4-27d: Sovfoto/Eastfoto   4-28a: Getty Images, Inc.   4-28b: © Morton Beebe, S.F./CORBIS   4-32: © Staffan Widstrand/CORBIS   4-33: © Galen Rowell/CORBIS   4-34: © Wolfgang Kaehler/CORBIS   4-35: Sovfoto/ Eastfoto   4-36: © David Samuel Robbins/CORBIS   4-B01-01: © Natalie Fobes/CORBIS   4-B01-02: Getty Images, Inc.   4-B02-01: Dean Conger/ CORBIS   4-B03-02: Eddy Van Wessel/Getty Images, Inc.   4-B06-01: © Ralph White/CORBIS   4-B07-01a: © Charles O'Rear/CORBIS   4-B07-01b: © Daniel Laine/CORBIS   4-B08-01: © Wolfgang Kaehler/CORBIS   4-B09-01a: © Gerard Degeorge/CORBIS   4-B09-01b: © Diego Lezama Orezzoli/CORBIS

Chapter 5

5-2a: Getty Images, Inc.   5-3: © Yann Arthus-Bertrand/CORBIS   5-6: O. Alamany & E. Vicens/CORBIS   5-7: Corbis Digital Stock   5-10: SuperStock, Inc.   5-11: Jeffrey L. Rotman/CORBIS   5-14a: Sandro Vannini/CORBIS   5-14b: Fulvio Roiter/CORBIS   5-17: Corbis Digital Stock   5-20: Farhang Rouhani   5-21: Mehmet Biber /Photo Researchers, Inc.   5-22: Robert Holmes/CORBIS   5-23: Tibor Bognar/Corbis/Stock Market   5-26: K.M. Westermann/CORBIS   5-28: Sallie A. Marston   5-30: Michel Bourque/Valan Photos   5-31: Michael Bonine   5-34: AP/Wide World Photos   5-36: Peter Turnley/CORBIS   5-37a: Courtney Kealy/Getty Images, Inc.   5-37b: Barry Iverson/Woodfin Camp & Associates   5-39: Robert Azzi/Woodfin Camp & Associates   5-40: Wolfgang Kaehler/CORBIS   5-41: Earl & Nazima Kowall/CORBIS   5-42a: Scott Peterson/Getty Images, Inc.   5-42b: Houman Sadr   5-43: Michael Bonine   5-44a: Michael Bonine   5-44b: Christine Osborne/CORBIS   5-44c: Robert Holmes/CORBIS   5-45: © Yann Arthus-Bertrand/CORBIS   5-46: SuperStock, Inc.   5-47: Juan Echeverria/CORBIS   5-B01-01: Bill Lyons   5-B01-02: Dave Bartruff/CORBIS   5-B01-03: Richard T. Nowitz/CORBIS   5-B03-02: David H. Wells/CORBIS   5-B04-01: Steve McCurry/National Geographic Society   5-B04-02: Seghilani/SIPA Press   5-B05-01a: Michael Bonine   5-B05-01b: Michael Bonine   5-B05-02: Zeynep Sumen/Getty Images, Inc.   5-B06-01: Georg Fischer/Bilderberg/Aurora & Quanta Productions   5-B06-02: Charles & Josette Lenars/CORBIS   5-B07-01: Dean Conger/CORBIS   5-B07-02: Owen Franken/CORBIS   5-B07-03: NASA/Johnson Space Center   5-B08-01: Susan Meiselas/Magnum Photos, Inc.

Chapter 6

6-2a: Courtesy of Tom Van Sant, Geosphere Project/Planetary Visions/Science Photo Library/Photo Researchers, Inc.   6-5a: Winifred Wisniewski; Frank Lane Picture Agency/CORBIS   6-5b: Sharna Balfour; Gallo Images/CORBIS   6-5c: Nik Wheeler/CORBIS   6-7a: Karl Ammann/CORBIS   6-7b: Brian Vikander/CORBIS   6-7c: Michael & Patricia Fogden/CORBIS   6-10: Dave Hamman; Gallo Images/CORBIS   6-11a: © Yann Arthus-Bertrand/COR-

Chapter 7

Chapter 8

Chapter 9

Chapter 10

Chapter 11

Inc.   11-19: Peter Ginter/Material World   11-20: G. Pangare/F.A.O. Food and Agriculture Organization of the United Nations   11-21: FAO Photo/G. Bizzarri   11-22a: © Catherine Karnow/CORBIS   11-22b: © Roman Soumar/CORBIS   11-23a: FAO Photo/G. Bizzarri   11-23b: G. Bizzarri/F.A.O. Food and Agriculture Organization of the United Nations   11-24: © Liba Taylorr/CORBIS   11-26a: © Lindsay Hebberd/CORBIS   11-26b: Harald Sund/Getty Images, Inc.   11-26c: G. Bizzarri/F.A.O. Food and Agriculture Organization of the United Nations   11-28: © David Cumming; Eye Ubiquitous/CORBIS   11-29: Daniel O'Leary/Panos Pictures   11-31: G. Bizzarri/F.A.O. Food and Agriculture Organization of the United Nations   11-32a: © Martin Jones/CORBIS   11-32b: © Catherine Karnow/CORBIS   11-34: Andrea Pistolesi/Getty Images, Inc.   11-35: Andrea Pistolesi/Getty Images, Inc.   11-36: © Hanan Isachar/CORBIS   11-38: © Yann Arthus-Bertrand/CORBIS   11-39: G. Grepin/F.A.O. Food and Agriculture Organization of the United Nations   11-40: Suraj N. Sharma/Dinodia Picture Agency   11-B01-01: © Tom Brakefield/CORBIS   11-B01-02: Frans Lemmens/Getty Images, Inc.   11-B02-01: G. Bizzarri/F.A.O. Food and Agriculture Organization of the United Nations   11-B03-01: CORBIS   11-B03-02: Stephen Dupont/CORBIS   11-B04-01: Andrea Pistolesi/Getty Images, Inc.   11-B05-01: © Catherine Karnow/CORBIS   11-B06-01: Namas Bhojani   11-B07-01: G. Bizzarri/F.A.O. Food and Agriculture Organization of the United Nations   11-B08-01: Namas Bhojani

Chapter 12

12-2a: Tom Van Sant/Geosphere Project/Photo Researchers, Inc.   12-4: Vic Sievey/Eye Ubiquitous/CORBIS   12-5a: Penny Tweedie/CORBIS   12-5b: Patrick Ward/CORBIS   12-6a: Robert Dowling/CORBIS   12-6b: Wolfgang Kaehler/CORBIS   12-7a: Bill Ross/CORBIS   12-7b: Douglas Peebles/CORBIS   12-9a: Howard Davies/CORBIS   12-9b: L. Clarke/CORBIS   12-9c: Tom Brakefield/CORBIS   12-14: Michael S. Yamashita/CORBIS   12-16a: Penny Tweedie/CORBIS   12-16b: Agence France Presse/Maya Vidon/CORBIS   12-17: Patrick Ward/CORBIS   12-18: Peter Menzel/Material World   12-19: Neil Rabinowitz/CORBIS   12-20a: AP/Wide World Photos   12-23a: Paul A. Souders/CORBIS   12-23b: Paul Souders/CORBIS   12-23c: Neil MacLeod/CORBIS   12-24a: Yann-Arthus Bertrand/CORBIS   12-24b: Yann Arthus-Bertrand/CORBIS   12-25: Heldur Netocny/Panos Pictures   12-26: Wayne Lawler/Ecoscene/CORBIS   12-27: Earl and Nazima Kowall/CORBIS   12-28: Galen Rowell/CORBIS   12-B01-01: Carl Purcell/Photo Researchers, Inc.   12-B01-02: Jeff Hunter/Getty Images, Inc.   12-B02-02: Kevin Fleming/CORBIS   12-B03-01: Bettmann/CORBIS   12-B03-02: © 1995 Miller/Greenpeace   12-B03-03: James Davis/Eye Ubiquitous/CORBIS   12-B04-01: Patrick Ward/CORBIS

Chapter 13

13-1: Courtesy of Steven G. Eick, Visual Insights   13-2a: Paul Quayle/Panos Pictures   13-2b: Pat & Tom Leeson/Photo Researchers, Inc.   13-2c: Noel Quido/Getty Images, Inc.   13-2e: AP/Wide World Photos   13-2f: Photo Researchers, Inc.   13-4: AP/Wide World Photos   13-11: AP/Wide World Photos   13-12b: David & Peter Turnley/CORBIS   13-12c: Alison Wright/CORBIS   13-13: Juda Ngwenya/Getty Images, Inc.   13-14: Agence France Presse/CORBIS   13-15: Mark Lewis/Getty Images, Inc.   13-16: Michael S. Yamashita/CORBIS   13-17a: Peter Menzel/Material World   13-17b: Shawn G. Henry   13-19: Michael S. Yamashita/CORBIS   13-20: Kevin R. Morris/CORBIS   13-21: Macduff Everton/CORBIS

Index

ANTARCTICA

Hawaii (U.S.)

MEXICO

GUATEMALA
EL SALVADOR
CHILE

ARGENTINA
PERU
ECUADOR
COSTA
RICA
BELIZE
HONDURAS

URUGUAY
PANAMA
NICARAGUA

UNITED STATES

CANAD

BOLIVIA
COLOMBIA

PARAGUAY
JAMAICA
CUBA

BRAZIL
VENEZUELA
HAITI
THE
BAHAMAS

GUYANA
DOMINICAN
REPUBLIC

SURINAME
FRENCH GUIANA
(France)

World States, 2002

See page 12 of the text for an explanation of this projection.

Source: Map projection, Buckminster Fuller Institute and Dymaxion Map Design, Santa Barbara, CA. The word Dymaxion and the Fuller Projection Dymaxion™ Map design are trademarks of the Buckminster Fuller Institute, Santa Barbara, California,